INTRODUCTION TO
ENVIRONMENTAL
SCIENCE
Second Edition

Joseph M. Moran
Michael D. Morgan
James H. Wiersma
University of Wisconsin, Green Bay

INTRODUCTION TO ENVIRONMENTAL SCIENCE

Second Edition

W. H. Freeman and Company
New York

Library of Congress Cataloging-in-Publication Data

Moran, Joseph M.
    Introduction to environmental science.

    1. Human ecology.   2. Environmental protection.
3. Environmental policy.   I. Morgan, Michael D.
II. Wiersma, James H.   III. Title.
GF41.M67   1985   333.7   85-13052
ISBN 0-7167-1684-4

Printed in the United States of America

1 2 3 4 5 6 7 8 9 0   RRD   4 3 2 1 0 8 9 8 7 6

C O N T E N T S

PREFACE

Environmental science is a dynamic field of study. Since publication of the first edition of *Introduction to Environmental Science,* scientists have learned more about the components of the environment and how they interact. They have gathered more data on the ecological relationships of endangered species, the impact of toxic materials on human health, the dispersal of pollutants in the atmosphere, the accumulation of persistent chemicals in aquatic food webs, and the control of agricultural pests with minimum application of pesticides. Equipped with this better understanding, scientists are now refining models of environmental systems and predicting more accurately the consequences of environmental change—both natural and human-related.

Since the first edition, much has happened as well in the arena of environmental issues. Some issues have gained in importance while others have gradually receded from the public eye. Today, acid rain and management of hazardous wastes are receiving considerable attention from the scientific community and the media. On the other hand, construction of numerous wastewater treatment plants has significantly lessened the impact of some water pollutants on aquatic ecosystems, and so water pollution has faded somewhat from public awareness.

While progress has been made in understanding and mitigating some environmental problems, others require more research and remedial action. With steady improvement in the quality of outdoor air, interest is shifting to the quality of indoor air. Although the quality of surface waters (lakes and rivers) has improved somewhat, the quality of groundwater continues to be threatened by improper disposal of toxic chemicals. Furthermore, resolution of some environmental issues is perhaps less urgent now than it appeared only a few years ago. Atmospheric scientists have discovered that in spite of dire warnings, the ozone shield is still intact. Conservation efforts and the global oil surplus have given us a little more time to plan for the inevitable depletion of traditional energy resources (fossil fuels).

Because of this dynamic nature of environmental science and issues, it is a challenge to write an up-to-date textbook on these subjects. In this edition, we meet this challenge much as we did in the first, by emphasizing scientific principles and the natural functioning of the environment. Although the importance of certain environmental issues may wax and wane with time, the underlying principles that govern the function of physical and biological systems do not change. Hence, the readers' grounding in basic concepts provides them with the interpretive skills and flexibility needed to analyze new environmental issues as they arise.

Organization and Content

Introduction to Environmental Science remains both a basic science text and a text on environmental issues. This is reflected in the organization of the text's 20 chapters: The basic principles of ecology are introduced in Part I and then applied to a wide variety of environmental concerns in Parts II and III.

Part I (Chapters 2–5) surveys the fundamental principles that govern the functioning of the environment. What is the natural flow of energy and materials through the environment? How do organisms and ecosystems respond to change? Why and how do populations grow? Readers will gain an understanding of these and other points that they can apply to their comprehension of more specific issues later in the book.

Part II (Chapters 6–15) explores dominant issues of environmental quality and management: water and air pollution; exploitation of the earth's rock, mineral, and fuel resources; waste disposal; endangered species; and conflicts in land use.

Part III (Chapters 16–20) focuses on problems at the core of most environmental issues: growing human population and shortages in food and energy resources.

As in the first edition, this organization provides instructors with flexibility in course format. The major topics covered in Chapters 6 through 20 may be taught in any order and any chapter may be omitted without loss of continuity. Hence, this text is appropriate for introductory, semester-long, college-level courses on the environment that emphasize basic science and its application to environmental problems and issues.

While we have retained the philosophical approach and chapter organization of the first edition, we have made significant changes in content—many of which were suggested in a survey of users of the first edition. Changes include improved readability, updated topic coverage, more statistical background information, use of two-color art, inclusion of many new photographs and line drawings that are tied directly to the narrative, and use of new examples and case studies. Boxed material that provides a deeper scientific explanation of certain topics is now clearer and more understandable. Many new boxes have been added that illustrate the social, political, economic, or personal aspects of environmental conflicts and their possible resolution.

Because so much of the reader's comprehension of environmental problems rests on an understanding of ecological principles, we have extensively revised and expanded, by about 30 percent, the basic ecology chapters (Chapters 2–5). This discussion now includes a greater in-depth analysis of such concepts as ecological succession, adaptation, and evolution. The chapter on population growth and regulation better reflects the increasing realization among ecologists that populations are regulated by a variety of factors and that the relative importance of those factors varies with species, habitat, and time. In addition, we have taken a new approach to world ecosystems by describing them in terms of rates of productivity and patterns of nutrient cycling.

A major change in content and approach is a special emphasis on the economic and policymaking dimensions of environmental issues. With the help of Professor Richard J. Tobin, a political scientist from the State University of New York at Buffalo, Chapter 1 was extensively rewritten and expanded to introduce the reader to the environment and the different views of environmental problems taken by scientists, economists, and public policymakers. Later chapters consider how economics and politics affect the resolution of specific environmental problems. By covering the economic and political arenas, the second edition provides a more complete and realistic treatment of environmental issues.

With these changes, we have made every attempt to analyze environmental problems as objectively as possible. As in the first edition, we do not editorialize, but we encourage readers to form their own opinions on controversial issues. New sections on policymaking (especially Chapter 1) make readers aware of how individual opinion can influence the formulation of environmental policy. Furthermore, we introduce readers to effective strategies for communicating their particular environmental concerns to appropriate policymakers (Chapter 20).

Features

We have included several features in our book that make it an effective teaching and learning text.

Pedagogical Aids

Each chapter ends with conclusions, summary statements, review questions, suggestions for group and individual projects, and an annotated bibliography. Both metric and British units of measure appear throughout the text and in many tables and illustrations. There are appendixes of scientific conversion factors, geologic time, and powers of ten notation.

Boxes

Where a deeper scientific explanation of certain topics—for example, energy laws and the nature of electromagnetic radiation—seems desirable, we have included complementary information set off from the text in boxes. Where specific examples will illuminate the social, political, economic, or personal aspects of such environmental conflicts as eutrophication of Chesapeake Bay and the development of barrier islands for human habitation, brief case studies are included in box form.

Illustrations

To bring more realism to the book we have included an unusually large number of high-quality photographs and line drawings. More than 450 illustrations—photographs, drawings, maps, and graphs—illustrate and clarify important points.

Glossary

All important terms are italicized and defined at first use in the text. They appear again in the glossary, which, with more than 420 entries, functions as a minidictionary of environmental topics and terms.

Supplementary Materials

The text is accompanied by an Instructor's Manual that contains learning objectives for each chapter, test questions (each with a parenthetical reference to the page in the book on which it is answered), topics of current concern for discussion or research, and a list of recommended slides and films that instructors might wish to obtain for classroom use.

Acknowledgments

In preparing the second edition of this text, we profited greatly from our interactions with the talented and dedicated personnel at W. H. Freeman and Company. We thank especially Linda Chaput, Peter Dougherty, Jim Dodd, Jim Maurer, Susan Moran, and Margaret Mason. David Barkan helped obtain more than 100 new photographs for this edition.

We are particularly indebted to our development editor, John Hendry, for his creative thinking and imagination. From the very beginning, John gave the project direction so that the product of our collaboration is a logical and cohesive text.

We are grateful to Richard Tobin for his contributions to Chapters 1, 11, and 20 and for his constructive criticism of other sections of the manuscript. Professor Tobin helped us to successfully integrate the political and economic dimensions of environmental issues.

It is always an imposing challenge to communicate basic science to nonscience students. In this regard we have benefited greatly from the talents of our copyeditor, Marsha Mirski. Marsha successfully transformed our diverse writing styles into a smooth flowing and highly readable book.

Our very special thanks and appreciation go to Nancy Lambert, Jeanne Broeren, Joy Phillips, and Hope Mercier for their prompt and accurate typing of the manuscript. Their steady dispositions even when deadlines loomed served as an inspiration.

To the students of Environmental Science 102, your curiosity and concerns about the environment are the reasons for this book.

To our families and friends, your years of patience and understanding cannot be adequately acknowledged. Indeed, without your support and sacrifices there would be no book.

In revising the first edition, we were fortunate to have the constructive criticism of many reviewers. We are especially grateful for those of Eckhart Dersch, Michigan State University; Boyce A. Drummond, Illinois State University; Morris W. Firebaugh, University of Wisconsin, Parkside; Ira W. Geer, SUNY, Brockport; Garrett Hardin, University of California, Santa Barbara; John P. Harley, Eastern Kentucky University; Donald G. Kaufman, Miami University; David Pimentel, Cornell University; Henry A. Regier, University of Toronto; William J. Rowland, Indiana University; K. M. Stewart, SUNY, Buffalo; Ronald W. Tank, Lawrence University; Edward J. Tarbuck, Illinois Central College; Kenneth J. Van Dellen, Macomb Community College; and Thomas R. Wentworth, North Carolina State University, Raleigh.

We wish to also acknowledge the contributions of reviewers of the book's first edition. They are Richard Anthes, Pennsylvania State University; Gary Barrett, Miami University; Russell F. Christman,

University of North Carolina, Chapel Hill; Dean Earl Cook, Texas A&M University; Eckhard Dersch, Michigan State University; John M. Fowler, National Science Teachers Association; Garrett Hardin, University of California, Santa Barbara; John P. Harley, Eastern Kentucky University; Marilyn Houck, Pennsylvania State University; Richard E. Pieper, University of Southern California; David Pimentel, Cornell University; Clayton H. Reitan, Northern Illinois University; Wayne M. Wendland, University of Illinois at Urbana-Champaign; Susan Uhl Wilson, Miami-Dade Community College.

We are especially indebted to John F. Reed for his continuous interest and encouragement. We are also grateful to other colleagues at the University of Wisconsin-Green Bay, particularly to Robert W. Howe, Charles A. Ihrke, Robert W. Lanz, Charles R. Rhyner, Dorothea B. Sager, Paul E. Sager, Leander J. Schwartz, Ronald H. Starkey, Ronald D. Stieglitz, and Richard B. Stiehl for their valuable contributions.

July 1985

Joseph M. Moran
Michael D. Morgan
James H. Wiersma

INTRODUCTION TO ENVIRONMENTAL SCIENCE

Second Edition

The environment is a complex of both living and nonliving things as exemplified by this scene in Colorado. Attempts to protect environments such as this from development involve scientific, economic, and political considerations. Indeed, as we see in this opening chapter, experts in these disciplines approach environmental issues in quite different ways. (Richard Dunoff, The Stock Market.)

CHAPTER **I**

THINKING ABOUT THE ENVIRONMENT

The environment is usually taken to mean the assemblage of all the external factors or conditions that influence living things in any way. Thus the environment includes all living and nonliving things (such as air, soil, and water) that influence organisms. Since environmental science is the study of all the mechanisms of environmental processes, it is considered to be a pure science. But environmental science also is an applied science, because it examines environmental problems with the goal of contributing to their solution.

However, to determine exactly what constitutes an environmental problem, we must consider several factors. For example, what some persons consider to be a problem, and what they are willing to do about it, largely depends on their personal values, goals, and interests, which vary for different persons and groups. Furthermore, what some persons may consider to be a serious problem, others may see as a welcome solution to a quite different problem. For example, consider how persons might react to an announcement that operations at a local industrial plant will expand. The plant manager and most of the employees will greet the news with enthusiasm. For them, it is a signal that the sluggish economy has at last turned around, and their jobs are once again secure. The Chamber of Commerce will claim that the plant expansion will create new jobs and pour money into the local economy. Thus

The second and third main sections of this chapter were written by Professor Richard Tobin.

many persons will view the plant expansion as a sign of prosperity. However, others will view it as yet another affront to the quality of the environment. Those persons may argue that the costs of added air and water pollution outweigh the benefits to the local economy. Clearly, some, but not all, persons will view the plant expansion as an environmental problem.

However, even when most of us agree that a particular situation constitutes a serious environmental problem, economic constraints may prevent us from doing much about it. For example, public tax coffers are not bottomless, and funding for environmental quality control must compete with funding for education, capital construction programs (such as highways and public housing), national defense, and welfare programs (including Social Security and all the other projects and programs that are undertaken by our federal, state, and local governments). And because the funds that are allocated for environmental concerns are limited, the expenditure of $10 million on, say, flood control, means that much less money is available to spend on agricultural research or the control of air pollution. Hence choices must be made concerning which problems will be tackled and how they will be approached.

As a result of all the conflicting interests, choices must be made, which means that environmental problems cannot be viewed as simply scientific or technical problems. They must be considered as economic and political problems as well. Thus we

Figure 1.1 *The various components of the environment are interdependent—so much so that a natural occurrence or a human activity can have far-reaching and often unexpected consequences. The shell of this brown pelican egg was too thin to support the weight of its nesting parent. Such defective eggs were tragically common during the 1960s and 1970s, when DDT—widely used to control mosquitoes and other insect pests—so disrupted the reproductive biology of certain fish-eating birds that, for a time, the very survival of those species of birds was in doubt. The process by which this occurred is called* food-web accumulation; *it is discussed in Chapter 2. (Photo by Joseph R. Jehl, Jr.)*

begin this chapter by examining some of the scientific tools that enable us to conceptualize and understand the workings of the environment. Then we discuss economic aspects of environmental issues, such as how the costs of dealing with environmental problems are determined and weighed against the benefits that accrue. And, finally, we consider the various political factors that initiate and shape public policy concerning the environment.

Environmental Problems and Science: Discovering What Can Be Done

The environment is complex both in terms of its great diversity of components and its interrelationships among those components. A multitude of species of plants and animals inhabit the globe in many different types of habitats. And air, water, and land, which are the nonliving constituents, have properties (such as temperature and chemical composition) that vary greatly from one place to another.

The various components of the environment are not isolated from one another but, rather, are strongly interdependent (Figure 1.1). For example, green plants acquire essential resources from the physical realm (water and mineral nutrients from the soil and carbon dioxide from the atmosphere). Predatory animals depend on other animals (their prey) as a source of food. And we harvest food from the land and the sea and mine minerals and fuels from rock in the earth's crust.

The linkages that exist among components of the environment are avenues along which energy and materials are transported. For example, the sun's energy is taken up by a plant and is converted into food energy, which is harvested by an animal, who eats the plant. Then, when an organism dies or is consumed by another organism, its chemical constituents are cycled into other components of the environment including the soil. Soil particles are washed by heavy rains into nearby streams or are picked up by strong winds and transported great distances.

Those linkages among components of the environment are not arbitrary and capricious acts of nature, but rather, are considered to be *systematic* because they can be described in terms of logical principles. Hence we can view the environment as a complex of interdependent systems, in which each system is understood to be an assemblage of components (living and nonliving) that function and interact in a regular and predictable manner.

For most of us, the term *system* is familiar in reference to the human body, in which the respiratory, digestive, and circulatory systems function separately and together in a methodical manner. For example, the respiratory system takes oxygen from the air and releases carbon dioxide. The digestive system breaks down food particles so they can be absorbed by the body. And the circulatory system transports nutrients and oxygen throughout the body. Although each system is separate, each one is interdependent on the others. To remain alive, the cells that compose the body must receive nutrients from the digestive system and oxygen from the respiratory system by means of the circulatory system. And the subsystems in which cells are organized—tissues (such as muscle tissue) and organs (such as the heart, liver, and eyes)—function properly only as a result of regulation by the nervous and hormonal systems.

In much the same way, the earth is not comprised of one homogeneous environment. It is a mosaic of different climates, soil types, and plant and animal assemblages. Hence it is useful to view the environment as being subdivided into a number of ecosystems (Figure 1.2). (An *ecosystem* is composed of all the associated organisms and physical features that exist within a specific geographic area, for example, deserts, tropical rainforests, and tundras.) We have much more to say about ecosystems throughout this book.

An environmental problem usually stems from some disturbance of the normal functioning of an ecosystem, that is, a disruption of the usual operation (or interaction) of physical and biological processes. For example, a mountain lake that is noted for its clear water and fine trout fishing might become choked with aquatic weeds and devoid of fish a few years after a housing development opens near the lake shore. And a tropical forest might be cleared to provide land for agriculture but, within a decade, the soil's fertility may decline to the point at which further cultivation is futile.

Since environmental science is concerned with the natural functioning of ecosystems, it also encompasses the problems that arise when ecosystems are disturbed. Because the environment is complex and composed of a myriad of interacting systems, environmental science is a broad endeavor that draws on the contributions of many sciences; that is, it is *interdisciplinary*.

Each of the traditional scientific disciplines contributes to our understanding of environmental functions and, as a result, to our understanding of environmental problems (Figure 1.3). Physics, the most fundamental of the sciences, is essentially the study of matter and energy, so it is concerned with the laws that govern the cycling of materials and energy within and between ecosystems. Chemistry is concerned with the composition of materials and their reactions, which enables us to study the origins and fate of air and water pollutants. Biology focuses on the functioning of organisms and their interactions, which enables us to understand how organisms respond to pollutants. Meteorology is the study of the atmosphere, weather, and climate, which enables us to discover how air pollutants are transported and distributed. Geology deals with the earth's structure and composition (soil, rocks, minerals, and fuel), which provides us with a perspective on our resources. These sciences and other domains of knowledge facilitate our understanding of the environment and our impact on it.

The Scientific Method

Scientists from a variety of disciplines gather information and draw conclusions about the workings of the environment by applying the *scientific method*, which is a systematic form of inquiry that involves observation, speculation, and reasoning. Our ultimate objective in using the scientific method is to reach an understanding of the physical and biological phenomena that go on all around us.

We may view the scientific method as a sequence of steps in which scientists: (1) identify a specific question; (2) propose an answer to the question in the form of an "educated guess;" (3) state the educated guess in such a way that it can be tested, that is, formulate a hypothesis; (4) predict what the

(a)

(b)

(c)

Figure 1.2 *Three ecosystems. (a) A relatively simple ecosystem. It has a single dominant plant species, a dozen or so other plant species (weeds), perhaps 200 animal species, most of which are invertebrates (insects, worms, and so forth), and a variety of microorganisms. (A. M. Wettach, Black Star.) (b) A very complex ecosystem. This one has hundreds of plant species and thousands of animal species, most of which are insects, and myriad microorganisms. (© John Nance, Magnum.) (c) Often the only immediately noticeable organisms in this ecosystem are members of its dominant animal species. Would you characterize it as a simple ecosystem or as a complex one? (© Andy Levin, Black Star.)*

(a)

(b)

(c)

Figure 1.3 *Environmental science is interdisciplinary; that is, it draws on the contributions of many sciences. (a) A meteorologist sends up a weather balloon. The balloon is equipped with an instrument package that tells him the temperature profile of the air, through which the instruments pass. This information helps the environmental scientist assess the likelihood of an adverse air-pollution situation in the region. (© Steve Northup, Black Star.) (b) A geologist explores for mineral resources. The development of those resources may entail mining, which could adversely affect land and water quality—a basic concern of the environmental scientist. (Illustrators Stock Photos.) (c) An aquatic biologist takes water samples to test for the presence of toxic materials. The widespread distribution of toxic substances is a major concern of environmental science. (Illustrators Stock Photos.)*

consequences would be if the hypothesis were correct; (5) test the hypothesis by checking to see if the prediction is correct; and (6) revise or restate the hypothesis if the prediction is wrong.

In actual practice, scientists often do not follow these steps exactly. And in some cases, the steps are not always followed in this order. Furthermore, some steps (such as 3 and 4) are often combined. Nor is the scientific method a formula for creativity, since it does not provide the key idea, the hunch, or the educated guess that forms the basis of the hypothesis. Rather, it is a method or technique that can be used to test the validity or worth of a creative key idea, however and wherever it originates.

The scientific method can be illustrated by an investigation that was done concerning an acid-rain problem in the lakes in the Adirondack Mountains of New York State. Those lakes had long been noted for their abundance of game fish and other aquatic life, and that good fishing had helped to make the Adirondacks a popular area for sportsmen and other vacationers. However, beginning in the early 1970s, the populations of fish in the Adirondack lakes began to decline. Thus local businesses became concerned that the lack of fishing would hurt the region's recreational industry, and conservationists worried that the lakes had become yet another casualty of the pollution problem.

Biologists initially proposed that toxic materials (poisons) that were entering the lakes were killing the aquatic life. But it was not known what kind of toxins were in the lakes and where they were coming from. As a result, tests were done on samples of the lake water, which revealed that the waters did not contain hazardous concentrations of toxins but, rather, that the waters were abnormally acidic. (Laboratory studies have shown that excessively acidic waters are lethal to young fish.) And that observation prompted scientists to hypothesize that acidic rainwater was causing the lake waters to become more acidic.

It is well known that rainwater normally is slightly acidic, because rainwater dissolves some of the carbon dioxide in the air, which forms a very weak—and harmless—solution of carbonic acid. However, laboratory tests of rainwater samples that were collected in the vicinity of the Adirondack lakes were at least 100 times more acidic than normal. Thus to determine why the rainwater was so acidic, further laboratory studies were done. And those tests identified sulfuric and nitric acids in the rainwater, which the investigators knew were formed as a result of oxides of sulfur and nitrogen—common industrial air pollutants—that were dissolved in rainwater. They were also aware that the Adirondack Mountains are downwind of some major industrial sources (such as coal-fired power plants) of those air pollutants.

By using this reasoning, it was determined that the loss of fish in Adirondack Mountain lakes was, at least circumstantially, linked to industrial air pollution. And the hypothesis that excessively acidic rainwater led to the fish kills is generally accepted, because it is consistent with our knowledge concerning the tolerance of fish to acidic waters, chemical reactions that involve rainwater and air pollutants, and the type of air pollutants that are transported into the Adirondack region.

As in our example concerning acid rain, a hypothesis merely serves as a working assumption that eventually may be accepted, rejected or modified. And, in fact, inquiry, creative thinking, and imagination are stifled when hypotheses are considered to be immutable. A new hypothesis (or an old, resurrected one) may be hotly debated within the scientific community. And sometimes, on a particularly controversial issue, disagreement among scientists is widely publicized, which may confuse the general public. Thus, in some cases, the prevailing reaction may be, "Well, if the so-called experts can't agree among themselves, who am I to believe? And is there really a problem at all?" However, debate and disagreement are common steps in the process of reaching an understanding. If a hypothesis survives the scrutiny and skepticism of scientists, the chances are that it is on target.

Scientific Models

Models can facilitate the application of the scientific method and can help us to understand complex situations by reducing them to their essential elements. A *model* is defined as an approximate representation or simulation of a real situation. It eliminates all but the essential variables, or characteristics, that can change. For example, to learn how to improve the fuel efficiency of automobiles, we can observe a model automobile in a wind tunnel. (An automobile can be designed to reduce its frictional, or air, resistance, which can increase its

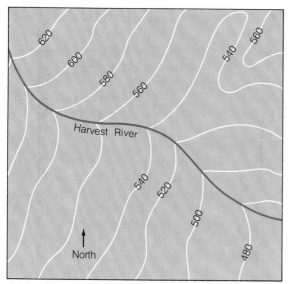

Contour interval 20 feet

⅝ inch = 2000 feet

(a)

Figure 1.4 *Two kinds of models. (a) A portion of a topographic map, which is a type of graphic model. If you were moving from left to right, would you be going uphill or downhill? (b) A portion of a physical model of the Mississippi River. (U.S. Geodetic Survey and the U.S. Army Corps of Engineers.)*

(b)

fuel efficiency.) Thus the shape of the model automobile is the critical variable, which becomes the focus of study. Other variables, such as the color of the automobile or whether it is equipped with a little mannequin inside, is irrelevant to the experiment and can be ignored.

Sometimes models are used to organize information. Because they are not cluttered with extraneous and distracting details, they may provide important insights concerning how things interact, or they may trigger creative thinking about complex phenomena. Models can also be used to make predictions. For example, in a model in which many important variables exist, one variable may be perturbed in order to assess its effect on the other variables.

Depending on their particular functions, scientific models may be classified as conceptual, graphical, physical, or numerical. *Conceptual models* describe the general relationships among components of a system. For example, a food web (Chapter 2) is a conceptual model that portrays the dependency of one group of organisms on another. *Graphical*, or *pictorial models*, compile and display data in a form that can be readily useful. For example, a topo-

graphic map displays the contours of land elevation, so that we can comprehend the characteristics of a general terrain at a glance (Figure 1.4). And a weather map, which integrates simultaneous weather observations at thousands of locations into a coherent representation of the state of the atmosphere, is another type of graphical model. (Some conceptual models, such as the above-mentioned food web, can also be presented graphically.) The U.S. Army Corps of Engineers operates a miniature *physical model* of the Mississippi River on 89 hectares (220 acres)* of land near Clinton, Mississippi. That model, which is also shown in Figure 1.4, enables engineers to forecast changes in the characteristics of the river's flow, including flooding, that occurs in response to changes that take place in the river's watershed.

Modeling has been greatly aided in recent decades by electronic computers, which accommodate enormous quantities of data and perform very rapid calculations. For example, L. M. Branscomb, who is a vice-president and chief scientist at IBM, observes that "The task of the scientist in the Western tra-

* Conversion factors are listed in Appendix I.

dition is to build the simplest possible model of nature and then test it against nature itself. . . . Science is a model-building process, and the computer is a model-building machine."[†] Typically, computers are programmed with *numerical models* that consist of one or more mathematical equations that describe, for example, the behavior of a particular physical or biological system. Then, the variables in the numerical model are manipulated, individually or in groups, to assess the impact on the system.

Computerized numerical models of the atmosphere have been used to forecast the weather since the 1950s. More recently, they have been used to predict the climatic effects of rising levels of carbon dioxide in the atmosphere. For example, as we see in Chapters 2 and 10, atmospheric carbon dioxide is rising, mostly because of the burning of coal as a fuel. As a result, we should expect warmer air temperatures, because carbon dioxide slows the loss of the earth's heat to space.

To determine how much warmer our air temperature will become, we must follow three steps. First, we must make a computerized numerical model of the atmosphere, which depicts the present worldwide air-temperature distribution, given the current level of atmospheric carbon dioxide. Then, holding all other variables in the model constant, we must elevate the carbon dioxide concentration, so the numerical model will compute a new worldwide temperature pattern. And finally, we must subtract the initial temperature distribution from the final temperature distribution. The net amount of warming that is calculated can be attributed to the elevated carbon dioxide concentration.

It is important to bear in mind that models are only approximations of reality, and that, as such, they are subject to error. For example, one potential difficulty with numerical models is the accuracy of their component equations. Usually, the equations are only approximations of the way that things really work in nature, and they may not account for all the relevant variables. For example, a computerized numerical model can be used to predict the change in the concentration of a water pollutant in a river as it is transported downstream. However, the resulting prediction may be unrealistic, because the model will fail to account for chemical reactions that involve, for example, the pollutant and other substances in the water that may enhance dilution.

Environmental Problems and Economics: Weighing the Costs

Earlier, we noted that environmental scientists attempt to understand ecological processes and to solve problems that adversely affect environmental stability. Although economists examine similar kinds of issues, they adopt a different approach to them. Economists start by assuming that such resources as money and raw materials are scarce, that persons act to maximize their self-interests, and that those persons want to obtain the greatest possible satisfaction for themselves. Thus one interest of economists is to determine how society's resources can be allocated most efficiently among all competing demands. For example, an *efficient economy* is considered to be operating when resources are being used in the best possible way to satisfy the largest possible number of consumer demands. An *inefficient economy* is considered to be operating when resources are not being used to their greatest advantage, so that if they were more productively employed, they would release more resources that could be used to satisfy other demands. Thus unemployment suggests an inefficient economy since labor—a resource—is not being used effectively; that is, persons who could produce something are not producing anything. Similarly, a manufacturer who uses 100 units of energy to make one product is considered to be inefficient if a technological change would make it possible to produce the same product with only 80 units of energy. However, no country's economy is wholly efficient, and all societies have opportunities to improve their economic efficiency and the use of their scarce resources.

Economists are interested in environmental issues only in terms of the extent to which they affect efficiency. Thus environmental scientists and economists might disagree about what constitutes an environmental problem. For the scientist, environmental problems ordinarily result from environmental disturbances. However, economists might not view those disturbances as problems unless they have an adverse affect on the economy. For example, al-

† Edward B. Fiske, "Computers in the Groves of Academe." *The New York Times Magazine*, May 13, 1984, p. 42.

though scientists have largely confirmed their hypotheses that acid rain damages crops, forests, and aquatic resources (such as fish), scientists can only speculate about how acid rain is formed and the precise relationship that exists between emissions of air pollution and the acid deposition that follows. As a result, an economist might argue that until those explanations are confirmed, it may be premature and, therefore, economically inefficient to spend money on uncertain remedies. Thus economists might recommend that resources be directed to more productive use, such as expenditures for new industrial equipment, which might have pollution controls that have no appreciable affect on acid rain.

In contrast to the possible disagreement about acid rain, scientists and economists are likely to reach easy agreement about the existence of problems in other areas. For example, both are likely to oppose policies that favor the rapid consumption of expensive virgin natural resources, which discourage recyclable materials, such as bottles, newsprint, and aluminum.

Once economists decide that an environmental problem exists and that an opportunity exists to improve economic efficiency, they will want to know how the persons, corporations, or industries that are responsible for the problem can be induced to change their behavior to improve efficiency. The range of possible techniques that can be used to produce change is broad, but the initial choice usually involves questions about the relative role of market forces versus government regulation.

Free Markets

Free, competitive *markets*, as many economists have asserted, are inherently efficient. In such markets, resources are allocated and prices are established on the basis of individual, voluntary exchanges among producers of goods and consumers. Thus in a free market system, persons get what they are able and willing to pay for, and markets adjust to accommodate changing desires. For example, if only a few consumers want widgets, few will be produced. And when particular goods are popular, producers will have incentives to provide more of them. According to this view, supply, demand and price tend to reach a point of equilibrium, and the production of goods and services reflects the demand for them.

Critics of this laissez-faire, or market, approach correctly note that economists have idealized the role that markets play in the efficient allocation of a society's scarce resources. For example, for market forces to operate efficiently, a large number of wise buyers and sellers must have complete information about the price, quality, and availability of all the products that are available. But no such market exists anywhere. In all capitalist societies, governments affect market forces through welfare policies, the use of progressive taxes, and controls on the supply of money. Thus we frequently use the term *mixed-market economies* to describe systems, such as in the United States, that combine private, competitive enterprise with some government involvement.

In addition, critics of the market approach contend that it fails in at least two important respects in terms of resource management and environmental protection. First, not all of the consequences of economic activities are reflected in market transactions. Such unreflected consequences are called *externalities*, or spillovers. Thus when we are advantaged by what someone else does, we reap external benefits and enjoy *positive externalities*. For example, when we landscape our lawns, paint our houses, and repave our driveways, we increase not only the value of our own property but the overall quality of the neighborhood as well. Nearby residents thus benefit, even though they have not done anything to their own homes.

Although we have no reason to complain about positive externalities, the opposite is true for *negative externalities*. Those externalities, such as pollution, odors from garbage dumps, or scarred landscapes that result from strip mining, are the undesirable side effects of activities that do not involve us directly. When we experience negative externalities, we face what economists call *external costs*; that is, we bear the negative consequences of someone else's activities. For example, steel mills produce air and water pollution, which are two externalities that can be detrimental to thousands of people. Residents in the communities that surround those mills, and not the mills' owners, bear the costs and consequences of the pollution. Therefore, damages that are caused by the pollution are not reflected in the selling price of steel, and the owners have no economic reason to stop polluting the environment. As a result, the free market system actually encourages pollution. In other instances, free markets might encourage extensive clear-cutting of forests, grazing of animals on fragile ecosystems, and the destruction of habitats of endangered species.

A second criticism of the market approach is that important distinctions can be made between private and public goods, and, private interests are usually favored. Since one of the major factors that motivates an economic exchange is profit, a producer must believe that he or she will gain a profit by providing something to a consumer. For example, if a developer builds a house, he or she can gain a profit from whomever is willing to buy or rent it (if he or she could not, it would have little or no market value). Similarly, once the house is sold or rented, one less house is available in the market and, as long as more demand exists, the developer will have reason to build another house—a *private good*. Free markets, then, encourage the production of private goods.

In contrast, free markets discourage the production of *public goods* such as national parks, flood-control projects, mosquito abatement programs, police and fire services, and environmental protection (Figure 1.5). All public goods share two common characteristics: (1) When public goods are provided to one person, everyone else has access to them at no extra cost. As a result, the provider of public goods cannot limit the consumption of those goods. And (2) the supply of the public good is not diminished as a result of being consumed or used. Under those conditions, because public goods are "free," rational consumers have no reason to pay for them and private producers in a market system have no incentive to produce them. Although we have many public goods, they are usually supplied by the government.‡ And to finance public goods, the government invariably relies on compulsory taxation.

Since market forces offer both advantages and disadvantages in terms of solving problems, few economists are in favor of relying on them exclusively to solve environmental problems. Hence, most economists acknowledge that governments should have a role in protecting the environment.

‡ Two major exceptions are radio and television signals. Do you know why private producers are willing to provide them free of charge to listeners and viewers?

Government Regulation

One major alternative to a free market system is government regulation. Instead of allowing prices and market forces to affect behavior, regulation imposes command-and-control techniques and specifies exactly what must be done. Thus if government regulation of the environment is imposed, laws will be passed that will require (1) firms to reduce their consumption of energy, (2) operators of surface mines to restore mined lands to their original contour and vegetation, and (3) the prohibition of disruption of habitats of endangered species or the discharge of cancer-causing substances into water supplies.

Government regulation is by far the most prevalent method that is used in the United States to change the behavior of those who affect environmental quality (and the well-being of consumers, investors, airline passengers, and so on). However, despite widespread reliance on regulation, a large number of opponents to regulation exist, including many economists. Those economists have more complaints than we can adequately discuss here. In brief, however, those critics maintain that government regulation discourages efficiency, innovation, and improved productivity, because it does not rely on the profit motive. According to many who are subject to it, government regulation frequently ignores the costs of compliance and the variations that exist among different regions of the country. For example, operators of stationary sources of pollution (such as coal-fired power plants and food-processing plants) frequently complain that they must meet the same pollution standards, regardless of the environment's capacity to absorb their wastes. In this case, they argue, the government causes a misallocation of scarce resources and contributes to economic inefficiency, because some factories must install equipment to address a problem that does not exist in the area where their factory is located.

Thus we should recognize that neither regulation nor a free market is a completely satisfactory way to affect the behavior of either industries or individual citizens. Since economists recognize this, they often recommend approaches that combine government intervention with some reliance on economic incentives. Those approaches can include preferential tax treatment for companies that install pollution-control equipment and subsidies for users

Figure 1.5 *Two examples of public goods. Once fire protection and national parks are available to one user, they are available to all users at no additional cost to those users. Because public goods are essentially free to those who use them, private producers have no incentive to supply them. Thus, in most cases, they must be provided by government and financed by taxation. (Illustrators Stock Photos.)*

of recycled products or for farmers who let some of their land lie fallow.

One potentially controversial approach is to have the government auction a limited number of permits to pollute within a certain region. Then, when someone purchases a permit, he or she could sell or trade it to another person or company, just as with any other product. And the value of the permits would be related to their supply and demand. That system would be similar to ones that already exist. For example, the United States government now auctions leasing rights to mine coal on federal lands and drilling rights to extract oil in offshore areas. And at some of the nation's busiest airports, the government has allowed airlines to trade or sell their rights to use particular landing gates. Still another approach that mixes government intervention with market forces would require industries to pay emission or effluent fees, which would be based on either the amount of pollution they produce or the estimated value of the damage their pollution causes. Those fees would probably be included in the prices that consumers pay, but fees have the advantage of *internalizing*, or accurately reflecting, the true costs of pollution.

Although some of these approaches are already being used, if criticism of command-and-control techniques continues, we can expect to see an increasing reliance on economic incentives to control pollution. Unfortunately, it is extremely difficult to select an appropriate method of control and, once selected, it still would not tell us how much to spend on pollution control.

Costs, Risks, and Benefits— Some Analytic Tools

Economists are interested in more than just having an affect on behavior. Since one of their goals is to improve the allocation of society's resources, they also want to know how much of a change in behavior is necessary—or how large an expenditure is desirable to control pollution—in terms of economic efficiency. To answer these questions, economists use several analytic concepts, or frameworks.

One of the most popular concepts is the *cost–benefit analysis*, in which all the gains, or benefits, of a project are compared with all the corresponding losses, or costs, of that project. If the benefits exceed the costs, a project or activity is usu-

ally deemed to be desirable and worthwhile to pursue. However, despite its apparent simplicity (or perhaps because of it), cost–benefit analysis is a controversial approach. We give more attention to this method in Chapter 11.

Cost effectiveness is a concept that is closely related to cost–benefit analysis. However, in *cost effectiveness analysis*, judgments are not made about the desirability of a project. Instead, given a particular goal, an analyst who uses cost effectiveness tries to determine how the goal can be achieved for the lowest cost. For example, if our goal is to reduce the incidence of cancer, we need to decide whether we should ban the use of all known cancer-causing substances, spend more money on research to prevent cancer, develop anticancer immunizations, or reduce persons' exposure to carcinogens. In assessing the cost effectiveness of those alternatives, we would consider social, economic, and administrative costs as well as many others.

Another approach, *risk–benefit analysis*, weighs the risks of a product or activity against its benefits to determine whether the activity should be tolerated or allowed. Many of us frequently apply risk–benefit analysis without realizing it. For example, if we choose to live or work in a notoriously smoggy city, we make a judgment that the economic or other benefits of doing so outweigh the health hazards of breathing air that is often badly polluted. Thus if we eat cured meats, such as bacon, we make a judgment that the benefits of eating bacon vastly exceed the potential risks of exposure to the preservative that is in it, which is usually sodium nitrite. And when sodium nitrite interacts with certain chemical compounds, nitrosamines result, which are potent carcinogens. On the other hand, if we eat cured meats that contain no preservatives, we increase the risk of contracting botulism (poisoning that is caused by the ingestion of toxins that are produced by certain bacteria).

Risk–benefit analysis has a certain appeal to persons who are involved in the issues that are related to environmental quality. Since most pollution-causing activities produce benefits that we enjoy on a daily basis, somehow we must judge whether those benefits outweigh the associated risks, and risk–benefit analysis can help us to make those decisions.

Marginal costs provide another useful analytic concept. In economists' terms, those costs reflect

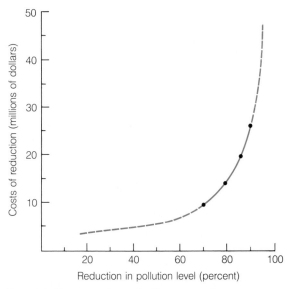

Figure 1.6 *The marginal cost of pollution control rises ever more steeply with each additional increment of control. In this example, a control level of 70 percent would cost $10 million, or approximately $143,000 for each percent of polluting substance that is removed. Increasing the control level to 80 percent would cost another $5 million, or $500,000 for each additional percent removed. And removing just 5 percent more would cost still another $5 million, or $1 million for each additional percent removed. The shape of the curve varies with actual conditions, but a steeply rising slope is typical of curves that plot the marginal costs of pollution control.*

the additional costs of producing an extra unit of output or one less unit of pollution. Thus marginal costs are especially important in the control of pollution. For example, the first increments of pollution are the cheapest and easiest to control. However, as an industry or municipality seeks to reduce further increments of pollution, the costs of control rise faster than the amount of pollution is reduced. Thus an industry might find that 70 percent of its emissions can be reduced at a cost of $10 million (Figure 1.6). However, to reduce 80 percent, it would cost $15 million; and to reduce 85 percent, it would cost another $5 million. Thus the industry has high marginal costs—the costs of reducing the last increments of pollution are significantly higher than the costs that are associated with the first increments.

Marginal costs can be an important consideration in deciding how much pollution should be controlled. For example, does it make economic sense for a government to require industries or municipalities to spend the last $5 million to reduce 5 percent of the emissions when another industry could spend the same amount of money to control 40 percent of its pollution? Consideration of marginal costs and the use of cost–benefit or risk–benefit analysis can help to provide an answer to that question.

Although economists employ many other concepts and analytic tools, the ones we have discussed provide an initial framework for understanding how economists view environmental issues. Thus if one theme characterizes this view, it is that decisions about the environment should be made rationally, with the goal of making the best use of our available resources. How likely we are to reach this goal depends on (1) our choice of decision-making processes, and (2) how responsive our political system is to both scientific and economic concerns about environmental problems.

Environmental Problems and Public Policies: Deciding What Actually Gets Done

When scientists discuss possible solutions to the problems of pollution, the exploitation of natural resources, or the problems that are associated with a rapidly growing human population, they frequently assert that governments must initiate new environmental control programs. Those calls for action reflect a belief that only governments have the authority and financial resources to develop and implement successful policies to protect the environment. Perhaps more importantly, however, such statements also tell us that environmental problems are issues of public policy. (For our purposes, we can say that *public policies* are those things that governments do or decide not to do.) Furthermore, those persons who request more government programs believe that the government's role in solving environmental problems is preeminent; that is, if our policies and agencies were more effective, then we would not have the problems (in such serious form) that we have today.

Given that situation, the government must have a central role in responding to environmental problems. Similarly, because environmental quality depends on what governments do, it is important for us to understand how public policymakers usually perceive those problems and make decisions concerning them. However, before discussing those topics, we should consider, from an environmental scientist's perspective, how we expect governments to act when they address the problems of pollution or allocate such resources as water, fuels, or minerals. Fortunately, even an elementary knowledge of environmental science provides us with several ways to assess a government's performance in terms of managing the environment.

First, if we want to reverse the processes of environmental degradation that are occurring, then public policies must address the root causes of pollution. And how policymakers view those root causes will affect the public policies they select. For example, if policymakers believe that pollution is caused by persons who litter or factories that grossly pollute the air and water, then only moderate changes in existing public policies are probably needed. For example, we can strengthen laws against littering or mandate the use of technological controls. In contrast, an alternative explanation might suggest that the root causes of pollution and resource exploitation can be traced to generally accepted patterns of consumption and the unwise use of finite natural resources. That explanation places the blame for pollution on a society that expects high and ever-increasing standards of living. If policymakers accept that explanation as being the cause of pollution, then fundamental shifts in existing public policies are probably necessary.

Second, we know that all parts of an ecosystem are interrelated, so if one part of an ecosystem is disrupted, other parts also will be affected. Some of those effects will be noticed immediately, but others will only become apparent gradually, after several years. Therefore, given that scientific principle, public policies should be formulated that reflect a holistic perspective, that is, considering a problem in its entirety. Thus at the very least, government officials should be required to consider the short- and long-term implications of whatever policy choices exist in terms of our resources and the environment.

Third, the need for a holistic perspective also suggests that different government agencies should thoroughly integrate their activities concerning resources and the environment to avoid contradictory policies.

Although these expectations seem to be reasonable, it is extremely difficult for the government to accommodate them. Policymakers are subject to constant and competing pressures from groups and individual citizens that have different values and goals. And it is impossible for a government to satisfy all groups, so public policies inevitably benefit some groups at the expense of others. For example, when a government allocates scarce water supplies in the arid parts of the western United States, some groups do not receive as much as they want and believe they should get. Groups that have certain advantages want to maintain their privileged position and, therefore, resist efforts to change the status quo, even though such a change might encourage economic efficiency. As a result, those groups that want a change in the status quo must persuade policymakers that change is necessary and that existing policies are unsatisfactory. First, however, policymakers must be persuaded that a problem exists and that the government has a responsibility to provide a solution.

Selecting Environmental Problems

Policymakers rarely employ a rational, comprehensive approach to problem solving. Thus even if most scientists agree that an environmental problem exists and that it should be remedied, that consensus does not necessarily extend to policymakers, who include elected officials and senior bureaucrats in the government's administrative agencies. Unless those persons view a situation as being a public problem, it will not receive their attention. Furthermore, once a problem has been defined, it does not always receive their attention, and not everything that receives attention produces a response. These points are important because they show us how issues can be misplaced in the political arena, which invariably affects the outcome.

Even if an environmental issue passes the first hurdle and is defined as being a public problem, it must still compete for a place on the government's agenda. Although many public problems are potential candidates for attention, policymakers focus on only a small number of them. Indeed, their agenda is limited; it is overloaded with problems that already require their full attention; and it reflects a preference for old, familiar, and recurring issues. For example, every year we can expect members of Congress to discuss taxes, the budget, the national debt, national defense, health care, interest rates, economic growth, unemployment, and foreign policy.

Therefore, any new problem that gains a position on the government's overloaded agenda must replace a problem that another group has already justified, which also deserves the government's attention. Then, if the agenda is changed, those other groups must lose their advantageous position, which explains why some groups oppose any attempts to change agendas and why changes are made very slowly.

Nevertheless, despite this opposition, changes in agendas do occur. For example, when hundreds of abandoned hazardous-waste sites were discovered throughout the United States in the late 1970s, the issue was placed quickly before state and federal officials (Figure 1.7). Thus the chances of having a new item added to an agenda are increased if the new problem is relatively easy to understand, can be decided by a yes or no response, or is the result of a crisis or catastrophe. New problems are also likely to gain status if their solution is not likely to be expensive, if a sense of urgency accompanies them, or if they attract widespread attention and produce widespread demands for change. Problems that do not share at least several of those attributes are unlikely to receive consideration. As a result, policymakers tend to delay or neglect consideration of new problems that are complex, are not easily understood, have uncertain short-term consequences, and are potentially costly to solve. Yet those features are exactly the ones that characterize many environmental problems.

Thus if a new issue gains recognition as a legitimate public problem, no guarantee exists that action will follow. For example, the prevention of cancer has been on the government's agenda for a long time. However, in 1977, the National Institute of Occupational Safety and Health acknowledged that it had not informed nearly 75,000 American workers, whose names and addresses had been collected, that they might have been exposed to cancer-causing substances. If they had been notified, many of those workers might have been saved or, at least, been given life-prolonging treatment. Despite that

Figure 1.7 *A hazardous-waste site in Niagara Falls, New York (Love Canal). Because the problem of leaking hazardous-waste sites was easily understood by an aroused public (which saw it as potentially catastrophic), this problem was added to the national political agenda in the 1970s. (Andy Levin, Black Star.)*

potential, the agency's director said that the workers had not been told because the agency did not have the funds or the legal authority to do so. If those workers had been informed of the hazards that they were exposed to and had been organized, the outcome might have been different. Organized groups almost always have an advantage in the political process over individual persons who share common concerns but who remain unorganized.

Responding to Environmental Problems

Underlying Political Considerations After a government decides that a problem requires a response, what usually happens? Earlier, we suggested that policymakers should adopt a holistic perspective in responding to problems that are related to resources and the environment. However, although that approach might be desirable, policymakers rarely employ it. Instead, policymakers—especially elected officials (such as governors, presidents, and state and federal legislators)—often make decisions on the basis of factors that are not related to the issue at hand, such as the environment. Many legislators vote for policies that reinforce their political ties (whether they are Democrats or Republicans or whether their vote will embarrass their opponents or be to their party's partisan advantage), not according to scientific hypotheses or economic concepts, such as efficiency or cost effectiveness.

However, other factors besides partisan advantage also affect policymakers' responses to environmental issues. Of those factors, two of the most important are elections and the basis of political representation in the United States. For example, most state legislators and all members of the United States House of Representatives face reelection campaigns every other year. Presidents have 4-year terms, as do most governors. Only United States senators have 6-year terms. For those politicians to be reelected, their short terms mean that they must continually demonstrate their accomplishments in a manner that quickly reflects favor on them. For many legislators, that requirement encourages them to support proposed laws that promise quick results. And, as a general rule, politicians universally favor programs that distribute benefits to the current generation of voters. In contrast, programs that might take many years to implement or that promise bene-

fits for future generations tend to be viewed as being much less desirable.

The basis of political representation also influences the kinds of policies that legislators support. For example, all state and federal legislators are elected to represent specific geographical areas, some of which are no more than a few square kilometers in size. And to be reelected, a legislator only needs to satisfy a majority of voters in his or her electoral district. As a result, he or she can safely ignore the concerns of all other districts—or even the concerns of the country as a whole. Therefore, in responding to public problems, a legislator often makes choices only in terms of the potential implications for his or her constituents, such as whether the proposal will increase the amount of money that will be distributed, the amount of construction that will be started in his or her district, or whether the proposal will create new jobs for some or cause others to lose them. These considerations are important, because voters often evaluate their senators and representatives on the basis of how much federal money those officials have procured for their state or district.

The consequences of narrowly based geographical representation should not be underestimated, because it encourages a parochial view of environmental problems at the expense of a regional, national, or international view. Indeed, many legislators are likely to support programs that will help their constituents and oppose programs that will not. For example, in voting on the Clean Air Act Amendments of 1977, many United States senators from midwestern states, which have large deposits of high-sulfur coal, consistently opposed provisions that would reduce the use of that coal.

Presidents also want to be reelected, and so they too must demonstrate quick successes, perhaps by introducing new ways to deal with existing problems. Moreover, presidents are traditionally assumed to have responsibility for the nation's economic health. Thus instead of promoting conservation and deferred consumption, presidential policies almost always favor a booming economy. If presidents reach that goal, then few persons notice how it was achieved. Presidents enjoy seeing increases in the gross national product (GNP), which measures the total value of all the goods and services that are produced. In computing the GNP, however, no distinction is made between environmentally sound and environmentally harmful activities, as long as they both contribute to economic growth. Presidents can, then, gain much public attention by manipulating the economy successfully. In contrast, they rarely gain public attention by exercising their administrative talents to ensure that existing programs are implemented properly. Thus presidents also realize that few political rewards exist for ensuring that laws actually achieve their goals.

Incremental Decision Making Now that we have some idea of the variables that affect policymakers, we can turn to the process that probably best describes the formulation of public policies—*incremental decision making*. The major assumption of incremental decision making is that existing approaches to public problems are satisfactory and preferred. Thus if problems arise with the implementation of the present approaches to problems, only small, or incremental, changes are needed. As a result, although many laws and public policies are claimed to be new or innovative, most are actually no more than revisions of existing practices.

For busy policymakers, incremental decision making has several appealing features. First, it assumes that the best guide to the future can be found in what is already being done, especially because existing policies are the products of political majorities, democratic political processes, and past decisions of agency officials. Second, the incremental approach is easy to use because it makes few demands on policymakers. As a rule, alternative solutions to problems are simply never examined, and those that are considered are usually limited to slight variations in the status quo. (As we noted earlier, for legislators, those variations tend to be assessed in terms of their short-range consequences on particular regions or electoral districts.) Third, when new problems arise, public officials usually conclude that existing approaches to similar problems can be applied. Such policymakers reject the belief that new problems require solutions that are different from existing ones.

Fourth, in contrast to decision-making processes that produce the most effective policy regardless of the public's reaction to it, incremental decision making allows elected officials to be responsive to public opinion and participation and to engage in bargaining and compromise. By compromising, the legislators can respond to competing demands and develop policies that are the most acceptable to as many well-organized groups as possible. Thus many legislators satisfy groups who voice competing demands by voting for vague or ambiguous laws. For example, in the Endangered Species Act, the United States government pledges to conserve "to the extent practicable" species that are facing extinction. And one goal of the Clean Air Act amendments is the protection and enhancement of the nation's air resources in order "to promote the public health and welfare and the productive capacity of its population."

As a result of the reluctance or inability of legislators to be more specific about their ultimate goals, problems are frequently created for officials in executive-branch agencies, who must interpret and implement the laws on a daily basis. Those officials are expected to develop precise regulations and guidelines without knowing precisely what the goals are that they are trying to achieve or their relative priorities. Should those officials interpret some environmental laws literally and possibly close down entire industries? Or should they adopt a more conservative approach to the control of pollution to protect jobs? Should they seek penalties from all those who violate environmental laws? Or should they focus their resources on the worst violators? In making these judgments, agency officials find that they are acting as policymakers in the circumstances for which public officials are elected.

For example, the congressional consideration of the Endangered Species Act shows how incremental decision making dominates the process of policymaking. That act was designed to protect endangered species, and it was passed in 1973. Then, subsequent amendments were passed in 1976, 1977, 1978, 1979, 1980, and 1982, which modified the original law. Similar examples can be provided for many other environmental laws as well. Thus instead of a series of well-coordinated environmental laws that reflect an integrated perspective, we have literally dozens of relatively uncoordinated federal environmental laws and hundreds of equally uncoordinated state laws. The fact that we have so many laws means that each resource or environmental problem is often treated as if it has little or no relationship to other environmental problems. Therefore, from a holistic perspective, those laws ignore the scientific principle that each part of the environment is interrelated and cannot be considered in isolation.

Why Environmental Policymaking Is Fragmented

Some understanding of how the Congress is organized helps to explain why we lack an integrated set of environmental laws. Briefly, whenever a proposed law is introduced to the United States Senate or House of Representatives, it is referred to a committee for consideration. And no single committee exists in either the House or the Senate that has complete responsibility for all environmental issues. In fact, approximately ten committees in each chamber handle these issues. In the Senate, proposals that concern the oceans or toxic chemicals may be sent to either the Committee on Environment and Public Works or the Committee on Commerce, Science, and Transportation. And despite their potential environmental implications, proposed laws that concern synthetic fuels, oil shale, and strip mining are not referred to the Committee on Environment and Public Works but, rather, to the Committee on Energy and Natural Resources. Moreover, since each chamber establishes its own committee structure, different committees in each chamber often consider the same environmental issues. For example, the Senate's Committee on the Environment and Public Works considers air-pollution legislation but, in the House that task is delegated to the Committee on Energy and Commerce.

Thus it should be evident that the committee structure in the Congress (and, likewise, in most state legislatures) fragments decision making and encourages a limited view of environmental problems. Although improved coordination among committees would be desirable, each committee jealously guards its own jurisdictions. And since the full Senate and House rarely overturn substantive recommendations of their committees, fragmentation occurs. Furthermore, both the Senate and the House have delegated the job of writing environmental laws to certain committees, but they allow entirely separate appropriations committees to decide how much money can be allocated to implement those laws, which causes even more fragmentation.

Given this situation, we might find it remarkable that any environmental laws emerge from the Congress. However, when they do, they consistently share several of the following features. We have already noted that lawmakers prefer incremental laws, which focus on one environmental problem at a time.

In addition, most federal environmental laws reflect a strong emphasis on regulatory as opposed to economic approaches and on engineering and technological solutions. As a result, our laws favor the development and use of such technology as catalytic converters for automobiles, scrubbers and electrostatic precipitators for power plants, high-temperature incinerators for toxic chemicals, and tertiary treatment plants for sewage. However well those controls may work, laws that require technological solutions to environmental problems shift the responsibility for pollution control to manufacturers and industrial concerns. And that type of reliance on such technological "fixes," or solutions, reflects an assumption that changes in individual or group behavior are largely unnecessary.

Environmental laws also reveal a mix of science, politics, and economics, but not necessarily in equal doses. Although science can show what is technically possible in any given scheme to reduce pollution, politics and economics temper what the laws actually require of polluters. For example, scientists can provide information about the harmful effects of different levels of pollution, but government officials must decide which of those levels (and which levels of pollution control) are acceptable. Since legal acceptability has no scientific meaning, those decisions invariably incorporate both political and economic considerations.

Furthermore, our environmental laws are subject to the same processes as all other laws. And opportunities are provided for dissatisfied groups or persons to challenge those laws in state or federal courts. For example, those persons who do not like the effects of a given law can argue that it is unconstitutional, that the agency responsible for implementing it has misinterpreted the law, or that it has exceeded its legal authority in administering the law. And to compound the problem, judges have become increasingly responsive to such complaints. For example, for many years, judges have deferred to the expertise of government agencies and have rarely overruled their decisions. However, during the last two decades, many judges have adopted an activist approach, and they have not hesitated to intervene in environmental decision making. Although that activism is well-intentioned, not everyone agrees that it should be encouraged. In many cases, judicial decisions have increased the effectiveness of several environmental laws and have facilitated citizens' access to the courts. At the same time, however, critics of judicial activism complain that many judges consistently ignore the political and economic implications of their decisions. In addition, although judges are expected to reach unbiased decisions, some evidence has become apparent which indicates that some federal trial judges are sympathetic to local economic and industrial concerns and often are reluctant to impose penalties on industries or municipalities that are persistent polluters.

However, polluters are not the only groups that initiate increased judicial activism. Frustrated with what they consider to be foot-dragging, environmental groups have frequently asked judges to force agencies to issue regulations or pollution standards that are required by law. For example, the Clean Air Act Amendments of 1970 require the Environmental Protection Agency (EPA) to issue standards for *hazardous air pollutants* (see Chapter 11). Through early 1984, however, the EPA had not issued standards for those pollutants unless they were challenged legally. Part of the explanation for the EPA's reluctance to act is that when Congress approved the law, it did not authorize the EPA to consider either marginal costs or cost–benefit analysis when it set standards for hazardous air pollutants.

Why Environmental Policy Implementation Is Fragmented

Our discussion of government policymakers indicates that they do not use a holistic perspective in their decision making. Furthermore, little indication exists that the government agencies that actually implement our resource and environmental laws even consider a holistic perspective.

If the United States government used a holistic perspective, it would coordinate its programs and policies that are intended to protect the environment and our natural resources. Although we can agree that such coordination would be desirable, the distribution of responsibility among federal agencies discourages that type of coordination—too many agencies share responsibility for the allocation of natural resources. And, as Figure 1.8 suggests, within the federal government's executive branch, the job of managing environmental protection activities is widely scattered and fragmented. For example, although the EPA has programs for pesticides, radiation, drinking water, solid wastes, toxic substances,

and air and water pollution, the agency must share responsibility for many of those programs with other state and federal agencies. One study, which was conducted in the mid-1970s, found more than 40 instances in which the EPA shared authority for environmental programs with other federal departments.

For those other departments and agencies, concern for the environment is often less important than their primary goals. For example, while the EPA is attempting to reduce the burning of coal, which contributes to air pollution, the Department of Energy is encouraging the use of coal to reduce our dependence on imported petroleum. Similarly, the Department of Agriculture has defended the use of some pesticides at the same time that the EPA has tried to discourage their use.

Although other agencies may not be pursuing environmentally harmful policies, they are expected to balance the allocation or development of resources with environmental protection. But how effectively they achieve that balance is often open to question. For example, during the first two years of the Reagan administration, many persons claimed that Secretary of the Interior James Watt favored the development of mineral resources on certain federally owned lands to the detriment of his department's other responsibilities. Those responsibilities include programs for the reclamation of surface-mined lands and the protection of endangered species, national parks, monuments, and wildlife refuges.

Still other federal agencies are expected to promote urban growth, the dredging of rivers and harbors, the filling of swamps and wetlands, the construction of dams, highways, and power plants, and, simultaneously, to minimize the environmental disruptions of their activities. To help ensure that these goals are reached, the federal government and many states require the preparation and publication of *environmental impact statements* (EISs). At the national level, those statements must be issued for all major federal actions "significantly affecting the quality of the human environment." Thus environmental impact statements must discuss the environmental effects of proposed actions, the environmental impacts that cannot be avoided if a project is completed, and alternatives to a proposed action.

The requirement that environmental impact statements be prepared has reduced the adverse environmental impact of many federal projects, but

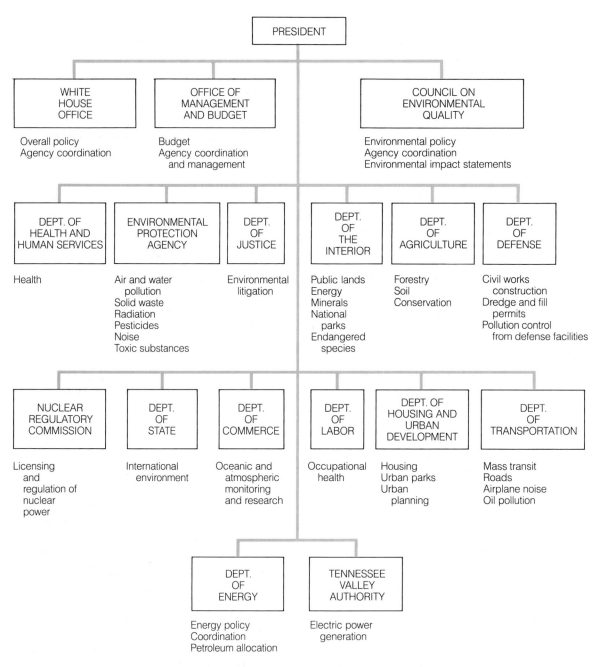

Figure 1.8 *The major agencies of the federal government's executive branch that have environmental responsibilities. (Slightly modified from the U.S. General's Accounting Office,* Environmental Protection: Agenda for the 1980s. *Washington, D.C.: U.S. Government Printing Office, 1982.)*

that process has experienced problems. For example, once a statement is published in draft form, interested groups and individual citizens can submit comments to the agency that is responsible for the preparation of the statement. In addition, for many environmental impact statements, the EPA is required to submit comments, but that task can be overwhelming. In a typical year, federal agencies prepare more than 1000 statements, which is far more than the EPA can review adequately. Then, the agency that prepared the statement must respond to any comments that are made, but it is not obligated to alter a project as a result of the comments. Also, environmental impact statements are subject to criticism, because they have led to litigation and delays in some projects, such as construction.

For other activities that do not require environmental impact statements, the EPA cannot do much to ensure that its sister agencies are sensitive to environmental values. For example, the EPA can express concern about practices that are detrimental to the environment and it can attempt to negotiate changes. And if those efforts fail, in theory, it can initiate legal action against an uncooperative agency. However, in reality, that rarely occurs. Since the Department of Justice handles environmental litigation for the federal government, the EPA must first convince that department that legal action is necessary. Furthermore, presidents actively discourage such legal action, because they want to avoid the open and potentially embarrassing conflict that would result if the EPA sued another government agency in court. Thus the lack of effective sanctions surely discourages any coordinating role that the EPA might wish to play.

The EPA has problems with cooperation and coordination not only with other federal departments but with the states as well. Most laws that are administererd by the EPA at the federal level state that the authority for actual implementation must be delegated to the individual states. For example, the EPA develops guidelines and regulations for laws concerning pesticides, solid wastes, drinking water, and air and water pollution, but the states apply and enforce those regulations. Thus if all states shared the same high-level concern for the environment, few problems would arise with this arrangement. However, with a few exceptions, the states' commitment to environmental protection historically has been limited. Some state governments sharply disagree with the intent of many federal laws and openly favor industrial growth at the expense of environmental protection. Other states compete with each other for new industrial development or are unable or unwilling to bear the costs of effective control programs. And still other states are reluctant to penalize industries or municipalities that cause pollu-

tion for fear that industrial plants will be closed, which possibly would cause the loss of hundreds of jobs.

Finally, we should note that the EPA is frequently criticized for its activities and, occasionally, for its leadership. For example, during the energy crises of the 1970s, many persons blamed the EPA's regulations for the energy shortages. And in his presidential campaign in 1980, Ronald Reagan complained that the EPA was not concerned with the economic plight of the coal and steel industries. Then, when Anne Gorsuch Burford headed the EPA (from 1981 to 1983), she allegedly filled many key positions in the agency with political appointees of questionable competence; she was accused of failing to enforce environmental laws; and she supposedly distributed funds to clean up abandoned hazardous-waste dumps on the basis of political considerations. Her refusal to cooperate with congressional committees that investigated those charges eventually led to her resignation. Although the consequences of the adversities that the EPA must face are not at all clear, we can suggest that any agency that experiences such a constant stream of both internal and external problems will be hesitant to regulate and will be much less assertive than it might otherwise be.

What Does All This Mean?

It is apparent that our government has not fared well in terms of our evaluation of its response to environmental problems. We might be disappointed by this fact, but we should not be surprised. In their efforts to be responsive to groups that have different values, all democratic governments attempt to achieve diverse and frequently competing goals. Concern for environmental quality and protection of our natural resources are only two of those goals, and they must compete with efforts to ensure a strong economy, an adequate national defense, sufficient energy supplies, a healthy and adequately fed population, appropriate care for the elderly and disabled, and so on (Figure 1.9). Few persons would disagree that these goals are reasonable or that the government should abandon them in favor of an exclusive concern for the environment. Moreover, in a democratic system of government, a close (but not exact) relationship usually exists between what persons

Figure 1.9 *Coal is cheap and plentiful in the United States, but it creates a great deal of air pollution. Should we use more coal or less? The question points up the conflict between two competing goals: the environmental advantages of cleaner air versus the economic advantages of cheap, abundant energy. A democratic government frequently must attempt to reconcile competing goals in its efforts to be responsive to citizen groups that have different values. (Chuck Rogers, Black Star.)*

want and what governments provide. As a result, the government's degree of attention to the environment is merely a reflection of public preferences and priorities. Viewed from another perspective, although many competing demands exist for public goods, not all those demands can be satisfied.

At least two other considerations should also temper our possible disappointment with the government's handling of environmental issues. First, we should understand that governments are expected to resolve society's most complex and difficult problems. As a result, governments find them-

selves saddled with issues that are highly controversial and often do not have any clear solutions. The debate over acid rain in the early 1980s provides a premier example. Except for an extraordinary program of energy conservation, all proposed solutions are likely to cost billions of dollars or involve thousands of decisions about appropriate marginal costs. Thus in the absence of indisputable data, which indicate exactly what can be done about acid rain, policymakers will be criticized for whatever they do or do not accomplish.

Second, whatever their shortcomings may be, our environmental laws and agencies have produced measurable improvements in the quality of our environment. As the chapters that follow demonstrate, progress can be seen in many areas. For example, our lakes, rivers, and streams are less polluted than in the past; it is easier to breathe in most metropolitan areas; and we have begun to clean up thousands of abandoned hazardous-waste dumps. Also, our endangered species program is one of the best in the world and, on a per-capita basis, few other nations spend as much as we do to protect the environment.

Our accomplishments may be impressive, but merely listing them can obscure some important questions. Therefore, as you read this book, ask yourself whether public policies in the United States and elsewhere are sufficiently responsive to the problems that exist concerning our resources and the environment. Also, try to determine how governments should respond if these problems get worse in the future. Think about whether we need more government, less government, or perhaps the same level of government but with different policies and priorities.

Unfortunately, ready answers to these questions do not exist. Some persons fear that our future environmental problems will be so severe that we will not be able to survive without vast increases in government authority. If those persons are correct, our civil liberties inevitably will be diminished, our standard of living will be forcibly reduced, and our sexual practices will be regulated. For example, Garrett Hardin, a biologist, suggests that the only way to limit population growth is to create a system of government-enforced polyandry, in which one woman would have several husbands at a time.

In contrast to those who anticipate the need for increased government authority, others contend that we already have too much government, that it creates unnecessary intrusions in our lives, and that public policies cause, rather than solve, environmental problems. Advocates of this view also believe that without continued economic growth, we cannot sustain or improve the world's standard of living or continue to pay for environmental protection.

Although questions remain concerning what the appropriate role of government should be, we can be certain that our behavior as individuals will affect the answers. If our life-styles and those of our parents have contributed to government as we know it today, then just as surely what we and our children do in the future will ultimately determine what governments can, must, or should do in response. You may want to consider this also as you read the rest of this textbook.

Conclusions

Environmental science is a problem-oriented, interdisciplinary field of study. Although environmental science primarily involves physical and biological processes that operate in the environment, we have seen that environmental issues have important economic and political dimensions as well. We have learned that natural scientists, economists, and policymakers differ in important ways in terms of their approaches to environmental issues. Scientists from many disciplines apply the scientific method in their attempts to understand the natural functioning of ecosystems and the causes of environmental problems. Economists view environmental issues in the context of economic efficiency and apply a variety of analytic tools in assessing the costs of dealing with environmental problems. And elected officials, government agency bureaucrats, and other public officials respond to pressures from special-interest groups and citizens in shaping public policy on the environment.

Summary Statements

The environment is the assemblage of external factors or conditions that influences living things in any way. Environmental science is an interdisciplinary endeavor that is concerned with both the natural functioning of ecosystems and the problems that arise from disturbances of environmental processes.

Our understanding of the environment is enhanced by the scientific method, which is a systematic form of inquiry that involves observation, speculation, and reasoning. The scientific method may include the use of models that reduce a complex situation to its essential components.

Scientists and economists do not always agree on what constitutes an environmental problem. In the economist's perspective, the disturbance of an ecosystem may be a problem only if that disturbance adversely effects economic efficiency.

Environmental protection and resource management are not encouraged by the free market system. Hence government regulation is the most prevalent method of dealing with environmental problems. Some economists argue for approaches that combine government intervention with some reliance on economic incentives.

Economists employ a variety of analytic tools (such as cost–benefit analysis) in assessing the costs of environmental protection in the context of economic efficiency.

Ideally, effective public policies on environmental issues must address the basic causes of environmental problems, reflect a holistic perspective, and be internally consistent.

Effective public policy on the environment must address the basic causes of pollution, and how policymakers view those causes will affect the public policies that they select.

Because the various components of the environment are interdependent, public policy on the environment should reflect a holistic perspective. Hence relevant government agencies should integrate their activities concerning the environment to avoid contradictory policies.

For action by policymakers, a problem first must be identified as a public problem and then it must be placed on the government's agenda. This is usually a slow process because of the many diverse and often competing issues that vie for the attention of policymakers.

Public policy is usually formulated through the process of incremental decision making, which assumes that existing approaches to public problems are satisfactory and only in need of small, or incremental, changes.

Public policymaking is fragmented because the legislative committee structure fragments decision making and encourages a limited view of environmental problems.

Environmental policy implementation is fragmented because of government's failure to adopt a holistic perspective on resources and environmental problems.

Selected Readings

BADEN, J., AND STROUP, R. C., (EDS.): *Bureaucracy vs. Environment.* Ann Arbor: University of Michigan, 1981. This book argues that government actions are more harmful than helpful to environmental quality.

COBB, R. W., AND ELDER, C. D.: *Participation in American Politics,* 2d ed. Baltimore: Johns Hopkins, 1983. An excellent study of how issues become problems and get on the government's agenda.

DOWNING, P. B.: *Environmental Economics and Policy.* Boston: Little, Brown, 1984. A useful introduction to environmental economics.

DYE, T. R.: *Understanding Public Policy,* 5th ed. Englewood Cliffs, N.J.: Prentice-Hall, 1984. A useful introduction to how governments handle public problems.

LAVE, L. B.: *The Strategy of Social Regulation: Decision Frameworks for Policy.* Washington, D.C.: Brookings, 1981. A leading economist discusses a different way to regulate.

OPHULS, W.: *Ecology and the Politics of Scarcity.* New York: W. H. Freeman, 1977. The best extant argument for increasing the government's authority to cope with resource and environmental problems.

WALKER, M.: *The Nature of Scientific Thought.* Englewood Cliffs, N.J.: Prentice-Hall, Spectrum Book, 1963. This book is concerned with the basic purpose (prediction of events) and procedure (construction and use of conceptual models) that are common to all the sciences.

WENNER, L. McS.: *One Environment Under Law.* Pacific Palisades, Calif.: Goodyear, 1976. A well-written introduction to the policies and politics of environmental protection.

P A R T   I

CONCEPTS OF ECOLOGY

To find practical answers to environmental problems, we must first understand how the environment works. In this part of the book, we consider some of the fundamental principles that govern nature's activities. In Chapter 2, we look at the flow of energy through food webs and the movement of materials through the environment. An understanding of these flows is essential to the solution of such problems as human hunger and the decline in the quality of our air and water. Because the environment is constantly changing, either naturally or as a consequence of human activities, we consider in Chapter 3 the response of organisms and ecosystems to such factors as water and air pollution, fire, weather, and agriculture. In Chapter 4, we look at how populations grow and at what controls their growth. Only by understanding population growth can we control the pests that attack our crops and livestock, save endangered species, and identify options to limit human population growth. In Chapter 5, we consider the characteristics of the diverse ecosystems of the earth and the points at which these ecosystems are vulnerable to human activity.

In succeeding sections of this book, our understanding of the basic principles that govern the working of the environment will help us to better understand the environmental consequences of human activities. As our increasing numbers and activities continue to have an impact on the earth, a better understanding of ecological principles becomes ever more important to our well-being.

This fox, carrying its prey, vividly illustrates one step in the flow of energy through an ecosystem. (Wilford L. Miller, National Audubon Society, Photo Researchers.)

CHAPTER 2

ECOSYSTEMS: THE FLOW OF ENERGY AND MATERIALS

In November of 1982, when scientists landed on Christmas Island, they could tell immediately that something was drastically wrong. Expecting to find hundreds of thousands of seabirds nesting on that large coral atoll in the central Pacific Ocean, instead, they discovered that the adult birds had deserted the island, leaving their young behind to die of starvation. When the scientists had visited the island in June of 1982, 8000 great frigate birds were on the island. Now, however, fewer than 100 adults remained on the island, and only six nestlings were there, all of which were starving. And those seabirds only represented one species. What had caused the adult birds to abandon their young and leave the island?

Scientists have good evidence that the cause of the problem on Christmas Island was El Niño, which is a natural phenomenon that occurs every 2 to 9 years on an irregular and unpredictable basis. El Niño is a warming of the surface waters in the tropical eastern Pacific, which causes fish to disperse to cooler waters and, in turn, causes the adult birds to fly off in search of new food sources elsewhere. Although that phenomenon resulted in complete breeding failure in 1982, the relatively long-lived adults were able to leave in search of available food for survival.

Through a complex web of events, El Niño (which means "the child" in Spanish because it usually occurs during the Christmas season off the coasts of Peru and Ecuador) can have a devastating impact on all forms of marine life. One of the first signs of

its appearance is a shifting of winds along the equator in the Pacific Ocean. The normal easterly winds reverse direction and drag a large mass of warm water eastward toward the South American coastline. And that top layer of warm water prevents the upwelling of nutrient-rich cold water from the ocean bottom to the surface. As a result, the growth of microscopic algae that normally flourish in the nutrient-rich upwelling areas diminishes sharply, and that decrease has further repercussions. For example, in 1983, the 20-fold decrease in algae that occurred along the South American coast reduced the number of anchovies (who feed heavily on the algae) to a record low level. Also, other fish, such as the jack mackerel, which feed on tiny animals, which, in turn, feed on the algae, also decreased in number as the algae population declined. Thus when their food resources disappeared, marine birds (such as frigate birds and terns) and marine mammals (such as fur seals and sea lions) also suffered large population declines as a result of breeding failures and the death of adult birds.

The 1982–1983 El Niño well illustrates the interdependence of organisms, their vulnerability to environmental change, and the linkages among components of the environment along which energy and materials flow. The wind and ocean currents reduced the amount of nutrients that were available in the surface waters, so that algae could not convert the sun's energy into food energy for growth and reproduction. In turn, fish, birds, and mammals could not reproduce successfully because they did not have the energy and nutrients that are supplied by the algae.

Human activities, such as the clearing of forests, draining of wetlands, construction of dams and highways, and the release of contaminants also disrupt the flow of energy and materials through the environment. Such movement is governed by certain principles that we frequently forget to consider. In our disregard of those principles, we simply try to make nature do more than it can do.

In this chapter, we examine the patterns of energy and material flow through the environment. We also consider the principles that govern such movements and see how those principles limit our abilities to exploit the environment as well as influence the disruptive impact that our activities often have on it.

Components of Ecosystems

The flow of materials and energy within the environment can be understood by examining the functioning of ecosystems. An *ecosystem* is defined as a functional unit of the environment, which comprises the interaction of all organisms and physical components within a given geographical area. Thus an ecosystem consists of both living *(biotic)* and nonliving *(abiotic)* components. The biotic community is composed of organisms (plants, animals, and microorganisms), each of which is considered to be either a producer or a consumer of food, or energy. By a process that is called *photosynthesis*, green plants manufacture their own food from existing supplies of water and carbon dioxide. They use sunlight as their source of energy to power photosynthesis. Therefore, plants are considered to be *producers*. Animals and most microorganisms cannot produce their own food. They must consume other organisms for energy and nutrition. Therefore, animals and most microorganisms are considered to be *consumers*.

Consumers fall into one of four classes on the basis of their food sources—herbivores, carnivores, omnivores, and detritus feeders. A consumer that eats only plants is called a *herbivore*, whereas a consumer that eats only animals is called a *carnivore*. A consumer that eats both plants and animals is called an *omnivore*. Thus a pheasant that feeds on

corn is a herbivore; a hawk that consumes a pheasant is a carnivore; and a human being who eats both corn and pheasants is an omnivore. *Detritus feeders*, or *decomposers*, include such organisms as bacteria, fungi (for example, molds and mushrooms), earthworms, termites, maggots, and clams. They feed on *detritus*—the freshly dead or partially decomposed remains of plants and animals—and acquire needed energy and nutrients from it. During that process, simpler decomposition products, including the decomposers' own wastes, are returned to the soil, air, and water, where they can be taken up again by green plants and recycled through the ecosystem.

The physical environment comprises the abiotic parts of an ecosystem. Those components include chemical substances that can be subdivided into two categories—*inorganic* and *organic substances*. Inorganic chemicals include water, oxygen, carbon dioxide, and minerals. Most organic substances are produced by organisms and include carbohydrates, fats, proteins, and vitamins. Besides chemical substances, abiotic components include such physical factors as heat, light, and wind, all of which are forms of energy.

We commonly picture an ecosystem as being a relatively undisturbed geographical area such as a desert, prairie, forest, lake, or stream, as shown in Figure 2.1. But even an area that has been changed greatly by human activity (for example, a cornfield) also functions as an ecosystem. A farmer may intend to grow only corn, but weeds, which are also producers, spring up as well. Herbivorous insects eat portions of the corn plants, and birds and carnivorous insects feed on the corn-eating insects. Bacteria, earthworms, and other decomposers thrive in the soil.

Even an urban area can be considered an ecosystem, albeit a highly modified one. Apart from pigeons, rats, and pets, human beings are the major consumers in cities and, unlike other ecosystems, urban ecosystems have few producers and decomposers. Food must be brought into an urban ecosystem, and wastes (garbage and sewage) must either be processed by special treatment plants or hauled out of the city to a dumping site, where they are broken down by decomposers or stored. Unfortunately, some cities still dump untreated wastes into nearby rivers and harbors, thereby befouling those waterways before the wastes can be decomposed.

(b)

(a)

Figure 2.1 *(a) A coral reef in the Fiji Islands. (Carl Roessler, Animals Animals.) (b) A sagebrush desert in Oregon. (Zig Leszczynski, Earth Scenes.) (c) Pronghorn on open grassland. (© Allen D. Cruickshank, National Audubon Society, Photo Researchers.) (d) A shallow lake that is dominated by bulrush and water lilies. (©Walter Dawn, National Audubon Society, Photo Researchers.)*

(c)

(d)

The Flow of Energy

Motion and activity within an ecosystem are possible as a result of the flow, or transfer, of energy. *Energy* is defined as the ability to do work or to produce change. An example from the physical world is the energy that drives the wind; an example from the biological world is the food energy that enables us to carry out our daily activities, such as walking and thinking. In this section, we focus primarily on how energy flows through and sustains organisms. In later chapters, we explore how energy sustains the water cycle, the weather, and landscape evolution. For a general description of energy and how nature regulates its flow, see Box 2.1.

BOX 2.1

THE FLOW OF ENERGY

Energy is a deceptively familiar word. We all speak of food energy, atomic energy, abundant energy, cheap energy, and so on. Moreover, we all know that our highly industrialized society would grind to a standstill without great quantities of energy. But energy is more than simply the force that turns the wheels of civilization. It pervades, indeed, powers, the whole universe—from the radioactive decay of atomic nuclei to the awesome power of hurricanes to the movement of planets, stars, and galaxies. Despite energy's pervasiveness and its seeming familiarity, it is an abstract and elusive concept. It cannot be seen, touched, tasted, or smelled. What, then, is it?

Physicists, who study this mysterious something that is responsible for everything that happens everywhere, usually define energy as the ability to do work or to produce change. *Work is done, change occurs, and energy is used whenever a leaf flutters, a baby cries, a truck climbs a hill, or a lightbulb glows. A hummingbird can flit about only because the food energy (sugar) in the nectar it sips is transformed biochemically into mechanical energy in the form of a blurred pair of beating wings. We can cook a meal on an electric stove only because, somewhere, a turbine is being driven by falling water or pressurized steam to produce electricity.*

Energy exists, then, in a variety of forms. Those forms include heat, light, chemical (food or fuel), kinetic, electrical, and nuclear energy. Every form of energy can be transformed into another form of energy. Scientists who observe those transformations always find that no new energy is created in the process and that no energy is ever destroyed. That observation underlies the first law of thermodynamics, also known as the law of the conservation of energy: Energy can neither be created nor destroyed. This law holds for all systems, living and nonliving.

Let us restate the law of conservation as it relates to living organisms: No organism can create its own food supply. *Every organism must fulfill its energy needs by relying on energy transformations within its ecosystem or adjacent ones. Thus plants must depend on light energy, and animals must rely on plants or other animals to obtain energy.*

A familiar example of an energy transformation occurs in an automobile engine. As gasoline burns in the cylinders, some of its chemical energy is transformed into motion (kinetic energy). A great deal of that chemical energy, however, is not converted into kinetic energy. Instead, it escapes through the engine walls in the form of heat or leaves the exhaust pipe in the form of heated combustion products: carbon dioxide, water vapor, and partially burned fuel. Because the engine obeys the first law of thermodynamics, no energy is created or destroyed in the process; energy input in fuel exactly equals energy output. In other words, only energy transformations *occur (although some of them are undesirable, or, in other words, do no work).*

Efficiency is a measure of the fraction or percentage of the total energy input of a process that is transformed into work or some other usable form of energy. It can be determined by the following formula:

$$\text{Percent efficiency} = \frac{\text{energy output as work} \times 100}{\text{total energy input}}$$

A typical gasoline-powered automobile engine typically has an efficiency of 20 to 25 percent. The

Food Webs

Most of us eat without thinking very much about where our food comes from or how it keeps us alive. We eat to obtain the energy and nutrients we need for sustenance and growth, and then we go about our business. However, every time we eat a meal, we form a link with other organisms, thereby playing a role in the continuous flow of energy through ecosystems.

All organisms, dead or alive, are potential sources of food energy for other organisms. That energy always travels in one direction only—from producers to consumers. In a clover field, for example, food energy is produced by a clover plant, which may be eaten by a rabbit, which, subsequently, may fall prey

---

efficiency of an incandescent light is much lower—approximately 5 percent. This means that only 5 percent of the electricity input to an incandescent light is transformed into light energy; all the rest is transformed into heat. (This is why a "burning" light bulb is hot to the touch.) Muscles, internal combustion engines, and other systems that convert chemical energy into work all operate at efficiencies lower than 50 percent. Devices that are designed solely to produce heat are generally more efficient. Home oil furnaces are approximately 65-percent efficient. Electric space heaters that do not need heat-wasting chimneys or vents are 100-percent efficient. (Note, however, that this 100-percent efficiency is attainable only in the conversion of electrical energy into heat by the space heater. At the other end of the electrical power loop, the efficiencies of the devices that transform chemical or kinetic energy into electrical energy are always much lower than 100 percent.)

It would be convenient if the biological and physical systems that we exploit approached efficiencies of 100 percent. But this is rarely possible, because energy's natural and unavoidable tendency is to spread out (become disorganized or disordered). The degree of disorder that occurs in any given system can be measured and expressed mathematically as entropy. *The higher the entropy value, the more disorder in the system, and so the less work the system's energy can do. For example, the chemical energy that goes into an automobile engine is more highly ordered energy than the mechanical and heat energy that comes out of it.*

Another example should help clarify the concept of disordered energy, or energy that can do little work. When we heat a bathtubful of water with an oil-fired heater, we convert the chemical energy in a small volume of fuel to heat energy and transfer it into a very much greater volume of water. The heat energy is now quite disorganized. It cannot drive a turbine: it cannot even heat a can of soup above the temperature of the bathwater itself. About all it can do is raise the temperature of the air in the bathroom by a few degrees.

Nor is there any way to transform this heat energy in the bathwater back into a more concentrated (ordered) form. For all practical purposes, then, it is lost forever. This is the inevitable consequence of the second law of thermodynamics, *which may be stated as follows:* In every energy transformation, some energy is always lost in the form of heat that is thereafter unavailable to do further useful work. *In other words, in all systems, energy "flows downhill" from more ordered to less ordered forms. And the most disordered form of all, which is the form that every bit of energy ultimately takes, is heat.*

One consequence of entropy, or the downhill flow of energy, is that organisms, which are highly organized systems, require a continual input of energy to maintain their order. For example, because our red blood cells break down at the rate of approximately 5 million per second, we have a never-ending need for energy to manufacture (organize) new cells.

Entropy has another important consequence for organisms and the ecosystems in which they live. As we shall see later in this chapter, many of the earth's materials are recycled again and again through biological and physical systems. But energy cannot be recycled through a system once it reaches the bottom of the "energy hill" in the form of heat. Fortunately for us and for all living systems, the energy forever lost as heat is being replaced continually by the inflow of solar energy, which is more ordered than heat energy and which can be transformed by plants into chemical energy through the process of photosynthesis.

to a fox. This type of pathway is called a *food chain*. Natural ecosystems consist of complicated networks of many interconnected food chains, called *food webs.* Figure 2.2 is a stylized food web; the arrows indicate the flow of energy, in the form of food, from one organism to another. Although it may seem complicated, the diagram is actually simplistic. Hundreds more species would have to be added to portray the actual complexity of almost any naturally occurring food web.

Each group of organisms along an energy pathway occupies a *trophic*, or feeding, *level.* All green

Figure 2.2 *A very simplified food web, made up of a network of interconnected food chains, such as the one at the right (plant to hawk).*

Weasel

Hawk

Fox

Snake

Sunlight

Toad

Rabbit

Mouse

Sparrow

Grasshopper

Producers

plants (producers) in an ecosystem belong to the first trophic level. Herbivores compose the second trophic level. Carnivores that eat herbivores make up the third trophic level. Carnivores that consume other carnivores comprise the fourth trophic level, and so on. Omnivores function at more than one level. For example, when a bear eats a pawful of berries, he is functioning at a lower trophic level than when he gulps down a fish.

The sun is the ultimate source of the energy that sustains organisms. On an annual basis, only approximately 1 percent of the sun's energy that strikes plant leaves actually enters food webs as a result of photosynthesis. The process of photosynthesis enables plants (for example, in a clover field) to use light energy from the sun to combine low-energy substances from their environment (carbon dioxide from the air and minerals and water from the soil) to produce sugar, which is a form of carbohydrate that contains a large amount of energy. Clover plants convert this sugar into other types of food—proteins, fats, and starch (another form of carbohydrate). Those foods not only serve as basic factors in constructing and repairing living *cells* (the basic structural and functional unit of plants and all other organisms), but they also serve as the energy source for performing the work of maintenance, growth, and reproduction.

Through photosynthesis plants also produce oxygen. Some of the oxygen produced by clover, for example, is used in various metabolic processes; the excess escapes into the surrounding environment. Figure 2.3 illustrates the sources of raw materials and the distribution of the products of photosynthesis within a plant. The process of photosynthesis is summarized by the following word equation:

Carbon dioxide + water + light energy →
sugar + oxygen

In a clover field, rabbits, mice, and other herbivores feed on plants to obtain needed energy and nutrients. In turn, foxes, weasels, hawks, and other carnivores feed on the herbivores, gaining from them the energy and raw materials to sustain their ability to maintain their health, grow, and reproduce.

The energy flow in a clover field illustrates a *grazing food web*. In grazing food webs, carnivores feed directly on herbivores, which feed directly on

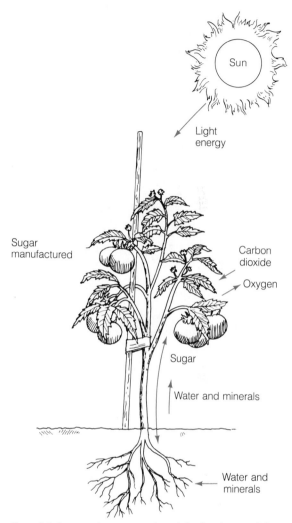

Figure 2.3 *Sources of raw materials and the distribution of the products of photosynthesis in a tomato plant. Sugars can move upward through the stem to be used for growth or stored in seeds or fruits, or they can move downward to the roots to be used for growth or storage.*

living plants. Thus living plants are the foundation of grazing food webs. Eventually, however, all plants and animals die, and their remains are the foundation of *detritus food webs*, which are often equally important but generally less conspicuous than grazing food webs. We can get a sense of how a detritus food web works by examining the energy flow within a marsh at the edge of a lake, as shown in Figure 2.4. When wetland plants and animals die and accumulate in the sediments at the bottom, their remains are broken down into increasingly smaller fragments by decomposers. Both the detritus and the

restrial ecosystems as well as in aquatic ecosystems, such as streams, rivers, and marshlands. In fact, as little as 10 percent of the leaves in a forest is eaten by herbivores; the rest dies and is funneled through detritus pathways. Decomposers in a forest include such organisms as bacteria, fungi, millipedes, and certain insect larvae. Those organisms, in turn, fall prey to such carnivores as centipedes, beetles, spiders, and rodents, which, subsequently, may be eaten by snakes and owls. In other types of ecosystems, however, energy flows predominantly through grazing food webs rather than detritus food webs. For example, in open water, where microscopic, free-floating plants (*phytoplankton*) are the chief producers, as much as 90 percent of the phytoplankton is eaten by microscopic animals (*zooplankton*) and thus enters the grazing food web, leaving only 10 percent to die and enter detritus pathways. Hence the pattern of energy flow will vary; depending on the type of ecosystem within which it exists.

We function primarily as herbivores in grazing food webs. Plant materials, such as cereals, vegetables, and fruits, account for approximately 64 percent of the food that is consumed in the United States. (Worldwide, that figure is about 89 percent.) Furthermore, even when we function as carnivores, most of our meat comes from herbivores (for example, beef cattle, chicken, and hogs). When we eat fish and shellfish (for example, crabs, clams, and oysters) or mushrooms, we are tapping food energy from detritus pathways, but those foods make up a very small portion of our diet (less than 1 percent in the United States).

It is important to remember that energy flows through food webs in one direction only—from lower to higher trophic levels. It never flows in the reverse direction—from carnivores to herbivores to green plants. Organisms at each trophic level depend on those at lower trophic levels for the energy to sustain themselves and reproduce. For example, we cannot convert energy directly from the sun into food energy. We depend on green plants to make that transformation for us.

Efficiency of Food Webs

Food webs constitute the sources of food energy, but nature limits the amount of food energy that is accessible to organisms within those webs. Not all food energy is transferred from one trophic level to

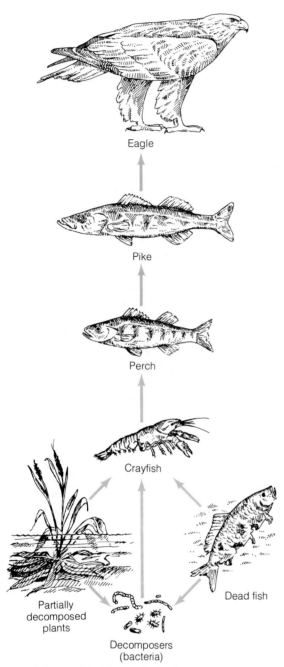

Eagle

Pike

Perch

Crayfish

Dead fish

Partially decomposed plants

Decomposers (bacteria)

Figure 2.4 *A simplified detritus food web.*

decomposers themselves are eaten by scavenging animals, such as crayfish, shrimp, and snails. Those animals then serve as food for small fish (such as perch), which are subsequently eaten by larger game fish (such as northern pike and walleyed pike) and fish-eating birds (such as eagles and herons).

Detritus pathways usually predominate in ter-the next. If we were to envision the amount of energy that is stored by the organisms at each trophic level, we would see a pyramid. The producers would form a broad base that contained most of the pyramid's energy, and each succeeding level would contain considerably less energy than the one below it. Figure 2.5 is an example of such an ecological pyramid.

A relationship within an ecological pyramid that is of considerable importance is *ecological efficiency*, which is determined by the amount of energy that is transferred from one trophic level to the next. It can be calculated by dividing consumer productivity by prey productivity. *Productivity* is defined as the accumulation of energy during a specified period of time at a given trophic level by means of growth and reproduction. It is usually determined by measuring the change in the *biomass* (the total weight or mass of all the organisms in an area) of one trophic level during a specified period of time. For example, the annual productivity of plants in a grassland can be estimated by harvesting the plants at the end of the growing season and then drying and weighing them to calculate the biomass that was produced during the growing season. Annual plant productivity on a North American prairie may be over 5000 kilograms per hectare (4450 pounds per acre). (See Appendix I for conversion factors for the

English and metric systems.) Some values for plant productvity in other ecosystems are given in Chapter 5.

The efficiency of energy transfer varies greatly depending on the types of organisms and ecosystems, but the efficiency is never very high. For many natural ecosystems, frequently less than 1 percent of the energy that exists at one trophic level becomes incorporated in the tissues of organisms that occupy the next, higher trophic level. In contrast, efficiencies of up to 20 percent have been achieved by some domestic livestock, which are raised in man-made environments.

For ease of calculation, ecologists often assume an ecological efficiency of 10 percent (that is, the *10-percent rule*) to estimate the amount of energy that is transferred through a food chain. For example, if we apply the 10-percent rule to the grain–beef–person food chain, we can predict that 100 kilograms (220 pounds) of grain will produce 10 kilograms (22 pounds) of beef, which, in turn, will produce only 1 kilogram (2.2 pounds) of persons. Hence only a relatively small amount of food energy is transferred from one trophic level to the next.

Since ecological efficiency is low and each organism requires a certain amount of food to survive, it is clear that only a limited number of organisms can survive at a particular trophic level. This fact applies to all organisms, including human beings: The earth cannot support unlimited consumers. Also, since the human population now totals nearly

Figure 2.5 *A pyramid of energy content, illustrating the low efficiency of energy transfer between trophic levels.*

Second carnivore — 3

First carnivore — 30

Herbivores — 200

Plants — 1000

Usable energy (kilocalories)

Energy flow

5 billion and, currently, the population is doubling nearly every 40 years, we cannot afford to ignore these natural limits in our struggle to eliminate hunger.

Causes of Low Efficiencies

If we are to have a fighting chance of feeding a growing human population and managing the earth's plant and animal life, we must determine where energy is lost between trophic levels and why the energy loss occurs.

Many steps occur between an organism's attempts to obtain food and the utilization of food for energy by that organism's cells. First, many organisms have characteristics that reduce their vulnerability to being eaten. Even plants can hinder herbivores. They cannot pull up their roots and run away, but they do possess various ways of warding off attack by their enemies. For example, needles and thorns on thistles and cacti discourage hungry herbivores, and many plants, including daisies, black-eyed Susans, oaks, and citrus trees, produce chemicals that repel or harm bacteria, fungi, insects, or other animals that attack them. Animals, too, possess characteristics that help them to escape predators. For example, prairie dogs find refuge in their burrows, skunks emit a pungent odor, fleet-footed antelope outrun their predators, and, as shown in Figure 2.6, walking sticks blend unnoticed into their surroundings. Although predators, in turn, have capabilities to overcome the defense mechanisms of their prey, their ability to find and capture it is limited. Hence often no more than 10 to 20 percent of the food present at one trophic level is actually harvested by the organisms at the next trophic level. Organisms that escape predators eventually die, and the energy that is present in their remains enters detritus pathways.

Once the prey is harvested by a predator, rarely is it all ingested. Although fur, feathers, and skeletons contain energy and nutrients, most carnivores do not eat these materials. And even if carnivores ingest these parts, most cannot digest them; nor can most herbivores digest the fibrous portions of plant tissue. Therefore, the energy contained in those materials cannot be utilized by most herbivores and carnivores; those indigestible materials, if they are consumed, are excreted and subsequently digested

Figure 2.6 *A walking stick, which often goes unnoticed by potential predators because of its resemblance to its surroundings. (Des Bartlett, Photo Researchers.)*

by detritus feeders. The portion of ingested food energy that is absorbed by the animal's digestive system is called *assimilated energy.*

Further energy losses occur when assimilated food undergoes *respiration.* That process, which "burns" the food energy in living cells, releases energy from digested food (usually in the form of sugars) to allow the organism to move, maintain itself, grow, and reproduce. Respiration occurs in all living organisms. However, under the best conditions, respiration is an inefficient process, because less than half of the energy that is present in the sugars is converted into work in an organism's system. Most of the rest is transformed into heat energy that cannot be used to perform work and is eventually lost to the environment, along with carbon dioxide and water—the other products of respiration. The process of respiration is summarized by the following word equation:

Sugar + oxygen →
$$\text{carbon dioxide} + \text{water} + \text{energy for work} + \text{heat energy}$$

Figure 2.7 summarizes the reasons for low ecological efficiencies between trophic levels. Those inefficiencies occur in grazing food webs as well as in detritus food webs.

Coping with Low Efficiencies

Mankind has developed several strategies to increase the production of food within the constraints of low ecological efficiencies. The success of these strategies can be demonstrated by the fact that the human population has been able to grow from less

than 5 million to nearly 5 billion—a 1000-fold increase—since the advent of agriculture, some 10,000 years ago.

Clearing of Forests for Raising of Crops   Almost all the food energy in a forest is composed of unpalatable or indigestible material, such as leaves and wood. Accordingly, for millennia, humans have cleared away forests and replaced them with crops, thereby channeling a greater percentage of photosynthetic energy into the growth of seeds and fruits—materials that are digestible for us as well as for domestic livestock. During the last 200 years, timberland in the United States has been reduced by some 20 percent (an area that is approximately the size of Texas). Worldwide, forest land has been reduced by 20 percent in just the past 30 years. Many of those forests were cleared to provide land for agriculture.

This means of increasing food production has its natural limits, however. Most forest land that is suitable for agriculture has already been converted to that use. The forests that remain must be maintained to sustain wildlife habitats and fulfill our needs for wood products and recreational areas. Therefore, as is always the case with environmental issues, hu-

Figure 2.7 Generalized flow scheme of energy transfer and losses between and within trophic levels. The relative sizes of the boxes will depend on the nature of the ecosystem, the trophic level, and the type of animal involved.

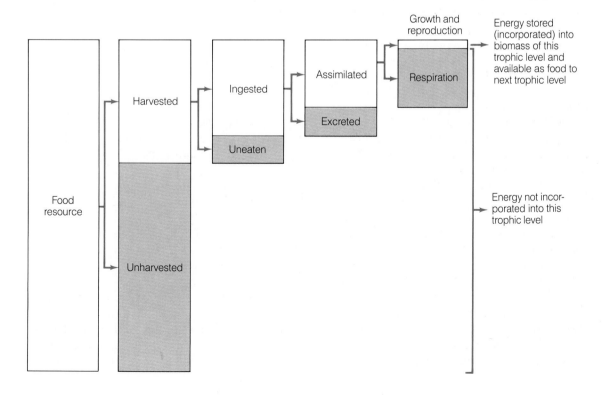

man inventiveness is met with constraints from nature and our own conflicting demands on available resources.

Attempting to Raise More Efficient Crops and Livestock As noted earlier, livestock have ecological efficiencies that are often many times higher than those found in natural ecosystems. This greater energy transfer is not a result of the inherent differences between domestic and wild animals but, rather, because of animal husbandry practices. For example, chickens no longer roam about the barnyard scratching the ground in search of seeds and chasing after insects. They are confined in climate-controlled houses, where ground-up grain that is supplemented with vitamins and minerals is brought to them. Thus, unlike wild animals in a natural ecosystem, they ingest virtually all the food that is present in their artificial ecosystem, because all of it is in the form of feed that is rationed to them in amounts they will eat. Because their food contains few indigestable materials, they assimilate (absorb from their digestive systems) a greater percentage of the food that they ingest. Because the feed is brought to them, they do not need to expend extra energy on movement. And because they are grown in climate-controlled houses, they spend little extra energy on maintaining a constant body temperature during cold weather. Hence a smaller percentage of assimilated energy is required for their respiration, thereby leaving more energy for them to grow and reproduce. Consequently, growing livestock in confinement usually enhances ecological efficiency, which, in this case, is calculated by the food energy that is required for growth and reproduction, divided by the food energy that is provided to the livestock in the feed. However, this increased ecological efficiency has been achieved at a cost—the consumption of fossil-fuels is increased enormously by hauling ground-up, enriched feed to the chickens, transporting wastes away, and heating and cooling the chicken houses.

Similar situations exist with crops. Although crop production per hectare has increased manyfold since the advent of agriculture, much of this improvement has occurred because of greater applications of fertilizers and pesticides, which often are accompanied by irrigation. All those amenities require a large investment in fossil fuels. In addition, plant breeders have developed higher-yielding varieties that flourish with increased fertilizer applica-

tions. But those new varieties do not convert a greater percentage of solar energy into greater amounts of sugars. Rather, they channel more of the sugars that are produced by photosynthesis into the production of the grain head (which is the prime human food source) and less of the sugars into the production of leaves, stems, and roots (which are usually unpalatable and indigestible for humans).

Harvesting Lower on Food Webs If we ate less meat and more fruits, vegetables, and cereals, the amount of food energy available to us would increase automatically. A gram of grain yields about the same amount of energy as a gram of meat. So, according to the 10-percent rule, approximately ten times more food energy would be available to us if we ate the grain we now feed to cattle instead of eating the beef. A strict vegetarian diet is unappealing to many Americans, however, since we are accustomed to a diet that contains approximately 40 percent of animal products. But in many overcrowded areas of the world, human beings have been forced to exist within shorter food chains. For example, the diet of Southeast Asians consists almost entirely of rice or wheat and some green vegetables; an occasional meal of fish for protein supplement accounts for only approximately 10 percent of their diet. Most of those persons need to take full advantage of existing food supplies and simply cannot afford to lose the energy between the trophic levels of herbivores and carnivores, which is about 90 percent, according to the 10-percent rule.

Reducing Competition Another way that we increase the food energy that is available to us is by reducing populations of other organisms that compete with us for food. For example, farmers apply pesticides to kill insects that feed on crops, thus increasing our share of available food. If our competitors get a bigger share (which they do, for example, during locust plagues) then fewer persons can be fed (see Figure 2.8). In those situations, famine and death may result unless food can be imported from regions that have surpluses of food. However, some pesticides kill valuable organisms as well, such as honey bees, which play a major role in food production as a result of their pollination activities. Some pesticides also kill the natural enemies of the pests that farmers are attempting to control, enabling the population of pests to grow despite the spraying of

Figure 2.8 *An invasion of locusts, an insect which has competed with human beings for food over the ages. (FAO Photo.)*

pesticides. We discuss the consequences of pesticide use further in Chapter 17.

Overharvesting In all our efforts to increase our food supply, we risk the hazards of overharvesting our sources. As mentioned earlier, many natural safeguards prevent predators from overharvesting their prey; consequently, predators rarely cause their prey to become extinct. Our ability to harvest our prey also would be quite limited if we lacked access to tools. Compared with many predators of equal size, we are generally rather slow afoot; we are rather poor at climbing up to, pouncing on, or digging out our prey; and we do not possess exceptional strength. But by learning to use tools and other technology, we can compensate for our physical limitations and increase our harvest of food energy. The cost of these advances, however, has not always been equal to our gains. In some instances, we have overharvested and brought about the demise of our prey, such as contributing to the extinction of the passenger pigeon, heath hen, and great auk by overhunting. We have also severely depleted the numbers of many commercially important types of fish, including the Atlantic salmon, cod, haddock, and herring, by overfishing. Unless we acknowledge and restrain our

technological capabilities for overharvesting our food webs, we will eradicate more species and will thereby lose valuable food sources.

In spite of our ingenuity in working around ecological inefficiencies, nature continues to restrict our ability to exploit available food resources. There is, after all, no such thing as a free lunch. In our efforts to make more food available to ourselves, we must try to foresee all the possible consequences of our actions. Then we can decide if, in the long run, we will be able to afford the losses that often accompany the gains. See Box 2.2, which suggests that the losses in food resources currently may be outrunning the gains in the world's most populous country.

The Movement of Materials

All organisms require certain chemical elements to survive. Nitrogen is a component of all the structural and functional proteins that sustain living tissues. Phosphorus is essential for such energy transformations as photosynthesis and respiration. Calcium is found in structures that provide strength, such as bones and shells in animals and cell walls in plants. In addition to those three nutrients, plants

BOX 2.2

CHINA'S PROGRESS HURTING LAND

China is causing widespread and serious damage to its natural environment in its effort to quadruple its economic output in the next 20 years, according to a survey of the published writings of more than a score of Chinese scientists.

Vast forest areas have been denuded, lakes and streams have been polluted and substantial soil erosion and loss of arable land have occurred, the survey reports, as China has expanded and upgraded its farms, factories and utilities.

The consensus among the scientists, writing in official scientific and political papers published in Chinese journals, is that the environmental deterioration poses a serious threat to their nation's physical well-being and hence to its social stability.

The survey was done by Dr. Vaclav Smil, professor of geography at the University of Manitoba in Winnipeg, Canada. It is believed to be the first comprehensive view of all aspects of the condition of China's natural resources.

Dr. Smil said in an interview that when taken separately, these accounts by reputable Chinese researchers "are worrisome enough," but when taken together, "the dimensions and implications of China's environmental degradation" were devastating.

In addition to the demands on trees for fuel use, huge forest areas have been cleared for agriculture. One survey showed that in Heilongjiang Province, which supplies nearly 50 percent of the nation's timber, the forested area was decreasing by 1980 at the rate of about 2 percent a year. Other areas such as the provinces of Sichuan and Yunnan show losses of forest cover of 30 and 45 percent in the last two decades.

Despite mass tree-planting programs initiated since 1950, the Chinese Ministry of Forestry last year estimated that no more than one-third of all saplings managed to survive, owing to careless planting, poor follow-up attention and a lack of scientific care that permitted the planting of many species in areas unsuitable for their survival.

Dr. Smil said studies showed that China, as a result of its "grain first" policy, was now producing about 70 percent more grain a year than it did in the late 1950s. But he said increasing population has literally eaten up the expanded output so that there has been no improvement in the already low food energy intake of the average Chinese diet.

"So taking the 'easiest way out' by expanding grain production," Dr. Smil said, "has not only destroyed forests but has extended to conversion of grasslands and filling of lakes, resulting in higher soil erosion rates and widespread ecosystemic disruption."

The most serious impact, he said, has been the loss of 30 percent of the country's farmland in the last two decades.

The study also reports that the denuding of forest watersheds under China's "grain first" policy has led to chemical pollution of lakes and streams, affecting fish and aquatic plant life. And leaks and seepages from irrigation canals are so widespread that only about 40 to 50 percent of the water is used.

Dr. Smil is currently writing a book on China's environment and is preparing a report on China's energy outlook for the World Bank, for which he is a consultant.

require 10 other elements; animals require approximately 12 other elements. The required elements common to both plants and animals include potassium, iron, sulfur, copper, and zinc. In addition, oxygen is essential for respiration, and carbon dioxide and water are necessary for photosynthesis. Since adequate amounts of those materials are needed for normal growth and development, it is apparent that their availability bears a direct relation to the amount of food and fiber produced by ecosystems. Hence we need to understand the sources of those materials and how they flow through ecosystems.

Although the earth receives a continuous and essentially unlimited supply of energy from the sun, it has no comparable source of materials. For all practical purposes, the quantity of materials on earth is fixed and finite. But for millions of years, those life-sustaining materials have been continually cycled and recycled within and among the earth's ecosystems. In this manner, organisms not only receive essential nutrients but play a major role in recycling them.

Although the energy that flows through a food web is eventually lost in the form of heat energy through respiration, nutrients continue to flow in cycles (See Figure 2.9). Those cycles follow two basic patterns—sedimentary cycles and gaseous cycles. In *sedimentary cycles*, the soil or rocks in the earth's crust provide the major reservoir (source) of the nutrients for organisms. In *gaseous cycles*, the atmosphere or the oceans provide the major reservoir of nutrients. A few of the many nutrient cycles that are essential for all life are discussed in the following sections. The *phosphorus cycle* exemplifies sedimentary cycles, whereas the *carbon*, *nitrogen*, and

oxygen cycles illustrate gaseous cycles. Another essential cycle, the *hydrologic cycle* (water cycle), is described in Chapter 6 in conjunction with our examination of water pollution and the management of aquatic resources.

The Phosphorus Cycle

Phosphorus is available to plants directly from the soil, which also supplies other major nutrients, such as calcium, magnesium, potassium, and sulfur. It is taken up from the soil, mainly in the form of phosphate (PO_4^{3-}), through the root system of a plant and is transported to the plant's growing parts. Figure 2.10 illustrates the central role of the soil in the earth's phosphorus cycle. Phosphorus is incorporated into a variety of organic compounds within the plant, including fats and nucleic acids. Furthermore, nutrients such as phosphate that are taken from the soil are concentrated in large amounts within the plant. Experiments have shown that the concentration of nutrients may be several thousand times greater in plants than in the soil around them. Thus plants not only provide energy for consumers; they also supply a concentrated source of essential nutrients. When plants and animals die, they are broken down by decomposers and their phosphorus is returned to the soil, where it can be taken up again

Figure 2.9 *Flow of energy and materials through an ecosystem. The blue lines indicate the flow of energy and the black lines the flow of materials. While nearly all the energy gained by photosynthesis is lost as heat energy through respiration of all organisms, many materials are eventually recycled. (Modified from G. M. Woodwell, "The Energy Cycle of the Biosphere." © Scientific American 223:64, 1970. All rights reserved.)*

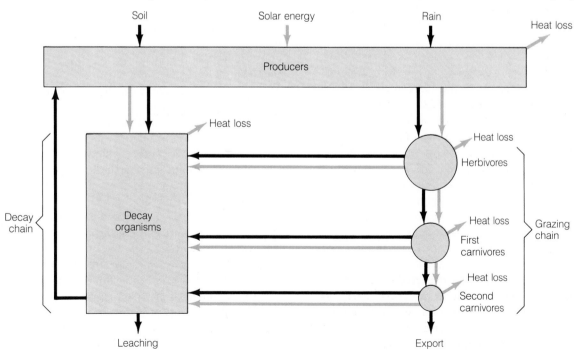

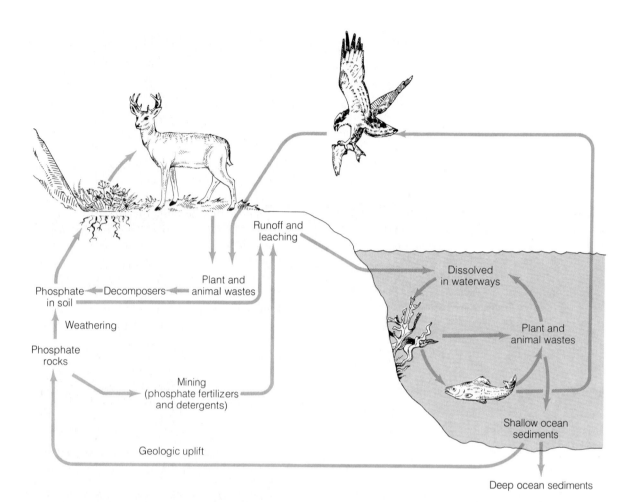

Figure 2.10 *The phosphorus cycle. Phosphorus is cycled into food webs from the soil; hence it is termed a* sedimentary *cycle.*

by plant roots. Thus decomposers play an essential role in the recycling of nutrients within an ecosystem and the maintenance of soil fertility. From this cycle, we can also see the basis for the old adage, "dust to dust."

Within undisturbed terrestrial ecosystems, nutrient cycling between soil and organisms is very nearly closed: that is, phosphorus and other nutrients are retained within the ecosystem and losses are minor. Ecosystems possess a variety of mechanisms for conserving nutrients. For example, plants serve as a protective covering for the soil; this covering and the plant's extensive network of roots help to retard soil erosion by rain and wind, thereby retaining the supply of nutrients within the ecosystem. Another natural mechanism for conserving nutrients is the partitioning of some nutrients into

long-term storage pools. For example, billions of tons of nutrients can be stored in wood and bark, where they are not subject to loss through surface runoff or seepage into the groundwater. Moreover, when a tree dies, its store of nutrients is released slowly to the soil over a long period of time, because it may take decades for a large log to decompose fully (Figure 2.11). Leaves, of course, comprise a short-term storage pool: except for evergreens, leaves live only one growing season, and they normally decompose within a year of falling off the tree. But even here, nutrient conservation may occur. Just before trees lose their leaves, some nutrients are transported back from the dying leaves to the woody tissues, where they are stored. As much as 30 percent of the phosphorus in leaves is resorbed before the leaves fall.

The retention of phosphorus within an ecosystem is critically important for organisms, because relatively little is available in the soil. Although phosphorus is added to the soil through the weath-

Figure 2.11 *Bracket fungi contribute to the slow decay of a fallen tree. (W. H. Hodge, © Peter Arnold, Inc.)*

ering of rocks that contain phosphorus, this process takes place very slowly. Thus the most effective means of providing an adequate supply of phosphorus in an undisturbed terrestrial ecosystem is through retention and recycling by organisms.

Still, no undisturbed terrestrial ecosystem is completely closed. Small amounts of phosphorus (which are usually bound to soil particles) are carried off to rivers and lakes. In those various aquatic ecosystems, some of the phosphorus is taken up by algae and rooted aquatic plants and again passed along food webs, including detritus pathways. Also, the phosphorus that enters lakes and rivers may sink to the sediment at the bottom. If the body of water is shallow, wave action can stir up the sediment, making the phosphorus available again for uptake by algae.

Eventually, small amounts of phosphorus may reach the oceans and be deposited in deltas and near-shore sediments. The remaining phosphorus is cycled through marine food webs or is deposited in deeper sediments, where it is removed from circulation until geologic processes reexpose them; the reappearance of those deposits may require hundreds of thousand or even millions of years. While phosphorus is being lost to aquatic ecosystems, uplifting and weathering of rocks that contain phosphorus slowly add phosphorus to the soil of terrestrial ecosystems. The best estimates that have been calculated indicate that, under natural conditions, additions of phosphorus to the ecosystem from weathering are approximately equal to the losses caused by erosion, but this conjecture remains to be verified experimentally.

The Carbon Cycle

Carbon, which is an essential component of all living things, is cycled into food webs from the atmosphere, as shown in Figure 2.12. Green plants obtain carbon dioxide (CO_2) from the air and, through photosynthesis, transform the carbon from CO_2 into sugar. Then the sugar is employed to manufacture a myriad of other organic compounds, including carbohydrates, fats, and proteins. Both producers and consumers transform a portion of the carbon in food

back into carbon dioxide as a by-product of respiration, which, subsequently, is released to the atmosphere. Carbon that is found in dead plants and animals also is returned eventually to the atmosphere through the respiration of detritus feeders.

Some dead plant and animal materials are buried in sediments before they can be broken down completely by decomposers. That process has been going on to a greater or lesser extent for hundreds of millions of years. But for us, it was particularly important during the Carboniferous Period, 280 to 345 million years ago (see Appendix II), when trillions of tons of organic materials were buried below the earth's surface. Much of those plant and animal

Figure 2.12 *The carbon cycle. The major carbon reservoir is the atmosphere, where the carbon is "stored" in the gas carbon dioxide (CO_2). For this reason, the carbon cycle is termed a gaseous cycle.*

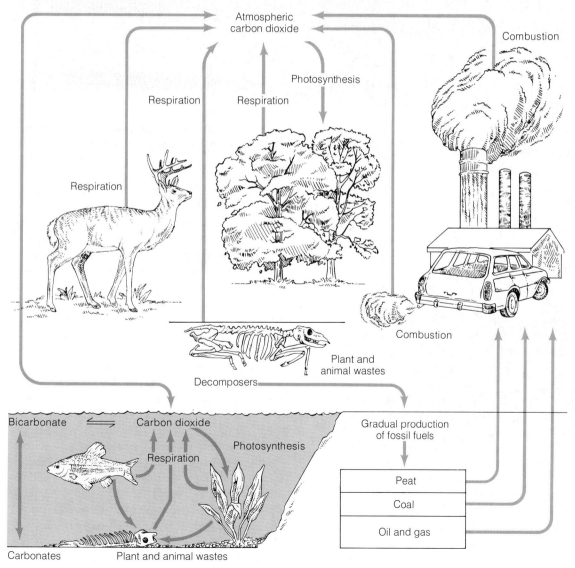

(a)

(b)

Figure 2.13(a) *A diorama, showing how a carboniferous swamp forest—similar to those that existed 280 to 345 million years ago—might have looked. The world's coal deposits are composed of the remains of such forests. (Courtesy of the Field Museum of Natural History.) (b) An exposed coal deposit that is about 2 meters (6 feet) thick, near Glendive, Montana. (Campbell MR 565, U.S. Geological Survey.)*

remains were transformed into coal, oil, and natural gas as a result of heat and compression through subsequent eons. Thus luxuriant swamp forests that inhabited the earth millions of years ago (Figure 2.13) were transformed into the coal reserves of today. In other situations, plant and animal remains have become buried and later transformed by complex and poorly understood processes into oil and natural gas. When coal, oil, and natural gas which are also known as *fossil fuels*, are burned, the energy that is released can be channeled to perform work. In the process, the stored carbon combines with oxygen in the air to form carbon dioxide, which enters the atmosphere.

If we consider the fact that it has taken natural processes millions of years to create the fossil fuels that are available in the world today, our use of them has lasted barely a moment in terms of geologic time. Some scientists contend that we will exhaust our available supplies of economically recoverable petroleum sometime within the next century—barely 200 years after petroleum was first used commercially as a source of energy. Although there is debate over when we will actually run out of economically recoverable petroleum, there is no doubt that we are consuming fossil fuels at a much greater rate than they are being formed. Hence if we are to continue our current standard of living, we must begin to develop alternative forms of energy. How this may be accomplished is explored further in Chapters 18 and 19.

The exchange of carbon dioxide between the atmosphere and the oceans creates another important aspect of the carbon cycle. That exchange occurs at the interface between the air and water, with carbon dioxide flowing in both directions. When atmospheric carbon dioxide levels increase above a certain level, the oceans absorb more of it than they release to the atmosphere. Conversely, when atmospheric carbon dioxide levels decrease, the oceans compensate by releasing more carbon dioxide to the atmosphere than is absorbed. Thus oceans help to maintain an equilibrium in the carbon dioxide content of the atmosphere.

The carbon cycle is also involved in the formation of limestone, dolomite, and carbonaceous shale. As a result of several rock-forming processes, carbon is incorporated into those components of bedrock. Subsequently, chemical and physical processes (*weathering*) may break down those substances, releasing carbon dioxide into the atmosphere. Although this source of carbon has little significance for us in the immediate future, the processes of formation and degradation of carbonaceous rock (which are called *rock cycling*) help to create a balance in the amount of carbon in the atmosphere, over a time scale of millions of years. For a more detailed discussion of rock cycling, see Chapter 12.

The Nitrogen Cycle

The nitrogen cycle is one of the most complex cycles. Although 79 percent of the atmosphere is composed of nitrogen gas (N_2), plants and animals cannot use the nitrogen gas directly to construct proteins and other compounds that contain nitrogen. In nature, nitrogen gas is changed into forms that plants can use by two processes—biological fixation and atmospheric fixation—which are collectively called *nitrogen fixation*. Both processes are essential parts of the nitrogen cycle (illustrated in Figure 2.14). *Biological fixation* results from the activities of specialized microorganisms that combine nitrogen gas with hydrogen gas (H_2) to form ammonia (NH_3). Some of those nitrogen-fixing microbes are specialized bacteria that live in the root nodules of leguminous plants, such as peas, beans, alfalfa, and clover (Figure 2.15). The ammonia produced by those bacteria is

then converted to other forms of nitrogen, which are absorbed by plant cells that surround the nodule and, subsequently, transport those forms of nitrogen to other parts of the plant. Biological fixation is also carried out by the blue-green algae that live free in the soil or in rivers, lakes, and oceans. The ammonia they produce can be taken up by plant roots or it can be converted by another group of bacteria to nitrate (NO_3^-), which also can be taken up by plant roots. Biological fixation contributes roughly 90 percent of the total nitrogen that is fixed each year.

The other natural process that makes nitrogen available to plants is *atmospheric fixation*, which occurs during thunderstorms. The high temperatures that are associated with lightning, shown in Figure 2.16, cause nitrogen gas to combine with oxygen gas in the atmosphere, eventually forming nitrate. That form of nitrogen is then captured by falling raindrops and carried to the earth's surface, where it can be taken up by plants.

Once within the plants, ammonia and nitrate are incorporated into a variety of organic compounds, including proteins and nucleic acids, and it is in the form of those compounds that nitrogen travels through food webs. When plants and animals die and decompose, the nitrogen is converted back into ammonia, and it is returned to the soil. Animal wastes, which are rich in urea, become part of another major recycling route. Organic nitrogen compounds, such as proteins and urea, are converted into ammonia by yet another specialized group of bacteria. Thus when ammonia returns to the soil, some of it may be changed into nitrate by microbial action, and both ammonia and nitrates can be taken up again by plant roots and recycled through food webs.

Ammonia and nitrate are lost from terrestrial ecosystems through two major pathways: denitrification and transport by water. *Denitrification* is a process that is carried out by another specialized group of bacteria. Those bacteria convert nitrate back into nitrogen gas, which escapes from the soil into the atmosphere. Transport by water is a simple mechanical process. Since both nitrate and ammonia are highly soluble in water, they are carried away readily by surface runoff or groundwater. Nitrate and ammonia normally remain dissolved in lakes, rivers, and oceans and can be taken up by aquatic plants and subsequently cycled through aquatic food webs. Ammonia does not move through the soil as readily

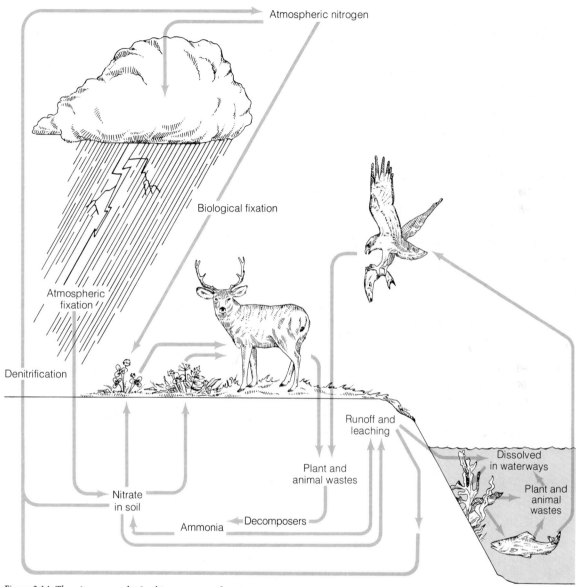

Atmospheric nitrogen

Biological fixation

Atmospheric
fixation

Denitrification

Nitrate
in soil

Ammonia

Decomposers

Plant and
animal wastes

Runoff and
leaching

Dissolved
in waterways

Plant and
animal
wastes

Figure 2.14 *The nitrogen cycle. In this gaseous cycle, nitrogen gas (N_2) enters the food web when it is converted, or fixed, into ammonia (NH_3) and nitrate (NO_3^-) and absorbed by plants.*

Figure 2.15 *Root nodules on soybean plants, which contain bacteria that transform nitrogen gas into ammonia. (U.S. Department of Agriculture.)*

Figure 2.16 *Lightning, an important factor in the nitrogen cycle. (National Oceanic and Atmospheric Administration.)*

as nitrate, since ammonia tends to adhere to soil particles.

The Oxygen Cycle

Oxygen is found almost everywhere on earth. As a gas, oxygen (O_2) makes up nearly 21 percent of the atmosphere; it is also found dissolved in surface waters and in the pore spaces of soils and sediments. Oxygen combines chemically with a multitude of other elements to form important substances, including water (H_2O); such gases as carbon monoxide (CO), carbon dioxide (CO_2), and sulfur dioxide (SO_2); plant nutrients, including nitrate (NO_3^-); organic substances, such as sugars, starch, and cellulose; and most rocks and minerals, including limestone

($CaCO_3$) and iron ore (Fe_2O_3). Clearly, in order to react with such a wide variety of materials, the oxygen cycle must be linked to other cycles of materials, including the carbon and nitrogen cycles. The oxygen cycle is so intricate and complex that we can outline only a few of the more important aspects here. Figure 2.17 illustrates some of those pathways in simplified form.

One important component of the oxygen cycle involves photosynthesis and respiration. As you will recall, in the process of photosynthesis, carbon dioxide combines with water in the presence of sunlight to produce sugar and oxygen. By contrast, oxygen combines with sugar through respiration, and one of the products is carbon dioxide. Furthermore, organic materials, such as wood, coal, or manure, will burn only in the presence of oxygen, and carbon dioxide is an end product. Indeed, a wood or coal fire and respiration are both examples of the energy-releasing process that is called *oxidation*. Of course, sugar "burns" in a more controlled fashion during respiration and at a much lower temperature than does wood or coal in a stove.

The equations for photosynthesis and respiration seem to suggest that each process is the reverse of the other. But, in reality, each process consists of a complex series of intermediate reactions, many of which occur exclusively in one process or the other.

Exchanges of oxygen between the atmosphere and the oceans make up another important aspect of the oxygen cycle. As with carbon dioxide in the carbon cycle, the exchange of oxygen between the oceans and the atmosphere is balanced, so the oxygen content of the atmosphere remains essentially constant. Currents carry dissolved oxygen into deeper waters to sustain aquatic life at great depths.

Earlier, we saw that oxygen enters the nitrogen cycle through atmospheric fixation, whereby oxygen gas combines with nitrogen gas to form nitrate. By means of denitrification, nitrate is transformed back into water and nitrogen gas, which escapes back into the atmosphere.

Food-Web Accumulation

Earlier, we saw close ties between the movement of energy and materials within an ecosystem. Plants provide both a source of energy and concentrated nutrients for all consumers. But in their normal

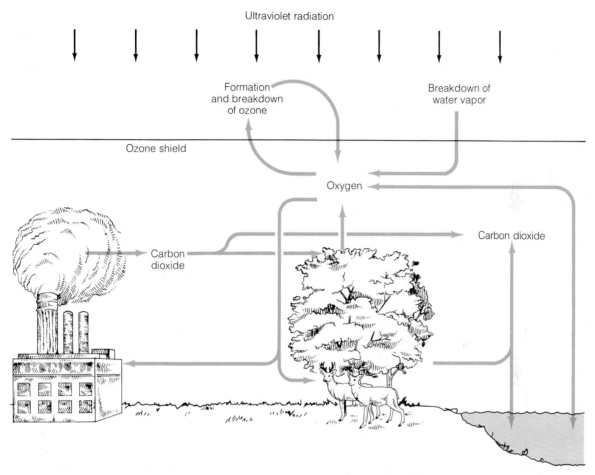

Figure 2.17 *The oxygen cycle.*

functioning, plants can also take up DDT, mercury, and various other toxic substances, which are *persistent;* that is, they are not broken down by chemical processes in soil, in water, or in organisms themselves. If those substances are not broken down by the plant, they are retained and concentrated in its tissues. Moreover, those chemicals can be passed from one trophic level to the next in ever-greater concentrations as one organism consumes another.

That concentration takes place as a consequence of the transfer inefficiencies that exist between trophic levels—the biomass declines at each successively higher level. Furthermore, a persistent toxic substance, such as DDT, is not readily broken down or excreted by organisms, so from one trophic level to another, the losses of it are small compared with the amounts that are transferred to the next trophic level. Because most of the DDT that is in-

gested is retained and most of the biomass is lost, the DDT becomes more concentrated at each higher trophic level, as illustrated in Figure 2.18.

That process of ever-increasing concentration is called *food-web accumulation,* or *bioaccumulation.* Bioaccumulation is especially pronounced in aquatic food webs, because they usually consist of four to six trophic levels rather than the two or three levels that are generally found in terrestrial ecosystems. Hence persistent contaminants that enter the lowest trophic level may become from a thousand times to over a million times more concentrated by the time they reach the uppermost trophic level. Consequently, fish-eating birds and persons that have a diet that consists principally of fish are at the greatest risk of consuming significant amounts of toxic materials.

We explore the problems of food-web accumulation further in the context of water pollution (Chapter 7), air pollution (Chapter 10), and hazardous and toxic materials (Chapter 13).

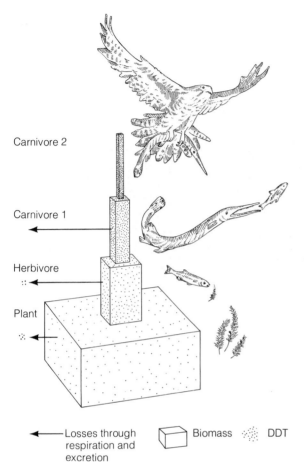

Carnivore 2

Carnivore 1

Herbivore

Plant

←— Losses through respiration and excretion

☐ Biomass ∴ DDT

Figure 2.18 *The accumulation of DDT in an aquatic food chain.*

Transfers between Ecosystems

We have seen that materials follow cycles not only within ecosystems but also between them. The major pathways for the flow of materials between ecosystems are air currents and waterways and, to a lesser extent, animal movements. Such transfers have several significant consequences.

First, an ecosystem can provide nutrients for another ecosystem. For example, the flow of phosphate and nitrate from a terrestrial to an aquatic ecosystem will enrich the waterways and enhance aquatic productivity. Moreover, decaying organic matter that is washed into the waterways from adjacent land areas provides an important food source for aquatic consumers in the rivers and tidal wetlands. Figures 2.10 and 2.14 illustrate that the flow between ecosystems can travel in both directions. Aquatic organisms, mainly fish and shellfish, serve as food

sources for animals that live primarily on land. And millions of seabirds feed on fish but nest on land near the shore. Those birds annually deposit onto the land hundreds of thousands of metric tons of wastes (guano) that are rich in nutrients.

Second, air currents and waterways disperse and dilute sudden surges of materials. For example, winds can spread gases and ashes from an erupting volcano over an entire hemisphere, thereby lessening the localized concentration of those materials. Although this dilution of materials may be beneficial locally, the global linkages among cycles also create problems. Anything emitted into the environment at one point will probably travel to other ecosystems. Thus scientists have found DDT in the tissues of penguins in Antarctica, far from any site where this pesticide has been used. Radioactive materials from atmospheric weapons testing have spread over the entire globe. And recently, scientists have found that the level of soot in Arctic air and snow is only three to four times lower than that found in a typical urban environment. Pollution that causes air filters to turn black in one day in New York City will cause filters to turn black within a week in the Arctic. Thus no part of the globe has been untouched by human-generated materials. We consider more specific implications of the global cycling of contaminants in Chapters 7, 10, 13, and 18.

Transfer Rates and Human Activity

Over time—at least until recently—the transfer rates of energy and materials among various components of the environment have become relatively stable and, at least for gaseous cycles, are close to equilibrium. (A *transfer rate* is defined as the amount of energy or material that moves from one component of the environment to another within a specified period of time). For example, if we take a closer look at the carbon cycle in Figure 2.19, we see that on a global basis, the flow of carbon dioxide between the atmosphere and terrestrial ecosystems, without human intervention, is essentially the same in both directions. In other words, approximately as much carbon dioxide is incorporated into plants through photosynthesis every year as is released by the respiration of plants and animals. A look at the patterns of flow between the oceans and the atmosphere also shows that, without human activities, the inflow of carbon equals its outflow.

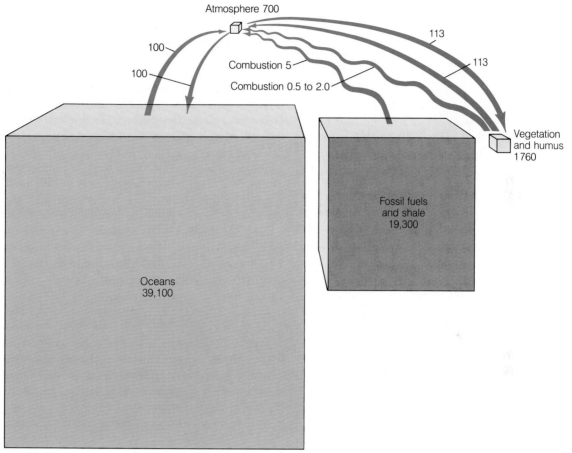

Figure 2.19 *The carbon cycle, showing the amount of carbon that is stored in various reservoirs. Straight lines show flow rates without human intervention between reservoirs. Wavy lines show flow rates that result from human activities. Values are given in gigatons (a gigaton is a billion tons). (After R. Revelle, "Carbon Dioxide and World Climate,"* © Scientific American *247:35–43, 1982. All rights reserved; and C. F. Baes, et al. "Carbon Dioxide and the Climate: The Uncontrolled Experiment."* American Scientist *65:310–320, 1977.)*

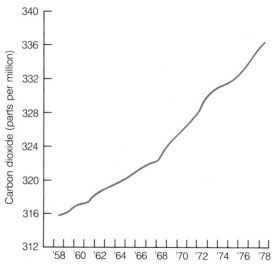

Figure 2.20 *Changes in atmospheric carbon dioxide levels due to increased consumption of fossil fuels and burning of vegetation. (Data from C. D. Keeling et al., Scripps Institute of Oceanography. After R. Revelle, "Carbon Dioxide and World Climate."* © Scientific American *247:35–43, 1982. All rights reserved.)*

In recent years, however, human activity has begun to disturb the equilibrium of transfer rates significantly. For example, as Figure 2.20 illustrates, atmospheric carbon dioxide levels increased by roughly 6 percent from 1958 to 1968, and they are expected to continue to rise. Several factors appear to be responsible for the changes in transfer rates that have caused these elevated concentrations of atmospheric carbon dioxide.

Since the dawn of the industrial revolution, tremendous quantities of carbon dioxide have been emitted into the atmosphere by the burning of coal, oil, and natural gas. As Figure 2.19 illustrates, currently at least 5 billion tons of carbon dioxide are being released each year as a result of the consumption of fossil fuels. In the late 1970s, scientists began to investigate the impact that human destruction of native forests and prairies has had on the carbon cycle. As we have discussed already, the woody parts of trees store large amounts of nutrients, including carbon. The primeval forests, which contain massive trees, are significant sources of carbon. When those forests were cleared and burned for agricultural and urban development in the nineteenth and twentieth centuries, huge amounts of carbon dioxide were released. That process continues today because of our increasing exploitation of tropical forests. Hence some scientists believe that the destruction of virgin vegetation has resulted in a net release of carbon dioxide from the earth's terrestrial ecosystems.

Recent studies indicate, however, that the regrowth of commercial timberlands may increase the absorption of carbon dioxide. Slowly, those regions are beginning to retain carbon dioxide, because they are now taking up more carbon dioxide as a result of photosynthesis than they are releasing through respiration. Thus it appears that the carbon dioxide being released by the destruction of tropical forests is now being offset partially by the increased amounts of carbon dioxide that are being taken up by the regrowth of temperate forests in such regions as the southeastern United States. While scientists continue to debate the magnitude of the net release of carbon dioxide from the clearing of forests, most estimates lie between 0.5 and 2 billion tons a year. The destruction of terrestrial vegetation is one important contributor to increased atmospheric carbon dioxide levels, but the burning of fossil fuels accounts for a release that is roughly three to five times greater.

Interestingly, only about half of the carbon dioxide released into the atmosphere during the past century is still there. Even though some compensating removal of it by the oceans must have occurred, obviously, the rate of human-induced release of carbon dioxide into the atmosphere has been greater than the oceans' ability to absorb it. Thus human intrusion in the carbon cycle has disturbed the transfer rates to the point that equilibrium flow no longer exists.

What are the consequences of this intrusion? If world consumption of fossil fuels and forests continue to accelerate, the carbon dioxide content of the atmosphere will rise approximately 20 percent by the year 2000, and it could double within 60 years. That increase will occur despite compensating transfers into the oceans. The added atmospheric carbon dioxide will probably warm the earth's climate by several degrees, but scientists cannot predict with confidence what the effect will be on local and regional levels. The complexities of climate are extensive and little is known about it at this time. We explore this predicament further in Chapter 10.

The disruptive effects of human activities on transfer rates also can be seen in the phosphorus cycle. Normally, the rate of phosphorus loss from an undisturbed ecosystem is low. But if the protective vegetative cover is removed, for example, when all the trees in a forest are cut down, the loss of nutrients in runoff increases tremendously. The phosphorus cycle is also disturbed when we mine deposits of phosphorus and incorporate it into fertilizers and detergents. Runoff of phosphorus from agricultural lands and its inefficient removal by municipal sewage treatment plants cause increased transfer of phosphorus to lakes and rivers.

Human activities also have altered the nitrogen cycle. Approximately 30 years ago, an industrial process, called the *Haber process*, was developed, which utilizes natural gas as an energy source to synthesize ammonia from atmospheric nitrogen. Today, more than 60 million metric tons of ammonia, synthesized by the Haber process, are added in various forms to the soil as fertilizer, and some of that ammonia runs off agricultural land into lakes and rivers. (That 60 million metric tons of ammonia roughly equals 30 percent of the annual production of ammonia by means of biological fixation in terrestrial ecosystems.) Unfortunately, nitrogen and phosphorus are entering waterways faster than they are being cycled out. Those higher-than-normal concentrations of phosphorus and nitrogen act as fertilizers, accelerating the growth of algae and rooted aquatic plants. And the accelerated plant growth, in turn, is indirectly responsible for the elimination of some species of fish in highly productive lakes. In Chapter 7, we explore the factors that control the cycling of nutrients and their impact on water quality.

Atmospheric oxygen levels do not seem to be endangered by human activity so far, but the corrosive effect of oxygen is visible on many of the products of civilization. For example, it crumbles our buildings and monuments and rusts our automobiles and farm equipment, reminding us that even they are not exempt from interaction with the planet's natural cycles.

Transfer Rates and Pollution

Materials and energy in their myriad forms are transferred continually among the components of ecosystems and among ecosystems themselves. In many instances, those transfer rates reach an equilibrium; that is, the flow into a reservoir essentially equals the flow out of it. But transfer rates may increase or decrease until they are no longer at equilibrium; that is, the flow into a reservoir is not equal to the flow out of it. Figure 2.21 represents schematically three basic patterns of flow into and out of a reservoir. Sometimes transfer rates change for natural reasons, as when tons of gases and particles are spewed into the atmosphere by an erupting volcano. Human activities also can change transfer rates, as we have seen in our discussions of the carbon, phosphorus, and nitrogen cycles.

Another type of change that can occur in an ecosystem is the introduction of substance into an ecosystem where it was not formerly found. Most instances that are significant to us involve such products of human technology as pesticides and radioactive materials.

From our considerations thus far, we cannot escape the fact that human activities disturb the natural flow within ecosystems. Since we are a part of the environment, it is inevitable that our activities will cause some changes in transfer rates of materials and energy within and between ecosystems.

If damage to human health or harm to plants or animals results from a change in transfer rates, then the overall process is called *pollution,* and the type of material or form of energy that is involved in bringing about that change is called a *pollutant.* Hence the increased flow of available phosphorus and nitrogen into waterways is a form of pollution.

Although we normally think of pollution as being a consequence of human activity, changes in transfer rates can occur independently of human activity.

Input equals output:
no change in water level
(equilibrium)

Output greater than input:
water level drops

Input greater than output:
water overflows

Figure 2.21 *Three possible patterns of transfer rates into and out of a reservoir.*

For example, volcanic activity, such as the eruption of Mount St. Helens in May 1980 (Figure 2.22), can devegetate thousands of hectares of land (a hectare equals approximately 2.5 acres). However, it is important to note that the effects of sporadic, non-human pollution are usually short term, lasting only a matter of days or at most a few years, whereas the continual impact of human activity may well have effects that last much longer. Already, vegetation has reappeared in the devastated area on Mount St. Helens (Figure 2.23). It will be many decades, however, before mature forests are reestablished.

Note that, by definition, pollution need not have a *direct* effect on organisms. For example, human

Figure 2.22 *This devastated slope in the Pine Creek area was covered with 10 to 50 centimeters (4 to 20 inches) of mud when Mount St. Helens erupted four months earlier. (Courtesy of Lawrence C. Bliss, Department of Botany, University of Washington.)*

activities have greatly increased the amount of carbon that is transferred from the earth to the atmosphere. Scientists believe that a continued increase in atmospheric carbon dioxide will cause a warming of the climate worldwide; and that change will, in turn, have detrimental effects on many species of plants and animals, including ourselves. In this example, carbon dioxide is an atmospheric pollutant, because it will affect indirectly the well-being of organisms by causing climatic change.

The major questions facing us include whether the disturbance is acceptable or unacceptable, where to draw the line to determine what is and what is not acceptable, and what exactly constitutes an unacceptable disturbance (pollution)? How seriously does the well-being of organisms have to be impaired before it becomes unacceptable to us? How many organisms must be seriously impaired or killed? Are we concerned about all organisms or just a few, including ourselves? If we are not concerned about

all organisms, what criteria do we use to decide which species we really are concerned about? There is no one answer to any of these questions. Viewpoints differ from those of hard-line industrialists, whose main concern is profit, to those of hard-line environmentalists, whose main concern is saving all of nature.

What are the consequences of changes in transfer rates for organisms, including ourselves? What are the limits of an organism's ability to adjust to change? What happens when those limits are exceeded? We explore these questions in Chapter 3.

Conclusions

In this chapter, we confront two related concepts that occur continually in environmental science: interrelationship and balance. We find that we cannot define an ecosystem without describing the interactions among the organisms and physical environment that compose it. We cannot describe the pathways taken by the flow of energy and materials without showing how each flow affects and inter-

Figure 2.23 *During the year after the Pine Creek area was covered with a layer of mud, precipitation and meltwater eroded away the mud. Some plants survived being buried and have slowly recovered. This photo was taken two years after the eruption. (Courtesy of Lawrence C. Bliss, Department of Botany, University of Washington.)*

acts with the others. And we really cannot describe environmental balance without referring to the dependency of each component or process in an ecosystem on the smooth working of all its interrelated parts.

These concepts are particularly important when we try to measure the effect that we have on the environment. If a raindrop strikes a pond, the effects of that disturbance will be felt in ever-widening circles. But unlike those natural ripples, which have a transient effect, man-made ripples may continue to disturb the environment for decades and perhaps centuries. One reason for studying environmental science is to make ourselves more sensitive to the effects of man-made ripples on the function of ecosystems and thus on our own future well-being.

Summary Statements

Most of our environmental problems are caused by disruptions of the flow of energy and materials through the environment.

Ecosystems contain both living and nonliving components. Organisms are distinguished by their major functions: producers (green plants) manufacture food, whereas consumers (animals, fungi, and certain types of bacteria) consume other organisms to obtain food. The physical environment comprises the nonliving portion of the ecosystem.

Within ecosystems, food energy moves only in one direction—from producers to consumers—in complex networks of pathways, called food webs. Two important types of food webs exist: grazing and detritus. The relative significance of each depends on the type of ecosystem in question.

For many reasons, only a small percentage of food energy is transferred from one trophic, or feeding, level to the next. Because of that inefficiency, any given geographical area can feed only a limited number of organisms.

With a knowledge of food-web dynamics, the quantity of food available to us can be increased by (1) reducing the number of organisms that compete for the same food, (2) converting forests and rangeland into cropland, (3) increasing the efficiency of energy use by livestock by improving animal husbandry practices, (4) growing crops that put more photosynthetic energy into edible parts, and (5) eating less meat and more fruits, vegetables, and cereals. All those efforts are limited by the energy inefficiencies that are inherent in food webs. Most efforts have also resulted in significant environmental degradation.

Although energy is supplied continually by the sun, the quantity of materials on earth is fixed and finite. The carbon, phosphorus, nitrogen, and oxygen cycles illustrate many of the pathways whereby necessary materials are recycled.

Toxic materials can enter food webs and become considerably more concentrated as they move into higher trophic levels.

Materials not only cycle within ecosystems, but also between them. Through such transfers, energy and nutrients travel from terrestrial ecosystems to aquatic ecosystems and back again. In addition, air currents and waterways help to disperse and dilute sudden surges of materials into an ecosystem, but that dispersal spreads toxic materials as well as necessary ones around the globe.

Over past millennia, the rates of energy and material transfer among various components of the environment have become relatively stable. In recent years, however, human activity has begun to disturb those rates significantly. Some of the consequences include a decline in air and water quality and a probable change in climate.

Pollution is any change in transfer rates of materials or energy that, either directly or indirectly, has an adverse effect on the well-being of organisms. Pollution can be either human-induced or natural.

Questions and Projects

1. In your own words, write a definition for each of the terms that are italicized in this chapter. Compare your definitions with those in the text.

2. What is a detritus food web? How does it differ from a grazing food web?

3. Visit several types of ecosystems in your region. For each, determine whether grazing or detritus food webs predominate. Describe the types of evidence that you used to make your determination.

4. The percentage of digestible food in the diet helps to determine the amount of energy that is transferred to the next trophic level. For each kilogram of food intake, which herbivore would obtain the most usable energy—a herbivore that eats seeds, wood, young foliage, or mature foliage? Explain your answer.

5. Why are ecological efficiencies in natural ecosystems commonly as low as 1 percent?

6. Describe the uses of fossil-fuel energy that are necessary to raise the ecological efficiencies of livestock up to 10 to 20 percent.

7. Describe the importance of low ecological efficiencies in evaluating mankind's attempts to feed an ever-growing human population.

8. Why is it that the diet of most Americans, Canadians, and Europeans contains significant quantities of meat, while most citizens of South America, Africa, and Asia eat very little meat? Do you expect those differences in diet to continue over the coming decades? Explain.

9. Describe several advantages that accrue to an animal that can occupy more than one trophic level.

10. Trace the route that phosphorus might follow as it cycles through a terrestrial ecosystem. Include at least four organisms in the cycle.

11. Trace the route that carbon might follow as it cycles through a terrestrial ecosystem. Include at least four organisms in the cycle.

12. How are the cycles that you constructed in Questions 10 and 11 similar? How do they differ?

13. How does the cyclical flow of phosphorus through an aquatic ecosystem differ from its cyclical flow through a terrestrial ecosystem? How are they similar?

14. Describe the roles of respiration and photosynthesis in both the carbon cycle and the oxygen cycle.

15. How does biological fixation of nitrogen differ from atmospheric fixation?

16. Microorganisms play many essential roles in the nitrogen cycle. Identify at least five points in the nitrogen cycle at which microbes are involved.

17. How can the rates of nutrient cycling within an ecosystem influence the rates of energy flow within the same ecosystem?

18. Describe the process of food-web accumulation. Describe the importance of setting limits on the quantities of persistent contaminants that are released into the environment in view of this phenomenon.

19. Millions of metric tons of ammonia are produced annually by the Haber process. What effects might this additional transfer of nitrogen from the atmosphere to the soil have on the environment?

20. A large amount of a poisonous chemical has been released in a lake. What criteria would you use to determine whether this pollution is acceptable or not? What kinds of evidence would be necessary to convince you that the pollution should be cleaned up and further incidences prevented?

21. Take a tour of your region and note examples of pollution. Does your community judge these incidences of pollution to be acceptable, or is something being done to control the pollution? What factors play a role in a community's decision to either ignore or to solve a pollution problem?

Selected Readings

ANNUAL REVIEWS, INC.: *Annual Review of Ecology and Systematics.* Published yearly. Palo Alto, Calif.: Annual Reviews, Inc. Each volume contains articles that review current research in energy and materials flow in the environment.

BARBER, R. T.: Biological consequences of El Niño. *Science* 222:1203–1210, 1983. An informative account of the impact of the 1982–1983 El Niño on algal productivity and the subsequent impact on higher trophic levels.

BELL, R. H. V.: "A Grazing Ecosystem in the Serengeti." *Scientific American* 226:86–93 (July), 1971. A description of how migrations of herds of zebra, wildebeest, and Thomson's gazelle are synchronized with the availability of specific food sources.

BOLIN, B., AND COOK, R. B.: *The Major Biogeochemical Cycles and Their Interactions.* New York: Wiley, 1983. A comprehensive examination of the major cycles and the impact of human influences.

BRILL, W. J.: "Biological Nitrogen Fixation." *Scientific American* 236:68–81 (March, 1977). An examination of how a few types of bacteria and blue-green algae are major contributors to the earth's nitrogen cycle and how they can be used to improve soil fertility.

DEL MORAL, R.: "Life Returns to Mount St. Helens." *Natural History* 90:36–46 (May), 1981. An account of devastating impact of the eruption on wildlife and vegetation and their initial recovery.

FRIDEN, E.: "The Chemical Elements of Life" *Scientific American* 227:52–60 (July, 1972). A good introduction to the role of chemical elements in organisms.

GOSZ, J. R., HOLMES, R. T., LIKENS, G. E., AND BORMANN, F. H.: "The Flow of Energy in a Forest Ecosystem." *Scientific American* 238:93–102 (March, 1978). A comprehensive examination of energy flow through food webs in a forest and adjacent stream.

REVELLE, R.: "Carbon Dioxide and World Climate." *Scientific American* 247:35–43 (August, 1982). An examination of human disruption of the carbon cycle and the potential impact on world climate.

SALATI, E., AND VOSE, P. B.: Amazon Basin: A system in equilibrium. *Science* 225:129–138, 1984. An examination of the nutrient and water cycles within the Amazon ecosystem and the potential consequences of large-scale deforestation on the current neo-equilibrium conditions.

SMITH, R. E.: *Ecology and Field Biology.* New York: Harper & Row, 1980. A well-written book, containing informative chapters on energy flow and nutrient cycling.

A minute change in elevation results in a major change in vegetation in Everglades National Park. This palm hummock is only a meter or so higher than the watery sawgrass prairie that surrounds it. (U.S. Department of the Interior, National Park Service. Photo by M. W. Williams.)

CHAPTER 3

ECOSYSTEMS AND ENVIRONMENTAL CHANGE

All environments change. The weather varies from hour to hour and day to day; the seasons come and go; rivers and lakes rise and fall; and forests and grasslands go up in flames. The very face of the planet changes as wind and water sculpts the land, volcanoes erupt, mountain chains rise, plains subside, glaciers flow, the seas advance and retreat, and the continents themselves drift over the face of the globe. Organisms die and are succeeded by their offspring; their populations wax, wane, and eventually disappear, to be replaced by those of other organisms; their remains build the soil and shape the seabed. Human beings, too, cause significant changes. We dam rivers, clear forests, reclaim farmland from the sea, cover the terrain with our agricultural and urban ecosystems, alter the rates of nutrient cycling, and foul the air, water, and land with our wastes.

To determine the effects that those changes have on organisms, including ourselves, we must first grasp the fact that each organism is a vastly complex entity whose survival depends on several intricate processes. First, all living organisms must obtain nutrients and water from their surroundings. Those nutrients must be processed, or metabolized, for maintenance, growth, and reproduction, and wastes, the by-products of metabolism, must be excreted back into the environment. Finally, all organisms receive and respond to stimuli from their surroundings. Thus for an organism to remain alive, each of those intricate processes must occur, and each must interact properly with the others. If even one process malfunc-

tions, the organism's survival is imperiled.

In most cases, organisms can adjust to the changes that occur in their environment, such as changes in the season, which initiate various physical and behavioral responses. For example, foxes and wolves adjust to the coming of winter by growing a thicker coat of fur; deer congregate in small, sheltered areas and greatly reduce their movement and foraging activities; chipmunks and groundhogs go into hibernation; and many birds and some insects migrate to warmer climates. We adjust to colder weather by wearing insulated clothing, staying indoors more often, and heating our homes. (Some of the more affluent persons follow the birds south.) The ability to adjust to change is essential to survival in an environment where change is inevitable.

However, some environmental changes may be so great or may occur so rapidly that some organisms cannot adjust to them. For example, massive fish kills as a result of dissolved oxygen depletion in streams and increased numbers of persons who suffer respiratory ailments during smoggy periods attest to the limitations that organisms have in adjusting to environmental change.

In this chapter, we examine the factors that determine an organism's ability to respond to and survive environmental change. We also consider the responses of ecosystems to change. An understanding of these processes helps us to realize the impacts of pollution and the problems of resource management, which are considered in subsequent chapters.

Limiting Factors

Many environmental factors change with time. Those factors include light levels; temperature; humidity; wind speed; availability of food, water, and mineral nutrients; pollutants; and the presence or absence of various species. As those factors change, they influence the well-being and survival of organisms. To understand the response of an organism to environmental change, we need to consider two laws—the law of tolerance and the law of the minimum.

The Law of Tolerance

An organism's ability to respond to an environmental change is illustrated by the *law of tolerance.* This principle states that for each physical factor in an environment, a minimum and a maximum limit exists, called *tolerance limits,* beyond which no members of a particular species can survive.

To illustrate the law of tolerance, let us return to the problems of phosphate enrichment and the excessive growth of algae and weeds in lakes, as discussed in Chapter 2. We can perform a simple experiment to demonstrate a cause-and-effect relationship between algal growth and phosphate concentration. If we place algae in an aquarium that is devoid of phosphorus but contains all other necessary nutrients, the algae will die; they—like all organisms—require a minimum amount of phosphorus to survive. If, little by little, we continue to add phosphorus to the aquarium, eventually a concentration will be reached that favors the survival of a few algae. That concentration represents the *lower tolerance limit* of phosphorus for algae (see Figure 3.1). If we continue to add phosphorus, the number of algae will increase until a maximum population is reached. The phosphorus concentration at that point is called the *optimum concentration.* Above the optimum concentration, additional phosphorus will become toxic to some algae, and the algal population will decline. At a still higher level, the phosphorus concentration will become lethal to even the most phosphorus-tolerant algae. The concentration above which no algae can survive is called the *upper tolerance level.*

In some respects, the law of tolerance supports the adage of moderation in all things. Algae require phosphorus to function properly (as do all organ-

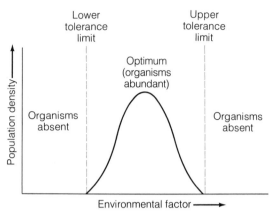

Figure 3.1 *The law of tolerance illustrates the ability of organisms to respond to environmental change.*

isms), yet excessive amounts can be fatal to them. In a similar way, we require many vitamins and minerals to maintain good health, but too much or too little of those nutrients can weaken, sicken, or even kill us. For example, an insufficient supply of vitamin A will lead to drying of the skin, abnormal bone formation, and night blindness, whereas an excess of vitamin A will produce gastrointestinal upset, dermatitis, loss of hair, and pain in the bones. Either too little or too much of any required factor, such as food energy, vitamins, minerals, water, heat, or oxygen, can jeopardize the survival of organisms and even an entire species.

The tolerance limits for each factor are determined by the genetic makeup of the organism (the information that is contained within its hereditary material). It is important to note that each individual in most populations of plants and animals is genetically unique. We are reminded of this point if we look around the classroom. Each person looks and behaves differently from all the others. (Even identical twins, who have the same genetic makeup, will differ slightly in appearance and behavior.) It also follows that if each person has a different genetic makeup, each person also has slightly different tolerance limits from others. For example, you may have a friend who catches a cold or the flu every winter and another friend who seemingly has never been sick a day in his or her life. Remember that a tolerance curve represents the overall response of a population composed of many individuals, each of which has tolerance limits that are somewhat different from those of the others. Thus as a population is exposed to any factor that is above or below the optimum concentration for that population, the ones that are least tolerant may die, while the more tolerant ones may experience no harm or discomfort.

If individuals within a population have different tolerance limits, what about differences in tolerance limits among species? Figure 3.2 illustrates some differences in heat tolerance among species.

The Law of the Minimum

Having observed the effect of increased phosphorus levels on algal growth in our aquarium experiment, and knowing that blooms, or dense mats, of algae

Figure 3.2 *Tolerance to temperature varies with species. Each bar represents the temperature range in which the animal can maintain a constant temperature without resorting to special temperature-regulating mechanisms. Note that the young have considerably narrower tolerance limits than the adults. (Data from G. W. Cox and M. D. Atkins,* Agricultural Ecology. *New York: W. H. Freeman, 1979; and H. Precht, J. Christophersen, H. Hensel, and W. Larcher,* Temperature and Life. *New York: Springer-Verlag, 1973.)*

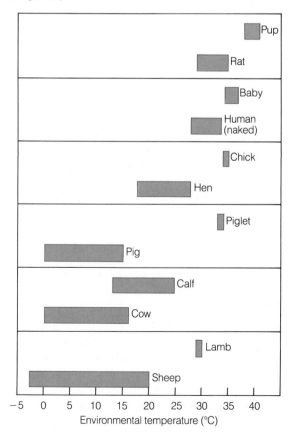

cover many ponds, lakes, and streams, we might conclude that the addition of phosphorus-rich sewage and agricultural runoff to waterways has raised their phosphorus concentrations to levels that approach the optimum level for algal growth. But, of course, other factors, including water temperature, light penetration, and concentration of other nutrients (such as nitrogen), influence algal growth as well. Thus one of those factors may have actually limited the growth of algae. Thus we shall now consider another ecological principle—the *law of the minimum*. This principle states that the growth and well-being of an organism ultimately is limited by that essential resource that is in lowest supply relative to what is required. That most deficient resource is frequently called the *limiting factor*. Hence if we visit a pristine mountain lake and we find only a few algae and rooted plants growing, further examination will show that the lake is also poor in nutrients. We might assume that the low availability of one nutrient, such as phosphorus, is the factor that is limiting algal growth. But some other factor, perhaps low water temperature, might be the limiting one. Thus more experimentation is needed to determine which factor is actually limiting the growth of algae in the lake.

Agriculture is an enterprise in which much effort is expended to reduce the impact of limiting factors on crop yields. For example, fertilizers overcome the growth-retarding effects of low soil fertility; irrigation eases the stresses of deficient soil moisture on plants; and pesticides reduce the yield-limiting impact of weeds, insects, and pathogens. In Chapter 17, we evaluate the efforts that are made to reduce the impact of limiting factors and thereby increase food production for a hungry world.

Determining Tolerance Limits

Determining the tolerance limits of a given organism for a particular environmental factor can be extremely difficult. One complication is that an organism usually has the genetic capacity to adjust to changes in its surroundings. Thus, for example, visitors to high mountains experience headaches, dizziness, and shortness of breath because the air at high altitudes contains less oxygen per unit volume. Those distressing symptoms indicate that the minimum tolerance level for oxygen is being approached. However, after 1 or 2 weeks of exposure to those same high altitudes, the symptoms usually disappear. We say that persons who make such adjustments have become *acclimatized* to the altitude. Their bodies adjust to the low oxygen concentrations by increasing the number of red blood cells (carriers of oxygen) in their bloodstreams. Furthermore, some persons are able to adjust better than others because of differences in their genetic constitutions. Still, a minimum limit does exist, and below that limit, no human being can adjust; thus at the summit of Mount Everest, an elevation of 8850 meters (29,028 feet), the oxygen level is so low that only those who are both genetically inclined and carefully acclimatized will not suffocate slowly. At higher altitudes, even hardy persons must be equipped with an auxiliary supply of oxygen.

Age is another factor that influences an organism's tolerance limits. In all species, the very young usually suffer the most harm from a stressful environment. A child whose diet is consistently deficient in protein is at risk of suffering permanent brain damage, whereas an adult on a similar diet probably would not suffer this effect, because his or her brain is already fully developed.

Despite these complications, we can determine the tolerance limits of an organism relatively simply if we can isolate one species and change only one factor at a time. However, this can only be done in a laboratory; in a natural, uncontrolled environment, many factors are operating and interacting simultaneously, often with unforeseen consequences. Some interactions create an effect that is called *synergism*. The interaction of two or more factors is said to be synergistic if the total effect is greater than the sum of the two effects acting independently. In other words, the whole may be greater than the sum of its parts. For example, sulfur dioxide fumes, a common air pollutant, attack the lungs and impair the respiratory tract. However, when laboratory animals are exposed to air that contains both sulfur dioxide and particulates (tiny particles of soot, ash, and so forth), their lungs sustain much greater damage than we would expect to occur if they were exposed to each pollutant separately and in sequence. Thus the interaction of sulfur dioxide with particulates is a synergistic interaction.

In contrast, two factors are said to be *antagonistic* if the total effect is less than the sum of the two factors acting independently. For example, nitrogen oxide is an air pollutant with effects that are similar to those of sulfur dioxide. But when nitrogen oxide and particulates interact, they somehow counteract each other, and their deleterious impact on the lungs is much less severe than we would expect. Thus the interaction of nitrogen oxide and particulates is an antagonistic interaction. Therefore, to predict confidently what the effect that a specific factor will have on an organism in a natural environment, we must first identify all the other factors that will interact with it and then try to determine the final outcome.

In the larger context of an ecosystem, the determination of limiting factors becomes formidably difficult. An ecosystem is composed of hundreds or thousands of species, each of which is influenced by both the physical environment and other species. Moreover, the number and kinds of species that are present and such physical factors as light levels, air temperature, and soil moisture fluctuate daily and seasonally.

Few simple cause-and-effect relationships exist in nature, so most predictions about the impact of an environmental change must be qualified. In most situations of interest, so many interactions are possible that it is difficult for scientists to identify them, let alone predict their outcome. And often that complexity leads to disagreement about the causes of (not to mention the solutions to) an environmental problem.

A case in point is the acid-rain controversy. In parts of the Adirondacks, northern New England, and southeastern Canada, lakes and streams are becoming acidic, and aquatic organisms are dying (Figure 3.3). Acid rain, which develops in the atmosphere as a result of interactions between sulfur dioxide and oxides of nitrogen (acid-rain precursors) and water vapor, is the suspected culprit. One proposed solution to the acid-rain problem has been to reduce the levels of sulfur dioxide in the atmosphere but, first, the major sources of sulfur dioxide in those regions must be identified.

The suspected source of sulfur dioxide is comprised of the coal-burning power plants that are located in and near the Ohio River Valley. Coal contains sulfur, and when coal burns, sulfur dioxide forms. Because prevailing winds generally move air from the Ohio River Valley toward New England and southeastern Canada, many scientists believe that those power plants are probably the main source of acid-rain precursors. But others, particularly those

Figure 3.3 *In an Adirondack stream, brook trout, confined in a wire cage, succumb to the effects of increased acidity of the stream water. (© Ted Spiegel, Black Star.)*

associated with the coal-mining and the electric-utility industries, point out that the relationships between acid rain and those sources of sulfur dioxide have not been established with scientific certainty. In other words, scientists have not actually traced sulfur dioxide from the smoke stacks in the Ohio River Valley to the East Coast, and they have not proved that those particular molecules of sulfur dioxide react with water vapor and, subsequently, fall to the earth in the form of acid rain. Given the complexities of the atmosphere, gaining this scientific evidence will be quite difficult.

An additional complication of acid rain stems from recent suggestions by soil scientists that soil-formation processes may also contribute to the acidification of lakes. The forested lands of eastern North America have been subject to cutting, burning, and subsequent cultivation. In many areas, agriculture has been abandoned and the land that had been cleared is now becoming reforested. As a consequence, soils are maturing and becoming more acidic, which, in turn, contributes to the acidificiation of

lakes. But again, little scientific data exist to quantify the actual contribution of soil maturation to lake acidification.

Lake acidification well illustrates the complexities of making cause-and-effect determinations about nature. The relative importance of natural soil acidification to water acidification versus the contribution to it by sulfur dioxide emissions remains unknown. But we do know that control of sulfur dioxide emissions will be expensive. Hence many industries are reluctant to initiate emission controls until the causes and effects are much better understood. In the meantime, the problem of lake acidification is becoming more severe.

Tolerance to Toxic Materials

To survive, all organisms must obtain energy, water, minerals, and air. But in the process, they are all exposed to substances that are *toxic* (poisonous). Some of those substances, such as mercury and lead, have always been in the environment in trace amounts. Today, however, industrial processes concentrate them to dangerous levels and then release them into the environment. Moreover, many toxic chemical products and by-products that are produced today were unknown a few decades ago—and dangerous amounts of those substances also enter the environment.

Determining Toxicity Levels Figure 3.4 illustrates a *dose-response curve,* which shows the response of an organism to increasing doses of a toxic substance. At very low levels of exposure, the body repairs all the damage, if any, that is caused by a toxic substance and recovery is complete. At higher levels of exposure, the body makes compensations. Although damage to the body may be ongoing, the symptoms may disappear. For example, industrial workers who are exposed to formaldehyde may stop feeling irritation in their lungs, but their continued exposure to formaldehyde will promote continued damage. As the dose increases even higher, adaptation will give way to symptoms that increase in severity. At higher doses, permanent disability will result. And, finally, at some point, a fatal dose will be reached.

The quantity that is needed to reach a particular point on the dose-response curve varies greatly among toxic agents. One measure of the danger that is associated with an agent is the LD_{50}, which is the dose of a substance that will kill 50 percent of a test population. That index is measured in units of poisonous substance per kilogram of body weight. For example, botulinus toxin–the very deadly bacterium-produced chemical that causes botulism (a form of food poisoning)—has an LD_{50} in adult human males of 0.000014 milligrams per kilogram. Thus if a dose of only 0.0014 milligrams (which is roughly equivalent to a few grains of table salt) is consumed by 100 adult human males who each weigh 100 kilograms (220 pounds), approximately 50 of them will die. Note again that within a population, the different genetic constitutions among persons render some of them less tolerant than others to particular factors—in this case, botulinus toxin. Table 3.1 illustrates the wide range of LD_{50} values among different chemical agents.

Lethal doses are not the only danger of toxic substances. Today, industry and government researchers are trying to determine the *threshold limit value* of many toxic substances: that is, the maximum concentration of a substance to which a person may be exposed for a long period of time without experiencing harmful effects. That information is especially important in industries where employees are frequently exposed to hazardous substances that escape as a result of manufacturing processes. If the threshold limit values are known and emissions are found to exist above the threshold, then either the

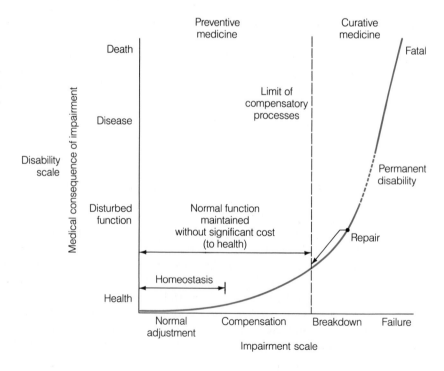

Figure 3.4 *A dose-response curve, illustrating the response of an organism to increasing doses of a toxic substance. (From T. F. Hatch, Changing objectives in occupational health.* Journal of the Association of American Industrial Hygenists *23:1–7, 1962. Reprinted with permission by the American Industrial Hygiene Association Journal.)*

Table 3.1 Approximate acute LD_{50} of a variety of chemical agents

AGENT	LD_{50} (mg/kg)	TOXICITY RATING	APPROXIMATE LETHAL DOSE FOR HUMANS
Ethyl Alcohol	10,000	Slightly Toxic	Between Pint and Quart
Sodium Chloride (Table Salt)	4,000	Moderately Toxic	Between Ounce and Pint
Morphine Sulfate	1,500		
DDT	100	Very Toxic	Between Teaspoonful and Ounce
Strychnine Sulfate	2		Few Drops
Nicotine	1		
d-Tubocurarine (Curare)	0.5	Super Toxic	One Drop
Dioxin (TCDD)	0.001		$\frac{1}{500}$th of a Drop
Botulinus Toxin	0.00001		$\frac{1}{50,000}$th of a Drop

Source: J. Doull, C.D. Klaasen, and M.O. Amdur, eds., *Casarett and Doull's Toxicology*, 2d ed. New York: MacMillan, 1980.

emissions can be reduced or work schedules or ventilation systems can be modified to lower the employees' exposure below the threshold level.

The process that is required to determine the toxicity of a chemical is complex. Toxicity of a substance depends in part on the age, sex, and general health of the person who is exposed to it; a dose that is lethal to one person may produce only mild symptoms in another. Furthermore, the length of time that a person is exposed to a toxic substance further complicates the determination of a toxic level. *Acute exposure* is defined as a single exposure that lasts from a few seconds to a few days. *Chronic exposure* is defined as a continuous or repeated exposure for several days, weeks, months, or even years. For example, acute exposure usually results from an accident, such as the release of chlorine gas from a ruptured railroad tank car. Acute exposures to lethal doses often make disaster headlines in newspapers, but chronic exposure to sublethal quantities of a toxic material presents a much greater hazard to public health. Chronic exposure rarely receives as much notoriety as acute exposure. Examples of chronic exposure are the frequent exposure of smokers to the toxins in cigarette smoke, the daily exposure of certain construction workers to airborne asbestos particles, and the continual exposure of city dwellers to low levels of sulfur dioxide that are produced by burning coal or fuel oil. Many deaths that are attributed to emphysema or cardiac arrest may actually be caused by a lifetime of exposure to sublethal amounts of air pollutants.

Carcinogens, Mutagens, and Teratogens We face daily exposure to a growing list of chemicals that take many forms—food additives, medicinals, pesticides, adhesives, sprays, cleaners, cigarette smoke, and insulating materials, to name a few. Particularly worrisome to many persons is the fact that some of those chemicals act as carcinogens, mutagens, or teratogens, all of which can have catastrophic effects on our health.

Table 3.2 lists some environmental agents that act as carcinogens, mutagens, or teratogens in human beings. Note that some agents can be harmful in more than one way. In addition to those chemicals, high-energy radiation, such as x-rays, can act as carcinogens, mutagens, and teratogens.

Carcinogens are environmental agents that cause cancer. The incidence of cancer has risen sharply in recent years. Today, in the United States, cancer is second only to heart disease as a cause of death; more than 400,000 persons die from cancer each year.

Mutagens are environmental agents that cause *mutations*, or random, heritable changes, in *chromosomes*, which are the cellular structures that contain hereditary information. Although some mutations may be beneficial, most are harmful because a random change has occurred in the natural functioning of an organism. For example, suppose that you take a fine Swiss watch, remove its back, grab a screw driver, and randomly make a change in the watch's mechanism. Do you think that the watch will run better or worse after you have completed the random change? In comparison with a watch, a

Table 3.2 Some environmental agents that are hazardous to human health

CARCINOGENS	TERATOGENS	MUTAGENS
Asbestos Fibers	Methyl Mercury	Mercury Compounds
Nickel Compounds	Lead Compounds	Lead Compounds
Chromium Compounds	Alcohol	Benzo(a)pyrene
Nitrosamines	Diethylstilbestrol (DES)	Mustard Gas
Vinyl Chloride	Thalidomide	Ultraviolet Light
Benzo(a)pyrene	Rubella (German Measles)	X-rays
Tar in Cigarette Smoke	X-rays	
X-rays		
Ultraviolet Light		

living organism is an immeasurably more intricate entity that is comprised of numerous finely tuned, smoothly running, interacting processes. Thus it is easy to understand why mutations (which are also random events) are much more likely to be deleterious than beneficial.

Teratogens are environmental agents that cause abnormalities in fetuses during their interuterine period. Persons frequently associate teratogens with monstrous deformities, such as the shortening or absence of arms or legs. Such was the impact from thalidomide—a sedative that was taken by thousands of pregnant women in West Germany during the early 1960s. But usually the effects of teratogens are less obvious. For example, rubella (German measles) infection during the first trimester of pregnancy may produce cardiac defects and deafness. And sometimes, the deleterious effect of a teratogen does not appear until many years after the mother has been exposed to it. For example, DES (diethylstilbestrol) was taken by pregnant women in the United States for more than 30 years to prevent miscarriages. Only in the past decade has the legacy of that drug become evident. Many daughters of women who took DES have difficulty getting pregnant and carrying a fetus to full term. Many sons of those women have a low sperm count. In addition, the risk of vaginal or testicular cancer appears to be significantly greater in children whose mothers took DES.

Since the turn of the century, human mortality from a wide variety of diseases, including pneumonia, influenza, tuberculosis, and diphtheria, have been reduced greatly (Figure 3.5). As a result, the average life expectancy at birth in the United States has increased from 47 years in 1900 to 75 years in 1983. That greater longevity has provided more time for the manifestation of symptoms that are caused by exposure to toxic and hazardous materials. Thus, as Figure 3.5 also illustrates, cancer has become a more significant cause of mortality. In Chapter 13, we consider more fully the risks that are associated with exposure to toxic and hazardous materials, and we evaluate the efforts that are being made to reduce those risks.

Ecological Succession

Sometimes a habitat is so disturbed that the tolerance limits of virtually all its organisms are exceeded. As a result, little life remains on the site. Such disturbances include agricultural plowing, clearing of forests, construction (dams, highways, towns, and so forth), industrial pollution, fire, wind, and volcanic lava flows.

Interestingly, when the disturbance ends, many plants and animals quickly invade the barren site. Successful invaders are those that can survive the stressful conditions of bare soil, rock, or sand. That first stage in recolonization is called the *pioneer stage,* and the successful invading species are called *pioneer species.* Over a period of months and years, those pioneer species are gradually replaced by other types of colonizers. The process of replacement, or succession, then continues until the assemblage of organisms is relatively stable; that is, able to maintain itself on the site (barring another major environmental disturbance). The relatively stable stage

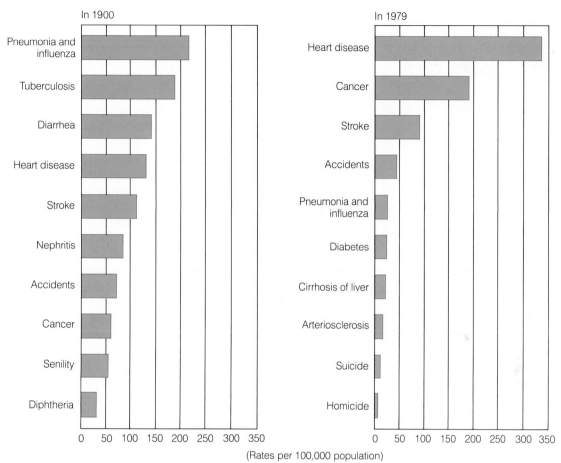

In 1900

In 1979

(Rates per 100,000 population)

Figure 3.5 *The leading causes of death among Americans. In 1900, the two leading causes of death were communicable diseases; by 1979, neither was communicable. Instead, heart disease and cancer now lead as causes of death. Both of these are influenced by such environmental factors as smoking and diet and thus illustrate the law of tolerance. (After the National Center for Health Statistics, U.S. Department of Health, Education, and Welfare, 1982.)*

of recolonization is called the *climax stage,* and the whole series of changes that occur in which one ecosystem is replaced by another until a relatively stable ecosystem is established (that is, one that is best adapted to that environment) is called *ecological succession.* Ecological succession can be seen as a series of recovery stages, through which a severely disrupted ecosystem evolves following a major disturbance.

The type of climax vegetation that emerges depends on the interaction of many factors, including the type of soil, the amount of precipitation, the

temperature, the drainage conditions, and the altitude. For example, the natural climax ecosystem of the Great Plains is grasslands (which have been converted to wheat fields), primarily because grasses are more tolerant than trees of the relatively dry climate and because frequent thunderstorms ignite fires that kill woody vegetation and promote the growth of grasses.

Ecologists recognize two basic types of succession: primary and secondary. Which one takes place depends on the soil conditions at the beginning of the process.

Primary Succession

Primary succession occurs on such surfaces as solidified volcanic lava, bare rock, or sand dunes, where no soil exists. Thus most plants cannot enter those barren areas until a soil forms in which they can grow. Soil development on bare rock often begins

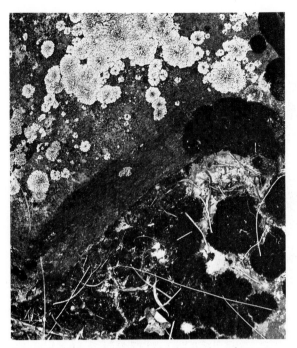

Figure 3.6 *Lichens and mosses, such as those shown here, contribute to the slow process of soil building. (M. L. Brisson.)*

with the establishment of mosses and lichens (such as those shown in Figure 3.6) in small cracks and depressions. There they trap small bits of debris and slowly grow outward. Concurrently, acids are produced by those organisms and water freezes and thaws in the tiny rock crevices, which break down the rock very slowly. Gradually, the soil depth increases, which leads to greater soil moisture and the inclusion of more organic matter, which, in turn, eventually leads to an invasion of grasses and herbs. With further soil development, shrubs may appear. If an extended period of dry weather occurs, the plants growing in those shallow pockets of soil may be killed and reinvasion of those pockets will have to begin again. Hence the initial stages of primary succession are extremely slow. Thousands of years may be needed for a cover of soil to develop that is deep enough to support shrubs and trees.

Once soil has been formed, however, the subsequent stages of primary succession are usually similar to those of secondary succession in comparable environments. *Secondary succession* occurs when soil is present but disturbances, such as agriculture or fire, have removed the natural vegetation. Because the soil usually remains after such disturb-ances have ended, revegetation normally begins within a few weeks.

Secondary Succession

The succession that occurs in abandoned agricultural fields, which is known as *old-field succession*, nicely illustrates secondary succession. For example, in the Piedmont region of North Carolina, European settlers cleared away the native oak–hickory forests and farmed the land. In subsequent years, the soil became depleted of its nutrients because of improper farming practices, thereby reducing the soil's fertility. As a result, after several generations, the farmers had to abandon their fields and migrate westward. We can reconstruct the successional events that occurred on those uncultivated farmlands by examining fields that have been abandoned for different lengths of time.

Abandoned fields in the Piedmont region of North Carolina did not revert immediately to the original oak–hickory forests, even though oak and hickory trees nearby were a source of seeds. Rather, a series of assemblages of plants and animals successively occupied the fields before an oak–hickory forest was reestablished. Figure 3.7 shows three stages in old-field succession. In the year following abandonment, a field is colonized by a mixture of pioneer species that includes crabgrass, pigweed, and horseweed. The next year, the field is dominated by asters, which are succeeded, the following year, by a nearly pure stand of a grass that is commonly called broom sedge. The pace of succession then slows; the broom sedge persists for as long as 20 years, and during that time the field is gradually invaded by pine seedlings. Eventually, the pines mature and shade out the broom sedge. Later, oak and hickory seedlings slowly establish themselves beneath the pine trees, and as the pines grow old and die, they are replaced by the hardwoods—the oaks and hickories. Eventually (some 150 to 200 years after abandonment) the climax oak–hickory forest is reestab-

Figure 3.7 *Various stages of old-field succession. (a) A marginal farm such as this one is often abandoned after a few years of cultivation. (b) This field is in the broom-sedge stage; the adjacent land is in the pine stage, and the broom sedge is being burned back. (c) The climax stage of old-field succession in this area is an oak–hickory forest. (Tennessee Valley Authority.)*

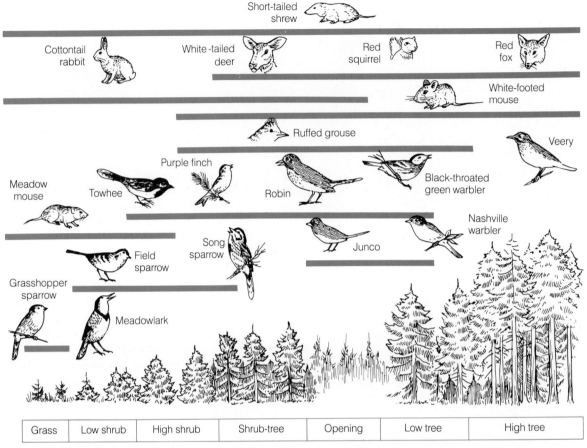

| Grass | Low shrub | High shrub | Shrub-tree | Opening | Low tree | High tree |

Figure 3.8 *Wildlife succession in central New York. Note that some species appear and others disappear as the vegetation changes; a few species are common to many habitats. Horizontal bars indicate the range of habitats for indicated species.* (From R. L. Smith, Ecology and Field Biology. *Courtesy of Harper & Row Publishers, New York. Copyright © 1980 by Robert Leo Smith.*)

lished. As the vegetation changes, the types of shelter and food that are available for animals change, too. Consequently, the numbers and types of animals also change as succession proceeds. Figure 3.8 illustrates the changes in wildlife that occur in old-field succession in New York.

Old-field succession and other successional patterns demonstrate that change in the composition of plants and animal species does take place, even if the area is no longer disturbed. Although ecologists do not yet understand the actual mechanisms by which one species replaces another, several mechanisms have been proposed (see Box 3.1).

We have examined one example of succession toward a climax ecosystem. But recall that the na-

ture of the climax ecosystem varies from one region to another, depending on climate, soil type, topography, and other environmental factors. It is a consequence of those interactions that the North American landscape varies from the majestic rain forests of the Pacific Northwest to the rolling grasslands of the prairie states and provinces to stands of scrub oak and pine along the coastal plain of the Southeast and from the tropical rain forests of southern Mexico to the arctic tundra of the Northwest Territory.

Effects of Human Activities on Succession

Many types of disturbances can destroy a climax ecosystem. Natural disturbances include fire, flood, drought, hurricane-force winds, and insect infestations; forestry and agriculture are probably the most significant of the stresses that are induced by humans. Around the globe, billions of hectares of climax forest and grasslands have been cleared and plowed under for cultivation. Thus those natural

BOX 3.1

HOW DO SPECIES REPLACE
ONE ANOTHER?

*Three mechanisms of species succession have
been proposed to describe the way species replace
one another in succession—the facilitation model,
the tolerance model, and the inhibition model.*

*The facilitation model was proposed initially
in the early 1900s, when the concept of ecological
succession was first scientifically formalized.
Essentially, the model states that the organisms
in each stage of succession modify the physical
environment in such a way that the site becomes
less suitable for themselves and more suitable
for another group of species. Thus as a result of
their effects, pioneer species facilitate colonization
of an area by later species. This model is based
on the common observations that plants and animals modify their physical environment. As
succession proceeds, the soil is enriched by humus
and nutrients. As a weedy field gradually reverts
to a forest, the climate beneath the developing
tree canopy changes from that formerly present in
the open field: fluctuations in air temperature
are reduced, and more moisture is retained in the
soil and in the air beneath the canopy. Recently,
however, a careful analysis has shown that although facilitation may occur in the early stages
of primary succession, when pioneer species first
begin to build up the soil, little experimental
evidence exists to verify that where a soil already
exists, pioneer species actually make the site more
favorable for later species.*

*The inhibition model and the tolerance
model both assume that any species that are able
to survive on the site as adults can establish
themselves early during succession. Because of
their ability to reproduce rapidly, produce many
seeds, disperse widely, and germinate and grow
readily, most of the early colonizers will likely be
pioneer species, but some of the species that
dominate later successional stages may also become established.*

*The tolerance model proposes that although
species that are dominant in later successional
stages will grow more slowly, they can tolerate the
difficulties encountered in an environment that is
created by the pioneer species that dominate the*
*site. As the slower growing, more tolerant species
that are characteristic of later stages of succession
continue to grow and mature, the faster growing,
less tolerant pioneer species will die out, and
dominance will pass on to the more tolerant
species of later stages. This sequence continues
until no species can replace the already established
ones. Note that in this model, the dominant
pioneer species neither inhibit nor facilitate the
establishment of later species.*

*In the inhibition model, the early occupants
modify the site so that it becomes less favorable
for colonization by both earlier and later successional species. Thus unless they are damaged
or killed, the established residents will exclude all
newcomers. When a plant dies and a spot opens
for colonization, the replacement plant may be
from the same stage or a later successional stage,
depending on the conditions of the site and the
characteristics of the species available for replacement at that time. If the new colonizer is a
member of the same successional stage, the existing successional stage will continue. If, however,
the newcomer is from a later stage and is able
to tolerate the environmental conditions, succession will progress to that later stage.*

*At this time, no consensus exists among
ecologists as to which model (inhibition or tolerance) better explains how secondary succession
actually takes place. Perhaps both processes occur
at the same time. As is usual in ecological studies,
several factors complicate the picture. For example,
even sites that appear quite uniform, such as a
recently abandoned cornfield, usually possess
considerable spatial heterogeneity with respect to
slope direction and angle, soil fertility, soil
moisture, and distance and direction from potential seed sources. Another complication is that
species vary in their ability to germinate and
establish themselves. Then, too, there is the element of chance, such as the direction and magnitude of the wind when seeds are dispersed. Thus
the initial mix of seeds that germinate on differing
sites will likely be quite variable. Finally, the
mechanisms which explain species replacement in
early successional stages may not be applicable
to later stages. Much remains to be learned about
the process of species replacement.*

ecosystems have been replaced by a few grasses, such as corn, wheat, oats, rye, and barley, and some legumes, such as soybeans and alfalfa. On a smaller but growing scale, forests are being cleared and replaced with single-species tree plantations. For example, in the southeastern United States row after row of pine trees now stand where climax oak–hickory forests once grew. In tropical areas, complex climax rainforest ecosystems, with their hundreds of plant species and thousands of animal species, including insects, are being chopped down and replaced by much simpler ecosystems, such as banana, rubber, and cocoa plantations. Other major disturbances that are caused by humans include air pollution, water pollution, and strip mining.

Although succession tends to result in an ecosystem that is most suited to the local environment, it can be halted at any stage before climax is reached. Natural succession can be interrupted and a developmental stage can be maintained, if that is desired

by its human overseers. For example, cultivation of farmland prevents old-field succession by continually disrupting the soil, and, in the southeastern United States, controlled burning is done to prevent oaks and hickories from invading pine plantations. When those climax species are allowed to invade a pine plantation, pine-tree production is reduced, subsequent pine regeneration becomes difficult, and the presence of oak and hickory trees increases the fuel that is available for a destructive wildfire. To overcome those difficulties, forest managers periodically set controlled ground fires, such as the one shown in Figure 3.9, to prevent the fire-sensitive climax species from invading the area. (Pines are fire-resistant and usually remain undamaged.)

As succession takes place, each succeeding stage is characterized by different flow patterns of materials and energy; thus, in certain fundamental ways, later stages of succession may differ considerably from early stages (as Table 3.3 indicates). Thus several consequences result from the large-scale human activities that replace climax stages with pioneer or developmental stages.

Figure 3.9 *Controlled burning of a pine stand, which prevents the invasion of fire-sensitive climax species and the buildup of plant litter that could fuel a destructive wildfire. (U.S. Department of Agriculture, Soil Conservation Service.)*

Table 3.3 Changes that occur during succession

	DEVELOP-MENTAL STAGES	MATURE STAGES
Net Biomass Production (Yield)	High	Low
Mineral Cycling	Open	Closed
Nutrient Exchange Rate	Rapid	Slow
Number of Species	Low	High
Food Chains	Linear	Weblike

Increased Productivity Increased production of food is the major reason that human beings replace climax communities with pioneer stages. For example, a corn field is more productive per hectare than a beech–maple forest. Furthermore, as described in Chapter 2, crop plants direct more of their energy into the production of seeds and fruits that are edible either directly by us or by our livestock. The situation is much the same in forestry. Many successional trees are more productive than climax species, and some are more desirable for high-demand wood products, such as lumber for construction. Commercially desirable successional species include pines in the Southeast and Douglas fir in the Northwest. Successional habitats also are the preferred homes of many popular game species, such as white-tailed deer, elk, moose, bobwhite quail, and pheasants.

Increased Nutrient Cycling As Table 3.3 indicates, several changes occur in nutrient cycling processes after a climax ecosystem is disturbed. Besides interrupting nutrient uptake and retention by plants, the loss of vegetative cover also raises the temperature and increases the aeration of the soil surface. Those changes speed the breakdown of *humus*, which is composed of organic materials that are generally resistant to decay and remain after the major portion of animal and plant residues have decayed. As a consequence, more nutrients are released into the soil. Hence a temporary surge in available nutrients often occurs after a disturbance. Concurrently, both the loss of humus and the loss of plant cover lessen the site's ability to retain water. Humus acts as a sponge that soaks up water and releases it slowly. The removal of plant cover creates a cessation of *transpiration*, which is the loss of water vapor from the pores of the plants' leaves into the atmosphere. Perhaps as much as 98 percent of the water taken up by plant roots is lost from an ecosystem as a result of transpiration. Since transpiration reduces soil moisture, thereby increasing storage space in the soil for rainfall, stream flow does not occur until the soil storage space is refilled by precipitation. Hence a loss of vegetative cover and the transpiration that is associated with it causes greater runoff, because the soil moisture remains higher and the soil cannot store as much of the rainfall. The combination of greater runoff and free soil nutrients can (and often does) result in a significant loss of nutrients from a disturbed ecosystem. Therefore, nutrient cycles in a pioneer stage are said to be *open*. As succession proceeds, the nutrient cycles become more *closed*.

That significant difference in nutrient cycles has been well documented by investigations that have been carried out at the Hubbard Brook Experimental Forest in New Hampshire since the mid-1960s. In one study, all the trees in a small watershed that were larger than 2 centimeters (1 inch) in diameter were cut, and subsequent regrowth was inhibited artificially for 2 years by periodic applications of herbicides. As Figure 3.10 indicates, nutrient losses were substantial. In the second year following the clearing of trees, stream-water concentrations of nitrate measured 56 times higher than before deforestation, which exceeded the levels that were recommended for drinking water. Clearly, that important plant nutrient was being lost from the soil through runoff. As Figure 3.10 shows, during the same period of time, substantial increases in the concentrations of other plant nutrients also were measured in the average stream water (for example, 16 times as much potassium, 4 times as much calcium and magnesium, and 2 times as much sodium were measured).

The Hubbard Brook data illustrate the potential magnitude of nutrient losses from a disturbed ecosystem, particularly if it remains disturbed for several years. Usually, however, new plant growth appears within only a few weeks after a disturbance. Recent studies have shown that recolonization by pioneer species quickly reduces nutrient losses by reversing those processes that produced the greater nutrient losses in the first place.

We already know that early successional species tend to grow rapidly and to produce a high yield. They also rapidly take up and retain soil nutrients. (Interestingly, the tissues of many successional spe-

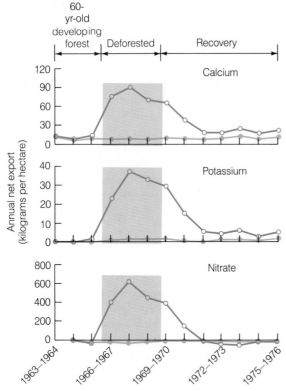

Figure 3.10 *Deforestation can significantly increase the concentration of nutrients found in stream water, indicated by open circles. Subsequent revegetation reduces their loss and thereby lowers their stream-water concentration. Solid circles represent stream flow from a similar forest that was not cut.* (Adapted from F. H. Bormann and G. E. Likens, Pattern and Process in a Forested Ecosystem. *New York: Springer-Verlag, 1979.)*

Figure 3.11 *Following deforestation, regrowth of vegetation increases the storage of nutrients. This causes a decline in nutrients lost from the ecosystem.* (Adapted from F. H. Bormann and G. E. Likens, Pattern and Process in a Forested Ecosystem. *New York: Springer-Verlag, 1979.)*

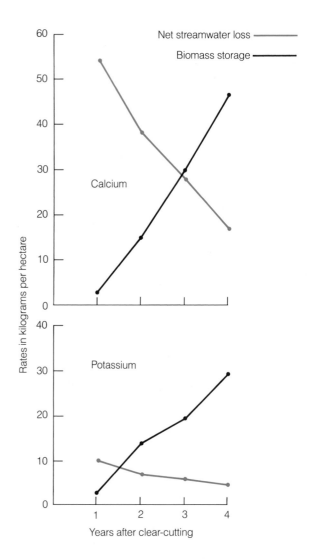

cies have been found to contain even greater concentrations of nutrients than those of climax species.) Thus successional plants conserve substantial pools of nutrients. Figure 3.11 shows the relationship between the increased storage of nutrients in vegetation and their reduced loss to stream water. Recolonizing vegetation also shades the ground, thereby reducing temperatures and airflow at the surface, which, in turn, slows the further release of nutrients from the decomposition of humus. Moreover, transpiration by the reestablished vegetation also begins to lessen water runoff. At Hubbard Brook,

transpiration reduced annual stream-water output by as much as 25 percent. Thus we see that if a site is not disturbed on a continual basis, colonization by early successional species can restore the cycling processes quickly to a pattern that approaches their original transfer rates.

Reduced Number of Species Another difference between pioneer stages and later successional stages is the number of species present. Usually, fewer species are present in early successional stages than were present in the climax stage. Associated with this trend is a tendency toward simple linear food chains in contrast to the more complex food webs that characterize more mature ecosystems. Those differences are particularly evident in agricultural regions, where climax forests and prairies have been replaced by a comparatively few species of crop plants, livestock, and associated weeds and pests (rodents and insects). However, that reduction in species diversity does not occur without consequences.

The traditional view is that the greater the number of species and the number of linkages between them, the greater will be the stability of the ecosystem. Intuitively, this conclusion makes sense. If we consider a very simple ecosystem, in which mice are the only important source of food for a population of foxes, and if disease drastically reduces the population of mice, we might suspect that some of the foxes would have to either migrate to another region in search of food or else die of starvation. Thus in a very simplified food web, our intuition tells us that energy flow may fluctuate tremendously, depending on the well-being of a few species.

What happens in a more complex ecosystem, where several species, such as mice, chipmunks, rabbits, and various ground-dwelling birds, are available as prey for the foxes? In that situation, if one species of prey becomes scarce, the foxes could turn to a more abundant population of prey. Therefore, as a consequence, we assume that energy flow from one trophic level to the next remains roughly unchanged. In turn, the more complex ecosystem remains stable despite internal changes (assuming that not all of the species of prey become rare at the same time). We can also find parallels in commerce. For example, a city or country that has a variety of natural resources, manufacturing processes, and export markets will be more stable economically in the long run than a city or country whose economy is based on only one or a few products.

But is our intuition actually confirmed when ecologists carefully examine the relationships between diversity and stability in natural ecosystems? Initial observations that led to the view that increased diversity leads to increased stability have turned out to be fragmentary at best and have not held up under closer scrutiny and experimentation. One major obstacle to a better understanding of the problem is the lack of controlled experiments. Natural ecosystems, even in pioneer stages, are so complex and variable that it is virtually impossible to find two sites that are identical, except in terms of their diversity of species. Another stumbling block

is that ecologists cannot run long-term experiments in which species diversity is the *only* variable in the ecosystem that changes. Today, most ecologists believe that no simple relationship exists between the diversity and the stability of an ecological system. Indeed, it may well be that a simple ecosystem is just as stable (or as unstable) as a complex ecosystem.

Changes in Species Composition A major problem with converting a landscape to an agricultural area is the continual invasion of it by unwanted pioneer species, which we call *weeds* (if they are plants) and *pests* (if they are insects or other animals). Pioneer plants are particularly hardy species. Their ability to grow rapidly and to take up soil nutrients makes them formidable competitors with crop plants. They can produce large numbers of hardy seeds that are readily dispersed by wind or animals (see Figure 3.12). Thus weedy plants can reestablish themselves during the next growing season in both the same fields and adjacent ones. Likewise, pioneer animals (mostly rodents and insects) have an innate capacity for proliferation and wide dispersion and, as a result, often inflict heavy crop damage. Thus in preparing the way for a profitable crop, a farmer is also providing opportunities for invasions by large numbers of animals and weeds that are well adjusted to the rigors of life in cultivated fields. We derive little or nothing in the way of food, clothing, or shelter from those plants and animals, but we spend a great deal of time, money, and effort trying to eradicate them.

In contrast, while pioneer organisms are proliferating in so many areas of the world, those species that require climax conditions for survival are disappearing. In dozens of cases where plant and animal species have become extinct, or nearly so, habitat destruction is the most significant contributing factor. The fate of Attwater's prairie chicken, shown in Figure 3.13, illustrates the problem. That species once thrived in the lush tall-grass prairie along the gulf coast in southwestern Louisiana and southeastern Texas. But the plowshare turned the prairie into cropland, thus reducing the prairie chicken's range by 90 percent and its population by 99 percent. Today, Attwater's prairie chicken is found in only a few isolated areas in southeastern Texas. Habitat loss contributes to the threatened extinction of many animal species, including the San Joaquin kit fox, the Florida panther, the Mexican grizzly bear, the

(a)

Figure 3.12 *The plumes of milkweed seeds (a) and the hooks of burdock (b) illustrate two means whereby seeds of weedy plants are widely dispersed. (Frank E. Toman, Taurus Photo; Jerome Wexler, National Audubon Society, Photo Researchers.)*

(b)

Figure 3.13 *Attwater's prairie chicken, which is in danger of extinction. (U.S. Department of the Interior, Fish and Wildlife Service. Photo by Luther C. Goldman.)*

African elephant, the Asiatic lion, and the Arabian gazelle. Plant species that are threatened by habitat loss include the Tennessee purple coneflower, persistent trillium, and several species of orchids. In Chapter 14, we discuss further the problems encountered in trying to preserve endangered species.

Limits to Succession

Nature's healing process takes a long time. Even under normal circumstances, 150 to 200 years are necessary for a climax oak–hickory forest to become established on abandoned farmland in the southeastern United States. Even more time may be required for restoration after a severe disturbance has occurred. In some instances, a disturbance is so severe that succession does not occur at all. Many such events are caused by human removal of vegetation and long-term, improper land-use practices.

Throughout most semiarid regions of the world, improper irrigation practices, which are often associated with poor drainage conditions, lead to an accumulation of salts at the soil surface, which is called *salinization*. If the problem is not corrected, increasing soil salinity leads to declining crop yields. In

more severe cases, farming becomes impossible. Today, millions of hectares of farmland have been abandoned throughout the world—from the Middle East to South America to western United States. Many believe that such agricultural failures contributed to the collapse of some ancient civilizations, such as those that occupied the Tigris-Euphrates and the Indus valleys. Few plants can survive highly saline soils and those regions have become barren. In fact, some man-made deserts in the Middle East have persisted for centuries; they are dramatic examples that succession may not heal all wounds to the landscape. As described in Chapter 17, soil salinization continues today in semiarid regions throughout the world.

In other instances, improper land-use practices have contributed to a complete loss of the soil cover. For example, along the east coast of the Mediterranean Sea, the mountainous landscape was once covered with the famous cedars of Lebanon. After those forests were cut down, some 5300 years ago, farmers moved in and attempted to cultivate the mountain slopes. Today, only a tiny remnant of the cedars remains, the soil is gone, and the bare limestone slopes bear mute testimony that over the centuries, succession has not restored the landscape to another cedar forest (Figure 3.14). Today, as a consequence of attempts to feed a rapidly growing human population, the soil in most areas of the world is being eroded

Figure 3.14 *The barren slopes of Lebanon well illustrate the limits of succession in restoring the landscape where human intervention results in a significant soil loss. This grove of some 400 trees is all that remains of the fabled Cedars of Lebanon. (Anne E. Hubbard, Photo Researchers.)*

away faster than it is being replaced by natural soil-forming practices.

Time may be the most significant single factor that influences our management of many natural resources. Natural succession is not a viable means of restoring much of the earth's disturbed land to its original state within our lifetimes. To speed up restoration, we must employ such management practices as adding fertilizer to the land and planting hybrid plants that do well under the poor growing conditions that are characteristic of many disturbed areas.

But even those techniques meet with natural limitations: soil building remains a slow process; and most trees take several decades to reach maturity. Slow as it is, ecological succession is the natural response of ecosystems as a way of recovering from an environmental disturbance. In some cases, our only hope for restoring the environment may lie in lowering our expectations and allowing succession to occur at its own rate. For a more detailed discussion of problems and progress in land restoration that follow mining, see Chapter 12.

Our evaluation of the consequences of reversing succession—that is, replacing climax stages with pioneer stages or early successional stages—provides a mixed picture. In agriculture, forestry, and wildlife management, the benefits that mankind has derived from reversing succession are considerable. Yet poor land-use practices have accelerated soil erosion, contributed to the pollution of waterways, and led to the loss of valuable genetic resources (as a consequence of species extinction). Without greater worldwide implementation of effective land-use practices, the benefits of reversing succession will diminish as more of the landscape becomes increasingly less productive.

Adaptation to Environmental Change

Organisms do not necessarily die when their environment changes. Most organisms have the capacity to adjust, at least in part, to environmental changes. Those adjustments are called *adaptations*. For example, our bodies can adjust to cooler temperatures by shivering in response to declining skin temperature. Shivering occurs as a result of increased muscle activity that is maintained by energy that is derived from the process of respiration. A by-product

of that process is heat energy, which helps to maintain normal body temperature. Chipmunks, ground squirrels, and woodchucks have the capacity to adjust to the onset of winter because they can hibernate. *Hibernation* involves the lowering of body temperature to just a few degrees above freezing and the slowing of heart and breathing rates. That sharp decrease in body activity diminishes an organism's energy needs, enabling it to rely on stored body fat. Thus an organism can live through the winter without having to leave its den to search for food under hostile weather conditions.

Some understanding of the process of evolution is required to answer (1) how adaptations to environmental change come about, (2) why some organisms have adapted better than others, (3) what limits an organism's ability to adjust to changes, and (4) what types of adaptation that occur.

Evolution: The Driving Force of Adaptation

Evolution is defined as a change in the genetic makeup of a population with time, and it entails a two-step process. The first step is the production of genetic variability. The second step is the ordering of that variability by natural selection. (Some evolutionary biologists argue that other agents besides natural selection work on the products of genetic variability, but most are agreed that natural selection is by far the most important agent.) We now consider how those two steps work together to bring about evolution.

Production of Genetic Variability We have already noted that in most populations of plants and animals, each organism is genetically unique. The human population now numbers nearly 5 billion persons. Discounting the relatively few pairs of identical twins, that is a lot of genetic variability! But even more remarkable is that billions of insects, such as grasshoppers, can be found in just a few square kilometers. What is the source of all this variability?

Ultimately, all new genetic materials arise by mutations. Mutations are random events; where they occur in the chromosomes and when they occur is a matter of chance. As a variety of genetic combinations become available through mutations, an almost endless variety of genetic combinations becomes possible through sexual reproduction. In the formation of *gametes* (eggs and sperms), the genetic

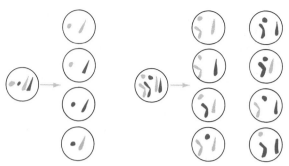

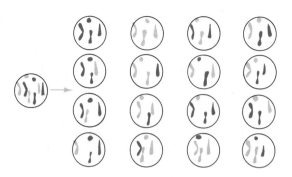

Figure 3.15 *Schematic representation of the possible distribution of chromosomes during the formation of gametes, using two, three, and four pairs of chromosomes. The dark chromosomes were originally of paternal origin and the light chromosomes, of maternal origin. Two pairs of chromosomes permit only four possible different distributions, but four pairs permit 16 possible different distributions, and the number of possible combinations doubles with the addition of every chromosome pair. Thus the chromosome number of our own species (23) makes possible over 8 million chromosome combinations in each of our gametes.*

combinations are, in effect, reshuffled. Every cell of the body, except the gametes, contains two sets of chromosomes—one set is inherited from each parent. When the gametes are formed, the number of chromosomes in each cell is reduced by half. Each gamete receives one complete set of the chromosomes that originally existed in pairs. (One member of each pair was inherited from the father and the other member was inherited from the mother.) As Figure 3.15 illustrates, however, the individual chromosomes are randomly distributed when the gametes are formed. Taking the simplest example, an organism that has two pairs of chromosomes can produce four possible chromosome combinations in its gametes. The two chromosomes in the gametes may both be of paternal origin or both of maternal origin, or one may be inherited from each parent. The number of possible chromosome combinations increases greatly as the number of pairs of chromosome pairs increases. Since we possess 23 sets of chromosomes, the number of possible chromosome combinations in the gametes of just one person is 8,388,608.

Variation can also occur when pieces of paired chromosomes are interchanged as they are distributed to the gametes (Figure 3.16). That process of interchange—called *crossing-over*—is also a random event.

A final source of variation occurs with the union

of an egg and a sperm. In humans, the number of combinations in the cell that is formed by the union of a sperm and an egg is 70,368,744,177,664. And this number probably increases by millions when crossing-over is taken into account. Furthermore, these numbers represent the possible combinations of only one man and one woman, considering their unique genetic makeup. When the entire human race is considered (each person having a unique genetic constitution) the number of possible genetic combinations is truly mind-boggling.

It is important to remember that the variation that is produced by mutation and sexual reproduction is random. That variation is not caused by, and is not related to, the present needs of organisms or the nature of their environment.

Natural Selection We have now arrived at the second step of evolution—natural selection. In a population of genetically unique organisms, some will have sets of *genes* (a portion of a chromosome that codes for a particular characteristic) that are better

Figure 3.16 *The crossing-over of segments between like chromosomes greatly increases genetic variability within a population.*

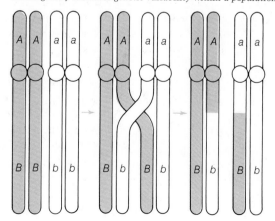

suited to their current environmental conditions. Such organisms have a greater probability of surviving and of having offspring than other members of the population. Moreover, it is likely that those offspring will inherit the same genes and that they also will have a greater probability of surviving and having offspring. Thus if the environment remains generally unchanged, eventually most members of the population will possess the beneficial set of genes. That sequence of events, in which a particular gene, or set of genes, is "selected" by the environment, is called *natural selection*. The result of natural selection is that a population contains a greater number of organisms that are better suited to a given environment.

Natural selection is illustrated by the classic example of the peppered moth, which lives in the forests of England. Before 1850, most peppered moths were light in color; only occasionally was a dark variant observed. In the industrial town of Manchester, for example, the first dark specimen was caught in 1849. However, by 1895, the dark variety made up some 98 percent of the peppered-moth population in the Manchester area.

It was not clear at that time what caused this dramatic shift in the predominant color of that species of moths. Thus in the 1930s, the following hypothesis was proposed, based on the fact that the peppered moth tends to rest during the day on tree trunks and rocks. Before the Industrial Revolution, those surfaces were mostly light in color, and often they were covered with light-colored lichens. Against the light-colored surfaces, the lighter moths were much less visible than the darker ones to bird predators (Figure 3.17). As a result, most of the dark moths were selected out of the population. However, with the growth of industry, soot from factory chimneys blackened the rocks and tree trunks and killed the lichens. At that point, the light moths, rather than the dark ones, were at a selectional disadvantage. Because they showed up more clearly against the soot-darkened surfaces on which they rested, they were selected against (eaten by birds) in far greater numbers than were the dark variants. As a result, over time, the dark variants increased in the population at the expense of the light ones, and eventually the dark variants predominated.

Figure 3.17 *Light and dark forms of peppered moths at rest on tree trunks. (a) The light variant is much less visible than the dark variant on this lichen-covered tree. (b) It is much easier to spot the light moth on this soot-covered tree. (From the experiments of H. B. D. Kettlewell, Oxford University, England.)*

(a)

(b)

This hypothesis seemed plausible to some biologists, but it was disputed by others. Finally, in the 1950s, H. B. D. Kettlewell of Oxford University put the matter to an experimental test. He released roughly equal numbers of the light and the dark varieties of peppered moths in a rural section of England, where the tree trunks had not been blackened by industrial pollution and where the light moths outnumbered the dark ones by almost 20 to 1. Kettlewell and his colleagues found that birds did, indeed, prey on the moths by sight and, also, that of the 190 moths that were observed to be eaten by birds, 164 were dark, while only 26 were light.

Next, Kettlewell repeated the experiment under the opposite environmental conditions. In another part of England, where soot had discolored the tree trunks and the dark variants naturally predominated, he released approximately equal numbers of light and dark moths. Once again, birds ate the moths, but that time it was observed that they ate roughly three times as many light moths as dark ones.

In the years since Kettlewell's experiments, air-pollution controls have been implemented in England. Today, there is less airborne soot, and tree trunks are becoming lighter. Thus light coloration in peppered moths should once again be increasingly selected for, and, just as we would expect, the light variants are beginning to predominate again.

Several important points in this example need emphasis. First, environmental change did not produce the dark form of the moth; it was already present in the population and probably was the result of a mutation. Natural selection, in response to environmental changes, can only occur with the genetic variations that are already present in a population. It favors only those members that already possess a trait that makes them better suited to the changing environment.

Second, the period of time that is required for an almost complete shift in a predominating factor in the population varies according to reproductive capabilities. The change in peppered moths occurred relatively quickly because moths, like most insects, reproduce large numbers of offspring several times a year. (Table 3.4 presents some reproductive capabilities for comparison.) The dark forms were able to multiply rapidly when bird predation was no longer an important limiting factor on their numbers. Also, the time that is necessary for a generation to mature and reproduce is a major determinant of the length

Table 3.4 Some measures of reproductive capabilities among different species

	MEAN LENGTH OF GENERATION*	TIME REQUIRED TO DOUBLE POPULATION
Rice Weevil	6.2 Weeks	0.9 Weeks
Brown Rat	31.1 Weeks	6.8 Weeks
Human Beings (1984 World Rate)	15–20 Years	40 Years

Source: After E. P. Odum, *Fundamentals of Ecology*. Philadelphia: Saunders, 1971.
*Mean length of generation is the average time between birth of individual members of a species and birth of their first offspring.

of time that is required for a population to adjust to an environmental change. For example, some bacteria can double their population in 20 minutes. Hence if some members of a bacterial population are resistant to a particular antibiotic, that population can adapt itself to the presence of the drug within hours or a few days. In marked contrast, long-lived organisms that reproduce slowly, such as ourselves, may require hundreds or thousands of years to undergo an evolution of comparable magnitude.

Some Types of Adaptation

The products of evolution are *adaptations*, which are genetically controlled characteristics that increase an organism's ability to survive long enough to produce offspring. Adaptations are structural, physiological, or behavioral. For example, adaptations to cold temperatures include the growth of fur (structural), reduction of the flow of warm blood to the skin (physiological), and the ability to fly south for the winter (behavioral). Adaptations include not only relatively static characteristics, such as our opposable thumbs, but also the capacity to adjust to an environmental change, such as our ability to shiver when our skin temperature is lowered.

Essentially every organism is a complex package that contains an immense number of interacting adaptations. For example, organisms adapt in innumerable ways to obtain energy, nutrients, and oxygen from their environment, to retain nutrients and water, to avoid becoming a meal for other organisms, to ward off disease-causing organisms, to compete with other organisms for limited resources, and to adjust to both physical and biological changes in their environment. To illustrate the fascinating array of adaptations that organisms possess, we consider some defensive adaptations and some types of learning.

Defensive Adaptations Many types of adaptations reduce an organism's vulnerability to attack, such as an organism's ability to produce defensive chemicals. One familiar chemical defense is the pungent chemical that is released by skunks when they are agitated. A less familiar example is the noxious chemical that is sprayed by the bombardier beetle when it is attacked by ants. That spray, which is emitted from the tip of the abdomen, can be aimed so precisely that the beetle can repel an ant from attacking any of its six legs.

Many chemical defenses are more subtle. Monarch butterflies, for example, contain a chemical that induces vomiting when it is ingested (Figure 3.18). Such a chemical is adaptive, because a predator that suffers from eating a monarch butterfly is less likely

---

BOX 3.2

THE SILENT BATTLE

Hundreds of gypsy moth caterpillars cluster on the tender green leaves of a beech tree, their ravening jaws chewing through every leaf in sight and stripping the bough completely bare. But there is a mystery here; only a few trees in the grove show signs of depredation by insects. Most of them are virtually untouched.

Scientists used to think that bug populations were controlled only by the weather and by their own predators, diseases, and parasites. But intriguing new findings show that many trees and other plants, far from being passive victims, fight back with an arsenal of deadly chemicals that can make foliage almost impossible to digest, or so toxic that it kills insects outright.

Some plants contain as many as eight toxins; others can change their poisons from one year to the next. Recently, scientists in Washington state and New Hampshire have discovered a fascinating dimension of this battle: trees can even sense when nearby members of their species have been attacked by insects, and this stimulates them to produce a flood of toxic chemicals even before a bug has taken its first bite.

According to David Rhoades, an ecologist at the University of Washington, trees react to insect infestations by setting off a silent alarm. An airborne chemical wafts from the attacked trees to others nearby, and induces the untouched trees to respond by increasing the concentration of chemicals that protect them from rampaging bugs.

Rhoades discovered this startling form of interaction quite by accident, in a Pacific Northwest tree called the Sitka willow. He was trying to find out whether willows could increase the toxins in their leaves when infested by insects. Rhoades infested one set of willows with western tent caterpillars, and protected an adjoining stand with collars around the tree trunks to prevent insects from crawling up. He collected leaves from both groups of trees every two or three days, and tested them for toxicity by feeding them to captive caterpillars. The attacked trees responded with astonishing speed: In eleven and a half days their leaves became unpalatable to the caterpillars. Quite unexpectedly, however, three days later the insects began turning up their noses at leaves from the uninfested trees as well.

When Rhoades reported his discoveries last year, Jack Schultz, an ecologist at Dartmouth, and his assistant Ian Baldwin started a series of experiments to test for similar reactions among eastern varieties of deciduous trees. They soon confirmed that other types, including poplars, sugar maples, and red oaks, communicate just as effectively.

While scientists do not yet know the details of tree communication, they have a good idea of the staggering array of poisons that plants produce to kill bugs. Most of these poisons are chemicals called secondary metabolites, some of which serve no direct function in a plant's growth, reproduction, or photosynthesis. Many secondary compounds seem perfectly innocuous: the fresh scents of herbs and evergreens come from volatile

to attack and eat a monarch butterfly the next time it sees one. As mentioned in Chapter 2, many plants possess chemicals that reduce attack by herbivores and microorganisms. Box 3.2 describes this chemical arsenal.

Another array of defenses involve the deceptive appearance of organisms. Many animals blend so well into their surroundings that they are nearly undetectable. We have already seen how differences in coloration make peppered moths more or less vulnerable to predation by birds. Natural selection favors *protective coloration* in many populations in which a highly visible individual is in danger of being eaten. Many desert animals, particularly lizards, are light-colored and blend with the light-colored background of desert soils. Other animals change colors with the seasons. For example, the Arctic hare is white during the winter snow season, but in the spring it sheds its white fur and becomes brown, thus blending better with the tundra vegetation.

Rather than match the color of the general background, some organisms *mimic*, or closely re-

---

chemicals called terpenes, which discourage bugs by means yet unknown, and most plant-derived drugs are made from secondary metabolites. Caffeine and nicotine belong to a class of secondary compounds called alkaloids, some of which disrupt nerve transmissions and immobilize bugs.

Trees often combine toxins with large amounts of less deadly chemicals that interfere with insect digestion. A ubiquitous group of chemicals called phenols has been found in almost all trees scientists have examined, including oaks, poplars, birches, and sugar maples, as well as many smaller plants and even marine algae. The most common phenols, called tannins, can make up as much as 80 percent of the dry weight of a leaf.

Tannins work on a bug's stomach the same way they tan leather; they bind with protein molecules, which makes it difficult for enzymes to break them down. Tannin is just the ticket to keep leather shoes from rotting, but scientists suspect it can wreak havoc with a bug by preventing the insect from extracting important nutrients from its leafy food.

Interfering with digestion is not enough to foil all foragers, so some trees increase the complexity of their defense systems by varying the amounts of toxins and phenols on different branches or leaves. Tom Whitham, a biologist at Northern Arizona University, has found that some branches of a cottonwood are 72 times as resistant to attacking aphids as adjacent branches. Naturally, many aphids converge and dig in on nontoxic leaves; but then the tree reacts by dropping those *leaves, thus shedding many of the tiny parasites. In effect, aphids arriving on a cottonwood face a heads-I-win-tails-you-lose dilemma: if they land on unprotected branches, they may be shed by the tree; but if they land on resistant foliage, they are confronted by some kind of toxin. "Such maneuvers," says Whitham, "put the bugs in a squeeze play."*

Ecologists who study how groups of insects and plants interact now agree that plant defenses are an important factor in keeping a lid on bug populations. Rhoades goes even further, saying that communication among trees, as well as the surge of chemicals they manufacture in response to attack, controls the rise and fall of many pests with precision. He thinks that recurring outbreaks of such insects as tent caterpillars and gypsy moths result when trees let down their guard for a few years. The trees remobilize after an attack, knocking out the bugs yet again.

"It's a lot like a continually escalating arms race," muses Whitham. For each defense a tree devises, an insect evolves a counter-adaptation—forcing the plant to come up with something new or perish. Since insects have short life spans, they evolve more quickly than their hosts. The burden of trickery—forcing insects to expose their presence to predators, varying toxins from year to year—is on the trees. "The whole point for the trees," says Whitham, "is to keep a pest evolutionarily zigging when it should be zagging."

Excerpted from Shannon Brownlee, "The Silent Battle," Discover (September), 1983.

(a) (b)

Figure 3.18 *A freshly eaten monarch butterfly (a) induces vomiting in a blue jay (b). After this experience, the blue jay will not be likely to prey on palatable, but similar-looking, viceroy butterflies. (Lincoln Brower.)*

semble, inanimate objects that are commonly found in the habitat. Walking sticks, for instance, look so much like sticks and leafhoppers look so much like leaves that predators often are unable to tell the difference and thus pass up many nutritious meals. A particularly interesting type of *mimicry* is that found in species that appear to be very similar to another species. The monarch butterfly and the viceroy butterfly, for example, are almost identical in appearance. Although the monarch butterfly can induce vomiting in its predator, the viceroy cannot. Nevertheless, laboratory experiments have shown that birds which have had bad experiences with the monarch butterfly will reject the viceroy butterfly. Thus the viceroy butterfly suffers little predation because predators cannot distinguish it from its unpleasant look-alike.

Although many species are deceptive in their appearance, a few boldly advertise their presence. Those species have evolved bright colors and bold patterns that contrast sharply with their surroundings. At first thought, we might question the survival value of such bold coloration, but further observations show that most of those animals either taste bad, smell bad, sting, or produce toxic chemicals. Hence after one or two unpleasant encounters with those animals, a predator will usually avoid further contact with them. Their conspicuous appearance allows the potential predator to recognize and avoid them easily. Thus although a few organisms may be eaten, a given population as a whole is benefited because predators stop feeding on other members of the species that has caused them problems for the remainder of the predators' lifetimes.

Learning as Adaptation Thus far we have discussed evolutionary adaptations to environmental change, that is, adaptations that have developed, or evolved, over many generations. But innumerable changes in an organism's environment take place within its lifetime. Indeed, some of those changes can occur within days, minutes, or even seconds. Fortunately for them, animals have evolved a behavioral process that enables them to adapt quickly to many relatively sudden changes in their environment. That ability, which is itself a product of evolution, is what we call *learning*.

Learning may also be defined as a tendency to acquire relatively permanent changes in behavior, which occur as a result of experience. An organism that can learn to adjust its behavior to changes in its environment has a great advantage over one that simply repeats the same behavior patterns regardless of its environmental conditions.

All multicellular animals, including lowly worms

and insects, are capable of learning certain simple responses to environmental stimuli. Indeed, it appears that even single-celled organisms, such as amoebas, can learn to ignore certain stimuli by what is perhaps the simplest learning process—*habituation* (learning to ignore unimportant stimuli). More complex animals rely heavily on three other learning processes: trial-and-error learning, insight learning, and imitation learning.

Trial-and-error learning often takes place as an animal develops its feeding habits. For example, a young squirrel that is just beginning to feed on the seeds contained in hazelnuts may spend a great deal of time gnawing and attempting to crack the nut. Sooner or later, the squirrel will hit on a method that will crack it open efficiently. For a short while thereafter, the squirrel will still gnaw randomly on each hazelnut, but soon it will begin to use the more efficient nutcracking method. By trial and error, the squirrel learns how to open the nut on its first attempt.

Insight learning involves "seeing through" a problem, that is, the ability to perceive the relationships that are essential to solve a problem. For example, if we place a hungry chimpanzee in a cage where boxes are scattered about the floor and bananas are hanging out of reach from the ceiling, the chimp may appear to be stumped at first. But then, rather suddenly, he will collect the boxes, stack them beneath the bananas, and climb up on them to reach his prize. In trial-and-error learning, what is learned cannot be generalized from one situation to another. In contrast, a solution that is achieved as a result of insight can be applied to new situations, because it involves an understanding of the nature of the problem, not just one specific problem or another. Once our hungry chimp has used boxes to reach the bananas, on another occasion he might reach apples (rather than bananas) by stacking flat stones or books (rather than boxes).

The insight learners, par excellence, are, of course, human beings. The whole edifice of science and technology, for example, is for the most part built on the insightful solution of problems that are both large and small.

Learning can be passed on to other members of the population by various means of communication, including sight, sound, and chemical substances. Communication, broadly defined, occurs when an animal performs an act that is perceived by, and alters the behavior of, another organism. Through *imitation learning*, an animal acquires a type of behavior by observing that behavior in another animal. This type of learning is very important in behavioral development, particularly in the development of such mammals as carnivores and primates, including human beings.

Communication allows a learned trait to spread very rapidly throughout a population. In contrast, a genetic trait can be transmitted only from parents to offspring, and so its spread through a population necessarily takes at least several generations. Adaptation by learning is particularly important for slowly reproducing organisms, such as ourselves.

Perhaps, most importantly, human learning is *cumulative*. Thanks to our ability to encode information—at first only in our own memories, then in writing, and now also in the artificial memories of machines—each generation has the benefit of the prodigious store of learning gained by those who lived before. For example, to calculate our share of last night's restaurant check, we did not have to reinvent arithmetic; and to drive home from the restaurant, we did not have to reinvent the wheel and the internal combustion engine.

It is important to remember, however, that the ability to learn is itself dependent on genetic makeup. In fact, an animal's ability to learn is ultimately determined by its genetic makeup. Figure 3.19 illustrates a raccoon's inability to solve a problem through insight; that inability is a genetically determined limitation.

The importance of adaptations through learning varies among species. In fairly simple animals, such as worms and oysters, learning is relatively unimportant. As the brain becomes more complex, adaptations through learning become more significant. Our own species represents the current peak in brain development on this planet, and adaptations that occur as a result of learning are critically important for us. Indeed, virtually all the behaviors that make us distinctively human are wholly or partly learned behaviors.

Two Important Human Adaptations Human beings are generalists. Although we are genetically adapted to perform a wide variety of activities, we excel at few. We can run and swim, but we are neither the fastest nor the slowest at those activities. Many animals far exceed us in physical prowess and agility.

Figure 3.19 *Lack of insight learning in a raccoon. Because its leash is looped around a stake, the raccoon cannot reach the food dish. A chimpanzee or a human being would immediately "see through" the problem, go back around the stake, and then go to the dish, but the raccoon must find the solution by trial and error.*

Nonetheless, we do, in fact, possess a few specialized adaptations which, along with our general abilities, set us apart from other forms of life.

One of our species' few specialized structural adaptations is the strong *opposable thumb*. The old-world monkeys and the apes also have an opposable thumb but, in them, it is very short compared with the other fingers, it is relatively weak, and its range of motion is more limited than that of our own thumb. Because our thumb moves freely in opposition to the other fingers and is relatively long in comparison with thumbs of other primates, we can manipulate objects in ways that other species cannot.

The human hand is capable of two major forms of grip—the power grip and the precision grip. In the *power grip*, an object is held between the palm and fingers. With that grip, we can apply great force to the grasped object (for example, a club, hammer, rope, tennis racquet, or whatever). In the *precision grip*, we can grasp the object (for example, a screwdriver or a scalpel) between one or more fingers and the thumb in its fully opposed position. We have developed the precision grip far beyond anything that is found in the other primates. It is that grip which makes possible the dextrous and delicate manipu-lations that are necessary to thread a needle, assemble a machine, dissect a frog, or play a violin.

The most significant of our adaptations is our enlarged forebrain (cerebrum), which is the locus of insight learning. As you can see in Figure 3.20, the forebrain is much larger in humans relative to other parts of the brain than it is in other animals. The development of the cerebrum, which is a genetic adaptation—like the development of the opposable thumb—has enabled us to add mightily to our adaptations through learning. First, we were able to use clubs, spears, and bow and arrows; later we developed firearms to kill our prey. Those weapons more than adequately substitute for a carnivore's fangs. And although we never evolved armor that would protect us from aggressors, as the armadillo did, through our technology we provided ourselves with metal shields, helmets, and breastplates. Second, although we were never the fastest or the strongest animal, by using the problem-solving ability that is centered in our forebrains, we have learned to domesticate other animals to supplement our speed and power. We have even learned to build machines that are faster and stronger than any animal, such as armored tanks. And third, we developed another skill that is related to insight learning, which is our ability to contemplate the past and, especially, the future, which has enabled us to give direction to our activities, both socially and technologically. Thus not only have we developed the technological means to protect ourselves and enhance our culture, but

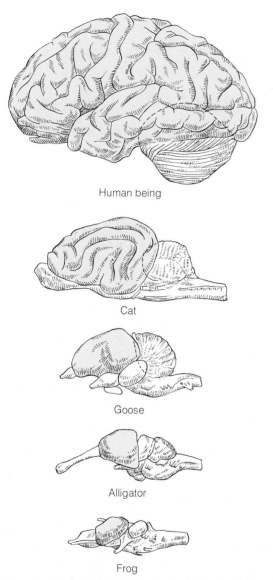

Figure 3.20 *The relative size of the forebrain (the locus of insight learning) in human beings and in four other animals. The forebrain (cerebrum) for each organism is shaded.*

we can visualize ways of applying them and imagine, in a limited way at least, the consequences of our actions.

It is to our capacity for learning, more than any other adaptation, that we owe our success as a species. That ability has enabled us to eliminate most of our predators, to at least partially control many of our major competitors, to live in almost all habitats on earth, and even to visit the moon. In short,

it has given us an ever-increasing understanding of—and control over—some segments of our environment.

Limits to Adaptation

It is tempting to believe that physical and technological adaptations will solve all our environmental problems. Perhaps we will develop a type of technology that will cleanse our air of all contaminants; perhaps our lungs eventually will become more resistant to air pollutants. But nature imposes strict limits on adaptations—limits to which all organisms are subject. We have already mentioned some major natural limits of adaptation. Let us review them here: First, each organism's capacity for adjustment to a particular change is limited by its genetic makeup; an organism cannot adjust to all environmental changes. Second, genetic limitations include adaptations that result from learning. Some species cannot learn certain types of behavior that would be quite adaptive for them; and some members of a species learn better and faster than others. Third, a changing environment selects only for characteristics that are already present in the population. And fourth, even if a heritable adaptive trait is present, the rate at which a population as a whole inherits the trait is limited by the population's reproductive capabilities (both the number of offspring per generation and the length of time that is required for each generation to mature and reproduce).

Our technological adaptations are limited as well. The development and large-scale implementation of most technological advances are very expensive and time-consuming. And often, as we point out many times in this book, our advances in technology create as many problems as they solve.

Feedback Controls

Throughout this chapter we have seen that individuals, populations, and ecosystems respond to environmental change. Those responses can take place because parts of a system (for example, the parts of a body) and its environment are interconnected by a flow of signals or information. If a change occurs in the output of the system and that change is somehow made known (fed back) to the sensor which

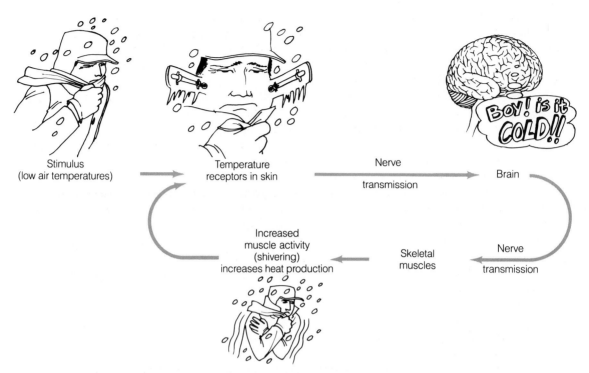

Stimulus
(low air temperatures)

Temperature
receptors in skin

Nerve
transmission

Brain

Increased
muscle activity
(shivering)
increases heat production

Skeletal
muscles

Nerve
transmission

Figure 3.21 *Shivering is a negative feedback response to declin-ing skin temperature.*

initiated the sequence of events, then the intercon-nected flow of information creates a control mech-anism, which is called a *feedback control.* The feed-back may be either positive or negative.

Negative feedback enables a system to adjust to an environmental change by counteracting the effects of the change. For example, shivering is the consequence of a negative feedback system, which acts to increase body heat production in response to declining air temperature. As Figure 3.21 illustrates, declining air temperature leads to a lowering of skin temperature. Temperature sensors in the skin re-spond to lower skin temperature by sending a mes-sage (nerve impulse) to the temperature control cen-ter of the brain (hypothalamus), where the message is interpreted. A second message (nerve impulse) is then transmitted to the skeletal muscles, which stimulates them to begin contracting—a response that we call shivering. As the skin warms up (as a consequence of the heat that is produced by shiv-ering), temperature sensors in the skin feed that new information back to the brain, from which another

message is forwarded to the muscles, instructing them to slow down or to cease contracting. Note that shivering fits the definition of feedback control, be-cause the change in output (increased heat produc-tion) is made known (fed back) to the temperature receptors in the skin that first initiated the sequence of events that led to shivering. If the environment becomes too warm, negative feedback may initiate sweating.

Negative feedback occurs when a change in a system induces an effect that reduces the change. Negative feedback, therefore, contributes to *homeo-stasis*—the tendency of a system toward mainte-nance of nearly constant internal conditions in the face of a changing environment. In addition to tem-perature regulation, other negative feedback con-trols in organisms deal with salt and water balance, blood composition, heart and breathing rates, and defenses against invading microorganisms. In the dose-response curve illustrated in Figure 3.4, nega-tive feedback is illustrated by the body's ability to repair damage that is caused by an exposure to very low levels of a toxic substance. Behavioral changes can also result from negative feedback and contrib-ute to homeostasis. For example, dressing warmly and seeking shelter are behavioral adjustments to lowered air temperatures.

Negative feedback and homeostasis also may be exhibited at the ecosystem level. Following the removal of a vegetative cover, the availability of many resources increases, because soil moisture and nutrients become more available. As a result, the disturbed area responds with a burst in plant activity that is often more productive than the vegetation was before the disturbance occurred. Moreover, within just a few years, the vegetation restores the nutrient cycling rates to those that approach predisturbance levels. Thus postdisturbance conditions promote negative feedback (succession), which counteracts the impact of the disturbance and may eventually return the area to predisturbance conditions—an example of homeostasis in ecosystems.

Ecosystem disturbance can also illustrate *positive feedback*, which is feedback that reinforces rather than counteracts a change. We have seen that initially, after a disturbance, removal of vegetation increases the temperature and aeration of the humus layer of soil. Those conditions, in turn, accelerate decomposition of the upper humus layer. As the humus decomposes, temperature and aeration conditions change in deeper layers so that decomposition and further loss of humus is accelerated. If revegetation (negative feedback) does not occur, positive feedback eventually will lead to a complete loss of humus and probably to severe soil erosion. In response to change, positive feedback will induce an effect that enhances the change, driving the system further and further away from homeostasis.

A similar situation exists in thermoregulation. Once our tolerance limit to cold is exceeded, the system goes out of control. For example, when the temperature of the body core (the abdominal, chest, and brain cavities) falls only 3°C (5.7°F) from a normal core temperature of 37°C (98°F), the negative feedback mechanisms become greatly impaired. If the core temperature falls to 29°C (85°F), the tolerance limits have been exceeded, and internal thermoregulation is no longer effective. As a result, positive feedback mechanisms take over and the core becomes progressively colder. Death ensues at core temperatures that are between 24° and 26°C (75° and 79°F). Survival is possible only if somebody can aid the victim by preventing further heat loss, by getting the victim close to a heat source, or both.

In some instances, positive feedback can be beneficial to the organism. For example, the formation of a blood clot in response to an injury that causes bleeding entails a complex sequence of events. Once clot formation is initiated, positive feedback tends to ensure that a complete clot will form, which stops further bleeding.

In later chapters, many examples are given of natural as well as human-originated negative feedback. In some instances, organisms, populations, ecosystems, or all of those entities are adjusting to environmental change. We also give many examples, particularly of human origin, in which the natural homeostatic controls have been disrupted, positive feedback has taken over, and the result has been death, species extinction, and ecosystem destruction.

Conclusions

In this chapter, we have examined some of the ways in which organisms, populations, and ecosystems respond to changes in the environment. We have seen that environmental change initiates a continual process of selection, which determines the prevailing characteristics of a population or ecosystem by favoring the continuing existence of some organisms over others. The net effect of environmental change, then, is a change in the type of organisms that can be found in an area.

In response to human activities, however, the environment itself is also changing. Some observers foresee a period of greater environmental change than we have yet experienced, as world population soars and human beings continue to strive for a higher standard of living. So far, our technological and cultural adaptations—the collective expression of our capacity for insight learning—have made us the most successful species in the history of the earth. Yet with those same adaptations, we have fouled the planet and gained the ability to destroy ourselves and most life on earth. We have the ability to contemplate the future and imagine the consequences of our actions. The question remains as to how responsibly, and how successfully, we will use this ability in years to come.

Summary Statements

The environment is changing continually. Some organisms are able to adjust to changes while others succumb. Tolerance limit curves help us understand the influence of environmental change on organisms, including ourselves.

The variables of age, ability to adjust to change, synergism, and antagonism all make it difficult to determine a specific organism's tolerance limits to an environmental change. Because so many interactions are possible among organisms and their environment, few simple cause-and-effect relationships exist in nature.

Today, a major and growing problem is the release of toxic substances into the environment. Those materials pose a significant threat to vegetation, wildlife, and persons. Efforts to establish standards for safe exposure levels of poisons, carcinogens, mutagens, and teratogens have been hindered by the same problems that arise in determining an organism's tolerance limits for other environmental factors.

If an essential factor in the environment lies outside the tolerance limits of an organism, then that organism either must be able to avoid the stressing conditions or it will die.

When many plants and animals in an ecosystem are killed by a disturbance, the functioning of that ecosystem is severely disrupted. If the stress is removed, the damaged ecosystem may or may not eventually return to its original structure and composition through a sequence of changes, which are called ecological succession. As populations of plants and animals succeed each other, the physical environment also changes. Many types of disturbance can destroy a climax ecosystem, and succession can be halted at any stage before climax is reached. Agriculture and forest management are important examples of the reversal of natural succession by human beings.

Early and developmental stages of succession differ from mature stages in several ways. The early stages are more productive, but they are also more subject to nutrient loss. Although the early stages exhibit less species diversity than mature stages, biologists do not yet understand the relationship between species diversity and ecosystem stability. Because many climax ecosystems have disappeared and others are disappearing quickly, weedy species and many species of game animals are more common today. But many other species of plants and animals, which require climax habitats, are threatened with extinction.

In most situations, succession to climax often requires many decades or centuries and, occasionally, a disturbance is so severe that the original climax ecosystem may never be reestablished.

Most organisms have the capacity to adjust, at least in part, to environmental change. That ability to adjust arises as a consequence of evolution, which consists of two steps: (1) the production of genetic variability, and (2) the ordering of that variability by natural selection. The products of evolution are adaptations.

Each organism's ability to adapt to an environmental change is determined, therefore, by its genetic makeup. A changing environment selects for (favors) only those characteristics that already exist in the population. If a trait is already present, the rate at which a population can adapt to any given change is limited by the population's reproductive capabilities.

An especially important means of adaptation is learning, because it enables an organism to adapt to new conditions very quickly—sometimes in a few seconds or less—and because learned adaptations can be communicated rather quickly throughout a population. Insight learning has enabled us to become the most successful species on earth. Yet that capability also has given us the potential to destroy ourselves and all other life on the planet. It remains to be seen whether we shall be sufficiently successful in the future to ensure our species' survival by using our insight learning and other adaptations.

Both positive and negative feedback mechanisms operate in the environment. Negative feedback tends to keep an organism, population, or ecosystem at nearly constant conditions. In contrast, positive feedback tends to take an organism or ecosystem further and further away from homeostatic conditions.

Questions and Projects

1. Write a definition in your own words for each of the terms that are italicized in this chapter.

2. What characteristics distinguish a living entity from a nonliving entity?

3. Describe how the law of tolerance and the law of the minimum are needed to understand the response of organisms to environmental changes.

4. How does the process of acclimatization make it more difficult to predict the impact of an environmental change?

5. How does the possibility of unknown synergistic effects influence evaluations that deal with the threat of environmental contaminants to human health?

6. What do LD_{50} values mean to you as just one person in a human population of nearly 5 billion people?

7. Although much less spectacular, chronic exposure to toxic materials is a greater threat to human health than acute exposure. Explain.

8. What are the differences in the effects of carcinogens, mutagens, and teratogens? Which one poses the greatest threat to the human species?

9. What are the major differences between primary and secondary succession?

10. Describe the mechanisms whereby pioneer species quickly slow the high rates of nutrient losses that normally follow the removal of the vegetative cover.

11. Cite examples from your area where improper land-use practices have created conditions in which ecological succession may not restore the landscape.

12. For your area, list plants and animals that are associated with pioneer communities and those that are usually found in climax communities. Is one group valued more than another? If so, for what reasons?

13. Contact the local office of your state's Department of Natural Resources or Conservation to find out where ecological succession is being used for vegetation or wildlife management. If possible, visit several of those sites and assess the success of the management practices.

14. Describe the various sources of genetic variability for a population.

15. After Kettlewell had run his peppered-moth experiment in a pollution-free area of England and found what he was looking for, why did he consider it necessary to rerun the experiment in a different polluted area?

16. By our activities, we have selected for bacteria that are resistant to antibiotics, insects that are resistant to insecticides, and larger cows that produce more milk. Compare and contrast those man-made selections with natural selection.

17. What is the difference between trial-and-error learning and insight learning? Which type would you say is more important for human beings?

18. List three specialized adaptations in human beings. Explain how those adaptations have contributed to the success of the species.

19. List some of the natural environmental changes that occur in your area daily, seasonally, and annually. How are plants, animals, and persons adapted to survive those changes?

20. Which group is better able to adapt to environmental change: human beings or insects, such as flies and mosquitos? Defend your choice.

21. What are the natural limitations on a population's ability to adapt to environmental change? How does the rate of environmental change influence a population's ability to adapt?

22. What is the major difference between negative feedback and positive feedback? Describe how those processes may be used to manage a relatively stable environment.

Selected Readings

AMES, B. N.: Dietary carcinogens and anticarcinogens. *Science* 221:1256–1263, 1983. A revealing look at natural carcinogens and mutagens as well as many natural anticarcinogens and antimutagens that are commonly present in the human diet.

ANNUAL REVIEWS, INC.: *Annual Reviews of Ecology and Systematics.* Published yearly. Palo Alto, Calif.: Annual Reviews, Inc. Each volume commonly contains articles that review current research on the response of organisms to environmental change.

BISHOP, J. A., AND COOK, L. M.: "Moths, Melanism and Clean Air." *Scientific American* 232:90–99 (January), 1975. The processes that shaped the selection for wing color of the peppered moth in England are examined.

BORMAN, F. H., AND LIKENS, G. E.: *Pattern and Process in a Forested Ecosystem.* New York: Springer-Verlag, 1979. An analysis and summary of the on-going Hubbard Brook studies.

CLARKE, B.: "The Causes of Biological Diversity." *Scientific American* 233:50–60 (August), 1975. The role of natural selection in maintaining a diversity of genetic traits is explored.

KRUG, E. C., AND FRINK, C. R.: Acid rain on acid soil: A new perspective. *Science* 221:520–525, 1983. An examination of the role of the natural processes of soil formation on the acidification of watersheds in eastern North America.

ROSENTHAL, G. A.: "A Seed-Eating Beetle's Adaptations to a Poisonous Seed." *Scientific American* 249:164–172 (November), 1983. A fascinating look at how one insect species can feed on seeds that are highly toxic to other insects.

SCIENTIFIC AMERICAN: *Evolution. Scientific American* 239 (September), 1978. A special issue devoted to many aspects of evolution and adaptation.

SCIENTIFIC AMERICAN: *The Dynamic Earth. Scientific American* 249 (September), 1983. A special issue that discusses various aspects of change in and on the planet Earth over time.

SMITH, R. L.: *Ecology and Field Biology.* New York: Harper & Row, 1980. A well-written text, containing several chapters on succession and adaptations of organisms to abiotic factors.

VITOUSEK, P. M., GOSZ, J. R., GRIER, C. C., MELILLO, J. M., REINERS, W. A., AND TODD, R. L.: Nitrate losses from disturbed ecosystems. *Science* 204:469–474 (May 4), 1979. A study of response mechanisms of forest ecosystems to disturbance.

WALDBOTT, G. L.: *Health Effects of Environmental Pollutants,* 2d ed. St. Louis: C.V. Mosby, 1978. A well-illustrated account of the basic effects of toxic and hazardous substances on human health. Plants and domestic livestock are considered to a lesser degree. The focus is primarily on air pollutants.

A mixed herd of zebra, waterbuck, and impala in South Luangwa National Park in eastern Zambia. The population sizes of these grazing animals are determined by many factors, whose relative importance varies from place to place and even from time to time in a particular area. (© Mark N. Brultan, National Audubon Society, Photo Researchers.)

CHAPTER 4   POPULATION: GROWTH AND REGULATION

We share this planet with more than 1.6 million known* species of plants and animals. Countless groups of individual members of species occupy the same geographical areas in what we call *populations.* Although we sometimes behave as if the human species were the only significant one on earth, our well-being rests on a balance among all populations of living things.

Populations that become too large can interfere with our well-being. Some examples include the overgrowth of algae, which fouls water, a swarm of locusts, which devastates crops, a large rat population in an urban area, which threatens human health by distributing rat-borne fleas that transmit typhus and bubonic plague, or large flocks of blackbirds and starlings which endanger airplanes, damage crops, and carry fungi that cause respiratory disease in human beings (Figure 4.1).

Populations that become too small can have an equally disruptive, and sometimes formidably complex, impact on human populations as well as on other populations in an ecosystem. It is obvious enough that a reduced population of beef cattle or wheat that occurs in Canada or Russia can affect the diets of Canadians or Russians (and possibly of Ghanians and Japanese, too). But the complex reverberations of population losses in nature are often less direct and less easily described. For example, a

Figure 4.1 *A large flock of blackbirds can significantly damage a grain crop. (Karl H. Maslowski, Photo Researchers.)*

---

* Some biologists estimate that as many as 10 million species exist, and that upward of 80 percent of them are arthropods (the animal phylum that includes insects, spiders, and mites).

significant drop in a rabbit population will cause foxes, owls, and hawks to feed more heavily on mice and ground squirrels or, failing that, to starve or migrate. Furthermore, a decline in the number of foxes (or owls or hawks) can trigger a surge in the population growth of rabbits, mice, and ground squirrels. Increased numbers of those animals, in turn, may damage vegetation, so that soil erosion is accelerated as well. Hence the effects of a population shift are by no means limited to the first species that experiences a change in its population. They are also felt, by way of food webs, throughout an entire ecosystem.

Because our interrelationships with other populations can affect us so profoundly, we attempt to regulate the size and quality of many of them. We control the growth of cattle and sheep populations when we raise them. We regulate game animals (such as deer and rabbits) and waterfowl (such as geese and ducks) in the name of wildlife management. One goal of such efforts is to keep the size of a game population within the capacity of its habitat for supporting them. For example, if a deer population grows too large for a particular habitat, the deer will over-browse and thereby damage vegetation, reducing the habitat's capacity to support the herd. Sadly, however, widespread disregard exists for the maintenance of ecosystems in favor of short-run goals. Many hunters have little regard for the ecological concerns in sustaining wildlife for any amount of time. To improve their chances for a kill, they pressure wildlife managers to allow deer herds to proliferate. Many persons in the tourist industry, too, favor uncontrolled growth of deer herds because deer in the wild attract visitors, who then spend their money in nearby resort areas. Thus not only must wildlife managers grapple with difficult wildlife population problems, but they must contend with those pressures as well.

Thus fluctuations in various plant and animal populations cause well-founded concerns, but, more worrisome is the problem of our own explosive growth worldwide. This ongoing increase, together with the soaring demands of the world's affluent and developing societies, is stressing the capacity of all the ecosystems on earth. In response to our unrelenting demands for food and timber, we have produced vast regions of greatly simplified ecosystems, which have created new conditions that favor population explosions of such organisms as insects, rats, and other pests. (A pest is defined as any species whose population is large enough to interfere significantly with the well-being of humans.) As a result of continued efforts to create those artificial environments, we are disrupting the natural balance among many species by polluting and destroying their habitats, and hundreds, perhaps thousands, of species are being pushed to the brink of extinction.

In this chapter, we examine how populations grow, what factors promote and inhibit their growth, and the implication of these principles for the management of biological resources, such as wildlife and fisheries. The same principles apply to the growth and regulation of all plant and animal populations, including our own. But human population growth is also greatly affected by social and economic factors, and so we defer our consideration of the growth and regulation of our own species until Chapter 16.

Population Growth

Populations are dynamic. One may be growing while another is declining, but all populations fluctuate between growth and decline. Rarely is a population *stationary*, that is, not changing in size. The growth rate of a population depends on the net difference between additions to it and subtractions from it. Additions are made by *natality* (the production of new individuals by birth, hatching, or germination) and *immigration* (migration into the population from elsewhere). Subtractions occur as a result of *mortality* (deaths) and *emigration* (migration out of the population). If the additions exceed the subtractions, the population will grow; if the subtractions exceed the additions, it will shrink.

Natality varies tremendously from one species to another. For example, codfish produce as many as 9 million eggs per female in one season. In marked contrast, female salmon lay fewer than 1000 eggs. Birds lay far fewer eggs than fish—from 1 (in the albatross) to 15 (in the quail) eggs per clutch. Although most bird species lay one clutch each year, some (such as robins and bluebirds) may produce two or three clutches. Small mammals, such as mice, may have a litter of four to six young as often as four times a year, but larger mammals characteristically have only one or two offspring each year. In general, the number of offspring is small in those species whose young require the most care.

Mortality rates, or chances of survival, also vary

considerably from one species to another. For example, in most species of fish, fewer than 5 percent of the eggs that are produced actually develop into young and survive for a year. In most bird species, fewer than one-quarter of the eggs that are laid produce young that survive the first season. Those species that receive parental care have a greater chance of surviving, but the young of all species (as pointed out in Chapter 3) are quite vulnerable to hostile weather, inadequate nutrition, predators, and disease. In most species, once a member survives its risk-laden first year, the chances that it will survive for the normal lifespan of the species remains more or less constant.

Exponential Growth Curve

When a population enters a new region that contains abundant resources, initially its growth follows the pattern illustrated in Figure 4.2a. It grows slowly at first but, soon, its growth rate increases rapidly. Such a pattern illustrates the exponential nature of population growth.

Exponential, or *geometric*, *growth* occurs when a factor such as a population increases by a constant

Figure 4.2 *(a) An exponential, or geometric, growth curve. In a population that is growing exponentially, the rate of increase remains constant, but the growth in the number of individuals accelerates even more rapidly as the size of the population increases. (b) A linear, or arithmetic, growth curve. In a population that is growing arithmetically, the size of the population increases by a constant amount in a given period of time. Such a curve, when plotted on graph paper, follows a straight line, as shown.*

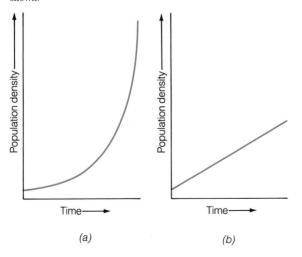

(a)   (b)

Table 4.1 An illustration of exponential growth involving wheat grains on a chess board

SELECTED SQUARES ON CHESSBOARD	NUMBER OF GRAINS OF WHEAT REQUIRED
1	1
2	2
3	4
4	8
5	16
10	512
20	524,288
40	549,755,809,568
64	9,223,372,036,854,775,808

percentage of the whole during a specific period of time. For example, assume that you have a savings account that is growing (accruing interest) at the rate of 6 percent per year. Next year, you will receive interest on the principal, as well as on last year's interest. Thus the total amount that you receive in interest will be greater than 6 percent. Most of us are more familiar with *linear*, or *arithmetic*, *growth* (Figure 4-2b), which is an increase of a consistent amount during a specific period of time. For example, if a youth grows in height at a constant rate of 2 centimeters per year, he or she is growing arithmetically. The differing shape of the curves in Figure 4.2 imply some highly significant differences between exponential and linear growth. To properly assess the impact of population growth of any species, including our own, we must understand those important differences.

One of the differences between exponential and linear growth is illustrated here by the following old legend: Many years ago a clever citizen presented a beautiful chess set to his king. The king was so pleased that he asked how he could reward his subject. In reply, the man asked the king to give him 1 grain of wheat for the first square on the board, 2 grains of wheat for the second square, 4 grains for the third square, 8 grains for the fourth square, and so on; that is, with each square, the number of grains would be doubled (increased by 100 percent). Thinking that this was a small price, the king readily agreed to pay the citizen for the chess set. Table 4.1 shows what happened. At first, the number of grains needed for each succeeding square appeared to increase slowly; but (just as the exponential growth curve in

Figure 4.2 illustrates) the number of grains required for the next succeeding square soon began to grow rapidly. The tenth square required 512 grains, the fifteenth 16,384 grains, and the fortieth, more than 549 billion. Needless to say, the king ran out of wheat long before the sixty-fourth square was reached. Obviously, the king was thinking about the more familiar linear growth when he agreed to the price. If he had agreed to pay at a linear growth rate, he would have had to add only one extra grain for each square, or (1 + 2 + 3 + 4 + · · · + 64). Thus, the total cost to the king would have been less than a bushel of grain. The cost for all sixty-four squares in terms of exponential growth would have equalled the world's present annual production of wheat for the next 2000 years. Thus exponential growth increases by huge proportions in a relatively short period of time. We consider another consequence of exponential growth in the next section.

Because populations also grow exponentially, any population has the ability to generate an enormous population explosion. For example, a single breeding pair of houseflies could have more than 6,000,000,000,000 descendants between April and August if all their eggs hatched and if all the young survived to reproduce. But, obviously, this problem does not occur—we are not buried under a deluge of flies or of any other species. No population can continue to grow exponentially for long because, at some point, some factor in the environment will limit it and its population growth will be retarded.

Sigmoid Growth Curve

The *sigmoid* (S-shaped) *growth curve* (Figure 4.3) illustrates one pattern of growth that can result if an environmental factor begins to limit population growth. In this case, the population begins to follow an exponential curve, but when it reaches a particular size (the inflection point), a limiting factor in the environment will cause the growth to decelerate. Eventually, the population will level off below the *carrying capacity* of the habitat—that is, the population size that can be supported by the total resources in the habitat. Populations rarely follow such a simple curve, however. The population often overshoots the carrying capacity and, when some factor becomes limiting, instead of leveling off, the population crashes (falls very quickly) below the carrying capacity and may then oscillate around it.

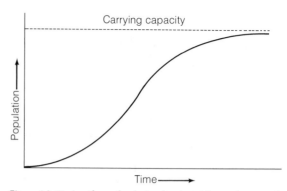

Figure 4.3 *During the early phase of a sigmoid growth curve, the population shows an accelerating rate of growth. At a population density, called the* **inflection point**, *the environment becomes limiting and the growth rate begins to decelerate. This deceleration continues as the population approaches the carrying capacity, at which no further growth will occur.*

To determine why populations tend to overshoot the carrying capacity of their habitats, another significant feature of exponential growth must be considered. That feature can be illustrated by the following modern-day version of an old French riddle. Assume that you are the proprietor of a plush resort, overlooking a beautiful lake. One day your landscape gardener informs you that an algal bloom is growing on the far side of the lake. The gardener has determined that the algal bloom is doubling in size (increasing 100 percent) each day, and that it will take 30 days for the bloom to cover the entire lake. He asks you what to do about it. You tell him that the bloom is too small to worry about now and that you will decide what to do when the pond is half covered. In that way, you believe that you will have plenty of time to stop the algal growth. To answer the riddle, you must determine which day you will have to make the decision. The answer is the twenty-ninth day. Since the bloom is growing exponentially, by the time it covers half the pond, you will have just one day to control it. Hence exponential growth has another deceptive characteristic—it is explosive and can reach the carrying capacity within a surprisingly short period of time.

This feature of exponential growth can cause a population to overshoot its carrying capacity because, in many cases, a population initially grows faster than its control agents. For example, if predators are important to control the numbers of a rapidly increasing prey population, the time that is needed for the predator population to reproduce or for outside predators to immigrate into the area might take some time. Thus the prey population will continue to grow and may overshoot the carrying capacity before the number of predators needed to control them becomes available.

In most cases, a growing population contains a large percentage of young members who are not yet making full demands on their habitat for food, water, space, and other resources. When those young members mature and begin to require an adult share of the resources, the habitat cannot supply them sufficiently. Meanwhile, the population has continued to grow exponentially and finds itself above the carrying capacity.

In these examples, a lag time becomes apparent, which often exists before various agents can effectively control a rapidly growing population. In the meantime, as the riddle illustrates, a population that is growing exponentially can quickly approach and will often overshoot the carrying capacity of the habitat. In response, an increased portion of the population will die as a result of predators, parasites, or insufficient resources, and some may migrate in search of more favorable conditions. As a consequence, the population is reduced to a size below the carrying capacity. This process may then be repeated, and the long-range effect is an oscillation around the carrying capacity.

Reproductive Strategies

Generally, the actual extent to which a population will fluctuate in size depends on how the population allocates its available resources. Ecologists speculate that each population, in effect, employs a strategy to balance the use of resources that are needed for current reproduction against the probability of survival in order to reproduce in the future. Two quite different, basic types of *reproductive strategies* are recognized, with many intermediate types. At one extreme are the *r-strategists*,[†] which opt for maximizing their current rate of reproduction at the expense of surviving long enough to reproduce in the future. Those organisms produce many offspring in a short period of time, and because they allocate most of their resources to reproduction, they are

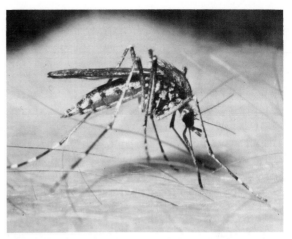

Figure 4.4 *The mosquito is a familiar r-strategist. (Hans Pfletschinger, Peter Arnold, Inc.)*

usually small and short-lived (Figure 4.4). This strategy allows the population to be opportunistic—that is, to reproduce rapidly when the conditions are favorable or a new habitat becomes available. Unfavorable habitat conditions, however, can easily cause a crash in the population. Hence populations of r-strategists exhibit large fluctuations in size—that is, they generally follow "boom-and-bust" cycles, as illustrated in Figure 4.5.

Algae are r-strategists. As noted in Chapter 3, a seemingly clear lake may become peagreen in only

Figure 4.5 *Variations in number of algae in the bay at Green Bay, Wisconsin. Because they are small, short-lived, and reproduce rapidly, algae are quite sensitive to environmental changes. Hence their numbers vary by more than 30-fold within a year.*

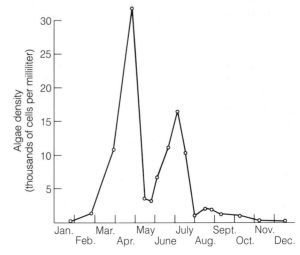

---

† The letters "r" and "K" (which appear later in this discussion) come from a standard mathematical equation that is used to describe population growth in a habitat with limited resources.

a few days when a miniscule population of algae reproduces at an explosive rate. Often, however, they disappear almost as quickly as they appeared, because sometimes the population will use up the available nutrient supply and then crash. Other times, a change in water temperature or a period of several days without sunshine will contribute to the crash.

Other r-strategists include insects, rodents, and annual plants. Because annual plants live only 1 year, they allocate most of their energy to seed production rather than to plant growth. Although they rarely grow much more than a meter in height, each annual plant commonly produces hundreds of seeds. Because r-strategists usually disperse widely and are good colonizers, they are common in the early successional stages of terrestrial ecosystems.

At the other extreme of reproductive strategies are the *K-strategists,* which opt to increase their probability of surviving to reproduce in the future rather than maximize their current rate of reproduction. Accordingly, K-strategists allocate far more energy to producing larger and longer-lived adults. Because they funnel less energy into reproduction, they produce fewer offspring, but usually their offspring are larger and require considerably more parental care. As Figure 4.6 indicates, populations of K-strategists are usually more stable than those of r-strategists. For large organisms, daily variations in the environment have a relatively minor impact on the size of a population. Even if the climate or resources (such as food, water, and shelter) become

limiting and diminish natality during any given year, many adults will survive to reproduce the following year. For example, the reproductive period of sheep extends from the age of 1 year until their deaths at 10 years (90 percent of the lifespan), and normally sheep produce one or two offspring each year. Reproductive characteristics such as these tend to reduce the degree of variation in the size of the population even if changes occur in the habitat.

Trees have reproductive strategies that are similar to those of large animals. Usually, they do not produce seeds until they are 15 to 20 years old, at which time they are 5 to 10 meters tall (approximately 15 to 30 feet) and have a well-developed root system. Once sexual reproduction begins, however, the trees produce seeds for many decades. Because seeds from trees are often contained in fruits, such as acorns and hickory nuts, which are larger and more obvious than the seeds and fruits that are produced by annual plants, it may appear that trees produce more seeds than annuals. But the veritable jungle of weeds that is found growing in an abandoned lot or another disturbed area shows that on any given unit of land, a dense growth of small annual plants will produce thousands more seeds than a stand of larger, but fewer, trees. The lower reproductive capacities of trees, along with a more limited ability to disperse, make K-strategists poorer colonizers than r-strategists. Hence K-strategists are more common in later successional stages of terrestrial ecosystems.

In crop cultivation, we obviously select for r-strategists. Our crop plants are small annual plants that put most of their energy into the production of seeds and fruits. Ironically, however, by selecting those plants, we also create conditions that favor pest species, which are also r-strategists. In raising livestock, animal husbandry practices represent a trade-off between both ends of the spectrum of reproductive strategy. Although cattle, swine, and sheep are K-strategists and thus are less productive than r-strategists, their larger size makes them easier to manage, and they are considered to be far more desirable as food sources than insects or rodents.

Many endangered species are extreme K-strategists. For example, the gorilla produces only one infant every 5 or 6 years; and the California condor produces only one chick every 2 years. Although K-strategists are more efficient than r-strategists in raising young, they have limits to their survival rates. Illness, predation, and accidents take their toll on

Figure 4.6 *Growth of a sheep population in the more than 100 years after their introduction onto the island of Tasmania. Because sheep are large, long-lived, and reproduce slowly, the sheep population is relatively insensitive to environmental change. Hence in contrast to the algae in Figure 4.5, the sheep population usually showed less than 25 percent in year-to-year variation in number after the population leveled off near the carrying capacity. The dotted line is the hypothetical sigmoid curve, about which the population fluctuates. (From J. Davidson, On the growth of sheep in Tasmania.* Transactions of the Royal Society of South Australia 62:342–346, 1938.)

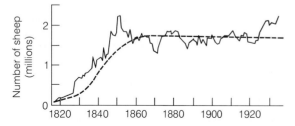

Table 4.2 Contrasting characteristics of r-strategists and K-strategists

r-STRATEGIST	K-STRATEGIST
Adults Small	Adults Larger
Many Small Young	Fewer, Larger Young
Little or No Care of Young	More Care of Young
Early Maturity	Later Maturity
Short Life	Longer Life

all populations, but their impact is particularly severe on populations with low reproductive capabilities. The gorilla, like many other endangererd species that are K-strategists, exists today in perilously small numbers and only in the most favorable habitats. Some ecologists suggest that such species as the gorilla and condor have pushed the K-strategy too far and that they are approaching extinction regardless of human activities. There is no doubt, however, that humans are accelerating the demise of those species by destroying their habitats. The characteristics and management of endangered species is discussed further in Chapter 14. The contrasting characteristics of r-strategists and K-strategists are summarized in Table 4.2

Population Regulation: Biological Factors

A population cannot continue to grow indefinitely. Sooner or later, some factor will become limiting and it will cease to grow. Frequently, the size of the population will actually decline as a result of one or more factors. Ecologists have found that both biological and physical factors play a role in limiting the size of a population. Biological factors consist of interactions between populations and include predation, parasitism, and competition. Physical factors include fire and weather stresses, such as hurricanes, floods, droughts, and temperature extremes.

Factors that influence the size of a population are also viewed as being either density-dependent or density-independent. The controlling influence of a *density-dependent factor* becomes stronger as the density of a population (the number of individuals per unit area) increases. Biological factors are usually considered to be density-dependent. For exam-

ple, in competition, the more the demand (the population size) exceeds the supply (the available resources), the more significant competition becomes in regulating the population.

Physical forces are frequently described as being *density-independent factors*. Unlike a density-dependent factor, the effect of a physical factor does not vary much as the population size changes. For example, a severe frost probably will kill all the annual plants in a field, regardless of whether 10 or 1000 of them exist.

Often both types of factors interact to influence the size of a population. For example, a late spring frost may kill many oak flowers, which may cause a failure in the acorn crop. As a result, the following winter, the squirrel population will face severe competition for the meager food resources that are available, and many will starve.

In the following sections, we consider more fully the intricate influences of density-dependent and density-independent factors on population growth. We first consider how various biological factors may be involved in controlling the size of any given population.

Ecologists have verified that several types of interactions take place among species. As Table 4.3 illustrates, some of those interactions play a role in regulating population size by having a negative impact on one population (such as reducing the size of the population by predation and parasitism) or on both populations (as a result of competition). Some interactions, however, actually benefit both populations. In many cases, the survival of each interacting species depends on the presence of the other. That kind of relationship is called *mutualism* (see Box 4.1). Our focus here, however, is on the role of

Table 4.3 Types of two-species interactions

TYPE OF INTERACTION	GENERAL NATURE OF INTERACTION
Predation	The Predator Benefits at the Expense of the Prey
Parasitism	The Parasite Benefits at the Expense of the Host
Competition	Both Species Are Inhibited by the Presence of the Other
Mutualism	Both Species Are Favored by the Presence of the Other

BOX 4.1

MUTUALISM: SPECIES HELPING EACH OTHER

Mutualism—an interaction between species that is beneficial to both—occurs widely and is important to many populations. Such an interaction can benefit a species in several ways. Those ways include (1) nutritional, either the digestion of food for the partner or supply of nutrients or energy by synthesis, (2) protection, either from enemies or from environmental variability, and (3) transport, such as the dispersal of pollen or of materials from unsuitable to suitable environments. Interactions between ruminants (such as cows, sheep, goats, and deer) and the microbes that inhabit their stomachs well illustrate nutritional mutualism. Because they are herbivores, a ruminant's diet consists mainly of cellulose. But like most animals, ruminants do not produce the necessary enzymes to break down cellulose. Unlike most animals, however, their stomachs do contain microbes that can digest cellulose. Moreover, the microbes can also synthesize proteins from the ammonia and urea that are present in the stomach. Some of those microbes are subsequently digested by the ruminant, thus providing it with protein. They can also synthesize all of the B-complex vitamins and vitamin K. Hence microbial activity enables ruminants to gain additional energy and nutrients. The microbes, in turn, benefit from free food and good housing—the warm, wet interior of the stomach provides a very favorable environment for their growth and reproduction.

We, too, have a complex assemblage of mutualistic bacteria residing in our guts. Unlike those of ruminants, our microbes cannot digest cellulose, but they do synthesize the B-complex vitamins and vitamin K.

Protection from enemies is provided by many organisms that, in turn, may be provided with food, protection, housing, or all of these. In East Africa, for instance, certain ants live in the swollen thorns of acacia trees. The thorns provide an ideal shelter and a balanced and almost complete diet for all stages of ant development. The ants, in turn, serve as protectors. As soon as a branch is touched by a browsing animal or some other organism, the ants feel the vibrations and respond by pouring out of their holes in the thorns and racing toward the ends of the branches. Along the way, they release a repulsive odor that usually drives the enemy away.

Ants also are involved in protective mutualisms with sap-feeding insects, such as aphids. Plant sap often contains too much sugar for the aphids to utilize. As a result, aphids commonly extrude a sugary solution, called honey dew, which the ants use as a major source of food. In return, ants protect the aphids by attacking an approaching predator and driving it away.

Flowering plants and their pollinators well illustrate transport mutualisms. A wide variety of animals (such as butterflies, moths, bees, flies, beetles, hummingbirds, and bats) visit flowers to obtain food. Flowers produce pollen, which is rich in protein, and nectar, which is a nutritious sugary fluid. During feeding, those animals often brush against the flower and their body is dusted with pollen. When the animal visits another flower, the pollen from the first flower may be deposited inadvertently on the next flower. That pollination process may then lead to fertilization and, consequently, to seed and fruit production. Several advantages accrue to flowering plants from pollination by animals. Because the animals move directly from one flower to another, less pollen is wasted than for those flowering plants that depend on the random movements of the wind for pollination. Moreover, the food-seeking activities of the animals increases the probability that pollination will occur over that of wind pollination, when the plants are relatively few in number and are widely scattered.

How do mutualistic interactions form? For example, are the mutualistic microbes that live in our guts the descendants of microbes that were at one time intestinal parasites of our ancestors? Are mutualistic interactions stable, or do they frequently break down? For example, instead of protecting their partners, do aphid-tending ants sometimes turn on the aphids and devour them? Ecologists still have much to learn about these fascinating population interactions.

negative interactions, that is predation, parasitism, and competition, in regulating population size.

Predation

Predation is an interaction in which one organism (the predator) derives its sustenance by killing and eating another (its prey). The role of predation in keeping the predator species alive is clear. It is more difficult to determine the effect that predation has on the prey population. Most studies of the effect of predation have focused on wild game and pests. Those studies indicate that the significance of predators in regulating the population of their prey depends on the types of predators, the types of prey, the ratio of predators to prey, and the conditions of the habitat.

Predation is a major agent in limiting the population size of small animals (such as insects and rabbits). In fact, *biological control*, the regulation of pests by natural enemies, including predators and parasites, is an important alternative to pesticides. A classic example of pest control by a predator involves ladybug beetles and the cottony-cushion scale, an insect that infests citrus plants (Figure 4.7). In the early 1870s, cottony-cushion scale insects were introduced accidentally into California from Australia. They spread rapidly, and within a few years, the infestation of that insect threatened the citrus industry. A resourceful government scientist journeyed to Australia to search out the natural predators of the scale. He turned up a species of ladybug beetle, which, when introduced into California citrus orchards, quickly brought the scale under control. The graph in Figure 4.8 illustrates the ladybug's effect on the scale's population. Overall, the ladybug has kept the scale population down, so it causes little damage to citrus crops. (Ironically, during the late 1940s, the scale population exploded when applications of the pesticide DDT decimated the ladybugs.)

In the United States, other insect pests that are regulated by biological control include bark beetles and sawflies on ponderosa pine, spotted aphids on alfalfa, and codling moths on apple trees. Today, scientists can point to nearly a hundred examples

Figure 4.7 *Ladybugs are used to combat cottony-cushion scale insects on citrus trees, which is an example of biological control. (Florida Department of Agriculture, Division of Plant Industry, Gainesville, Florida.)*

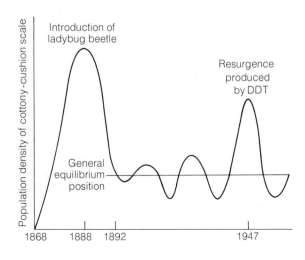

Figure 4.8 *The effect of ladybug beetles on population size of cottony-cushion scale. (After V. Stern et al., "The Integrated Control Concept." Hilgardia 29:81–101, 1959.)*

around the world in which insect pests have been at least partly controlled by predators.

Although predation is often important in controlling the populations of insects, the role of predators in regulating populations of large animals is much less clear. Despite many field studies, few generalizations can be made. In one region, predators may limit a particular prey population; in another, the impact of predation on the same population may be negligible. For example, a study of wildebeest herds on the Serengeti Plains in Tanzania revealed that large predators, such as lions, cheetahs, and hyenas, had little impact on the wildebeest population. Apparently, the major factor that was controlling the numbers of wildebeest was the separation of offspring from their mothers. This type of separation occurs frequently when the large herds migrate and, subsequently, the offspring die from starvation.

In marked contrast, other studies indicate that large predators do play a major role in controlling the size of prey populations. For example, lions in Kruger National Park in South Africa have dramatically reduced the numbers of such prey species as wildebeest and zebra (Figure 4.9). Two factors account for the differences in the role of predators between the Serengeti and Kruger National Park. One factor is the number of lions per prey animal, which was much greater in Kruger Park than in Serengeti. In Serengeti, the predator-to-prey ratio was 1 lion to 1000 prey animals; in Kruger National Park, it was 1 lion to only 110 prey animals. All other things being equal, then, one would expect lion predation to have a much greater impact in Kruger Park than in Serengeti. Habitat conditions comprise the other factor. A 5-year period of unusually heavy rainfall in Kruger Park caused the large aggregations of wildebeest and zebra, which are common during drier years, to break up and disperse into small groups over a large area of land. That distribution pattern allowed the young lions, which were forced out of their prides, to hunt without competition from adult lions. Hence the total lion population increased by

Figure 4.9 *Large predators—such as these lions feeding on a zebra carcass—can, at times, be important in controlling the numbers of a prey population. At other times, their impact on prey numbers is negligible. (David C. Fritts, Animals Animals.)*

about 60 percent in 2½ years, which caused more hunting of wild game herds. The impact of predation subsequently lessened when the weather returned to normal, the temporary water holes dried up, and the herds began to concentrate again around permanent water holes.

The complex interactions between predators and prey are well illustrated by the wolf and moose populations that reside on Isle Royale in Lake Superior. A particularly important finding of this on-going study, which began in the early 1950s, is that the nature of interactions between predators and prey change surprisingly often (see Box 4.2).

Hence we see that habitat conditions may influence the importance of predation as a regulator of prey population size. Also, the amount of predation may vary from one place to another as well as over time within a particular area. In the past, predators were killed frequently on the assumption that the game populations would then prosper. But the evidence that exists, which indicates a variable influence of predators on prey populations, makes it difficult for those in game management to make decisions about the need for predator control. Moreover, wildlife managers must contend also with pressures from special-interest groups, such as hunters, who think that killing predators will save more wildlife for them, and environmentalists, who point out that many predators are endangered species and argue that perhaps all predator species should be protected.

The role of large predators in controlling the numbers of prey is not yet fully understood, but it is clear that predators do contribute to the stability of prey populations in other ways. For example, large predators tend to keep prey populations healthy by killing off the weak, diseased, and old members. Those animals are normally the first to be overtaken when their herd is migrating or being pursued, because they tend to lag behind. Thus some predator species often act more as scavengers than as actual attackers. They contribute to the general well-being of the herd by eating animals that are suffering from lethal diseases—animals that would have died soon anyway. If those diseased animals lived longer, they would probably spread their disease to the remaining members of the herd.

In general, as we have seen, many prey species possess adaptations that help them to escape predation. But not all species are so equipped. Many

insects, for example, do not run or hide from predators. Aphids merely sit on leaves, sucking up plant juices (as shown in Figure 4.10). Hence they are easy prey. But such insects compensate for their inherent vulnerability by producing offspring by the hundreds and thousands. Thus, although predators consume them in great numbers, their inherent high natality rate maintains the size of their population. This balancing effect demonstrates the fundamental concept that population size results from the interaction of many factors. Some factors are genetically inherent to the population itself, while other factors stem from the nature of the ecosystem and the other populations that reside within them.

Parasitism

Parasitism is an interaction in which one organism (the parasite) obtains its nourishment by living within or on another living organism (the host). Parasites do not kill their hosts to get their nourishment, as predators do, but they may eventually kill their hosts inadvertently in the process of obtaining their meals. When a host dies, the parasite frequently also dies, because a parasite population that infests and weakens a host population to the point of extinction will lose its own habitat in the process. However, as a result of natural selection, less virulent parasites and more disease-resistant hosts are often produced. Those changes are examples of the process known as *coevolution*.

A classic example of the coevolution of a parasite and a host is illustrated by the short-lived reduction of European rabbits in Australia. After the rabbits were introduced in Victoria, Australia, in 1859, their numbers quickly exploded and they rapidly spread across the continent. By 1928, they had invaded nearly two-thirds of the Australian continent, ruining much of the grasslands that supported the sheep, Australia's main industry. By 1950, poisons, predators, and fences had proved to be ineffective against them. But a solution was found—a virus that causes myxomatosis, which is a disease that is related to smallpox. That virus, which (like all viruses) is a parasite, was discovered in South American rabbits that were resistant to it. European rabbits, however, were not resistant, and they quickly died of myxomatosis when they were infected.

The virus was introduced in Australia in 1951,

BOX 4.2

WOLVES AND MOOSE ON ISLE ROYALE: AN EVERCHANGING STRUGGLE FOR SURVIVAL

Rolf Peterson is the wolf man of Isle Royale National Park. Mr. Peterson has spent much of the past 13 years howling through a megaphone at the wolves here (they howl back, tipping him off to their whereabouts), lugging tons of old moose bones (he studies them to determine the animal's age and state of health when it died) and painstakingly picking apart dozens of wolf "scats," the biological term for carnivore droppings (to see what the wolves have been eating lately).

All of this is in the name of science, and it has helped bring knowledge about wolves into the twentieth century. As a biology professor at Northern Michigan University, the wiry, 34-year-old Mr. Peterson is continuing a research project begun 25 years ago by the noted Purdue University naturalist Durward L. Allen.

When the research on this island's wolf–moose relationship started, Canis lupus still had its "big bad wolf" image of the Middle Ages. It was generally believed, for example, that this island's thriving moose colony was a goner after the first wolves trotted 15 miles across the ice from Canada in 1949. Instead, after living high on the moose for over three decades, it was the wolves that suddenly began starving a few years ago, even though hundreds of moose were left. Today there are 23 wolves on this 210-square-mile island and about 900 moose.

The sudden decline of the wolves occurred just when Mr. Peterson thought he was finally beginning to understand wolves and moose. Over the years, it had become clear that the two could coexist. Scientists theorized that, with nobody shooting at either animal, wolf and moose populations would strike a delicate natural balance. That didn't happen. The number of moose soared and plunged wildly, and the number of wolves also fluctuated, to a lesser extent.

So biologists began thinking of the situation as a seesaw of survival. And the great "wolf crash" underscored the seesaw's volatility. Now Mr. Peterson is trying to sort out how the seesaw is affected by snowfall patterns, a nasty little tapeworm, the mysterious intricacies of wolf society, and the eating habits of moose.

The equation here almost entirely excludes humans. The local wolves avoid people, and rangers have placed the island's trails and campsites away from wolves' favorite areas.

One of the most startling research findings is that wolf ferocity doesn't count for much. "I find myself cheering for the wolves," Mr. Peterson says, "because I so seldom see a kill." The research here indicates that wolves can kill only one of every two dozen or so moose they attack.

The moose is one tough critter. "There are some moose that wolves just can't kill," Mr. Peterson says. At 8 feet tall and 900 pounds, a healthy moose can kick like an Oriental boxer in any direction, with any leg, with enough force to crush a wolf's—or even a human's—skull. A 100-pound wolf, even attacking with a pack, seldom is any match for a fighting machine like that.

(Bull moose also have antlers, which they grow each year, but they don't fight wolves with them. Antlers are for impressing the ladies during the mating season and for jousting with other bulls.)

Just how tough are moose to kill? On a winter flight in 1979, Mr. Peterson watched an old bull with big white cataracts wander around in circles on the Lake Superior ice. When a pack of 11 wolves attacked, the moose backed up against a small, brushy island and, lashing out with his front hoofs, fought them off for two nights. (Wolves apparently prefer to attack at night.) On the third night, the moose fell off the ice and drowned.

Understandably, wolves are careful about which moose they attack, usually picking out calves or moose past their prime. As moose grow older, arthritis and a tapeworm that creates stifling cysts in their lungs make them more vulnerable to wolves. That wolves are so selective, avoiding the strongest animals, has come as something of a surprise to naturalists who had been steeped in Wild West lore about wandering packs of wolves killing anything they came across and leaving the carcasses to rot.

Knowing about this selectivity may be important, experts say, if wildlife managers are to broaden their efforts to preserve wolves in the United States, where they are an endangered, if unloved, species. Some wolf advocates suggest that wolves eventually be set loose again in Yellowstone and other national parks. They hope that wildlife "management" by wolves would help strengthen herds of deer, elk, and bison by culling out the weak animals, as it apparently has improved the moose stock here.

On an early-morning flight last February 14, Mr. Peterson spotted some gory evidence of that process. Six wolves were curled up in inky spots on the ice of a small inland lake, sleeping off their feast of the night before. In the blood-splashed forest snow nearby lay a fresh moose carcass.

It is months later now, and the reddish-bearded Mr. Peterson clambers over boulders and fallen trees and rushes through thickets to visit the site in a remote, trackless part of the island. (There aren't any roads on Isle Royale, only primitive trails at best, and the only way to get here is by boat or plane.)

All that remains of the moose is a huge mat of straw-like brown and gray fur, a mass of chewed-up twigs that was in the stomach, and the skull, the jaw, and some other bones scattered nearby. The wolves ate everything else—hide, hoofs, and all. Mr. Peterson determines from tooth wear that the moose, the 1728th carcass examined since the research began, was a cow at least 10 years old. Few moose survive past 14, and the average life span is about 8 years.

The moose probably tried to run when the wolves attacked, Mr. Peterson speculates. Moose that stand and fight usually survive. That gives the wolves a chance to slash away with their fangs on her rear. The moose then apparently backed up against two old jack pine to try to defend herself, the typical strategy.

"It works every time—except the last time,"

Mr. Peterson observes. He points out splintered pine seedlings that still testify to the fight she put up.

Two weeks later, wolves that Mr. Peterson recognized as the same pack were feeding on another moose carcass. Suddenly, while Mr. Peterson spied from above, a rival pack of six wolves bounded over the ridge and attacked. A breakneck, mile-long chase along a stony ridge ended when the attackers caught and briefly scuffled with one of the fleeing wolves. Instead of killing it, for some reason the attackers let it go this time.

The skirmish was part of a war over hunting territories that erupted 2 years ago, Mr. Peterson believes. Its seeds may go back a decade or so. In the early 1970s, deep snows made it hard for moose to find good browse—the trees and shrubs they nibble in winter. This not only made them easier pickings for the wolves but also created a whole generation of constitutionally weakened moose.

That encouraged the wolves to prosper, to split into five breeding packs, and to push their numbers in 1980 to a record 50, double the 1960s average. (Packs are social organizations with rigidly enforced pecking orders. The top male and female permit only themselves to breed, limiting pack growth.)

By then, the wolves had whittled the moose down to about 600 from a mid-1970s peak of 1000. But among the moose that remained, there weren't enough that wolves could kill. Facing starvation, the wolves began fighting and killing each other. By 1982, there were only 14 wolves left. Since then, the wolves have recovered to 23 in three packs.

Although the moose have soared back to over 900, Mr. Peterson believes that the wolf population will stay about where it is for several years because a string of mild winters has created a generation of strong, well-fed moose that can fight off the wolves. Mr. Peterson says he needs at least 3 more years to test that theory.

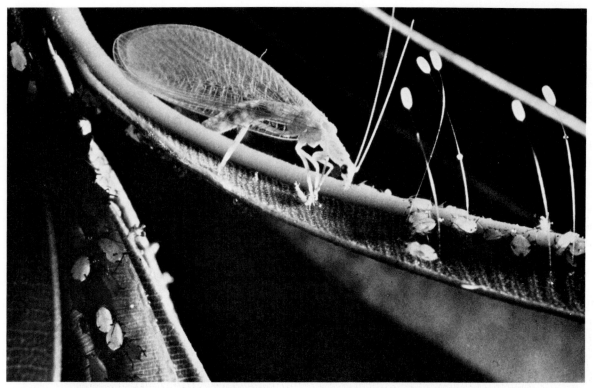

Figure 4.10 *Aphids are "easy pickings" for the predatory lace-wing insect. (Grace Thompson, National Audubon Society, Photo Researchers.)*

and hopes soared as myxomatosis spread quickly through the rabbit populations of southeastern Australia, killing hundreds of thousands of the animal pests. The rapid dispersal of the virus was accomplished by mosquitos, which picked it up when they bit infected rabbits and, subsequently, spread it among healthy rabbits by biting them also. As the years passed, however, fewer and fewer of the remaining rabbits succumbed to the virus. Today, the rabbits are a problem again.

In their attempt to determine what happened, investigators found that, through natural selection, the virus had lost its "punch" and that the rabbits had become less susceptible to myxomatosis. The rabbits that were infected by the most virulent strains died almost immediately, which had two effects. First, it inhibited the spread of the most virulent forms of the virus, and, second, it drastically reduced the reproduction rate in the most susceptible rabbits. The most virulent forms were not spread adequately enough, because the mosquitos (the carriers) were not attracted to dead animals. Hence if the virus

killed its host quickly, less time would be available for a mosquito to bite a host that was infected with the virulent strain of virus and to transmit it to another rabbit. Meanwhile, the rabbits that were infected with the less virulent viral strains lived longer, and the mosquitos transmitted more of the less virulent strains. Because they were more readily transmitted, those strains increased in the viral population at the expense of the more virulent strains. At the same time, natural selection continued to favor the rabbits that had some resistance to the virus. Since both rabbits and viruses reproduce rapidly, within a few years, selection had occurred for both, favoring a less virulent strain of virus and rabbits with greater resistance. The result was a stable interaction between the relatively nonvirulent parasite and the relatively resistant host. Once that balance was reached, the virus was no longer a significant regulating factor, and the rabbit population began to increase virtually unchecked.

Thus we see that parasite–host relationships tend to evolve toward a compromise, wherein the host can usually tolerate a widespread, low-grade infection. As a result, the parasite is not an important regulator of the size of the host population. But conditions can change. If habitat conditions for the host population become less favorable, the consequent stress may render some hosts less tolerant of the parasite, thereby increasing host mortality; or a mutation in a few parasites may lead to a parasite population that is more virulent, which also will increase host mortality.

Coevolution between a parasite and a host is by no means inevitable, however, particularly if the host is a long-lived species. For example, let us consider what happened when chestnut blight was introduced into the United States. Before the turn of the

century, the American chestnut tree was a dominant member of the Appalachian forests. In 1904, Asiatic chestnut trees were brought to New York, carrying a parasitic fungus to which they were resistant. American chestnuts, however, had no resistance to the new parasite. The fungus quickly spread, and by the early 1950s, the American chestnut had been virtually eliminated from the Appalachian forest ecosystem. In a few places, the roots of destroyed trees are still alive and continue to send up new shoots. But those, too, will be killed eventually by the fungus. Another example is Dutch elm disease, whose effects are illustrated in Figure 4.11. This disease also demonstrates how a new parasite can essentially eradicate a nonresistant, long-lived host. Partly because it usually takes 20 years or longer for a tree to produce its first seed, the possibility of

Figure 4.11 *The effects of Dutch elm disease, which is an example of the destruction that can follow introduction of a new parasite. (a) Elm-lined Gillett Avenue in Waukegan, Illinois, as it appeared in the summer of 1962. (b) Gillett Avenue in 1969, after the elms were destroyed by Dutch elm disease. (Elm Research Institute, Harrisville, New Hampshire.)*

(a)

(b)

coevolution between a very susceptible tree species and a new, virulent parasite is much reduced. Hence, instead of coevolution, extinction of the host might well result. Moreover, if the parasite can survive only on that particular host, then the parasite would also become extinct.

Parasitism well illustrates the nature of density-dependent regulation of population size. The greater the density of the host population, the more likely it is that an uninfected member will come into contact with an infected member. Thus crowding increases the probability that a parasite will spread throughout a host population and kill a greater proportion of it. Moreover, if a growing host population exceeds the carrying capacity of its habitat, competition for limited resources may force some members into marginal areas. There, weakened by a scarcity of resources or an inhospitable climate (density-independent factors), they may become particularly susceptible to predators or to infection by parasites and, consequently, succumb.

Human beings, too, have always been affected by parasites. In fact, human history has been shaped partially by the impact of parasitism. In Chapter 16, we examine the historical role of parasitism in limiting the human population.

Competition

The resources that are necessary for life (such as food, water, and space) are finite, and sometimes the demand for them exceeds the supply. The interactions among organisms to secure a resource that is in short supply are collectively called *competition*. Recall that competition often illustrates the interaction of density-dependent and density-independent factors. The degree of competition depends on both the availability of resources, which often is determined by the weather, and the sizes of the populations that are dependent on them.

Ecologists recognize two types of competition: intraspecific and interspecific. *Intraspecific competition* occurs among members of a single species, whereas *interspecific competition* takes place among populations of different species.

Intraspecific Competition among Animals Intraspecific competition among animals often occurs as a result of certain types of social behavior that regulate population size. Many animals, such as song-birds, hawks, muskrats, and Alaskan fur seals, to name a few, establish territories, which they defend against intruders of their own species. For example, during the breeding season, each male robin establishes and defends the territory within which he and his mate build their nest. The male robin's song informs other robins of his territorial limits. If another male robin enters his established territory, he will attempt to drive the intruder out (Figure 4.12). That sort of aggressive defense, which is an example of *territorial behavior*, separates animals and tends to ensure an adequate supply of resources for members within each territory. Thus territorial behavior favors population growth by partitioning resources in such a way that the members that can occupy and defend a territory will more likely have sufficient resources.

However, the same behavior pattern also tends to limit populations, because the number of breeding pairs cannot exceed the number of available territories. Although the size of each breeding territory may shrink as a population increases, a minimum size exists for every species. The territory of red-winged blackbirds is 0.3 hectares (0.7 acres). In contrast, the bald eagle has a territory of 250 hectares (617 acres). Once the minimum is reached, those animals that do not have a territory of their own are forced to emigrate to marginal areas. Since the areas beyond the prime territories have limited resources, many of those outcasts die. Some succumb because of inadequate food, water, or shelter. Others, in their weakened condition, fall victim to predators and disease. Poor habitat conditions also reduce the number of offspring that are produced by the outcasts. Occasionally, an outcast will replace a member of the breeding population that has died, but outcasts cannot force their way into a fully occupied habitat as a rule. Territorial behavior, therefore, appears to regulate population size by limiting the number of animals that can breed in a favorable habitat.

Social hierarchies, or peck orders, comprise another behavior pattern that may play a role in competition. In some species, including baboons, wolves, ring-necked pheasants, and chickens, encounters between animals result in dominant–submissive relationships, which involve the entire population. One animal may dominate all the other animals in the population, whereas another, though submissive to the first one, dominates the remaining animals, and

Figure 4.12 *Two male robins fighting over territory. (Phillip Strobridge, National Audubon Society, Photo Researchers.)*

so on. (This example is very much oversimplified; in nature, patterns of dominant–submissive relationships are far more complex.)

Once a peck order has been established, greater stability and order prevail within the population. Because each animal "knows his place," less fighting occurs. Therefore, animals do not injure one another, and the population as a whole remains healthier than it would be if they were continually struggling for dominance. Thus social hierarchies foster the growth of a population by maintaining its overall health, but they also limit it as well, since low-status animals often fail to breed. During times of stress, submissive animals are the first to be deprived of adequate food, water, and shelter. Even in the best of times, low-status animals may be so harassed by dominant animals that they are driven out into marginal habitats, where they may die.

Interspecific Competition among Animals Interspecific competition occurs among animals of different species that compete for the same resources. The results of this type of competition vary, depending on many factors. Laboratory experiments suggest that if two species need resources that are too similar, one will cause the extinction of the other. Such *competitive exclusion* occurs, for example, when two species of *Paramecium* (unicellular animals) are introduced into a tube that contains a fixed amount of bacterial food. Although both species thrive when they are grown separately, *Paramecium caudatum* cannot survive when grown in the same culture with *Paramecium aurelia* (Figure 4.13). But does competitive exclusion occur in nature where habitat conditions are much more complex than the simplified conditions used in laboratory experiments?

Direct evidence of competitive exclusion in the field is scarce, even though it is inferred in many cases. In fact, the evidence suggests that many cases of localized extinction that were once thought to be a result of competition were probably caused by changes in the habitat. As an example of such an occurrence, let us consider a recent reexamination of the effect of the European cabbage butterfly on its American cousin. Following the arrival of the European species in New England in the 1860s, local butterfly collectors began to notice that the native butterfly species was disappearing from towns and nearby countrysides. They blamed the extinction of the local species on competition from the European invader. Today, however, in Vermont, adults of both species still occur in large numbers in the same hay fields, mate in the same areas, and during the summer, lay their eggs on many of the same species of food plants. Despite this large overlap in resource

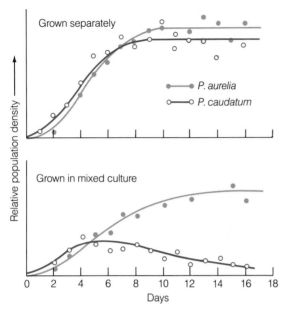

Figure 4.13 Paramecium caudatum *and* Paramecium aurelia *thrive when grown separately in a controlled environment with a fixed food supply. When grown together, however, the* Paramecium caudatum *population becomes extinct—an example of competitive exclusion. (Adapted from G. F. Gause, Experimental analysis of Vito Volterra's mathematical theory of the struggle for existence.* Science *79:16–17, 1934. © 1934 by AAAS.)*

utilization, which creates a great potential for competition, recent studies show that no evidence of competition exists between the two species. However, if competition did exist, each species has a refugium from the other. Whereas both species are found in hay fields and roadsides, only the native species flies in the wooded areas. Moreover, although the larvae of both can feed on some of the same plants, only the larvae of the European species can feed on a weedy mustard species, which was also introduced from Europe and has successfully invaded hay fields and roadsides. Significantly, that mustard plant is lethal to the larvae of the native butterfly.

If those two species can seemingly coexist today, why did the native species become extinct from towns in the last half of the nineteenth century? No one knows for sure, but our improved understanding of the ecological relationships of the native butterfly provides an explanation that differs from that of competitive exclusion. In May and early June, the only abundant plant on which the eggs and larvae of the native species can develop is a species that is restricted to woodlands. Moreover, the major summer larval food plant can persist only where the land is cultivated each year. In nineteenth century New England, woodlands were cut steadily to supply railroads and manufacturing industries. Cleared land was often used as pastures, but they were abandoned later when the cities expanded. As a result, suitable sites for the foodplants—on which the larvae of the native butterfly could feed—declined dramatically. Thus the localized extinction of the native butterfly in the urban areas may well have resulted from a loss of food plants rather than from competition with the introduced species.

Other studies have shown how closely related species are able to coexist in the same region. In a classic study, the noted ecologist R.H. MacArthur studied the feeding behavior of five species of Maine warblers. He observed that, although all five species feed on basically the same type of insects, which are found in spruce trees, competition is greatly diminished by differences in feeding habits. As shown in Figure 4.14, the hunting activities of each species are confined to a specific part of a tree. Although feeding zones overlap to some extent, other behavioral patterns further reduce competition. For example, one kind of warbler captures insects on top of spruce needles, while another hunts the insects hidden underneath the needles. Also, because the five species all have different nesting times, their times of greatest food need differ as well. Hence those five species are able to coexist with a minimum of competition.

Many factors reduce competition among species. One of those factors is the difference between the hunting periods of predators. For example, hawks and owls both feed on similar types of animals, but, because hawks capture their prey during the day and owls are nocturnal, they usually do not compete for the same prey. Other feeding habits disperse species throughout an entire ecosystem, thereby reducing competition. For example, some species of birds feed on the ground, while others seek food in trees and shrubs; and a few species capture their prey in the air. Structural adaptations also reduce competition between species. For example, the downy woodpecker and the hairy woodpecker occupy similar habitats in North America. As you can see in Figure 4.15, those species are similar in appearance, but the hairy woodpecker is larger and has a larger bill. Those structural differences allow the hairy woodpecker to eat larger insects and to reach insects that are hidden

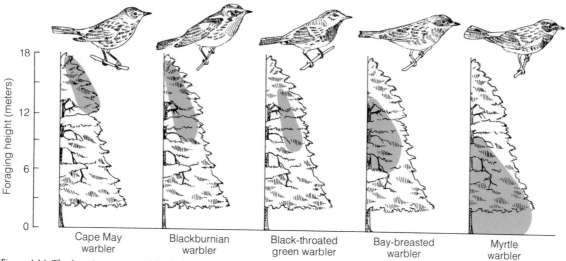

Figure 4.14 *The foraging areas used by five species of warblers in spruce forests in Maine. (After R. MacArthur, Population ecology of some warblers in northeastern coniferous forests.* Ecology *36:533–536, 1958.)*

Figure 4.15 *Differences in the structural adaptations of the downy woodpecker (a) and the hairy woodpecker (b) enable them to exploit somewhat different food sources. This reduces competition between the two species. (Henry C. Johnson, National Audubon Society, Photo Researchers; Karl H. Maslowski, Photo Researchers.)*

(a)

(b)

deeper within tree trunks. With respect to nesting sites, we find again that different species of birds use different resources. For example, some birds nest on the ground, while others nest on a tree branch or in the tree trunk. A few species use a burrow in the ground; others employ a hole in the side of a cliff.

Thus we see that numerous mechanisms allow animals to make a living without competing extensively with their neighbors. If a region possesses many types of ecosystems, then a species is more likely to be able to avoid competition. Although the result of competition, which is popularly referred to as the "survival of the fittest," conjures up images of animals fighting tooth and nail for survival, such violent encounters are rare. After all, as we describe in Chapter 3, the most fit are those whose young make up a greater percentage of the next generation. When competition is lessened because of adaptations that reduce confrontation, more energy can be directed to reproduction and rearing of young. The term *fittest*, therefore, often applies best to those organisms that avoid competition.

Intraspecific Competition among Plants Intraspecific competition plays a significant role in controlling the size of plant populations. Generally, most seeds fall close to the parent plant. When those seeds germinate, they produce a dense growth of seedlings. Unlike animals, plants are unable to make compensating adjustments in their spacing to reduce competition, and so the whole mass of seedlings competes for sunlight, water, and soil nutrients in the same space.

As Figure 4.16 indicates, the impact of competition depends on the density of seedlings, (which is density-dependent regulation). At extremely low densities, plants may grow vigorously, because little competition exists. As densities increase, the biomass yield for each unit of area also increases. At moderate densities, plants respond to increased competition by a reduction in growth rate, which leads to a reduction in plant size and in the number of seeds, the size of seeds, or both. Despite increasing plant densities, the biomass for each unit of area continues to increase. Eventually, however, it approaches a constant value, which is equivalent to the carrying capacity. At high densities, competition becomes so severe that some plants die. The mortality associated with severe competition has been termed *self-thinning*. The yield for each unit of

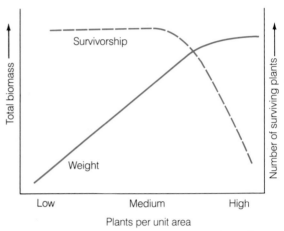

Figure 4.16 *As plant density increases, total biomass production increases until a maximum production level is reached. Increasing plant density also eventually results in lowered survivorship.*

area remains little changed from the yield that is associated with medium densities, which again illustrates the concept of carrying capacity. Such relationships are important for natural plant communities, and they are crucial for agricultural and forestry planning. Experienced farmers and foresters know that by increasing the planting density, their yield for each unit of area will only increase up to the carrying capacity of the site. A further increase in planting density will only initiate self-thinning. It will not improve the total yield of crops.

Interspecific Competition among Plants In contrast with our evidence on animals, competitive exclusion is common in the plant kingdom. Take, for example, the competitive interactions of bluebunch wheatgrass and cheatgrass on the rangelands of the Columbia River Basin in Washington. Before the advent of domestic grazing, that rangeland was dominated by bluebunch wheatgrass, along with a rich diversity of other plants. But heavy overgrazing destroyed the vegetation, creating a habitat that was open to invasion. Cheatgrass, a native of Europe, soon dominated several million hectares that were previously occupied by bluebunch wheatgrass. Although ranchers preferred to have their cattle graze on wheatgrass, it has been almost impossible to reestablish the growth of wheatgrass and most attempts have failed—often even where grazing has been eliminated.

Field and laboratory experiments indicate that

cheatgrass can exclude wheatgrass for the following reasons. The seeds of both species germinate at approximately the same time—during the fall—when the soil is moist. Then during the winter, wheatgrass becomes dormant while the cheatgrass continues to grow. As a result, the extensive root system of cheatgrass grows deeper and gains control of the soil before the wheatgrass seedlings can become better established. Also, cheatgrass matures 4 to 6 weeks earlier than wheatgrass, so it withdraws most of the water from the soil before the wheatgrass has a chance to take it. During the summer, when it is dry, wheatgrass seedlings usually succumb, because the cheatgrass has already used up the soil moisture that was available. Hence in places where little rain falls during the summer, cheatgrass excludes wheatgrass by out-competing it for soil moisture. In areas where plentiful rain falls in the summer, however, cheatgrass does not compete as well, and it is found only on recently disturbed sites.

Another type of competition in plants occurs when one plant species produces chemicals that inhibit the germination or growth of another species. That type of competition occurs on abandoned farmland during secondary succession. Recall from Chapter 3 that the grass broom sedge invades farmland within 3 years after it is abandoned, quickly replaces the pioneer weeds that are present, and then remains there for 15 to 20 years, often in almost pure stands. Investigators have found that the roots and shoots of broom sedge contain chemicals that inhibit the growth of other plant seedlings. Thus when broom sedge enters a field, it releases those chemicals into the soil, retarding the growth of pioneer weeds. After the broom sedge has become established, its decaying parts continue to release growth-inhibiting chemicals into the soil. Since the presence of chemicals retards any possible invasion by other plants, the rate of secondary succession is slowed down considerably.

Although competitive exclusion sometimes does eliminate a plant species from a particular habitat (where it might otherwise survive), abundant observational data suggest that plants, like animals, have evolved distinct strategies for using resources that also reduce competition. Such strategies may involve temporal or spatial differences. For example, if we visit a forest in the early spring, we will find the forest floor carpeted with a spectacular array of spring wild flowers. Those spring ephemerals renew their growth 3 to 5 weeks before the trees leaf out. The ephemerals flower, set seed, and many die before the tree canopy closes. Because they bloom early in the spring and their growth cycle is rapid, the plants have access to sunlight before the tree canopy can close and shade them.

The fact that plants grow to differing heights also reduces the competition for space among plants. In forests, a vertical stratification of vegetation occurs, as shown in Figure 4.17. In mature forests in the southern Appalachians, tall trees (such as beech, sugar maple, basswood, and tulip trees) constitute the upper canopy. Beneath that layer are smaller trees (such as dogwood, magnolia, and ironwood) that seldom or never reach the upper canopy. The shrubs

Figure 4.17 *Vertical stratification in a forest. Large trees are approximately 30 meters tall, small trees are 5 to 10 meters tall, shrubs are 1 to 3 meters tall, wildflowers are 10 to 50 centimeters tall, and mosses are 2 to 5 centimeters tall.*

(such as spicebush, witchhazel, and pawpaw) grow beneath the smaller trees; below the shrubs, flourish wild flowers and a ground layer of mosses, lichens, and blue-green algae. In tropical rain forests, which have three to five overlapping strata of trees, vertical stratification is most apparent. Because of the number of layers of trees in those forests, little light penetrates to the forest floor, and such vegetation as shrubs, herbs, and ferns is sparse compared with that of a temperate forest. As a result, most herbivores in a tropical rain forest live in the tree canopies.

By using various depths of the soil, plants can enhance coexistence with other plants. For example, ecologists have found that three species of annual weeds can coexist successfully in agricultural fields that have been abandoned recently in east-central Illinois. In most regions, only one or two species usually dominate. However, each of those three annuals—smartweed, Indian mallow, and bristly foxtail—possesses a combination of characteristics that allows it to coexist with the others. As shown in Figure 4.18, smartweed exploits the deepest region of the soil, thereby ensuring itself a continuous supply of water. Indian mallow exploits the middle region, which usually has less soil moisture available near the end of the growing season, but it compensates for the reduced moisture by maintaining a high rate of photosynthesis, even when its leaves are partially wilted. The roots of bristly foxtail occupy the upper region, which is low in moisture during most of the growing season. Yet that plant compensates by rapidly recovering from water stress after it rains and, like Indian mallow, it has a relatively high rate of photosynthesis, even when it is partially wilted.

Population Regulation: Physical Factors

Recall that physical factors can act as density-independent agents in influencing population size. The most dramatic factors are catastrophic events, such as hurricanes, tornadoes, floods, and fires, that can wipe everything out. Less dramatic are the year-to-year variations in the seasonal patterns of temperature and precipitation, as well as the length of the frost-free period. Those weather components can affect population size both directly and indirectly. For example, an extended period of hot, dry weather can

Figure 4.18 *Partition of soil volume by three plant species. (After N. K. Wieland and F. A. Bazzaz, Physiological ecology of three codominant successional annuals. Ecology 56:681–688, 1975.)*

produce habitat conditions that exceed the tolerance limits of many aphids in a hay field. Hence the aphid population will decline dramatically.

Extreme weather may also lower population size indirectly by reducing food resources. For example, pregnant female kangaroo rats in the Mojave desert need an adequate supply of herbaceous (nonwoody) plants in January and February to provide them with water, vitamins, and energy. If winter rainfall is too scant to stimulate sufficient growth of those plants, kangaroo rat natality suffers.

The impact of a highly variable environment on any given species usually depends on the sensitivity of that species. As we noted earlier in this chapter, species that are small and short-lived, which reproduce rapidly, usually are the most sensitive to extremes in the weather. Hence insects and rodents often show dramatic fluctuations in population numbers. Large, long-lived, slowly reproducing organisms, such as large game animals and trees, are less sensitive to variable environmental conditions and hence usually exhibit only slight changes in population size.

Although short-term, weather-related stresses may temporarily reduce the size of a population, it will usually recover. For example, when deep snow prevents bobwhite quail from scratching for food, their populations decline drastically. But during the following spring and summer, the quail's reproductive success is often unusually high—probably as a result of reduced intraspecific competition. Thus by fall, their population is restored to near-normal numbers.

The population size of plant species, too, can be severely reduced in the short run by such stresses as drought, frost, or fire. But through a variety of mechanisms, most plant populations avoid extinction. For example, the seeds of many species remain viable for many years in the soil despite such stresses as drought or fire, which would kill the parent plants. Other plant species die back yearly into underground structures (for example, tulip bulbs and the underground stems of many grasses), thus avoiding exposure during seasons when stressful conditions occur. By such mechanisms, some plant species are able to withstand stressful conditions for several years or longer. Such populations persist in an area because they are well adapted to the major short-term stresses that frequently occur.

The Effects of Underpopulation

Overpopulation is a serious problem, but underpopulation also poses dangers. Consider what can happen to bobwhite quail when their population becomes abnormally small. Except when feeding, a covey of bobwhite quail usually forms a compact circle (as shown in Figure 4.19). During cold weather, they draw closer together in their circle, thereby decreasing their heat loss by exposing less body surface to the cold air. Field observations show that quail that belong to large coveys usually survive severe winter temperatures, while single birds or small coveys often succumb. This mechanism is another example of the often close relationship between density-dependent and density-independent factors—the impact of cold weather on the population size of quail depends in part on the number of birds in a covey.

Note that the birds face outward from the circle. A covey, in effect, watches in all directions for the approach of predators. If a predator approaches them, the alarmed birds produce a phenomenon that is known as the confusion effect. They fly up suddenly in all directions, thereby so disorienting the predator that it often fails to capture even a single bird. A covey with too few birds to form a circle loses this advantage.

The use of circling as a protective strategy is exhibited in other species as well. For example, adult male pronghorn form a protective ring around the females and young and stand their ground against predators. Musk oxen also form such protective circles. Westward bound pioneers "circled up" their wagons in similar fashion, as a protective measure against raiders (Figure 4.20). In each case, a group that is too small to make a tight circle cannot gain the protective effect. In fact, if a herd of pronghorn has fewer than 15 members, the herd will attempt to run away from a predator rather than try to form a circle and stand its ground.

For predators that hunt in groups, such as wolves, lions, and hyenas, underpopulation can impair their ability to stalk, attack, and kill their prey. For example, moose and caribou can repel an attack by one or two wolves, but they fare less well against wolves that run in a pack, attacking from all sides (Figure 4.21). Lionesses have a less obvious means of cooperating in a hunt. They spread out, and each one

Figure 4.19 *Gathered in a circle, all facing outward, quail protect themselves against cold and predators. (Jack A. Stanford, Missouri Conservation Commission.)*

Figure 4.20 *Pioneers "circled up" their wagons to protect themselves from intruders. (The Bettmann Archive, Inc.)*

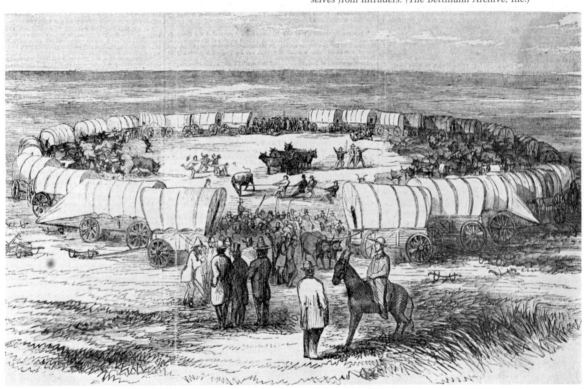

Figure 4.21 *Hunting as a pack, wolves have a greater chance of killing large prey. (Jim Brandenburg, Woodfin Camp and Assoc, Animals Animals.)*

stalks the prey from a different direction, often growling. When the presence of a lioness is detected, the prey animals may bound off and come within striking range of another lion, which it had not seen originally. Clearly, when a predator population is reduced to the point at which its cooperative hunting suffers, the survival of the remaining members is threatened.

As the size of the population declines, *inbreed-*

ing often results. Inbreeding is defined as mating between close relatives. Frequently, inbreeding facilitates the buildup of deleterious characteristics within a population. The hazards of inbreeding are well illustrated by recent studies of the cheetah. Although it ranged over Africa, Asia, Europe, and North America as recently as 20,000 years ago, the cheetah, today, is restricted to isolated populations in South Africa and East Africa, with numbers as low as 1500 to 5000.

Investigators recently found that cheetahs possess dramatically lower levels of genetic variation than has been found in other cats and mammals in general. Concurrently, they found that the sperm count (sperm per unit volume of ejaculate) was 10 times lower than it is in related cat species. Moreover, 70 percent of the sperm were structurally aberrant. Investigators believe that the lack of genetic variability and the sperm abnormalities are a result of inbreeding. What do these features portend for the cheetah's future?

Currently, the cheetah is a highly specialized and successful hunter. For the present, at least, population data suggest that their rate of increase is low, if not actually declining. But the greatest concern is the cheetah's apparent level of vulnerability to environmental change. Being so specialized and possessing so little genetic variability, the cheetah could have little chance of adapting to a significant change in its environment.

Another concern about small populations is that they may be unable to make up for normal losses caused by predation, disease, accidents, aging, or natural disasters. If it cannot compensate, the population is doomed to extinction unless natural immigration or the artificial introduction of animals from another population adds to its numbers.

The results of undercrowding are of particular interest to those concerned with preserving endangered species. We consider the consequences of undercrowding in Chapter 14, where we evaluate efforts to save endangered species.

How Are Populations Actually Regulated?

Today, ecologists are in the midst of a major debate over which factor, or factors, are the most significant in determining population size. For example,

some ecologists hold that competition is the predominant regulator of population size, whereas other ecologists find that populations rarely build up to levels at which competition becomes important. In other instances, random physical events, such as hurricanes, floods, and early fall frosts, are very important influences on population size, whereas biological factors are rather insignificant. What conclusions can we draw from these debates?

Recent field and experimental observations show that the natural world is a more complex and uncertain place than most ecologists once believed it to be. Out of the bewildering complexity of ecological systems, ecologists realize that the size of a population at any one time can be the result of the interaction of several factors, and the relative importance of any one factor depends on the nature of the species and the ecosystem. Moreover, the relative importance of a factor will often change from one time to another.

Interspecific competition is likely to be more important in certain circumstances—for example, as a population regulator at higher trophic levels in food webs rather than at lower levels, for species-rich ecosystems rather than for species-poor ecosystems, and in ecosystems that are subject to only modest physical disturbances.

Physical factors are likely to have a greater influence in other circumstances—for example, as a population regulator for r-strategists rather than K-strategists, and in temperate environments, where weather is more variable, rather than in a tropical rain forest, where the weather is relatively more stable.

Changes that occur from 1 year to another can shift the relative importance of population control factors. Recall that lions killed more wildebeest and zebra in Kruger National Park when those prey animals were widely dispersed and that the dispersion, in turn, was a result of increased rainfall. When rainfall subsequently returned to normal levels, the prey became concentrated into a few herds, and the relative importance of predation in regulating the population of prey declined.

Continuing research will further clarify the relative roles of competition, predation, parasitism, catastrophic events, and weather variability on population size and how the relative importance of those factors varies with species, types of habitat, and time.

Conclusions

The well-being of any population, including our own, depends on its physical environment and its interactions with other populations. In the long run, these density-independent and density-dependent factors influence the size of a population and tend to keep it in balance with its habitat. Although the fluctuations in size may be quite large, population growth rarely becomes out of control for long, nor do most populations become extinct.

In recent decades, human intrusion into ecosystems has dramatically altered population-regulating mechanisms. Many ecosystems have been greatly simplified, and in many cases, new predators, parasites, and competitors have been introduced. The result has been the extinction or near extinction of many other species. In other cases, controlling agents, such as predators or competitors, have been removed. The result has been the explosion of populations of some species to such numbers that they become pests. Because our well-being rests on a balance among other populations, we must learn to consider the impact that our activities have on our fellow inhabitants of this planet.

Summary Statements

The size of a population at any given time is determined by the difference between additions to it (natality and immigration) and subtractions from it (mortality and emigration). Both natality and mortality vary with environmental conditions and the type of organism that is being considered.

Populations exhibit exponential growth. In contrast to linear growth, exponential growth generates huge numbers in a very short period of time. Moreover, because of exponential growth, a population can reach a habitat's carrying capacity in a surprisingly short period of time.

The size of a population usually fluctuates around the habitat's carrying capacity for that species. The range of fluctuations depends on the habitat conditions and the nature of the species. Species that are r-strategists generally experience larger fluctuations than those that are K-strategists.

Both biological and physical factors play a role in regulating population size. Biological factors include predation, parasitism, and competition. Physical factors include fire and weather-related stresses, such as floods, droughts, hurricanes, and temperature extremes.

Predation, parasitism, and competition are density-dependent factors. Their regulating impact becomes stronger as the density of the population increases. In contrast, fire and weather-related stresses are density-independent factors. Their impact on population size does not vary greatly as the size of the population increases. In most cases, both density-dependent and density-independent factors play a role in determining the size of a population.

Predation is often important in regulating the population of small animals, such as insects. The impact of predators on populations of large animals appears to vary considerably with time and often depends on habitat conditions.

The establishment of a new parasite-host relationship often has a devastating impact on the host population, particularly if the host is a long-lived species. If both the parasite and the host reproduce relatively rapidly, a stable interaction may soon result as a result of natural selection, that is, by the development of a less virulent parasite and a more immune host. Such an interaction is known as coevolution.

Territorial behavior and peck orders are examples of intraspecific competition, or competition among members of the same species, that may regulate population size. If interspecific competition, or competition between species, is reduced, many similar animals can coexist in the same habitat. Factors that reduce interspecific competition include differences in timing of hunting activities, location of feeding activities, and types of food.

Self-thinning illustrates the impact of severe intraspecific competition of plants. Although competitive exclusion is a more common consequence of interspecific competition between plant species than it is of such competition between animal species, vertical stratification and differences in timing of growth reduce competition among some plant species.

Short-term, weather-related stresses may temporarily reduce population size, but populations usually recover quickly from those occurrences. K-strategists are usually less sensitive to variable environmental stresses than are r-strategists.

Inability to ward off predators, inbreeding, and inability to make up normal losses caused by accidents, aging, and interactions with other species are some of the dangers posed by underpopulation.

The relative importance of biological and physical factors in regulating a population's size depends, in part, on the nature of the habitat and the nature of that population as well as on the characteristics of other populations that reside in, or move through, the ecosystem.

Questions and Projects

1. In your own words, write a definition for each of the terms that are italicized in this chapter. Compare your definitions with those in the text.

2. Describe the differences between exponential growth and linear growth. Why should a person in wildlife management or pest management be aware of those differences?

3. Why do some populations overshoot the capacity of their habitat to support them? What might be their impact on the habitat during the time of overpopulation?

4. Contrast the reproductive strategies of an r-strategist with those of a K-strategist. Why is an r-strategist more likely to be a better colonizer than a K-strategist?

5. Why are endangered species more likely to be extreme K-strategists than extreme r-strategists?

6. Describe how the importance of predation in regulating prey populations will vary with the type of prey and habitat conditions.

7. Although some predators may not be the primary agents in controlling the size of prey populations, they do contribute to the stability of those populations. Explain.

8. What circumstances will normally favor coevolution of a parasite and its host? Under what conditions will coevolution probably not occur?

9. Describe how territorial behavior and peck orders function in population regulation. Identify animals in your area that demonstrate that type of social interaction.

10. Compare and contrast the means whereby plants and animals reduce interspecific competition.

11. Visit a park or natural area in your region, and observe the many species of plants and animals in the ecosystem. Identify and describe some of the adaptations that reduce competition for resources among those species.

12. Using examples of plants or animals from nearby ecosystems, describe how predation, parasitism, and competition act as density-dependent regulating mechanisms.

13. Can density-dependent regulation of population size also be described as a negative feedback system? Defend your answer.

14. Why are short-term changes in the weather of relatively little significance in the long-term regulation of population size?

15. What conditions foster inbreeding? How is inbreeding a threat for the continued existence of a population?

16. Some scientists argue that the valiant efforts that are being made to save certain nearly extinct species are doomed to failure, because the populations involved are too small. What is the basis for that assertion?

17. Describe how physical and biological factors may interact to control the population size of a species. Cite examples from your area.

18. The population size of many r-strategists fluctuates widely. Can those large fluctuations be reconciled with the concept of the "balance of nature"?

19. Check with the local office of your state Department of Natural Resources, Department of Conservation, or Department of Agriculture to learn whether it is involved in projects to control pests or overpopulation of wildlife. If it is, determine which population control measures are being used. How successful are those measures in reducing the overpopulation?

20. Some ecologists argue that the days are gone when someone can state that one particular population regulator is the predominant means by which population numbers are controlled. What is the basis for that assertion?

Selected Readings

ANNUAL REVIEWS, INC.: *Annual Review of Ecology and Systematics.* Published yearly. Palo Alto, Calif.: Annual Reviews, Inc. Each volume contains up-to-date reviews of research areas in population ecology.

BERGERUD, A. T.: "Prey Switching in a Simple Ecosystem." *Scientific American* 249:130–141 (December), 1983. A study of a prey–species' population crash and the subsequent impact on its major predator and other prey populations.

BERTRAM, B. C. R.: "The Social System of Lions." *Scientific American* 232:54–61 (May), 1975. An account of how behavior plays a role in the population ecology of lions.

GILBERT, L. E.: "The Coevolution of a Butterfly and a Vine." *Scientific American* 247:110–121 (August), 1982. An interesting account of co-evolution between a parasite and its host.

LEWIN, R.: Predators and hurricanes change ecology. *Science* 221:737–740, 1983. A fascinating review of the recent controversies over the relative importance of predation, competition, and the physical environment in community organization.

MACK, R. N.: "Invaders at Home on the Range." *Natural History* 93:40–46 (February), 1984. An interesting account of the conditions that allowed cheatgrass to successfully invade and take over much of the intermountain west.

MAY, R. M.: Parasitic infections as regulators of animal populations. *American Scientist* 71:36–45, 1983. An interesting review of the dynamic relationships (including coevolution) between parasites and their host populations.

O'BRIEN, S. J., ROELKE, M. E., MARKER, L., NEWMAN A., WINKLER, C. A., MELTZER, D., COLLY, L., EVERMANN, J. F., BUSH, M., and WILDT, D. E.: Genetic basis for species vulnerability in the cheetah. *Science* 227:1428–1434, 1985. An investigation into the cost of genetic uniformity in a small population.

PUTNAM, A. R.: Allelopathic chemicals: Nature's herbicides in action. *Chemical and Engineering News* 61:34–45 (4 April), 1983. An informative account of how plants wage chemical warfare with one another.

SCHOENER, T. W.: The controversy over interspecific competition. *American Scientist* 70:586–596, 1982. A review of the importance of interspecific competition in nature.

SMITH, R. L.: *Ecology and Field Biology,* 3rd ed. New York: Harper & Row, 1980. A well-written text, containing several chapters on population ecology.

SMUTS, G. L.: Interrelations between predators, prey, and their environment. *Bioscience* 28:316–320, 1978. A review of the influence of habitat conditions on predator-prey relationships.

This swamp forest is a wetland transition between an upland forest (terrestrial ecosystem) and a lake (aquatic ecosystem). (U.S. Department of the Interior, National Park Service. Photo by George Grant.)

CHAPTER 5   THE EARTH'S MAJOR ECOSYSTEMS

Barefoot, clad in shorts and T-shirts, we can float down a gently flowing river in a canoe, enjoying the warmth of the sun and the gentle breeze that ripples the surface of the water.

By walking through a grove of majestic redwoods, we can enjoy the cool stillness and the shafts of sunlight that slant through the green canopy, which is high overhead, to dapple on the shady forest floor.

Propelling ourselves with rubber flippers on our feet and breathing air that hisses from pressure tanks on our backs, we can startle dozens of brilliantly colored fish that dart away from us as we inspect the fantastic underwater environment of a coral reef.

Shielded by white clothing and headgear, we might emerge from a Land Rover, gasping at the shock of the ovenlike heat and the dazzling light that assaults every living thing in the desert, when the summer sun is high in the sky.

Encased in down-filled parkas and leggings, felt-lined boots, and goggled facemasks, we might struggle forward into the Antarctic wind—a wind so cold that it would freeze our skin in seconds, and kill us in minutes, if we braved it unclothed.

Our small planet is covered with an astounding variety of ecosystems. Each one owes its uniqueness to a singular combination of myriad organisms and physical conditions. Thus each ecosystem houses a different set of resources, whose exploitation generates a different group of problems for its inhabit-

ants. Because each of these ecosystems is unique, the solutions to those problems are also necessarily unique.

Every ecosystem on earth has, by now, been modified to some extent by human activity. The most severely modified ecosystems are those that have been most heavily populated or purposefully altered to meet human needs. The once great forests of eastern North America, Europe, and China, for example, were originally cleared for agriculture, villages, and cities; and the man-made ecosystems that now dominate those areas comprise a significant percentage of the world's urban and industrial centers. Most of the world's grasslands, including the Great Plains of North America, the pampas of Argentina, and the steppes of Russia, have been greatly modified by grazing and cultivation.

Today, as the human population continues to grow, efforts must reach farther and farther afield to meet the demands for resources. Ecosystems that were once considered to be remote are now bearing more of the brunt of human exploitive activities. The North American tundra and continental shelves are being altered as their oil, natural gas, and minerals are being removed. Tropical rain forests of Central and South America are being cleared to meet demands for timber and food. At the same time, we continue to disturb those ecosystems that already have been changed significantly by human activity.

Terrestrial Ecosystems

Limiting Factors: Climate

Blizzards, heat waves, hurricanes, droughts, and floods all remind us that terrestrial environments can be hostile places to live. If we have any choice in deciding where to live, we often consider the climate as a major determining factor. *Climate* is defined as the weather conditions of an area, which are averaged over a period of years, considering the weather extremes that are likely to occur. We speak of harsh climates and mild climates but without proper shelter and access to water, all climates are hazardous to human life. Indeed, the two most common limiting factors for all terrestrial organisms are air temperature and moisture.

Perhaps the greatest threat to life on land is desiccation. Air, particularly moving air (wind), dehydrates every surface it touches. For example, it speeds the evaporation of water from the outer surface of the body and from the surface of the lungs. To retard dehydration, animals are covered with hair, scales, feathers, or skin that has an outermost layer of hardened, dead cells. They conserve their internal supply of water by excreting concentrated urine and relatively dry feces. Bark protects tree trunks and branches from drying out; and leaves are covered with a waxy layer that is relatively impervious to water. However, wind also provides several benefits to terrestrial life. The continual movement of air maintains a uniform concentration of oxygen and carbon dioxide which are vital to life. Moreover, the wind plays a vital role in the hydrologic cycle (as described in Chapter 6) and in the dilution of air pollutants (as described in Chapter 10).

Air temperatures fluctuate both diurnally and seasonally in terrestrial environments. But along the coastlines of oceans and large lakes, the water moderates the fluctuations, which reduces the amplitude of temperature change. However, in the continental interior, temperatures are particularly variable. In the midwestern United States, for example, the temperature may vary by as much as 10 to 15°C (18 to 27°F) between day and night; it may vary by perhaps 70°C (126°F) between summer and winter. Thus to survive seasonal extremes, many animals either migrate or hibernate, and most plants become dormant. In Chapter 3, we described other adaptations that organisms make in response to temperature changes.

Temperature and precipitation are the major parameters of a region's climate. Because temperature and precipitation are such significant limiting factors, a strong correlation exists between those climatic factors and the type of vegetation that is present in a given area. Figure 5.1 illustrates the general relationship between the types of world vegetation and the variables of temperature and precipitation.

Because air temperatures gradually decline poleward from the equator, the type of dominant vegetation also changes with latitude (Figure 5.2). Precipitation gradients also can be found. For example, annual precipitation increases from west to east across the Great Plains to the East Coast and Gulf Coast. The response of vegetation to that gradient is also illustrated in Figure 5.2. Another important variable for terrestrial life in temperate latitudes is the *photoperiod*, or relative length of day and night. As Figure 5.3 illustrates, the length of the day does not change much with the seasons near the equator, but differences in the length of the day become significant with the seasons as one travels poleward. At midlatitudes, the changing length of the days controls the timing of many essential functions of organisms. For example, in many plants, the length of the day determines the breaking of dormancy, the timing of flowering, and the onset of dormancy. In animals, the photoperiod is a primary control for the onset of breeding activities, molting (in birds), migration, and hibernation. As we noted in Chapter 3, an organism's ability to adapt to a change in weather depends on when the stress occurs. For example, an extended outbreak of abnormally cold temperatures is much more detrimental to mammals in early autumn that it is in midwinter, because the change in the photoperiod during early autumn is not sufficient to trigger the onset of the processes that enable mammals to adapt to winter conditions.

Climate, then, is one basic determinant of terrestrial ecosystems. Another is soil.

Limiting Factors: Soil

Soil is a dynamic habitat whose processes are vital to the proper functioning of terrestrial ecosystems. As you will recall from Chapter 2, soil is the major source of essential mineral nutrients and water for

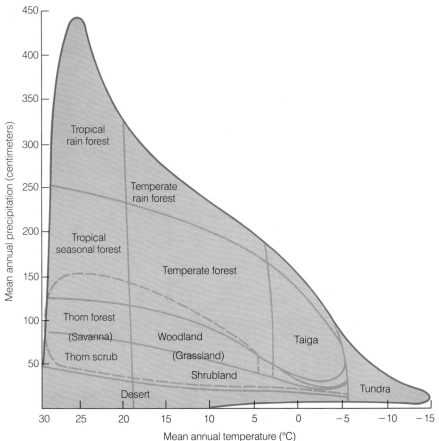

Figure 5.1 *The relationships between climatic variables (temperature and moisture) and vegetation. The dashed line encloses a wide range of environments in which either grassland or one of the communities dominated by woody plants forms the prevailing vegetation, depending on the particular region. (After R.H. Whittaker,* Communities and Ecosystems, *2d ed. New York: Macmillan, 1975.)*

plants, and it is the home of most terrestrial decomposers. Soil also contains humus and broken-down rock, which is usually in the form of sand, silt, and clay. The pore spaces in soil contain varying amounts of air and water. The quantity of available mineral nutrients, the water-holding capacity, and the aeration of the soil play major roles in determining soil fertility, and those factors, in turn, govern plant productivity and the number of consumers a terrestrial ecosystem can support.

A vertical profile of most soils reveals a sequence of distinct layers, called *soil horizons*, which occur upward from undisturbed bedrock or sediment to surface vegetation. Although no single set of soil horizons can be selected to illustrate the variety of soil types, Figure 5.4 illustrates the set of horizons that are commonly found in the forests of eastern United States. Horizons are classified on the basis of physical and chemical characteristics of the soil. Those characteristics, in turn, are determined—for each soil type—by the degree of leaching to which each layer is subjected. (*Leaching* is defined as the process by which downward seeping water dissolves, transports, and redeposits soluble soil constituents.)

Four horizons can be identified in many soil profiles: the O-horizon, A-horizon, B-horizon, and C-horizon. The *O-horizon* is the surface layer; it is composed of fresh or partially decomposed organic matter *(litter)*. The topsoil, which is called the *A-horizon*, is the major locus of mineral nutrients for plants and, therefore, usually contains most of the plant roots. The A-horizon also accumulates humus. Some of the soluble soil constituents of the A-ho-

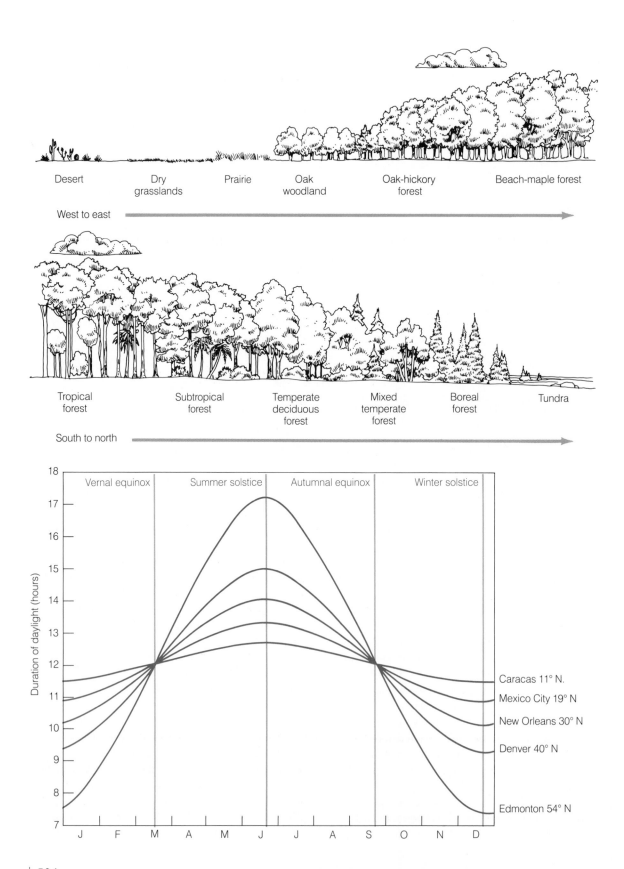

Desert — Dry grasslands — Prairie — Oak woodland — Oak-hickory forest — Beach-maple forest

West to east

Tropical forest — Subtropical forest — Temperate deciduous forest — Mixed temperate forest — Boreal forest — Tundra

South to north

Vernal equinox | Summer solstice | Autumnal equinox | Winter solstice

Duration of daylight (hours)

Caracas 11° N.
Mexico City 19° N
New Orleans 30° N
Denver 40° N
Edmonton 54° N

J F M A M J J A S O N D

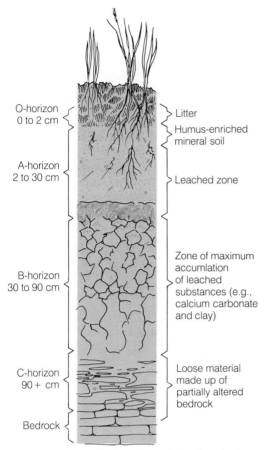

O-horizon
0 to 2 cm

Litter
Humus-enriched
mineral soil

A-horizon
2 to 30 cm

Leached zone

B-horizon
30 to 90 cm

Zone of maximum
accumlation
of leached
substances (e.g.,
calcium carbonate
and clay)

C-horizon
90 + cm

Loose material
made up of
partially altered
bedrock

Bedrock

Figure 5.4 *A soil profile that is typical of soils in the forests of eastern United States. Depth of horizons vary with site conditions.*

Figure 5.2 *Gradients of vegetation in North America from west to east and south to north. The west to east vegetation gradient is determined primarily by a moisture gradient, whereas the south to north vegetation gradient is determined primarily by a temperature gradient. (After R. L. Smith, Ecology and Field Biology. New York: Harper & Row, 1980. © 1980 by Robert Leo Smith. Courtesy of Harper & Row.)*

Figure 5.3 *The difference in day length between winter and summer increases significantly as one travels poleward. Note that day length is equal everywhere at the time of the vernal and autumnal equinoxes.*

rizon, such as calcium carbonate, are transported from it by leaching to the subsoil, which is called the *B-horizon*. Because the B-horizon contains less organic matter, it is less fertile than the A-horizon. The B-horizon may also be a zone of clay accumulation. Below the B-horizon lies the *C-horizon*, which is a mineral layer that is made up of the partially decomposed, underlying bedrock or sediment. Because its fertility is low, the C-horizon is penetrated by few plant roots.

Some soils form from the breakdown of the underlying bedrock. Others develop from sediments that have been deposited on the bedrock. Sources of sediments include windblown deposits, water-borne deposits, volcanic ash, and glacial drift, which are deposits left by a melting glacier.

Five factors play a role in the development of a soil profile and its properties, such as moisture-holding capacity, degree of aeration, and nutrient availability: (1) the composition of parent materials, (2) the climate (prevailing moisture and temperature), (3) the topography (the degree of water drainage), (4) the types of plants and animals, and (5) the time involved.

Two examples of how those factors interact to influence soil development follow. Prairie soils, such as those in Iowa and Illinois, are exceptionally fertile. Those soils are derived from *loess*, which is a dusty, wind-borne parent material that has a high moisture-holding capacity and is rich in plant nutrients. It was deposited in that region between 14,000 and 20,000 years ago from sources throughout the central United States. Also, it was commonly deposited to a depth of several meters. These soils, along with moderate precipitation and warm summer temperatures, promote the growth of grasses and other nonwoody plants, such as prairie clovers, sunflowers, and goldenrods. Many of those plants characteristically send roots several meters into the soil—a growth pattern that is aided by the deep loess. Thus when the roots die and begin to decay, the soil humus layer becomes quite deep. Because these soils are relatively young, leaching has removed relatively few nutrients from them. Hence we see that the highly productive prairie soils of Iowa and Illinois result from a particular combination of factors—parent materials, climate, organisms, and time.

In contrast to those prairie soils, some tropical rain forest soils in the Amazon Basin are quite infertile. A major contributing factor to the low nutrient levels in tropical rain forests is the climate. High temperatures and frequent heavy rains accelerate the rate of decomposer activity, so dead plants and animals are broken down rapidly. However, few nutrients are released directly into the soil. Rather, for the most part, decomposition is carried out by fungi that grow in live plant roots and extend into the soil. As those fungi break down the dead plants and animals, the mineral nutrients they release are taken up and transported directly into plant roots. Because of the heavy precipitation, the few nutrients that are released directly into the soil are quickly carried away by heavy rains. Hence few plant nutrients are present in the soil.

Another contributing factor to low soil fertility in the Amazon Basin is the geologic stability of that region. Relatively little large-scale geological activity, such as volcanic and glacial activity that would provide new sources of nutrient-rich parent materials for soil building, has occurred for millions of years. Although some deposition of sediments has occurred, most soils are derived from sandstones (parent material) that contain few nutrients. That geologic stability, coupled with a warm, wet climate, has produced deep, highly-leached, nutrient-poor soils.

To help them organize what they have learned about soils, soil scientists have developed a classification system that is based on the chemical and physical properties of soil. Table 5.1 presents some of the major groups of soil. As can be seen from the table, prairie soils are classified as Mollisols and the tropical rain forest soils are classified as Oxisols.

Knowing that climate plays a major role in soil formation, we would expect to find a relationship between the characteristics of a regional climate and the soil groups found in that region. Such relationships are illustrated in the world climate and soil maps, shown in Figures 5.5 and 5.6, respectively.

Biomes of the World

All the continents are covered by very large communities of plants and animals, called *biomes*, which are named for the type of plants that dominate the landscape. For example, most of the eastern United States is covered by a deciduous forest, whose trees lose their leaves in autumn, whereas the midsection of the continent is covered by vast grasslands.

Figure 5.7 shows the geographical distribution

Table 5.1 Major soil groups and their general properties

GROUP	PROPERTIES
Aridisols	Desert Soils; Little Development of Soil Horizons, Little Accumulation of Organic Matter
Mollisols	Grassland Soils; Moderate Development of Soil Horizons, Black, Organic-Rich A-Horizon, High in Plant Nutrients
Alfisols	Deciduous Forest Soils; Moderate Development of Soil Horizons, Brown or Grey-Brown Topsoil, Moderate Accumulation of Organic Matter, Moderately Rich in Plant Nutrients
Ultisols	Forest Soils of Humid, Warm Regions; High Development of Soil Horizons, Low Accumulation of Organic Matter, Yellow and Red Topsoil, Low in Plant Nutrients. Generally on Older Landscapes or Younger, but Highly Weathered Parent Material
Spodosols	Soils under Conifers, such as Those in Boreal Forests and Pine Barrens; Acidic, Rather Thick Humus Layer Overlying White, Highly Leached, Sandy A-Horizon; B-Horizon Enriched in Organic Matter and Iron Compounds. Generally Low in Plant Nutrients
Oxisols	Soils under Tropical Rain Forests. Highly Weathered Soils, Little Horizon Differentiation; Little Organic Matter within A-Horizon; Highly Leached of Plant Nutrients; High in Iron and Aluminum

Source: *Soil Taxonomy: A Basic System of Soil Classification for Making and Interpreting Soil Surveys.* U.S. Soil Conservation Service, 1975.

of the major biomes of the world. As we noted earlier, the presence of a particular type of vegetation is determined mainly by the climate. Hence the spatial pattern of biomes that is shown in Figure 5.7 exhibits many similarities to the spatial pattern of climates that is shown in Figure 5.5. Moreover, because vegetation and soil development also influence each other, the biome spatial pattern corresponds, to a great degree, with the world pattern of soil groups, which is illustrated in Figure 5.6.

The type of vegetative cover that is found within a biome varies considerably. The interaction of such factors as soil, altitude, slope exposure (for example, a south-facing slope is warmer and drier than a north-facing slope), drainage patterns, and local climate produce a myriad of different climax communities within a biome. Moreover, because of disturbances that originate from humans, other species, and physical factors, such as fires, hurricanes, and volcanic activity, a region will exhibit communities that exist in various stages of succession. Hence the map that is shown in Figure 5.7 is an oversimplification of the complex mosaic of ecosystems that actually exists within biomes.

Table 5.2 lists the major biomes and some of their distinguishing characteristics. Figure 5.8 illustrates the appearance of several of those major biomes. In the following sections, we compare and contrast two of the most important characteristics of biomes—their productivity and their rates of nutrient cycling.

Productivity

Because temperature and precipitation strongly influence productivity, plant production varies greatly among biomes, as shown in Table 5.3. Abundant

Table 5.3 Estimates of plant productivity for various ecosystems (numbers indicate grams of biomass per square meter per year)

TYPE OF ECOSYSTEM	MEAN PLANT PRODUCTIVITY
Desert Scrub	71
Tundra	144
Temperate Grassland	500
Savanna	700
Boreal Forest	800
Temperate Deciduous Forest	1200
Tropical Rain Forest	2000

Source: Adapted from R. Whittaker and G. Likens, Carbon in the Biota, in G. Woodwell and E. Pecan, eds., *Carbon and the Biosphere*. Washington, D.C.: Technical Information Center, USAEC, 1973.

Table 5.2 Earth's major biomes and their characteristics

BIOME	DOMINANT GROWTH FORM	REPRESENTATIVE PLANTS (MOSTLY NORTHERN HEMISPHERE)	CLIMATE
Forests			
Tropical Rain Forests	Trees, Broad-leaved Evergreen	Many Species of Evergreen, Broad-leaved Trees, Vines, Epiphytes	125 to 1250 cm Annual Rain; No Dry Period; Temperatures from 18° to 35°C
Tropical Deciduous Forests	Trees, Both Evergreen and Deciduous	Mahogany, Rubber Tree, Papaya, Coconut Palm	Marked Dry Season, Generally Lower Precipitation
Temperate Deciduous Forests	Trees, Broad-Leaved Deciduous	Maple, Beech, Oak, Hickory, Basswood, Chestnut, Elm, Sycamore, Ash	60 to 225 cm Precipitation; Droughts Rare; Some Snow; Temperatures from −30° to 38°C
Northern Coniferous (Boreal) Forests	Trees, Needle-Leaved	Evergreen Conifers (Spruce, Fir, Pine), Blueberry, Oxalis	35 to 600 cm Precipitation; Evenly Distributed; Much Snow; Temperatures from −54° to 21°C; Short Growing Season
Reduced Forests			
Scrubland Chaparral	Shrubs, Sclerophyll Evergreen	Live Oak, Deerbrush, Manzanita, Buckbush, Chamise	25 to 90 cm Precipitation; Nearly All During Cool Season; Temperatures from 2° to 40°C
Grasslands			
Tropical Savannah	Grass (and Trees)	Tall Grasses, Thorny Trees, Sedges	25 to 90 cm Precipitation During Warm Season; Thunderstorms; Dry During Cool Season; Temperatures from 13° to 40°C
Temperate Grasslands	Grass	Bluestem, Indian Grass, Gramma Grass, Buffalo Grass, Bluebunch Wheatgrass	30 to 200 cm Precipitation; Evenly Distributed or High in Summer; Snow; Temperatures from −46° to 60°C
Tundra	Diverse Small Plants	Lichens, Mosses, Dwarf Shrubs, Grass, Sedges, Forbs	10 to 50 cm Precipitation; Snow Drifts and Areas Blown Free of Snow; Temperatures from −57° to 16°C
Deserts			
Warm	Shrubs, Succulents	Creosote Bush, Ocotillo, Cacti, Joshua Tree, Century Plant, Bur Sage (in U.S.A.)	0 to 25 cm Precipitation; Very Irregular; Long Dry Seasons; Temperatures from 2° to 57°C with High Diurnal Fluctuations
Cold	Shrubs	Sagebrush, Saltbush, Shadescale, Winterfat, Greasewood (in U.S.A.)	5 to 20 cm Precipitation; Most in Winter; Some Snow; Long Dry Season; Temperatures from −40° to 42°C with Diurnal Fluctuation

Source: Vegetation data from W. A. Jensen and F. B. Salisbury, *Botany: An Ecological Approach*. Belmont, Calif.: Wadsworth, 1972. Climatic data from W. D. Billings, *Plants, Man, and the Ecosystem*, 2d ed. Belmont, Calif.: Wadsworth, 1970.

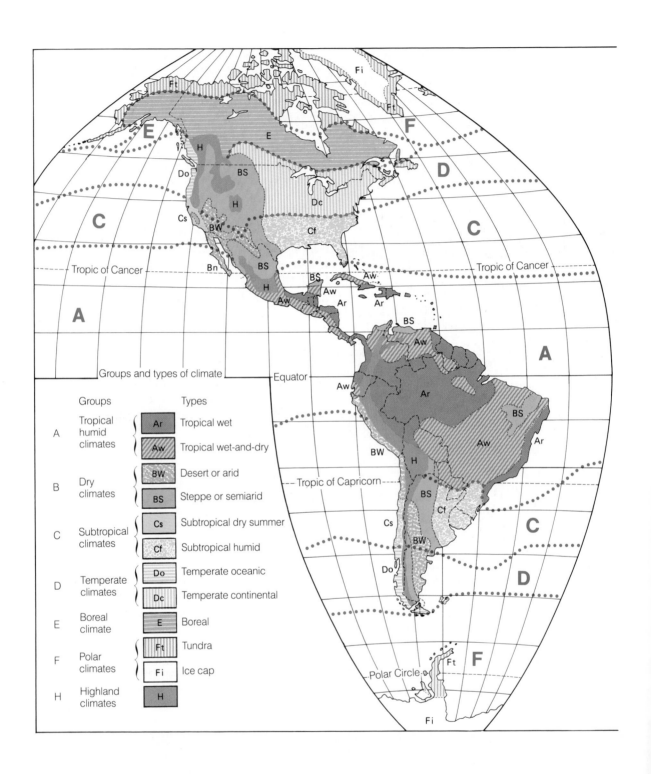

Figure 5.5 *Distribution of major climate types of the world.*
(After G. T. Trewartha, A. H. Robinson, and E. H. Hammond,
Elements of Geography. © McGraw-Hill Book Company, New
York, 1967.)

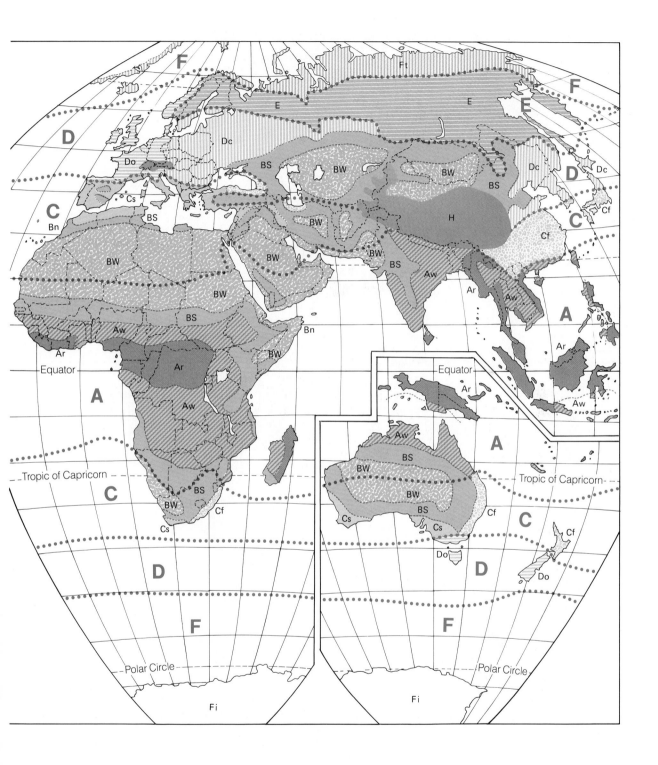

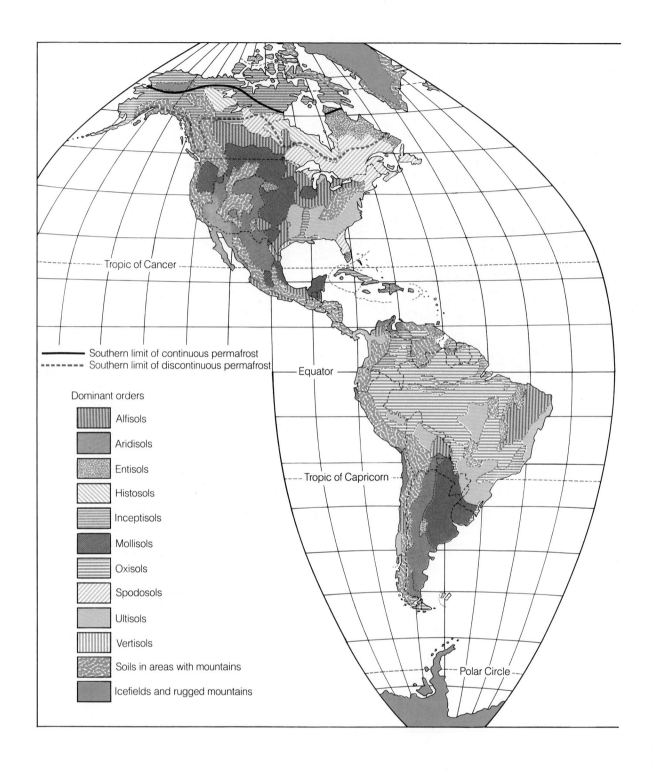

Figure 5.6 *Major soil types of the world. (U.S. Soil Conservation Service.)*

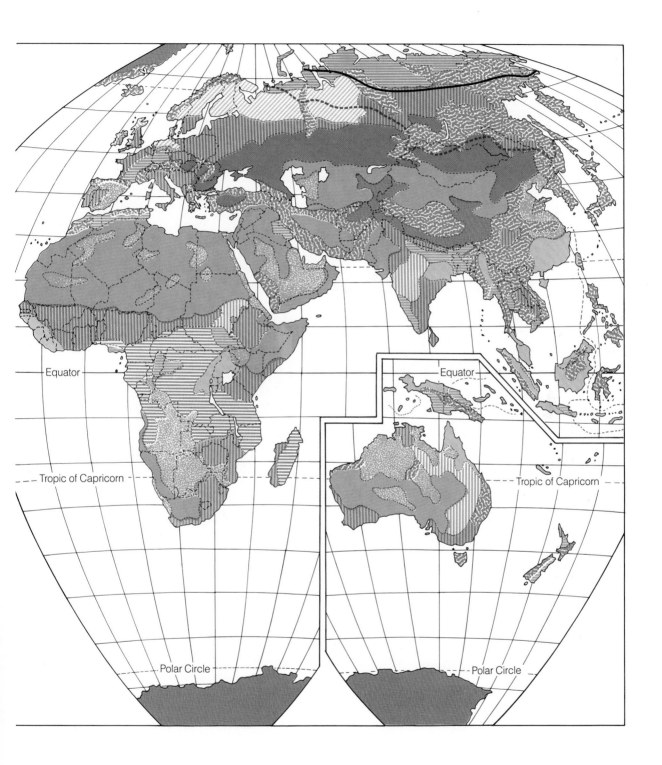

Equator

Tropic of Capricorn

Polar Circle

Equator

Tropic of Capricorn

Polar Circle

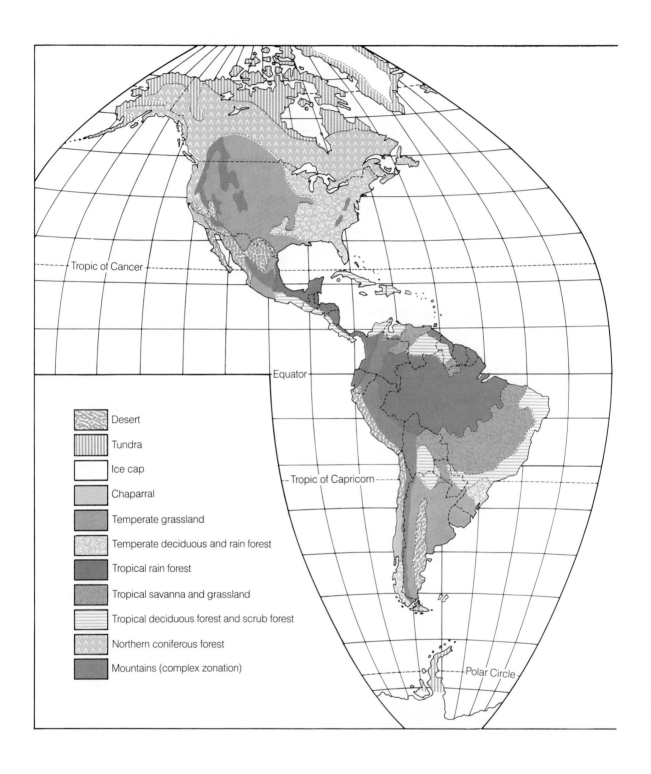

Tropic of Cancer

Equator

Tropic of Capricorn

Polar Circle

Desert

Tundra

Ice cap

Chaparral

Temperate grassland

Temperate deciduous and rain forest

Tropical rain forest

Tropical savanna and grassland

Tropical deciduous forest and scrub forest

Northern coniferous forest

Mountains (complex zonation)

Figure 5.7 *The major biomes of the world. (After P. R. Ehrlich, A. H. Ehrlich, and J. R. Holdren,* Ecoscience: Population, Resources and Environment. *New York: W. H. Freeman, 1977.)*

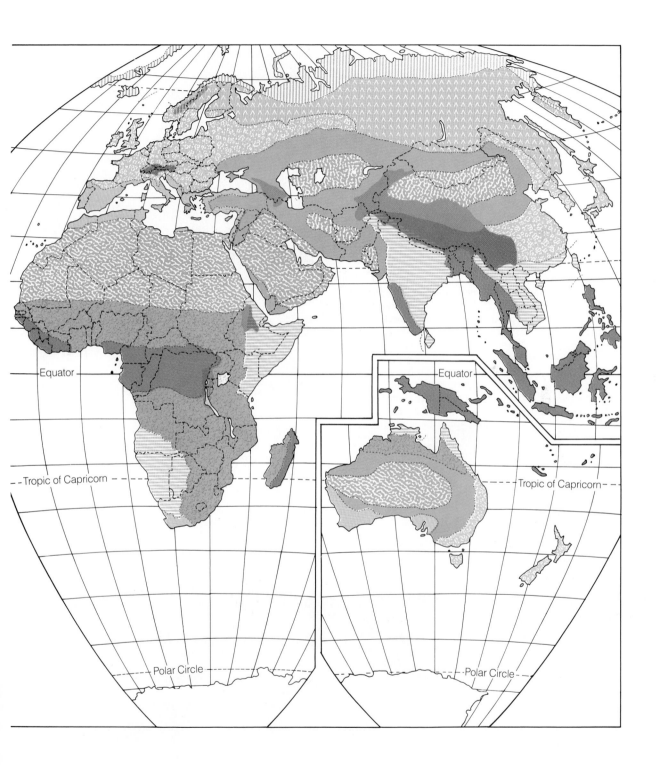

Equator

Tropic of Capricorn

Polar Circle

Equator

Tropic of Capricorn

Polar Circle

(a)

Figure 5.8 *(a) Cacti are characteristic of the Sonoran Desert. (Peter B. Kaplan, Photo Researchers, Inc.) (b) Trees along a waterway, a common configuration in grasslands. (U.S. Department of the Interior, National Park Service.) (c) Tropical savannas are characterized by a continuous cover of grasses interrupted by scattered trees and shrubs. (R. D. Estes, Photo Researchers.) (d) The chaparral, such as this one in California, is dominated by evergreen shrubs. (W. H. Hodge, Peter Arnold, Inc.) (e) Spruce and fir are the major trees in the boreal forest. (Charlie Oh, National Audubon Society, Photo Researchers.) (f) A road through a deciduous forest in the eastern United States. (J. M. Conrader, National Audubon Society, Photo Researchers.)*

(b)

(c)

(d)

(e)

(f)

moisture, favorable temperatures, and a year-round growing season combine to make tropical rain forests the most productive terrestrial biome on an annual basis. Figure 5.9 illustrates the luxuriant plant growth that is commonly found in tropical rain forests.

Tropical rain forests are also the earth's most diverse ecosystems. A two-hectare (five-acre) area can contain over 100 species of trees, whereas the same area in an eastern deciduous forest typically contains fewer than 25 species of trees. Trees of all heights are found beneath the upper canopy, and epiphytes are abundant. The luxuriant canopy growth permits little light to reach the forest floor, so the ground flora is generally sparse, except for areas in which the canopy has been broken by a fallen tree. Consequently, most of the ecosystem's animals live in the trees. An abundance of fruits supports a great diversity of animals, particularly insects and birds. Although temperatures are favorable year round for plant growth, many tropical rain forests experience a short dry period, during which many trees lose their leaves. With the advent of the rainy season and renewal of plant growth, most animals give birth to their young.

Figure 5.9 *A luxuriant tropical rain forest. (K. Weidmann, © Earth Scenes.)*

Figure 5.10 *Sparse precipitation limits plant production in this sagebrush desert. Hence much of the ground is bare of vegetation. (John Zuiner, Peter Arnold, Inc.)*

Moving poleward from the tropics, the length of the growing season and temperatures during that season decline. Such changes contribute to significant reductions in annual productivity. The productivity of the temperate deciduous forests is some 40 percent lower than that of the tropical rain forests. Moving farther poleward, we find that annual productivity of the boreal forest is 30 percent lower than that of the temperate deciduous forests. In temperate grasslands, the growing season is shortened by frosts in spring and autumn as well as by droughts in the summer. Those summer dry periods are frequently so severe that prairie plants cease to grow and become dormant for a significant portion of the growing season.

At the end of the productivity spectrum are the desert and the tundra biomes. The extreme impact of limiting factors on productivity is illustrated in those regions. Precipitation is the obvious limiting factor in deserts (Figure 5.10). Not only is precipitation sparse—less than 25 centimeters (10 inches) annually—but rainfall is also sporadic, often occurring as brief, intense downpours in thunderstorms. The plants and animals that live in desert biomes are adapted to burst into activity when those infrequent deluges occur. When sufficient precipitation falls, for example, the seeds of many desert plants germinate rapidly, producing seedlings that flower and set seed within a span of only a few weeks. The

appearance of that new plant growth, in turn, triggers reproductive behavior in such desert herbivores as grasshoppers and rabbits. However, once the water has evaporated or has been absorbed by the plants, activity slows down and production comes to a virtual halt again.

Productivity in the tundra illustrates the impact of light precipitation, low growing-season temperatures, and a short growing season (60 days or less). Vegetation grows slowly and consists of low plants, such as grasses, sedges, and such dwarf woody plants as willows and birches (Figure 5.11). Few animals

---

BOX 5.1

ICEBOUND OASIS

Although they are cold and snow-covered, the Arctic islands in Canada's Northwest Territories are as dry as any desert in hot southern regions. Less than 7 inches of precipitation falls annually on most of the Arctic archipelago north of mainland Canada—the same amount that reaches arid Phoenix, Arizona. No wonder that early visitors described these islands as a vast wasteland, though one of rock and snow, not sand.

And yet, just as in hot deserts, the Arctic, too, has oases that shift nature's odds to favor life. Some are in sheltered valleys that calm winter's bitter gales and extend the short growing season a few extra days. Others are meadows with an abundant supply of water which can support large numbers of animals. Still others are special areas of ocean water, where upwellings or currents provide critical sources of food.

Polar Bear Pass is such a place. It is located on Bathurst Island, midway in a group of islands— one as large as Great Britain, another as big as Manitoba—that extends into the Arctic Ocean, 550 miles from the North Pole. Just over 6000 square miles in size, Bathurst is connected most of the year to other lands by solid ice. Polar Bear Pass cuts it almost in half, a broad valley that stretches between inlets on the island's east and west coasts. The hill-and-valley topography of this passageway takes the teeth out of the harsh Arctic climate and provides plentiful fresh water in summer.

Scientists describe the pass as a "thermal" oasis, created largely by the hills that surround the valley. Elsewhere in the Arctic, the sun's rays are reflected by white snow. At the pass, dust from these hills, carried by prevailing winds in

spring, settles on the snow, absorbing the sun's heat. That melts the snow, slowly at first, then more quickly. By early June when the island's spring melt is underway, vegetation is exposed and shallow pools and ponds are formed.

As the land surface warms, there is a stirring of the tiny creatures that have over-wintered in mud, mosses, and plant debris. Insects and other invertebrates begin again the cycle of life. They have only the brief summer to reproduce. For another season, Polar Bear Pass is transformed into a fertile nursery, while the surrounding hills and uplands remain swathed in cold and winds reminiscent of winter.

Tundra ponds, formed by porous dams of moss and other vegetation, regulate the flow of meltwater during the growing season and thus irrigate broad, luxuriant meadows of sedge spotted with creeping Arctic willow. Warmth and water make Polar Bear Pass the most extensively vegetated part of Bathurst. There are more than 65 species of vascular plants, 112 mosses, 182 lichens.

Rich in insect life, the tundra ponds are prime feeding areas for many birds. Thirty species breed in the pass. Insect eaters include sandpipers and plovers, ptarmigan, snow buntings, Lapland longspurs, and the young of geese and eider ducks, which hatch when the swarms of insects also are emerging. Brant and snow geese, in particular, benefit from the thermal oasis, for they will not nest if the spring melt is delayed. With a plentiful food supply, these birds remain in the wetlands until the fledging of their young.

Red-throated loons carry fish from the sea to their nesting sites by the inland ponds. Eggs and young of all species are ravaged by jaegers, aggressive birds of prey, and by the big glaucous gulls,

(such as caribou, musk oxen, Arctic fox, Arctic hare, lemmings, and ptarmigans) make the tundra their home year-round. In the summer, however, large numbers of migratory birds, mostly geese and ducks, arrive, nest, and raise their young before migrating toward the equator again. Box 5.1 describes some fascinating dynamics among tundra organisms and their harsh environment.

Nutrient Cycling

A fundamental distinction in nutrient cycling patterns among terrestrial biomes occurs between oli-

---

though carrion also suits them. The number of nesting snowy owls and rough-legged hawks fluctuates according to the periodic rise and fall of the lemming population. Many other migrating birds enroute to distant breeding grounds stop at the oasis. In all, some 53 species of birds have been recorded at the pass.

A place like Polar Bear Pass gives animals an edge in the constant push to get on with life, to get such essential tasks as breeding over with in the short time before the long winter comes again. For birds, the timing is critical. While they are preoccupied with nesting, predators take advantage of this season of easily obtained meals. Weasels and Arctic foxes make regular forays for eggs and young as well as for adult birds.

The foxes are a hardy lot, and an excellent example of the superbly adapted Arctic mammal. With thick fur, a compact body, short snout, and small ears to minimize heat loss from extremities, Arctic foxes are active throughout the year. They mate in winter and pups are born in late May and early June. Dens, usually located where there is good drainage so they will be dry, may be hundreds of years old.

The bears, from which the pass gets its name, travel through the lowlands from March to November, either looking for denning sites inland or moving from one coast to another in search of seals. Peary caribou, in seriously decreasing numbers, also cross and linger in Polar Bear Pass twice a year, moving 80 miles from wintering areas on southern Bathurst to calving and summering areas further north.

Even in the oasis, however, an unexpected change in the weather can bring disaster. Many caribou died during the months following a heavy wet snowfall in the first week of September 1973.

Because of strong winds and bitter cold, the snow sealed the ground-hugging vegetation as effectively as a layer of concrete. Some of the caribou that survived left Bathurst and traveled over sea ice to neighboring islands.

Arctic animals live in environmental conditions that are extreme and sustain no weaklings. Seasonal changes are relentless; unprepared animals may perish. Shifts in the weather may be abrupt and severe; they can forestall or interrupt breeding seasons or even wipe out large numbers of animals if food is unavailable. But, despite natural catastrophes, in the long run populations survive. It is when natural disasters coincide with man-made pressures that animals are in big trouble.

Like some other key ecological areas in the north, the pass is also coveted by miners and oil drillers. That has elevated it from an unusual biological phenomenon to a hot political issue. Since 1968, when the United Nations began to examine the biological productivity of ecosystems all over the planet, Canadian scientists have identified 151 special ecological areas in the far north, many of which can be considered oases. Some are already in existing parks and reserves, but nothing was done to preserve any of the rest—until late last year. Then, Canada's Department of Indian and Northern Affairs proposed a tract of 1017 square miles, including Polar Bear Pass, a national wildlife area. Exactly what kind of protection that entails is still being hammered out, but conservationists are hoping the designation will mean that a remarkable Eden in a barren desert can also be insulated from the bitter winds of northern politics.

Excerpted from Ron Vontobel, "Icebound Oasis." © International Wildlife (July–August), 1983.

Figure 5.11 *Poor growing conditions restrict Alaskan tundra vegetation to low plants, such as grasses, sedges, and dwarf willows. (Charlie Ott, Photo Researchers.)*

gotrophic and eutrophic ecosystems. The basic distinguishing factor is soil fertility. *Oligotrophic ecosystems* contain soils that are low in fertility, whereas *eutrophic ecosystems* contain soils that are high in fertility. Oligotrophic ecosystems include tropical rain forests that grow on old, highly-leached soils (Oxisols); pine forests that grow on sandy, acidic, well-leached soils (Spodosols); and chaparral (a shrubland that grows in a moderately dry, temperate, maritime region, with little or no summer rain) that grows on shallow, coarse soils. Eutrophic ecosystems include grasslands that grow on soils with a great deal of humus and nutrients (Mollisols), deciduous forests that grow on soils with an intermediate to high availability of nutrients (Alfisols), and tropical rain forests that grow on relatively young soils, which are derived from nutrient-rich volcanic rocks.

Because soils in oligotrophic ecosystems contain few plant nutrients and also often have a low nutrient storage capacity, plant uptake and storage is the major mechanism by which those essential nutrients are conserved. Thus several nutrient-conserving adaptations have been identified. Perhaps the

most important adaptation is the mat of humus and fine roots that are located at or near the soil surface in pine forests and tropical rain forests. As the leaves fall to the mat, they are quickly enveloped by a mass of fine rootlets. Those rootlets, which we have mentioned previously in terms of the special fungi that are associated with them, reduce nutrient leaching by efficiently taking up the nutrients from decomposing leaves and twigs. Very few nutrients are stored in the mineral soil in oligotrophic ecosystems.

Many plants that are found in oligotrophic ecosystems have evergreen leaves—leaves that remain on the plant longer than one growing season (Figure 5.12). Evergreen leaves conserve nutrients in several ways. (1) They utilize nutrients efficiently, because nutrients that are invested in a leaf are used for photosynthesis for several growing seasons rather than just one. (2) When they die, after 3 to 4 years, a large proportion of the nutrients they contain are transported back into the stem before the leaf falls. (3) They frequently contain chemicals that reduce herbivore attacks (some deciduous leaves contain such chemicals, too). If leaves are consumed, then plants must take up more nutrients from the soil to produce new leaves. Those toxic chemicals thus help to conserve nutrients for the plant. (4) When evergreen leaves, which contain toxic chemicals, do fall,

Figure 5.12 *By conserving nutrients in their leaves, evergreen plants, such as manzanita, enhance productivity in oligotrophic ecosystems. (John Christopher, Photo Researchers.)*

those chemicals slow the breakdown of dead leaves by decomposers, thereby increasing the likelihood that the nutrients will be taken up again by the vegetation. Thanks to these effective nutrient-conserving mechanisms, productivity in oligotrophic ecosystems is frequently much greater than would be predicted on the basis of soil fertility alone.

An important difference between oligotrophic and eutrophic ecosystems is exhibited by their responses to a disturbance. A particular concern is the impact of clearing oligotrophic rain forests to make way for intensive agriculture. It was long assumed that the luxuriant forests are indicative of fertile soils that can support high-yielding crops. But many soils under tropical rain forests are not fertile. What, then, is the impact of removing the vegetation from an oligotrophic ecosystem?

Because most of the nutrients in an oligotrophic ecosystem are incorporated in the leaves, stems, and root mat, most of the ecosystem's nutrients are removed when the vegetation is cleared and the root mat is destroyed by cultivation. Moreover, because those soils often have little nutrient-holding capacity, the application of fertilizers does little to improve crop yields. As a result, productivity usually falls off quite rapidly and, within a few years after the original forest vegetation is cleared, agricultural yields fall to the point at which further cultivation

becomes futile. For more information about the problems that are threatening tropical rain forests, see Box 5.2.

In contrast, most of the nutrients in a eutrophic ecosystem are in the soil, and the nutrient-conserving mechanisms of plants are less important. Thus the loss of the vegetative cover does not greatly diminish the nutrient content of the ecosystem. Furthermore, because those soils have substantial nutrient-storage capacity, they respond well to the application of fertilizers. Thus, following a disturbance, a eutrophic ecosystem retains much of its productive capacity, and agricultural activities in it are usually successful, barring severe erosion.

If the disturbance is short-lived, such as a fire, both types of ecosystems usually experience rapid revegetation. Often, colonization is particularly rapid in oligotrophic ecosystems after a fire, because many species of shrubs and trees that grow in such ecosystems resprout from underground stems and roots that are not harmed by the fire. In eutrophic ecosystems, however, recolonization takes longer because the new vegetation is initiated, in most cases, by seeds, which take longer to become established. Revegetation that occurs from resprouting plants is quicker because sprouts already have a well-developed root system. Nutrients contained in the ashes of a fire also spur rapid regrowth, but in oligotrophic ecosystems, that burst in regrowth soon slows, because the ash nutrients are depleted and the plants become dependent again on soil that is poor in nutrients.

Convergent Evolution

Tundra, boreal forests, deciduous forests, grasslands, and deserts are all found on every continent except Antarctica. Tundra and boreal forests form nearly concentric bands around the North Pole, as you can see in Figure 5.7. Deciduous forests grow in eastern North America, western Europe, and northeastern China. Extensive grasslands are present in the Great Plains of North America, the pampas of South America, the steppes of Eurasia, and the African veld. The world's major deserts include the Sonoran in North America, the Sahara in North Africa, the Gobi in Mongolia, the Atacama along the west coast of South America, and the great desert that makes up the interior of Australia.

Within each type of biome, selection pressures

BOX 5.2

DAMMING RIVERS IN THE TROPICAL RAIN FORESTS: POTENTIAL ECOLOGICAL CATASTROPHES

Tropical rain forests lie mainly within nations that are struggling to improve the standard of living for their citizens. To achieve their goals, some nations have adopted ambitious plans to tap the hydroelectric potential of their waterways. Brazil, for example, hopes to build some 25 dams in the vast Amazon River Basin. What will be the ecological ramifications of such development? Because so little is known about tropical rain forests, few well-grounded predictions can be given. But the lessons from dams that have already been constructed provide us with reasons for great concern. The following describes some of the ecological impacts in tropical rain forests that have resulted from dam construction.

Because of engineering and logistical problems, only a few dams have been built in rain forest areas anywhere in the world, and their history has not been a happy one. The first one, completed in 1964, created Lake Brokopondo in the former Dutch colony of Suriname just north of Brazil. Five hundred seventy square miles of dense, virgin rain forest were flooded for this reservoir. As the trees decomposed, they produced hydrogen sulfide and the resultant stink brought complaints from people many miles downwind. For 2 years, workers at the dam had to wear gas masks. The worst aspect of the decomposing vegetation was that it made the water more acidic and corroded the dam's expensive cooling system. The cost of extra maintenance and of repairing and replacing damaged equipment was estimated in 1977 to have totaled $4 million, or more than 7 percent of the total cost of the project.

Once the dam was built, water hyacinth, which up to then had been rare in the Suriname River, began to spread over the surface of the

lake, impeding navigation and getting entangled in machinery. A lake in which decaying trees are releasing nutrients is an attractive habitat for waterweeds. They appear first on the edges of the lake, fed by the nutrient-rich water lapping up and down on the shores. Eventually they break loose and float into the main part of the lake, forming dense mats and anchoring in the branches and on the trunks of still standing trees. Within a year a 50-square-mile carpet of water hyacinth was floating on Lake Brokopondo, and by 1966 almost half the reservoir was covered by the weed. A floating fern, Ceratopteris, *covered another 170 square miles.*

Decomposition and waterweeds have also caused problems elsewhere. The newest dam in the Amazon Basin, Curua Una near Santarem, two hours by plane from Tucurui [the site of a large dam on the Tocantins River in the Amazon Basin] began operating in 1977. Although Curua Una's reservoir is only one twenty-fifth the size of Tucurui's, people 40 miles away complained of the sulfurous smell for months after the uncleared reservoir began filling. Once the stench died down the real troubles started. By 1982, the steel casings of its turbines had been corroded by the acidic water; replacement casings made of stainless steel cost more than $5 million.

Within a short time about half of Curua Una's lake was covered with floating mats of water hyacinth and sedge. One consequence was that many of the fish in the lake died. Water hyacinth is unpalatable or even toxic to most fish. In addition, it absorbs available nutrients and blocks the sun from penetrating the depths of the lake. As a result, other plants, and the fish that depend on them, have little chance of survival.

At Jupia, a dam on the Parana River outside the Amazon region, the sheer weight of the mass of water hyacinths growing on the lake's surface has snapped steel cables. Special filters designed to protect the turbines from waterweeds have been so badly clogged that, on occasion, the dam has had to be shut down. Floating on the water's sur-

face or stranded on the shore, waterweeds also provide food, oxygen, and breeding sites for the carriers of two serious diseases, malaria and schistosomiasis.

Malaria is already widespread in Amazonia, and its incidence in Tucurui was serious even before the influx of workers to build the dam. In 1976, one person out of every five examined there was found to have the disease. The dam will greatly increase the number of suitable breeding sites for mosquitoes; during periods of reservoir drawdown, as much as 340 square miles of ideal mosquito breeding grounds will be exposed. If the period of low water is longer than 2 weeks, there will be time for an entire generation of mosquitoes to breed.

Schistosomiasis is caused by a parasitic flatworm and is transmitted to humans by a few genera of aquatic snails that thrive in slow-moving, weedy tropical waters. The parasite's larvae swim from the snail and penetrate the skin of people swimming, washing, or just standing in infested water. Once in the bloodstream, the larvae spend time in the liver and then, depending on the species, take up residence around the bladder or intestines of the human host. Symptoms range from chronic diarrhea and urinary blood to more severe, life-threatening ailments of the liver, kidneys, spinal cord, and other organs. Available treatments are costly and carry potential dangers of their own.

Fourteen million Brazilians are infected with schistosomiasis, which is spreading in many parts of the tropics as a result of water-resource development. At present, the Tocantins River itself does not harbor the snail vectors in any numbers. But Robert Goodland, a World Bank ecologist who undertook an environmental assessment of Tucurui for Eletronorte in 1977, believes that a few years after the lake is filled, it "will provide ideal conditions for snail proliferation." Since the host snails occur north, south, east, and west of Tucurui, and since many of the dam workers are from the poor northeast of the country, where

schistosomiasis is widespread, an intensive program to monitor snail prevalence and human infection will be necessary—and even then it is not clear how the spread of the parasite can be prevented.

Deforestation of the surrounding uplands and river basin poses other threats to the smooth functioning of the dam. One is accelerated erosion, which could lead to increased siltation of the reservoir and its feeder rivers. Another is the likelihood of exaggerated wet and dry season conditions, since rainwater rushes off denuded lands rather than being absorbed for later percolation. The result could be a shortage of water to operate the turbines at full capacity during the dry season.

The prospect of a multimillion-dollar investment being undermined by siltation resulting from deforestation may seem fanciful, but it has already happened more than once, especially in smaller reservoirs. The Anchicaya Reservoir in Colombia silted up almost completely within 10 years of its inauguration. In the Philippines, the Ambuklao Reservoir, which accountants calculated would pay for itself in 60 years, is now expected to silt up after only 32 years of operation. Intensive agriculture in the watershed of Brazil's Rio Alto Pardo caused such severe runoff that, in 1977, two dams on the river were completely smashed.

Ecologists know little about the dynamics of tropical rain forests and the waterways that flow through them. Because many rain forests are oligotrophic, we should be concerned about their ability to withstand the intensive development associated with dam construction. As in the past, human beings are conducting large-scale ecological experiments by changing the landscape without having more than the vaguest idea of the consequences of those experiments (development projects).

Excerpted from Catherine Caufield, "Dam the Amazon, Full Steam Ahead." With permission from Natural History, vol. 92, no. 7; copyright The American Museum of Natural History, 1983.

are quite similar, regardless of geographic location. For example, grasslands everywhere provide little shelter from predators, and the small seeds that are produced by grasses and other herbaceous plants are a major source of food. Interestingly, we find that large flightless birds, whose powerful legs help them to outdistance their predators, have evolved separately on three continents—the ostrich in Africa, the emu in Australia, and the rhea in South America. Although they are different species, they have striking similarities in body form (Figure 5.13). Moreover, all three species derive most of their food from the small seeds and insects that abound in the grasslands. That process, whereby unrelated, isolated populations become more similar to each other as a result of independent adaptation to similar environmental factors, is called *convergent evolution.* Other examples of convergent evolution on grasslands include the pronghorn of North America, the pampas deer of South America, the wild horse of Asia, and the zebra of Africa. All are large, grazing herbivores, whose speed and herding behavior reduce their vulnerability to predators in open grasslands.

We can also find examples of convergent evolution in the plant kingdom. As a result of selection pressures that are caused by moisture stress, many plants in desert biomes have a similar growth form, which consists of large green stems that are often devoid of leaves. The barrellike green stems not only store water, but they also carry on photosynthesis. Although the plants that are shown in Figure 5.14 appear to be very similar, they are actually members of two quite different families. They have evolved similar growth forms under similar desert conditions, but on two different continents.

Human beings have taken advantage of both the products of and the conditions that produce convergent evolution in several ways. Most of the food production in the United States consists of the raising of crop plants that were introduced to North America from similar biomes in Europe and Asia. For example, the grasslands of the Great Plains have been turned into expansive fields of wheat, which was originally domesticated from wild varieties of wheat that were in the grasslands of the Tigris-Euphrates Valley in the Near East. Other Old World food plants that are now successfully being cultivated in North America include rye, barley, oats, rice, soybeans, peas, carrots, lettuce, apples, pears, and cherries.

The movement has not been all one way toward the New World, however. Maize (corn), which originated in Mexico, and the potato, which is native to the Andes Mountains of Bolivia and Peru, are now cultivated worldwide. Similarly, the Monterey pine, which is native to southern California, has been transplanted to Africa, Australia, and Europe, where it is now an important timber tree.

Figure 5.13 *The ostrich, emu, and rhea illustrate convergent evolution in similar grassland ecosystems in different parts of the world. (a) The ostrich (Courtesy of the Field Museum of Natural History, Chicago.) (b) The emu. (Smithsonian Institution.) (c) The rhea (U.S. National Zoological Park.)*

(a)   (b)   (c)

(a)

(b)

Figure 5.14 *Because they have evolved in similar climates, the saguaro cactus of the Sonoran Desert (a) and the great euphorbia of East Africa deserts (b) have a very similar appearance. They actually belong to two different families. (Leonard Lee Rue III, © Earth Scenes; W. H. Hodge, Peter Arnold, Inc.)*

A few introductions of game animals also have been successful. A notable example is the introduction of the ring-neck pheasant to the central Midwest, where it is a major upland game bird. In addition, breeds of domesticated animals, such as cattle, chickens, sheep, and swine, are now found on every continent except Antarctica.

Although the introduction of some species from, for example, a biome on one continent to a similar biome on another continent has been invaluable to humankind, all too often, the introduction of new species has seriously disrupted some ecosystems and has caused severe economic loss. A classic example of this problem is the fungus that causes chestnut blight, which was introduced to the United States in the 1900s from China. In its native home, the fungus and its host, the Chinese chestnut tree, have coevolved to the point at which the fungus does little harm to the chestnut trees. But in the eastern United States, the fungus was a new parasite to which the American chestnut trees were quite susceptible. As we described in Chapter 4, when the fungus was accidentally introduced on nursery stock, it quickly spread, killing millions of chestnut trees. Today, the American chestnut tree has all but disappeared from the forests of eastern United States.

When Old World crops were introduced into the New World, their weedy counterparts accompanied them. Most of the weeds (approximately 20 species) that cause the worst problems for North American farmers are native to Eurasia. Even the familiar dandelion, so common in our lawns, is from Europe. Millions of dollars are spent each year to reduce the impact of introduced weeds on crop yield.

Although many pest species were introduced accidentally, some importations were intentional. In Chapter 4, we noted the impact of the English rabbit in Australia. Intentional introductions into the United States include the house sparrow, European starling, and carp.

Whether it is intentional or accidental, frequently an alien species is introduced without its natural enemies (predators, parasites, competitors, or all of them), and the alien species has a severe impact on the native species. In fact, introduced species have placed many native species on the endangered-species list. We discuss this consequence more thoroughly in Chapter 14. Table 5.4 lists some of the injurious plants and animals that have been introduced into the United States.

Altitudinal and Latitudinal Zones

If we were to climb a mountain from its base to its summit, we would encounter ecosystems that are similar to those that exist between the equator and the North and South Poles. Figure 5.15 illustrates the series of zones we would cross on our ascent in the Colorado Rocky Mountains. We would begin our journey in a sagebrush desert community, and as we climbed, we would move successively through a grassland, a Douglas fir forest, a spruce fir forest, and, crossing the timberline, we would enter an Al-

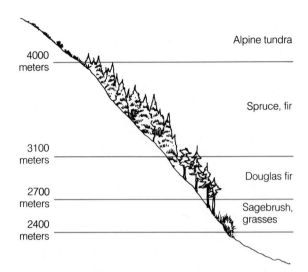

Figure 5.15 *Vegetation zones on a mountain slope, which develop as a response to change in moisture and temperature. In the central Colorado Rocky Mountains, a hiker would encounter a series of zones similar to those in the illustration.*

Table 5.4 Some harmful animals and plants introduced into the United States

NAME	ORIGIN	HOW INTRODUCED	TYPE OF DAMAGE
Mammals Nutria (Coypu) (a Giant Rodent)	Argentina	Intentionally Imported; Escaped Captivity (1940)	Alteration of Marsh Ecology; Damage to Levees and Earth Dams; Crop Destruction
Birds European Starling	Europe	Intentionally Imported Because a Group Wanted to Import All Birds Mentioned in Shakespeare's Plays (1890)	Noise; Displacement of Native Songbirds; Crop Damage; Transmission of Swine Diseases; Airport Interference
House Sparrow	England	Released Intentionally by Brooklyn Institute to Control Insect Pests (1853)	Crop Damage; Displacement of Native Songbirds
Fish Carp	Germany	Intentionally Released Because German Immigrants Wanted a Reminder of Home (1877)	Displacement of Native Fish; Uprooting of Water Plants, Causing Loss of Waterfowl Populations
Insects Gypsy Moth	Europe	Accidentally Escaped (1869)	Defoliation of Many Commercially Important Trees
Japanese Beetle	Japan	Accidentally Imported on Irises or Azaleas (1911)	Defoliation of More than 250 Species of Trees and Other Plants, Including Many of Commercial Importance
Plants Water Hyacinth	Central America	Intentionally Imported in Horticultural Exhibit (1884)	Clogs Waterways and Drainage Canals; Disrupts Aquatic Ecosystems
Dutch Elm Disease 1. *Cerastomella ulmi* (a Fungus; the Disease Agent)	Europe	Accidentally Imported in Infected Elm Timber Used for Veneers (1930)	Destruction of Millions of Elms; Disturbance of Forest Ecology
2. Bark-Beetle (the Disease Carrier)	Europe	Accidentally Imported in Unbarked Elm Timber (1909)	

Source: D. W. Ehrenfeld, *Biological Conservation.* New York: Holt, Rinehart and Winston, 1970.

pine tundra. We would also note changes in climate: the air temperature would decrease at higher altitudes, and precipitation would probably increase. In effect, mountains telescope biotic and climatic zones (that is, the same biotic–climatic zones that exist in several thousand meters of altitude are found in several thousand kilometers of latitude). Every 300 meters (1000 feet) that we ascend would roughly correspond to a northward advance of 500 kilometers (310 miles).

Ecotones

Biomes usually do not have readily discernible boundaries. Rather, a transition zone exists between ecosystems, which is called an *ecotone*. An ecotone contains species of both ecosystems and, often, its own unique species as well. Therefore, that zone often exhibits a greater diversity of plants and animals than either of its adjacent ecosystems (Figure 5.16), and that increased diversity is called the *edge effect*. Human settlement has greatly increased the amount of edge habitat on the planet. For example, a forest that was once continuous for many square kilometers may now be reduced to small isolated woodlots that are surrounded by pastures and fields (the situation today in much of the land in the eastern deciduous forest biome). Also areas that were once sweeping grasslands are now fields, dotted with farm houses and small towns, where trees have been planted (a common land pattern in the prairie states and provinces).

Ecotones vary in width. They may be as narrow as the boundary between a woods and a field. Others may consist of wide transitional belts between biomes, where one biome gradually blends into the other. Such a transition belt is the taiga, which lies between the boreal forest and the tundra across northern Eurasia and North America. Across that broad zone, changes in climate occur quite gradually, so the vegetation also changes gradually. If we were to walk north across the ecotone from the boreal forest to the tundra, we would encounter spruce stands that become progressively less dense and more stunted. Further on, patches of tundra vegetation would become more common, and spruce stands would become smaller and less frequent, as illustrated in Figure 5.17. Finally, we would reach a point where no trees are present—the tundra.

Ecotones, therefore, are not only boundaries between biotic communities, but they are also separations between major climatic zones. Hence even a small variation in climate will have its most immediate effects in ecotonal areas and will cause a geographical shift in the ecotone zone. For example, recent droughts in the Sahel savanna region of Africa have expanded the Sahara Desert southward into the savanna, and they have had a profound impact on its inhabitants (see Chapter 17).

The impact of climatic variability in ecotonal areas can be exacerbated by improper land management. The expansion of the Sahara into the Sahel has been accelerated by the overgrazing of livestock herds, which has expanded along with the rapidly growing human population in that area. The United States Agency for International Development estimates that along the southern fringes of the Sahara, 650,000 square kilometers (about 251,000 square miles) of land that was once suitable for agriculture or intensive grazing have become wasteland during the past 50 years. The severity of the problem was confirmed by a special worldwide conference that was convened by the United Nations in 1977 in Nairobi, Kenya. That conference brought a new word to the public's attention—desertification. *Desertification* is defined as the degradation of terrestrial ecosystems as a result of human activity. It reduces the potential for crop production and the land's capacity to support livestock. It also increases environmental deterioration by accelerating erosion by water and wind, and it usually lowers the residents' standard of living. The term is primarily applied to events that take place in arid and semiarid regions.

Aquatic Ecosystems

Limiting Factors: Dissolved Oxygen and Sunlight

Aquatic habitats differ from terrestrial habitats in several important respects. Whereas the main limiting factors that affect terrestrial organisms are moisture and air temperature, the most important limiting factors that affect aquatic animals and plants are dissolved oxygen and sunlight. In the atmosphere, oxygen is mixed with other gases, and its concentration is quite constant. But in aquatic ecosystems, it is dissolved in water, where its concen-

Figure 5.17 *Near the limits of tree growth in Alaska, stunted, widely distributed black spruce trees intermingle with bogs. This vegetation pattern marks the northern reaches of the broad ecotone between the boreal forest and the tundra. (Jen and Des Bartlett, Photo Researchers.)*

Figure 5.16 *The edge between a forest and a field contains a wide diversity of animals and plants.*

tration varies considerably. *Dissolved oxygen* concentrations in fresh water average approximately 0.0010 percent (also expressed as 10 *parts per million*, or 10 ppm) by weight, which is 150 times lower than the concentration of oxygen in an equivalent volume of air. The concentration of atmospheric oxygen is quite constant, but the amounts of dissolved oxygen in aquatic ecosystems vary widely, depending on the factors that influence the input and output of oxygen.

Oxygen enters an aquatic system through the air–water interface and by the photosynthetic activities of aquatic plants. Thus the amount of dissolved oxygen that is present in an ecosystem depends on the rate at which those processes occur. For example, the turbulence that occurs in waterfalls, rapids (Figure 5.18), and dam spillways as well as wave activity that occurs in open water increase the rate of oxygen transfer from the air to the water (unless the water is already saturated with oxygen). The profile

of a waterway also affects the transfer of oxygen, because a wide, shallow section of a river has a larger surface area for oxygen transfer than a narrow, deep segment. Also, the amount of oxygen that is produced per unit of area as a result of photosynthesis is directly related to the density of aquatic plants.

Dissolved oxygen leaves the water through respiration (by decomposers, zooplankton, shellfish, and fish) and through the air–water interface. Temperature influences the amount of dissolved oxygen that is retained by water because oxygen is less soluble in warm water (as shown in Figure 5.19) and also because warm water enhances decomposer activity and, therefore, the rate at which oxygen is removed through their respiration.

In certain circumstances, large populations of decomposers will remove nearly all the dissolved oxygen in *surface waters* (such as lakes, rivers, and streams) through respiration. That is most likely to occur during late summer, when high water temperatures and low stream flow reduce the level of dissolved oxygen even more. When the dissolved oxygen level falls below 3 to 5 ppm, many aquatic organisms die.

Figure 5.18 *Turbulent rapids add significant quantities of oxygen to waterways. (U.S. Department of the Interior, Bureau of Reclamation.)*

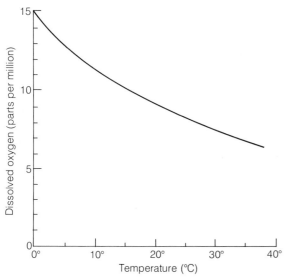

Figure 5.19 *The amount of dissolved oxygen that water can hold at saturation declines as water temperature rises.*

The water temperature of aquatic ecosystems does not vary much either daily or seasonally in contrast to the air temperature of terrestrial ecosystems. Furthermore, water temperature changes more slowly than does air temperature, because water has a considerably higher *specific heat* than air; that is, much larger amounts of heat energy must be added to, or taken from, water to raise or lower its temperature. Thus because water temperatures are less subject to change, aquatic organisms are not selected for wide temperature tolerance. Consequently, small changes in water temperatures are a greater threat to the survival of aquatic organisms than comparable changes in air temperature are to terrestrial organisms.

Sunlight is a limiting factor for aquatic plant life because, generally, less than 1 percent can penetrate beyond a depth of about 30 meters (100 feet) below the water surface. Thus photosynthesis is essentially confined to that area, which is called the *euphotic zone*. The depth of the euphotic zone may be significantly reduced by *suspended materials* (mostly silt, clay, and algae), which prevent sunlight from penetrating to greater depths.

The limiting factors discussed in this section are generally applicable to all aquatic ecosystems—

ponds, lakes, marshes, rivers, and seas. Each type of aquatic ecosystem, however, has certain unique properties. We turn now to a survey of the major types of aquatic ecosystems and their distinctive characteristics.

Freshwater Ecosystems

Lakes vary greatly in their productivity and rates of nutrient cycling. Using the same terminology that is applied to terrestrial ecosystems, we can classify lakes according to their productivity. *Oligotrophic lakes* are characterized by slow rates of nutrient cycling, low productivity, and a relatively high diversity of aquatic organisms. An example of an oligotrophic lake is shown in Figure 5.20. *Eutrophic lakes* have rapid rates of nutrient cycling, high productivity, and relatively few species of aquatic organisms. Oligotrophic and eutrophic lakes represent the two extremes of a continuous range of productivity in lakes.

The quantity of living organisms that a lake can support depends on the rate at which limiting nutrients are cycled. Sources of nutrients are the surrounding land area, which drains into the lake, the

Figure 5.20 *A high-altitude lake near Beartooth Pass, Wyoming. It is characteristically oligotrophic; that is, its water quality is high and its productivity is low. (Harvey Lloyd, Peter Arnold, Inc.)*

bottom sediments, and the lake organisms themselves. Some lakes are eutrophic from the time of their formation because of the influx of high concentrations of nutrients from their drainage basins. Most lakes, however, were originally oligotrophic and have become more eutrophic over many thousands of years. That process of aging through nutrient enrichment is called *eutrophication.*

Eutrophication proceeds as aquatic plants grow more extensively and increase in density, which occurs when greater quantities of mineral nutrients become available. For example, in shallow near-shore waters, rooted aquatic plants grow luxuriantly. With the onset of warm weather, the population of algae may increase suddenly, causing the water to become the color of pea soup or green paint within a few days. Those high algal concentrations, or blooms, are usually composed of only a few species of algae—most of them members of the blue-green group. Because they usually are too large to be eaten by zooplankton, those blue-green algae constitute a dead-end food chain, that is, a food chain that ends at the first trophic level. Nevertheless, photosynthesis by algae adds oxygen to surface waters. But after a bloom develops, the algae die, and most settle to the bottom, where they are acted on by decomposers. With such a large food supply available, the decomposers (mostly bacteria) multiply and consume large amounts of dissolved oxygen. As a result, the colder, deeper water of highly eutrophic lakes becomes depleted of dissolved oxygen during the summer. When that happens, less tolerant cold-water fish may die.

As large masses of rooted aquatic plants and algae die, they contribute to the gradual buildup of the bottom sediment. The lake gradually fills in, and cold-water fish die, while warm-water fish (such as perch, bluegills, bass, and pike) multiply and eventually predominate. At the same time, greater portions of the ever-shallower lake are invaded by aquatic plants that take root in the shallow water (such as cattails, bulrushes, and water lilies). Eventually, the entire lake becomes covered with rooted aquatic plants. Thus the lake becomes a marsh.

The successional progression that changes a lake into a marsh usually takes thousands of years. As noted earlier, however, human activities have greatly accelerated that aging process in many lakes. Accelerated soil erosion and the dumping of wastes that are rich in plant nutrients have speeded up the filling-in process. That accelerated aging, called *cultural eutrophication,* is discussed in more detail in Chapter 7.

Although the level of productivity varies greatly among lakes, most rivers exhibit low productivity. For example, near a river's source, the channel is narrow, water flow is swift, and rapids and waterfalls are common. Few organisms are equipped for survival under such turbulent conditions. Downstream, rapids disappear, and the current gradually slows as the river broadens. Yet even there, algae and rooted plants generally are not plentiful enough to meet the demands of river consumers. Hence, by necessity, a river ecosystem is open. That is, most consumers are detritus feeders that rely on the input of decaying organic matter from adjacent land areas or lakes that feed the river. Runoff also contributes nutrients that are needed to sustain plant productivity.

Rivers and streams have long been the traditional dumps of our society. Many persons seem to believe that waterways can dilute any type of wastes without affecting our well-being. But water's power of dilution is limited, and hundreds of kilometers of waterways in North America are now essentially devoid of desirable aquatic life. For example, trout or bass are being replaced in many rivers by "trash fish," such as carp and gar. Although many nations have done much to upgrade the water quality of their rivers, the continual growth of urban-industrial complexes and the development of intensive agriculture along rivers will continue to stress those aquatic ecosystems.

Aquatic Ecotones

Along the margins of lakes and bays, we commonly find a wetland transition zone between the uplands and an open body of water. That ecotone may be a *marsh* (which is a treeless form of wetland that is dominated by grasses) or a *swamp* (which is a form of wetland that is dominated by shrubs or trees). With time, the whole lake or bay may fill in to form a marsh or a swamp. Examples of that process include the Great Dismal Swamp of Virginia and North Carolina and the Everglades of Florida (a marsh). Freshwater marshes are especially widespread in the formerly glaciated lowlands of North America. Marshes and swamps act as sponges, since they hold

Figure 5.21 *Intensive agriculture on drained wetland soils immediately south of Lake Isotokpoga, Florida. The grove of cypress trees that fringe the margin of the lake are all that remains of the cypress swamp that once dominated this area. (U.S. Department of Agriculture, Soil Conservation Service.)*

large volumes of water. Thus they regulate streams that flow through them, absorbing water at times of flooding and releasing it at times of low river discharge. Because nutrients and detritus accumulate in those wetlands, they are very productive and support a wide variety of wildlife, including water birds and small fish.

Wetlands are disappearing in many regions because of human intervention. If artificially drained by ditching, some wetlands yield mucky soils that are highly productive when they are used for intensive cash-crop agriculture. Figure 5.21 shows a former wetland in Florida under cultivation. Many marshes have been filled in and used for industrial parks, shopping centers, and housing developments.

An *estuary* is an ecotone that is created by mixing fresh and salt water. Some examples are river mouths, tidal marshes, and bays. In an estuary, water is subjected to daily tidal oscillations. Hence organisms that reside in an estuary must have wide tolerances, because they are subject to frequent changes in temperature, salinity, and concentrations of sus-

pended sediment. Because of their unique combination of physical features, estuaries are among the most productive ecosystems on earth.

Many estuaries are located at the endpoint of rivers and, therefore, receive water that is rich in nutrients, which promotes plant growth. River water also delivers large quantities of organic matter. Once those materials are contained within the estuary, patterns of water circulation, which are called *nutrient traps* (Figure 5.22), retain and recirculate the detritus and other nutrients. Also, because estuaries are relatively shallow, sunlight usually penetrates to the bottom, so the entire depth of an estuary is in the euphotic zone.

That combination of physical features supports luxuriant plant growth (both phytoplankton and sea

Figure 5.22 *Circulation in an estuary, mixing lighter fresh water with denser seawater to form a nutrient trap. (After E. Odum,* Fundamentals of Ecology, *3rd ed. Philadelphia: W. B. Saunders, 1971.)*

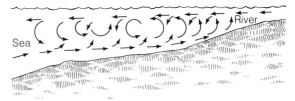

Figure 5.23 *These houses near Fort Lauderdale occupy a former estuary. (© Foto Georg Gerster, from Rapho/Photo Researchers.)*

fish and shellfish catch off the East Coast of the United States is composed of species that spent part of their life cycles in estuaries. The unique environmental conditions of the estuary favor development and protection of young: food is abundant, and the low salinity and shallowness of the water serve as barriers to many ocean predators.

The high productivity of estuaries illustrates the advantages of natural energy subsidies. The energy of water that flows downhill also transports nutrients and detritus to estuaries from land surfaces. Waves and currents retain the detritus and other nutrients in the estuary and keep the water stirred up, thereby cycling those materials to plants and detritus feeders. Even modern agriculture, with its massive fossil-fuel energy subsidies, has not been able to exceed (at least over an extensive area) the productivity that results from the natural energy subsidies of estuaries.

As with freshwater wetlands, more than 75 percent of the estuaries in the United States have been moderately to severely modified by human activity. Leading causes of damage include cultural eutrophication, dredging to maintain navigation channels, and filling in to make sites for industrial parks, shopping centers, and marinas. Figure 5.23 gives dramatic evidence of the encroachment of developments into those areas. This subject is discussed further in Chapter 15.

and marsh grasses) as well as large populations of detritus feeders. Animals that occupy the lower trophic levels include zooplankton, oysters, mussels, crabs, horseshoe crabs, shrimp, and snails. Those consumers, in turn, constitute a copious source of food for consumers at higher trophic levels, particularly fish and birds. In fact, about two-thirds of the

Figure 5.24 *A cross section of marine habitat zones.*

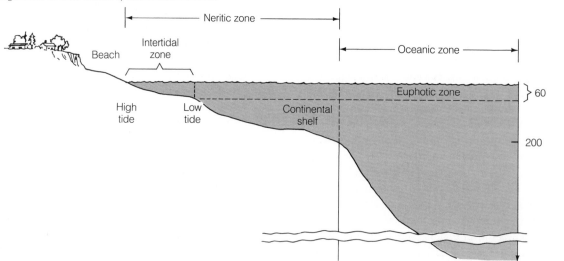

Table 5.5 Fish production of the oceans

TYPE OF AREA	PERCENT OF OCEAN	AREA (IN SQUARE KILOMETERS)	AVERAGE NUMBER OF TROPHIC LEVELS (APPROXIMATE)	ANNUAL FISH PRODUCTION (IN MILLIONS OF METRIC TONS)
Open Ocean	90.0	326,000,000	5	1.6
Coastal Zone	9.9	36,000,000	3	120.0
Coastal Upwelling Areas	0.1	360,000	1.5	120.0
Total Annual Fish Production				241.6

Source: After J. Ryther, Photosynthesis and fish production in the sea. *Science* 166:72–76, 1969.

Marine Ecosystems

Seaward from the high-tide mark, the *marine* (salt-water) environment can be divided into two basic zones, or types of ecosystems, as shown in Figure 5.24. The *neritic zone* extends from the high-tide mark to the edge of the continental shelf, where the water depth averages approximately 200 meters (660 feet). Beyond the 200-meter depth, the *oceanic zone* is found.

Although the seas cover nearly three-quarters of the world's surface, they comprise a surprisingly uniform habitat compared with the ones on land. However, productivity differs significantly from one portion of the ocean to another, as indicated in Table 5.5. Those differences stem primarily from differences in two limiting factors—sunlight and nutrients.

While the neritic zone accounts for only approximately 10 percent of the ocean's surface, it is the principal site of commercial fish harvests. Sea life is abundant and very diverse throughout the depths of this zone. Together with estuaries, the coastal zone accounts for approximately 50 percent of the productivity of the sea. Because they are the terminus of land drainage, coastal waters are particularly rich in nutrients. Also, because the continental shelf is relatively shallow, a significant portion of the water lies in the euphotic zone. Hence light and mineral nutrients are sufficient to promote productivity in the coastal zone.

Most of the remaining fish production in the sea takes place in *upwelling* areas. Most of the world's upwelling fisheries are found along the west coasts of continents, at the outer edge of the neritic zone. Upwellings are the result of the drag of prevailing winds on surface waters. The prevailing winds transport the surface layer of water offshore, away from the continents, thereby causing nutrient-rich bottom waters to rise (well up) to the surface. The nutrient enrichment of surface waters spurs algal productivity, which, in turn, enhances fish productivity. Although upwelling regions constitute less than 1 percent of the ocean's surface, they account for almost 50 percent of the productivity in the sea (see Figure 5.25, which shows the distribution of the world's fisheries).

Some of the major fish that are used as food sources are found in or near the neritic zone (for example, the Peruvian anchovy, Atlantic cod, mackerel, Atlantic herring, Alaskan walleye, pollack, halibut, flounder, and salmon). Those fish support huge concentrations of birds that nest near shore but feed from the sea. Common types of seabirds include cormorants, petrels, shearwaters, and sea ducks.

The *oceanic zone* covers the deep water that stretches beyond the edge of the continental shelf. Although enormous in area, open oceans are quite unproductive. Simply put, the raw materials for photosynthesis are not present simultaneously in sufficient quantities. Although ample light exists near the surface, nutrients are particularly limiting in the euphotic zone. Hence open oceans are biological deserts.

Although desertlike, the open seas do have scattered oases of abundant sea life. Phytoplankton form the base of food webs in the open oceans, where trillions of tons of those microscopic organisms are

produced. Herbivorous zooplankton graze on the phytoplankton and, in turn, are preyed on by carnivorous zooplankton, which are then fed on by small fishes and squids; those, in turn, fall prey to a wide range of larger carnivores. The carnivorous fish (such as tuna, dolphins, porpoises, swordfish, and sharks) are some of the largest and most superbly designed animals ever to inhabit the earth. The scattered nature of food sources accounts for the powerful swimming ability of those carnivores. Their streamlined, torpedolike shape enhances speed and efficiency of movement. The coloring of many marine organisms is quite similar, which acts as a camouflage when seen from above. Their dark blue-grey backs blend in with the color of the water. To predators below, their white undersides are very similar to the silvery

appearance of the surface. Denizens of the open oceans have developed such remarkable adaptations as sonar, extreme olfactory sensitivity, and complex senses of orientation and homing.

The sea is one of our few remaining frontiers and, until recently, the impact of human activity has been relatively minor. Today, however, things are changing. Overfishing has greatly reduced the stocks of several kinds of fish, including the anchovy, the California sardine, and the Norwegian herring. Oil spills from tankers and drilling rigs and the dumping of domestic and industrial toxic wastes threaten marine life at the local scale. Many of our reserve deposits of petroleum and minerals are on the continental shelves; if they are not extracted carefully and responsibly, marine and estuarine life

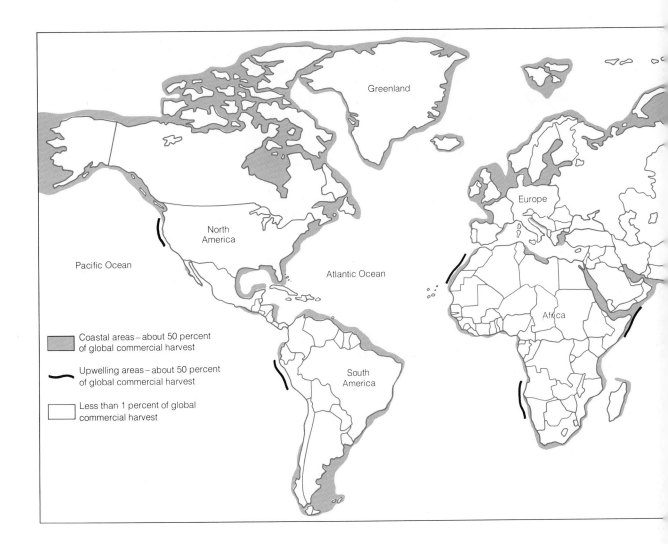

Figure 5.26 *A sea otter preparing to eat a sea urchin. (Scott Ransom, Taurus Photos.)*

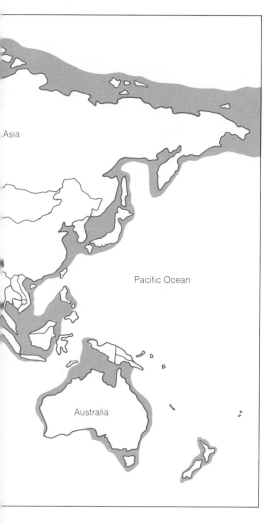

can be damaged gravely. Marine species, like terrestrial species, are interdependent, and their relationships can be disturbed dramatically by human activity.

A striking example is provided by three species in a food chain on the California coast. There the sea otter preys on sea urchins (Figure 5.26) that eat kelp (a giant brown algae). When sea otters were all but exterminated by fur traders in the nineteenth century, the sea urchin populations exploded and reduced the extensive kelp beds to small scattered patches. Once their food source was reduced, the sea urchins began to starve and die off. That subsequent reduction of the sea urchin population allowed the kelp to make a comeback. But because the otter population had not recovered much from the overhunting, they could not effectively control the sea urchin population. Thus the kelp beds did not regain their

Figure 5.25 *The distribution of the world's fisheries. (After* Patterns and Perspectives in Environmental Science. *Washington, D.C.: National Science Foundation, 1972.)*

former abundance. That ecosystem was disturbed further by the increased loads of sewage that were being poured into the ocean. The sea urchins ate the sewage and, with that added food source, the sea urchin population exploded, even though the kelp beds were greatly reduced. As a result, the sea urchins finished off the kelp beds in areas where enough sewage was present to increase the sea urchin population. Thus human intervention reduced those areas that once supported luxuriant kelp forests to biological deserts.

With legal protection, sea otter populations are slowly making a comeback. However, since sea otters also eat abalone, abalone fishermen do not want the sea otter to be protected. Therefore, those in the fishing industry should be made aware that sea otters, kelp, sea urchins, and abalone can cohabit the Pacific coastline successfully, as they did for thousands of years before humans intruded.

Figure 5.27 *These birds, which feed in the water and nest in trees, illustrate some of the linkages between terrestrial and aquatic ecosystems. (© Allan D. Cruickshank, National Audubon Society, Photo Researchers.)*

Linkages between Land and Water

Many important linkages exist between terrestrial and aquatic ecosystems: (1) Aquatic ecosystems are enriched by the movement of nutrients, such as phosphorus and nitrogen, from the land. (2) Decaying organic matter that is washed in from adjacent land areas provides a great deal of food for aquatic consumers in rivers and estuaries. (3) The erosion of soil builds up sediments in lakes and slow-moving rivers. (4) The rates at which nutrients, detritus, and soil particles flow through ecosystems are influenced significantly by human activities. Every time the landscape is disturbed, those materials are transferred at a greater rate from the land to the waterways.

The flow of materials and energy, however, is not restricted to just one direction. Aquatic life—mainly fish and shellfish—provides food for animals that are land dwellers. For example, large concentrations of seabirds feed on fish but nest on land near the shore (Figure 5.27). In addition, the wastes deposited by those birds on the land help to maintain the cycle of nutrients, such as phosphorus and nitrogen, from the sea to the land.

Hence those two diverse types of ecosystems are closely linked. Any change that occurs in one is likely to produce changes in the other.

Conclusions

In our brief survey of ecosystems, we have caught only a glimpse of the rich diversity and interactive complexity of life that exists on this small planet. But we have seen enough to know that human activities threaten to decimate that rich diversity and, therefore, seriously disrupt the roles that are played by organisms in the functioning of ecosystems. We as a species must begin to assess more accurately the degree to which our enterprises disrupt the earth's ecosystems. We must come to realize that those disruptions ultimately have a serious impact on the quality of human life. Therefore, we must determine how much deterioration we are willing to tolerate.

Summary Statements

The terrestrial environment can be a hostile place to live. Temperature and moisture are two common limiting factors. A strong correlation exists between a region's climate and the type of vegetation it supports.

Soil helps to govern plant productivity and the number of consumers that an ecosystem can support. The development of a soil profile and its properties are determined by five factors: (1) parent material, (2) climate, (3) topography, (4) types of plants and animals, and (5) time.

As a result of such factors as climate, soil, topography, and human and physical disturbances, the earth's land masses are covered by a complex mosaic of biotic communities.

Plant productivity varies greatly among ecosystems. Abundant moisture, warm temperatures, and, essentially, a year-round growing season combine to make tropical rain forests the most productive biome. At the low end of the productivity spectrum, plant production is limited by sparse, sporadic precipitation (in deserts) and by light precipitation, low temperatures, and a short growing season (in tundra).

A fundamental distinction occurs between the nutrient cycling patterns of oligotrophic ecosystems (based on low-fertility soils) and eutrophic ecosystems (based on high-fertility soils) among terrestrial biomes. Oligotrophic ecosystems possess many nutrient-conserving mechanisms. Because oligotrophic ecosystems store most of their plant nutrients in the plants themselves, whereas eutrophic ecosystems store most of those nutrients in the soil, oligotrophic ecosystems are much less likely to recover from severe disturbances (such as clearing for agriculture) than are eutrophic ecosystems.

Many biome types are found on all continents of the world except Antarctica. As a result of convergent evolution, similar but unrelated species are found within the same type of biome, but on different continents. The transplanting of species from a biome on one continent to a similar biome on another continent has been both beneficial and detrimental.

Steep temperature and precipitation gradients on mountainsides produce zonations of ecosystems that are similar to those encountered when moving from south to north in North America. Ecotones are transition zones between two different types of ecosystems: they usually display a high diversity of organisms from both ecosystems. Climatic variability usually has its greatest impact in ecotones.

The concentration of dissolved oxygen in water and the penetration of sunlight into water are limiting factors for aquatic organisms. A rapid and marked change in water temperature can put severe stress on aquatic organisms.

Lakes age through nutrient enrichment. Natural aging processes are accelerated by human activities. Oligotrophic and eutrophic lakes represent extremes in lake productivity.

Because the quantities of algae and rooted plants that grow in a river are insufficient to meet the demands of river consumers, those animals feed on the decaying organic matter that is washed in from adjacent land areas or lakes.

Marshes act as reservoirs for nutrients and water and help to regulate stream flow. Many marshes have been drained for agriculture or filled in for commercial development.

Estuaries are coastal ecosystems in which fresh and salt water mix. Because they receive natural energy subsidies and have a unique combination of physical features, estuaries are among the most productive ecosystems on earth.

Productivity in the oceans depends primarily on light penetration of water and nutrient availability. Coastal waters and areas of upwelling account for more than 99 percent of the oceans' fish production. Although enormous in area, open oceans are biological deserts. The oceans constitute one of our few remaining frontiers, and they will increasingly bear the brunt of increased exploitation of fish and mineral, particularly petroleum, resources.

Questions and Projects

1. In your own words, write a definition for each of the terms that are italicized in this chapter. Compare your definitions with those in the text.

2. How are the plants and animals adapted to the climate of the biome in which you live?

3. Describe the characteristics of the four horizons commonly present in a soil profile. How do those characteristics affect the number and type of plants that can grow in the soil?

4. What factors have most influenced soil development in your area?

5. What factors account for the great difference in plant productivity between tundra and deserts on the one hand and tropical rain forest ecosystems on the other?

6. Distinguish between terrestrial oligotrophic ecosystems and terrestrial eutrophic ecosystems with respect to cycling rates and their ability to recover from disturbances.

7. Why do oligotrophic ecosystems often appear to be more productive than they actually are?

8. Describe the selective advantage of evergreen leaves for plants that live in oligotrophic ecosystems.

9. Visit an arboretum, botanical garden, or a zoo, and describe the examples of convergent evolution that you find.

10. Compare altitudinal ecological zones with latitudinal ecological zones.

11. Why are ecotones particularly vulnerable to disruption by climatic variability?

12. Visit a lake. Would you describe it as an oligotrophic or a eutrophic lake? What factors are contributing to its nutrient enrichment? Is it aging slowly or rapidly?

13. Although wetlands are highly productive, regulate water flow, and remove water pollutants, most wetlands have been destroyed by human activities. Suggest reasons for this paradox.

14. Describe how natural energy subsidies contribute to the productivity of estuaries.

15. Why is the open ocean sometimes referred to as a biological desert?

16. Compare and contrast the characteristics of terrestrial oligotrophic and eutrophic ecosystems with those of aquatic oligotrophic and eutrophic ecosystems.

17. Whether an organism lives in a terrestrial or an aquatic habitat, its physical environment can be quite hostile at times. Compare and contrast the advantages and disadvantages for animals that live in those two distinct environments.

18. Which ecosystems have the greatest species diversity? Which have the least? What accounts for the differences in diversity?

19. Although all organisms require essentially the same resources from their physical environment to survive, certain resources are more plentiful in some ecosystems than in others. Give examples of ecosystems in which each of the following is plentiful as well as of ecosystems in which one or more are often limiting: light, water, oxygen, soil nutrients (such as nitrate and phosphate), and space.

20. Many ecosystems are quite interdependent. In terms of the flow of energy and nutrients, describe the interdependence of the plants and animals in a deciduous forest, in a river that flows through the forest, and in an estuary at the mouth of the river. How would those relationships be affected if the forest were cleared for agriculture or if a large industrial city were located on the river?

21. Which biomes do you believe will be most severely affected by human activities during the next 20 years? Defend that choice.

22. In which biome do you live? What other types of ecosystems are present in your region? What environmental factors are responsible for the geographical patterns of ecosystems that you find? You may wish to arrange a field trip.

Selected Readings

BOERNER, R. E. J.: Fire and nutrient cycling in temperate ecosystems. *Bioscience* 32:187–192, 1982. A description of nutrient cycling patterns in oligotrophic and eutrophic terrestrial ecosystems.

CLOUDSLEY-THOMPSON, J. L.: *Terrestrial Environments.* New York: Halsted Press, 1975. An overview of the many biomes on earth. The adaptations of organisms to the particular physical characteristics of the biome, concentrating on animals, are described.

CURRY-LINDAHL, K.: *Wildlife of the Prairies and Plains.* New York: Harry N. Abrams, Inc., 1981. A well-illustrated look at prairies and plains on several continents.

EMSLEY, M.: *Rain Forests and Cloud Forests.* New York: Harry N. Abrams, Inc., 1979. A well-illustrated view of the lush vegetation and diverse animals of the rain forest and of the dwarf variant, known as the cloud forest.

JENSEN, A. C.: *Wildlife of the Oceans.* New York: Harry N. Abrams, Inc., 1979. A fascinating look at the plant and animal life that lives in and around the world's oceans.

JORDAN, C. F.: Amazon rain forests. *American Scientist* 70:394–401, 1982. An examination of the nature of nutrient cycling in Amazon rain forests and its implications for forest management.

LAWS, R. M.: The ecology of the southern ocean. *American Scientist* 73:26–40, 1985. A fascinating look at the interactions of organisms that occupy the Antarctic ecosystem.

SALAT, E., AND VOSE, P. B.: Amazon Basin: A system in equilibrium. *Science* 225:129–137, 1984. A study of the water cycle and nutrient balance within the Amazon Basin and the implications of continued large-scale deforestation.

SCIENTIFIC AMERICAN: *Life in the Sea: Readings from Scientific American.* New York: W. H. Freeman, 1982. An informative and well-illustrated look at the organisms in the sea and the conditions in which they live.

SMITH, R. L.: *Ecology and Field Biology*, 3rd ed. New York: Harper & Row, 1980. A basic, well-written ecology textbook, containing several chapters that survey aquatic and terrestrial habitats.

SUTTON, A., AND SUTTON, M.: *Wildlife of the Forests.* New York: Harry N. Abrams, Inc., 1979. A look at the plant and animal life of forests from equatorial rain forests to boreal forests.

WAGNER, F. H.: *Wildlife of the Deserts.* New York: Harry N. Abrams, Inc., 1980. An informative, interesting examination of the animal and plant life of the desert.

PART II

ENVIRONMENTAL QUALITY AND MANAGEMENT

In Part I, we examined some fundamental principles governing natural processes. From our discussion, it is evident that both we as a species and our environment are governed by certain natural laws. For example, our surroundings have a limited capacity to absorb waste; when this limit is exceeded, natural functions are impaired or even destroyed. Likewise, when foreign matter, in the form of pollutants, enters our bodies our health may suffer.

In this part, we examine in detail the natural processes that occur in the major components of our physical environment: the waterways, the atmosphere, and the land. Then, we study how our own activities disrupt those natural processes. Finally, we discuss management techniques, including feasible alternatives to those now being practiced, and we assess our progress toward restoring areas that are degraded and preserving environmental quality in regions that are relatively undisturbed.

The cauldron left by the eruption of Mt. Mazama contains 2000-foot-deep Crater Lake. Wizard Island, a volcanic formation, is at the extreme right. (Oregon Department of Transportation.)

CHAPTER 6   THE WATER CYCLE

Earth is sometimes referred to as the watery planet because almost three-quarters of its surface is covered by water. It is not surprising, then, that water has a multitude of important functions. For example, approximately 70 percent of our body weight is water, which is the medium in which all biochemical reactions—the chemistry of life—take place. Also, water is just as crucial for modern industry as it is for each of us individually. Its use in transportation, power generation, food production and processing, manufacturing, and waste disposal are absolutely basic. Those activities simply could not occur without water.

However, less than 3 percent of the water on earth is fresh water that is usable for domestic, agricultural, and many industrial purposes. Most of the water on earth is salt water. Moreover, most of the earth's freshwater supply exists in the form of ice and snow in polar glaciers. The quantity of fresh water that is available depends on the global water cycle. The processes that are involved in that water cycle determine the natural limits to our freshwater supply and the mechanisms by which water transports materials.

Because it is essential to us in so many ways, we sometimes naively expect water to exceed its natural capabilities. For example, in recent years, we assumed that the water we used would cleanse itself of added wastes. Although an aquatic ecosystem can rid itself of some wastes if it is given enough time, many substances, such as certain pesticides and heavy metals, do not break down once they enter an aquatic environment. The retention of those pollutants in the water is a direct threat to us as well as to the many organisms that depend on an uncontaminated aquatic environment for their survival. Moreover, water pollution also threatens us through those organisms, because many are essential sources of food for millions of persons. Although detritus feeders, such as crayfish, provide a free water-cleansing service by feeding on detritus from natural sources and human-related activities, in a polluted environment, many of those organisms perish, which results in a buildup of contaminants and an increase in the numbers of less-desirable organisms.

Our concerns about water include both its quality and its quantity. It is important for us to be aware if the supply is going to diminish as the human population continues to grow. We can envision the effects of a long-term shortage of water by considering the hardships that are endured during a short-term drought. For example, the summer heat wave of 1980 affected large areas of the United States, especially in the southern Mississippi Valley and Southwest. The combination of heat and drought contributed to the loss of more than 1200 human lives and an estimated $20 billion in crops and livestock. Only three years later, in 1983, searing heat and dry weather created perhaps the worst drought east of the Rocky Mountains since the dust bowl days of the 1930s. More than 220 people died as a result of the heat, and corn production declined by 25 percent. (Those

Figure 6.1 *Clair Engle Lake, a water reservoir in California Central Valley, filled to approximately one-tenth its storage capacity during the drought of 1976–1977. Note the extensive exposed shoreline. (U.S. Department of the Interior, Bureau of Reclamation.)*

A Historical Overview

Most major areas of human settlement are located adjacent to rivers, lakes, or oceans. That settlement pattern is not coincidental; it demonstrates the utter dependency that civilizations have on adequate water supplies. A brief historical survey reveals the role of water in the development of civilization and in the evolution of modern society.

As agriculture was developed, the advantages of irrigation were discovered. For example, well-watered crops yielded larger, more reliable harvests. Later, when surplus food became available as the result of irrigated land, trade among tribes became possible. Those early commercial transactions were instrumental to the origins of early civilizations (for example, Mesopotamia, between the Tigris and Euphrates rivers, which is now Iraq; Egypt, in the Nile Valley, and the Indus culture, along the Indus River, which is in Pakistan). As civilizations developed and cities were built, finding sources of communal water became more complicated. At first, public water was

reductions do not include the impact of voluntary acreage-reduction programs that were sponsored by the federal government.) Residents of the East, too, suffered when reservoir levels remained abnormally low for several years during the late 1960s. Although the actual time of onset of a drought remains largely unpredictable, we know that virtually all parts of the United States will experience droughts from time to time (see Figure 6.1).

In this chapter, we examine the essential links in our natural water-delivery system. In Chapters 7 and 8, we concentrate on water quality: water pollution and water management, respectively.

drawn from nearby rivers or shallow wells. Later, as demands for water increased, governments supervised the construction of elaborate aqueduct systems to transport it from remote sources to the growing cities. The system of aqueducts constructed by the Romans is perhaps the most familiar of those systems. A few of those aqueducts are still being used today (Figure 6.2).

Eventually, ships were built and transportation by water proved to be less expensive and more effective than overland transportation. Those ships were able to carry many more goods with less effort than camels or horse caravans could accommodate. Because many northern European rivers were deeper and thus navigable by ships, those European civilizations eventually gained supremacy over the Mediterranean region. Not only could they transport supplies and products more easily, but they also had

easy access to water power—water wheels provided power for sawmills, iron forges, and flour mills—which enabled them to process more raw materials.

The harnessing of water power was a key to the onset of the Industrial Revolution. Then, industrialization received an added impetus when pressurized steam was recognized as a source of power. Steam-driven machinery was built that could be used at great distances from where sources of large volumes of flowing water could be found. Moreover, steam power and steam engines made railroad transportation possible. Subsequently, the invention of the versatile reciprocating steam engine revolutionized manufacturing and transportation. Although only a few steam engines are still in operation today, the successor to that remarkable engine, the steam turbine, is used in most electric-power generating plants worldwide. With the advent of electricity, large volumes of water were needed to condense the steam in power-generating plants. Thus even when the source of power shifted to nuclear and fossil fuels, the role of water was not diminished. From primi-

Figure 6.2 A segment of the Roman aqueduct system at Segovia, Spain, which is still in use today. The elevated section of the aqueduct, built of granite blocks, is approximately 1 kilometer (2700 feet) long. (Kenneth E. Maxwell, California State University, Long Beach.)

tive times until the present, water has been absolutely essential to civilization.

As the Industrial Revolution progressed, many new manufacturing processes required the use of water. For example, papermaking, steel manufacturing, petroleum exploitation and refining, food canning, and many mining operations are just a few of the processes that required (and still require) large quantities of water. Clearly, then, we are as dependent on a readily accessible water supply now as were our primitive ancestors. In fact, our need for water has increased 1000-fold, and most of that increase has occurred during the last hundred years.

Figure 6.3 indicates the extent to which modern society relies on water. The human body only requires 2 liters (2 quarts) of water a day to prevent fatal dehydration, but each American uses 630 liters (160 gallons) a day for use around the home. The quantity of water that is used by industries (excluding water that is used for the generation of electricity) is much greater than the amount that is supplied to cities and municipalities. Industry uses the equiv-

alent of 950 liters (250 gallons) of water per person every day. However, even those water withdrawals appear small when compared with the volumes of water that flow through agricultural irrigation systems. Every day, the irrigation of crops in the United States requires the equivalent of 2500 liters (650 gallons) of water per person. We should remember, however, that the supply of water is not guaranteed. In the past, persistent drought contributed to the decline of several civilizations. The Mycenaean civilization in Greece (1200 B.C.), the Harrappan civilization in India (1700 B.C.), and the American Indian Mill Creek culture (1400 A.D.), in what is now Iowa, all succumbed as a result of drought. Even our recent experiences have taught us that it is a mistake to think of our water supply as an unlimited resource.

The Water Cycle

We saw in Chapter 2 that ecosystems are dependent on the continual cycling of materials and energy. Water plays a key role in all of those cycles in various ways, which are discussed in the following sections.

Water occurs in solid, liquid, and vapor forms, and it is distributed among oceanic, atmospheric, and terrestrial reservoirs (Table 6.1). The total volume of water on earth has remained nearly constant for several hundred million years. Each year a very small amount of water vapor is decomposed by sunlight to hydrogen and oxygen in the upper atmosphere. The hydrogen escapes into space. That loss of water is compensated for by the addition of water vapor through volcanic emissions.

Air–Water Interaction

The ceaseless flow of water among oceanic, atmospheric, and terrestrial reservoirs, which is called the *water cycle*, or *hydrologic cycle*, is illustrated in Figure 6.4. That cycle is sustained by energy from the sun. The heat that is derived from solar energy causes water to vaporize from the sea and the land, and that vapor condenses as clouds, from which rain and snow fall to earth, to resupply the rivers, which then flow back to the sea. The endlessness of the water cycle is expressed in a verse from Ecclesiastes:

Figure 6.3 *Trends in freshwater withdrawals by various groups in the United States. (National Water Summary, 1983—Hydrologic Events and Issues. U.S. Geological Survey Water Supply Paper 2250, 1984.)*

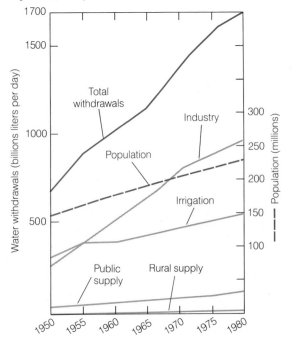

Table 6.1 The distribution of the earth's water reservoirs

MODE OF STORAGE	LOCATIONS	WATER VOLUME (IN CUBIC METERS)	PERCENTAGE OF TOTAL WATER
Surface	Freshwater Lakes	1.3×10^{14}	0.006
	Saline Lakes and Inland Seas	1.0×10^{14}	0.008
	Average in Stream Channels	1.3×10^{12}	0.0001
Subsurface	Water in Zone of Aeration	6.7×10^{13}	0.005
	Groundwater within Depth of Half a Mile	4.2×10^{15}	0.31
	Groundwater at Great Depths	4.2×10^{15}	0.31
Other	Icecaps and Glaciers	2.9×10^{16}	2.15
	Atmosphere (at Sea Level)	1.3×10^{13}	0.001
	World Ocean	1.3×10^{18}	97.2
Totals		1.4×10^{18}	100

Source: Data from U.S. Geological Survey.

All the rivers run into the sea, yet the sea is never full; unto the place from whence the rivers come thither they return again.

Water is transported from the earth's surface into the atmosphere by means of evaporation, transpiration, and sublimation. *Evaporation* is defined as the process by which water changes from a liquid to a vapor at a temperature that is below the boiling point of water. Water evaporates from all open bodies of water as well as from the wetted surfaces of plant leaves, plant stems, and the soil. Direct evaporation of water from the oceans is by far the largest source of atmospheric water vapor.

Transpiration, as we saw in Chapter 3, is the loss of water vapor from plants. On land, transpiration is considerable. It is often more important than direct evaporation from lakes, streams, and the soil surface. In fact, a single hectare (2.5 acres) of growing corn typically transpires 35,000 liters (8800 gallons)

Figure 6.4 *The water cycle. The world's water is distributed among its oceanic, terrestrial, and atmospheric reservoirs. Most of it (some 13 million trillion cubic meters) is found in the ocean (see Table 6.1).*

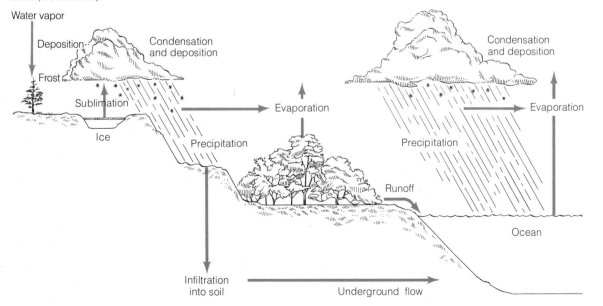

of water each day. Usually, measurements of water loss from the land include both evaporation and transpiration, which is collectively called *evapotranspiration.*

Sublimation is defined as the process by which water changes directly from a solid to a vapor without passing through an intervening liquid phase. The gradual disappearance of snowbanks during periods when the temperature remains well below freezing is an example of sublimation. Energy for evaporation, transpiration, and sublimation is supplied directly by solar radiation.

Water is returned to the land and the sea from the atmosphere by means of condensation, deposition, and precipitation. *Condensation* is defined as the process by which water changes from a vapor to a liquid (in the form of droplets). *Deposition* is defined as the process by which water changes directly from a vapor into a solid (ice crystals). In the atmosphere, water droplets and ice crystals that are produced by condensation and deposition are visible as clouds. On the land, condensation or deposition of water is visible as dew or frost, respectively. *Precipitation* is defined as the portion of atmospheric water that returns from clouds to the earth's surface in the form of rain, snow, ice pellets, and hail. At the earth's surface, most of the precipitation eventually vaporizes and is returned to the atmosphere.

Water is purified by the evaporation–condensation and the sublimation–deposition sequences. When it vaporizes, the suspended and soluble substances, such as sea salts and microorganisms are left behind. Thus the condensate, which is almost entirely free of those substances, forms freshwater precipitation.

Humidity, Evaporation, and Condensation

"It's not the heat, it's the humidity," is a familiar statement, by which we correctly ascribe the discomfort we feel on a hot, muggy day to the content of water vapor in the air. The water-vapor concentration in air is an important determinant of our physical comfort, and we know from experience that it is quite variable. In this section, we examine some ways in which the water-vapor content of air can be measured.

Simply stated, *vapor pressure* is the pressure exerted solely by the water vapor that is present in a sample of air. Vapor pressure is the result of the exchange of water molecules occurring at the interface between air and water or between air and ice. At the air–water interface, some water molecules escape from the surface of the water and become vapor, while other molecules escape from the vapor and return to the water surface as liquid. *Net evaporation* occurs if more water enters the air as vapor than returns to the liquid state. *Net condensation* occurs if more water enters the liquid state than enters the atmosphere as vapor.

The same type of two-way exchange occurs at the interface between ice and air, except that water molecules do not escape as readily from an ice surface as they do from a liquid surface. Net sublimation occurs when more water enters the atmosphere as vapor than returns in the form of ice; and net deposition takes place when less water becomes vapor than returns as ice.

When water molecules enter the air as vapor, the molecules disperse very rapidly, spreading throughout the air. In the process, water vapor mixes with the other atmospheric gases and thereby contributes to the total pressure exerted by air. Basic laws describing the behavior of gases indicate that we may consider the contribution of water vapor to the total pressure separately from the other gases in air. The partial pressure that is exerted by water vapor alone is a very small fraction of the total air pressure.

An upper limit exists, which determines the maximum concentration of water vapor that can remain in the air. When air contains its maximum water-vapor concentration, it is said to be *saturated.* The *saturation vapor pressure* increases as the air temperature increases. Thus more water vapor can be held by warm air than by cold air. That temperature dependence of the saturation vapor pressure is not surprising, because temperature regulates the motion of all molecules, including the escape of water molecules from a water surface. The higher the temperature, the greater the molecular activity, and, therefore, the more rapid the escape of water molecules from the surface of water as vapor.

We can also describe the water-vapor concentration in air by the mass of water vapor per mass of dry air (grams of water vapor per kilogram of dry air) (Figure 6.5). This measure of water-vapor concentration is called the *mixing ratio,* since it specifies how much water vapor is mixed with other gases in air. The *saturation mixing ratio* is the maximum concentration that water vapor can reach in air. As with saturation vapor pressure, the saturation mixing ratio is proportional to the temperature.

Relative humidity is perhaps the most familiar way of describing the amount of water vapor in the air. At a particular temperature, the relative humidity is determined by comparing the actual concentration of water vapor in the air with the concentration of water vapor at its point of saturation. Relative humidity usually is expressed as a percentage and can be calculated approximately from either the vapor pressure or the mixing ratio using the following equations:

Percent relative humidity

$$= \frac{\text{vapor pressure}}{\text{saturation vapor pressure}} \times 100$$

or

Percent relative humidity

$$= \frac{\text{mixing ratio}}{\text{saturation mixing ratio}} \times 100$$

To calculate the relative humidity for a muggy summer afternoon, we can measure the temperature and the vapor content of the air. Thus suppose that we make some measurements and find that the temperature is 32°C and that the vapor content is 20 grams of water vapor per kilogram of dry air. We can see in Figure 6.5 that the saturation mixing ratio of air at 32°C is 30 grams per kilogram. By using the mixing ratio equation, we can determine that the relative humidity is 67 percent. Thus our 1 kilogram air sample is capable of holding 10 grams more of water vapor before it becomes saturated at that temperature.

Achieving Saturation

The movement of water from the atmosphere back to the earth's surface (precipitation) depends on events that occur in the atmosphere, which produce conditions that are near saturation. (At saturation, the relative humidity is 100 percent.) But it is possible for air to achieve *supersaturation* as well, that is, for the relative humidity to exceed 100 percent. As the relative humidity approaches or exceeds satu-

ration, condensation or deposition of water vapor becomes increasingly probable. Since clouds are composed of condensation (water droplets) or deposition (ice crystals) products, or both, the possibility that clouds will form becomes more likely.

Precipitation cannot form without clouds, which, in turn, cannot form unless high relative humidity exists, which will initiate condensation or deposition. The relative humidity will increase when (1) the air is cooled, which reduces its capacity to hold water vapor, or (2) more water vapor is added to the air. The more common way for air to approach saturation is by cooling.

One way that air will cool (and its relative humidity will increase) is when it expands. In fact, expansional cooling is the most important means of cloud formation in the atmosphere. *Expansional cooling* occurs when the pressure on a volume of air is decreased, such as when air ascends within the

Figure 6.5 *The saturation mixing ratio (solid line) rises as temperature increases and is a measure of the weight of water that dry air can hold at its point of saturation. The dashed line indicates that at 32°C, 1 kilogram of dry air requires 30 grams of water vapor for saturation.*

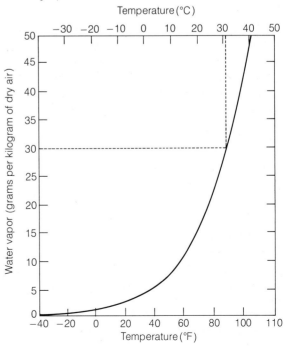

atmosphere. Rising air expands, just as a helium-filled balloon expands (and eventually bursts), as it drifts skyward. For example, even on a hot day, air that is released from a tire will feel cool, because the compressed air expands as it escapes from the tire. Conversely, when air is compressed, its temperature rises (and its relative humidity decreases). For example, the cylinder wall of a bicycle pump will become warm as air is pumped (compressed) into a tire. Thus *compressional warming* occurs when air descends (and compresses somewhat) within the atmosphere. Therefore, a rising air mass cools, which leads to an increase in relative humidity; a descending air mass grows warmer, which leads to a decrease in the relative humidity.

Four major processes lead to expansional cooling in the atmosphere: (1) lifting by an advancing warm front, (2) lifting by an advancing cold front, (3) convection cell development, and (4) orographic lifting (Figure 6.6).

Clouds and often precipitation are triggered by frontal uplift, which occurs when masses of warm and cold air meet. If the leading edge of a moving air mass that is warm and humid (a *warm front*) encounters a cold air mass, the warm air will ride over the cold air because the warm air is less dense. This uplift results in expansional cooling and may lead to cloud formation and subsequent precipitation (Figure 6.6a).

Because cold air is denser than warm air, the leading edge of a moving mass of cold air (a *cold front*) slides under a warm air mass, thereby displacing warm air upward. Here, too, the rising air experiences expansional cooling, and the result may be cloud formation (Figure 6.6b).

Convection can also lead to the formation of clouds (and precipitation). *Convection* cells develop when the sun heats the earth's surface and the warm surface heats the air that is in contact with it. The warm air rises, expands, and cools. Eventually, the

Figure 6.6 *The four mechanisms whereby relative humidity increases, leading to cloud formation: (a) warm frontal lifting, (b) cold frontal lifting, (c) convective cell circulation, and (d) orographic lifting.*

(a)

(b)

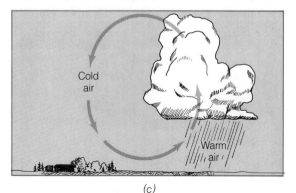

(c)

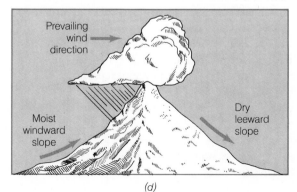

(d)

air becomes cool and dense enough so that it begins to sink back toward the surface, thereby forming a convective cell circulation (Figure 6.6c). Clouds may form where convective air currents move in an upward direction, but clouds dissipate where air currents move downward. In general, the higher the altitude that the convection cell reaches, the greater the expansional cooling and the greater the likelihood that clouds and precipitation will form. Convective currents that reach to great altitudes can spawn severe thunderstorms.

As winds sweep across the earth, irregularities in *topography*—the physical relief of the land—force moving air to rise and fall. *Orographic lifting* refers to the rising motion of air that is induced by topography. If the ascent of air is great enough, its expansional cooling and compressional warming will affect the patterns of cloud formation and precipitation. For example, a mountain range that is perpendicular to the direction of the prevailing wind will form a natural barrier, which will result in heavier precipitation on the windward side of the range (the side facing the wind). As air is forced to rise, it will expand and cool, which will increase the relative humidity, condensation may occur, and, eventually, precipitation (called *orographic rainfall*) may develop. On the leeward side (the side away from the wind), however, air will descend and warm, which will reduce the relative humidity, so that cloud formation and precipitation will probably not occur. Hence the mountain range encompasses two contrasting climatic zones: a moist climate on the windward side and a dry one on the leeward side (Figure 6.6d).

Orographically induced disparity in precipitation is especially apparent from west to east across Washington and Oregon, where the north–south Cascade Range intercepts the prevailing moist air that flows from the Pacific. As a result of the location of those mountains, exceptionally rainy conditions are prevalent in the western portions of those states; arid conditions prevail in the eastern portions. Figure 6.7 shows two contrasting landscapes from Oregon, which exhibit this disparity. On Mount Waialeale, Hawaii, the contrast is spectacular. Annual rainfall varies from 11,700 millimeters (460 inches, which is the heaviest in the world) on one side of the mountain to only 460 millimeters (19 inches) on the other side. In every case, orographic rainfall results in markedly different plant and animal communities on opposing slopes of a mountain barrier. It also has a direct impact on the domestic water supply, the type of crops that can be grown, and the type of human shelter that must be built. For example, the city of Denver, which is located on the leeward side of the front range of the Colorado Rocky Mountains, must obtain some of its water from the wet side by means of a 37-kilometer (23-mile) tunnel through the mountains.

Precipitation Processes

In their myriad forms, *clouds* are visible manifestations of condensation and deposition of water vapor within the atmosphere. They are composed of tiny water droplets, ice crystals, or both.

Condensation and deposition take place on minute particles—*condensation nuclei*—that are suspended in the atmosphere. Those nuclei are the products of both natural and human activity. For example, forest fires, volcanic eruptions, wind erosion of soil, saltwater spray, and effluents of domestic and industrial chimneys provide a continual supply of those particles.

Hygroscopic nuclei are condensation nuclei that have a particularly strong affinity for water molecules. Condensation begins on those nuclei when the relative humidity is below 100 percent. In fact, magnesium chloride, a constituent of seawater, is a hygroscopic substance that initiates condensation when the relative humidity is as low as 70 percent.

However, just because clouds develop, no guarantee exists that it will rain or snow. In fact, most clouds do not yield any precipitation, because a special combination of circumstances, which are not yet completely understood, are required in clouds before precipitation will occur.

The water droplets, ice crystals, or both, which make up clouds, are so minute that they remain suspended indefinitely unless they vaporize or somehow grow considerably larger. Updrafts that occur within clouds usually are strong enough to prevent them from falling toward the earth's surface. And, even if droplets or ice crystals descend from a cloud, their velocities are so slow that they travel only a short distance. Then they vaporize in the unsaturated air that exists beneath the cloud. Hence for precipitation to occur, the cloud particles must grow large enough to counter updrafts and then survive a descent to the earth's surface in the form of raindrops or snowflakes. It takes about 1 million cloud droplets in the 5- to 10-micrometer (a micrometer equals 10^{-6} meters or 1/one-millionth of a meter) range to form a single raindrop (approximately 1 millimeter in radius). A great deal of experimentation has been done to determine how this growth takes place.

Cloud physicists have discovered that condensation alone cannot result in the precipitation of raindrops or snowflakes. Thus they have identified two important processes whereby cloud particles will grow large enough to precipitate: the collision–coalescence process and the Bergeron process.

The *collision–coalescence process* occurs in some *warm clouds*, which exist at temperatures that are above the freezing point of water. For precipitation to form in those clouds, some relatively large droplets (larger than 20 micrometers in radius) must be present, which may be produced, for example, by "giant" hygroscopic sea-salt nuclei. When they fall, those large droplets fall more rapidly than the smaller droplets that compose the cloud. Furthermore, as

(a)

Figure 6.7 *(a) A rain forest in Western Oregon's Silver Falls State Park. (b) A desert in eastern Oregon. A mountain range between the two areas results in differences in precipitation, which, in turn, account for differences in vegetation. (Oregon Department of Transportation.)*

(b)

they fall, large droplets intercept and coalesce with smaller water droplets that occur in their paths. Collision and coalescence is repeated perhaps a million times before the droplet is large enough to fall out of the cloud in the form of a raindrop.

However, it is unusual for clouds to contain droplets that are large enough to initiate the collision–coalescence process. In fact, more precipitation originates through the *Bergeron process*. That process, named for the Scandinavian meteorologist who first described it (around 1930), applies to *cold clouds*, which exist at temperatures that are below 0°C (32°F). Also, the coexistence of water vapor, ice crystals, and *supercooled water* droplets (water that remains in the liquid state despite the fact that it is below the freezing point) is required.

The Bergeron process depends on the saturation vapor-pressure difference between ice and water, which triggers the growth of ice crystals at the expense of supercooled water droplets. As we noted

earlier, water molecules escape (vaporize) from liquid water more readily than from solid ice at temperatures that are below freezing. Hence water evaporates from supercooled water droplets and deposits on ice crystals, which then begin to grow.

As ice crystals grow larger and thus heavier, they begin to fall at an accelerating rate. Then as they fall, they collide and coalesce with supercooled water droplets and smaller ice crystals, which they encounter in their paths, thereby growing still larger. Eventually, the ice crystals become heavy enough to counter updrafts and then fall out of the cloud. If the air temperature is below freezing all the way, or even most of the way, to the ground, the crystals reach the earth's surface in the form of snow. If the air temperature beneath the cloud is above the freezing point, the crystals melt and fall as rain.

Once a raindrop or a snowflake leaves a cloud, it enters a hostile environment in which either evaporation or sublimation occurs. In general, if the length of time that is required for the precipitation to reach the ground is extensive and if the air in its path is dry, more water is returned to the atmosphere as vapor. In Figure 6.8, rain falling from a

Figure 6.8 *During a thunderstorm, not all the raindrops leaving the cloud reach the ground. Because the drops fall through unsaturated air, some evaporate. (National Center for Atmospheric Research.)*

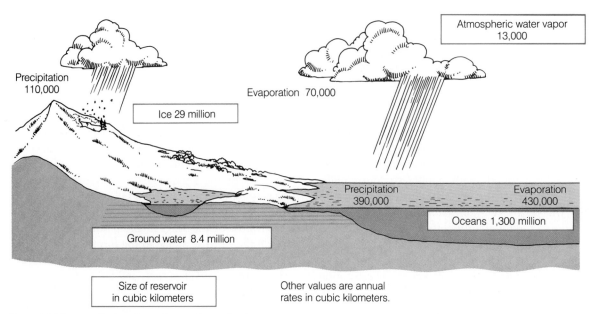

Figure 6.9 *The water budget shows that, each year, land areas receive 40,000 cubic kilometers more precipitation than is lost through evaporation and transpiration from land. The disproportionate share comes from the oceans. (R. P. Ambroggi, "Water,"* in Economic Development, *A Scientific American Book, New York: W. H. Freeman, 1980.)*

distant thunderstorm enters drier (unsaturated) air. As a result, some of the raindrops evaporate as they descend. Partly for this reason, more rainfall reaches the ground in highlands than in the surrounding lowlands.

The Water Budget

When water returns from the atmosphere to the land and the sea in the form of precipitation, an essential subcycle in the global water cycle is completed. To learn more about the water cycle, we can compare the flow of water as it moves into and out of terrestrial reservoirs with the flow of water as it moves into and out of the sea. The balance sheet for the total amount of input and output of water that moves both to and from the various global reservoirs is called the *water budget* (Figure 6.9). Each year on the continents, the total amount of precipitation exceeds evapotranspiration by approximately one-third. But at sea, the annual amount of evaporation exceeds the amount of precipitation. Hence over the course

of a year, the water budget indicates a net gain of water on land and a net loss of water from the oceans, where the excess on land is approximately equal to the deficit for the oceans. The land, however, is not getting soggier, any more than the world's oceans are drying up, because the excess precipitation on the land drips, seeps, and flows back to the sea.

The Earth's Water Reservoirs

Once precipitation reaches the earth's surface, it follows one or more of the various routes shown in Figure 6.10. On its journey from the land to the sea, water is often stored for varying periods of time in one of several kinds of reservoirs (snowfields, the ground, rivers and lakes, or glacial ice). All those reservoirs are potential sources of fresh water for agriculture or domestic use. To understand better how contamination and overexploitation affect those freshwater resources, we now examine their characteristics and functions.

Groundwater

More than 97 percent of the unfrozen resources of fresh water in the United States is contained in underground reservoirs. Groundwater that lies within

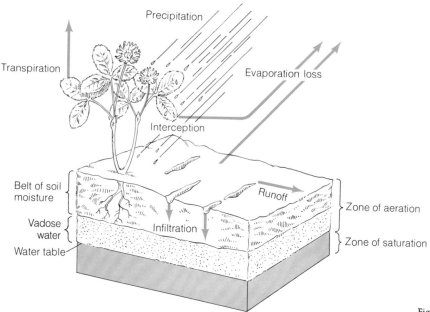

Figure 6.10 *The routes of precipitation.*

1000 meters (3300 feet) of the earth's surface is considered to be economically recoverable. Its volume is more than nine times that of the Great Lakes. At present, approximately 50 percent of the United States population obtains its water supply from groundwater sources. As shown in Figure 6.11, irrigation accounts for approximately 68 percent of groundwater withdrawals. Only 13 percent is drawn off for public water supplies, which is the next highest use. Because the quality of much surface water is deteriorated, it is all the more important that we maintain the quality of groundwater. In many areas, however, human activities threaten both the quantity and the quality of groundwater.

The Flow of Groundwater *Groundwater* collects underground. Not all the precipitation that reaches the earth's surface, however, finds its way into underground reservoirs (Figure 6.10). A portion of it evaporates, while the remainder runs off into streams or rivers. Precipitation that does reach the earth's surface and that seeps into the ground is called *infiltration*. The relative amounts of water that infiltrate into the ground depend on the total amount of precipitation as well as the topography, the climate, and the physical properties (e.g., coarseness or fineness) of the soil.

Figure 6.11 *Groundwater withdrawals by various groups in the United States (National Water Summary, 1983—Hydrologic Events and Issues. U.S. Geological Survey Water Supply Paper 2250, 1984.)*

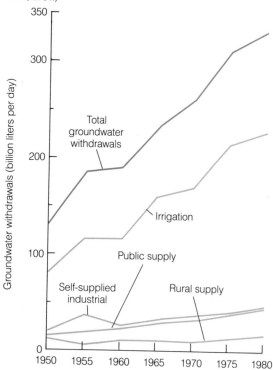

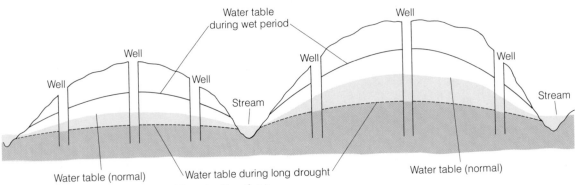

Figure 6.12 *The elevation of the water table varies. Note that it tends to follow topography and that it changes during extended wet or dry periods.*

Infiltrating water typically accumulates in two zones within the upper soil and rock layers of the earth's crust. In the uppermost band, the *zone of aeration*, pore spaces (that is, the spaces between soil particles and rocks) contain both water droplets and air. The water that is required by most land plants is supplied by the moisture that is held in the uppermost division of that zone, which is called the *belt of soil moisture*. Most of the moisture in that belt is lost to the atmosphere as a result of evaporation and transpiration. The rest of it adheres tightly to soil particles and, therefore, is not available to plants.

Under the influence of gravity, moisture that remains in the soil is gradually drawn down through the lower portion of the zone of aeration (the vadose zone) and accumulates in the *zone of saturation*. Pore spaces and fractures in the rocks and soil that are in the zone of saturation are completely filled with water. Thus the zone of saturation constitutes the *groundwater reservoir*, and the surface that separates it from the zone of aeration is called the *water table*. To withdraw groundwater, wells must be dug or drilled deeply enough to penetrate the water table and enter the zone of saturation.

The water table usually parallels the overlying topography, as illustrated in Figure 6.12. Its level changes in response to extended wet or dry periods. In most areas, the zone of saturation is less than 30 meters (100 feet) below the surface, but in some areas, especially in arid regions, wells must be drilled several hundred meters before the water table is intercepted. When groundwater is withdrawn at rates that exceed natural or artificial recharge, the water table drops, and the groundwater is eventually depleted.

The property of rock and soil that allows it to transmit water in the zones of aeration and saturation is called *permeability*. The degree of permeability that exists is determined by the volume of pore space (porosity) as well as by the interconnections between pores. Layers of sediment and bedrock that have high permeability and transmit water are called *aquifers*. In general, layers of sand and gravel are good aquifers, whereas clay and most crystalline rocks, such as granite, that have low permeability (unless they are highly fractured) are poor aquifers.

In some regions, aquifers are sandwiched between two layers of impermeable rock, as illustrated in Figure 6.13. If the aquifer exists along an incline and a well is dug at a point where the water is under pressure, such as at a point that is near the bottom of the incline, water will flow freely from the well. Free-flowing wells, which are called *artesian wells*, are of particular value because they do not require an energy source and expensive pumps.

Groundwater flows naturally through the permeable layers toward various points of discharge, such as rivers, lakes, and seas. That movement is extremely slow, however. A speed of 15 meters (50 feet) per year is typical, but the speed varies widely from one location to another. That slow movement allows groundwater to flow in smooth, continuous paths with little mixing, as shown in Figure 6.14. Some of the water that seeps in from the soil surface flows straight downward, and some of it flows toward either side, in this case, supplying water to a stream and a marsh. Also, marshes, springs, and some streams and lakes are formed when the water table

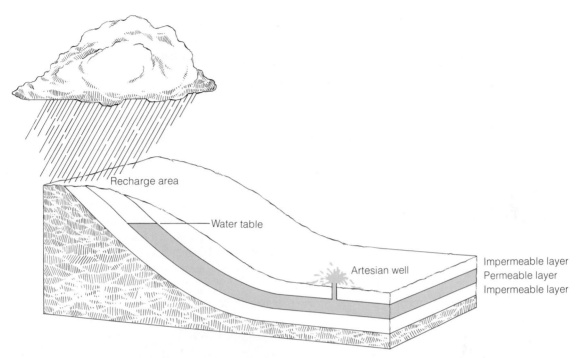

Figure 6.13 *A cross section of the types of rock strata and conditions required for artesian wells. At some point, the aquifer must be at a higher level than the top of the well for water to flow freely.*

Figure 6.14 *Patterns of groundwater flow.*

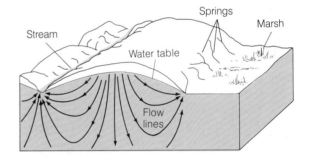

either intercepts or is higher than the land surface. In those situations, no zone of aeration exists. During long dry spells in which the water table is much lower than a stream bed, the stream may supply water directly to the groundwater reservoir. Such streams may dry up completely unless periods of heavy rains occur.

When groundwater is pumped out, the vacated volume forms a *cone of depression* in the water table (Figure 6.15), which does not fill quickly, because groundwater flows slowly. Furthermore, if wells are too close to each other, heavy withdrawal from one well will cause an increase in the size of the cone of depression, sometimes to the point that adjacent wells run dry.

Figure 6.15 *A cone of depression in the water table results when groundwater is withdrawn. Heavy pumping causes shallow wells in the vicinity to go dry.*

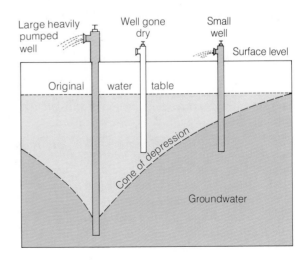

Another consequence of the slow, smooth movement of groundwater is that mixing and dilution of pollutants is severely inhibited once they enter groundwater aquifers. The United States Geological Survey reports that contaminants may remain in some subsurface aquifers for hundreds of years, fouling the water supply for generations to come. We consider some specific groundwater pollution problems in Chapter 7.

Groundwater Depletion Groundwater reservoirs are depleted if the rate of withdrawal exceeds the rate

---

BOX 6.1

EBBING OF THE OGALLALA: THE GREAT WATERING HOLE BENEATH THE GREAT PLAINS IS GOING DRY

There are two documentary images of the Great Plains. The first is a black-and-white photograph of the '30s Dust Bowl, with windblown homesteaders treading the cracked earth. The second: a glossy color shot of the same land 40 years later, showing the lush checkerboard farms of America's breadbasket. Now, as if through a strange reversal in time, the second image threatens to fade into the first. For in another 40 years, the territory could backslide into dust and dispair. The Ogallala Aquifer, the vast underground reservoir of water that transformed much of the Great Plains into one of the richest agricultural areas in the world, is being sucked dry.

The aquifer is a California-size deposit of water-laden sand, silt, and gravel. It ranges in thickness from 1000 ft in Nebraska, where two-thirds of its waters lie, to a few inches in parts of Texas. Although it was first tapped in the 1930s, it has been extensively exploited only since the development of high-capacity pumps after World War II. The Ogallala's estimated quadrillion gallons of water, the equivalent of Lake Huron, have irrigated farms in South Dakota, Nebraska, Wyoming, Kansas, Colorado, Oklahoma, Texas, and New Mexico, changing a region of subsistence farming into a $15 billion-a-year agricultural center.

For the past three decades, farmers have pumped water out of the Ogallala as if it were inexhaustible. Nowadays they disperse it prodigally through huge center-pivot irrigation sprinklers, which moisten circular swaths a quarter-mile in diameter. The annual overdraft (the

amount of water not replenished) is nearly equal to the yearly flow of the Colorado River. Like all aquifers, the Ogallala depends on rain water for recharging, and only a trickle of the annual local rainfall ever reaches it. Gradually built up over millions of years, the aquifer is being drained

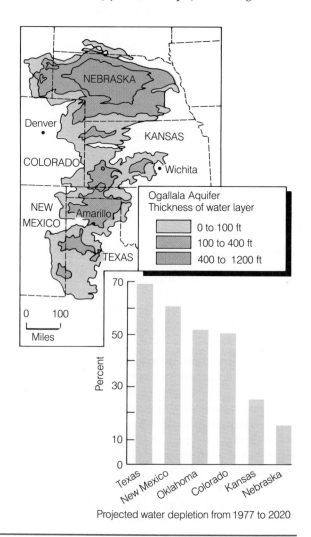

Projected water depletion from 1977 to 2020

of natural recharge (see Box 6.1). Such a situation now prevails in the High Plains region of western Texas. In the vicinity of Lubbock, Texas, over 14,000 square kilometers (5500 square miles) are irrigated by water that is pumped from the ground. Because evapotranspiration rates are high and precipitation rates are low in that area, little surface water is available for groundwater recharge. As a result, withdrawal is occurring 50 times faster than natural recharge. The water table is also dropping dramatically in the Houston, Texas area (see Figures 6.16 and 6.17). In both those regions, the withdrawal of

---

in a fraction of that time. The question is no longer if the Ogallala will run dry, but when.

W. B. Criswell has been raising cotton on his 1700-acre farm in Idalou, Texas, since 1955. Cotton is called the camel of crops because it requires little water, yet Criswell is now in trouble. His water table has dropped 100 ft since he started farming. Nine years ago, he paid $4 an acre to water his cotton; today he pays more than $45. "It's like a disease," he says. "You just accept it and go on." Gerald Wiechman farms 6000 acres and feeds 2500 head of steer near Scott City in western Kansas. When his farm's first well started pumping, it tapped water at 54 ft. Today he has to go 130 ft. "I've got another 20 years, maybe," he reckons. On the High Plains of eastern Colorado, the water level has dropped as much as 40 ft since the 1960s. In parts of Oklahoma it has dipped that much in 4 years. Texas, the thirstiest of the eight states, has consumed 23 percent of its Ogallala reserves since World War II.

According to a major study just completed by Camp Dresser & McKee, a Boston engineering firm, 5.1 million acres of irrigated land (an area the size of Massachusetts) in six Great Plains states will dry up by the year 2020. If current trends continue, Kansas will lose 1.6 million irrigated acres, Texas 1.2 million, Colorado 260,000, New Mexico 224,000, Oklahoma 330,000. Yet this drastic estimate, declares Herbert Grubb of the Texas department of water resources, is "20 percent too optimistic."

"When the water goes," says W. E. Medlock, a stoic, third-generation farmer from Lubbock, Texas, who has lost 47 of his 73 wells in 10 years, "we'll just go back to dry-land farming." To the farmers of the Great Plains, those words summon up visions of The Grapes of Wrath. *Dry-land farming means larger farms with lower yields, fewer workers, and probably higher prices in the supermarkets. Cattlemen know that less water means less corn and therefore smaller herds. Grubb calls such farming the "Russian roulette" of agriculture. Over a 10-year period, he says, dry-land farming will yield two strong harvests, four average ones, and four "busts."*

Although the cause of the trouble is obvious, the cure is not. Indeed, there may be no fundamental solution to the ebbing of the Ogallala. "We can prolong the supply," concludes John Weeks, a U.S. Geological Survey engineer who heads a 5-year U.S.G.S. study of the situation, "but we are mining a limited resource, like gold, and we can't solve the ultimate depletion problem."

Conservation may forestall the end. Farmers can simply use less water. They are already converting from profitable but water-thirsty corn to water-thrifty crops, such as wheat, sorghum, and cotton. James Mitchell, a cotton farmer from Wolfforth, Texas, has installed an experimental center-pivot sprinkler that, instead of spraying outward, gently drops water directly into the planted furrows, thereby reducing evaporation. Sophisticated laser-guided land graders can now almost perfectly flatten the terrain so that water is not wasted in runoff. Electrodes planted in the fields can measure soil wetness and determine exactly when water is needed. Today, these techniques are rarities, but they may soon be routine. As Kansas cattle feeder Harold Burnett puts it: "Water misers" will last longer. But even the stingiest will go under if neighbors are wasteful and the whole aquifer dries up.

Excerpted from Richard Stenogl, "Ebbing of the Ogallala: The Great Watering Hole Beneath the Great Plains Is Going Dry." Time Magazine 119:98–99 (May 10), 1982. Copyright 1982. Time, Inc. All rights reserved. Reprinted by permission from TIME.

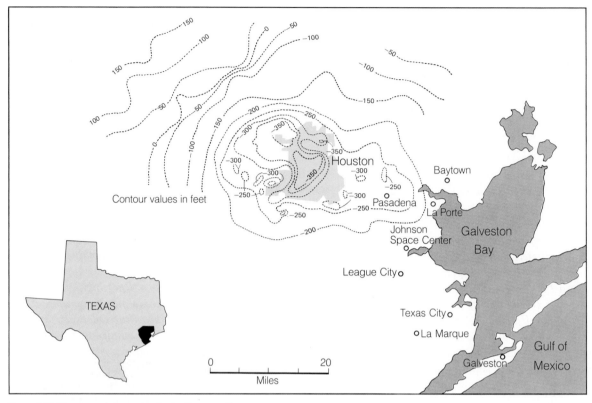

Figure 6.16 *The cone of depression in the Houston, Texas, area in 1982 is shown by the approximate water levels in wells. (The data shown are for the Evangeline aquifer, which underlies another aquifer.) (Department of Interior. U.S. Geological Survey Open File Report 82-559.)*

water is appropriately called *groundwater mining*, since the water reservoir is being depleted. Also, as the water table is falling, the pumping costs are increasing by approximately $0.70 per hectare for each additional meter ($0.10 per acre per foot) that the water must be raised. Those increased costs are slowly forcing farmers to stop raising crops (such as alfalfa and barley) that require a great deal of water and to start raising less profitable dry-land crops (such as cotton). Although efforts to conserve water may ease the problem somewhat, the region must soon begin to make some painful economic adjustments. A Texas Agricultural Experiment Station study of the Lubbock area predicts that by the year 2015, irrigated acreage will have to be decreased by 95 percent. Between now and then, agricultural production is ex-

pected to decline by 70 percent in that area, and, undoubtedly, other businesses in the Lubbock area will also suffer. Water mining occurs in other areas as well (see Figure 6.18), including the Mojave Desert of California; the cold-desert regions of Nevada, Utah, and Oregon; western Kansas and Nebraska; and eastern Colorado. In New York and Pennsylvania, groundwater depletion in the Upper Susquehanna region is reducing stream flows during periods of subnormal rainfall.

In regions where fresh groundwater is located adjacent to salty groundwater, overwithdrawal of fresh water can cause the salt water to migrate toward the fresh groundwater reservoir (Figure 6.19). The result is that fresh water is replaced by salt water in the aquifer, and wells suddenly begin delivering salt water. That phenomenon, which is called *salt-water intrusion*, is especially prevalent in flat coastal plains, such as in Florida and southeastern Georgia. Salt-water intrusion is also possible inland, where un-

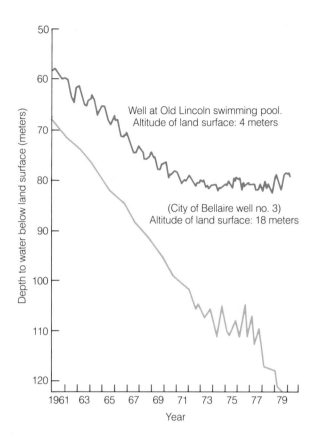

Well at Old Lincoln swimming pool.
Altitude of land surface: 4 meters

(City of Bellaire well no. 3)
Altitude of land surface: 18 meters

Figure 6.17 *Changes in water levels in two wells in the Houston area. Some wells in the area are beginning to recover as more water is used from surface sources. Water levels in other wells continue to decline because overwithdrawal continues. (U.S. Geological Survey Open File Report 82-431, p. 17.)*

Figure 6.18 *Regions in the lower 48 states where the water table or the artesian water level declined by at least 13 meters (40 feet) since predevelopment. Smaller, but significant, declines occurred in many other parts of the country (see, for example, Box 6.1). (National Water Summary, 1983—Hydrologic Events and Issues. U.S. Geological Survey Water Supply Paper 2250, 1984.)*

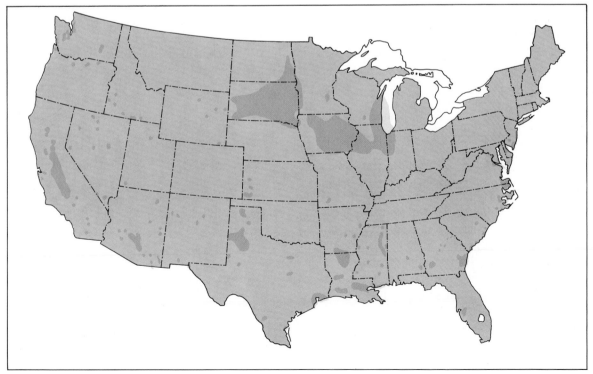

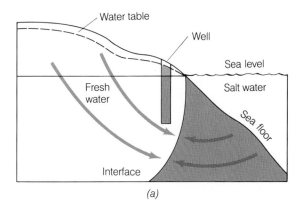

(a)

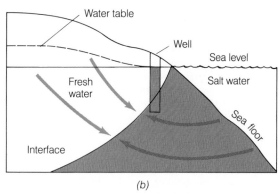

(b)

Figure 6.19 *(a) Cross-sectional diagrams of the interface between fresh and salt groundwater along a coastal region. (b) The fresh-water–saltwater interface moves inland as groundwater over-withdrawal proceeds. Note that the well, which formerly furnished fresh water, now supplies salt water.*

derground brines (water that contains more salt than ocean water) are located next to fresh water.

Another problem, which is related to groundwater withdrawal that exceeds recharge, is *ground subsidence*. Sometimes, when groundwater is removed, the earth materials that constitute aquifers become compacted, which causes the overlying land to subside. Ground subsidence often accompanies heavy withdrawal of groundwater, and it is occurring in several regions at this time (Figure 6.18). Multimillion-dollar losses can result from ground subsidence, since it often causes structural damage to buildings and pipe networks, changes in the direction of flow in sewers and canals, and increased tidal flooding along seacoasts. For example, Houston

and Mexico City have already experienced such problems. A particularly dramatic example of subsidence has occurred southeast of Phoenix, Arizona, where some 310 square kilometers (120 square miles) of land have sunk more than 2 meters (7 feet). Railroads, highways, and homes were damaged when cracks developed in the ground that were 3 meters deep, 3 meters wide, and 300 meters long (10 × 10 × 1000 feet). Further damage is expected as urbanization expands into the fissured area and as the falling water table causes new fissures to develop. In California's San Joaquin Valley, at least 30 percent of the land has subsided at least 0.3 meter (1 foot), which is enough to disrupt irrigation wells. The damaged wells must be replaced, typically at a cost of $35,000 each.

Streams and Rivers

Water that does not infiltrate into the ground or evaporate remains on the surface of the earth (as *surface water*) and becomes runoff. The runoff then travels as streams and rivers (the major pathways taken by the runoff component of the water cycle) in its journey from the land to the sea. It enters the stream and river channels by overland movement following a rainfall or snowmelt. Groundwater seepage, springs, and precipitation that falls directly into the stream also contribute to stream flow. The land area that delivers the water, sediment, and dissolved substances to a common point is called a *drainage basin*, or *watershed*. The geographical region of a drainage basin is defined by the river and the tributaries that drain it. The major drainage basins of the contiguous United States are shown in Figure 6.20. The climate, vegetation, and topography of a drainage basin, along with upsets that are caused by natural processes and human activities, affect the quantity and quality of river water. Those factors are discussed more fully in Chapter 7. Of particular concern to us here is the variety of ways in which human beings exploit river resources. Although both groundwater and rivers provide fresh water for drinking water and irrigation, rivers also play a valuable role in transportation and electrical power generation.

Gravity causes river water to flow downhill on its seaward journey. In areas where the gradient of

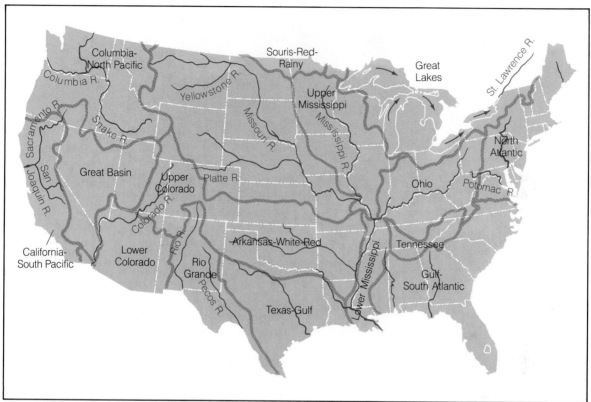

Figure 6.20 *The major drainage regions of the lower 48 states and selected major rivers in the various basins. The arrow in the Great Lakes indicates the direction of flow through them.*

a channel is steep, water flow is rapid and usually turbulent; where a channel slope levels off, the rate of flow slackens. As the water moves downhill, its potential energy is converted to *kinetic energy* (the energy of motion). A portion of that kinetic energy will erode and reshape the channel along the river's course. Additional energy is used to transport the material that is loosened from the channel bed and banks as well as other substances that enter the river, such as domestic and industrial wastes.

Because the amount of kinetic energy that exists in a river at any given time depends on the rate of water flow as well as on the quantity of water that is present, the amount of stream discharge is a better measure of the potential energy of a river than the speed at which the water is traveling. *Discharge*

is defined as the volume of water that passes a fixed point along the river's course within a given unit of time. It is usually expressed in terms of cubic meters (or cubic feet) of water per second. When the stream discharge is great, such as during a spring snowmelt in middle latitudes, channel erosion is accelerated. If the gradient is steep, a river can transport stones and even large boulders. Floods, such as that shown in Figure 6.21, are familiar examples of what happens during periods of great discharge and high sediment transport. When the quantity of river discharge lessens, however, such as when water is diverted for irrigation or is dammed, the transport of materials decreases and suspended sediments settle out.

The primary factor that determines the amount of a river's discharge is the weather in the drainage basin. The discharge usually declines from the rainy season to the dry season and, in some areas, it may

Figure 6.21 *Floodwaters invading the inhabited portions of the Snohomish Valley of western Washington. (U.S. Army Corps of Engineers, Seattle District.)*

fluctuate greatly even during a single day. For example, rapid discharge oscillations are especially characteristic of the arroyos of the American Southwest. Arroyos are normally dry streambeds until an afternoon thunderstorm causes a dramatic, short-lived "gulley washing" discharge. In contrast, some regions have rivers that flow out of heavily watered mountains, only to dry up completely in the desert plains beyond, never reaching the ocean.

Rivers are the lifeblood of many cities and areas, especially where irrigation water from rivers permits agriculture. For example, Egypt is said to be a gift of the Nile. In the southwestern United States, the entire economy is based on an adequate supply of water from the Colorado River. Thus it would seem logical to keep the growth of communities and concomitant economies within boundaries that would not overexploit the natural capacity of the river. However, a strong temptation exists to expand those communities, especially during extended wet periods, when more than enough water appears to be available. In the southwestern United States, we have already overcommitted the water supply, and battle lines over who is to use the inadequate water supply have been drawn between regions (Arizona versus California) and between agricultural and urban interests.

It is difficult to establish guidelines for predicting the amount of water that is available to us in heavily water-reliant areas. During dry periods, when runoff into rivers is near zero, the discharge in rivers originates from groundwater and water that is released slowly from wetlands. That *base flow* of a river, as it is called, is considered to be the dependable water supply. Therefore, a wise policy would be for us to budget water uses for a given region in accordance with the river's dependable supply. Such policies are becoming more common as a result of necessity. Still, we must be prepared for the fact that during extended dry periods, even base flow levels can decline, which require that we implement extensive measures to conserve water.

Materials that are washed into a river from the land, together with the sediment that is eroded from

the channel, constitute the river's *load*. The river transports its load in suspension, in solution, and along the riverbed by the thrust of the current. Soil erosion contributes suspended silt, clay, sand, gravel, and bits of detritus along with dissolved substances, including plant nutrients. The types of substances that are dissolved in a river vary with the climate, the rock and soil composition, and the human activities in the basin. A significant amount of the dissolved river-water components are contributed by groundwater, whereas an insignificant amount is contributed by air pollutants as a result of rainfall. An added burden has been placed on rivers in the form of such human waste materials as sewage and mine tailings (waste rock).

Most rivers have a tendency to flood. And rather than follow a straight course, river channels meander back and forth within the bounds of a broad *floodplain*, as shown in Figure 6.22. Figure 6.23 shows a cross section of a river and its floodplain. During times of abnormally high discharge, the river channel cannot accommodate all of the water, and the excess spills over the riverbanks and rapidly spreads over flat floodplains. The extent of the area that becomes flooded depends on the volume contained in the discharge. Planners and engineers use the expected frequency of flooding to designate specific parts of a floodplain. For example, the 100-year floodplain is that part of a flood-prone region that is subject to flooding once every 100 years on the average. Unfortunately, human settlement is often extensive in river floodplains. People are lured to them by productive soils, a reliable water supply, an inexpensive means of transport, the fishing and rec-

Figure 6.22 *This aerial view shows the meandering channel of Big Creek in east–central Louisiana. (U.S. Army Corps of Engineers.)*

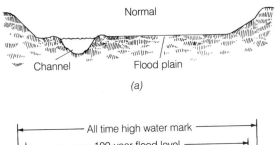

Normal

Channel   Flood plain

(a)

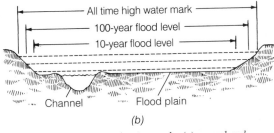

All time high water mark

100-year flood level

10-year flood level

Channel   Flood plain

(b)

Figure 6.23 *Cross section of a river under (a) normal and (b) flooded conditions. The land area that is flooded is defined as the floodplain. The greatest flood that is expected on the average of once every 100 years defines the 100-year floodplain.*

reational potential, and for esthetic reasons. This serious problem is discussed in Chapter 15.

Lakes

Depressions in the landscape allow lakes to exist. The water that fills those depressions comes from runoff or groundwater or both. Thus some lakes have rivers that run into and out of them, some have only outlets, others have only inlets, and some have neither inlets nor outlets.

A lake can be the result of any one of a variety of events. Lake basins are formed by geologic processes: earthquakes (such as Lake Nyasa in East Africa), volcanic activity (such as Crater Lake, Oregon, shown in Figure 6.24), and glaciation (such as the Great Lakes). Abandoned rock quarries and other surface-mining operations are often sites of water accumulation, and, of course, we create artificial lakes, for various reasons, when we construct dams across rivers and streams.

Figure 6.24 *Crater Lake in southern Oregon, which was formed by the collapse of a volcano. (Oregon Department of Transportation.)*

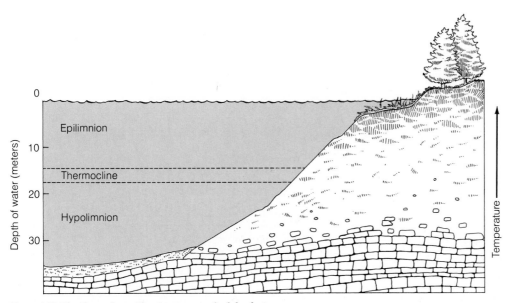

Figure 6.25 *The thermal stratification pattern of a lake during the summer.*

Although lakes are found in many different climates and contain waters that range widely in terms of salinity and acidity, they all share one characteristic: They are relatively short-lived in terms of geologic time. As we saw in Chapter 5, lakes experience a gradual aging process and eventually become filled with sediments that are eroded from the watershed as well as undecayed plant and animal matter. Lakes that are maintained by springs or groundwater seepage may disappear if the regional water table falls. A climatic shift that results in increased evaporation, decreased rainfall, or both may reduce a lake's water volume or even cause it to dry up. For example, about 10,000 years ago, a major climatic change from moist and cool to arid and warm conditions resulted in the disappearance of numerous lakes in the western region of the United States, primarily California, Nevada, and Utah. Remnants of some of those lakes exist today; the Great Salt Lake in Utah is what remains of the prehistoric Lake Bonneville. Many others, such as the Harney-Malheur Lake in Oregon, have nearly dried up, leaving wetlands. Salt flats were often left behind when those lakes dried up.

In temperate latitudes, lakes undergo an annual mixing cycle. Depending on their relative sizes and depths, lakes tend to become thermally stratified into two layers during the summer (see Chapter 5). Under bright summer sunshine, surface waters warm and become less dense (lower in mass per unit volume) than the colder water below. The result is a stable layer of less-dense water that floats on a layer of cooler, denser water. Little vertical mixing occurs between the two layers. The upper layer of a stratified lake is called the *epilimnion;* the lower layer is called the *hypolimnion;* and the narrow transition zone between the two is referred to as the *thermocline.* Figure 6.25 shows a cross section of a stratified lake.

When a lake is stratified, the dissolved oxygen supply of the epilimnion is replenished as usual by transfer through the air–water interface and by photosynthesis of algae. But the dissolved oxygen supply of the hypolimnion may be reduced by the decomposition of detritus that settles into the hypolimnion of the lake. Some fish (such as lake trout, whitefish, and cisco) must stay in the cold hypolimnetic waters, because they cannot tolerate the seasonally warm waters of the epilimnion. Those fish also require large amounts of dissolved oxygen and thus could not survive if the dissolved oxygen supply in the hypolimnion were not replenished. Thus

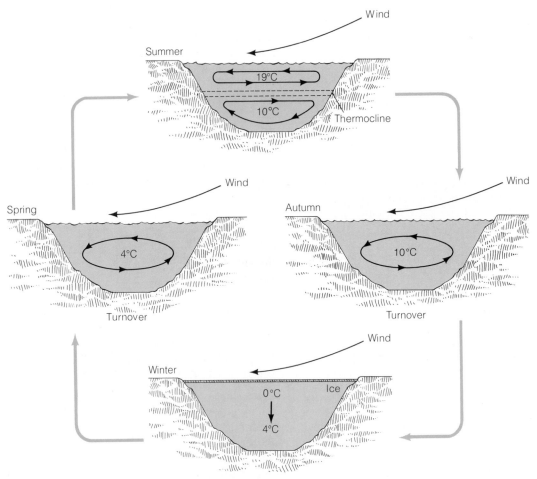

Figure 6.26 *The seasonal mixing patterns of a lake in temperate regions is determined by its temperature profile. Deep lakes undergo complete mixing in the spring and the fall; that turnover replenishes the dissolved oxygen of their deeper waters.*

we need to understand how oxygen replenishment occurs in those lakes.

Lakes do not remain stratified year-round. In autumn, the surface water cools. Eventually, the temperatures and, therefore, the densities of the two layers become equal. Assisted by the force of wind, water mixes from the top to the bottom and that mixing replenishes the supply of dissolved oxygen in the lower waters (Figure 6.26). The mixing period is called the *fall turnover.*

During winter months in the high and midlatitudes, lakes are covered with ice, and the water temperature may vary from 0°C (32°F) just below the ice to near 4°C (39°F) near the bottom. With the arrival of spring, the ice melts and surface waters warm. But because water becomes more dense as its temperature rises from 0°C to 4°C (39°F), the temperature at which water reaches its maximum density (Figure 6.27), the warmer but denser surface water sinks. That mixing period is called the *spring turnover.* Twice a year, then, in the spring and the fall, the surface and bottom waters are recirculated in temperate lakes.

Depending on its depth, a tropical or subtropical lake may experience only one turnover period—during the winter, when the surface of the lake cools. Lake turnovers are essential to replenish the dissolved oxygen supply of bottom waters. They thus help to ensure the survival of fish that require cold water and high concentrations of dissolved oxygen. Furthermore, recirculation brings nutrients (primarily nitrogen and phosphorus) from bottom waters in contact with the sediments to surface waters, thereby increasing algal productivity.

Glaciers

Glaciers, which are really rivers of ice, represent another pathway taken by the runoff component of the hydrologic cycle in its journey to the sea. Today, glaciers constitute the largest freshwater reservoir, although that resource goes unused, because it is virtually inaccessible. Water that exists in the form of glacial ice covers approximately 10 percent of the earth's land surface and is confined primarily to Antarctica (85 percent), Greenland (11 percent), and some high mountain valleys in various parts of the world (4 percent). If all of that ice were to melt, the level of the seas would rise by approximately 60 meters (200 feet), inundating many of the world's major cities.

In the earth's history as a whole, the presence of any glacial ice at all has been a rare event. Yet during the past 2 million years, glaciers have expanded and receded many times. As recently as 18,000 years ago, more than 30 percent of the earth's land surface was covered with ice. In North America, an ice sheet that was perhaps 3 kilometers (2 miles) thick covered much of Canada and the northern states, as shown in Figure 6.28. At that time, sea level was about 100 meters (330 feet) lower than it is now, because so much water was in the form of ice.

In the northern United States, the Ice Age left us a valuable legacy. Hundreds of thousands of the lakes, swamps, and bogs that dot the landscapes of the upper Midwest owe their origin to the gouging action of glaciers. Under the tremendous forces of the creeping glaciers, the landscape was carved, scraped, and pushed into new topographic patterns. Excavated materials were left behind as extensive deposits of sand and gravel layers. Some of those layers now function as aquifers, while others are mined for sand and gravel. Glaciation was indirectly responsible for the rich soils of the Great Plains and central Midwest because, as the glaciers retreated, extensive loess deposits were left behind (Chapter 5). The loess was the parent material of the rich soils of the Great Plains and Midwest.

Figure 6.27 *The effect of temperature on the density of water; maximum density is reached at 4°C.*

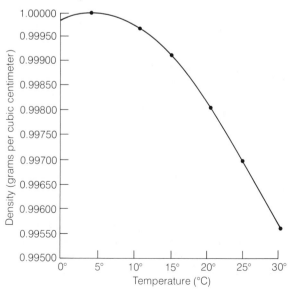

The Oceans

The sea, by far the earth's largest reservoir of water, is the ultimate receptacle of terrestrial water, whether it trickles into it from melting glacial ice, slowly seeps through permeable rock and soil, or rushes down the courses of stream channels. Viewed from space, the ocean is the most prominent feature of the earth's surface. It covers 71 percent of the total area, to an average depth of 4 kilometers (2.5 miles). Also, the ocean plays an extremely important role in the environment.

Oceans circulate heat energy through persistent currents, which redistribute the solar energy that is absorbed by the surface waters. In the Atlantic, the warm Gulf Stream flows poleward and the cold Labrador currents flow southward. The circulation of surface ocean currents is controlled by a coupling of winds with surface waters. That link results in a

Figure 6.28 *The extent of glaciation in North America 18,000 years ago.*

clockwise movement of surface waters in the northern hemisphere and a counterclockwise movement of surface waters in the southern hemisphere.

The sea is an essential reservoir in most of the cycles of material and energy flow on earth. Materials that are cycled into the ocean waters may remain there for thousands or even millions of years before being cycled out again. Thus the sea is the final destination for much of the waterborne and airborne pollutants that are released as a result of human activity. Because the sea also supports life forms that are essential food sources for millions of persons, we are finding it necessary to identify the substances that enter the oceans and pose potential hazards to marine ecosystems. Also, because we are rapidly depleting the terrestrial supplies of some fuel and mineral resources, we are growing increasingly interested in the resources contained in seawater as well as in the rock and sediment that make up the oceans' basins (see Chapter 12). Until now, the mineral resources of oceans have remained relatively unexploited.

Seawater contains large quantities of dissolved salts, the most abundant of which are sodium chloride and magnesium sulfate. Sea salt is derived primarily from the weathering and erosion of soil and rock, the products of which are dissolved and carried to the sea by rivers. The salinity of seawater renders it undrinkable; in fact, drinking salt water actually causes dehydration and an increased craving for water. The salinity of seawater is remarkably uniform. It averages 35,000 parts per million (ppm), or 35 parts per thousand—approximately 70 times the maximum concentration of salt that is allowed in our public supplies of drinking water. When evaporation is excessive or sea ice is formed, the level of salinity may be as high as 38,000 ppm; in particularly rainy climates or when melting ice or stream runoff dilutes seawater, it may be as low as 32,000 ppm. Seawater contains numerous dissolved materials, but only a few exceed a trace percentage, as Table 6.2 shows.

The salinity of seawater virtually precludes its use as a direct source of fresh water, not only for drinking but for cleaning and irrigation as well. In Chapter 8, we see that the desalination of ocean water for domestic, industrial, and agricultural purposes is technically feasible but too costly in terms of dollars and energy consumption. Although seawater is potentially the largest single source of several raw materials besides fresh water, the only ones that are economically feasible to extract are sodium chloride (table salt), bromine (which is used primarily in the production of an antiknock agent in leaded gasoline), and magnesium (which is used as metal, in metal alloys, and in pharmaceuticals).

Conclusions

Although the oceans contain the largest reservoir of water on earth, the high level of salinity in seawater renders it unusable for domestic, agricultural, and most industrial processes. The water cycle, however, processes seawater naturally through evaporation and condensation, making it available to us as fresh water, which we tap as surface water and groundwater. Although our freshwater supply is renewed each year, the world's volume of fresh water is essentially fixed at approximately 41,000 cubic kilometers (9800 cubic miles). Therefore, as the world's human population soars, the amount of fresh water that is available for each person declines. Furthermore, the distribution of fresh water is uneven; many of the world's population centers are located in regions that have limited freshwater resources. It is evident that we must conserve freshwater resources to ensure that we will have sufficient water of acceptable quality in the future.

To protect the quality of our fresh water, we must understand how it moves through the water cycle. All parts of the water cycle interact, and the functioning of the entire system results in both our freshwater supply and a habitat for freshwater organisms. Contaminants that enter the cycle at any point eventually enter all parts of the system. Natural cleansing processes cannot keep pace when the system becomes overloaded. The types and properties of pollutants that enter our water supply, their effects on aquatic life, and methods of pollution abatement and freshwater conservation are explored in Chapters 7 and 8.

Table 6.2 Concentration of the major substances dissolved in seawater

SUBSTANCE	PERCENTAGE	MILLIGRAMS PER LITER (PARTS PER MILLION)
Chloride	1.9	19,000
Sodium	1.06	10,600
Sulfate	0.27	2,700
Magnesium	0.13	1,300
Calcium	0.041	410
Potassium	0.039	390
Bicarbonate	0.01	100
Bromide	0.0067	67
Strontium	0.0008	8
Fluoride	0.00013	1.3
Iodide	0.000006	0.06

Summary Statements

The water cycle is the continual circulation of water among the atmospheric, land, and oceanic reservoirs.

Air is said to be saturated when it is holding all the water vapor it can tolerate. The degree of saturation is indicated by the relative humidity. Warm air can hold considerably more water vapor than cold air. As air parcels warm, however, they become less saturated. As air parcels cool, they approach saturation and may lose water through precipitation processes.

Expansional cooling in the atmosphere causes air parcels to approach saturation. Expansional cooling is caused by (1) lifting by an advancing warm front, (2) lifting by an advancing cold front, (3) convection cell development, and (4) orographic lifting.

Precipitation processes are initiated by condensation nuclei, especially hygroscopic nuclei. Precipitation forms in clouds through the collision–coalescence process and Bergeron ice-crystal process.

Topographic barriers, such as mountains, increase orographic precipitation on the windward side and reduce it on the leeward side.

Water flows from land to sea by means of infiltration (groundwater flow) and runoff (rivers and streams).

Groundwater flows through aquifers, which are composed of materials that are both porous and permeable. Free-flowing, or artesian, wells are possible in areas where groundwater is confined and under pressure.

Groundwater undergoes little mixing compared to water that exists in rivers and lakes and, therefore, the dilution of pollutants in that water is minimal.

Overuse of groundwater, called water mining, can lead to ground subsidence and saltwater intrusion and, eventually, the loss of the water supply.

Rivers drain a defined area, called a drainage basin, or watershed. The discharge (volume of flow per unit of time) of rivers varies with precipitation and activities within the drainage basin. Rivers are important agents in the erosion of the landscape.

Lakes stratify in the summer because of density differences that are caused by temperature differences. The water in most lakes in temperate regions tends to mix completely (turn over) twice each year, whereas the water in tropical lakes may mix only once.

Glacial ice represents the largest reservoir of fresh water, but it is unused because it is virtually inaccessible.

Seawater is highly saline and, therefore, unusable for drinking or for irrigation. Salt, bromine, and magnesium are the only substances that are now extracted from seawater on a commercial basis.

Evaporation of seawater by the sun and subsequent condensation are the natural processes that renew the earth's freshwater supply.

Questions and Projects

1. In your own words, write a definition for each of the terms that are italicized in this chapter. Compare your definitions with those in the text.

2. How has water played a role in the location and historical development of your community?

3. From what specific reservoir of the water cycle do you obtain your water supply?

4. Give one common example of each of the following water-transfer processes: sublimation, deposition, evaporation, and condensation. Which processes are most important in your region?

5. Usually, little difference exists between the quantity of water vapor in the air over the Southwest and that in the air over the Northeast. However, large differences do exist in the amount of precipitation that falls on the two regions. What is the fundamental reason for that difference?

6. Cite specific attempts in your area to modify the water cycle for energy, water supply, or flood prevention.

7. In what parts of the United States would you expect high evaporation rates? Explain your answer. What happens to the dissolved salts in a lake that experiences high evaporation rates? What happens to dissolved salts in irrigation water?

8. The residence time for water in a water reservoir (for example, a lake, groundwater, river, or glacier) is the average amount of time that is required for a volume of water to cycle through a reservoir. In which reservoirs does water have long residence time?

9. What is the average annual rainfall in your region? What fraction of that amount evaporates?

10. Cite specific factors that would affect the level of the water table in a region.

11. How is groundwater used in your region? What problems can arise from overuse of groundwater?

12. How might paved roadways, parking lots, and rooftops affect groundwater levels in an area?

13. Explain what is meant by the term *floodplain*. Is there a developed floodplain in your region?

14. What problems are created by sediment transport and deposition in rivers in your region? Cite specific examples and explain what is being done to minimize or alleviate those problems. Your local

U.S. Soil Conservation Office will have publications or information on this problem.

15. Why is distillation a purification process? What is the source of the energy that fuels natural distillation?

16. Using the temperature–density relationship of water, explain how lakes in temperate regions may undergo complete mixing (turnover) twice a year.

17. Explain how oxygen from the air enters and mixes into a stratified lake in the summer, the fall, and the spring.

18. Explain why relative humidity is often lower indoors during winter months than during summer.

19. A region experienced little change in its water-table level when only small private wells were used. After irrigation became an accepted practice, water-table levels began to fall. Describe some of the changes that the region can anticipate.

Selected Readings

AMBROGGI, R. P.: "Underground Reservoirs to Control the Water Cycle." *Scientific American* 236:21–27 (May), 1977. A discussion of methods for increasing groundwater supplies.

AMBROGGI, R. P.: "Water." In *Economic Development*. A Scientific American Book. New York: W. H. Freeman, 1980. Chapter 4 presents a discussion on the use of water on a global scale and its economic and physical limitations.

DUNNE, T., AND LEOPOLD, L. B.: *Water in Environmental Planning*. New York: W. H. Freeman, 1978. A textbook that examines methods for preventing water-related problems.

FREEZE, R. A., AND CHERRY, J. A.: *Groundwater*. Englewood Cliffs, N.J.: Prentice Hall, 1979. A textbook for persons who are interested in a more advanced treatment of groundwater quality and movement.

LEET, D. L., JUDSON, S., AND KAUFMAN, M. E.: *Physical Geology*, 6th ed. Englewood Cliffs, N.J.: Prentice Hall, 1982. A textbook that provides additional information on the water cycle and its associated problems.

LEOPOLD, L. B.: *Water*. New York: W. H. Freeman, 1974. A primer on water cycling and water use.

MURRAY, R. C., AND REEVES, E. B.: *Estimated use of water in the United States in 1975*. U.S. Geological Survey, Circular 765, 1977. A statistical report on total water uses in the United States.

PRESS, F., AND SIEVER, R.: *Earth*, 3rd ed. New York: W. H. Freeman, 1982. A textbook that covers the hydrologic cycle and related processes in detail.

OBERLANDER, T. M., AND MULLER, R. A.: *Essentials of Physical Geography Today*. New York: Random House, 1982. A readable, comprehensive, and well-illustrated textbook.

STRAHLER, A. N.: *Physical Geology*. New York: Harper & Row, 1981. A textbook with several chapters devoted to the work of running water and groundwater.

WETZEL, R. G.: *Limnology*. Philadelphia: Saunders, 1983. A textbook that covers the physical, chemical, and biological aspects of aquatic ecosystem function.

A dead bird, one of many victims of water pollution. (M. L. Brisson.)

CHAPTER 7   WATER POLLUTION

Debris bobs along the course of a river and clutters the shoreline. Oil slicks slither and shimmer, and refuse and dead fish foul the air and water. Signs here and there along the shore warn, "Danger—Water Polluted—No Fishing or Swimming." Such sights and smells were common along the rivers and lakes near our nation's cities in the 1960s and early 1970s. In our quest for industrial–technological and economic growth, we allowed our cities and industries to foul many aquatic habitats. Our rivers were not always in such deplorable condition. The consequences have been major, but since 1975, much progress has been made in stemming the release of pollutants that gravely damage aquatic ecosystems and turn sparkling surface waters into murky eyesores. In this chapter, we survey what is known about the sources of water contaminants and their effects on surface water and groundwater.

A Historical Overview

In the late 1800s, when our cities were growing rapidly in response to industrialization, horse manure and garbage in the streets became a deplorable problem. None too frequently, garbage collectors cleaned the streets and dumped the refuse into the nearest river. In New York City, platforms were specially constructed so that citizens could dump garbage directly into the river. Although river-dumping was discontinued in 1872, the method that was substi-

tuted for it—barging the wastes a short distance from shore before dumping them into the ocean—polluted city beaches, to which droves of New Yorkers flocked during the summer (Figure 7.1).

Such dumping, along with a complete lack of sanitary practices, which we now take for granted, frequently led to epidemics of waterborne diseases throughout the world. City and county folk alike experienced and feared outbreaks of the so-called filth diseases, such as typhoid fever, cholera, infectious hepatitis, and dysentery. Epidemics were usually traced to water or food that had been contaminated with human feces. Many city water supplies were drawn from surface water supplies, which were easily contaminated. The transmission of waterborne diseases then went unchecked, since supplies of surface water were not treated. On farms, the barnyard, pigsty, chicken coop, privy, and cesspool often were located close to the open well that supplied the drinking water. In cities, efforts to control those diseases began with the filtering of public water supplies through sand. Later, beginning in 1908, chlorine was added to municipal water before it entered the water mains. By the early 1960s, waterborne diseases had become relatively rare in the United States, although they still plague the less-developed countries of the world, as we see in Chapter 8.

However, as biological pollution (in the form of various waterborne diseases) was diminishing during the first half of this century, chemical pollution was increasing. As our technology advanced, becom-

Figure 7.1 *A political cartoon from the late 1800s, depicting water pollution problems in New York City. (Bettman Archive, Inc.)*

ing more sophisticated, the types of wastes that were discharged into our waterways became even more varied. Today, our water resources are being assaulted by such chemicals as commercial fertilizers, pesticides, detergents, trace quantities of metals, acidic mine drainage, radioactive substances, and a wide variety of industrial chemicals.

Today, in this country, industrial wastes and other chemical pollutants are the most worrisome threat to our waters. Primary long-range concerns are focused on food, which is being contaminated with potential carcinogens that are a threat to human health. Another concern is that pollutants may disrupt aquatic ecosystems by interrupting the life cycle of certain organisms. The long-range effects of chemical pollution are poorly understood at this time. The visible short-range effects, however, are clear

enough—fish kills that follow accidental oil and chemical spills, unusable beaches, and lakes that are choked with aquatic plants.

It is encouraging that the contamination of our water no longer continues unchecked. Most industries and municipalities are now acting in compliance with legislation that requires certain minimum treatment standards for effluents. But some industries and municipalities continue to battle with regulatory agencies concerning those standards. For example, they disagree with the amount of treatment that is required to control, for example, oxygen-demanding substances and plant nutrient levels. Moreover, guidelines for controlling low-level discharges of certain toxic chemicals have not been established, and government regulatory agencies are not progressing very rapidly toward decisions. Frustrated by what they consider to be glacial progress in efforts to cope with waterborne pollution, many persons and citizens' groups have turned to the courts for relief and are suing polluters, regulatory agencies, or both. Legal battles to control the levels of pollutants in discharge waters will continue in the foreseeable future.

Drainage-Basin Activities

Most water-pollution problems stem from land-based activities within drainage basins (see Figure 7.2) rather than from water-based activities, such as shipping, boating, and swimming. As we discussed in Chapter 6, a *drainage basin* is a geographic region that is drained by a river or stream. We can identify three general types of uses within drainage basins: (1) areas that are designated for wildlife and or recreation, which are left natural and relatively undisturbed, (2) areas that are designated for rural–agricultural activities, including cultivation, and (3) areas that accommodate urban–industrial activities. Each use introduces a different mix of contaminants into surface water and groundwater.

Natural Pollutants

Natural areas, such as forests, marshes, and grasslands, generally contribute only small amounts of materials to waterways. Surface waters that flow through those ecosystems contain characteristic,

Figure 7.2 *A few of the many activities in drainage basins that can affect the quality of surface water. In natural drainage basins, flowing water (1) slowly dissolves and erodes rock, and wetlands (2) allow sediment to settle from water. In urban–industrial basins, municipal sewage treatment plants (3) fail to remove all the wastes added during use; storm sewer water (4) contains wastes that are washed from city streets and lots; industrial wastewater (5) contains a wide array of different water pollutants; acid water flows from mines and strip-mined land (6); heated water flows from power plants (7); industrial gases are washed out of the air, forming acid rain (8); the transportation of oil (9) results in spills, including many minor spills during unloading; and faulty landfill sites (10) contaminate groundwater. In agricultural drainage basins, crop spraying (11) adds pesticides and herbicides to rain and runoff; fertilizers that are applied improperly (12) are dissolved in runoff; and animal wastes (13) are washed from farmland.*

"normal" concentrations of dissolved substances. The composition of sediments varies considerably and, therefore, the water that flows over and through them also varies widely in terms of dissolved mineral content. For example, the headwaters of the Wolf River in Wisconsin have a low concentration of dissolved substances. Typical values are: calcium, 6 ppm; magnesium, 7 ppm; and bicarbonate, 42 ppm, because the bedrock in the Wolf River consists mainly of insoluble granite. Downstream, however, the granite gives way to more soluble dolomite (a rock that is composed of calcium and magnesium carbonate). Hence as the river flows along its course, concentrations of the dissolved material increase to calcium, 13 ppm; magnesium, 25 ppm; and bicarbonate, 170 ppm. Although those elevated levels cause no problems for aquatic life, from the human standpoint, higher levels cause the water to become hard, which means that more soap or detergent must be used for cleaning. Later, the detergent may become a pollution problem.

However, not all water that comes from natural areas is considered to be of high quality. For example, streams that originate or pass through swamps or marshes are usually somewhat acidic and discolored, because they pick up organic acids that are generated by decomposer organisms. Rivers that flow from melting glaciers become choked with finely ground-up rock, which causes intestinal disorders in humans and prevents the existence of most aquatic life.

Point and Nonpoint Sources of Pollution

Agricultural areas degrade water quality in several ways. For example, excessive soil erosion will increase the load of sediment in a river. Pesticides, fertilizers, and animal wastes that are washed from fields and orchards will run off into streams or seep into the groundwater. Since the contaminants are contributed from throughout an area rather than from one or more concentrated sources (point sources), such as drainpipes, they are called *nonpoint sources* of pollutants. In contrast to *point-source* pollutants, which are concentrated wastes from sewage treatment plants and industry, pollutants from nonpoint sources are especially difficult and expensive to control, because their concentrations are relatively low and their volumes are enormous.

Urban–industrial drainage basins contain the greatest number and variety of point sources of pollutants. Although the amount of untreated wastes that is dumped directly into waterways from such areas has decreased, sewer systems are still important contributors to surface-water pollution. For example, although most American cities now have adequate sewer systems, they do not all have adequate waste-treatment facilities. Thus some of the wastes from such diverse sources as hospitals, meat-packing plants, electroplating industries, paper mills, and petroleum-related industries find their way into surface waters. Moreover, not all industries have met their federally mandated deadlines to clean up the water that is discharged directly back into the environment.

Sewer System Design

The basic design of sewer systems contributes to the problem of water quality in urban areas. In cities, buildings and paved areas render a large part of the urban surface impermeable to rainwater and snowmelt, thereby increasing the volume of runoff. To prevent flooding, large *storm sewer pipes* that lie under the city streets channel that runoff to the nearest river, lake, or ocean. Thus during a rainstorm, the air, streets, and ground surface are washed, and many pollutants (such as hydraulic fluid, oil, radiator coolant, road salt, and pet droppings) are carried into surface waters by the storm sewer system. The components of a city sewer system, including the storm sewer pipes, are shown in Figure 7.3.

A second, smaller system of sewer pipes, called the *sanitary sewer*, carries waste water, or *effluents*, from homes and commercial areas to treatment plants, where it is treated and discharged into nearby surface water. (As we shall see later, the quality of the treated water depends on the sophistication of the treatment plant.) When both storm sewer pipes and sanitary sewer pipes service an area, the system is called a *separated sewer system*.

Another type of sewer system employs a single pipe to transport both runoff and sewage to a treatment plant. That type of system is called a *combined sewer*. In areas where combined sewers are used, the treatment plant receives only domestic wastes during dry weather. However, when it rains

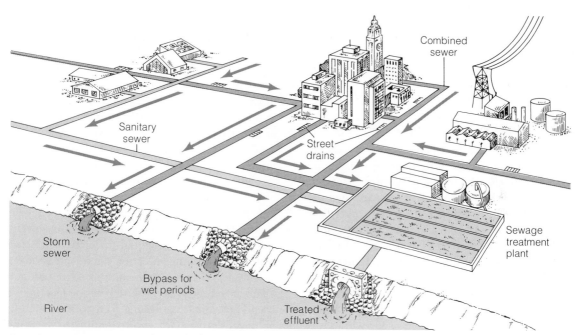

Figure 7.3 *A city sewer system, showing both separated and combined sewer systems.*

heavily, the volume of water that flows through the combined sewer often exceeds—by as much as 100 times—the amount of water that can be processed by the treatment plant. As a result, only a small fraction of the sewage water is treated, and the overflow, which contains raw sewage, is bypassed and discharged directly into the surface waters. Today, parts of many major cities, such as Boston, are still served by combined sewer systems. Every time a treatment plant bypasses raw, untreated sewage, the health of the city's residents and their downstream neighbors is threatened.

The wastewater of some industries is treated by municipal treatment plants, but sometimes it is so toxic that municipal treatment plants will not accept it. Industries that do route their wastewater to municipal treatment plants are required to limit the level of potential toxicity by following pretreatment procedures on their own premises. But not all factories are equipped properly, and even responsible industries have occasional accidental discharges. Furthermore, as a result of the tightening of air- and

water-quality standards, many industries and municipalities have been forced to dispose of most solid wastes in landfills rather than processing them as effluents. That process usually protects surface waters, but those pollutants will contaminate groundwater if the landfills are improperly sited, constructed, or operated.

Surface-Water Pollution

Any threat to the quality of water that is used for drinking and recreation quickly angers the persons that are affected. In this section, we discuss the eight basic categories of water pollutants and their effects on humans and aquatic ecosystems. Later in the chapter, we discuss some of the types of activities and pollutants that pose special problems to groundwater resources.

Until recently, our concerns about water pollution were focused mainly on disease-producing organisms that threatened human health. Today, we also worry about the effects of polluted water on aquatic organisms and the stability of aquatic ecosystems. Also, we recognize that disease-producing organisms as well as plant nutrients, pesticides, heavy metals, oil, sediments, radioactive wastes, and heat (thermal pollution) can have a negative impact on aquatic ecosystems. We now consider the particular risks that are associated with specific types of pollutants.

Infectious Agents

Water transmits disease when it contains waterborne *pathogens*, or disease-producing organisms. Those pathogens, which can be viruses, bacteria, protozoa (unicellular animals), or parasitic worms, cause such diseases as dysentery, typhoid fever, and cholera (which are caused by bacteria) and infectious hepatitis (which is caused by a virus). Table 7.1 shows some important characteristics of the more common waterborne diseases of humans. In the United States, the *chlorination* of water supplies has greatly reduced the transmission of disease through drinking water. But for a large segment of the world, waterborne pathogens in drinking water are still a serious health hazard (Figure 7.4). In fact, as the United Nations Environmental Program has pointed out, the number of persons who contend with unsafe water supplies is greater today than it was in 1970. Increases in population have allowed that trend to continue despite the fact that a greater proportion of the population in developing countries is now provided with safe water supplies.

Pathogens enter the water mainly through the feces and urine of infected persons and animals. Infected body wastes can enter water in a variety of ways: in seepage from malfunctioning cesspools and septic tanks; in unchlorinated sewage, when sewer systems become overloaded or treatment plants malfunction; in waste discharges from pleasure boats; in untreated discharges from meat-processing plants; or directly from swimmers, hikers, and so forth. If drinking water supplies that are drawn from contaminated surface waters are not treated properly, an epidemic may occur. Indeed, when even a single case of a virulent disease, such as typhoid fever, occurs in a community, the potential for an epidemic exists. Thus during such periods, personal hygiene and cleanliness are essential if the disease is to be kept from spreading. Infection can be spread directly as a result of drinking or swimming in contaminated water or indirectly as a result of eating food that has been contaminated through food webs. Swimming or other direct contact with contaminated water leads to an increased incidence of skin, ear, nose, and upper-respiratory-tract infection.

Detecting the presence of specific pathogens, especially viruses, is a time-consuming, costly, and difficult process. Thus microbiologists usually analyze the water for a more readily identifiable group of microorganisms—the *coliform bacteria*. Since those organisms are normally present in the intestinal tract of humans and animals, large numbers of coliform bacteria in a water sample indicate recent

Table 7.1 Some common waterborne diseases transmitted through drinking water and food

DISEASE	TYPE OF ORGANISM	SYMPTOMS AND COMMENTS
Cholera	Bacterium	Severe Vomiting, Diarrhea, and Dehydration; Often Fatal if Untreated; Primary Cases Waterborne; Secondary Cases Carried by Contact with Food and Flies
Typhoid Fever	Bacterium	Severe Vomiting, Diarrhea, Inflamed Intestine, Enlarged Spleen; Often Fatal if Untreated; Primarily Transmitted by Water and Food
Bacterial Dysentery	Several Species of Bacteria	Diarrhea; Rarely Fatal; Transmitted Through Water Contaminated with Fecal Material or by Direct Contact Through Milk, Food, and Flies
Typhoid Fever	Several Species of Bacteria	Severe Vomiting and Diarrhea; Sometimes Fatal; Transmitted Through Water or Food Contaminated with Fecal Material
Infectious Hepatitis	Virus	Yellow Jaundiced Skin, Enlarged Liver, Vomiting and Abdominal Pain; Often Permanent Liver Damage; Transmitted Through Water and Food, Including Shellfish Foods
Amoebic Dysentery	Protozoan	Diarrhea, Possibly Prolonged; Transmitted Through Food, Including Shellfish

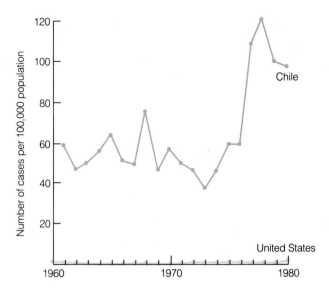

Figure 7.4 *Comparison of the number of cases per 100,000 of typhoid fever in the United States and Chile from 1960 through 1980. (United States: Annual Summary, 1981. U.S. Department of Health and Human Services, Vol. 30, No. 54, October 1982.; Chile: Pan American Health Organization Epidemiological Bulletin, Vol. 4, No. 1, 1983.)*

contamination by untreated feces. It is assumed that if coliform bacteria are present in the sample, it is likely that infectious pathogens are also present. In addition, coliform bacteria can act as pathogens in certain cases. Hence when coliform bacteria exceed 2.2 organisms per 100 milliliters of drinking water, a municipality must either chlorinate the water more heavily or seek alternative sources of water. The Environmental Protection Agency has set an upper limit of 200 coliform bacteria per 100 milliliters of water for recreational areas. If that limit is exceeded, the contaminated area and contingent beaches are usually closed.

The chlorination of public water supplies has virtually eliminated epidemics of often-fatal waterborne diseases in developed countries, including the United States. Figure 7.4 indicates how the incidence of one such disease—typhoid fever—has declined to almost zero in this country. However, the methods for controlling it do fail occasionally, as the data in Table 7.2 indicate. Moreover (as we shall see later in the chapter), chlorination itself may create other pollution problems. For this reason as well as for economic considerations, many European nations do not chlorinate sewage effluents unless direct evidence exists (such as a case of typhoid fever) that an outbreak of a waterborne disease is possible. Thus, although we now take the protection of our water-treatment systems for granted, the potential for waterborne epidemics always exists. Domestic livestock and wildlife also can be infected by waterborne diseases.

Oxygen-Demanding Wastes

Most aquatic organisms acquire their oxygen from the supply that is dissolved in the water. However, that supply quickly diminishes when organic wastes decompose in the water. Recall from Chapter 5 that the amount of dissolved oxygen in surface water is dependent on the processes that add oxygen (turbulence and photosynthesis) and those that remove oxygen (primarily respiration by decomposers). When organic materials are added to surface waters, the proliferation of oxygen-consuming decomposers, mainly bacteria and fungi, is encouraged. Those decomposers reduce the oxygen supply and, consequently, members of aquatic communities, espe-

Table 7.2 Waterborne disease outbreaks from various water supply systems in the United States

YEAR	COMMUNITY*	NONCOMMUNITY†	PRIVATE‡	TOTAL OUTBREAKS	TOTAL INDIVIDUALS
1971	5	10	4	19	5,182
1972	10	18	2	30	1,650
1973	5	16	3	24	1,784
1974	11	10	5	26	8,363
1975	6	16	2	24	10,879
1976	9	23	3	35	5,068
1977	12	19	3	34	3,860
1978	10	18	4	32	11,435
1979	23	14	4	41	9,720
1980	23	22	5	50	20,008
1981	14	16	2	32	4,430

Source: Center for Disease Control: Water-Related Disease Outbreaks Annual Summary, September 1982.
*Mostly public water supplies.
†Motels, trailer parks, and so forth, with own well.
‡Private wells.

cially fish and shellfish, become deprived of adequate oxygen and perish.

In an undisturbed aquatic ecosystem, the quantity of organic material (detritus) is small and, therefore, the amount of oxygen that is utilized by decomposers is also limited. As a result, under natural conditions, the concentration of dissolved oxygen usually remains relatively constant and at a level that is higher than 5 ppm—the level that is usually considered to be critical for the survival of the most sensitive fish. That balance represents the smooth functioning of a natural purification process. When we say that a river can cleanse itself, we are referring to its natural, but limited, capacity to decompose and remove organic wastes, without depleting the dissolved oxygen lower than the 5 ppm level.

Most industrial and municipal wastes, however, contain high concentrations of organic substances. Their presence encourages the growth of decomposers, and those organisms consume large quantities of dissolved oxygen. (Some inorganic industrial wastes, too, can react chemically with, and thereby remove, dissolved oxygen.) When the presence of oxygen-demanding materials ultimately results in the loss of organisms through oxygen deprivation within aquatic ecosystems, they are, by definition,

pollutants. Industrial and municipal wastes are short-term pollutants, because they are almost completely decomposed within several weeks after being introduced to surface waters. However, in cold water, the rate of decomposition is slowed considerably. Concentrated organic wastes cause the oxygen supply in the water to become depleted before decomposition is complete. (Most waste water also contains small amounts of organic materials that resist decomposition.) If organic-waste pollution is halted, oxygen is replenished in the affected area in a matter of a few weeks in rivers and within a year in the hypolimnion of lakes.

The amount of dissolved oxygen that is needed by decomposers to break down organic materials in a given volume of water is called the *biochemical oxygen demand* (BOD). Thus BOD is a measure of the level of organic contamination in wastewater. Human wastes are a major source of BOD. Sewage-laden wastewater that enters a sanitary sewer has an average BOD level of 250 ppm. But that water is initially likely to contain only about 8 ppm of oxygen, so all of its dissolved oxygen is quickly depleted through microbial decomposition of sewage. In fact, the decomposition of the daily wastes of just one person would require all the dissolved oxygen in 9000 liters (2200 gallons) of water if no oxygen were added to it.

When water is used in the processing of organic materials, such as vegetables, fruits, paper, meat, and dairy products, some of those materials, which are also oxygen-demanding substances, are added to the water. The levels of BOD in the water that is used in those processing activities vary widely, but they are usually much higher than those that occur in domestic sewage. In fact, some concentrated industrial wastes have BOD levels that are higher than 30,000 ppm. Other high-level BOD wastes include runoff from livestock feedlots and organic sediments that are dredged from polluted harbors and canals.

When effluents that have high levels of BOD are released into a stream, dissolved oxygen levels downstream follow a characteristic pattern, called an *oxygen sag curve* (Figure 7.5). At the point of discharge, bacteria begin to consume the organic material. The bacterial populations grow in direct relation to the amount of organic material present, and they consume dissolved oxygen faster than it can be replenished. As the material moves downstream, increasingly more of it is decomposed, so the BOD level becomes smaller and smaller. The dissolved oxygen deficit caused by the decomposers is slowly replaced by oxygen transfer from the air into the water and by the photosynthetic activity of aquatic plants. Eventually, the rate of oxygen replacement exceeds the rate of removal, and the oxygen level in the stream is restored to its original level. Organic-waste discharges have their greatest impact on aquatic life during warm summer months, when the stream flow is low (there is less dilution and turbulence). Also, the warm water holds less dissolved oxygen, and it enhances microbial activity.

The responses of aquatic organisms in rivers to large quantities of organic wastes have been well-documented. Upstream from the discharge point, a river can support a wide variety of fish, algae, and other organisms, but in the section of the river where oxygen levels approach zero, only a few types of sludge worms survive. As oxygen levels recover further downstream, species of rough fish (such as carp, gar, and catfish) that can tolerate low oxygen levels begin to appear. At some further distance downstream, a more diverse and desirable community of fish can live. But a single overloading discharge of organic wastes can eliminate even that aquatic community for some time. Moreover, if the discharge points are distributed along the length of the river, multiple oxygen sags will occur, and if those discharge points are spaced too close together, the entire downstream stretch of the river will suffer severe oxygen depletion.

If a complete loss of oxygen occurs in the stream, a change in the type of decomposer bacteria that are present in the water will also occur—from *aerobic decomposers* (those that require oxygen) to *anaerobic decomposers* (those that require the absence of oxygen). The products of aerobic and anaerobic decomposers also differ. The products of decay by aerobic decomposers, which are mainly carbon dioxide, water, nitrate, and sulfate are not usually harmful, whereas those generated by anaerobic decomposers, such as methane, ammonia, and hydrogen sulfide (which is recognized by its rotten-egg smell), are hazardous gases. Under anaerobic conditions, waterways become a putrid, turbid, decaying mess, and bubbles of methane and hydrogen sulfide rise to the surface. Under those degraded conditions, even sludge worms and fungi cannot survive, and the only living organisms left are the anaerobic bacteria.

Plant Nutrients and Cultural Eutrophication

If excessive quantities of plant nutrients are discharged into lakes, ocean bays, and rivers from any one of many possible sources of human activity, the natural aging processes of those waterways is accelerated (see the discussion of eutrophication in Chapter 5). Accelerated eutrophication that results from human activity is called *cultural eutrophication*. Cultural eutrophication of freshwater resources is one of the most serious problems facing us today in terms of water quality. It jeopardizes the use of water for drinking, recreation, sport and commercial fishing, agriculture, and industry. Its effects on the ecosystem in the Chesapeake Bay are described in Box 7.1.

Like terrestrial plants, aquatic plants require nitrogen, phosphorus, potassium, and other mineral nutrients. In aquatic systems, the two nutrients that most commonly act as limiting factors are phosphorus, in the form of phosphate (PO_4^{3-}), and nitrogen, in the form of either nitrate (NO_3^-) or ammonia (NH_3). Thus when levels of limiting nutrients in-

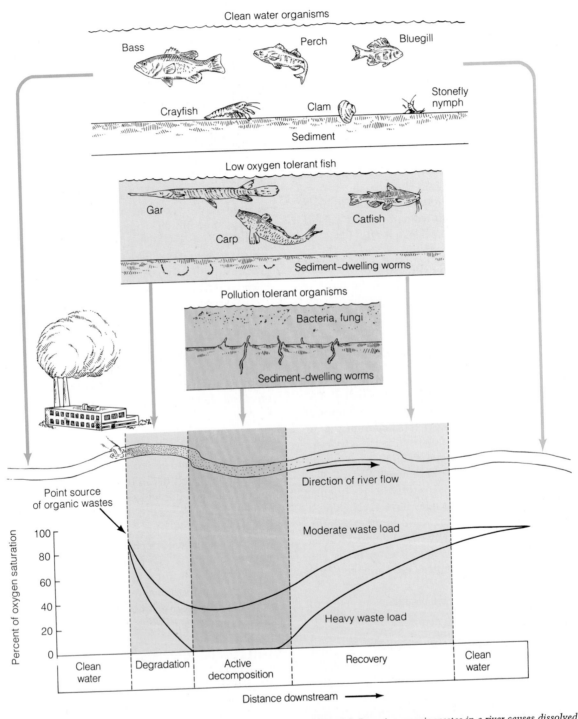

Figure 7.5 *Dumping organic wastes in a river causes dissolved oxygen levels to fall. The level of dissolved oxygen affects the number and types of organisms that inhabit various sections of the river.*

BOX 7.1

DYING OR EVOLVING, THE CHESAPEAKE BAY SHOWS SIGNS OF STRESS

Cambridge, Md.—Twenty-five years ago, when Ray Stevens was growing up in this town on the Eastern Shore of the Chesapeake Bay, he often went wading in the clear, knee-deep waters to catch crabs. Making his way through the lush, waving grasses on the bay's bottom, he carried a long pole with a net on the end and dragged along a galvanized metal washtub tied by a string to his waist.

"The crabs would scamper out of the grasses, and I'd flip them into the tub and never break stride," Mr. Stevens says. "I remember the water's being so clear I could see my tennis shoes."

Mr. Stevens now is 40 years old, and he still lives here, but the water at the mouth of the Choptank River is brown and turbid. The crabs have moved out to deeper zones. "The grasses are gone," Mr. Stevens says sadly. "There just aren't any. We don't know why."

During Mr. Stevens's lifetime, submerged grasses have vanished from many parts of the upper and middle bay. Scientists believe that they are beginning to understand why. The answers lie in assaults on the Chesapeake, many of them caused by humans, that are changing its ecology. Some environmentalists suspect the bay is dying. Most experts think the situation isn't that dire; they believe the bay is merely evolving.

The grasses are among the obvious casualties. Less obvious but no less important to the character of the bay's ecosystem are the effects on many other forms of life—plankton and larvae, crabs and oysters, striped bass and shad, canvasback ducks, herons, ospreys, Canada geese, whistling swans, and even bald eagles—that depend for nourishment on the environment those grasses provided.

For centuries, the Chesapeake has been what H. L. Mencken described as a "great big outdoor protein factory." It is the largest most bountiful estuary in the United States. Year after year it yields tens of thousands of tons of seafood valued

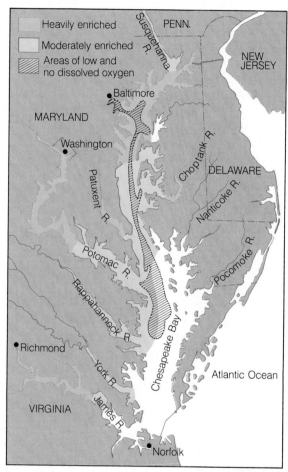

Map of Chesapeake Bay showing areas that are moderately or heavily enriched with plant nutrients and areas of low or no dissolved oxygen.

in the millions of dollars.

The bay actually is a "drowned" river basin, formed about 15,000 years ago when glacial ice melted. This generally shallow, 200-mile-long cornucopia of aquatic life is sustained and stimulated by a constant mixing of salt water from the Atlantic with fresh water from the nine large rivers that empty into it.

But ever since the first European settlers arrived and began to clear the land, the rivers have been delivering more than water. "Mud goes into the bay every time it rains," says Jerome Williams, a professor of physical oceanography at the U.S.

Naval Academy. "Fewer trees, more construction, more parking lots, more runoff—just about everything man does stirs up the water. . . . If you want a pristine bay, then forget about people."

The rivers also dump nutrients into the Chesapeake, especially nitrogen and phosphorus from sewage-treatment plants and from the fertilizer runoff of farms. These nutrients stimulate the growth of free-floating algae blooms in the bay, giving the water a murky quality and cutting off sunlight to plants and grasses below. When the algae blooms die, they consume large amounts of dissolved oxygen needed by plants and fish.*

"As it looks now, the prime culprit is light limitation caused by nutrient enrichment," says Thomas DeMoss, the deputy director of the Environmental Protection Agency's six-year, $27 million research program on water quality in the bay.

The EPA will report the program's findings to Congress later this week. The report will show correlations between the buildup of nutrients in northern parts of the bay and the virtual disappearance of grasses and other submerged vegetation in the same area.

It also will describe the accumulation of toxic materials on the bay floor, which experts say appears less serious than expected.

About 100 scientists and dozens of federal and state officials participated in the EPA project. Participants say that what they learned about the bay confirms some fundamental themes of nature: that everything is related to everything else, that stresses produce changes, that nature always finds ways to restore equilibrium to a system, that adapting is part of surviving.

"The bay will always be in balance," says Virginia R. Pankow, an oceanographer for the Army Corps of Engineers. "It will always have water and life. The water may be polluted, and the only fish may be carp, but it will always be in balance."

The EPA project also suggests the kinds of steep technical and political obstacles still to be overcome in trying to manage the resources of the bay. The human population of the bay area is expected to increase to about 15 million from about 8 million in the next 20 years, so the bay probably will be under greater stress and stress management should become more complicated.

Research on nutrients shows annual increases in concentrations of nitrogen and phosphorus over the past 30 years. Since the mid-1960s, the concentrations have roughly doubled in the upper half of the bay. The Susquehanna, Potomac, and James rivers are significant contributors of these nutrients.

The best evidence of stress on the bay's environment is the disappearance of the underwater grasses. Grace Brush, a plant geographer at Johns Hopkins University, analyzed about 25 core samples of sediment taken from the bottom of the bay to determine if the phenomenon might have occurred before in natural cycles.

Using the sediment layers to construct a "profile" of plant populations dating back centuries before the arrival of Europeans, Mrs. Brush concludes that the "wholesale extinction" of grasses that occurred in many parts of the bay about 15 years ago "is a unique event within the last 1000 years."

Some fishery experts note a corresponding decline in populations of striped bass, or rockfish, and other species that spawn in fresh water and feed on organisms that thrive in bay grasses. The U.S. Fish and Wildlife Service also associates the decrease in grasses to departures from the bay of redhead and pintail ducks that dive in the water for food. Meanwhile, however, the populations of menhaden, bluefish, and other species that spawn in salt water and that feed around algae blooms have been increasing in the bay.

"Some people would say that's a degradation," says the EPA's Mr. DeMoss. "But that's a value judgment. It's not for us to say whether this is good or bad, desirable or undesirable. All we can say is that it's a change."

*Readers should recognize this as an oversimplification.

crease, some aquatic organisms respond by increasing in number and size. In the shallow near-shore waters that receive added nutrients through runoff, rooted aquatic plants grow luxuriantly, which hinders swimming and boating and even reduces shoreline property values (see Figure 7.6). Algae blooms during the summer give the water the appearance of green soup. Those algae, primarily the blue-green species, release foul-smelling and unpleasant-tasting substances into the water, which adversely affects its recreational value. Furthermore, the cost for removing the algae and the compounds they release causes the cost of water that is supplied to municipalities and industries to increase. In shallow waters, some algae attach to the rocks, which gives them a green, slimy appearance. And, finally, oxygen depletion that results from the microbial decay of dead algal masses often causes fish kills in rivers and in the epilimnion of lakes. Lakes that receive overloads of nutrients also lose oxygen from the hypolimnion when the decaying algae settle to the bottom. Thus oxygen-depleted lakes also lose their populations of cold-water species of fish, such as lake trout, cisco, and whitefish.

Several sources of nutrients contribute to cultural eutrophication. Domestic sewage is an impor-

tant point source. Approximately 50 percent of the phosphorus that is contained in sewage comes from detergents (in areas where such detergents have not been banned). A modern treatment plant that provides secondary treatment of sewage (described in Chapter 8) may remove only as little as 10 percent of the nitrogen compounds and 30 percent of the phosphorus compounds that are present. Thus effluents that have been processed through those treatment plants may contain approximately 20 ppm nitrogen and 9 ppm phosphorus following treatment. Scientists estimate that excessive algal growth will occur when the phosphorus levels exceed 0.10 ppm (if phosphorus is limiting) or when the nitrogen levels exceed 0.3 ppm (if nitrogen is limiting).

Urban runoff is another significant source of plant nutrients. Studies in several cities across the country indicate that storm water runoff contains several parts per million of both nitrogen and phosphorus. Those materials evidently enter the runoff from lawn fertilizers, animal fecal material, dust, leached leaves, and combustion products. Table 7.3 lists the amounts of various pollutants in different sources of runoff.

Industrial inputs of nutrients vary widely. Paper mills contribute enormous quantities of oxygen-demanding substances but release low concentrations of plant nutrients. In contrast, industrial facilities that have large surface areas that need to be cleaned, such as creameries or car washes, use great amounts of phosphorus-containing cleaning agents. Phosphate mining is another major source, which, in the United States, has caused the most problems in Florida.

Figure 7.6 *The shallow zone in eutrophic lakes is often infested with dense growths of aquatic weeds. Shown here is (a) a harvester removing aquatic weeds from a Wisconsin lake, and (b) the harvested weeds being loaded for land disposal. (University of Wisconsin, Madison, Department of Agricultural Journalism.)*

(a)

(b)

Table 7.3 Land use and amounts of various water pollutants washed from surface (kilograms per hectare per year)

LAND USE		BOD	SUSPENDED SOLIDS	NITROGEN	PHOSPHORUS
Urban	Range	53–82	728–4794	3.2–18	1.0–5.0
	Average	75	1700	8.5	2.0
Pasture	Range	6–17	11.8–840	2.5–8.5	0.24–0.66
	Average	11	840*	5.3	0.30
Cultivated	Range	4–31	286–4200	15.0–37.0	0.18–1.62
	Average	18	4200*	26.0	1.05
Woodland	Range	4–7	45–132	2.4–5.1	0.01–0.86
	Average	5	98	3.1	0.10

Source: M. P. Wanielistra, Y. A. Yousef, and W. M. McLellon, *Journal Water Pollution Control* 49:441, 1977.
*Data limited; therefore, high value is used.

Studies indicate that nutrient runoff from properly managed farmland is only slightly greater than that from natural areas. However, in areas where large amounts of fertilizers are applied and drainage conditions promote runoff and erosion, the quantities of nutrients that are washed from farmland are significant. For example, in California's Central Valley, the nitrate concentration in water that seeps from irrigated fields approaches 50 ppm on occasion. In northern states, such as Minnesota, Wisconsin, and New York, the practice of spreading animal manure over frozen ground in the winter adds nutrients to runoff that occurs in the spring. Feedlots that drain directly into streams are another important source of plant nutrients. In contrast, groundwater and precipitation are usually minor sources of nutrients. Also, the importance of each type of nutrient source within a drainage basin varies from one drainage basin to another, depending on the relative proportion of nutrients that are contributed from each specific source.

Toxic Substances

Aquatic organisms are affected by oxygen-consuming pollutants and added plant nutrients because those substances directly or indirectly alter the amounts of oxygen available to them. *Toxic substances,* in contrast, affect organisms adversely, and sometimes fatally, because they disrupt the metabolism of organisms as a result of ingestion or contact. An organism's exposure to a toxic substance may be *acute* (as a result of a single heavy dose) or *chronic* (as a result of repeated, smaller doses over a long period of time).

During the past two decades, we have become more aware of the effects of toxic substances in aquatic environments. Although large industrial spills are reported on the evening news because of their immediate threat to human health and safety, we have also become more sensitized to an equally serious, if less dramatic problem: the escape of small amounts of persistent toxic chemicals that are released into the aquatic environment, and their subsequent movement through food webs, some of which may ultimately have a far-reaching effect on organisms.

Bioaccumulation of Toxic Substances Just as nutrients pass from one trophic level to the next, so also do some persistent toxic substances, for example, DDT and methyl mercury, pass along through food webs. As we saw in Chapter 2, if those substances are not excreted or broken down biologically by an organism, they are retained in its tissues. Over a period of time, if the organism continues to ingest that contaminated food, the chemical becomes more concentrated in the organism. Thus persistent pollutants that enter the food web become considerably more concentrated by the time they reach the highest trophic level. As a result, the organisms that feed at the highest trophic levels are exposed to the highest doses of toxic substances. Accumulation is especially pronounced in aquatic food webs, because they usually consist of four to six trophic levels; in comparison, terrestrial food webs usually only have two or three levels. Since humans normally eat organisms from the upper trophic levels of aquatic food webs, we run the risk of ingesting large amounts of toxic substances. For example, the mercury poisoning of hundreds of people in Minamata, Japan, dramatically illustrated the food-web accumulation of

a toxic substance and its effects on the human central nervous system (Box 7.2).

News of the Minamata tragedy triggered immediate concern about the potential health hazards of mercury in the United States. Testing was done, and fish from some rivers that receive industrial

---

BOX 7.2

MINAMATA—THE TOXIC BAY

People have long put up with pollution from local industries that are their economic lifeblood. Such was the case in the small coastal village of Minamata, Japan, where fishermen tolerated pollution problems from local industries for several decades. In the early 1950s, animals in the Minamata Bay region began acting strangely. Birds fell from their perches and flew into trees and buildings. Cats walked with an awkward gait; some became enraged, foamed at the mouth, and ran in circles until they died or were killed. Soon ominous symptoms appeared in the fishermen and their families; numbness, tingling sensations in the hands and feet, trembling, headaches, blurred vision, impaired speech. For the unluckiest victims, milder symptoms were followed by violent thrashing, paralysis, and even death. After some investigation it became clear that something was ravaging the villagers' brains and central nervous systems. The symptoms pointed to mercury poisoning.

Investigators found that fish and shellfish from Minamata Bay, which made up most of the villagers' diet, were contaminated with mercury compounds, including mercuric chloride. They also found these mercury compounds in the effluent that flowed from the nearby chemical plant of the Chisso Corporation into Minamata Bay. And they learned that the plant had greatly increased its use of mercuric chloride not long before the residents of Minamata began falling ill.

By 1959, researchers had demonstrated that certain microorganisms in bay sediments were converting the mercuric chloride and other toxic

mercury compounds into an even more toxic compound—methyl mercury. That chemical readily moves through food chains and accumulates at higher trophic levels.

How the surviving victims fared after the epidemic of Minamata disease, as the affliction had come to be called, is as tragic a story as that of the epidemic itself. The Japanese government did not officially recognize the cause of the disease until 1968, about 15 years after the first symptoms appeared. As late as 1970, government investigations recognized only 121 of the more severe cases as Minamata disease; of those, 46 proved fatal. And for years the Chisso Corporation itself managed to evade responsibility for what had happened. Not until 1973, after a 4-year struggle in the Japanese courts, was the company required to pay monetary damages—$3,530,000—to just 112 victims and their families. By 1975, more detailed health surveys had recorded another 3500 cases of the disease. But payments to so many more victims threatened Chisso's financial health. To avoid making them, the corporation established independent new companies from profitable subsidiaries. This move left the victims little hope of financial compensation if the parent company, stripped of its financial resources, filed for bankruptcy.

But monetary compensation, however generous, could not have undone the irreparable damage to the minds and bodies of many victims. On that day in 1973 when Chisso was ordered to make its first compensation payments, Shinobu Sakamoto, a 16-year-old girl whose muscular control had been gravely impaired, reacted grimly to news of the decision. "Money will not cure the disease," she said as she hobbled to school. "I want them to restore my body."

wastes, especially those downstream from chlorine manufacturing plants, were found to have levels of mercury that were much higher than the 1.0 ppm standard set by the United States Food and Drug Administration (FDA). Consequently, either fishing was banned in those rivers or warnings were posted, which stated that no more than one meal per week prepared from fish caught in those areas should be eaten. Since that time, industrial plants that use mercury have made major changes in their operations, so that, by now, the amount of mercury that is discharged is substantially reduced. That action has caused mercury levels in fish and shellfish to subside somewhat. But in some waterways, the levels remain higher than the 1.0 ppm limit set by the FDA, so commercial fishing cannot be resumed in those areas. The Minamata tragedy alerted the world to the dangers of mercury poisoning as a result of bioaccumulation in food webs. Had it not occurred, persons in the United States and many other countries may well have contracted that terrible disease.

Certain chemical properties that lead to bioaccumulation can be identified. For example, substances that are highly insoluble in water but soluble in fat tend to bioaccumulate. Also, the toxic compound must be chemically stable, that is, it must be able to persist in the environment for months to years. In order to persist, compounds must be resistant to biodegradation by organisms, reaction with water or moisture (hydrolysis), or degradation by sunlight (undergo photolysis).

Heavy Metals Mercury—the pollutant that was responsible for the Minamata tragedy—is only one member of a troublesome group of metals, which are called *heavy metals*. Chemicals that contain arsenic, cadmium, chromium, copper, lead, nickel, silver, zinc, thallium, vanadium, platinum, and gold are classified as heavy metals and are considered to be hazardous. Some of those compounds—arsenic, cadmium, nickel, and some forms of chromium (chromates)—are considered to be carcinogens as well. Moreover, most of the heavy metals act as enzyme inhibitors, which disrupt the metabolic processes of organisms. The heavy metals are of special concern because, since they are elements, they cannot be broken down to nontoxic forms. Thus once aquatic ecosystems are contaminated by those chemicals, they remain a potential threat for many years. Their danger is reduced only by either dilution in the en-

vironment or by being covered or buried, for example, by sediment.

Organic Chemicals Certain organic chemicals are particularly dangerous threats to the environment because they, too, bioaccumulate through food webs. The chemicals in this category that cause the most concern include pesticides that are applied to cultivated fields and forests. In the United States, farmers and foresters use enormous quantities of pesticides to control weeds, insects, rodents, and disease-producing fungi. Some 1800 different toxic chemicals are used as active ingredients in pesticides, and those chemicals are combined in more than 40,000 different pesticide formulations! Pesticides enter lakes and streams through runoff, through the effluents of pesticide manufacturing plants, from drifting spray mists during application, through the washing of spraying equipment, and by accidental discharges. Moreover, pesticides adhere to soil particles. When those particles are washed into streams, they may form contaminated sediments.

Of all the pesticides that enter surface waters and accumulate in food webs, one class of compounds, the *chlorinated hydrocarbons*, poses the greatest risk to our species and the environment in general. The Environmental Protection Agency (EPA) has banned the widespread use of several chlorinated hydrocarbons, for example, DDT, aldrin, dieldrin, heptachlor, chlordane, and toxaphene. Those chemicals can only be used to treat pest outbreaks on an emergency basis and then only with EPA approval.

Certain chlorinated industrial chemicals, PCBs (polychlorinated biphenyls), and the extremely toxic chemical, dioxin, (a chlorinated hydrocarbon), also bioaccumulate. The movement of PCBs through aquatic environments is well-documented. The complex interactions that occur when PCBs are introduced into aquatic ecosystems provide us with insight into the process of bioaccumulation as well as the complexities of reducing environmental contamination.

Because PCBs are fire-resistant and stable at high temperatures, they are used as insulating materials in electric capacitors and transformers, in hydraulic fluids, to a limited extent as heat-exchanging liquids, and in the manufacture of some plastics. Since 1977, the manufacture of PCBs has been discontinued in this country, but the problems associated with them have not disappeared. It is estimated that 300 million kilograms of the 540 million kilograms of PCBs that have already been manufactured are still in use and pose a continual threat if they are not handled properly. As a result of their use and eventual disposal, PCBs can enter aquatic ecosystems in a variety of ways. For example, when PCB-containing paper and plastics are burned, the PCBs vaporize and condense on dust particles. Those dust particles then either settle to the ground or onto surface waters or are washed out of the air by precipitation. PCBs also gain entry into aquatic ecosystems by means of runoff, evaporation of accidental spills, treatment of plant effluents, and leakage from improperly constructed or operated municipal and industrial waste-disposal sites.

PCBs accumulate at the upper levels of food webs. The significance of their bioaccumulation is illustrated by the level of contamination in Lake Michigan, which is one of the world's largest reservoirs of fresh water (Figure 7.7). That type of contamination occurs in the following way. First, seemingly insignificant amounts of PCBs in sediments and water become incorporated into bottom-dwelling organisms and algae. Then those slightly contaminated organisms are consumed by small fish, such as alewives, young salmon, and trout, which are then consumed by larger salmon and trout. The PCB levels in the larger fish increase as the fish grow older, until the levels of PCBs in their bodies make them unsafe for human consumption. To protect the public from the PCB hazard, the EPA has banned

Figure 7.7 *The accumulation of PCBs in Lake Michigan trout and salmon. The PCBs in these fish are over a million times more concentrated than those in the water.*

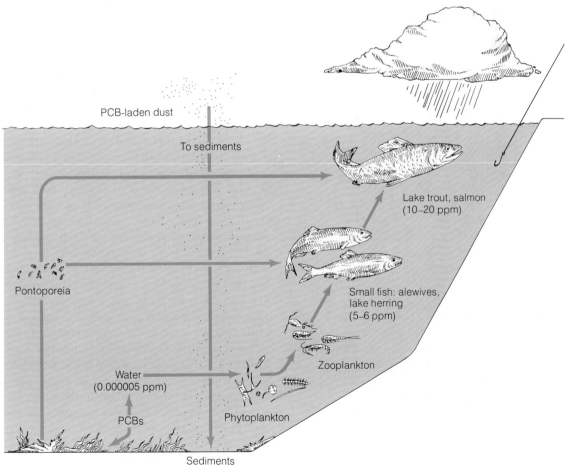

the commercial sale of Lake Michigan's highly prized salmon and trout, and sportsmen have been warned not to eat large fish and not to eat more than one meal per week of medium-sized trout or salmon that is caught in Lake Michigan. The FDA considers 2.0 ppm to be the safe limit in fish. Most large predator fish in Lake Michigan exceed those levels, as shown in Figure 7.8. Fortunately, because PCBs are now being used and disposed of more responsibly, the levels of PCBs are steadily declining in Lake Mich-

igan. (Disposal techniques for hazardous chemicals are discussed in Chapter 13.)

PCBs accumulate in humans in the same way that they accumulate in other animals. Samples of human fatty tissue contain approximately 1 ppm, because we ingest small amounts of PCBs in our food every day. Fish from the Great Lakes, New York's Hudson River, and the Upper Mississippi River, to mention just a few sites, contain high levels of PCBs. As a result, those who eat fish regularly from those sources are believed to be endangering their health.

Many studies have been conducted on the health effects of PCBs. Tests on fatty tissues from persons who are exposed to PCBs at work show PCB levels up to 700 times higher than normal levels. That group shows some increased incidence of chloracne (a skin disruption that is similar to acne) and impaired liver function. In Yusko, Japan, many persons who ate rice oil that had been accidentally contaminated with PCBs ingested an estimated 0.5 to 2.0 grams of PCBs per day during a 4-month period. At least 1057 persons suffered symptoms of chloracne, headaches, nausea, diarrhea, and rash. (Some evidence also exists that other, more toxic substances may have been present as a contaminant in the PCBs that contaminated the rice oil, and those substances may have been partially responsible for the observed symptoms.)

PCBs can be transferred to infants in the milk of nursing mothers who consume PCB-containing foods, especially fish. Because infants and young children are the most susceptible to most toxins, including this one, and because breast milk is often the infant's sole source of nourishment, health officials recommend that young girls, who are in their prereproductive years, and women, who are in their reproductive years, minimize their consumption of fish that are taken from contaminated areas.

The discovery that heavy metals and PCBs bioaccumulate has caused scientists to suspect that other chemicals, heretofore not detected, also may be contaminating food supplies, especially aquatic food resources. Recent studies indicate, for example, that dioxin and other closely related compounds accumulate in food webs. Although the dioxin levels are much lower (approximately 1 million times lower than the levels of PCBs in the fish from Lake Michigan), dioxin is many times more toxic than PCBs.

Figure 7.8 *Levels of PCBs in filets from two Lake Michigan fish species that are important food sources for humans: (a) coho salmon and (b) lake trout. Coho salmon typically live 4 years while lake trout can live much longer. Note the much higher levels of PCBs in larger (older) lake trout. (Data from the Wisconsin Department of Natural Resources.)*

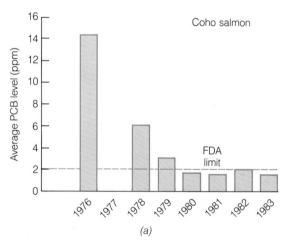

(a)

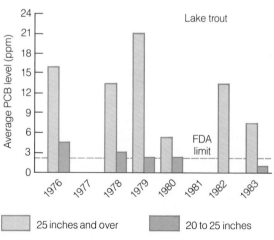

(b)

For example, a single dose of 6 micrograms (6 millionths of a gram) per kilogram of body weight will impair the immune system (disease-fighting system) in mice.

Today, we know that bioaccumulation is a natural concentration mechanism, which must be considered in the overall management of toxic chemicals. Much more scientific study will be required to identify all the chemicals that are potential hazards and to find the technical information that we need to determine how to control them.

Other Toxic Chemicals Some chemicals that are used in industry pose direct hazards for both humans and aquatic life. Some examples are ammonia, cyanides, sulfides, fluorides, strong acids, and alkalis. (Acid water in the form of acid rain is discussed in Chapter 10, and acid mine drainage is covered in Chapter 12.) Most of those chemicals are used in concentrated forms, which are directly toxic to organisms but indispensable to industry. For example, compounds of cyanide, such as potassium cyanide, are essential for electroplating electronic components with silver and gold. Other toxic chemicals, such as phenols, are used in the manufacture of plastics. Accidental discharges of those highly toxic wastes, which are usually the result of human error, occasionally wipe out aquatic communities for kilometers downstream. The subtle long-term effects of those toxic chemicals, which occur at concentrations that are lower than the tolerance limit for aquatic organisms, include lowered rates of reproduction, reduced growth rates, and abnormal behavior. Most sublethal effects of toxic substances on aquatic life are difficult to document, however, and further research is necessary to identify them accurately.

Oil Spills Oil pollution is an ever-present threat to our surface waters, especially to the oceanic and river waterways that are used to transport "black gold." During the past decade, several dozen major oil spills have occurred worldwide, and one of those incidents, the grounding and breakup of the *Amoco Cadiz*, on March 16, 1978, was the worst spill in maritime history. That supertanker, which was carrying some 223,000 metric tons of crude oil, encountered heavy seas in the North Atlantic. Its steering mechanism failed, and it went aground 2 kilometers (1.2 miles) from the northwest coast of France, rupturing its storage compartments. The entire cargo escaped during the following 2 weeks. Approximately 320 kilometers (200 miles) of the French coastline experienced heavy deposits of oil. In total, about 15,000 square kilometers (5800 square miles) of ocean experienced oiling.

Oil that is spilled into the ocean is dispersed by several routes. The insoluble fraction, which is by far the largest fraction, is lighter than water and gradually spreads and thins to form an ever-widening oil slick on the water's surface. The most volatile (and toxic) components in crude oil (typically 30 percent) immediately begin to evaporate into the atmosphere. In areas that experience wave action, the oil is whipped into an oil–water emulsion, which contains 50 to 70 percent water. If the spill occurs near the shore, the waves transport the oil to the intertidal zone and beaches, which become coated with oil (see Figure 7.9). Wave action and currents also distribute the emulsion throughout the upper layer of the water, where bacteria break down most of the oil's components. However, that biochemical process is slowed considerably in cold water.

Oil that is not rapidly degraded by bacterial action collects to form floating tar balls that range from the size of marbles to the size of baseballs or it sinks. Crude oil (which is untreated, from the ground) usually contains a fraction that is heavier than water. That type of oil sinks to the ocean floor and coats and kills the animals at the bottom. Oil-coated, coarse-grained sands that exist on beaches or on the ocean floor in high-energy environments—where strong waves or currents prevail—clean up quite rapidly. However, sheltered areas that contain fine-grained sediments remain contaminated for several years. Therefore, aerobic, high-energy environments tend to disperse and degrade oil rapidly, whereas low-energy, anaerobic conditions allow it to persist.

The ecological and economic consequences of oil spills are not easy to assess. Near-shore oil spills that do not disperse rapidly, however, are particularly damaging, because they affect many vulnerable marine organisms. Layers of oil form coats on birds and aquatic furry animals (for example, sea otters and seals), thereby destroying their natural insulation and buoyancy. Most of those victims die of exposure or drown. Evidence also exists that environ-

Figure 7.9 *This oil-coated beach was the result of the grounding of the oil tanker Amoco Cadiz. The beach is located on France's Normandy Beach area along the English Channel. (© Martin Rogers, Woodfin Camp & Associates.)*

of attention, but they account for only 20 percent of the estimated 3.6 million metric tons of oil that is spilled or dumped into oceans each year (Table 7.4). The types of oil that are spilled range from thick, crude oil and asphalt to the lighter, refined products, including heating oil, diesel fuel, and gasoline. The major causes of oil spills include collisions and groundings of tankers and barges, oil-well blowouts, and ruptures or leakages of oil storage tanks and pipelines. The U.S. Coast Guard estimates of the amount of oil that is spilled annually in the United States waters varies widely. Reports show, however, that oil is spilled in all the nation's waterways. Only isolated surface waters that have no contact with petroleum in their drainage basins are free from the threat of an oil spill.

Pollution by Sediments

When raindrops strike bare ground, they dislodge soil particles. Those particles then travel through runoff into streams, lakes, or oceans and are deposited there as *sediment*. Coarse sediments, such as sand, settle out of the water quite rapidly, but fine particles, such as clays, can remain in suspension for months, making the water turbid for the duration. Although rivers have always transported enormous quantities of sediment to the sea, their sediment loads today are greater than ever. Soils that are laid bare by crop cultivation, timber cutting, strip

mental stresses, such as low temperatures, act synergistically with oil because, in several oil-spill incidents, even light coatings of oil have led to the death of seabirds.

The effects of oil on the water surface generally last for a few days to a few weeks. However, oil that remains on the ocean floor is thought to have the greatest long-term impact on marine ecosystems. Initially for a short period of time after a spill, shellfish die, but for a longer period of time, when the pollution levels decrease, shellfish in the contaminated area remain offensive to smell and to taste. Thus commercial fishermen lose their catches for several seasons. The long-term effects of low levels of oil are difficult to assess, but, to date, no major effects have been identified.

Spectacular accidental spills draw a great deal

Table 7.4 Estimated inputs of petroleum hydrocarbons in the world's oceans during the early 1980s

INPUT SOURCE	QUANTITY (MILLIONS OF TONS PER YEAR)
Coastal Refineries	0.02
Urban Runoff	0.12
Atmosphere	0.15
Offshore Production	0.18
Municipal and Industrial	0.41
Natural Seeps	0.54
Transportation	0.73
River Runoff	1.45
Total	3.6

Source: B. Hileman, Offshore oil drilling. Reprinted with permission from *Environmental Science and Technology* 15(11):1260, © 1981, ACS.

Figure 7.10 *Erosion from construction sites is a major source of sediment-caused water pollution. (M. L. Brisson.)*

mining, overgrazing, road building, and other construction activities are subject to high rates of erosion. It has been estimated that construction sites (see Figure 7.10) contribute 10 times more sediment per unit of area than cultivated land, 200 times more than undisturbed grassland, and 2000 times more than undisturbed forested land. Drainage basins, where strip mining has occurred, also experience erosion rates that are comparable to rates that are associated with unprotected crop cultivation. Overall, in the United States, 85 percent of the soil erosion is estimated to originate from cropland. (The rates of erosion in the United States are shown in Figure 7.11.) Although soil losses that are as high as 1 metric ton per hectare (3 tons per acre) per year are viewed as being acceptable in terms of soil fertility, those losses are not acceptable from a water-quality standpoint.

Figure 7.11 *Relative rates of erosion in the United States. (U.S. Department of Agriculture, Soil Conservation Service.)*

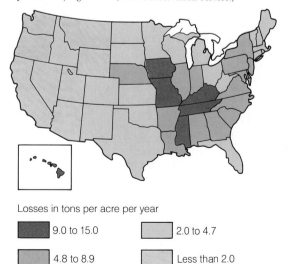

Losses in tons per acre per year

■ 9.0 to 15.0	2.0 to 4.7
4.8 to 8.9	Less than 2.0

Figure 7.12 *Silt deposits can fill a water reservoir completely if soils are left unprotected. (U.S. Department of Agriculture.)*

The loss of valuable agricultural soils is not the only serious result of erosion; many other water-related problems are associated with the wearing down of soil as well. For example, eroded soil particles eventually fill lakes, reservoirs, navigation channels, harbors, and river channels (Figure 7.12). As a result, the excessive sediments greatly reduce the attractiveness of lakes and reservoirs, which causes them to be less suitable for water-based recreational activities. They also impede navigation, cover bottom-dwelling organisms, eliminate valuable fish-spawning areas, and reduce the light penetration that is necessary for photosynthesis. In addition, soils that are eroded from farmlands sweep nutrients in the form of nitrogen and phosphorus into surface waters. Thus reservoirs that receive large sediment deposits are usually eutrophic, because the shallow water and high rate of nutrient input result in high rates of nutrient cycling. The ecological

changes that have occurred in the Chesapeake Bay, which were caused partly by the influx of sediments and nutrients, were described earlier in Box 7.1.

Although $15 billion has been spent on erosion control in this country since the Dust Bowl days of the 1930s, soil erosion is still one of our most pervasive environmental problems. It is expensive to dredge and transport sediments from reservoirs and stream channels. Each year, the United States Army Corps of Engineers dredges about 300 million cubic meters (400 million cubic yards) of sediment from our waterways (which is enough to cover 360 square kilometers, or 90,000 acres, of land to a depth of 1 meter) at a cost of approximately $300 million. Thus if we consider both the economics and our efforts at soil conservation, preventive measures make more sense than cleaning up after the fact. But those measures are difficult to implement, because the persons who must initiate and pay for the conservation measures often reap few economic benefits. One way of controlling erosion would be to persuade farmers to use control techniques in cultivating their land (some of those techniques are discussed in Chapter 17). Other strategies that can be used to reduce the amount of sediment that reaches our surface waters include (1) minimizing the time during which construction and mining sites are left unprotected by vegetative cover, (2) reducing the size of those areas as much as possible, (3) building sediment-retention basins, and (4) directing water through swamps and marshes, which slows its flow and thereby allows sediments to settle out.

Thermal Pollution

Efficient large-scale generation of electricity requires enormous quantities of water for cooling—water that is usually drawn from oceans (bays and inlets), major rivers, and large lakes. Power plants that burn coal or oil or that use nuclear fuel generate great amounts of waste heat, some of which is removed by circulating cool water around and through hot electricity-generating machinery. Thus the heat is transferred to the water, raising its temperature. When that water is discharged, it can have an adverse effect on aquatic ecosystems. For example, organisms that live in water that receive the waste heat from power plants are killed when the temperature of their aquatic environment rises to levels that are higher than they can tolerate. Waste heat that disrupts an aquatic ecosystem is called *thermal pollution*.

In a body of water that is thermally polluted by a discharge from a power plant, the heated water is confined to a mixing zone, which is called a *thermal plume*. It is usually located downstream from the point of discharge (Figure 7.13). Fish can be killed in that zone if they encounter temperatures that are higher than their upper temperature tolerance limits.

Thermally polluted water can also stress aquatic life indirectly. For example, animals, including fish, oysters, and clams, whose body temperatures are nearly the same as that of their surroundings, are particularly vulnerable. When the water temperature rises, their body temperatures also rise, causing an increase in their respiration rates, which, in turn, causes an increase in their oxygen demands. But, as we saw earlier, at higher temperatures, water con-

Figure 7.13 *An infrared photo of the thermal plume from a power plant located on Lake Michigan. The plume extends approximately 1500 meters (4900 feet) offshore. The photo was taken during relatively calm seas. (Professor T. Green, Marine Studies Center, University of Wisconsin–Madison.)*

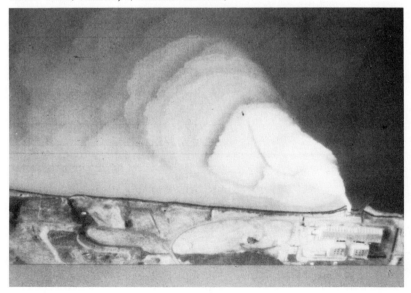

tains less dissolved oxygen, so thermal pollution causes those animals to suffer oxygen stress. Thermal pollution is greatest during the summer, when the demand for power is greatest in most areas of the United States, and when organisms are the most vulnerable to a rise in water temperature. Moreover, toxic chemicals and diseases also pose a greater threat when the water that contains them reaches elevated temperatures.

Some fish can survive the elevated temperatures in a thermal plume. Many game fish, however, cannot, because they have lower tolerance limits for temperature and higher requirements for oxygen. Thus thermal plumes become inhabited with fish that can tolerate higher temperatures, which are often undesirable species, such as carp.

If a power plant shuts down quickly, causing the mixing zone to disappear temporarily, fish that have become acclimatized to the warm waters in that zone experience *thermal shock*, because the water tem-

peratures fall suddenly. Thermal shock, which results from either sudden cooling or sudden warming, is often fatal to fish. Thus when power plants shut down, they should do so over a period of a day or two, so the fish that live in its plume have sufficient time to adjust to the change in temperature. In general, marine fish, especially those located in tropical or subtropical regions, have lower tolerance limits than freshwater species have to temperature changes and thermal shocks.

In regions where the discharge of heated water would badly disrupt aquatic ecosystems, evaporative cooling towers (such as the one shown in Figure 7.14) or cooling ponds are constructed. Those alternative methods of dissipating heat require less water and eliminate the detrimental effects of waste heat that is discharged into rivers, lakes, and marine environments. However, the cooling towers can have negative effects on the environment as well. For example, in cold climates, the resulting water vapor may increase driving hazards by fogging nearby highways and icing roads.

Although heated discharge water creates some ecological problems, several potential uses for it have been suggested, which could be beneficial. For example, *aquaculture*, or fish farming, has promise in some locations, where warm water can be cycled

Figure 7.14 *Large evaporative cooling towers, such as those shown at the Three Mile Island Nuclear Power Plant near Middletown, Pennsylvania, are used to disperse heat into the air rather than into surface water, where thermal pollution is a problem. Only one set of cooling towers is in use because an accident damaged one of the nuclear reactors. (© Bill Pierce, Woodfin Camp & Associates.)*

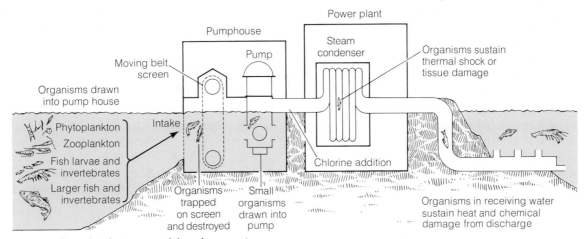

Figure 7.15 *Effects of cooling-water withdrawal on aquatic eco-systems.*

through pens to speed the growth of catfish, shrimp, and oysters. In Japan, various types of aquaculture enterprises have been successful. In the United States, some operating catfish farms exist, and other types of aquacultural enterprises are in developmental stages. Heated discharge water can be used for warm-water irrigation to extend the growing season in frost-prone regions. Heated water might also be used to heat buildings and swimming areas, de-ice canals and their lock systems, and provide heat for various industrial processes (see Chapter 18).

Unfortunately, economic considerations dictate that waste heat must be used in the immediate vicinity of its source. Thus if waste heat is used for heating buildings or as a heat source for an industrial process, the power plant must be located near the urban area. The major problem with this possibility is that fossil-fueled and nuclear-fueled electric power plants, which are by far the largest contributors of waste heat, have other adverse effects on nearby environments (see Chapters 10, 11 and 18). Whether the beneficial uses of recovered waste heat outweigh the increased risks of living near a power plant has yet to be decided by our society.

The withdrawal and use of cooling water that is discharged from power plants also causes other problems that are not directly related to heat discharge, as shown in Figure 7.15. Many fish die on the intake screens that are installed to prevent fish from being pumped through the small pipes in power plants. Moreover, toxic levels of chlorine are periodically flushed through the cooling system (usually for about 20 minutes every 4 hours) to kill and remove bacterial slimes that build up inside the pipes and reduce plant efficiency. That chlorine remains concentrated in the discharge water for a short period of time, where it can be toxic to aquatic organisms. Another problem is that in moderate-size lakes, the pumping of large volumes of cold, nutrient-rich hypolimnetic water to the surface increases the rate of nutrient cycling and speeds up the eutrophication process.

Radioactive Substances

Radioactive substances are another type of pollutant that can affect aquatic ecosystems. Those substances are either produced or used at a number of different sites, such as nuclear power plants, nuclear-weapons production sites, hospitals, and research laboratories. Although the Nuclear Regulatory Commission has strict requirements concerning the amounts of radioactive substances that can be discharged into sewers or the environment, the major concern is the possibility of accidental discharges at sites where high-level radioactive substances are stored. For example, at Hanford, Washington, 16 leaks occurred between 1958 and 1973. Since that time, better double-walled storage tanks, more reserve storage capacity, and a liquid-waste solidification program have reduced both the volume of liquid wastes and the chances of an accidental release.

In general, strict controls on the release of radioactive substances have prevented the cycling of radioactive substances through aquatic ecosystems, thus preventing contamination and bioaccumulation problems. In particular, the ban on above-ground nuclear weapons testing has reduced the amount of radioactivity that cycles through ecosystems. If those controls fail, some radioactive substances, such as strontium or iodine, would cycle and concentrate in human bones and the thyroid gland, respectively.

Chapters 13 and 18 discuss further the problems of radioactive waste disposal and the effects of radiation from radioactive substances.

Groundwater Pollution

The same drainage-basin activities that pollute surface waters can also contaminate groundwater. Landfills, agriculture, septic tanks, industrial waste lagoons, underground injection wells (in which liquid industrial wastes are pumped under pressure into

Figure 7.16 *Common activities that lead to groundwater contamination: leaky underground storage tanks, surface lagoons, leachate from landfills, and seepage through mine tailings.*

Underground storage tanks

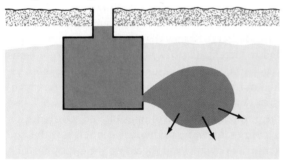

Pit, pond, or lagoon

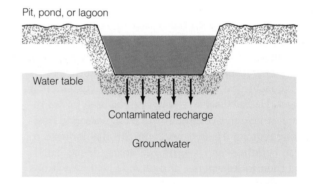

Landfills

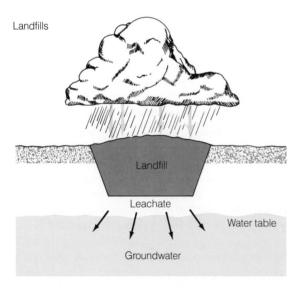

Mining

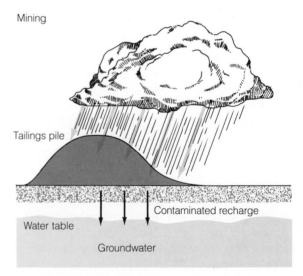

permeable rock layers), underground storage and petroleum and natural gas production can all lead to groundwater pollution (see Figure 7.16). Because approximately one-half of all United States' residents rely on groundwater—most of it used directly from the ground without any treatment—for their primary water supply, groundwater pollution can be a very serious problem.

The quality of groundwater varies, depending on its source and the interactions it has with the soil and rock materials through which it flows. For example, groundwater that is drawn from surface sand aquifiers is relatively "soft," because sand is virtually insoluble in water and thus contributes little dissolved material to it. More frequently, however, water is "hard," that is, it contains calcium and magnesium, which were dissolved in the water when it passed through layers of dolomite or limestone or both. Hard water can be a problem in residential use, because it requires large quantities of soap or detergent to achieve proper cleaning action. Water-softening detergents, which contain phosphorus, are used widely in hard-water areas, but as we have seen, phosphorus stimulates the overgrowth of plants in surface waters, where it is a limiting factor. Moreover, at high temperatures, the calcium in hard water precipitates out of solution, which causes hot-water pipes to clog and a buildup of scale to occur in industrial boilers. Other naturally occurring substances in groundwater that cause problems are iron and manganese. Those dissolved metals form the red-orange and black stains that are often seen on bathroom fixtures and impart a metallic taste to the drinking water. Hydrogen sulfide is also found in groundwater, which causes the odor of rotten eggs.

As water enters the ground, it is filtered naturally through soil, and in some soils, that process quite effectively removes many substances, including suspended solids, bacteria, and some viruses. Some chemicals, such as phosphates, bind chemically to the surface of soil particles and thus are removed. In fact, infiltration through soil is so effective as a purifying mechanism that the locations of septic tanks are carefully chosen to take advantage of it (Figure 7.17). In some areas industrial and municipal wastes are sprayed on the ground surface so that they will filter through the soil, become purified in the process, and recharge the groundwater

reservoir. Spray irrigation and ponding (shown in Figure 7.18) both employ infiltration to purify wastewater and recharge groundwater supplies at the same time. In the United States, approximately 600 communities, mostly in arid and semiarid areas, reuse municipal treatment plant effluents after they have been treated in this way.

Some wastewater systems that are designed to take advantage of the soils' natural filtering capacity work well. Others fail because proper filtration only occurs under certain soil drainage conditions. Filtration is impaired by high water tables, soils with low permeability, near-surface bedrock, and extremely sandy soils (soils with extremely high permeability). In areas where any of those conditions exist, wastewater is not in contact with the soil long enough for it to be purified adequately before reaching the groundwater. The bedrock of the Yucatan Peninsula of Mexico, for example, is composed of highly permeable limestone. Pollutants from cesspools and accumulated animal wastes seep into cracks and crevasses, which act as conduits into the wells of the region. Contaminated drinking water has caused infectious hepatitis to reach epidemic proportions in the Yucatan, and dysentery now ranks as the leading cause of death in that region. In certain areas of the United States, where only a thin layer of soil overlies bedrock, contamination problems also occur. Groundwater under areas in which a thin layer of soil covers the dolomite bedrock (such as in the area illustrated in Figure 7.19) is particularly susceptible to contamination.

The importance of choosing appropriate soil conditions for the sites of septic disposal systems is revealed by the statistics on this practice. In the United States, approximately 20 million single-family housing units (40 to 50 million persons) use on-site wastewater disposal. Those systems represent the single largest source (about 4 trillion liters per year) of wastewater that is discharged directly into the ground. Also, they are the most frequently cited sources of bacterial contamination of groundwater, especially the contamination of private wells. The problem is particularly acute in heavily settled suburban areas that are served by septic disposal systems. The high density of septic systems, together with a high density of private wells, greatly increases the danger.

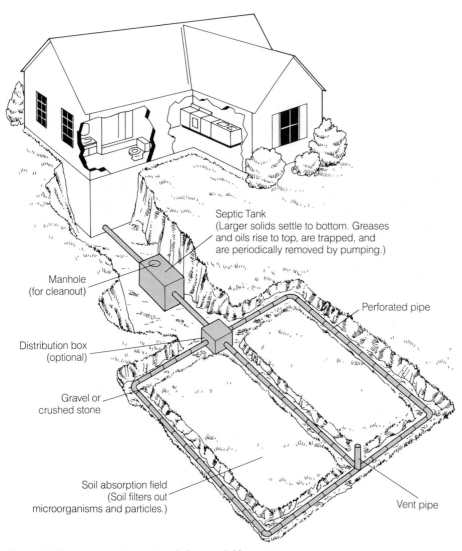

Septic Tank
(Larger solids settle to bottom. Greases and oils rise to top, are trapped, and are periodically removed by pumping.)

Manhole
(for cleanout)

Distribution box
(optional)

Gravel or
crushed stone

Soil absorption field
(Soil filters out
microorganisms and particles.)

Perforated pipe

Vent pipe

Figure 7.17 *Cross section of a septic tank drainage field system, which is used for domestic waste disposal in rural and suburban areas.*

Furthermore, the inadequacies of old disposal systems, both septic tank systems and cesspools, are creating pollution problems today. Cesspools are simply big holes that are dug into the ground, into which household liquid wastes are emptied. In the past, many systems were sited in unsuitable soils and even directly over fractured limestones, and dolomites. Water from disposal systems that are located directly on top of such rock finds its way into wells, and systems placed in clay soils quickly be-

Figure 7.18 (facing page, top) *This artificial recharge basin near Bushland, Texas, is used to purify runoff and to recharge groundwater supplies in the southern Great Plains. (USDA Photo by Fred S. Witte.)*

Figure 7.19 (facing page, bottom) *An area where a thin layer of soil covers dolomite bedrock, making the groundwater below particularly susceptible to contamination. (M. L. Brisson.)*

come plugged and rapidly overflow into nearby surface waters. Older septic systems, even when properly sited, often become overloaded, because many of them were designed to handle much smaller volumes of wastewater than those produced in modern households. Dishwashers, automatic washing machines, and garbage disposal units, as well as increased personal use of water as a result of today's greater emphasis on personal hygiene have resulted in the generation of much more wastewater per household than there was when many septic systems were installed.

We have seen that, under proper conditions, soil can cleanse water that moves through it by filtering out suspended materials and by adsorbing (chemically binding) certain dissolved chemicals. But some kinds of water pollutants are not retained by soil particles, and such substances can percolate directly into the groundwater. Some of the substances that cause trouble are nitrate (NO_3^-), road salts (calcium chloride or sodium chloride), petroleum products (gasoline and oil), odorous chemicals in water that seeps from landfills, and some agents that are used to remove grease deposits from household plumbing (such as trichloroethylene).

Nitrate is a particularly troublesome substance. It seeps into the ground from feedlots, septic tanks, and fertilized fields. It is highly soluble in water and is not adsorbed by soil particles. Nitrate is dangerous when it gets into well water. In concentrations above 45 ppm, it can induce spontaneous abortion in cattle. It is also linked to methemoglobinemia (which is commonly called *blue-baby disease*), an often fatal condition in newborn infants that results from oxygen depletion in the blood. Nitrate concentrations can build up in the groundwater in regions where agricultural fertilizers are used heavily. In the Central Valley of California, concentrations of nitrate have reached levels that are considered to be a public health hazard, and doctors there recommend that only pure bottled water be used in preparing infants' formulas.

A recently recognized threat to groundwater is trichloroethylene, which is a carcinogen in mice, and its related compounds, which are used to remove grease from the drains in homes and in industry as a degreasing solvent. The chemical is not destroyed in the septic tank systems serving homes in rural areas, nor is it retained by soil particles once

it is discharged. The result has been that trichloroethylene and its related compounds are found frequently in public water supplies. In Nassau County and Suffolk County, New York, several community water-supply wells (23 and 13, respectively) have been closed, which has affected the lives of approximately 2 million persons. In January, 1980, 39 public wells, which supplied 400,000 persons, were closed in 13 cities in the San Gabriel Valley. EPA data show serious contamination of the well water in those areas by organic chemicals, even though the locations from which samples were taken were not selected on the basis of potential contamination.

The Threat of Underground Fuel Storage

An increasingly common contamination problem is the leaking of underground fuel storage tanks. Service stations, farms, and many industries and institutions store fuel in steel tanks—thousands of them underground—which are subject to corrosion and thus to leakage. In 1983, the EPA warned that each year some 42 million liters (11 million gallons) of gasoline escape from 75,000 to 100,000 leaking storage tanks in this country and that the number of leaking tanks is increasing.

Because soils do not retain gasoline and other petroleum-based fuels, and because it is not required by law in this country that storage facilities be monitored for leakage, slow leaks usually remain undetected until someone discovers that a well is contaminated. However, by that time, the problem is already very serious, because a very small amount of gasoline can contaminate a great deal of water. In fact, an EPA spokesman (who testified before the Senate Environment and Public Works Committee's Toxic Substances Subcommittee in 1983) said that a gasoline leak of just 1 gallon a day could seriously contaminate the water supply of 50,000 persons.

The Threat of Solid Wastes

Waste that is disposed of by municipalities, industrial plants, and mines also can contaminate groundwater. (This topic is discussed in more detail in Chapter 13.) Some 200,000 sources of solid wastes exist in the United States, where they are either treated or buried. Water seeps through those sites and dissolves (leaches) soluble chemicals, thereby becoming polluted. Furthermore, many disposal sites that contain hazardous wastes are very old and pose serious problems because they were not sited or engineered properly. In the mid-1980s, the Federal Office of Technology Assessment identified 15,000 problem sites that need to be cleaned up at an estimated cost of $10 to $40 billion. In most of those cases, groundwater contamination is the most serious problem.

The quality of water that enters the groundwater reservoir depends largely on the quality of the water that percolates through the recharge zone. Water that passes through a contaminated recharge zone can contaminate a groundwater reservoir for hundreds or even thousands of years, because little dilution or pollutant removal occurs in aquifers. Thus the key element in protecting groundwater quality is to prevent pollution. This means that we must control the location and operation of human activities in recharge zones: waste-disposal sites of industries, housing developments, the use of chemical sprays to control pests, and so on. Protecting groundwater quality is a complex problem that most states are just beginning to wrestle with.

Conclusions

Our water pollution problems are complex. The sources of pollution are many, the types are varied, and the effects are often poorly understood. Instances of catastrophic water pollution can distract our attention from the subtle effects of low-level water contamination, which is considered by many scientists to be the more important problem. Those complexities make the control of water pollution difficult. The variety and multitude of pollution sources challenge the ingenuity of water-pollution control engineers and regional planners. Moreover, the actual effects of low concentrations of many chemical wastes on human health have not been established, and those uncertainties make industries, governments, and the public reluctant to pay for costly pollution-abatement measures. Nevertheless, every effort must be made to control sources—especially point sources—of pollution, because once pollutants are released, they are virtually impossible to retrieve. In the next chapter, we survey the ongoing efforts to clean up our waterways and to evaluate their overall effectiveness.

Summary Statements

Until recently, waterborne diseases comprised our primary concern about water pollution. The chlorination of water supplies has controlled that danger in the United States and in other developed nations, but waterborne diseases still afflict the under-developed countries.

The presence of coliform bacteria is used as an indicator of waterborne pathogens. The threat of waterborne disease is always present; to minimize it, public water supplies and wastewater effluents from municipal sewage systems are chlorinated.

The decomposition of organic wastes in water by bacteria and fungi removes dissolved oxygen from the water. The concentration of organic material in water is measured by the biochemical oxygen demand (BOD). Surface waters that have reduced levels of dissolved oxygen are inhabited by pollution-tolerant organisms, most of which are considered to be undesirable by humans.

Municipal treatment plant effluents, some industrial effluents, and runoff from urban and agricultural areas contain significant quantities of plant nutrients. Phosphorus and nitrogen compounds are the nutrients in those sources that are most often responsible for cultural eutrophication. Eutrophication reduces the quality of water, making it less able to support desirable aquatic life and less suitable for municipal and industrial use.

Many chemicals are toxic to aquatic life. Chemicals that are resistant to biodegradation are stable in water; fat-soluble chemicals are an especially serious problem, because they bioaccumulate in food webs. PCBs and mercury are examples of substances that have accumulated in fish to levels that are higher than FDA limits.

Oil spills kill most of the organisms that become coated with oil. Contaminated areas in high-energy environments—those with heavy wave action or fast-flowing currents—cleanse themselves quite rapidly, whereas low-energy environments remain contaminated for several years or longer. The chronic (long-term) effects of compounds in oil that are resistant to breakdown are poorly understood.

Heated-water discharges may exceed the temperature tolerance limits of some aquatic organisms and, therefore, must be treated by cooling towers or cooled in ponds. The withdrawal of cooling water from surface waters for power plants may kill fish on intake screens.

Erosion-caused sediments fill reservoirs, lakes, harbors, and navigation channels. Suspended sediments impede photosynthesis and carry nutrients into bodies of water; as they settle out, they cover bottom-dwelling organisms and destroy fish-spawning areas.

Human activities that produce or use radioactive materials have strict restrictions in terms of disposal and, therefore, they are seldom a problem in the surface waters of the United States.

The quality of groundwater is controlled mainly by purification processes—biodegradation by soil organisms, filtering, and adsorption—that occur as water infiltrates soil. Areas that have thin layers of soil over fractured bedrock are especially susceptible to groundwater contamination. Defective or improperly sited septic-tank systems in suburban and rural areas endanger groundwater quality; so do faulty landfills, wastewater ponds, and fuel-storage tanks. To protect groundwater quality in some areas, it may be necessary to restrict practices that lead to groundwater pollution, such as heavy fertilization of fields and the use of certain chemicals, such as trichloroethylene, in recharge zones.

Questions and Projects

1. In your own words, write a definition for each of the terms that are italicized in this chapter. Compare your definitions with those in the text.

2. Develop a short history of water-pollution problems in your area.

3. List the various types of water pollutants that are discussed in this chapter. Give an example and a possible source for each.

4. What are pathogenic organisms? How do scientists determine whether they may be present in water?

5. Describe the sequence of events that leads to the lowering of dissolved oxygen levels in streams that receive organic wastes.

6. What does it mean to say that a stream has the capacity to cleanse itself?

7. How do the chemical products of aerobic decomposition differ from those of anaerobic decomposition?

8. Which of the following substances will decompose in bodies of water or soil water: chromium-plating wastes, mercury, bread-baking wastes, meat-packing wastes, and PCBs?

9. How does natural eutrophication differ from cultural eutrophication? What water bodies in your area have suffered from cultural eutrophication? What are the major sources of the nutrients that promote cultural eutrophication?

10. Cultural eutrophication often leads to excessive populations (blooms) of blue-green algae. List three reasons why those blooms are undesirable.

11. Is fishing prohibited in any lakes or rivers in your region? Are people warned to restrict their consumption of fish or shellfish from any of those bodies of water? What contaminants are responsible for the warnings?

12. Explain how a contaminant in dilute concentration can become a problem through bioaccumulation.

13. What characteristics of a chemical allow it to bioaccumulate?

14. How can a large oil spill affect the economics of a region? Would you expect the economic effects to be permanent?

15. What properties of oil make it lethal to birds?

16. Where are oil spills most likely in your region? What aquatic ecosystems would be disrupted?

17. What water-related problems develop as the result of severe soil erosion? List the various types of land use in order of increasing susceptibility to soil erosion.

18. Distinguish between point and nonpoint sources of pollutants.

19. What types of systems are used to dispose of household wastes in the rural areas of your state? How are they regulated?

20. Do the rock and soil in your region have properties that are conducive to groundwater pollution? What are they? Where are they located?

21. What types of pollutants are most often responsible for groundwater contamination in urban–industrial areas? Suburban areas? Agricultural areas?

22. What single water-pollution problem has been debated the most in your community?

Selected Readings

COUNCIL ON ENVIRONMENTAL QUALITY: *Annual reports, 1970–1984.* Washington, D.C.: U.S. Government Printing Office. A series of 14 annual reports that include data on water pollution problems, sources, and cleanup.

COUNCIL ON ENVIRONMENTAL QUALITY: *Contamination of groundwater by toxic organic chemicals.* Washington, D.C.: U.S. Government Printing Office, 1981. Overview of types of pollutants found in groundwater and their sources.

CONSERVATION FOUNDATION: *State of the environment*, 1984. Washington, D.C., 1984. The Conservation Foundation. A report by a nonprofit organization that includes a chapter on current trends in water use and water pollution of both surface and groundwater.

EHRLICH, P. R., EHRLICH, A. H., AND HOLDREN, J. P.: *Ecoscience: Population, Resources, Environment.* New York: W. H. Freeman, 1977. An in-depth survey of water supply and pollution problems is presented in Chapters 6 and 10.

GUNDLACH, E. R., BOEHM, P. D., MARCHAND, M., ATLAS, R. M., WARD, D. M., AND WOLF, D. A.: Fate of Amoco Cadiz oil. *Science* 221:122–129, 1983. Summary of environmental effects of the largest and most intensely studied oil spill in history.

HILEMAN, B.: Offshore oil drilling. *Environmental Science and Technology* 15:1259–1263, 1981. Survey of effects of oil drilling and controversies surrounding extraction of oil from the continental shelf areas.

HOLGATE, M. W., KASSAS, M., AND WHITE, G. F., EDS.: *The World Environment 1972–1982.* Dublin: Tycooly International, 1982. Chapters 3 and 4 deal with global pollution problems of marine and freshwater ecosystems.

HUTCHINSON, G. E.: Eutrophication. *American Scientist* 61:267–279 (May–June), 1973. The scientific basis of eutrophication problems.

KESWICH, B. H., AND GERBA, C. P.: Viruses in groundwater. *Environmental Science and Technology* 14:1290–1297(November), 1980. Survival and migration of viruses as related to health problems are surveyed.

KRENKEL, P. A., AND NOVOTNY, V.: *Water Quality Management.* New York: Academic Press, 1980. A textbook that includes a thorough discussion of the physical, chemical, and biological aspects of water pollution.

MARSHALL, E.: The "lost" mercury at Oak Ridge. *Science* 221:130–132, 1983. A report on the uncovering of large losses of mercury from the Department of Energy's Oak Ridge, Tennessee, plant.

SMITH, W. E., AND SMITH, A. M.: *Minamata.* New York: Holt, Rinehart and Winston, 1975. A well-illustrated account of the tragedy at Minamata, Japan.

U.S. WATER RESOURCES COUNCIL.: *The Nation's Water Resources, 1975–2000.* Washington, D.C.: U.S. Government Printing Office, 1978. Volume 2 deals with water quality, quantity, and related land issues.

WETZEL, R. G.: *Limnology.* Philadelphia: Saunders, 1983. A textbook that deals with the scientific aspects of freshwater ecosystems.

Reservoirs created by dams are one of the controversial methods used to enhance water supply. (Illustrators Stock Photos.)

CHAPTER **8**

MANAGING AQUATIC RESOURCES

In July, 1977, the city of Chicago won a lawsuit against the city of Milwaukee, its neighbor, which is 130 kilometers (80 miles) to the north. The decision required that Milwaukee cease its occasional discharge of raw sewage into rivers that empty into Lake Michigan, which is Chicago's source of drinking water. To comply with that decision, a new sewer system, which will separate runoff from sewage, or a system that can treat both, is needed to prevent raw sewage from entering Lake Michigan during rainy periods. Since the court order, Milwaukee has embarked on a $1.7 billion, decade-long program to build a huge tunnel system under the city, which will store excess sewage and runoff until the system can accommodate it. The city is also upgrading its existing treatment plants. The alternative, to replace the old combined sewers, mostly in the downtown area, was considered to be too expensive and disruptive.

By 1983, however, the mayor of Milwaukee realized that polluted runoff from the agricultural land upstream was still entering Lake Michigan from rivers that were running through Milwaukee, thereby counteracting his city's herculean efforts. "We can spend $2 billion here," he contended, "and still have a polluted stream."

This situation illustrates the kinds of conflicts that arise when a region's water resources are used by different groups for different purposes. Examples of such conflicts abound. Groups that are interested in sport and commercial fishing want to keep the

rivers free of chemicals that make fish and shellfish unsafe to eat or spoil their taste, whereas those who are involved in industry depend on the rivers to degrade and dilute wastes and generally balk at installing costly pollution-control measures. Environmental groups strive to preserve or even create wetland habitats for waterfowl, while farmers dig ditches to drain wetlands, thereby converting them into tillable lands. The U.S. Army Corps of Engineers deepens channels and harbors to accommodate increased traffic of barges and ships and spreads the sediment that is dredged up over shoreline marshes, destroying much of the wildlife in those areas. In arid regions, cities and agriculture compete for scant water supplies, one interest group often diverting the entire flow of a mountain stream away from another interest group. In some locations, the political clout of persons in highly populated areas is strong enough to wrestle water away altogether from persons in sparsely populated areas. For example, some insight into the struggles that Phoenix, Arizona, has experienced in trying to gain access to more water rights are presented in Box 8.1. Once restrictions are placed on the quantity of water that may be drawn from surface or groundwater resources—or when minimum limits are placed on the quality of water that can be returned to surface waters—confrontations and lawsuits often follow, because those who use the water tend to assume that they have the right to use public resources as they see fit.

Today, developed nations are entering a new era of water management. The uncontrolled exploitation of water resources has become a thing of the past, and countries must now work their way through the complex socioeconomic and ecological problems that are associated with resource allocation. Fortunately, new technology—some of it in existence, and some of it under development—will help us to solve some of our water-management problems. Formerly, water-resource development consisted only of hydraulic engineering efforts, such as the building of dams, irrigation projects, and aqueducts, which often disregarded the ecological disruption they caused. Today, water management is a much more complicated business. Not only must the quantity of water be fairly regulated, but complex water-quality problems must be addressed at the same time.

Solving the broad range of water-resource problems that we are currently facing requires the expertise of many specialists—regional planners, civil engineers, biologists, chemists, economists, and recreation specialists—as well as ideas and support from special-interest groups. As we saw in Chapter 7, our water problems extend far beyond the fundamental problem of ensuring adequate freshwater supplies. In this chapter, we examine some of the ways in which technology can help us to resolve conflicts between the quality and the quantity of water needed

---

BOX 8.1

STATE FEARS BEING LEFT HIGH AND DRY

Phoenix—At Lake Havasu, on the border with California, giant pumps lift water from the Colorado River up an 800-foot mountainside and into the first stretch of a concrete aqueduct that unfolds across the remote reaches of Arizona's high desert. When it is finished, the aqueduct will travel 300 miles, an unbroken ribbon of concrete and pipe designed to transport billions of gallons of Colorado River water to thirsty Phoenix and Tucson.

By any measure, the Central Arizona Project is a monument to man's determination in attempting to reshape the facts of nature. In good years, when the river runs full, the project should enable Arizona to make use of its full share of the Colorado's water, as set down in a 1922 Federal compact, apportioning the flow of the river among the seven states of the basin.

The project is among the largest nonmilitary public works programs under way in the nation, an undertaking that has already consumed more than $1 billion of increasingly scarce Federal tax dollars. By the time the network of pumping stations, canals, storage reservoirs, and new flood control dams above Phoenix is completed, the project could cost $2 billion more.

The project, authorized by Congress 15 years ago, is expected to bring water to Phoenix by the end of 1985. But overall, it is barely 40 percent finished, and among some here, questions are being raised about when the entire network of dams, reservoirs, and canals will be completed. Amid a new and stingier mood in Washington, a state task force is now studying ways to accelerate construction by providing local financing.

Wesley E. Steiner, who has been Arizona's chief water official since 1969, wonders whether Congress or the Administration are willing to put that kind of money in. In fact, there are fewer Federal funds available to the Central Arizona Project this current fiscal year than there were a year ago. And in recent public statements, Interior Secretary James G. Watt has suggested that Arizona may already be getting a disproportionate share of reclamation money.

"Even if we could sustain past funding levels of about $150 million, it will take 25 years to finish this project," Mr. Steiner said, "and who is going to let Arizona constitute that kind of draft on the Federal budget for that long!"

The project is 6 years behind its original target date of 1995. For a lot of people in Arizona, where water has the emotional and political viscosity of blood, further delays are viewed with horror. Indeed, among the state's political leadership, there is universal agreement that the Central Arizona Project remains an absolute—and nonnegotiable—necessity.

by enhancing the quality of water and extending its supply.

Legislative Responses to Water-Resource Problems

We can trace the development of our water-resource problems in the United States by examining the legislative history of water-quality management and planning. In general, individual states have not taken much action to control water pollution (see Figure 8.1). The first action to improve water quality was taken by the federal government in 1899, when the Rivers and Harbors Act which is better known as the Refuse Act, was passed. One of the provisions of that act prohibited the dumping of trash into rivers and harbors. However, no provisions were made to prohibit discharge of sewage and runoff into navigable waterways. Unfortunately, that law was soon forgotten, and it was not used to prosecute violators until the 1970s.

The federal government's next response to the country's growing water-pollution problems was the passage of a series of water-pollution laws between 1948 and 1970. However, because of political pressures, those efforts resulted in weak laws that were difficult to regulate and enforce. Vexed by the failure

---

By the year 2025, planners anticipate that the population of the state will have tripled, with most of the newcomers drawn to Phoenix and Tucson, those urban islands in the desert. In Tucson, the pumping of groundwater supplies from beneath the city is causing water tables to fall 3 to 8 feet a year. A tough groundwater management law passed by the state mandates strict conservation measures, but state officials concede they will be helpless to reduce the overdraft without supplemental water from the CAP, as the project is known here.

Meanwhile, competition will continue to sharpen. Several Indian tribes in the state have successfully pressed their own prior claims, including some 310,000 acre feet of the project's proposed annual flow, which in dry years could account for more than one-third of the water available throughout the entire diversion system.

All this has impelled state officials this year to press Washington to accelerate rather than scale back funding for the project. At the same time, the task force now under way was set up by Governor Bruce E. Babbitt, a Democrat and a leading regional voice on water issues, to examine how the state might raise its own revenues to speed construction of some of the flood control dams or distribution systems designed to bring water to irrigators. Among the ideas is legislation that would permit the state to issue bonds towards

its share of the construction.

Cost-sharing is not a new notion. For more than a year, the Reagan Administration has been studying a plan that could force private water users in the West to pay up to 35 percent of the cost of new reclamation projects. In a period of record deficits, even this Administration—which vowed to restore water development projects cut by Jimmy Carter—simply cannot afford to bankroll huge reclamation projects. Whether they are right or wrong, many outside the West perceive reclamation as synonymous with pork-barrel politics.

Most Westerners disagree, typically arguing that through hydroelectric revenues and the purchase of water by irrigators, cities and industry, reclamation projects pay at least part of their cost. "That's more than you can say for the Washington subway system," said Mr. Steiner. The subway is often cited in the West as an example of Eastern pork-barrel—an expensive public works project that doesn't return a dime to the Treasury.

Still, most Westerners have come, however reluctantly, to accept the idea of cost-sharing, just as they concede the mega-projects like the CAP are probably the last of their kind.

Figure 8.1 *Fish kills are common scenes along rivers that receive water pollutants. (Allain Nogues/Sygma.)*

of earlier legislation—and feeling increasing pressure from concerned voters—Congress passed the Water Pollution Control Act of 1972. That act and its amendments, now called the Clean Water Act, empowers the federal government to set minimum water-quality standards for rivers and streams (although individual states may set more stringent standards if they so desire). According to the 1972 act, it is illegal to discharge any pollutant into a waterway unless a permit is first obtained from the state. This permit system, the National Pollution Discharge Elimination System (NPDES), is the basic mechanism for controlling the level of pollutants from point sources. The permit sets specific limits for pollutants on a facility-by-facility basis. The act also establishes a goal of zero discharge into "navigable" waters by 1985. As of 1984, 36 of the states assumed the responsibility for enforcing compliance; the EPA is policing permit holders in the other states.

The Clean Water Act empowers the EPA or a state agency to impose limitations on effluents, deadlines for compliance, and fines for industries and municipalities that do not meet the requirements. Violations of conditions that are specified in a permit can result in fines of up to $10,000 per day. Repeated intentional violations are punishable by fines of up to $50,000 per day and a prison term of up to 2 years.

The 1972 act attempts to distribute more equitably the economic burdens of cleaning up the water. According to the provisions of that act, government experts, after consulting industry representatives, set standards that are appropriate for each industry. For example, limitations on effluents from steel mills are different from those for paper mills. Also, industries were given until July 1, 1977, to install the "best practical treatment" methods for their effluents. Following that regulation, industries and municipalities were given until July 1, 1983, to comply with a more restrictive regulation, which requires that they employ the "best available treatment" technology to treat effluents. The national goal was to clean up all the nation's surface waters, "wherever attainable," so they would be in fishable and swimmable condition by July, 1983.

Municipal discharges into surface waters have been controlled primarily as a result of federal technical and financial assistance, which has enabled municipalities to build and upgrade sewage treatment plants. Although some progress has been made toward the goals set in 1972, many formerly heavily polluted areas are still not fishable and swimmable.

The 1972 Water Pollution Control Act is one of the most ecologically enlightened and most powerful pieces of environmental legislation ever passed in the United States. It recognizes that some wastes concentrate in food webs and that emission of those materials must be reduced. In addition, it recognizes that nonpoint sources of pollution are major problems in some watersheds. But the need to achieve compliance is still questioned in some areas, especially along the seacoasts, where the authorities in some cities insist that minimally treated wastewater will not damage marine ecosystems. They refuse to accept the added expense of fully treating their wastewater by secondary treatment, (which is explained later). Those types of problems are discussed in Box 8.2. Furthermore, the compliance can

---

BOX 8.2

RAW SEWAGE FOULS HARBORS
11 YEARS AFTER POLLUTION LAW

Boston, July 24—Eleven years after Congress approved legislation to clean up the nation's waters, the Atlantic coast's two largest metropolitan areas, Boston and New York, are still releasing billions of gallons of untreated sewage into their harbors each year.

And the Federal agency charged with enforcing the Federal Water Pollution Control Act, the Environmental Protection Agency, says it has no idea how many coastal cities are complying with the law, which ordered all sewer systems emptying into marine waters to provide at least secondary treatment of sewage wastes.

"Nobody knows," says Wen Huang, a senior sanitary engineer with the agency's Office of Water Program Operations in Washington.

In Boston, the apparent threat to public health has made a state court judge determined to find and enforce a solution.

Experts say that repairing the Metropolitan Boston system and raising its treatment to the secondary level could cost $1.5 billion, and would come at a time when the Federal share of such capital improvements is scheduled to drop from 75 percent or more to 55 percent, as it will in 1985.

Boston, which discharged 12 billion gallons of untreated sewage into its own waters last year, is not alone among coastal cities with old and patchwork systems that are prone to failure or function at lower levels of treatment than required by Federal law. But public apathy began turning to disgust last summer when the sewage invaded public beaches.

For the solicitor of the City of Quincy, just south of Boston's harbor, the final insult came one morning when the young lawyer went down to run along the shore. It was low tide, and in the pale dawn he saw what he took to be a scattered gleaming of jellyfish, all down the beach.

To his revulsion, William B. Goldman discovered that "they weren't jellyfish. They were little patties of human waste, and patches of grease." From that experience has come a lawsuit dramatizing the gross pollution of Boston's rivers and harbor, and the inability of government at all

levels to do anything about it. Quincy has sued the Metropolitan District Commission, which operates the sewage plants, and Justice Paul G. Garrity of Superior Court is hearing the case.

New York City has made more progress, but problems still remain. The city, which has spent $2.2 billion in the last decade to upgrade nine of its plants, had to close all public beaches in Brooklyn, Queens, and Staten Island two years ago when one of two old plants in Brooklyn failed.

The city system still releases 200 million gallons of raw sewage into its harbor every day, the same amount as it did 10 years ago, says Andrew McCarthy, a spokesman for the city's Department of Environmental Protection. That practice is expected to end by the summer of 1987, when two new plants begin operation. But when the volume is too great for the pipes, the system will continue to lose 10 to 15 percent of the 1.3 billion gallons of liquids and solids it receives each day.

The Boston and New York City systems are among more than 200 sewage systems on the Atlantic and Pacific coasts that have asked the Environmental Protection Agency to waive the stringent secondary treatment requirements of the 1972 law.

In 1977, Congress modified the law, now known as the Clean Water Act, to allow such waivers, and the EPA is just now getting around to approving or denying those requests. The Federal agency has made decisions in 55 cases, but only one, a denial in the case of a system emptying into Puget Sound, in Washington state, is considered final.

All requests for waivers from the North and Middle Atlantic Coasts, including Boston's and New York's, have been denied, but are still open to appeal.

When the deadline for waiver applications expired last December, an estimated 223 systems had not been heard from and Paul Pan, acting director of the environmental agency's Marine Discharge Evaluation Branch, said he did not know if those systems were complying. He said he did not believe anyone else in the agency knew, either.

More improvement is thought to have been made among inland systems emptying into rivers

because Congress did not extend them the waiver opportunity.

While much progress has been made in cleaning up rivers, there is still a considerable distance to go, particularly by municipalities, in reaching the law's goals of fishable, swimmable waters. Major remaining problems include contamination by toxic chemicals and pollution from urban and rural surface runoff.

The history and structural problems of the Boston and New York systems are the same, and are shared by other old American coastal cities. Installed largely in the nineteenth century, and extended as the population grew, the pipes and tunnels were built to receive both sewage and stormwater. The two mixed in the pipes, which poured them into the rivers and harbors, to be carried away by the outgoing tide.

In the metropolitan Boston area, the system evolved as 5000 miles of pipe owned by 43 separate cities and towns. They feed into 200 miles of pipes and tunnels operated by the Metropolitan District Commission, a state agency, which did not build its first sewage treatment plant until 1953, on Nut Island in the harbor.

The second plant, on Deer Island, was built in 1968. Both were built only to provide primary treatment, the separation, reduction, and chlorination of the solids and liquids in the sewage and not secondary treatment, which further reduces the demand for oxygen by treated wastes after they were released. The joint capacity of the two systems is less than the volume carried by the pipes when it rains.

"It's not physically possible to separate the storm and sewer systems now," says David Standley, an engineer and former state and Federal environmental officer here. "You'd have to dig up the City of Boston and much of Cambridge and Somerville to do it."

There are also illegal connections, and infiltration by goundwater, and when the volume is too great for the piping network, the overflow discharges through more than 100 outfalls along the rivers and harbor front.

"Every year, there are about 6 billion gallons of raw sewage that feed out through the combined sewer outfalls into the waterways," said Stephen P. Burgay, of the Metropolitan District Commis-

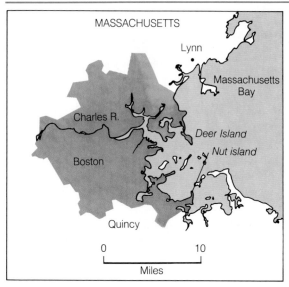

Sewage treatment plants on Deer and Nut islands often over-flow. (New York Times, July 25, 1983.)

sion. That has been the case since about the turn of the century, and "for all practical purposes, the inner harbor is biologically dead," Mr. Burgay said. The Charles River is closed to swimmers.

When the volume in the pipes is still too great for the capacity of the plants, which are now old and in disrepair, the raw sewage flows out through bypasses into the harbor, at one point only 1400 feet from the Quincy beach. On Mother's Day this year, when a coupling split on one of the pumps at the Deer Island plant, 60 million gallons of untreated sewage poured out of the bypass.

The bottom two floors of the plant filled with raw sewage, burying the cause of the problem at a depth of 20 feet. For three days, a team of police frogmen from the Metropolitan District Commission dove repeatedly in a maze of machinery, through two tunnels and beneath a grate, groping with their hands in the blindness of the sewage to find and identify what had broken.

They found it, but for 3 weeks afterward, they say, they suffered from dysentery and other infections, and have yet to receive any official recognition from the Commonwealth.

The failings of the Metropolitan Boston sewer system are visible and have long been known. But with so many local and state govern-ment offices involved in the 1950s and 1960s, and with the advent of the EPA in the 1970s, government action toward a consolidated solution never progressed beyond a series of studies, reports and negotiations which ended in stalemate.

"To the making of reports, there is no end," Charles M. Haar, Louis D. Brandeis Professor of Law at Harvard University who is working with the court in the Boston case, declared last week. Waving at boxes piled in his office, he said, "All you've got here is the cutting down of trees— all these reports."

He later toured the Deer Island plant, whose manager has told him the battered facility was so engorged with leaked diesel oil that its stacks burst frequently into flame, threatening to burn the whole plant down.

The Environmental Protection Agency took 3½ years to consider Boston's application for a waiver, rejecting it only at the end of last month, as the matter began to heat up in court.

It is thought that the City of Quincy's attorneys took care to file their suit against the Metropolitan District Commission before Superior Court Justice Paul G. Garrity. No relation to the Federal judge who took control of the school system in the Boston desegregation case, Justice Garrity is nonetheless well known here for his approaches to jurisprudence. He took the unprece-dented act of placing this city's public housing system in receivership for its failings, appointing the court's own agent to run it.

In his initial ruling, late last month, Justice Garrity found that "Boston Harbor is polluted, that the current and potential impact of that pollution upon the health, welfare, and safety of persons who live and work nearby Boston Harbor and who use it for commercial, recreational and other purposes is staggering."

Professor Haar agreed to be appointed as the court's special master in the case, with just 30 days to wade through the complexities of the problem and make recommendations to Justice Garrity. "I think there is a chance here," he said, "to set a kind of pattern about a very crucial na-tional problem, which is the quality of life."

Figure 8.2 *Matthews Beach, on Lake Washington near Seattle, before and after pollution abatement by surrounding communities. (Municipality of Metropolitan Seattle.)*

be undermined if the federal government reduces the availability of funds to help municipalities. A state or federal administration can also flout the law by underfunding the judicial system, which is responsible for enforcing legislation, or by failing to set the standards to comply with the legislation. By closely monitoring our country's aquatic ecosystems during the next decade, we will be able to determine whether the Clean Water Act is sufficient to return our waterways to the fishable and swimmable conditions that are mandated.

Strict enforcement of the stringent Clean Water Act has reversed a long, steady trend of deteriorating water quality. In 1982, 21 of 33 states reporting indicated that the quality of their water was improving, which was primarily due to increased removal of conventional pollutants—suspended solids and biochemical oxygen-demanding substances (BOD)—from point sources. For example, the Willamette River in Oregon and the Detroit River in Michigan have recovered to a degree that astonishes persons who are closely associated with the problem. Furthermore, salmon have reappeared in the Connecticut and Penobscot Rivers of New England, and polluted beaches have been reopened for swimming (Figure 8.2). Most such gains are the result of aggressive pollution abatement efforts of local, state, and federal governments as well as local citizens' groups.

Overall, however, water quality has not been improved as much as many had hoped for in the decade following the passage of the Clean Water Act. Not enough surface waters have been cleaned up, even though legislation has been passed to require cleanup. The most frequently cited reasons for violations of water-quality standards are improper oper-

ation and management of wastewater-treatment facilities, inadequate treatment of industrial wastewater prior to its discharge into a publicly owned treatment plant, and delays in the design, funding, and construction of water-treatment facilities. Thirty states have cited localized impairment of water by toxic chemicals and heavy metals. And nonpoint sources of water pollutants were cited as the major reason for deteriorating water quality by 20 percent of the states that reported in 1982.

To answer the recurring question of whether the nation's water quality is improving or deteriorating, the U.S. Geological Survey established the National Stream-Quality Network (NASQAN) in 1972. One goal of that network was to detect changes in stream quality over time. Altogether, 504 sampling stations were located at the outlets of major drainage basins, and the data collected during the period from 1974 to 1981 were analyzed for statistically significant trends. A partial listing of that analysis (which is presented in Table 8.1) shows that in a significant number of places, the water quality

Table 8.1 Trends in selected water quality constituents and properties at NASQAN stations, 1974–1981

| CONSTITUENTS AND PROPERTIES | NUMBER OF STATIONS WITH | | | |
	INCREASING TRENDS	NO CHANGE	DECREASING TRENDS	TOTAL STATIONS
Nitrate-Nitrite	76	203	25	304
Ammonia	31	221	30	282
Total Organic Carbon	36	230	13	279
Phosphorus	39	232	30	301
Suspended Sediment	44	204	41	289
Conductivity	69	193	43	305
Turbidity	42	199	18	259
Fecal Coliform Bacteria	19	216	34	269
Phytoplankton (Algae)	22	234	44	300
Dissolved Trace Metals:				
Arsenic	68	228	11	307
Cadmium	32	264	7	303
Chromium	12	152	2	166
Copper	6	83	6	95
Lead	5	232	76	313
Manganese	30	250	19	299
Mercury	8	194	2	204

Source: Excerpted from Table 4, U.S. Geological Survey, Open File Report 83–533, 1983.

has deteriorated despite efforts to improve it. Only a few new efforts are underway to reverse these trends in declining water quality.

A whole new set of concerns about water pollution surfaced in 1974, when 66 industrial chemicals, some of them known carcinogens, were identified in the drinking water in New Orleans. Those findings hastened the passage of the Safe Drinking Water Act of 1974, which charges the EPA with the responsibility for setting and enforcing minimum national drinking-water standards. That act includes

Table 8.2 National interim primary drinking-water regulations (Data in milligrams per liter† unless otherwise specified. tu = turbidity)

CONSTITUENT	MAXIMUM CONCENTRATION
Arsenic	0.05
Barium	1
Cadmium	0.010
Chromium	0.05
Lead	0.05
Mercury	0.002
Nitrate (as N)	10
Selenium	0.01
Silver	0.05
Fluoride	1.4–2.4*
Turbidity	1–5 tu
Coliform Bacteria	$1/100$ ml (mean)
Endrin (Pesticide)	0.0002
Lindane (Pesticide)	0.004
Methoxychlor (Pesticide)	0.1
Toxaphene (Pesticide)	0.005
2,4-D (Herbicide)	0.1
2,4,5-TP Silvex (Herbicide)	0.01
Total Trihalomethanes [the Sum of the Concentrations of Bromodichloromethane, Dibromochloromethane, Tribromomethane (Bromoform), and Trichloromethane (Chloroform)]	0.10
Radionuclides:	
Radium 226 and 228 (Combined)	5 pCi/L‡
Gross Alpha Particle Activity	15 pCi/L
Gross Beta Particle Activity	4 mrem/year

Source: U.S. Environmental Protection Agency, 1982. U.S. Code of Federal Regulations, Title 40, Parts 100 to 149, revised as of July 1, 1982, pp. 315–318.
*Maximum fluoride levels vary according to the amount of water consumed by persons in a region; warm areas consume more and thus allowable fluoride levels are lower.
†Milligrams per liter is the equivalent of parts per million (ppm).
‡pCi/L (picocuries per liter) and mrem are measures of the level of radioactivity. One picocurie releases 0.037 radioactive particles per second.

Table 8.3 National secondary drinking-water regulations

CONSTITUENT	MAXIMUM LEVEL
Chloride	250 mg/L*
Color	15 color units
Copper	1 mg/L
Corrosivity	noncorrosive
Dissolved Solids	500 mg/L
Foaming Agents	0.5 mg/L
Iron	0.3 mg/L
Manganese	0.05 mg/L
Odor	3 (Threshold Odor Number)
pH	6.5–8.5 units
Sulfate	250 mg/L
Zinc	5 mg/L

Source: U.S. Environmental Protection Agency, 1982. U.S. Code of Federal Regulations, Title 40, Parts 100 to 149, revised as of July 1, 1982, p. 374.
*Milligrams per liter is the equivalent of parts per million (ppm).

public water systems that have at least 15 service connections or that regularly serve at least 25 persons.

One of the major provisions in the Safe Drinking Water Act establishes maximum levels for specific substances so that they have "no known or anticipated adverse effects on the health of persons" and which "allow a margin of safety." Those primary standards include the presence of toxic chemicals, pathogenic microbes, radioactive substances, and turbidity. Although the act has been in force since December, 1974, only the 17 chemicals (8 heavy metals, 6 pesticides, trihalomethanes, nitrate, and fluoride) and radioactivity, which are listed in Table 8.2, are included in the current regulations. To date, more than 700 different chemicals (several of which are known carcinogens or mutagens) have been identified in the various drinking-water supplies across the United States. The levels of those chemicals have been very low, and the adverse health effects of low levels of pollutants have not been demonstrated. The only recent addition to the list of regulated chemicals in drinking water by the EPA has been the trihalomethanes. Later in this chapter, we consider how the EPA is attempting to meet the mandate of "no adverse health effects" in the regulation of water standards.

Secondary standards for the quality of drinking water have also been established and are directed primarily at the esthetic qualities of water. Those levels (listed in Table 8.3) are only guidelines and are not federally enforceable.

Another major provision of the Safe Drinking Water Act requires that states establish programs to prevent the underground injection of wastes that endanger the quality of drinking water that is taken from groundwater sources. That provision protects "sole-source aquifers," so a community can petition the EPA to designate an aquifer as its sole source of drinking water. Thus if an aquifer is so designated, the agency may then delay or halt federal assistance to projects that would degrade the quality of water in it. By January, 1984, 17 communities had received a "sole-source" designation.

Table 8.4 Federal legislation concerned with the control of toxic substances in water

YEAR	LEGISLATION	AREAS OF CONCERN
1954	Atomic Energy Act	Regulates Low-Level Radioactive Waste Discharges into Water
1970	National Environmental Policy Act	Requires Filing of an Environmental Impact Statement for All Projects that Affect the Environment
1972	The Water Pollution Control Act (Clean Water Act)	Regulates Discharges from Industry and Municipal Sewage Treatment Plants
1972	Marine Protection Research and Sanctuaries Act	Regulates Ocean Dumping
1972	Ports and Waterways Safety Act	Regulates Oil Transport
1972	Insecticide, Fungicide, and Rodenticide Act	Classifies Pesticides and Regulates Their Use
1974	Safe Drinking Water Act	Sets Limits on Chemical and Radiation Levels in Drinking Water
1974	Hazardous Materials Transportation Act	Regulates the Transport of Hazardous Materials on Land
1976	Resource Conservation and Recovery Act	Regulates the Treatment, Storage, and Disposal of Hazardous Wastes
1976	Toxic Substances Control Act	Regulates the Use of Dangerous Chemical Substances
1980	Comprehensive Environmental Response Compensation and Liability Act ("Superfund")	Provides Funds for Cleaning Up Chemical Dump Sites that Endanger the Public

In addition to the Water Pollution Control Act of 1972 and the Safe Drinking Act of 1974, which together have primary authority for protecting public water supplies, nine other federal laws help to prevent water pollution. Those laws are summarized in Table 8.4.

Water-Quality Control

Water-quality goals are not the same for all groups of consumers. But once community goals are established, various technical, land-use, and legal strategies can be used to persuade—or compel—persons or groups to cooperate in efforts to attain those goals. In many instances, the level of improvement is dictated by how much money a community is willing to spend on water-quality improvements. Let us now examine some methods for improving water quality.

As water travels through an urban–industrial system, it is pumped from a source, treated, distributed, used, collected, treated again, and then discharged into the nearest appropriate surface water (river, lake, bay, or ocean). Figure 8.3 illustrates the urban–industrial water cycle. That cycle has two treatment stages: one prior to use and one after use. Although treatment of wastewater is designed to protect the quality of water in aquatic ecosystems, the standards deemed sufficient for that purpose are lower than those for drinking water. Thus surface waters that are treated before they are used for drinking water must meet more stringent requirements. We discuss wastewater treatment first, because it also employs the techniques that are used to treat public water supplies.

Wastewater Treatment

Wastewater treatment plants provide the principal means of improving water quality in point-source discharges. They are designed primarily to remove BOD and suspended solids. A modern treatment plant for a small town with a population of 5000 might cost $2.2 million (in 1983 dollars); one for a city with a population of 250,000 might cost $60 million to $70 million (in 1983 dollars) assuming that a secondary level of treatment were provided (which is discussed later). Thus sewage treatment plants are expensive to build and to operate; typically 3 to 8 percent of property taxes go toward the financing of

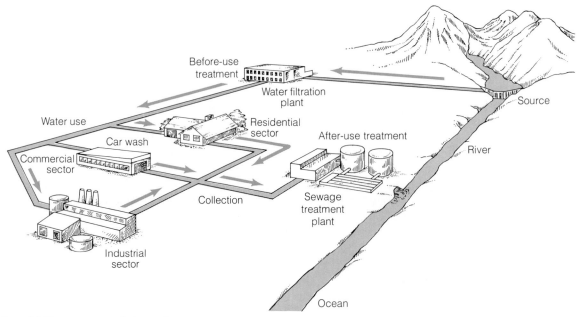

Figure 8.3 *The water-use cycle in an urban area.*

water-treatment facilities. In addition, an "average" household pays about $100 per year in user charges.

Wastewater treatment plants remove from municipal sewage a wide variety of substances that persons dump into sewers. Those substances fall into three general categories, each of which is treated differently: (1) insoluble materials, such as grease, oils, fats, sticks, and beverage cans are mechanically screened or skimmed off the surface; (2) suspended substances, such as human wastes, paper, ground-up garbage, and so forth, are removed by diverting the wastewater into holding basins, where those substances slowly settle out; and (3) dissolved and colloidal pollutants, such as sugar, starch, and milk, are removed by decomposers that feed on organic wastes, and the microorganisms themselves subsequently settle out in settling basins.

Activated sludge treatment is the most common technique used to treat municipal wastes and industrial wastes, for example, from food and paper industries. That technique combines natural biological processes (aerobic decomposition) and mechanical processes to remove and break down waterborne wastes. Activated sludge treatment plants take up

less space than other treatment methods, which is an important consideration in view of the high land values in cities.

A simplified diagram of an activated sludge treatment plant is shown in Figure 8.4. The sewage flows by gravity or is pumped into the treatment plant. As it enters, it is typically 99.9 percent water and only 0.1 percent impurities. In the first step, the sewage passes through a screen that removes debris, such as rags, sticks, cans, and so forth. It then flows into a small *grit tank*, where fast-settling dense materials, such as sand and pebbles, are removed to protect pumps and other mechanical equipment. The materials removed in those first two steps are disposed of usually at a sanitary landfill (Chapter 13). Once the material that settles rapidly is removed, the sewage enters the *primary settling tank*, where its velocity is reduced. Within 2 hours, approximately 45 percent of the organic suspended solids settle as *sludge* to the bottom of the tank. That process of freeing wastewater of its suspended solids is called *clarification*. The settled sludge is then pumped to sludge-handling equipment.

After being partially clarified in the primary tank, the sewage enters an *aeration tank*, where it remains for 6 to 8 hours. At that point, compressed

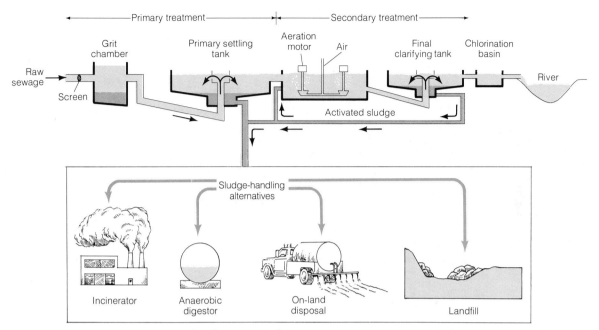

Figure 8.4 *An activated sludge treatment plant for sewage. Such a plant removes approximately 90 percent of the organic material that enters it in the raw sewage.*

air is pumped continually through the sewage to provide oxygen for aerobic bacteria, which degrade some of the suspended material and most of the dissolved organic wastes. Then, from the aeration tank, the wastewater flows into a final clarifying tank, where the bacterial cells in the water form clumps, which are called *floc*. Floc, along with most other suspended material, settles out in the final clarifying, or settling, tank, and some of it is recycled back into the aeration tank to provide "seed" bacteria to maintain the bacterial colony there. Because recycled bacteria are well adapted to conditions in the aeration tank, they multiply quickly. Therefore, they are especially efficient at removing organic wastes. Unrecycled floc is pumped to sludge-handling equipment. The water that leaves the final settling tank is quite clear and contains only about 10 percent of the organic material that was present originally in the incoming sewage.

Primary, Secondary, and Tertiary Treatment Sewage treatment is classified as primary, secondary, or tertiary, according to the degree to which wastewater is processed. *Primary treatment* takes sewage

only through the primary settling stage, which removes approximately 35 percent of the BOD and 60 percent of the suspended solids. In *secondary treatment*, the sewage is subjected to the additional step of biological (bacterial) degradation that takes place in the aeration tank. That step brings the total BOD and suspended solids removed to approximately 90 percent (see Figure 8.5). But even after secondary treatment, some 25 to 75 percent of the phosphorus, 90 percent of the nitrogen, and approximately 12 percent of the organic materials that are resistant to biological breakdown remain in the treated effluents. If those substances aggravate water-pollution problems in the receiving surface water, advanced secondary treatment procedures, which are often called tertiary treatment, are required.

Tertiary treatment is accomplished by a variety of procedures that have different treatment objectives. For example, if cultural eutrophication is a problem, treatment plants must be designed to reduce phosphorus levels. Thus the EPA requires more efficient removal of phosphorus by treatment plants that empty into bodies of water, such as the Great Lakes, where phosphorus is the limiting nutrient for algal growth. Under those circumstances phosphorus is removed most commonly by adding chemicals that form precipitates in the wastewater, which coagulate and then settle (Figure 8.5, step C). For ex-

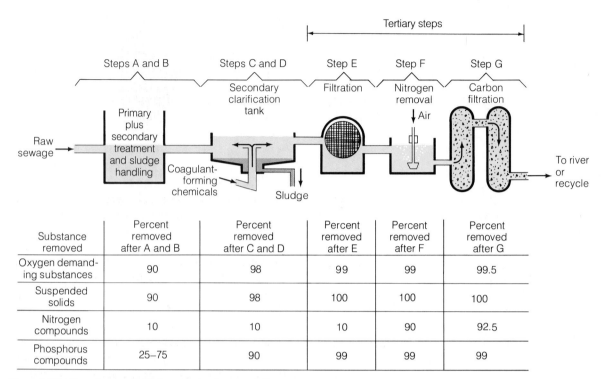

Substance removed	Percent removed after A and B	Percent removed after C and D	Percent removed after E	Percent removed after F	Percent removed after G
Oxygen demand-ing substances	90	98	99	99	99.5
Suspended solids	90	98	100	100	100
Nitrogen compounds	10	10	10	90	92.5
Phosphorus compounds	25–75	90	99	99	99

Figure 8.5 *Tertiary steps available for water treatment and the resulting improvement in water quality. (After the U.S. Environmental Protection Agency.)*

ample, alum (aluminum sulfate), ferric chloride, and lime (calcium hydroxide) remove phosphorus by forming precipitates that attract soluble phosphorus. The precipitates, as they settle, also drag along small suspended organic particles, which further increases the total BOD and suspended solids removed to approximately 98 percent. As much as 99 percent of the phosphorus is removed from wastewater if the water is filtered after the precipitate has settled (Figure 8.5, step E).

In some instances, it has been more economical to avoid cultural eutrophication of lakes by diverting partially treated water around lakes rather than providing a high level of treatment. For example, Lake Washington, near Seattle, formerly received nearly 80 million liters (20 million gallons) of treated domestic wastes per day. Those wastes contained at least 50 percent of the original nutrient load, so aquatic weeds and algae thrived in the lake. To overcome that problem, a pipeline was installed to funnel the wastes around Lake Washington to a secondary treatment plant on Puget Sound, which does

not provide expensive tertiary treatment. Treated effluents from that plant are discharged into the offshore tidal flow, which carries the remaining nutrients out to sea. Because treated effluents are no longer discharged into Lake Washington, the quality of water in the lake has been improved vastly. For example, algae populations are much smaller and, consequently, the water is discernibly clearer.

Although nitrogen is not removed normally from sewage effluents, it is generally a good idea to remove it when it reaches high levels (especially when it exists in the form of ammonia, which is toxic to fish) or in areas where it may be a limiting factor, such as in some lakes and estuaries. For example, to preserve Lake Tahoe's pristine nature, Nevada citizens agreed to add nitrogen-removal equipment (Figure 8.5, step F) to their sewage-treatment facilities. Nitrogen removal may also be required to prevent high nitrate levels when treatment-plant effluents are used to recharge groundwater.

Removal of Persistent Chemicals Even after all those measures for treating water have been taken, some chemicals usually still remain in the water. Examples of such persistent materials include sulfonatedlignins from paper-mill effluents, which impart a brown-stained appearance, and phenols from steel plants. However, most of those substances can be removed by passing the water through beds of powdered or granulated activated carbon (Figure 8.5, step G). When that technique, which is called *carbon adsorption*, is used after clarification and filtration, the water is usually of such high quality that it is fit for human consumption immediately after it is chlorinated. In late 1984, Denver, Colorado, put into service a 4-million-liter (1-million-gallon) per day wastewater recycling plant that employs carbon filtration, among other techniques, to recycle wastewater into the drinking-water supply. The town of South Lake Tahoe empties its carbon-treated wastewater into Lake Tahoe, which supports trout—a type of fish that is especially sensitive to water pollutants. In general, only affluent communities voluntarily install costly tertiary treatment equipment; less affluent communities install it only if they are legally required to do so to meet water-quality standards.

Most large industrial plants have their own wastewater treatment facilities that discharge into the nearest suitable surface water. Industries that produce organic wastes (such as paper mills and food processing plants) rely on the same techniques that are employed in primary and secondary stages of conventional sewage treatment. If nondegradable wastes are absent from their wastewater, industries dispense with primary and secondary treatment and use only one or more of the tertiary steps. For example, toxic metals are often removed through the addition of chemicals that induce coagulation and settling. Also, many new technologies are available to industries to reduce the quantity of particular chemical pollutants.

Most technologies, however, do leave some residual level of contaminants in the discharge water. To further reduce residual contaminant levels, more costly procedures must be employed—and, usually, it costs as much to remove approximately the last 10 percent of the residual pollutants as it does to remove the first 90 percent (Chapter 1). Thus the state or the EPA specifies the required level of treatment, and that is based on the limitations of available technology and the cost of the treatment process.

Chlorination The next-to-last step in municipal systems and some industrial treatment processes in the United States is *chlorination*. Chlorine can be used to kill pathogenic organisms that may have survived the earlier treatment steps. Although this step is necessary to prevent the spread of waterborne diseases, scientists have discovered that chlorination also results in the formation of small amounts of *chloroorganic compounds*, some of which are known carcinogens. The most prevalent of those compounds comprise the *trihalomethanes*, of which chloroform is the most frequently encountered. The extent of danger that extremely low concentrations of those chemicals presents is yet to be determined. (The levels of trihalomethanes are controlled in drinking water; see Table 8.2).

The discovery that chlorination results in the formation of chloroorganic substances poses a serious conflict for policymakers. Although chlorination is seen by some scientists as necessary to control waterborne diseases, the practice of chlorination adds significant quantities of chloroorganic compounds to aquatic ecosystems. Thus some treatment plants in the United States now only chlorinate wastewater during the summer months, when recreational activities associated with water are at their peak. In many European countries, wastewater treatment plants do not chlorinate treated water, but they maintain the capability to do so in the event of an outbreak of waterborne diseases. Because of this dilemma, new methods of disinfection are currently being sought. (We discuss the implications of those findings for water-supply treatment in the following section.)

The Clean Water Act of 1972 requires that all communities that are served by sewage treatment plants treat their wastewater through the secondary treatment stage. (Coastal cities can apply for a waiver.) Some communities are satisfied with meeting EPA requirements, but others choose to bear the cost of further treatment in the interest of producing cleaner water. The decision is determined, in part, by the fact that each step beyond secondary treatment increases treatment costs by roughly 20 percent.

Disposal The final step in the sewage treatment process is *disposal*. Most communities discharge their treated wastewater into nearby waterways. As we discussed in Chapter 7, however, approximately 1200 treatment plants dispose of their treated water effluents on land to utilize the natural filtering action of soil. Typically, soil filtration removes approximately 98 percent of the BOD, 95 to 99 percent of the phosphorus, and 50 percent of the nitrogen that remain after secondary treatment. The disposal of wastewater onto the land from secondary treatment plants cycles moisture and nutrients for crop production, thus turning a disposal problem into an asset. Generally, the wastewater is delivered to the crops by conventional irrigation methods, such as fixed or moving sprinkling systems. For example, in Muskegon, Michigan, over 110 million liters (28 million gallons) per day of treated wastewater irrigates approximately 2240 hectares (5540 acres) of corn. And Muskegon has plans to nearly double the capacity of its facility. Other cities, especially those in arid and semiarid regions, spread treated water on land to recharge the groundwater, again taking advantage of the soil's capacity to filter the water. Coastal cities use the technique to prevent saltwater intrusion (see Chapter 7).

Sludge and Dredge Spoil Disposal

The treatment process itself creates another major problem—the disposal of accumulated sludge. Sewage *sludge* is defined as the mixture of liquids and solids that are produced by domestic wastewater treatment plants—and the quantities of sludge that are generated are enormous. For example, a conservative estimate of the amount of sludge that is disposed of from the New York–New Jersey coastal cities approximates 16 million metric tons per year. In 1980, approximately 40 percent of this country's sludge was hauled to landfills, 20 percent was worked into agricultural lands, 25 percent was incinerated, and the remaining 15 percent was dumped into the oceans.

Sewage sludge has characteristics that can be used to advantage, as shown in Figure 8.6. Because it contains significant quantities of nitrogen and phosphorus, it can be used to fertilize crops. For example, the sludge from Milwaukee's treatment plant that is used in this way saves farmers some $180 per hectare ($75 per acre) each year (most of which is fertilizer costs). Sludge also acts as a soil conditioner, because it increases the water-retention capacity of the soil. And, since its protein, fat, and caloric content are similar to organic detritus, it has potential as a cattle or poultry feed supplement. Moreover, many sewage sludges can be used as fuel when sufficiently dried.

The many possible uses for sludge are limited, however, because sludge tends to concentrate toxic chemicals in wastewater. Those chemicals, in turn, vary with the types of industries that use the wastewater-disposal system. The most troublesome inorganic chemicals are the heavy metals (cadmium, copper, lead, mercury, zinc, and so forth); the organic chemicals that pose the greatest hazards are PCBs. Some communities reduce the levels of such substances by requiring industries to pretreat their wastewater before sending it to their treatment plants.

Land application of sewage sludge makes sense, because it recycles nutrients and reduces disposal costs. Much research is being done on that method of disposal, and many states now have guidelines that specify the maximum slope (steepness), the type of soil, and the rates of application that are acceptable for sludge disposal. The principal concern is to keep the amount of metals at nontoxic levels in crops. Although metals from sludge do not accumulate in seeds, they do accumulate in other plant tissues. Usually, cadmium levels are the biggest worry, because cadmium is a carcinogen.

Some treatment plants use anaerobic bacteria to digest sludge. The bacteria then produce methane gas, which is either sold as fuel or used at the treatment plant itself as an energy source. Anaerobic digestion of sludge fell into disfavor in the early 1970s because fuel prices were low and serious failures had occurred in the process from time to time. But it is coming back into favor as a partial solution to the problem of higher energy costs. Anaerobic digestion is also now more reliable, because the conditions that lead to failure are now better understood and thus can, in most instances, be prevented. Sludges that contain high levels of toxic substances, however, are not suitable for anaerobic digestion because the anaerobic bacteria are poisoned by them. For this reason, such sludges must be landfilled (see Chapter 13) or incinerated. Modern incinerators can recover heat from sludge, provided that the sludge is dry enough and there is a nearby use for the heat.

Figure 8.6 *Sewage sludge contains valuable nutrients, especially nitrogen that can in some cases be applied to land. (a) Sewage sludge is shown being applied to one of approximately 500 sites used by the Milwaukee, Wisconsin Metropolitan Sewerage District. (b) The sludge is injected under the surface using injectors. (Milwaukee Metropolitan Sewerage District.)*

(a)

(b)

Until December 31, 1981, no federal limits existed for ocean dumping of sludge, but the Marine Protection and Sanctuaries Act now strictly limits ocean dumping. In 1980, approximately 10 million metric tons per year of wastes (mostly sludge and some acid wastes and demolition debris) were dumped in several ocean locations. However, cities that used this method are now supposed to find one or more land-based alternatives, such as landfilling, incineration, or land application. But, in 1984, in New York and New Jersey, nine sewage districts were still dumping their sludge into the ocean—six in New Jersey, and three in New York. All nine plants dump their sludge at a site that is located 19 kilometers (12 miles) from the New York–New Jersey shoreline. All six of the New Jersey plants and two of the New York plants are under court order to phase out dumping. But in early 1984, New York City still had no terminal date for dumping its sludge in the ocean, because a court ruling required that the EPA first determine the effects of land disposal versus those of ocean disposal. Once that study is completed, the EPA will review New York City's permit application for sludge dumping. The EPA then could issue an interim permit to continue dumping at a designated ocean site, or it could order a phase-out program and set a date for termination.

At the time the Marine Protection and Sanctuaries Act was passed, most scientists were opposed to ocean dumping, but few were opposed to land disposal. Nevertheless, at present, it is not known whether ocean disposal of sludge has a greater or lesser environmental impact than land-based alternatives. Disposal in shallow ocean water is generally thought to have significant negative impacts (for example, oxygen depletion and covering of spawning areas), which have been observed close to New York City. But some scientists speculate that in deeper water, sludge could benefit marine detritus food webs. They note that sludge has been successfully incorporated into the diets of cattle and poultry and that dumping has not significantly disrupted the deeper off-shore aquatic ecosystems.

A recent report by a group of the world's leading marine scientists, which was sponsored by the United Nations Environmental Programme, concluded that the harmful effects of pollution from sewage and sludge disposal at sea can be seen in semienclosed marine environments, such as the Mediterranean Sea, the Gulf of Mexico, the North Sea, and the Baltic Sea. But the report also stated that the fertilizing value of sewage disposal at sea, if it were controlled adequately and the disposal sites were selected properly, might outweigh any harmful effects of toxic substances in sewage.

Another problem is created by the ocean dumping of 40 million cubic meters (52 million cubic yards) of dredge spoils that are removed each year from this country's harbors and navigation channels. Some of that material is contaminated with PCBs, heavy metals, and other industrial chemicals. The problem is that nobody knows the level that various pollutants can reach in the dredge spoils before causing adverse environmental effects. The Army Corps of Engineers is responsible for determining those levels. Thus in 1984, about 70 ocean sites were receiving dredge spoils, whereas, the heavily polluted dredge spoils were usually disposed of by the more costly process of piping or hauling them to approved land-disposal sites.

Water-Supply Treatment

Approximately 240,000 public water supply systems exist in the United States. Of those systems, 30,000 to 40,000 are municipal systems, most of which were installed decades ago, when water was relatively clean and treatment was designed only to prevent the transmission of waterborne diseases. The remainder are used regularly by the public, including approximately 200,000 hookups in service stations, motels, and so forth, which are not connected to larger regulated municipal supplies.

As a result of the degradation of surface-water and groundwater quality over the years, additional precautions are necessary to ensure that supplies of drinking water are safe. The Safe Drinking Water Act of 1974 requires that public supplies of drinking water meet standards for physical, chemical, biological, and radioactive components (see Table 8.2). The law allows citizens to sue their municipalities if those criteria are not met.

Although contaminants in drinking water can be controlled by established treatment procedures, the degree of treatment hinges on the initial water quality. Thus since most groundwater is sufficiently pure, it usually requires only chlorination before it is distributed. Surface water is generally stored for several days in a quiescent reservoir to allow the suspended materials to settle out and the dissolved oxygen content to increase. Those procedures improve water clarity and taste.

However, if surface waters contain substances that are slow to settle out, such as algae, bacteria, or suspended clay particles, additional treatment steps are needed before the water can be chlorinated. Normally, sand filters are used to remove algae and bacteria. Suspended materials are removed through coagulation and settling procedures that include the use of chemicals, such as alum and lime, which are added to the water to form insoluble precipitates. The precipitates then form a floc, which gathers the suspended materials as they settle. In addition, some surface-water and groundwater supplies are exceptionally hard. The calcium and magnesium that cause hardness in water can also be removed by the addition of alum and lime. Generally, it is less expensive to soften water at a central treatment facility than to try to soften it in individual homes by means of commercial water softeners. Water that is purified by coagulation procedures may still contain taste- and odor-causing chemicals, natural organic chemicals, and trace quantities of industrial chemicals.

The practice of chlorination to eliminate pathogens in drinking water must be monitored carefully in water supplies that contain dissolved organic chemicals. In some cases, small amounts of chloroorganic or bromoorganic compounds may be formed when those surface waters are chlorinated. In addition, those compounds may find their way into water supplies from industrial discharges. Some of those substances, such as chloroform, are known carcinogens. Collectively, they are called *haloforms*. In an effort to minimize human exposure to haloforms, the EPA has set a limit of 100 ppb (parts per billion) for the total of all of them, including chloroform. Water supplies that exceed this level must be treated, usually by activated carbon filtration, which was mentioned earlier as a tertiary treatment method. Activated charcoal filtration adds $7 to $30 annually to the water bill of an average household in areas where treatment is required.

To date, we still do not know if a definite connection exists between haloform chemicals in drinking water and increased rates of cancer. But the connection may well exist, so the EPA carries out its mandate to protect the public health by requiring that haloforms be removed by carbon filtration.

Nonpoint Sources of Pollution

Pollutants from nonpoint sources are much more subtle and difficult to control than are those from point sources, because the concentrations are relatively low and the volumes are enormous. Nutrients, sediments, and trace quantities of toxic metals and pathogenic organisms are the most common offenders. Requiring control strategies is controversial, because the persons who must control the level of pollutants often do not benefit directly from their efforts. Furthermore, since one person's contribution to control nonpoint pollution is small, especially when it is compared to pollution from most point sources, those persons may feel that their cleanup efforts do little to improve the situation.

Many lakes across the United States have eutrophication problems that are entirely due to pollution from nonpoint sources. In fact, 94 percent of New York's 4000 lakes are affected by eutrophication from nonpoint sources. Excessive nutrients and sediments comprise the most common types of nonpoint pollution. Recall from Chapter 3 that the loss of vegetative cover commonly increases the rate of flow of nutrients and sediments into nearby waterways. When raindrops fall on bare soil, their impact dislodges some of the soil particles and puts them into suspension, so they become a part of the runoff along with the nutrients that are attached to them. As Table 8.5 indicates, the greatest nonpoint pollution occurs where the soil is exposed (general agriculture cropland and developing urban land). Thus the best way to reduce nonpoint pollution is to protect the soil from raindrops and flowing water.

Construction sites constitute major sources of nonpoint pollution in urban areas. Exposed ground should be covered as soon as possible by mulching newly seeded areas (Figure 8.7) and by sodding waterways. In the meantime, retention basins should be built to trap the sediments that are present in the runoff from open construction sites. Some cities have passed ordinances that require builders to prepare plans for the control of erosion and runoff before new developments can be started. In some areas, developers must submit their plans to county Soil and Water Conservation Districts for approval.

Table 8.5 Quantity of materials in runoff under various land uses (kilograms per hectare per year)

	SUSPENDED SOLIDS	TOTAL PHOSPHORUS	TOTAL NITROGEN
Rural Land Use			
General Agriculture	3–5600	0.10–9.1	0.6–42
Cropland	20–5100	0.20–4.6	4.3–31
Improved Pasture	30–80	0.10–0.5	3.2–14
Forest Woodland	1–820	0.02–0.67	1.0–6.3
Idle/Perennial	7–820	0.02–0.67	0.5–6.0
Urban Land Use			
General Urban	200–4800	0.3–4.8	6.2–18
Residential	620–2300	0.4–1.3	5.0–7.3
Commercial	50–830	0.1–0.9	1.9–11
Industrial	450–1700	0.9–4.1	1.9–14
Developing Urban	27,500	23.0	63.0

Some communities can improve the quality of their stormwater runoff by taking advantage of the "free services" that are provided by natural ecosystems. For example, at La Belle, Florida, storm-water runoff is diverted to marshes, swamps, and weedy edges of lakes, where vegetation filters out the suspended solids and retains some of the nutrients. (Lakes that are used in this manner may experience some increases in eutrophication.) Stormwater is also ponded there, so it can soak into the ground and recharge the groundwater supplies.

Efforts to clean up the sources of nonpoint, urban pollutants are also beneficial. For example, frequent sweeping of streets reduces the levels of nutrients, metals, and pathogenic organisms that collect

Figure 8.7 *Mulching is a technique used to reduce sediment entry into runoff. Here straw is spread on a road construction site to slow the erosive forces of falling raindrops. Mulch also helps to retain moisture, so it aids in establishing a new sod cover. (USDA-SCS Photo, Edward B. Trovillion.)*

Figure 8.8 *Runoff from feedlots is prevented by the use of detention basins such as those shown on this Cass Co., Michigan, farm. (USDA-SCS Photo, Paul W. Koch.)*

in urban runoff. Prompt collection of fallen leaves reduces the amount of phosphorus that is leached during fall rains. And reducing the use of road salt (calcium chloride) or sodium chloride to a minimum further lessens the damage (as a result of groundwater contamination and the pollution of some lakes and ponds) that is done by the presence of those dissolved salts in water.

Problems that result from nonpoint pollution in agricultural areas can be reduced by using conventional soil-conservation practices. Those practices include strip-cropping, contour plowing, keeping livestock out of streambeds, adding riprap (a disordered layer of good-sized rocks) to stream banks (where they are particularly vulnerable to erosion), and preventing barnyard and feedlot runoff (such as that illustrated in Figure 8.8) from entering streams by impounding it in small lagoons. (Some of those techniques are described in more detail in Chapter 17.) Modern practices of handling manure also help. For example, storing manure during the winter months (Figure 8.9) and working it into the soil in the early spring reduces contamination of runoff during the spring and helps to retain nutrients in

the soil. Those practices are especially important in the northern states, where the freezing of the soil during the winter prevents infiltration of water and causes the loss of nutrients in spring runoff.

Unfortunately, many farmers view soil-conservation practices as both bothersome and expensive frills. Today's farms use large machinery that requires a great deal of room to maneuver, and it is most efficient when it is used for large, unbroken areas of crops. It is difficult and time-consuming to cultivate narrow strips of cropland with such equipment. Generally speaking, farmers have few short-term economic incentives to reduce pollution from nonpoint sources, even though some cost-sharing programs are available through county Soil and Water Conservation Districts.

One recently developed farming technique that does offer an economic incentive is the soil-saving "no-till" planting method. That method, which is becoming more popular (Figure 8.10), eliminates plowing entirely. In the spring, when soil conditions are right, a farmer simultaneously sows seeds, adds fertilizers, and applies herbicides to kill competing vegetation. Because the method leaves crop residues on the surface, it eliminates bare soil. Thus the crop residues from the previous season break the fall of raindrops and slow running water, which greatly re-

Figure 8.9 *Animal manure retention facilities such as the one shown are now being used by farmers to store manure. By waiting until the soil is thawed or dry enough to apply the manure, farmers prevent losses of nutrients through runoff. (Courtesy A. O. Smith, Harvestore Products, Inc.)*

Figure 8.10 *Crop residues that are left on the surface substantially reduce soil erosion rates. The techniques used to accomplish this are called minimum tillage, or no-till. Soybeans grown by the no-till method are shown here. (USDA-SCS Photo, H. E. Alexander.)*

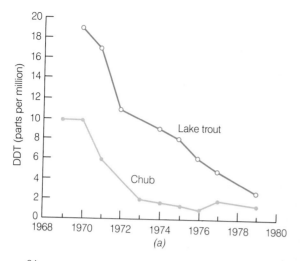

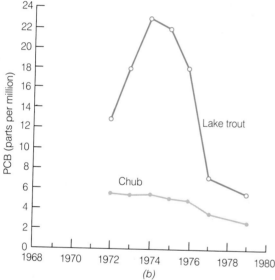

Figure 8.11 *The decline in (a) DDT and (b) PCBs in Lake Michigan fish following restriction of their use (in an average of three locations). (From National Pesticide Monitoring Program: Organochlorine Residues in Freshwater Fish, 1976–1979. U.S. Department of the Interior, U.S. Fish and Wildlife Service, Washington, D.C., 1983.)*

ample, in the past, the location of a cesspool or a septic tank was left to the whim of the landowner. Today, however, many states are beginning to permit the installation of septic tanks only in places where they are unlikely to contaminate groundwater or surface water. In areas where soil conditions are not suitable for septic tanks, landowners are required to install holding tanks. Also, periodically, the owners must pump those tanks out and have the wastewater hauled to treatment facilities—a procedure that is not always followed, because it is less expensive to pump the wastewater onto nearby land. If pollutants from septic tanks that are sited on improper soils create eutrophication problems in a lake, a sewage-collection system must be installed to serve the homes around the lake. However, such systems cost several thousands of dollars for each lot that is served.

Some nonpoint water-pollution problems can be solved by regulating the use of specific chemicals. For example, since the EPA banned the widespread use of DDT, the concentrations of that pesticide in aquatic ecosystems have declined greatly. Figure 8.11 shows the decline in the DDT levels that are found in fish from Lake Michigan and indicates that the PCB levels are also responding to the cessation of production of PCBs. However, PCB levels have declined more slowly than those of DDT because many PCB-containing devices, such as capacitors and transformers, remain in use. Moreover, because waste PCBs were dumped legally before their dangers were known (and are still being dumped illegally today), they continue to enter ecosystems from many unknown sources. Undoubtedly, other persistent toxic chemicals that can move through ecosystems and into groundwater will be regulated in the future.

Groundwater Protection

Groundwater is continually being recharged by precipitation that filters into the soil. As water slowly migrates downward (the rates vary from a few tenths of a meter to several tens of meters each year), it can carry a variety of substances with it. Because of that slow downward migration, contaminants can accumulate in the unsaturated zone that is located above the water table. However, the quality of water in the unsaturated zone is difficult to monitor, and it is seldom analyzed by state or federal authorities. Problem areas that are most often identified become

duces nonpoint sources of pollution. A closely related technique, which is called *minimum tillage*, is also gaining acceptance by farmers. That technique loosens the soil but leaves a substantial portion of the crop residues on the surface so that soil erosion is reduced.

Changes in land-use regulations for rural and suburban areas also influence water quality. For ex-

apparent as a result of testing samples of well water, but usually only after the saturated zone has been contaminated.

The first step in a proper approach to the management of groundwater resources is to identify groundwater flow patterns and recharge areas because, ultimately, human activities in recharge areas

Figure 8.12 *Groundwater recharge areas in the Houston region are separated from the heavy water-use areas. This exemplifies a thorny problem that faces groundwater managers: How to get people in one area to restrict their activities for the benefit of others. (U.S. Geological Survey, Open File Report 77-754.)*

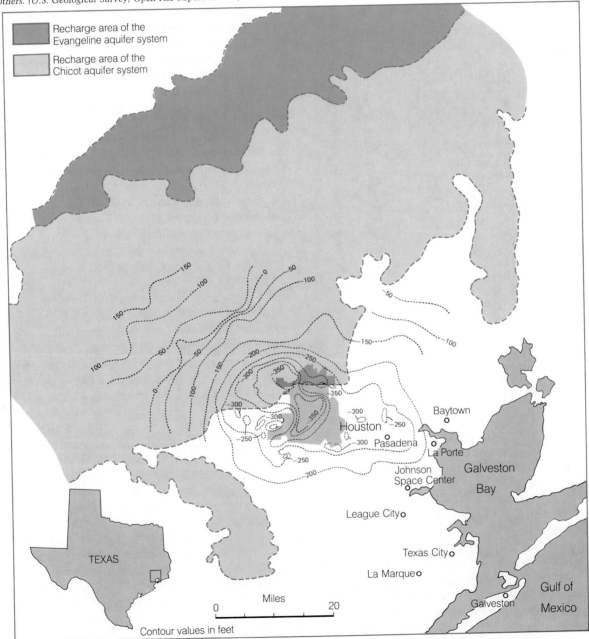

will determine the composition of future groundwater supplies. For example, the areas where groundwater is withdrawn for the city of Houston are widely separated from the areas of groundwater recharge (see Figure 8.12). That situation, which is quite common, points out one of the major difficulties in managing groundwater—the problem of convincing persons in a recharge area to restrict activities that will pollute the water in order to protect water quality for those who live in another area. The problem is complicated further by the fact that many landowners are mainly interested in short-term profits and believe that any restrictions on their activities infringe on their freedom, which puts them at a competitive disadvantage.

Water managers are beginning to recognize that, in some regions, certain practices must be curtailed if groundwater quality is to be protected. For example, pesticides must not be used on recharge areas where the soils do not slow down the downward migration of pesticides. Although pesticides are vital to agriculture, some of them—especially the water-soluble types—are now being found in groundwater, particularly in irrigated areas. Another problematic practice is the heavy use of fertilizer or the application of manure, which can lead to high concentrations of nitrate in groundwater. Those problems can be reduced, if not completely overcome, by lighter, less frequent applications of commercial fertilizers, manure, and pesticides. But many farmers are reluctant to switch to such practices because they are more time-consuming, may cut crop yields, and are more expensive. Moreover, many farmers today are not aware that their particular farming practices may contribute to future groundwater problems.

The contamination of groundwater reservoirs by toxic chemicals is another major problem. Accidental spills in transportation accidents, accidental discharges from plant equipment, and leakage from storage tanks occur every day. Furthermore, even though chemical companies, transportation companies, and state and local governments have done much to prevent spills and to clean them up once they occur, the costs are high. For example, in Beulah Station, a small community in southern Wisconsin, a railroad car went off the tracks and spilled 34,000 liters (9000 gallons) of phenol (carbolic acid), which contaminated an aquifer. As a result, several railroad cars of contaminated ground had to be

scooped up and hauled to a landfill. Cleanup also involved the cooperative effort of six federal and state agencies and the analysis of 800 water samples; 21 families in the vicinity had to use bottled water for 2½ years. Then a new $600,000 well was dug into a deeper aquifer, which was funded by state and federal grants. Eventually, a jury also awarded the families and the town a $500,000 settlement for expenses and hardships that had been incurred.

The chemical transportation industry has focused much attention on determining the type of equipment and procedures that are necessary for emergency use in the event of a spill. They have found that the best management technique requires the initiation of training programs that focus on spill prevention. But the only way to minimize contamination as a result of a spill is to contain it promptly and clean it up as quickly as possible. Illegally dumped chemicals present an even more costly and complex cleanup problem (see Chapter 13).

Toxic chemicals also enter groundwater as a result of use and indiscriminate disposal. For example, one such substance is trichloroethylene (TCE), which is a solvent in paint thinners, an engine-part degreaser, a dry-cleaning fluid, and a drain and septic-tank cleaner. In 1980, TCE, which is known to cause cancer in mice, contaminated 39 public wells in the San Gabriel Valley in California. Public health officials had to close those water supplies, which were a primary source of water for 400,000 residents. The pattern of pollution suggested multiple widespread sources. Septic tanks were one of the suspected sources, because TCE had been sold as a septic-tank cleaner. Because TCE is not biodegradable in septic tanks and is not retained on soil particles, it seeps into groundwater supplies. Petroleum fuels, especially gasoline, which is stored in buried tanks, pose a similar threat to groundwater. Underground storage tanks eventually rust through, and the fuels that leak from them migrate into the groundwater, because they, like TCE, are not retained by soil particles.

Mining activities contaminate groundwater in various ways. Some release toxic heavy metals; some increase the acidity of water; and some change groundwater levels and groundwater flow patterns. When groundwater levels are reduced in areas that have underground deposits of sulfide minerals, or when mining operations uncover such deposits, the sulfides are exposed to air. (Sulfide minerals are associated with the mining of coal, copper, nickel, zinc, and lead.) Exposure to air causes them to oxidize, and that process is catalyzed (accelerated) by certain strains of bacteria. Thus when water filters through those oxidized minerals, it becomes more acidic, which increases the solubility of metals in water that filters through the newly aerated deposits. Thus such water can have a harmful level of toxic materials.

Groundwater also can be contaminated by precipitation that seeps through stockpiles of bulk chemicals, such as road salt and fertilizers. At present, few states require that bulk stores of water-soluble chemicals be protected from the weather in buildings.

All sensible persons would agree that groundwater is a valuable resource that must be protected. But they often disagree about how it should be protected and who should pay the costs of the protective measures. For example, arguments arise on issues, such as whether restrictions should be placed on the kinds of fertilizers and pesticides farmers use and their rate and frequency of application. Also, it has not been determined how frequently and intensively groundwater should be sampled and monitored. As of 1984, the EPA had limits for only 17 specific chemicals (see Table 8.2) in groundwater that is used for drinking purposes. In the future, many of the other thousands of chemicals that are in use today will be found in groundwater unless we take action to protect our groundwater reservoirs.

High-quality water for human consumption, industry, and agriculture is an imperiled resource in many regions. Rough estimates indicate that 1 to 2 percent of groundwater is contaminated already, and once contaminated, it remains so for dozens, even hundreds of years. Purifying groundwater to acceptable health standards is generally not economical and often is not technically feasible.

Policies for sound water management should combine surface water and groundwater as a single resource. To provide adequate future water supplies, each region of the country must determine its own optimal conditions for the use of surface water, groundwater, and effluents, that is, the pattern that comes closest to meeting all user requirements without depleting or contaminating water resources.

Where water quality problems exist, a holistic approach should be initiated. Scientists can begin by compiling an inventory of point and nonpoint sources of pollution, either by estimating or actually measuring the quantities of pollutants from various sources. Such an inventory must be compiled for each water resource (lake, river, or aquifer) that may be polluted. Because each drainage basin or recharge area is used for different purposes, the mix of pollutants from each one is unique. Thus the quantities of various pollutants that enter each water resource from surface water, precipitation, groundwater, and each point source are calculated, and their relative percentage contributions are computed. Once those values are known, the available remedial options and their costs can be considered. Later, various cleanup plans, along with plans to finance them, are brought before meetings of people representing property owners, industries, municipalities, government agencies, environmental groups, and the voters, who must decide the degree of cleanup that will be done and the means of spreading the costs equitably.

Enhancing Freshwater Supplies

A highly respected international water resource specialist, Dr. R. P. Ambroggi, projects that, on a worldwide basis, plenty of water will be available for years to come. (Figure 8.14 shows the disposition of fresh water that is delivered by the water cycle.) He is also quick to point out, however, that 30 countries will face severe water shortages during the next 20 years if some action is not taken. Water-pressed regions have three choices: (1) they can increase their water supply by investing in dams or other measures that control a small part of the water cycle; (2) they can conserve water and use it to its maximum benefit; or (3) they can employ some combinations of those two approaches.

A case in point is the Canary Islands, where rates of water consumption now exceed the rate of renewal. Groundwater mining now makes up the deficit. But that practice cannot last longer than two more decades at present rates of use if current estimates of groundwater reserves are accurate. Eventually, the type of agriculture which is the islands' major consumer of water, will have to change. High-value crops that require less water will have to be cultivated if the islanders' standard of living is to be maintained. Water-supply experts also point out that certain regions of many countries cannot keep up

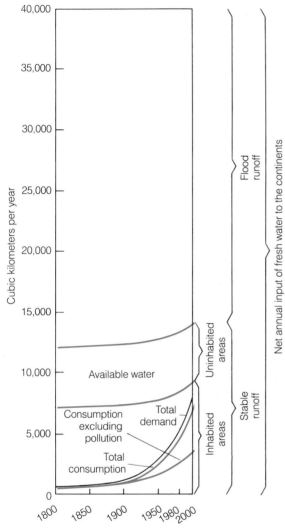

Figure 8.13 *Global water supply consists mainly of the base flow, or stable runoff, of rivers and streams. Of the net continental influx or 40,000 cubic kilometers per year, all but approximately 12,000 cubic kilometers runs directly to the sea in floods. A small fraction of the floodwater can now be captured by dams, so that by the end of the century the stable and regulated runoff will amount to 14,000 cubic kilometers. Of that amount, however, some 5000 cubic kilometers flows in sparsely inhabited regions, such as rain forests. Hence the total volume of water that is readily available at the end of the century will be approximately 9000 cubic kilometers per year. Actual consumption then will be approximately 3500 cubic kilometers, but water made unusable by pollution will effectively add another 3000. Total demand, which includes requirements for water that is not consumed, but still must be made available, will reach almost 7000 cubic kilometers per year. (R. P. Ambroggi, "Water," in* Economic Development, A Scientific American Book. *New York: W. H. Freeman, 1980.)*

with the water demand. For example, California now consumes 12 percent more water than is renewed by the water cycle in that state (the balance is imported from the Colorado River); and its demand for water continues to grow.

These two examples point up the need for long-range planning when devising water-supply strategies and evaluating their environmental impacts. Today, our primary way of increasing freshwater supplies are groundwater withdrawal, dam construction, and watershed transfers. We have already examined the implications of groundwater mining; next we turn our attention to dam construction and watershed transfers. Before doing so, however, we should have some understanding of the different legal aspects of the use of water resources.

Water Law

Water law is administered by the individual states. Nearly all the states east of the Mississippi are governed by *riparian* (bank or shoreline) *doctrine,* which states that the owner of land which adjoins a stream is entitled to receive the natural flow of that stream, undiminished in its quantity and quality. Also, riparian rights are transferred to the new owner when the land is sold. During times of drought, all riparian owners have equal rights to the reasonable use of water, although a priority use system may be imposed (domestic uses are usually given highest priority). Some states have enacted permit legislation that allows permit holders (riparian owners) to use water under prescribed conditions. For example, if stream flow is diminished during dry periods, the permit holder is not allowed to use any water. Also, the transfer of water from one watershed to another is prohibited.

In contrast, watercourse use in most western states is governed by *appropriation water laws.* Under those laws, only the owners of water rights may divert water from a watercourse, and then only for purposes that are deemed to be beneficial. Also, ownership of water rights is not tied to ownership of land that adjoins the watercourse, which is the case in areas where riparian doctrine rules. These laws came about after irrigation was introduced to the dry parts of western United States. They were instituted to prevent anyone from buying land upstream and then diverting the limited water supply away from dependent downstream users. Today, the law also protects the system whereby water is appropriated to owners of water rights. Thus the owners of the earliest acquired water rights have priority over those whose rights were obtained at a later date. In essence, an early water-right holder will get his full share of water during a drought, while owners of water rights that were acquired later may be forced to do without. In addition, various treaties and compacts affect the quantity of water that various states can withdraw from the rivers of a particular region.

Laws that govern the use of groundwater distinguish between two types of underground water resources: subterranean streams (water in the zone of saturation) and percolating water. Subterranean streams are generally subject to the same water laws as surface streams. Percolating water is owned by the landowner, who is allowed to withdraw enough of it for his own use, as long as the withdrawal does

not infringe on the water rights of others. Furthermore, landowners may not pollute groundwater.

Dams

Because precipitation is irregular, both seasonally and geographically, dams are built to conserve surface water and regulate its availability. Thus reservoirs that are created by dams collect water during wet periods and store it for use during drier periods by municipalities and industries. In many locations around the world, dams also provide the water for agriculture, which is dependent on irrigation. In turn, irrigation allows more land to be farmed, increases the yield per hectare, and often enables the farmer to grow more crops per year on the same land. Because dams reduce or eliminate irregularities in water supply, they stabilize the agricultural economy, especially in drought-prone regions.

But dams have other important functions as well. In mountainous areas, deep canyons are dammed to create the large, deep reservoirs that are needed to generate hydroelectric power. And many dams are built as flood-control structures. Such dams make the fertile floodplains of major rivers habitable and reduce or eliminate crop damage that occurs as a result of flooding. Dams also create recreational opportunities, such as boating, swimming, and fishing.

Figure 8.14 *Some of the multiple uses that must be managed when water is impounded. Note that withdrawal is not the same thing as consumption. Water that is withdrawn for agriculture is mostly consumed (lost through evapotranspiration), but sewage treatment is essentially a nonconsumptive use. Virtually all the water withdrawn for this purpose is returned to the stream and becomes available for reuse farther downstream. (After R. F. Ambroggi, "Water," in* Economic Development, A Scientific American Book. *New York: W. H. Freeman, 1980.)*

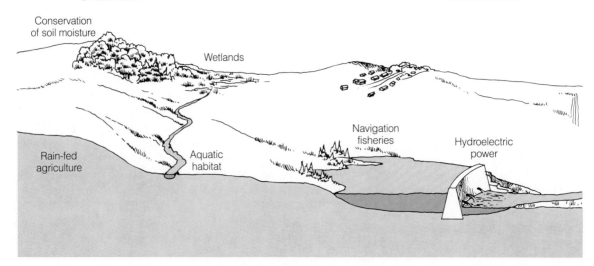

On-site uses

Streamflow uses

Conservation of soil moisture

Wetlands

Rain-fed agriculture

Aquatic habitat

Navigation fisheries

Hydroelectric power

Some of the uses of an artificial reservoir are shown in Figure 8.14.

The benefits derived from dams, however, must be weighed against the costs. For example, (1) dams trap nutrients along with sediments. The nutrients accelerate eutrophication in the reservoirs, which reduces the quality of the water. The sediments are retained in the reservoirs, and dams inevitably lose their value when the reservoirs fill with sediments, which is a process that takes anywhere from only a few years to a few centuries. (2) Dams interfere with the spawning migration of some fish, such as salmon, unless fish ladders (Figure 8.15) are provided. Those ladders enable such fish to cross dams and continue their migration upstream to their spawning grounds, where they complete their reproductive cycle. (3) Dams can also disrupt estuarine ecosystems. When dams are built across rivers that flow into estuaries, they can trap spring runoff that would otherwise dilute the seawater in an estuary. The consequent increase in the salinity of the seawater can prevent the reproduction of some types of fish and shellfish. (4) Dams reduce the flow of nutrients into estuaries, thereby limiting their productivity. (5) The flooding of land behind dams can destroy vast areas of valuable agricultural land, wildlife habitat, and scenic beauty. (6) The storage of water behind dams raises the water table, often waterlogging the soil on ad-

jacent land, thereby cutting crop or forestry production. And finally, (7) faulty construction or earthquakes can cause dams to fail, and water that floods from a breached dam can take a terrible toll in lives and property.

New dam projects are always controversial, because special-interest groups benefit from the construction, while other groups must bear the brunt of the costs. As a specific example, let us examine the major issues that are associated with the damming of the Tuolumne River, which drains part of the Sierra Nevada Mountains of northern California. The Tuolumne, which flows through Yosemite National Park, is already dammed in four places. Those dams deliver enough water for nearly 2 million Californians, generate enough electricity for 400,000 homes, and irrigate 92,000 hectares (230,000 acres) of fertile farmland. The city of San Francisco and the Turlock and Modesto irrigation districts plan a series of hydroelectric projects that would quiet rapids and reduce the flow in other segments of the river. The proposed hydroelectric projects would help to maintain electric utility rates in the region at a level that is among the lowest in the nation and add to the quantity of water that is available for irrigation. Opponents to the projects are naturalists, fishermen, white-water boaters, resort owners, and local residents, whose homes would be inundated by one

Withdrawal uses

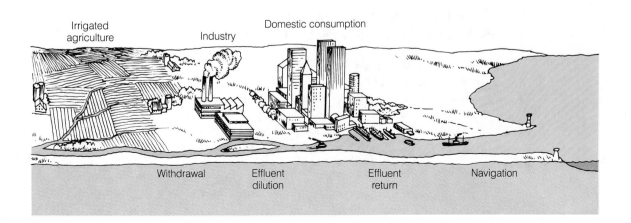

(a)

(b)

Figure 8.15 *(a) A fish ladder (foreground) at the Red Bluff Diversion Dam on the Sacramento River in California. (b) A king salmon, using a ladder to complete its migration and thus its reproductive cycle. (U.S. Department of the Interior, Bureau of Reclamation.)*

of the proposed projects. These persons ask, Why does this scenic wild river have to be dammed to provide for yet more growth?

The fate of the undeveloped sections of the Tuolumne has not yet been determined. Because 44 kilometers (27 miles) of it comprise a series of thundering white-water rapids that flow through a nearly mile-deep canyon, it may gain protection under proposed legislation. For example, one legislative effort is to include parts of the Tuolumne River under the provisions of the National Scenic and Wild Rivers Act, which would give those sections so designated protection from any sort of development. Another effort is to have the Tuolumne Canyon protected under the California Wilderness Act, but five other rivers in northern California—the American, Eel, Klamath, Smith, and Trinity—were denied wilderness protection in a mid-1980 federal court decision.

Watershed Transfers

One of the most common techniques for increasing a limited water supply is the transfer of water from a watershed that has an excess of water to a watershed that has a deficit of water. Several major cities in the United States, among them Los Angeles, Phoenix, and New York, rely on other watersheds in addition to their own. For example, southern California transports water from the Colorado River through a 714-kilometer (444-mile) aqueduct system. The route of the system and a portion of the aqueduct are shown in Figures 8.16 and 8.17. But, in 1985, as the result of a U.S. Supreme Court decision (*California* v. *Arizona,* March 9, 1964), Arizona will begin to use some of the water that was previously allocated to California. As a result of that decision, Arizona gained access to 2.8 million acre-feet—the amount of water that is needed to flood 1 acre of land to a depth of 1 foot—of water per year from the Colorado River, which was used previously by southern California. Arizona will use the water for its Central Arizona Project, a diversion system that is expected to deliver water from the Colorado River to Phoenix in 1985 and to Tucson in 1991 (see Box 8.1). The diverted water partially will replace groundwater withdrawals because groundwater levels have fallen 122 meters (400 feet) during the past 50 years in parts of Arizona. This impending reduction of water to southern California has northern Californians, especially those in the San Fran-

cisco–Sacramento River delta area, concerned about the effects of those increased diversions, because they fear that the water now needed by southern California will come from their rivers.

Those diversions and others along the Colorado River have created some other major problems. For example, as the water travels through the arid Southwest, over 10 percent of it evaporates, leaving the remainder more saline. Thus to meet drinking standards, it must be diluted with less salty water from northern California sources. The waters of the Colorado River are also used to irrigate crops in the Imperial and Coachella Valleys of southern California (see Figure 8.17), but only crop varieties with a high salt tolerance can be grown in those valleys.

Similar problems occur in Mexico. By the time the Colorado River flows into Mexico, the water is highly saline. Because of evaporation and the use of water for irrigation, water in the Colorado watershed returns to the Colorado River with a much higher salt content. Hence by the time river water flows into Mexico, it is too saline for human consumption and the raising of most crops. Furthermore, Mexico has no low-saline water with which to dilute the saline water. To alleviate that problem, both countries signed an agreement on August 30, 1973. To meet the water-quality terms of the agreement, the United States is engaged in a $350 million desalination project to reduce the salt content of the water that is delivered to Mexico for irrigation in the Mexicali Valley. That water costs United States citizens $300 per acre-foot to produce, which is three times the cost of irrigation water north of the desalination plant at Yuma, Arizona.

Disputes over water rights, especially in arid and semiarid areas, can be expected to intensify as demand exceeds the supply. Increased urban development, agricultural demands for increasing amounts of irrigation water, and the mining industry's demand for more water to process coal and oil-shale resources can only create a three-way tug-of-war over this precious resource, especially in the Colorado River watershed.

Other Enhancement Methods

Damming rivers, increasing groundwater withdrawal, or transferring of watershed surpluses may not be feasible in a particular region or may not suffice to meet the region's needs. Thus in recent years,

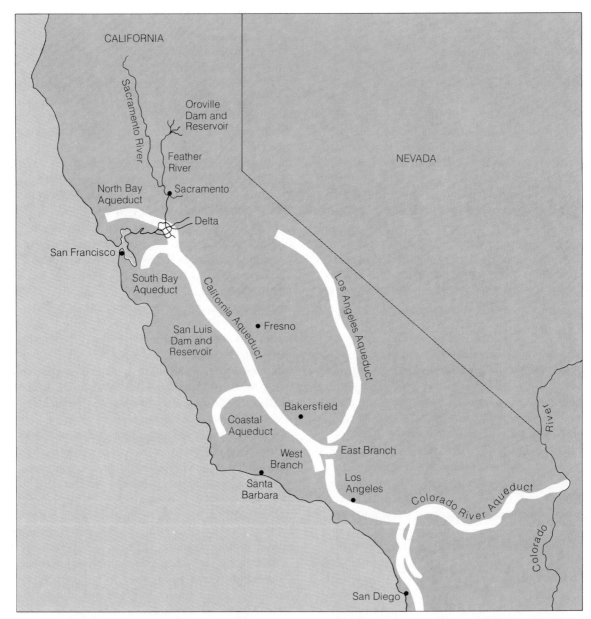

Figure 8.16 *Route of the Colorado River, Los Angeles, and California aqueduct systems. (The Metropolitan Water District of Southern California.)*

Figure 8.17 *A portion of an aqueduct (a) and fields (b) in California's Coachella Valley that are irrigated with Colorado River water. The ridged field (center) is being flushed of salts. (U.S. Department of the Interior, Bureau of Reclamation.)*

(a)

(b)

Direction of prevailing wind

Ground-based silver iodide generators

Figure 8.18 *One scheme that is used to increase precipitation utilizes silver iodide smoke (composed of countless freezing nuclei that produce ice crystals) that is released into the air to seed clouds.*

research has been directed toward other methods of fresh-water enhancement. Those techniques include cloud seeding, desalination, and various water conservation strategies.

Cloud Seeding Since World War II, much research has been done on cloud seeding—the injection of nucleating agents into clouds to trigger precipitation processes. (Precipitation processes were described in Chapter 6.) Usually, silver iodide, which catalyzes ice formation by providing an icelike template for ice-crystal formation, is injected into cold clouds that are deficient in ice-crystal nuclei (See Figure 8.18). Silver iodide crystals, which are freezing nuclei, produce ice crystals at −4°C (25°F) and below. The objective is to stimulate the Bergeron process. Sometimes dry-ice pellets are used instead of silver iodide crystals to seed cold clouds. *Dry ice*, or solid carbon dioxide, maintains a temperature of −80°C (−110°F) as it vaporizes. When it is introduced into a cloud, the dry-ice pellets lower the temperature of the su-

percooled water droplets, which causes them to freeze. Then those frozen droplets function as nuclei for the Bergeron process. As little as 1 gram of dry ice can produce 5 trillion nuclei at −10°C (14°F). Warm clouds are seeded with sea-salt crystals or other hygroscopic substances to stimulate the collision–coalescence process. The trick to cloud seeding is to stimulate the process of nucleation without overdoing it, because too many ice-forming surfaces or collision-inducing agents can hinder the process of precipitation formation.

To determine how effective those efforts are, we must recognize first that not all clouds can be successfully seeded. For example, cloud-seeding experiments that were conducted by scientists under favorable conditions in Colorado and Israel were judged to increase precipitation. Later, two sequential experiments in Florida, which were conducted to show that cloud-seeding experiments could be duplicated, failed to do so. In the first experiment, precipitation was enhanced by 20 percent, but the second experiment showed no significant increase in precipitation. Atmospheric scientists still do not clearly understand the exact conditions that trigger precipitation formation, but they continue to study those processes so that someday cloud seeding may become a tool that can be used to enhance the water supply.

Before cloud-seeding techniques can be used, potential conflicts concerning the effects of increased precipitation must be resolved. For example, consider the conflicts that would arise if the mountain snowpack in the Colorado River drainage basin were increased through cloud seeding. Such a project has been contemplated to alleviate the frequent shortages of water that occur in the Colorado River basin. (The water rights that have been granted to seven states and Mexico to the Colorado River's water presently amount to 110 percent of the average amount of water present; thus its water is overcommitted by 10 percent.) In an effort to counteract those shortages, experiments were conducted between 1970 and 1975, which showed that snowfall increases of 10 percent would be possible if cloud-seeding techniques were used. The additional water supply would benefit irrigation projects, communities that are dependent on the Colorado River for their water supply, hydroelectric generating facilities, and the exploitation of coal and oil-shale resources in the region.

But except for those who rely on the ski industry for a living, inhabitants of the mountains have much less to gain from cloud seeding. Making the winter snow cover thicker and longer lasting would shorten the growing season, reduce the amount of land that is available for grazing of cattle and sheep, and increase the danger of avalanches. In addition, tree growth would be slowed, road travel would become more difficult, and logging and mining seasons would be shortened. Indeed, if cloud seeding had been used in the winter of 1982–1983, it would have further aggravated the severe flooding problems that occurred along the Colorado River in the spring of 1983, which were caused by the melting of abnormally heavy late-spring snowfalls.

Moreover, rainmaking by cloud-seeding—if it is successful—may merely redistribute a fixed supply of rain; thus increased precipitation in one area may be offset by a decrease in another. Thus although rainmaking might, for example, benefit agriculture in the high plains of eastern Colorado, it might deprive wheat farmers of rain in adjacent states of Kansas and Nebraska. Clearly, cloud seeding poses problems that may be just as complex and severe as the ones that we are attempting to solve. Such conflicts have caused legal wrangles in the United States between adjoining counties and states and, in Canada, between provinces.

Desalination Since 97 percent of all the water on earth is in the oceans, it is tempting to think that efficient methods of *desalination*—the removal or reduction of dissolved salts from seawater—might solve some freshwater supply problems. Because seawater contains 35,000 ppm dissolved salts, and brackish groundwater can have even higher salt content, those sources are not suitable for drinking or irrigation. Dissolved salt content in drinking water should not exceed 500 ppm, and for most crops, the salt content in irrigation water should not exceed 700 ppm. Thus the salt content of seawater or brackish water must be reduced substantially before it can be used.

Many methods of desalination exist, but all require great amounts of energy and, therefore, are expensive. Only the richest countries can afford to desalinate water, and then only for domestic or industrial purposes. Using desalinated water for the conventional irrigation of crops is economically out of the question.

Conservation Conservation is the most practical means of extending available water supplies. Irrigation, which accounts for 80 percent of water use in the western United States, requires especially judicious conservation, since it is a *consumptive use,* that is, most of the water (more than 60 percent) that is used in irrigation is lost by evapotranspiration or infiltration and thus is not available for immediate reuse. (In contrast, water that is used by municipalities and industries is considered to be *nonconsumptive use,* since approximately 95 percent can be recovered for immediate reuse.) Besides applying only the amount of water that is absolutely necessary for a crop, farmers who are intent on conservation, particularly during critical shortages, may find it necessary to employ costly measures, such as using buried pipes to irrigate below the ground rather than using wasteful in-air spraying techniques or open-furrow irrigation. Furthermore, by lining irrigation canals with plastic sheets, seepage (Figure 8.19) losses can be prevented during transport. In areas where it is appropriate, farmers who anticipate droughts can grow more drought-resistant crops, such as cotton or wheat.

Water conservation is also possible in industry through recycling. For example, such industries as paper manufacturing are beginning to identify and employ processes that utilize a treatment-use-treatment cycle. Such processes drastically reduce the amount of water used and facilitate compliance with government water-quality effluent standards.

Wastewater and storm runoff can also be conserved by routing it into recharge basins, which allow the water to seep into the ground, rather than emptying it into rivers or coastal waters. Groundwater recharge is commonly used in coastal regions to prevent saltwater intrusion into freshwater aquifers.

In the home, the primary cause of unnecessary water loss is leaky faucets. Thus householders can conserve water by mending all leaks. Other steps to reduce the use of water include refraining from watering lawns, decreasing the frequency of toilet flushing and car washing, and taking short showers rather than baths. Although toilets that use approximately 12 liters (3 gallons) of water per flush rather than the normal 20 liters (5 gallons) are now available, the amount of water used by conventional toilets can be reduced by installing a water dam in the toilet's water tank or simply by putting a few bricks in it.

In Orlando, Florida, city officials have gone from door to door, offering to install water-saving devices on showers, sink faucets, and toilets. A survey that sampled 500 of the 33,000 homes that have installed those devices show that the consumption of water is down by 18 percent. Furthermore, officials have calculated that the $750,000 spent for the devices will save $10 million on sewage-treatment expansion costs.

Figure 8.19 *A plastic liner being installed in an irrigation canal to prevent water losses through seepage (U.S. Department of the Interior, Bureau of Reclamation.)*

Table 8.6 Estimated population served by drinking water supply and sanitation services in less-developed countries, 1970 and 1980

	DRINKING WATER SUPPLY			
	1970 POPULATION SERVED (MILLIONS)	PERCENT OF TOTAL POPULATION	1980 POPULATION SERVED (MILLIONS)	PERCENT OF TOTAL POPULATION
Rural	182	14	526	29
Urban	316	67	469	75
Total	498		995	
	SANITATION SERVICES			
Rural	134	11	213	13
Urban	337	71	372	53
	471		585	

Source: United Nations Environmental Programme, The World Environment, 1972–1982.

In areas where voluntary water conservation is insufficient to reduce water consumption, water departments may have to force reduced consumption. For example, restricting devices can be installed on household water meters to reduce the flow of water into a house to a mere trickle. Water companies may also decide to impose heavy fines on households that waste water during crisis periods.

Global Water Problems

Developed countries are primarily concerned with water pollution caused by BOD, nutrients, and toxic chemicals. But in the less-developed countries, the most serious water pollutants are still pathogens— organisms that transmit infectious diseases. Waterborne diseases are prevalent in less-developed countries because the drinking water is contaminated by pathogens and sanitation facilities are inadequate. Furthermore, the construction of water impoundments to improve local water supplies has, in some instances, increased the incidence of waterborne diseases by providing habitats for organisms that spread disease.

A glimpse into the magnitude of global water problems is gained by examining the statistics in Table 8.6. Although impressive gains have been made in supplying safe water to the residents of less-de-

veloped nations, more than two-thirds of the rural residents (1,200 million) and one-quarter of the urban residents (175 million) do not have a safe water supply. Furthermore, even though a massive effort was made during the 1970s to reverse this trend, only the regions in Asia and the Pacific Islands actually increased their safe water supplies at a rate that exceeded their rate of population growth. Trends in improving sanitation facilities are even more disheartening. About 1400 million rural and 330 million urban residents do not have sewers, latrines, or any other means of disposing of human wastes. Statistics gathered by the United Nations show that efforts to improve sanitation have not kept pace with population growth.

The less-developed nations cannot solve these problems without massive infusions of capital from the wealthier nations. And even if the capital is supplied, it will be decades before the necessary drinking water supplies are available and sanitary facilities are constructed.

Malaria, schistosomiasis, onchocerciasis, yellow fever, and encephalitis are the greatest scourges among the less-developed nations. All these diseases are transmitted by water-related vectors and thus are related to water-development projects. Thus we now look at some of the specific water-related disease problems those nations face.

Malaria remains a constant threat for about 290

million persons who live in sub-Saharan tropical Africa, and it is one of the most significant diseases in other tropical parts of the world. In Africa, an estimated 1 million children die each year from malaria, and 50 percent of the children contract the disease by the time they are 3 years old. Indeed, the threat of malaria in the less-developed countries was greater in 1980 than it had been a decade earlier.

Malaria is spread by the *Anopheles* mosquito. During the 1930s, gains against the disease were made by draining the mosquito's breeding sites. During the 1950s, countries that could afford pesticides, such as DDT, made further dramatic reductions in the incidence of the disease. But during the 1970s, that encouraging trend was reversed because the mosquitos became increasingly resistant to DDT and other insecticides, including malathion and dieldrin. At this point, scientists believe that the best long-term means of combating malaria is to develop a vaccine that could be administered to persons who live with those mosquitos. At present, no such agent is available, but research scientists are making some progress in that direction.

Several forms of *schistosomiasis* also affect some 200 million persons and threaten approximately 600 million more. For every person who shows disabling symptoms of the disease (such as liver fibrosis or bladder cancer), there are many who have only mild symptoms. This disease, which is virtually unknown in the United States, is particularly prevalent in Africa, Asia, and some of the West Indian Islands. In Egypt, schistosomiasis is the primary health problem. Since the disease is spread by a parasitic worm, which requires snails as an intermediate host, the construction of large hydroelectric power plants and irrigation projects has promoted the spread of schistosomiasis. Irrigation canals that are associated with such water projects are filled with water year-round, which increases the survival of the host snails as well as opportunities for the parasitic worm to penetrate the skin of persons who wade in the water. After the worm penetrates the skin, it lays eggs, which hatch in the human body. The developing worms then interfere with the flow of blood, lead to infections, and destroy tissues in the kidneys, lungs, liver, and other organs. Over a period of years, the victim becomes weakened and his or her ability to work is reduced. In severe cases, brain damage and even death

result after a number of years. The strong drugs that are used to treat the disease also produce severe side effects and have even led to death in some cases.

A particularly horrible disease is *onchocerciasis*, or "river blindness." That disease is widespread in parts of Guatemala, Mexico, Venezuela, Brazil, Columbia, Northern Yemen, and tropical Africa. West Africa and the Sahel region of Africa have the highest rates of incidence. Persons who contract the disease are infected by a parasitic worm that is spread by the blackfly. In contrast to the quiescent water conditions that are required for the propagation of schistosomiasis, blackflies require fast-moving water as part of their life cycle. Those conditions exist in rapids and below the outlets of dams. In the mid-1970s, 10 percent of the population of Africa's Volta River Basin was afflicted with the disease, and 70,000 victims were blinded by it. The high rates of blindness among young persons have had serious social and economic consequences. Thus efforts have been made to control the disease by treating the habitat (to kill the larval stages of the blackfly), which has greatly reduced the rate of incidence, but unless pesticide treatments continue, the disease will regain its foothold in those regions.

Conclusions

Worldwide, the hydrologic cycle furnishes humans with about three times as much water as we withdraw and about 14 times as much water as we consume. But we must still determine whether we have water supplies that are adequate in both quality and quantity both where and when we need it. These matters are up to us. We can develop new supplies, implement conservation strategies, reallocate existing supplies, and implement policies that will ensure high-quality water.

Today, we have the technical means to reduce greatly the level of pollutants in wastewater from point sources. In the United States, levels of removal are dictated by the states or the federal Environmental Protection Agency. Those levels, in turn, are based on the cost as well as the availability of pollution-abatement technology. The application of that technology has led to significant improvements in water quality in certain regions, but the quality of some surface waters has deteriorated.

Attempts to meet future demands for water in water-poor regions of the world pose many problems. Possible solutions will entail trade-offs in terms of environmental quality if those regions are to continue to grow. Less-developed nations must first address the problems posed by waterborne diseases, but many of them will also have to begin planning now to meet greater demands if they choose to raise their standards of living.

Summary Statements

The 1972 Water Pollution Control Act was the first really effective legislation to reverse a trend of deteriorating water quality. During the 1970s, other laws were also enacted that directly or indirectly relate to water-pollution control, such as the significant Safe Drinking Water Act of 1974.

The primary stage of sewage treatment removes insoluble and suspended materials from wastewater. Secondary sewage treatment steps, in effect, use bacteria that grow on dissolved organic materials in the wastewater and then "harvest" the bacteria to remove the waste materials. Tertiary treatment steps remove phosphorus compounds, nitrogen compounds, and toxic compounds. Those same techniques are employed in the treatment of industrial wastewaters. Sludges that are generated from treatment processes result in large volumes of solid wastes that cause a disposal problem. Alternative solutions are land spreading, landfilling, anaerobic digestion, and incineration.

Drinking water supplies are treated to remove suspended materials and are chlorinated to prevent the spread of waterborne diseases. The presence of small amounts of chloroorganic compounds in drinking water supplies presents a possible chronic health hazard. The technology for removing those substances exists, but its use adds to the cost of drinking water.

Nonpoint-source pollutants often add significantly to water quality problems in a watershed. Control of nonpoint sources is more difficult and costlier than control of point sources.

Agricultural nonpoint-source pollutants can be reduced by using the "no-till," or minimum till, planting method, which, by reducing runoff, reduces pollution.

Training in accident prevention is the best management technique for reducing the number of spills of harmful substances that can pollute surface waters and groundwaters.

If groundwater quality is to be protected over the long term, the use of fertilizers and pesticides must be limited in recharge areas with soils that do not retain those substances, which is a difficult prospect, indeed. Leaky underground fuel-storage tanks and the use of certain chemicals in septic-tank systems are serious groundwater problems that are beginning to be addressed.

Groundwater supplies are finite and should not be withdrawn at rates that exceed natural recharge rates. Artificial recharge can be used to augment natural recharge.

Water supplies can be enhanced by damming rivers, transferring water from one watershed to another, desalination, and precipitation enhancement. All those methods entail costs and benefits that must be weighed carefully. Desalination, for example, is so costly in terms of energy use that it is now economically feasible to use it only in certain areas, where critical shortages of drinking water now exist.

Water conservation practices can substantially lower the amount of water we use in homes, industries, and agriculture. Conservation efforts in agriculture will reap the largest benefits.

Waterborne diseases are still a constant threat in many less-developed nations, where water supplies are often contaminated with human and animal wastes.

Questions and Projects

1. In your own words, write a definition for each of the terms in this chapter. Compare your definitions with those in the text.

2. List several sources of nonpoint pollutants and propose a strategy to reduce the amount of pollution that can enter runoff from a watershed.

3. How does groundwater become contaminated? List some substances that threaten groundwater quality.

4. List the stages of the sewage treatment process (primary, secondary, or tertiary) that would remove the following materials: paper fibers, maple syrup, aluminum cans, table salt, phosphate in laundry detergent, dirt in wash water, and hand soap.

5. Visit your local sewage treatment plant. Find out how efficiently suspended solids, biochemical oxygen-demanding substances (BOD), phosphorus, and nitrogen are removed from wastewater and determine whether your local treatment plant meets designated effluent standards for those substances.

6. Why should industries be prohibited from dumping toxic substances into sanitary sewers?

7. Should all communities or industries, regardless of size, be required to treat their wastewater to the same degree? State your reasons for your answer.

8. Are sanitary sewers and storm drains completely separated in your city? If not, what happens during wet weather? If there are serious

problems, ask someone to speak to your class about the nature and frequency of the problems.

9. What is the environmental impact of effluents from your sewage treatment plant? Are the effluents being used for a useful purpose? What are some alternative uses?

10. Determine how sewage sludge is disposed of by your community. Are local officials exploring or implementing any new alternative disposal methods?

11. Use a topographic map to outline your watershed. Identify some water-pollution sources (both point and nonpoint) in the watershed. Draw up a plan to improve the quality of water in that watershed.

12. Contact your local water-supply utility to determine whether your water supply has been analyzed recently for organic chemicals, especially chloroorganics. If so, and if any have been found, which of them are of primary concern? How or where do they originate? Are treatment steps provided to remove them?

13. Evaluate the dependability of your community's water supply. What circumstances might disrupt or cut off supplies?

14. Which specific nonpoint source pollutants cause groundwater-quality problems in your area?

15. Develop a report on a water-quality improvement or rehabilitation project that has been completed for surface water. Pick a nearby example if you can. Your state's Office of Environmental Protection is a good source of information.

16. How is your state or community equipped to deal with accidental spills of toxic substances?

17. Evaluate a dam site in your region. What are the environmental and economic benefits and costs of the dam? Be sure to consider the land-use patterns that existed prior to the dam's construction.

18. Should large metropolitan areas be allowed to import water from more sparsely populated watersheds? Explain your answer.

19. In arid regions, who should have the right to limited water supplies—cities, industries, farmers of irrigated lands, or energy resource development interests? Set up teams in class to debate this issue.

20. What water-conservation measures, if any, are used in your region?

21. Devise and implement a water-conservation plan for your home. Use water-meter readings to measure the effectiveness of your plan.

22. Is the water in your state governed under riparian or appropriation water law?

Selected Readings

AMBROGGI, R. P.: "Underground Reservoirs to Control the Water Cycle." *Scientific American* 236:21–27 (May), 1977. A discussion of methods for increasing groundwater supplies.

AMBROGGI, R. P.: "Water." In *Economic Development.* A Scientific American Book. New York: W. H. Freeman, 1980. Chapter 4 presents a discussion on the use of water on a global scale and the economic and physical limitations.

CLARK, C.: *Flood.* Alexandria, Va.: Time-Life Books, 1982. A well-illustrated book, with much discussion on the impact of dams.

THE CONSERVATION FOUNDATION: *State of the Environment—1984.* Washington, D.C.: The Conservation Foundation, 1984. One chapter is devoted to analyzing trends in water quality during 1981.

COUNCIL ON ENVIRONMENTAL QUALITY: *Environmental Quality, 1983.* The President's Council on Environmental Quality, 14th Annual Report. This report includes a section on water quality.

CRITES, R. W.: Land use of wastewater and sludge. *Environmental Science and Technology* 18:140A–147A, 1984. Review of current technologies and treatment performance for those municipal wastes.

DUNNE, T., AND LEOPOLD, L. B.: *Water in Environmental Planning.* New York: W. H. Freeman, 1978. A textbook that examines methods for preventing water-related problems.

FREEZE, R. A., AND CHERRY, J. A.: *Groundwater.* Englewood Cliffs, N.J.: Prentice Hall, 1979. A textbook for persons who are interested in a more advanced treatment of groundwater quality and movement.

KRENKEL, P. A., AND NOVOTNY, V.: *Water Quality Management.* New York: Academic Press, 1980. A technical textbook that deals with water-quality problems. Most introductory sections are written at the layperson's level.

JOSEPHSON, J.: Restoration of aquifers. *Environmental Science and Technology* 17:347A–350A, 1983. Methods of identifying, containing, and reclaiming contaminated groundwater are discussed.

MILLER, S.: The marine environment. *Environmental Science and Technology* 16:645A–646A, 1983. A summary of the Report of the United Nations on ocean pollution problems and trends.

NAYLOR, L. M., AND LOEHR, R. C.: Increase in dietary cadmium as a result of application of sewage sludge to agricultural land. *Environmental Science and Technology* 15:881–886, 1981. Sludge management techniques that should protect public health.

PYE, V. I., PATRICK, R., AND QUARLES, J. R.: *Groundwater Water Contamination in the United States.* Philadelphia: University of Pennsylvania Press, 1983. An excellent overview and guidelines for dealing with groundwater pollution problems.

UNITED NATIONS ENVIRONMENT PROGRAMME: *The World Environment, 1972–1982.* Dublin: Tycooly International, 1982. Chapters 3, 4, and 10 deal with water-pollution issues on a global basis.

UNITED STATES GEOLOGICAL SURVEY: *National Water Summary, 1983— Hydrologic Events and Issues.* Washington, D.C.: U.S. Government Printing Office, 1984. Contains maps of every state, which show locations of problems that are related to water quality and quantity.

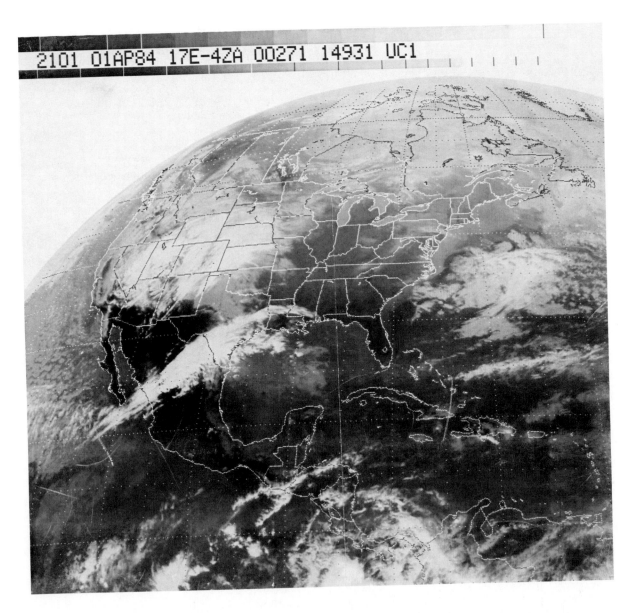

2101 01AP84 17E-4ZA 00271 14931 UC1

A satellite provides a unique perspective of the cloud patterns that are generated by the circulation of the earth's atmosphere. (From NOAA.)

CHAPTER 9

THE ATMOSPHERE, WEATHER, AND CLIMATE

The weather during January and February of 1983 was devastating for Californians. One storm after another swept ashore from the Pacific Ocean, bringing strong winds and heavy precipitation. Coastal areas bore the brunt of nature's fury when huge waves and rough surf caused severe beach erosion that damaged or destroyed more than 1600 beachfront homes and other buildings (Figure 9.1). Many hillside residences and roads were undermined or washed away by rivers of mud. Well inland, in the Sacramento and San Joaquin Valleys, floodwaters inundated 140,000 hectares (350,000 acres) of prime farmland, which ruined many fruit and vegetable crops. The total amount of damage that was caused by the storms was estimated at more than $500 million.

Fortunately, disastrous and disruptive extremes are not the usual pattern of weather. Although in any year we can expect (somewhere across the globe) drought, flood-producing rains, scorching heat, and bone-numbing cold waves, most storms are not severe and the rainfall and snowfall that they produce usually are welcomed. Furthermore, weather typically tends to be fair more often than it is stormy in most regions of the world.

Clearly, the importance of the weather in day-to-day activities and the potential impacts of severe weather on economic and societal issues provide enough incentive for us to learn about weather and climate. But other reasons exist as well, because weather and climate are involved in some major environmental issues: (1) Weather affects and is af-fected by air quality; (2) atmospheric circulation patterns are a primary consideration in the current international debate concerning acid rain; (3) supplies of energy, food, and fresh water are influenced by weather conditions; (4) possible threats to the atmosphere's ozone layer, which protects us from exposure to unhealthy doses of ultraviolet radiation, are a source of concern; and, (5) much speculation is voiced about the climatic future, especially in view of rising levels of atmospheric carbon dioxide. Before considering those issues, however, we must know something about the structure and composition of the atmosphere, the basic reasons for weather, and the role of climate in the functioning of ecosystems.

Weather is defined as the state of the atmosphere at some place and time, which is described in terms of such variables as temperature, cloudiness, precipitation, and wind. *Climate* is defined as weather at a particular region, which is averaged over some period of time plus weather extremes that occur during that same period. Climate is the ultimate environmental control that determines what crops can be cultivated, the long-term water supply, and average heating and cooling requirements for homes in any given area.

Origin and Evolution of the Atmosphere

Today's atmosphere is the product of a lengthy evolutionary process that began perhaps 4.8 billion years

Figure 9.1 *Beachfront homes are swept away or partially destroyed by heavy surf that occurred during winter storms along the California coast in January 1983. (Mark Leet, Sygma.)*

ago. The Earth and the entire solar system, it is believed, developed out of an immense cloud of dust and gases within the Milky Way galaxy. In the beginning, the Earth was an aggregate of dust and meteorites that were surrounded by a gaseous envelope of hydrogen and helium. No gaseous oxygen existed. Then, after millions of years, the Earth's mass grew by accretion as the planet swept up cosmic dust. The evolving planet was bombarded by meteorites that caused the surface to heat up, which drove off most of the atmosphere's original gases.

Then the earth became geologically active, and volcanoes spewed forth huge quantities of lava, ash, and a variety of gases, such as Mount St. Helens did in the spring of 1980. Evidence suggests that at that time (and as it is now), water vapor, carbon dioxide, and nitrogen were important constituents of volcanic emissions. Gaseous, or free, oxygen did not exist, although oxygen was bound to other elements in various chemical compounds, such as in carbon dioxide (CO_2). After millions of years, volcanic activity eventually produced an atmosphere that was rich in nitrogen and carbon dioxide. The breakdown

of water vapor (which is a compound of hydrogen and oxygen, whose chemical formula is identical to that of liquid water) contributed only minor amounts of oxygen to the early atmosphere. And radioactive decay of an isotope of potassium, which is embedded in the earth's crustal rock, added the inert gas argon to the evolving atmosphere. Then, with the formation of seas and the coming of life, important changes took place in the composition of the atmosphere.

Volcanic activity eventually subsided more than 3 billion years ago, and the Earth and its atmosphere gradually cooled. The process of cooling caused some of the water vapor to condense into clouds, which initiated torrential rains, which then gave rise to the first rivers, lakes, and seas. The first primitive forms of life appeared 2 to 3 billion years ago in those seas. Photosynthesis began when the first marine plants appeared, which contributed oxygen to the air. Then, through subsequent millions of years, photosynthesis released huge quantities of oxygen into the atmosphere, until it eventually became the second most abundant atmospheric gas.

While all those changes were taking place, the concentration of atmospheric carbon dioxide fell significantly. Photosynthesis removed some of the carbon dioxide, but the bulk of it was dissolved in ocean waters. Eventually, carbon dioxide made up only a tiny fraction of the atmospheric gases.

Those long evolutionary processes yielded our current atmosphere, which is a mixture of many different gases and suspended particles. Because the lower atmosphere undergoes continual mixing, those gases occur almost everywhere in about the same proportions up to an altitude of approximately 80 kilometers (50 miles). Hence we may travel anywhere over the earth's surface and be confident that we breathe essentially the same type of air. Higher than 80 kilometers (50 miles) those gases become stratified—the heavier gases comprise the lower layers and the lighter gases comprise the upper layers.

If we exclude water vapor, nitrogen usually constitutes 78.08 percent of the lower atmosphere (below 80 kilometers) by volume, and oxygen constitutes 20.95 percent. The next most abundant gases are argon (0.93 percent) and carbon dioxide (which averages 0.03 percent). The atmosphere also contains trace quantities of neon, helium, methane, krypton, hydrogen, xenon, ozone, water vapor, and several other gases. The volume of some of those trace gases (for example, carbon dioxide and water vapor) varies from one place to another and with time. Table 9.1 shows the relative proportions of the principal gases that exist in the lower atmosphere, exclusive of water vapor.

The atmosphere also contains minute liquid or solid particles, which are collectively called *aerosols*, most of which are found in the lower atmosphere near their source—the earth's surface. They originate as a result of forest fires and wind erosion of soil, as sea-salt crystals from ocean spray, and in volcanic eruptions, as well as from industrial and agricultural activities. Some particles enter the atmosphere in the form of meteoric dust.

Although it may be tempting to dismiss as unimportant those substances that make up only a small fraction of the atmosphere, the significance of an atmospheric gas or aerosol is not necessarily related to its relative abundance. For example, water vapor, carbon dioxide, and ozone occur in minute concentrations, but they are essential for life. Also, no more than approximately 4 percent by volume of water vapor occurs in the lowest kilometer of the atmosphere, not even in the warm, humid air over tropical oceans and rain forests. But without water vapor, no rain or snow would fall to replenish soil moisture, rivers, lakes, and seas. Carbon dioxide is an essential raw material for photosynthesis; and ozone (O_3), which is formed in the upper atmosphere by the action of solar energy on oxygen (O_2) molecules, shields all living things from being exposed to potentially lethal levels of ultraviolet radiation (UV) from the sun. The aerosol content of the atmosphere is also relatively small, yet those suspended particles participate in several important processes. For example, some aerosols act as nuclei for the development of clouds and precipitation (Chapter 6), while others influence the air temperature by interacting with sunlight.

These processes are discussed further in this chapter as well as in Chapter 10, because it is important not to underestimate the role of trace gases and aerosols in the atmosphere.

Table 9.1 The relative proportions of gases in the lower atmosphere (below 80 kilometers), excluding water vapor

GAS	PERCENT BY VOLUME	PARTS PER MILLION
Nitrogen	78.08	780,840.0
Oxygen	20.95	209,460.0
Argon	0.93	9,340.0
Carbon Dioxide	0.03	340.0
Neon	0.0018	18.0
Helium	0.00052	5.2
Methane	0.00015	1.5
Krypton	0.00010	1.0
Nitrous Oxide	0.00005	0.5
Hydrogen	0.00005	0.5
Ozone	0.000007	0.07
Xenon	0.000009	0.09

Pressure and Density Variations

Air pressure is defined as the weight of the atmosphere over a given unit area of the earth's surface. The average air pressure at sea level is approximately 1 kilogram per square centimeter (14.7 pounds per square inch). Thus the pressure of the atmosphere on the roof of a typical three-bedroom ranch-style house at sea level is approximately 2.1

million kilograms (4.6 million pounds), which is equivalent to the pressure that is caused by the combined weight of 1500 full-sized automobiles. Thus it is a wonder that the roof does not collapse under this enormous pressure. However, the reason it does not collapse is that the air pressure at any point inside and outside the house is the same in all directions—up, down, and sideways. Hence the air pressure within the house exactly counterbalances the external air pressure in all directions.

Gravity, which is the force that holds everything on the earth's surface, also holds atmospheric gases in a thin envelope around the planet. Gravity compresses the atmosphere toward the earth's surface so that air pressure decreases very rapidly with increasing altitude (Figure 9.2). Air density (which is the mass per unit volume) also diminishes with altitude. That is, the air becomes thinner at increasing altitudes, which makes sense, because air expands as the pressure on it decreases.

Although air pressure and air density decrease at higher altitudes, it is impossible to specify a particular altitude at which the Earth's atmosphere definitely ends. No point can be clearly identified as the beginning of "space." Rather, the vertical extent of the atmosphere is described in terms of the relative distribution of its mass. Thus although 99 percent of the atmosphere's mass lies between the earth's surface and an altitude of approximately 32 kilometers (20 miles), half of its mass is concentrated within 5.5 kilometers (3.4 miles). At altitudes that are higher than 80 kilometers (50 miles), the relative proportion of atmospheric gases changes markedly, and beyond approximately 950 kilometers (590 miles), the atmosphere merges with the highly rarefied interplanetary gases—hydrogen and helium.

From a different perspective, the atmosphere thins so rapidly with altitude that at an altitude of only 5.5 kilometers (3.4 miles), the air pressure is only one-half of the pressure that exists at sea level. That rapid fall in pressure means that we experience appreciable changes in air pressure as we travel up mountains and down into valleys (that is, surface air pressure varies significantly with the land's elevation). Thus the average air pressure at Denver, the mile-high city, is only 83 percent of the average air pressure at Boston, which is located just above sea level.

As we noted in Chapter 3, the expansion and thinning of air, which accompany lower air pres-

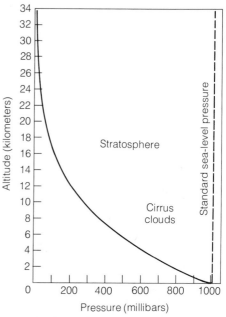

Figure 9.2 *Air pressure falls very rapidly with higher altitude. (From A. N. Strahler*, Physical Geography, *4th ed. Copyright © 1975. Reprinted by permission of John Wiley & Sons, Inc.)*

sures at higher altitudes, trigger physiological changes in humans that allow adjustments. For example, persons who visit mountainous areas often complain of dizziness, headaches, and shortness of breath as a result of the relatively low concentration of oxygen in the thin air. But after a week or two, those distressing symptoms usually disappear. Persons gradually adjust (or *acclimatize*) to low oxygen levels as the number of red blood cells (which are carriers of oxygen) in their bloodstreams rises to increase their oxygen transporting capacity. Limits to acclimatization do exist, however, and most people cannot adjust to pressures at altitudes that are higher than approximately 5500 meters (18,000 feet).

However, elevation is not the only factor that contributes to differences in surface air pressure from one place to another. In fact, atmospheric scientists are more interested in air-pressure variations that arise from factors other than the land's elevation. Hence the influence of elevation is usually eliminated from air-pressure readings at a weather station. This is done by adjusting air-pressure measurements to what the air pressure would be if the station were actually located at sea level. When that reduc-

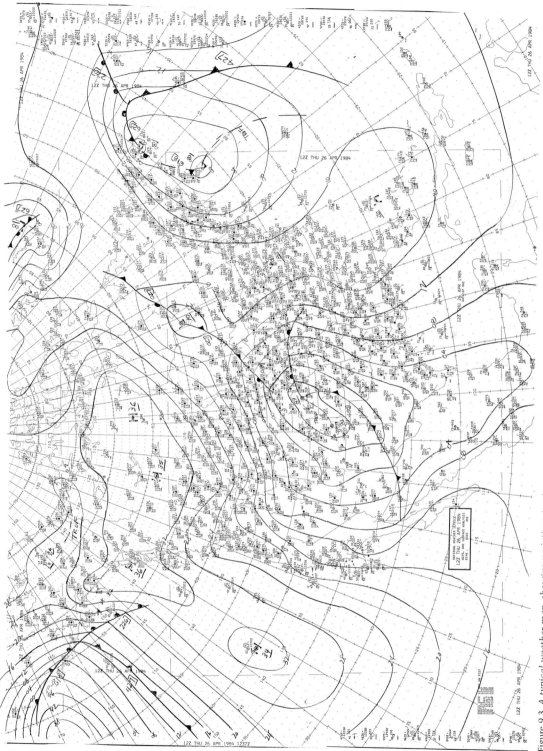

Figure 9.3 A typical weather map, showing variations in air pressure (reduced to sea level) from one place to another. Solid lines are isobars, which connect regions that report the same air pressure in millibars. (National Weather Service.)

tion to sea level is applied everywhere, we still find that air pressure varies from one place to another (Figure 9.3). Furthermore, in any region, the air pressure will fluctuate from day to day and even from hour to hour (Figure 9.4).

Television and radio weather casters usually report air-pressure readings in units of length (millimeters or inches), according to the type of instrument that monitors air pressure, which is called the *barometer*. It is more appropriate, however, to measure air pressure in *millibars* (mb), which is a metric unit of pressure. The average sea-level pressure in millibars is 1013.25 mb, and the usual worldwide range in air pressure that is reduced to sea level is 970 to 1040 mb.

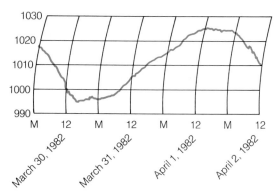

Figure 9.4 *Variation in air pressure in millibars (reduced to sea level) at Green Bay, Wisconsin, from March 30 through April 2, 1982. Note that significant changes in air pressure occur from one day to another and even from one hour to the next.*

BOX 9.1

HEAT AND TEMPERATURE

Temperature *is an approximate measure of the kinetic energy, or energy of motion, of molecules. The faster the molecules move, the higher the temperature. Heat, which is a closely related concept, is the total molecular energy of a given amount of a substance, such as 1 gram.*

For most scientific purposes, temperature is described in terms of the Celsius scale. Established by the Swedish astronomer Anders Celsius in 1742, the Celsius temperature scale has the numerical convenience of a 100-degree interval between the freezing and boiling points of pure water. (This scale is still sometimes called the centigrade *scale, but* Celsius scale *is now the preferred term.) The United States is virtually the only nation where the Fahrenheit temperature scale is still used in everyday affairs, including weather reports. Introduced in 1714 by a German scientist, Gabriel Daniel Fahrenheit, the Fahrenheit scale is numerically more cumbersome than the Celsius scale. If a thermometer that is graduated according to both scales is immersed in a glass that contains a mixture of ice and water, the Fahrenheit scale will read 32°, whereas the Celsius scale will read 0°. If it is immersed in boiling water at sea-level air pressure, the readings will be 212 Fahrenheit (°F) and 100 Celsius (°C).*

*Molecular activity is slower in cold sub-*stances than in hot substances. The temperature at which all molecular activity ceases is called absolute zero, or 0 degrees Kelvin, which corresponds to $-273.15°C$ $(-459.6°F)$. On the Kelvin scale, temperature is the number of degrees above absolute zero. Therefore, the Kelvin scale is a more direct measure of molecular activity, or heat energy, than the Fahrenheit and Celsius scales. (Since nothing can be colder than absolute zero, which corresponds to the complete cessation of molecular motion, no negative temperatures are recorded on the Kelvin scale.) A 1-degree interval on the Kelvin scale corresponds precisely to a 1-degree interval on the Celsius scale. The three scales are compared in Figure 1.

Temperature is a convenient way to describe relative heating, but we may also quantify heat energy directly. Meteorologists traditionally measure heat energy in units that are called calories. *A calorie is defined as the amount of heat that is needed to raise 1 gram of water 1 Celsius degree. Note that this definition specifies a substance (water), a mass of water (1 gram), and a temperature change (1 Celsius degree). (The "calorie" that is used to measure the energy content of food is actually 1000 heat calories, or 1 kilocalorie.) Heat energy is also expressed in another unit, the British thermal unit (Btu). A Btu is defined as the amount of heat that is required to raise the temperature of 1 pound of water*

Although spatial and temporal fluctuations in surface air pressure are relatively slight, they can accompany some important changes in weather. For example, in middle latitudes, the weather is controlled by a continuous procession of different air masses—some exert relatively high pressures and others exert relatively low pressures. An *air mass* is defined as a huge volume of air that covers hundreds of thousands of square kilometers, which is relatively uniform in terms of its temperature and its content of water vapor. As air masses move from one place to another, surface air pressures rise or fall and the weather changes. As a general rule, the weather becomes stormy when the air pressure falls, and it improves when the air pressure rises.

The Thermal Profile

Besides pressure and density, another important feature of the atmosphere is the distribution of heat. (For a review of the distinction between heat and temperature, see Box 9.1.)

On the basis of air temperature, we can subdivide the atmosphere vertically into a series of four concentric layers (Figure 9.5): the troposphere, stratosphere, mesosphere, and thermosphere. Most weather occurs in the lowermost layer, the *troposphere*, which extends from the earth's surface to an altitude that varies between 16 kilometers (10 miles) at the equator and 8 kilometers (5 miles) at the poles. The troposphere is normally (but not always) char-

---

1 Fahrenheit degree. Thus 1 Btu = 252 calories.

Two different objects at the same temperature probably do not contain the same amount of heat energy. For example, a bathtub that is full of hot water contains thousands of times more heat than the red-hot head of a burning match, but the temperature of the match head is many hundreds of degrees higher. Nor is the same quan-

tity of heat required to raise the temperature of equal amounts of two different substances by 1 degree. The amount of heat that is required to raise the temperature of 1 gram of a substance 1 Celsius degree is called its specific heat. *The specific heat of all substances is measured relative to the specific heat of water, which by design is 1.00 calorie per gram per degree Celsius (at 15°C). Water has the highest specific heat of any naturally occurring substance on earth. It takes only one-fifth as much heat to raise the temperature of a gram of sand by a given number of degrees as it does to bring about an equivalent temperature rise in a gram of water. Thus substances with low specific heats, on exposure to the same heat source, warm up more rapidly than those with high specific heats. The specific heats of some familiar substances, expressed in calories per gram per degree Celsius, are given below:*

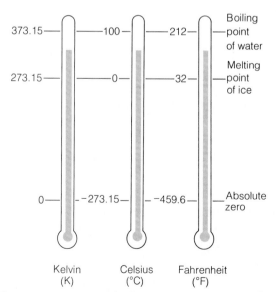

Figure 1 *A comparison of the three common temperature scales. Note that the scale intervals on the Celsius and the Kelvin scales are equal.*

Water	*1.000*
Ice	*0.478*
Wood	*0.420*
Aluminum	*0.214*
Brick	*0.200*
Granite	*0.192*
Sand	*0.188*
Gold	*0.031*
Air (at 0°C)	*0.240*

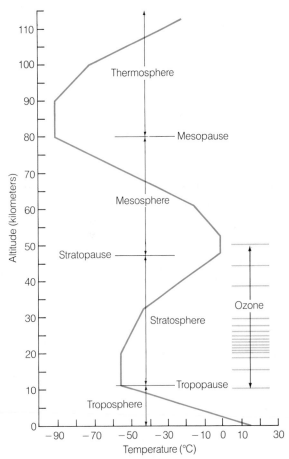

Figure 9.5 *The atmosphere is subdivided vertically into zones that are based on the profile of air temperature. The ozone concentration within the stratosphere is greatest in the areas where the horizontal lines are closest together. (After M. Neiburger et al.,* Understanding Our Atmospheric Environment. *New York: W. H. Freeman, 1973.)*

acterized by temperatures that decrease steadily with altitude. Thus air temperatures at the tops of mountains are usually lower than those in surrounding valleys. At the top of the troposphere is the *tropopause*, which is a transition zone between the troposphere and the next higher layer.

The next layer, which is called the *stratosphere*, extends up to an altitude of approximately 50 kilometers (30 miles). It is *isothermal* (of constant temperature) in its lower portion and is marked by a gradual increase in temperature at higher altitudes, in its middle and upper levels. Pilots of jet aircraft prefer to fly in the stratosphere, because that layer is above the weather and hence offers excellent visibility and generally smooth flying conditions. Since the early 1970s, however, scientists have been wor-

ried about the effects of pollutants that enter the stratosphere. Because little air exchange occurs between the stratosphere and the troposphere, pollutants that enter the stratosphere tend to remain there for long periods of time. Thus some aerosols, which have been thrown into the stratosphere by violent volcanic eruptions, will remain there for many years. Atmospheric scientists are particularly concerned about the possibility that certain pollutants may erode the life-protecting layer of ozone that is located in the stratosphere, and that threat has become a major environmental issue in recent years. We have more to say about this topic in Chapter 10.

The *stratopause* is the transition zone between the stratosphere and the next higher layer, the *mesosphere*, where temperatures are isothermal at first but then rapidly decrease with altitude. The mesosphere extends to the *mesopause*, which is approximately 80 kilometers (50 miles) above the earth's surface and is marked by the lowest temperatures (which average −90°C) in the atmosphere. Above that transition zone is the *thermosphere*, in which temperatures increase again at higher altitudes. Human activities so far appear to have had little direct impact on those outermost atmospheric layers.

The Dynamism of the Atmosphere

Weather is a composite of a wide variety of phenomena—spectacular lightning displays, devastating tornadoes and blizzards, feathery frosts, and gentle June breezes, to mention only a few. Although these phenomena are very different, each one is a manifestation of the dynamic nature of the atmosphere. The ceaseless flow of energy from the sun to the earth, from one place to another on earth, and, finally, from the earth back out into space maintains that dynamism. Let us now examine the characteristics of that energy flow.

Solar Energy Input

More than 99 percent of the energy that is involved in weather phenomena comes from the sun. The sun's energy travels through space at the speed of light (300,000 kilometers per second) in the form of *electromagnetic radiation*, about half of which is visible sunlight. Electromagnetic radiation, which is also called *radiant energy*, moves in the form of os-

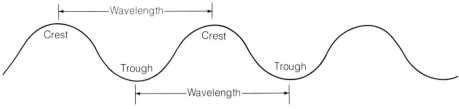

Figure 9.6 *The wavelength of an electromagnetic wave is the distance between successive crests or, equivalently, between successive troughs.*

cillating waves (Figure 9.6), which do not require an intervening physical medium. Thus radiant energy can travel through a vacuum. Many forms of electromagnetic radiation exist, which are all variations of the same basic phenomenon. They can be distinguished from one another by energy level, frequency, and wavelength. Familiar types of electromagnetic radiation include x-rays, visible light, microwaves, and radio waves. The *electromagnetic spectrum* is shown in Figure 9.7, and key points about radiation are summarized in Box 9.2.

Of the enormous amount of energy that is radiated by the sun (approximately 5.6×10^{27}[*] calories each minute), only about one-half of 1 billionth of that amount is intercepted by the earth. As the earth moves in space, that radiant energy is distributed over its surface. The rotation of the earth on its axis accounts for day-to-night variations in the

Figure 9.7 *The electromagnetic spectrum. The various types of electromagnetic radiation are arranged according to wavelength. (From M. Neiburger et al.,* Understanding Our Atmospheric Environment. *New York: W. H. Freeman, 1982, p. 58.)*

[*] Numbers expressed as powers of 10 are explained in Appendix III.

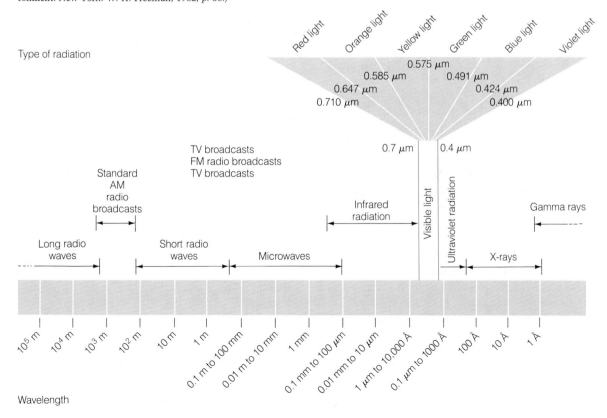

amount of energy that is received at a given place, while the orbiting of the earth around the sun is responsible for seasonal variations.

Throughout the course of a year (actually 364¼ days), the earth describes a slightly elliptical orbit about the sun, with the sun situated at one focus of the ellipse (Figure 9.8). Its orbital eccentricity (departure from a circle) is slight; at *perihelion* (when it is nearest to the sun), it is only approximately 3.4

percent closer to the sun than it is at *aphelion* (when it is farthest from the sun). Because the intensity of solar radiation diminishes rapidly with distance from the sun (as the inverse square of the distance traveled), the intensity of solar radiation that reaches the earth's surface varies approximately 7 percent about its mean value between perihelion (maximum) and aphelion (minimum).

Perihelion occurs on January 3 or 4; aphelion,

BOX 9.2

ELECTROMAGNETIC RADIATION

The world is continually bathed in electromagnetic radiation, which is so named because it exhibits both electrical and magnetic properties. Most of the electromagnetic energy that reaches the earth originates in the sun; only a vanishingly small amount comes from more distant stars and other celestial objects. The many types of electromagnetic energy together make up the **electromagnetic spectrum**, *which is illustrated in Figure 9.7. Light, which is visible radiation, represents only a very small portion of that spectrum. Other types of electromagnetic energy include radio waves, infrared radiation, ultraviolet radiation, and gamma radiation.*

Electromagnetic radiation travels in the form of waves that are usually described in terms of wavelength or frequency. The length of a wave, or **wavelength**, *is the distance from one crest to another or one trough to another, as shown in Figure 9.6. A wave's* **frequency** *is defined as the number of crests (or troughs) that pass a given point in a given period of time, usually 1 second. The passage of one complete wave is called a* **cycle**, *and a frequency of one cycle per second is termed 1* **hertz** *(Hz). The frequency of a wave is inversely proportional to its wavelength; that is, the higher the frequency, the shorter the wavelength. Some radio waves have frequencies of just a few hundred Hertz and wavelengths that are hundreds of kilometers long. Gamma rays, in contrast, have frequencies that are as high as 10^{24} (a trillion trillion) Hz and wavelengths that are*

as short as 10^{-14} meters (a hundred trillionth) of a meter.

Electromagnetic waves can travel through space as well as through gases, liquids, and solids. In a vacuum, all electromagnetic waves travel at their maximum speed—300,000 kilometers (186,000 miles) per second. All forms of electromagnetic radiation slow down when they pass through materials. Their speed depends on their wavelengths and the types of materials they are passing through. As electromagnetic radiation passes from one medium to another, it may be reflected or refracted (that is, bent) at the interface. That happens, for example, when solar radiation strikes the ocean surface: some is reflected and some becomes bent when it penetrates the water. Electromagnetic radiation may also be absorbed, such as when solar radiation is absorbed at the earth's surface and converted to heat.

Although the electromagnetic spectrum is continuous, it is convenient to assign different names to different segments of this energy spectrum, because we detect, measure, generate, and use different segments in different ways. But all types of electromagnetic radiation are similar in all respects except in terms of their wavelengths, frequencies, and energies. Wavelength and energy are functions of frequency; the higher the frequency, the higher the energy level. Also, the different types of electromagnetic radiation do not begin or end at precisely marked points along the spectrum. For example, red light shades into invisible infrared radiation (infrared = below red) beneath it on the frequency scale. At the other end of the visible portion of the electromagnetic spectrum, violet light shades into invisible ultra-

on July 3 or 4. Hence it is winter in North America when the earth is closest to the sun and it is summer when it is farthest from the sun. Hence it is not the eccentricity of the earth's orbit that accounts for the seasons. Seasons arise from the fact that the earth's equatorial plane is inclined at 23°27' to its orbital plane. That angle is called the *obliquity of the ecliptic.* As shown in Figure 9.9, at that angle, the most intense solar beam is focused on the southern hemi-

sphere for one-half of the year and on the northern hemisphere for the other half.

The intensity of the solar beam that is received at the top of the atmosphere is called the *solar constant.* The solar constant is defined as the rate at which solar radiation falls on a unit area of a plane surface, which is oriented perpendicular to the solar beam, when the earth is at its mean distance from the sun. Actually, the "constant" designation is

violet radiation (ultraviolet = beyond violet).

At the low-energy (low-frequency, long-wavelength) end of the electromagnetic spectrum are radio waves. *Their wavelengths range from hundreds of kilometers to a small fraction of a centimeter, and their frequencies can extend to a billion Hertz. Frequency modulation (FM) radio waves, for example, range from 88 million to 108 million Hz (Note, for example, the familiar 88 and 108 at opposite ends of the FM radio dial).*

Next comes the microwave *portion of the electromagnetic spectrum, which has wavelengths that range from 300 millimeters to around 0.1 millimeter. Some microwave frequencies are used for radio communication, for microwave cooking, and in weather-tracking systems (radar).*

Between microwaves and visible light is infrared radiation, *which is often called* infrared heat *or* radiant heat. *We cannot see it, but we can feel the heat it generates if it is sufficiently intense—as it is, for example, when it is emitted from a hot stove. Actually, small amounts of infrared radiation are emitted by every known object or material, no matter how cold it is. Naturally occurring infrared radiation (some of which comes directly from the sun but most of which is converted by the earth and atmosphere from solar radiation) warms the atmosphere and powers weather phenomena, as we see in this chapter.*

At its uppermost frequencies, infrared radiation shades off into the lowest-frequency visible radiation, which is red light. Wavelengths of visible light *are in the range of approximately 0.40 micrometers at the violet end to about 0.70 micrometers at the red end. (A* micrometer *is a*

millionth of a meter.) Light is essential to many activities of plants and animals. In plants, light provides the energy that is needed for photosynthesis; it also coordinates the opening of buds and flowers and the dropping of leaves. In animals, light regulates reproduction, hybernation, and migration, and, of course, makes vision possible.*

Beyond visible light on the electromagnetic spectrum, in the order of increasing frequencies, increasing energy levels, and decreasing wavelengths, are ultraviolet radiation, x-rays, *and* gamma radiation. *All three of those types of radiant energy occur naturally, and all can be produced artificially. All have medical uses: ultraviolet radiation is a potent germicide; x-rays are a powerful diagnostic tool; and both x-rays and gamma radiation are used in treating cancer patients.*

But those three highly energetic types of radiation may be dangerous as well as useful. For example, ultraviolet rays can cause irreparable damage to the retinal cells of the eye. One can be permanently blinded by looking at the sun (say, during a partial solar eclipse) unless a filter is used to block out ultraviolet rays. Also, overexposure to ultraviolet rays, x-rays, or gamma rays can cause sterilization, cancer, mutations, or damage to fetal tissue.

Fortunately for us, the atmosphere blocks out most incoming ultraviolet radiation and virtually all x-radiation and gamma radiation. Without that protective shield, all life on earth would be destroyed quickly. Some radioactive materials emit dangerous levels of x-rays and gamma rays, as do such devices as x-ray machines. Thus overexposure to high-energy electromagnetic radiation, whatever its origin, must be avoided.

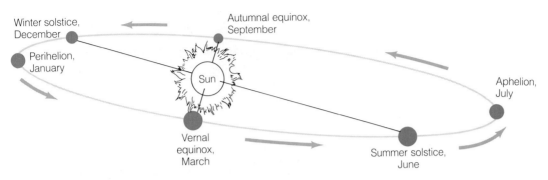

Figure 9.8 *The earth's orbit is an ellipse, with the sun located at a focus. The earth is closest to the sun at perihelion (January 3 or 4) and farthest from the sun at aphelion (July 3 or 4).*

misleading, since solar energy output fluctuates by a few tenths of a percent during the course of a year and exhibits some longer-term variations. Nonetheless, we can approximate the solar constant as 2.0 calories per square centimeter per minute, or 1367 watts per square meter.

Atmosphere–Radiation Interactions

As solar radiation travels through the atmosphere, it interacts with the atmosphere's constituent gases and aerosols. Some of it (about 19 percent) is absorbed by oxygen, ozone, water vapor, ice crystals, water droplets, and dust particles. As a result of absorption, that radiant energy is converted to heat energy, and the air is warmed to some extent. Some solar energy is reflected—primarily by clouds—back to space. Another portion is *scattered*, that is, dispersed in all directions. In fact, it is the scattering of the blue portion of visible sunlight by nitrogen and oxygen molecules that gives the sky its blue color. The portion of solar radiation that is not reflected, scattered back to space, or absorbed by gases and aerosols reaches the earth's surface. Solar radiation that is transmitted directly through the atmosphere to the earth's surface constitutes *direct insolation*, which is augmented by *diffuse insolation*—solar radiation that is scattered, reflected, or both, to the earth's surface.

The radiation (both direct and diffuse) that reaches the earth's surface is either reflected or absorbed by it. The amount that is reflected varies from one place to another, depending on the nature of the surface. More solar radiation is reflected by light surfaces than by dark surfaces. For example, skiers who have been sunburned on the slopes know that snow is very reflective. Also, smooth surfaces are more reflective than rough ones, and surfaces that are less perpendicular to the sun's rays are more reflective than surfaces that make an angle that is close to 90° with the incoming solar radiation. The measure of a surface's reflectivity is known as its *albedo*, which is the ratio of reflected radiation to incident radiation. For example, freshly fallen snow typically has an albedo that is between 75 and 95 percent; that is, 75 to 95 percent of the solar radiation that is incident on snow is reflected. At the other extreme, the albedo of a rough, dark surface, such as a blacktopped road or a green forest, may be as low as 5 percent. The albedos of some common surfaces are listed in Table 9.2. The portion of radiation (insolation) that is not reflected is absorbed by the earth's surface, thereby warming it.

The Solar Radiation Budget

Recent satellite measurements indicate that approximately 30 percent of the total incoming solar radiation is reflected or scattered by the *earth–atmosphere system*, which is the earth's surface and the atmosphere considered together, and is lost to space. That reflected light (which has been seen as "earthshine" by American astronauts on the moon) is known as the earth's *planetary albedo*. The remaining 70 percent of the solar radiation that is intercepted by the earth–atmosphere system is absorbed (that is, converted to heat) and, ultimately, is involved in the functioning of the environment.

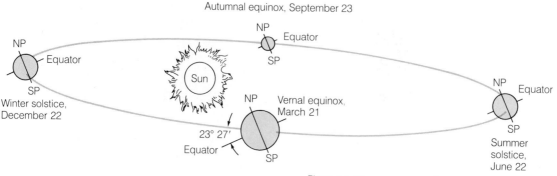

Figure 9.9 *The seasons occur because the earth's equatorial plane is inclined (at 23°27') to its orbital plane.*

Table 9.2 The reflectivity, albedo, of some common surface types for visible solar radiation in percent reflected

SURFACE	ALBEDO
Grass	16–26
Deciduous Forest	15–20
Coniferous Forest	5–15
Crops	15–25
Tundra	15–20
Desert	25–30
Blacktopped Road	5–10
Sea Ice	30–40
Fresh Snow	75–95
Old Snow	40–70
Glacier Ice	20–40
Water (High Sun)	3–10
Water (Low Sun)	10–100

Of the total amount of solar radiation that is intercepted by the earth–atmosphere system, only 19 percent is absorbed directly by the atmosphere; in other words, the atmosphere is relatively transparent to solar radiation. The remaining 51 percent of the solar radiation (remember, 30 percent was reflected) is absorbed by the earth's surface—most of it by ocean waters, which have a low albedo and cover nearly three-quarters of the globe. The oceans' relatively low reflectivity (and hence high absorption of solar radiation) is evident in the satellite image of the earth in Figure 9.10. Note how the oceans appear darker than the adjacent continental land masses. Figure 9.11 summarizes the distribution of solar radiation in the earth–atmosphere system.

From the global solar radiation budget, it follows that the earth's surface is the principal locus of solar heating. The earth's surface, in turn, continuously radiates heat back to the atmosphere, which eventually radiates it off to space. Hence the earth's surface is also the main source of heat for the lower atmosphere, which is evident in the usual vertical temperature profile of the troposphere (Figure 9.5). Normally, in the troposphere, air is warmest close to the earth's surface and its temperature falls at higher altitudes, that is, at a greater distance from the main source of heat.

The Infrared Response

If solar radiation were continually absorbed by the earth–atmosphere system without any compensating flow of heat out of the system, then air temperatures would rise steadily. In reality, however, little change occurs in global air temperature from one year to another because heat leaves the earth–atmosphere system in the form of infrared radiation. But why does the earth–atmosphere system radiate in the infrared portion of the electromagnetic spectrum? The explanation is found in one of the physical laws that govern electromagnetic radiation.

Wien's displacement law states that the higher the surface temperature of a radiating object (such as a hot stove, the sun, or the earth–atmosphere system), the shorter are the wavelengths of its radiated energy. Therefore, since the effective radiating temperature of the sun is about 6000°C (11,000°F), electromagnetic waves that are emitted by the sun are

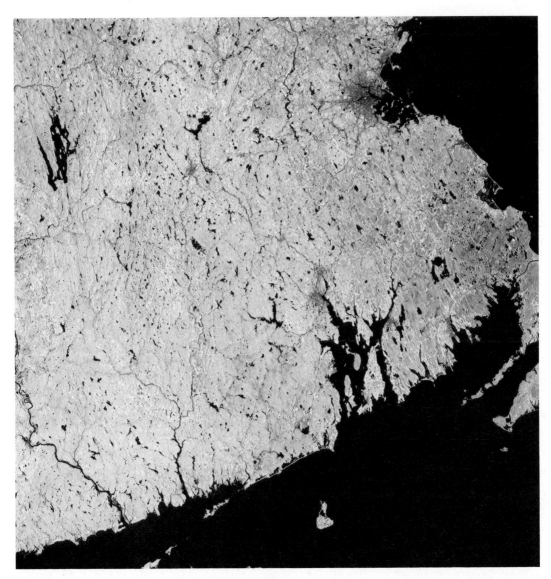

Figure 9.10 *A satellite view of the earth. Note how the highly absorptive oceans appear dark compared to the more reflective continents. (NASA.)*

relatively short. Figure 9.12 shows how the intensity of solar radiation varies according to wavelength. The greatest amount of energy is emitted by waves that are 0.48 micrometers long (in the green segment of the visible spectrum). The much cooler earth's surface, radiating at an average temperature of approximately 15°C (59°F), emits longer waves, as shown in Figure 9.13. In fact, the earth–atmosphere system emits radiation of maximum intensity at wave-

lengths that are approximately 10 micrometers long (in the infrared). Hence the earth–atmosphere system responds to solar heating by emitting comparatively long-wave infrared radiation.

Because solar radiation and terrestrial radiation peak in different portions of the electromagnetic spectrum, they have different properties and their interactions with the atmosphere also differ. We noted earlier that atmospheric constituents absorb only approximately 19 percent of the incoming solar radiation. But then the atmosphere absorbs a larger percentage of the infrared radiation that is emitted

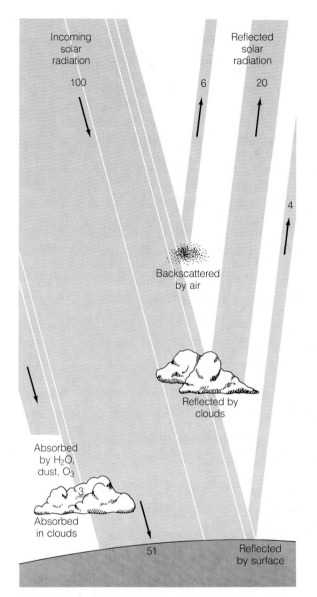

Figure 9.11 *The disposition of 100 units of solar radiation that interact with the atmosphere and interact at the earth's surface. (From A. P. Ingersoll, "The Atmosphere." © Scientific American 249(3):164, 1983. All rights reserved.)*

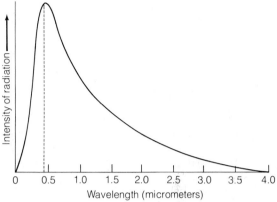

Figure 9.12 *The energy intensity of solar radiation as a function of wavelength. Peak intensity is emitted by waves that are 0.48 micrometers long, in the green region of the visible spectrum. (From H. R. Byers,* General Meteorology. *New York: McGraw-Hill, 1959, p. 19.)*

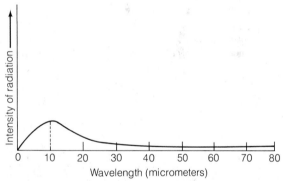

Figure 9.13 *The intensity of terrestrial radiation as a function of wavelength. Peak energy intensity occurs in the form of waves that are about 10 micrometers long, in the infrared region of the spectrum. (From H. R. Byers,* General Meteorology. *New York: McGraw-Hill, 1959, p. 19.)*

by the earth's surface. As shown in Figure 9.14, the atmosphere, in turn, reradiates some of that absorbed infrared back toward the earth's surface. That reradiation slows the escape of heat into space, and so the lower atmosphere has a higher average temperature than the upper atmosphere, which allows it to be more hospitable to life. In fact, thanks to this reradiation, the earth's average surface temper-

ature is more than 30°C (54°F) higher than it would be otherwise.

The atmospheric gases that absorb and reradiate infrared radiation, and thereby impede its loss to space, are primarily water vapor and carbon dioxide and, to a lesser extent, ozone, methane, and nitrous oxide. The percentage of infrared radiation that is absorbed by those gases varies with wavelength; the percent of absorption is very low in the wavelength bands around 8 and 11 micrometers (which include

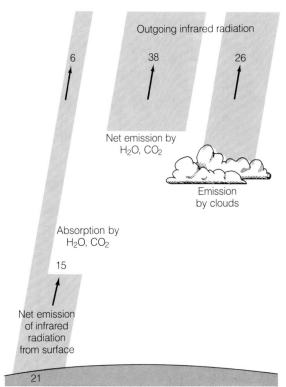

Figure 9.14 *Emission of infrared radiation by the earth's surface and atmosphere. Numbers are based on 100 units of incident solar radiation. (From A. P. Ingersoll, "The Atmosphere." © Scientific American 249(3):164, 1983. All rights reserved.)*

the wavelength of peak infrared intensity). Hence it is through those *atmospheric windows* that heat eventually escapes to space.

Because the glass that is installed in greenhouses slows the transmission of infrared radiation in much the same way as the atmosphere, the absorption and reradiation of infrared radiation by atmospheric gases has been called the *"greenhouse effect."* Actually, greenhouses keep plants warm primarily by preventing air movement and only to a lesser extent by trapping infrared radiation inside. Hence, strictly speaking, it is incorrect to apply the greenhouse analogy to the earth–atmosphere system. Still, reference to the "greenhouse effect" is so common in discussions of the atmosphere's heat balance that we use the term here, but always in quotations. A more appropriate name is the *atmospheric reabsorption effect.*

To illustrate the "greenhouse effect" in the atmosphere, we can compare the typical summer

weather of the American Southwest with that of the coast along the Gulf of Mexico. Both areas are at the same latitude and, therefore, receive approximately the same intensity of solar radiation. As a result, both areas commonly experience afternoon temperatures that are near 30°C (86°F). At night, however, the air temperatures in those areas may differ markedly. In the Southwest, relatively little water vapor exists in the air to impede the escape of infrared radiation into space. Thus considerable heat is lost, the earth–atmosphere system cools rapidly, and surface air temperatures may fall below 15°C (59°F) by dawn. Along the Gulf of Mexico, however, the air is more humid so it absorbs more infrared radiation. Because a portion of that heat is reradiated toward the earth's surface, the surface air temperature may fall only into the twenties in Celsius degrees (or the seventies in Fahrenheit degrees). Also, since clouds are largely composed of water droplets, which also produce a "greenhouse effect," the nights at any location are usually colder when the sky is clear than when the sky is cloud-covered.

So far, our discussion has focused on the flow of solar and terrestrial radiation within the earth–atmosphere system. But we have also seen that when radiation is absorbed, heating occurs. Radiant energy is never destroyed. It changes into heat according to the *law of energy conservation,* which states that energy is neither created nor destroyed but can change from one form to another (see Box 2.1). However, heating that is caused by the absorption of radiation is not uniform everywhere within the earth–atmosphere system. Indeed, imbalances of heating are the primary causes of the atmospheric processes that we call weather.

Heat Imbalances between the Atmosphere and the Earth's Surface

If we make separate measurements of the radiant (solar short-wave and terrestrial long-wave) energy distribution for the entire earth's surface and for the entire atmosphere over the course of a year, we will find an imbalance in the distribution of heat. In the atmosphere, the rate of cooling (due to infrared emission) is greater than the rate of warming (due to absorption of solar radiation). On the other hand, at the earth's surface, the rate of warming (due to absorption of solar radiation) exceeds the rate of cooling (due to infrared emission). That heating imbalance is evident in Figure 9.15, where we combine the disposition of incoming solar radiation with that of outgoing terrestrial infrared radiation.

The distribution of radiant energy appears to imply a net cooling of the atmosphere and a net warming of the earth's surface. But the atmosphere is not cooling relative to the earth's surface, so a transfer of heat must be taking place from the surface to the atmosphere on a global scale. That heat is transferred by two mechanisms, which are called sensible heating and latent heating. The flow of heat from the earth's surface to the atmosphere is brought about to a lesser extent by sensible heat transfer (approximately 23 percent) and to a greater extent by latent heat transfer (approximately 77 percent).

Sensible Heating

Sensible heat transfer is the more readily understood of the two heat-transfer mechanisms, and it actually involves two processes: conduction and convection. The word "sensible" is used because the

Figure 9.15 *The global disposition of solar radiation and terrestrial infrared radiation, based on 100 units of incident solar radiation. Note that the solar radiation that is absorbed at the surface (51 units) exceeds the infrared radiation that is emitted by the surface (21 units). The excess heat at the surface (30 units) is transported to the atmosphere by sensible heating (7 units) and latent heating (23 units). (From A. P. Ingersoll, "The Atmosphere."* © Scientific American *249(3):164, 1983. All rights reserved.)*

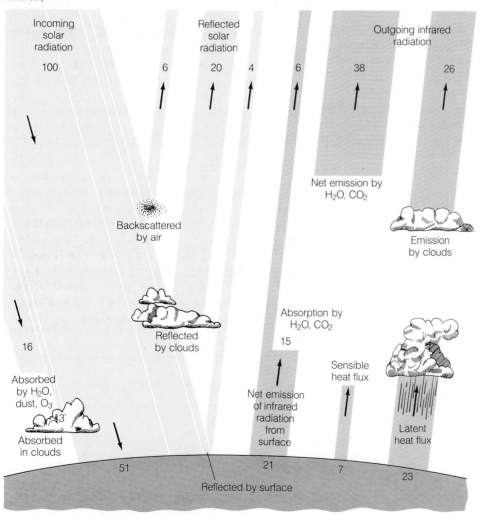

Incoming solar radiation

100

Reflected solar radiation

6   20   4   6   38   26

Outgoing infrared radiation

Backscattered by air

Net emission by H$_2$O, CO$_2$

Emission by clouds

Reflected by clouds

Absorption by H$_2$O, CO$_2$

15

Sensible heat flux

16

Absorbed by H$_2$O, dust, O$_3$

3

Absorbed in clouds

Net emission of infrared radiation from surface

Latent heat flux

51   21   7   23

Reflected by surface

redistribution of heat that occurs as a result of sensible heating can be monitored, or sensed, as the temperature changes.

With *conduction*, the kinetic energy of atoms or molecules (that is, heat) is transferred as a result of collisions between neighboring atoms or molecules. For example, a spoon heats up when it is placed into a steaming cup of coffee. In the same way, heat is conducted from the relatively warm earth's surface to the cooler overlying air. However, air cannot transfer much heat through conduction (air is a poor conductor of heat) and that process is only significant in a very thin layer of air (a few centimeters thick), which is in immediate contact with the earth's surface. The process of convection then transports that heat higher into the atmosphere.

Convection is the transport of heat within a fluid that occurs as a result of the fluid's movement. Convection often develops within the atmosphere as a result of differences in density that are triggered by the conduction of heat from the earth's surface. For example, as the lower portion of the atmosphere is heated by the ground, the heated air becomes lighter than the overlying cooler air and begins to rise. Then the surrounding air sweeps in, forcing the warmer air to rise. The result is a transfer of heat upward in the atmosphere by *convective currents* (Figure 9.16).

Latent Heating

Latent heat transfer is more subtle than sensible heat transfer. It arises from differences in molecular activity (energy), which is represented by the three physical phases of water: solid, liquid, and vapor. In the solid phase (ice), water molecules are relatively inactive and vibrate around fixed locations; hence an ice cube retains its shape. In the liquid phase, water molecules move around with greater freedom, so that liquid water takes the shape of its container. In the vapor phase, water molecules exhibit their maximum activity, which enables them to diffuse readily throughout the entire volume of a container. Hence a change in phase involves a change in the level of molecular activity, so during a change in phase, heat must either be added to, or released from, the substance that is undergoing the change.

As an illustration, let us follow the fate of an ice cube as it is heated. The specific heat (Box 9.1) of ice is approximately 0.5 calorie per gram per degree Celsius, which means that 0.5 calorie of heat must be supplied per gram for every degree of temperature rise. Once 0°C (32°F) is reached, an additional 80 calories of heat per gram (which is called the *heat of fusion*) must be supplied to break the forces that bind the molecules together in the ice phase. Thus the temperature of the water and ice remains at 0°C (32°F) until all the ice is melted. The specific heat of liquid water is 1.0 calorie per gram per Celsius degree, so from that point on, only 1.0 calorie is needed to raise the temperature of 1 gram

Figure 9.16 *Excess heat at the earth's surface is transported into the atmosphere by conduction, convective currents, and latent heat transfer. (After M. Neiburger et al.,* Understanding Our Atmospheric Environment. *New York: W. H. Freeman, 1973.)*

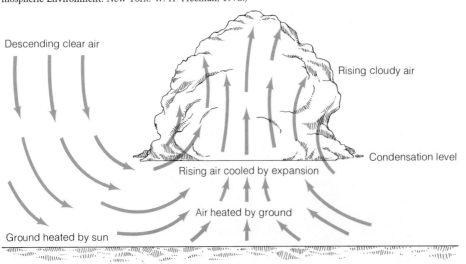

Descending clear air

Rising cloudy air

Condensation level

Rising air cooled by expansion

Air heated by ground

Ground heated by sun

Figure 9.17 *Rising convective currents can trigger the formation of clouds like these. (From NOAA.)*

of water 1 Celsius degree. However, a phase change from liquid water to water vapor requires the addition of much more heat. The *heat of vaporization* of water is temperature-dependent. It varies from approximately 600 calories to vaporize 1 gram at 0°C (32°F) to 540 calories to vaporize 1 gram at 100°C (212°F). If the sequence is reversed, that is, if the water vapor is cooled until it becomes liquid and then ice, the temperature falls and phase changes take place as equivalent amounts of heat are released.

If we apply this energy concept of phase changes to the earth–atmosphere system, we can illustrate the latent heat-transfer mechanism clearly. Thus as the earth's surface absorbs solar radiation, some of the heat is used to vaporize water from oceans, lakes,

rivers, soil, and vegetation. Conversely, when air rises and cools, some of its water vapor changes back to the liquid phase (forming water droplets) or the solid phase (forming ice crystals); that is, clouds develop within the atmosphere, heat is released, and the atmosphere is warmed. The heat that is required to change liquid water to vapor is supplied at the earth's surface, and that same heat is subsequently released by cloud formation within the atmosphere.

If convection extends deeply into the troposphere, convective clouds (Figure 9.17) billow upward to eventually form thunderstorm clouds. When that happens, sensible heat transfer combines with latent heat transfer to channel the heat from the earth's surface into the atmosphere. Hence an important heat-transfer function is served by thunderstorms, even though they may wash out your ball game or send you scurrying home from the beach.

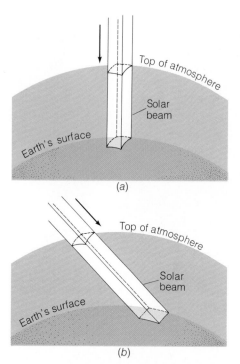

Figure 9.18 *Variations in the angle of incidence of solar radiation translate into more intense insolation per unit area of the earth's surface at (a) tropical latitudes than at (b) middle and higher latitudes.*

Because ocean waters cover so much of the earth's surface (70.8 percent), it is not surprising that latent heat transfer is more significant than sensible heat transfer on the global scale. But the ratio of sensible to latent heating varies from one region to another, depending on the amount of surface moisture that is available. Thus the ratio of latent heating to sensible heating is about 10:1 for oceans but only 1:2 for deserts. The drier the surface, the more important is sensible heating.

Also, frequently heat is transported from the atmosphere to the earth's surface, which is the reverse of the average global situation. That situation occurs, for example, when warm air masses travel over cold, snow-covered ground or relatively cool ocean water.

Earlier, we noted that nearly all weather is confined to the troposphere, which implies that sensible and latent heat transport operate primarily within the troposphere. The heat and temperature distribution above the troposphere is determined by radiative factors alone.

Heat Imbalances: Latitudinal Variation

Imbalances in the distribution of radiant energy occur horizontally, with latitude, as well as vertically (between the earth's surface and the atmosphere). Because the earth is nearly spherical in shape, parallel beams of incoming sunlight strike lower latitudes more directly than higher latitudes (Figure 9.18). Hence at higher latitudes, solar radiation spreads over a greater area and is less intense. The output of infrared radiation, on the other hand, varies less with latitude. As a result, over the course of a year at higher latitudes, the rate of infrared cooling exceeds the rate of warming by absorption of solar radiation. But in lower latitudes, the reverse is true: the rate of warming by the absorption of solar radiation is greater than the rate of infrared cooling. For many years, it was assumed that the circles marked at 38° latitude divided the areas in which net radiative cooling and net radiative warming occurred. But recent satellite measurements, shown in Figure 9.19, suggest that the division lies closer to 30° latitude.

Figure 9.19 *The variation by latitude of absorbed solar radiation (solid line) and outgoing long-wave radiation (dashed line), which was obtained by satellite measurement from June, 1976, to May, 1977. Cooling and warming rates are equal at about 28° N and 33° S. (From J. S. Winston et al., Earth-atmosphere radiation budget analysis derived from NOAA satellite data June 1974–February 1978. Washington, D.C.: NOAA Meteorological Satellite Laboratory, 1979.)*

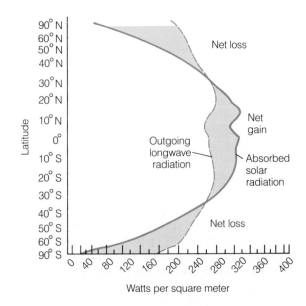

In any event, although the implication is that higher latitudes experience net cooling while tropical latitudes experience net warming, in fact, lower latitudes do not become increasingly warmer relative to higher latitudes. Hence the additional heat that is experienced in the tropics must be transported to middle and higher latitudes. In the northern hemisphere, that *poleward heat transport* is brought about primarily by the northward movement of warm air masses that form in lower latitudes and replace cold air masses that flow southward from higher latitudes. Approximately 50 percent of the heat that is transported poleward is sensible heat, whereas latent heat transfer—vaporization of water in lower latitudes and condensation in storms at higher latitudes—accounts for approximately 30 percent of the transported heat. The north–south exchange of cold and warm ocean waters is responsible for transporting the remaining 20 percent of heat toward the poles. Thus warm ocean currents flow poleward and cold ocean currents flow toward the tropics.

Weather: Response to Heat Imbalances

Weather redistributes heat within the earth–atmosphere system. The flow of cold and warm air masses from one place to another, vaporization and cloud formation, storms, and convective currents all serve to counter heat imbalances. Thus a cause-and-effect (or forcing-response) chain operates in the earth–atmosphere system, starting with the sun as the prime source of energy and resulting in weather.

We have seen that within the earth–atmosphere system, some solar radiant energy is converted to heat as a result of absorption and some of that heat, in turn, is radiated in the form of infrared radiation to the atmosphere and off to space. A portion of solar energy is also converted to kinetic energy, which is the energy of motion in weather systems. Kinetic energy is manifested in winds, convective currents, and the exchange of air masses. Weather systems do not last indefinitely, however, and, ultimately, the kinetic energy is dissipated as frictional heat. Figure

9.20 is a schematic diagram of the major energy transformations that operate within the earth–atmosphere system.

In summary, the sun drives the atmosphere in two ways: (1) imbalances in solar heating spur atmospheric circulation that redistributes heat, and (2) solar energy is the source of kinetic energy that is manifested in the circulation of the atmosphere.

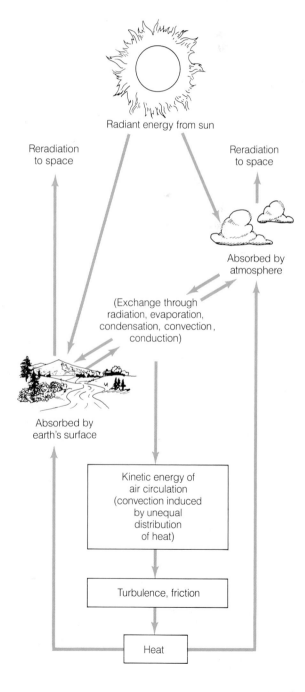

Figure 9.20 *The flow of energy within the earth–atmosphere system. (From A. Miller, et al.,* Elements of Meteorology. *Columbus, Ohio: Charles E. Merrill, 1983, p. 48.)*

Scales of Weather Systems

Atmospheric scientists view atmospheric circulation in terms of the weather systems, which are classified according to the spatial dimensions, or scale, of the phenomena. From the largest systems to the smallest, scales of weather are designated as being global, synoptic, meso, and micro. For example, the large-scale wind systems of the globe (polar easterlies, westerlies, and trade winds) and semipermanent centers of low and high pressure are features of *global-scale circulation* (Figure 9.21). *Synoptic-scale weather* is continental or oceanic in scale. For example, migrating high-pressure systems, storms, air masses, and fronts, which are usually highlighted on television and newspaper weather maps (Figure 9.3), are important components of that scale. *Mesoscale systems* include thunderstorms, hailstorms, and sea and lake breezes—phenomena that may influence the weather in one section of a city, while leaving other sections unaffected. The flow of air within a small environment, for example, the emission of smoke from a chimney, represents the small-est spatial subdivision of atmospheric motion, or *microscale weather.*

Climate and Climate Variability

As we saw in Chapter 3, long-term average weather conditions constitute a primary regulator (limiting factor) for organisms. Those weather conditions, which are averaged over an extended period of time, plus extremes in weather, constitute the *climate.* Climate is the principal factor that controls the geographical distribution of the various worldwide biomes and ecotones that we examined in Chapter 5. And climate is one of the critical factors in soil formation, food production, and energy demand for space heating and cooling. Clearly, then, climatic variations can have profound effects on the environment—and thus on humankind. However, before we can assess the possible consequences of such variations, we must have some understanding of climatic processes.

Some Principles of Climate

The globe is a mosaic of many types of climate, just as it is a mosaic of numerous types of ecosystems. A variety of controls interact to shape the climate of any particular region. The controls that exert a regular and predictable influence on the climate are latitude, proximity to large bodies of water, and elevation of the land. (1) Latitude is the principal climatic control because it determines the intensity of solar radiation and the duration of daylight. (2) Large bodies of water, such as seas and lakes, modify the air temperature over adjacent land areas because water has a relatively large specific heat. (3) Land elevation is important because it affects air temperature and the type and amount of precipitation that occurs.

In addition to those controls, the climate of any particular region is strongly influenced by the prevailing circulation of the atmosphere. That circulation determines the types of air masses that regularly develop over a region or move across it from other regions. Thus it is useful to elaborate on this climatic control, particularly as it relates to middle and higher latitudes.

The moisture and temperature characteristics of an air mass are determined by the type of surface over which the air mass develops and travels. For

Figure 9.21 *Global-scale wind belts and pressure systems.*

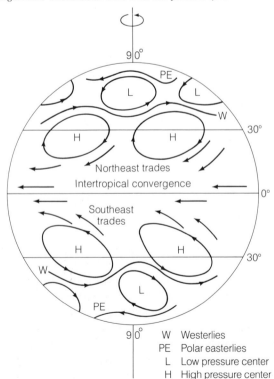

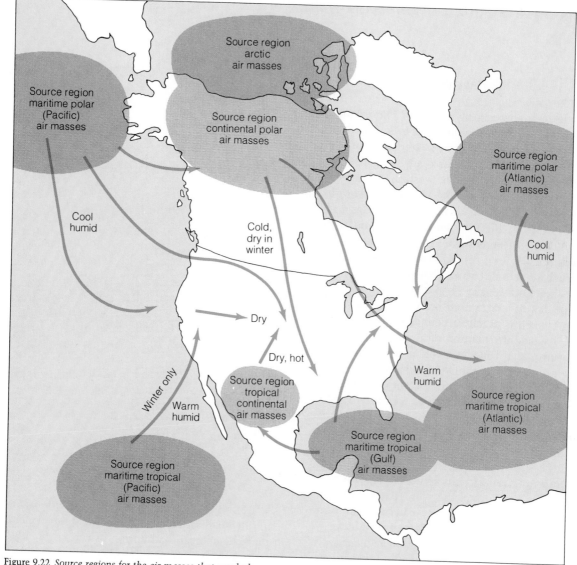

Figure 9.22 *Source regions for the air masses that regularly stream across North America. (From A. N. Strahler,* Introduction to Physical Geography. *New York: Wiley, 1970.)*

example, air that is situated for an extended period of time over cold, snow-covered ground (such as northern Canada in winter) becomes cold and dry. On the other hand, an air mass that forms over a wet, warm surface (such as the Gulf of Mexico) becomes warm and humid. Four basic types of air masses exist: warm and dry, warm and humid, cold and dry, and cold and humid. The source regions that initiate the various types of air masses that invade the United States regularly are identified in Figure 9.22. As a general rule, cold air masses that develop in the north push southward, while the warm air masses of the south stream northward. Recall that this air-mass exchange is the principal mechanism of poleward heat transport.

As air masses move from one place to another, their temperature and moisture content become modified to some extent. Thus in the winter, a cold air mass loses some of its punch as it plunges southeastward from Canada into the Central United States. Cold air masses tend to become modified much more rapidly than warm air masses. Hence a heat wave may spread from the Gulf Coast into southern Canada with about the same intensity, but a cold wave usually warms considerably by the time it reaches the deep South. Also, the thermal characteristics of air masses change with the seasons—warming during the summer and cooling during the winter.

The particular type of air mass that flows into a region that is located in a middle or higher latitude is determined by the westerly wind pattern in the midtroposphere. Partially for that reason, the upper-level westerlies are referred to as *steering winds.*

Aloft, the westerlies girdle the middle latitudes in wavelike patterns of troughs and ridges (Figure 9.23). Thus the westerlies actually consist of a north–south airflow that is superimposed on an overall west-to-east wind. In regions where the westerlies blow from north to south, cold air masses are steered southward. But in regions where the westerlies flow from south to north, warm air masses are funneled northward. Hence the type of air mass that invades a region depends on the location of that region with respect to the westerly wave pattern.

A particular wave pattern in the westerlies tends to persist for periods of time that vary from several days to weeks or longer. Then, abruptly (usually in a single day or less), a shift occurs to a new wave pattern, which is usually accompanied by a change in the regional distribution of air masses. For example, areas that have been cold and dry may suddenly turn mild and stormy, while areas that were warm and moist may abruptly become cool and dry. Thus the weather that occurs at a particular region that is in a middle latitude involves a sequence of spells (episodes) of weather that last for varying periods of time and are punctuated by abrupt periods of change.

Different wave patterns of westerly winds prevail at different times of the year. Hence, in many regions, different air masses dominate, depending on the season. For example, in portions of the Midwest, cold, dry air from Canada is the most frequent type of air mass in January, while warm, humid air from the Gulf of Mexico prevails in July. A change in climate, then, means a change in prevailing atmospheric circulation patterns and, at least in some regions, a change in the frequency of various types of air masses.

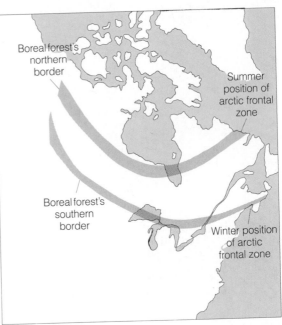

Figure 9.24 *The southern boundary of Canada's boreal forest corresponds to the average winter position of the leading edge of Arctic air. Its northern border corresponds to the average summer position of the leading edge of Arctic air. (After R. A. Bryson, Air masses, streamlines and the boreal forest. Geographical Bulletin 8:228–269, 1966. Reproduced by permission of the Minister of Supply and Services, Canada.)*

The influence of atmospheric circulation patterns and air-mass frequency on climate is evident in the geographical distribution of certain biomes. For example, the region of Canada that is dominated by cold, dry Arctic air corresponds closely with the region that is occupied by the boreal forest. As shown in Figure 9.24, the southern boundary of the boreal forest coincides with the average position of the leading edge of Arctic air during the winter, and the northern border of the forest closely corresponds to the average position of the leading edge of Arctic air during the summer. The geographical position of the grassland biome of North America can also be explained in terms of air-mass distribution. Hence climatic variations that alter the frequency of certain types of air masses in certain regions may also change the boundaries of biomes.

Normal Climate

Through international agreement by climatologists, normal values for temperature and precipitation (and other elements of weather) for each region are com-

Figure 9.23 *At middle and higher latitudes, the westerly winds encircle the globe in a wavelike pattern of ridges and troughs. To the east of a trough (and west of a ridge), warm air moves northeastward. To the west of a trough (and east of a ridge), cold air moves southeastward.*

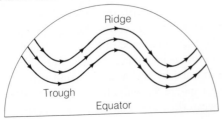

puted from weather instrument measurements and compiled over a 30-year period. Current climatic averages are based on the period from 1951 to 1980. The assumption is that these temperatures and precipitation values represent a reasonable guide for future weather expectations.

Problems arise, however, when we assume that the weather of this 30-year period actually constitutes normal climate. First, it turns out that climate can change significantly during periods of time that are much shorter than 30 years. In addition, a 30-year period gives us a very restricted view of the climatic record and the variability that climate can exhibit. For example, the weather during this century has been unusually mild when compared to the climatic record of the last several hundred years.

Some perspective on recent weather conditions can be gleaned by comparing weather over some period with the climatic record. For example, if we average the temperatures and the amount of precip-

itation over some specified period of time, say, for the month of January, the statistics can then be compared to the long-term average temperature and precipitation for the month of January. Typically, we will find a departure from the long-term average, which is known as an *anomaly*, that is, we will find that the recent January was colder or warmer or wetter or drier than the long-term average.

If climatic anomalies are computed for the same period of time at numerous regions across the nation, they form a pattern that is geographically nonuniform in both magnitude and direction (Figure 9.25). Also, an extreme anomaly in temperature or precipitation never occurs everywhere at the same time. For example, the United States is so large that anomalous cold or drought never grips the entire country at the same time.

The nonuniformity of climatic anomalies—in both magnitude and direction—is linked to the westerly wind pattern aloft. For example, suppose that a certain westerly wave pattern causes abnormally cold conditions in the eastern United States, exceptionally mild weather in the western United States,

Figure 9.25 *Monthly anomaly pattern (departures from long-term averages) of temperature (°F) for the United States in June, 1983. (NOAA data.)*

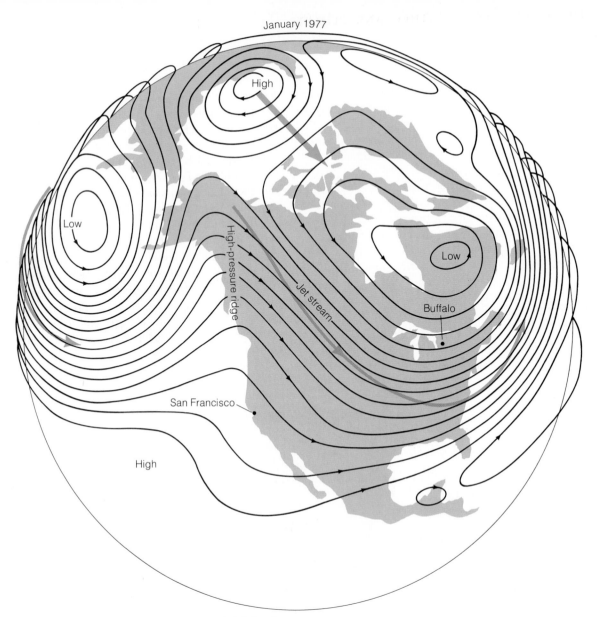

January 1977

High

Low

High-pressure ridge

Jet stream

Low

Buffalo

San Francisco

High

Figure 9.26 *The westerly wind pattern that prevailed during the winter of 1976–1977, bringing record cold to the East and record drought and heat to the West. (After T. Y. Canby, "The Year the Weather Went Wild."* National Geographic 152:799–829, 1977.)

and near-normal weather in the middle of the country. Now suppose that this pattern persisted through the course of a winter, as it did during the notorious winter of 1976–1977. This pattern is represented in Figure 9.26. During such winters, the average temperatures across the country mirror the characteristics of the dominating westerly wave pattern. The nonuniformity is represented by temperatures that are colder than normal in the east, warmer than normal in the west, and near normal in between.

The geographical nonuniformity of climatic anomalies has some implications on our demand for fuel for space heating. For example, since the amount of fuel that is consumed for heating our homes depends largely on outside air temperature, it would be useful if we could predict whether next winter will be colder or warmer than usual. Then we could plan strategies to conserve energy and develop appropriate ways to allocate fuel. However, it is not possible for meteorologists to predict with any degree of accuracy whether next winter in your community will be colder or warmer than usual. All we can say with confidence is that, across the nation, some places will be warmer than normal, some places will be near normal, and some places will be colder than normal.

Northern Hemispheric Temperature Trend

The geographical nonuniformity of climatic anomalies has some interesting consequences in terms of recent trends that have occurred in air temperatures across the northern hemisphere. For example, if we compute the average temperature for the northern hemisphere for each year, we must gather data on the temperature for hundreds of places. Figure 9.27 shows the average annual hemispheric temperature from 1880 to 1980. Note that a general warming trend began in the 1890s and continued until about 1938. Then, an oscillatory cooling trend set in and apparently ended by the late 1970s. Because of the geographical nonuniformity of climatic anomalies, however, the temperature trend that is shown in Figure 9.27 is not representative of all the regions across the hemisphere. On the contrary, while the average annual hemispheric temperature trended downward after 1938, in some places, the average annual temperatures rose, and in other regions the annual temperatures remained virtually unchanged. For example, in the United States, the cooling trend in the 1960s and 1970s was quite marked in the southeastern states, while a warming trend occurred in the Pacific Northwest.

Just as it is misleading to apply the direction of hemispheric climatic trends to all regions, it is also misleading to assume that the magnitude of climatic change is the same everywhere. In fact, a small change in average hemispheric temperature may translate into a much larger change in certain regions. For example, it appears that the slight fall in average hemispheric air temperature after 1938 was accompanied by a considerably greater fall in temperature in polar regions. Thus it follows that, in any meaningful assessment of the potential impact of hemispheric climatic trends on ecosystems, we must not consider small changes in average hemispheric temperature as being unimportant. Locally, hemispheric trends may be amplified significantly and even reversed, to the point of disrupting physical processes and the activities of plants and animals.

Responses of Ecosystems to Climatic Change

Plants and animals respond to changes in climate. If the climatic shift is severe, the tolerance limits of certain organisms may be exceeded. Then, if that occurs, the species that live in the region will change. Organisms vary greatly in their ability to disperse, become established, and reach equilibrium with a new climate. As a result, population interactions may be disrupted for years, and some organisms may succumb while others may migrate into or out of a region. Recall from Chapter 4 the severe disruption that often follows the establishment of new interactions between populations. Centuries may pass before a climax ecosystem can become established that is adapted to the new climate.

An extreme example of the kind of disruptions that can occur in an ecosystem comes from the Ice Age. Climatic deterioration 25,000 years ago triggered the formation of a huge sheet of ice that spread over most of Canada and, by 18,000 years ago, engulfed the northern tier states of the United States. In nonglaciated areas that were south of the ice sheet, climatic shifts caused marked changes in ecosystems. For example, tundra developed at the edge of the ice, and prairies were replaced by coniferous forests. In the American Southwest, moist conditions erased desert ecosystems, and in the East, deciduous forest species retreated to tiny refuges.

In more recent times, abrupt changes in climate

Figure 9.27 *The trend in mean annual temperature for the northern hemisphere from 1880 to 1980, expressed as departures from the 100-year mean temperature (in degrees Celsius). (From P. D. Jones et al., Variations in surface air temperatures: Part 1. Northern hemisphere, 1881–1980.* Monthly Weather Review *110:67, 1982.)*

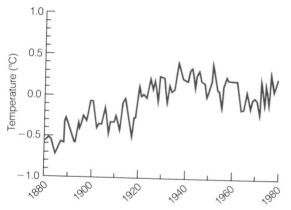

have also had a disruptive effect on agriculture. For example, in the 1300s, a succession of severe droughts may have forced the Pueblo Indians of Mesa Verde to abandon their southwestern Colorado homes. Later, during the Dust Bowl era of the 1930s, farms were abandoned by the scores on the western Great Plains.

Agriculture that is most vulnerable to climatic change is located in areas where the climate is just marginal for food production. Typically, in those regions, just enough rain falls, or the growing season is barely long enough, for crops to mature. Thus, even a small change in the wrong direction of these critical parameters can spell disaster. Apparently, this is what happened to the early European settlements in Greenland, which is described in Box 9.3.

The Uncertain Climatic Future

It is difficult to determine what the climatic future holds for us. Although weather forecasters now have many sophisticated tools, including satellites and high-speed electronic computers, long-range weather forecasters still cannot predict the weather for more than approximately 2 weeks in advance. The problem stems partially from the margin of error that is seemingly inherent in the worldwide network of weather monitoring instruments and from the lack of observational data from vast areas of the ocean. But a more imposing difficulty involves the complex way in which weather occurs. For one thing, the episodic pattern of weather requires that the forecaster predict not only the type of future episodes

---

BOX 9.3

CLIMATIC CHANGE AND THE GREENLAND TRAGEDY

Late in the ninth century, a lengthy episode of unusually mild climate began in the North Atlantic region. That warming enabled Viking explorers to probe the far northerly reaches of the Atlantic Ocean. Previously, severe cold and extensive drift ice had proved to be insurmountable obstacles to European navigators. By 930 AD, the Vikings established the first permanent settlement in Iceland, some 970 kilometers (600 miles) west of Norway and just south of the Arctic Circle.

Among Iceland's early inhabitants was Eric the Red, a troublesome person whose exploits eventually caused him to be banished from Iceland. He sailed west, and in 982 AD, he discovered a new land, which he named Greenland. Some historians think he called it Greenland to entice others to follow him. Then, as now, much of the new land was buried under a massive sheet of ice. But in the late tenth century, the climate was so mild in Greenland that some sheltered areas probably were, indeed, quite green.

In such a coastal place, on Greenland's southwest shore, Eric founded the first of three colonies (Figure 1). Although far from prosperous, it developed into a successful agrarian society.

Initially, it fared well, and its population climbed to roughly 3000. By the fourteenth century, however, the climate began to deteriorate rapidly. Drift ice again expanded in the North Atlantic, hampering and eventually halting navigation between Iceland and Greenland, which was cut off from the rest of the world. What happened to the Norse settlers in succeeding years can only be inferred from their graves and the ruins of their homes; there were no survivors.

In 1921, an expedition from Denmark examined the remains of the Greenland settlements. The expedition reported that the colony had lasted for 500 years and had suffered a slow, painful annihilation. Evidence shows that grazing land was buried under advancing lobes of glacial ice and that most farmland was rendered useless by permafrost. Near the end, the descendants of a once robust and hardy people were ravaged by famine; they became crippled, dwarflike, and diseased. Furthermore, some suggestion exists that the malnourished colonists were attacked by pirates when they turned to the sea for food.

The Greenland tragedy may be the only historical example of the extinction of a European society in North America. What lesson does it teach contemporary society? The lesson is that climate changes, and that it can change rapidly—sometimes with serious and even disastrous

but also their duration, which is still an impossible task, particularly as the forecast period lengthens.

Although climate forecasters do not know enough about the workings of the atmosphere to formulate reliable long-range forecasts, we can gain some idea of what the climatic future holds by studying the climatic records of the past. Those records reveal that the mild weather that occurred during the first half of this century was an unusual event—one that was unprecedented in many hundreds of years. The long-range record, then, appears to favor cooling weather in the future. On the other hand, recent studies suggest that rising levels of atmospheric carbon dioxide will cause pronounced global warming by the close of the century.

Which will it be, warming or cooling? Nobody

knows. Some climatologists are becoming less concerned about the trends in average temperature. They are more concerned about prospects for increasingly unreliable weather, that is, *climatic destabilization.* The first half of this century was not only relatively mild, but the weather was also remarkably stable and reliable. However, recent climatic trends suggest that an increased possibility now exists for hostile episodes of weather, that is, greater extremes in the weather, including record warmth, cold, drought, and excessive rain and snow.

Climatic destabilization could have a particularly severe impact on food production. Although modern agriculture can adjust to gradual changes in climate, it would be difficult to devise an agricultural strategy that would be flexible enough to cope

consequences. Nowhere are people more vulnerable to climatic shifts than in regions where the climate is just marginal for their survival. Those regions include areas where barely enough rain falls to

sustain crops and livestock and places where temperatures are so cold and the growing season so short that only a few hardy crops can be cultivated successfully. In such regions, even a slight deterioration of climate can make agriculture impossible. If the inhabitants have no alternative food source, and if they cannot migrate to more hospitable lands, their fate may be similar to that of the early Greenland people.

Figure 1 *Locations of the first Greenland settlements. (R. A. Bryson, and T. J. Murray,* Climates of Hunger. *Madison, Wisc.: The University of Wisconsin Press, 1977, p. 48. © The Regents of the University of Wisconsin System.)*

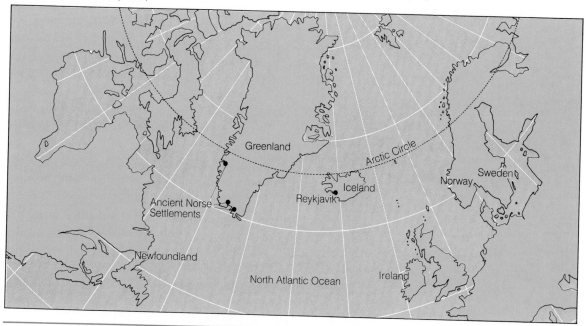

with successive years during which rainfall and temperatures might fluctuate from one extreme to another. Although food production and human population soared during the generally favorable climatic conditions of the first half of the twentieth century, climatic destabilization could reduce the production of food in some key areas of the globe. This important topic is discussed further in Chapter 17.

Conclusions

In this chapter, we have seen that weather is far from a capricious act of nature. Rather, weather, in its many forms, is a response to unequal heating and cooling within the earth–atmosphere system. Excess heat is transported from the earth's surface to the atmosphere and from the tropics to higher latitudes by migrating storms, evaporation and condensation, conduction and convection, and the exchange of cold and warm air masses.

Weather patterns that are averaged over long periods of time constitute the climate. The climate in middle latitudes is strongly influenced by the prevailing patterns that characterize global-scale westerly winds. When the prevailing patterns change, the climate also changes. And since the well-being of an ecosystem depends on climate, a climatic change could cause serious disruptions.

Until recently, average hemispheric temperatures were cooling slowly, and the occurrence of extreme weather appears to be more likely than it has for some time. The role that air pollution may play (if any) in climatic trends is among the topics we consider in our examination of atmospheric pollution in Chapter 10.

Summary Statements

The atmosphere is the product of a long evolutionary process that began billions of years ago and was linked to the development of continents, seas, and life.

The atmosphere is a mixture of gases in which solid and liquid particles are suspended.

The significance of an atmospheric gas or aerosol is not necessarily related to its relative concentration in the atmosphere.

Air pressure and density fall rapidly with increasing altitude, but the atmosphere has no clearly defined upper boundary; rather, it simply becomes progressively thinner as the distance from earth increases.

Differences in air-mass characteristics cause the surface air pressure (reduced to sea level) to differ from one place to another. Important weather changes can accompany relatively small changes in surface air pressure.

The atmosphere is subdivided into concentric layers on the basis of the vertical profile of temperature. The troposphere is the site of most weather events.

Infrared, visible, and ultraviolet radiation represent small segments of a continuous spectrum of electromagnetic radiation. Forms of electromagnetic radiation differ in wavelength, frequency, and energy level.

The earth's elliptical orbit, spherical shape, and tilted rotational axis cause incoming solar radiation to be distributed unevenly over the earth's surface and to change through the course of a year. The tilted rotational axis is the factor that causes seasonal variations.

Solar radiation that is not absorbed by the atmosphere or reflected and scattered by it to space reaches the earth's surface, where it is either reflected or absorbed, depending on the surface albedo.

The earth's surface radiates in the infrared portion of the electromagnetic spectrum and is the primary source of heat for the troposphere.

Water vapor, carbon dioxide, and other gases absorb and reradiate infrared radiation back toward the earth's surface, thereby moderating the temperature of the lower atmosphere.

Heat imbalances develop between the earth's surface and the atmosphere and between tropical and higher latitudes as a result of differences in rates of solar heating and infrared cooling. At the earth's surface, solar heating exceeds infrared cooling; within the atmosphere, the opposite is true. In lower latitudes, solar heating exceeds infrared cooling; in higher latitudes, the opposite is true.

Excess heat is transported from the earth's surface into the atmosphere through conduction, convection, and vaporization of water and its subsequent condensation within the atmosphere.

Excess heat is transported poleward from the tropics by means of air-mass exchange, storm systems, and ocean currents.

Vertical and poleward heat transport within the earth–atmosphere system is brought about by various weather systems that span a wide range of spatial scales.

The climate of a region is shaped by its latitude, proximity to large bodies of water, and elevation as well as the prevailing atmospheric circulation. Atmospheric circulation controls air-mass frequency.

The wavelike pattern of westerly winds in middle latitudes is responsible for the regional distribution of air masses and the episodic behavior of weather.

Anomalies and trends in climate are geographically nonuniform in both magnitude and direction.

Locally, hemispheric trends in climate may be significantly amplified and even reversed, thereby disrupting plant and animal distributions and physical processes.

In the perspective of the long-term climatic record, the weather of much of this century has been unusually mild and stable.

Climatic change can be rapid, which can have a potentially disruptive impact on ecosystems.

Questions and Projects

1. Explain how the evolution of the earth's atmosphere is tied to the development of oceans and the beginning of life on earth.

2. Present several examples of how "minor" constituents (gases and aerosols) of the atmosphere play important roles in the functioning of the environment.

3. Explain why temperature is an inexact way of describing the heat energy of different substances.

4. Is the air temperature on a winter day more likely to be higher if the ground is snow-covered or if it is bare? Explain your answer.

5. Identify the various ways in which heat is added to or removed from the atmosphere.

6. Although insolation reaches its maximum intensity at noon, surface air temperatures do not typically reach a maximum until several hours later. Why?

7. For the entire globe, why must incoming solar radiation balance outgoing infrared radiation?

8. How does the albedo of the moon compare with the planetary albedo of the earth? Why is there a difference?

9. Insolation diminishes as the inverse square of the distance traversed. If the mean earth–sun distance were *three* times what it is today, the solar constant would be reduced to what fraction of its present magnitude?

10. Speculate on what circumstances might cause a change in the "solar constant."

11. In your own words, explain the following statements: (a) The sun drives the atmosphere; (b) the atmosphere is heated from below.

12. Is there evidence from your area of the country that climate is becoming more extreme? Check with local newspapers, fuel companies, or agricultural suppliers.

13. How might climatic change (to warmer, colder, drier, or wetter conditions) influence the economy of your community?

14. Are weather data routinely collected in your community? If so, where and by whom? What factors are monitored (for example, temperature, precipitation, wind, and solar radiation)? Where are those data stored and how are they summarized?

15. Is it possible for a slight cooling trend in hemispheric average temperature (for example, 1°C per decade) to disrupt the functioning of certain ecosystems? Defend your answer.

16. In the long-term perspective, our current "normal" climate actually may be anomalous. Explain this possibility.

17. Why are convective air currents more likely to develop over a vegetated surface than over a snow-covered surface?

18. The bulk of solar radiation on earth is absorbed by the oceans, with lesser amounts being absorbed by air and land. In view of this distribution, speculate on the role that oceans might play in any large-scale climatic shift.

19. Reliable weather records in the United States do not extend back much further than 100 years. This brief record provides us with a restricted view of climatic variability. Can you think of any ways in which we can lengthen the record by reconstructing the climatic past?

20. What is the purpose of the large fans that are used in some Florida orange groves during calm, clear, cold winter nights?

21. The maximum poleward energy transport takes place at about 30° N latitude. Would you expect this latitude belt to be particularly stormy? Explain your answer.

22. What is the effect of spreading coal dust over the snow cover in late spring?

23. Design a house that will take advantage of the "greenhouse effect" during the winter heating season but will not overheat in the summer.

24. Describe the factors that explain the fact that although the earth is closest to the sun in January, it is a winter month in the northern hemisphere.

Selected Readings

BATTAN, L. J.: *Weather in your Life.* New York: W. H. Freeman, 1983. A well-illustrated and lucid introduction to weather, with special emphasis on the role of weather in societal concerns.

BRYSON, R. A., AND MURRAY, T. J.: *Climates of Hunger.* Madison, Wisc.: The University of Wisconsin Press, 1977. A description of how climatic change has an impact on society.

HUGHES, P.: Weather, climate, and the economy. *Weatherwise* 35(2):60–63, 1982. An outline of some of the ways in which weather extremes and climate have an impact on the economy.

INGERSOLL, A. P.: "The Atmosphere." *Scientific American* 249(3):162–174, 1983. A comprehensive and timely review of global radiative balance, the problems that limit weather forecasting, and the nature of climate variability.

KELLOGG, W. W., AND SCHWARE, R.: "Society, Science and Climate Change." *Foreign Affairs* 60(5):1076–1109, 1982. A climatologist and a political scientist speculate on the implications of future climatic change on agricultural productivity, ecology, and human health and disease in specific regions and nations of the earth.

LEHR, P. E., BURNETT, R. W., AND ZIM, H. S.: *Weather.* New York: Golden Press, 1975. An exceptionally well-illustrated and lucid description of weather phenomena.

LUTGENS, F. K., AND TARBUCK, E. J.: *The Atmosphere, An Introduction to Meteorology.* Englewood Cliffs, N.J.: Prentice Hall, 1982. A well-organized and clearly written introduction to the atmospheric sciences.

NATIONAL RESEARCH COUNCIL: *Solar Variability, Weather, and Climate.* Studies in Geophysics. Washington, D.C.: National Academy Press, 1982. A sophisticated summary report on possible linkages between variations in solar output and weather and climate.

NATIONAL RESEARCH COUNCIL: *Understanding Climatic Change, A Program for Action.* Washington, D.C.: National Academy of Sciences, 1975. A classic report of the U.S. Committee for the Global Atmospheric Research Program on current research concerning climatic variations and the need for further research on global climate.

SKINNER, B. J., ED.: *Climates Past and Present.* Los Altos, Calif.: William Kaufmann, 1981. A collection of articles from *American Scientist* on controls of climate, past climate, and the interaction of people and climate.

Industrial and other human activities threaten the quality of air.
(Dick Hanley, Photo Researchers.)

CHAPTER 10   AIR POLLUTION

Air pollution is not a new phenomena. Indeed, it is older than civilization itself. The first episode of air pollution probably occurred when early humans tried to make a fire in a poorly ventilated cave. Reference to polluted air appears as early as Genesis (19:28): "Abraham beheld the smoke of the country go up as the smoke of a furnace." About 400 BC, Hippocrates noted the pollution of city air. And in 1170, Maimonides, referring to Rome, wrote, "The relation between city air and country air may be compared to the relation between grossly contaminated, filthy air, and its clear, lucid counterpart."

London's air became heavily contaminated early in the thirteenth century, when soft coal was introduced as fuel. At that time, air quality deteriorated so much that several government commissions were appointed between 1285 and 1288 to solve the problem. The commissions were unsuccessful, but London's air improved in 1307, when Edward I banned the burning of coal and reinstated wood as the principal source of fuel. Prohibition of coal lasted into the sixteenth century, when the depletion of forests forced a shift back to the use of coal. Again, the quality of air steadily deteriorated, until abatement legislation was implemented in the 1950s.

The Industrial Revolution was the single greatest cause of chronic air pollution in Europe and North America. In the United States, in post-Civil War days, the cities swelled as a result of new industries and new immigrants to work for them. Thus by the turn of the century, the urban environment was becoming increasingly fouled by the fumes of foundries and steel mills. In those days, a city took pride in its smokestacks, such as those shown in Figure 10.1. They were considered to be a sign of a prosperous economy. Efforts to regulate air quality were meager, and little was known about the effects of air pollution on human health. In an attempt to placate the wheezing and coughing population, some physicians even argued that polluted air had medicinal value.

In fact, even today, concern over polluted air does not stem from disenchantment with the fruits of industrialism. Belching smokestacks still reassure many persons that the economy is healthy. But many of us are troubled by charges that polluted air adversely affects our health, agricultural productivity, and the weather. In this chapter, we examine several factors that are related to these concerns: for example, how polluted air affects the well-being of living things; how weather conditions influence levels of air pollution; and whether air pollution has an impact on weather and climate.

Types and Sources of Air Pollutants

Air pollutants are airborne gases and aerosols that occur in concentrations that threaten the well-being of living organisms or disrupt the orderly functioning of the environment. Interestingly, many of those substances are natural constituents of the atmo-

Figure 10.1 *Industrial smokestacks in Pittsburgh in 1906. (Carnegie Library of Pittsburgh.)*

sphere. For example, sulfur dioxide and carbon monoxide are normal components of the atmosphere, which only become pollutants when their concentrations approach or exceed the tolerance limits of organisms. On the other hand, some air pollutants exist that do not occur naturally in the atmosphere and may be hazardous in any concentrations. For example, asbestos fibers are known to cause cancer.

Air pollutants are products of both natural events and human activities. Natural sources of air pollu-

tants include forest fires, pollen dispersal, wind erosion of soil, organic decay, and volcanic eruptions. The single most important source of atmospheric pollutants that is attributed to human activity is the motor vehicle. According to the U.S. Environmental Protection Agency (EPA), transportation vehicles emit more than 100 million metric tons of the major air pollutants every year. Many industrial sources contribute as well. For example, pulp and paper mills, iron and steel mills, oil refineries, smelters, and chemical plants are prodigious producers of air pollutants. Additional sources of pollutants include fuel combustion by industrial and domestic furnaces, re-

fuse burning, and various agricultural activities (for example, crop dusting and cultivation). In the United States, almost 160 million metric tons of air contaminants are emitted into the atmosphere each year as the result of human activities—almost 0.7 metric ton per person (Table 10.1).

Some substances, which are called *primary air pollutants*, pollute the air immediately when they are emitted into the atmosphere. In addition, within the atmosphere, chemical reactions that involve primary air pollutants, which are both gases and aerosols, produce *secondary air pollutants*. Examples of secondary air pollutants are acid mists and smog. In some cases, the environmental impact of certain primary pollutants is less severe than that of the secondary pollutants they form. In this section, we survey the major air pollutants, noting their natural cycling within the environment and the contributions that occur as a result of human activity.

Oxides of Carbon

The burning of carbon-laden fossil fuels releases *carbon dioxide* (CO_2) into the atmosphere. Much of this carbon dioxide is absorbed by ocean water, some of it is taken up by vegetation through photosynthesis, and some remains in the atmosphere. Today, the concentration of carbon dioxide in the atmosphere is approximately 340 parts per million (ppm) and is rising at a rate of approximately 18 ppm every decade. The increasing rate of combustion of coal and oil has been primarily responsible for this occurrence, which (as we see later in this chapter) may eventually have an impact on global climate.

By far the most important natural source of atmospheric *carbon monoxide* (CO) is the combination of oxygen with methane (CH_4), which is a product of the anaerobic decay of vegetation. (Anaerobic decay takes place in the absence of oxygen.) However, carbon monoxide is removed from the atmosphere by the activities of certain soil microorganisms, so the net result is a harmless average concentration that is less than 0.15 ppm, in the northern hemisphere. The principal source of carbon monoxide that is caused by human activities is motor vehicle exhaust. Breathing carbon monoxide causes drowsiness, slows reflexes, and impairs judgment; at high concentrations, death ensues. However, carbon monoxide poses serious health hazards in areas where its concentrations build. Because it is odorless and tasteless, carbon monoxide defies direct human detection, especially where sources are concentrated—for example, in highway tunnels (Figure 10.2) and underground parking garages.

Hydrocarbons

Hydrocarbons [also called volatile organic compounds (VOC)] encompass a wide variety of chemical compounds that contain exclusively hydrogen and carbon. Of the hydrocarbons that occur naturally in the atmosphere, *methane* (CH_4) is present at highest concentrations (1 to 1.5 ppm). But even at relatively high concentrations, methane is nonreactive (that is, it does not interact chemically with other substances) and causes no ill health effects. Much more reactive are the *terpenes*, which are a variety of volatile (readily vaporized) hydrocarbons that are emitted by vegetation. Terpenes occur naturally in lower concentrations (less than 0.1 ppm). They are responsible for the aromas of pine, sandalwood, and eucalyptus trees. Terpenes also form particles that scatter sunlight, which produces the bluish haze that is often seen hanging over forests, such as the Great Smokies.

In urban areas, the principal source of reactive hydrocarbons is the incomplete combustion of gasoline in motor vehicles, which emits hundreds of different hydrocarbons. Because gasoline is very volatile, some hydrocarbons (perhaps as much as 15 percent of the total in some cities) enter the air during gasoline delivery and refueling operations at service stations.

Not much is understood about the natural cycling of hydrocarbons in the atmosphere. But we do know that the typically low concentrations of most

Table 10.1 Estimated emissions of the principal air pollutants in the United States during 1980 (million metric tons per year)

Total Suspended Particulates	7.8
Sulfur Oxides	23.7
Oxides of Nitrogen	20.7
Volatile Organic Compounds (Hydrocarbons)	21.8
Carbon Monoxide	85.4
Total	159.4

Source: 1982–1983, *Statistical Abstract of the U.S.*, U.S. Department of Commerce, Bureau of the Census.

Figure 10.2 *Carbon monoxide poses the greatest health hazard where sources are concentrated, such as in this highway tunnel. (Dan McCoy, Black Star.)*

hydrocarbons that occur in city air do not appear to pose environmental problems. The most serious health effects of hydrocarbons are caused by the products of reactions of hydrocarbons with other pollutants in the presence of sunlight. Those photochemical reaction products (secondary pollutants) include formaldehyde, ketones, and peroxyacetylnitrate (PAN), which are substances that irritate the eyes and damage the respiratory system. Also, some hydrocarbons, such as benzene (C_6H_6) (which is a component of many consumer goods, including rubber cement) and benzo(a)pyrene (a product of fossil-fuel combustion) are carcinogenic.

Oxides of Nitrogen

The activities of soil bacteria are responsible for the bulk of nitric oxide (NO) that is produced naturally and released to the atmosphere. Within the atmosphere, nitric oxide combines readily with oxygen to form nitrogen dioxide (NO_2) and, together, those two oxides of nitrogen are usually referred to as NO_x.

Although human activities contribute only about 10 percent of the atmosphere's total load of NO_x, those contributions tend to be much more concentrated than the atmospheric average. Scientists distinguish between two types of NO_x—thermal and fuel—depending on its mode of formation. Thermal NO_x forms when high combustion temperatures, such as those within internal combustion engines, cause nitrogen (N_2) and oxygen (O_2) in the air to combine. Fuel NO_x forms when nitrogen that is contained within a fuel, such as coal, is oxidized, that is, combines with oxygen in the air. For both types of NO_x, nitric oxide is generated at first and then, when it is vented and cooled, a portion of nitric oxide is converted to nitrogen dioxide. Approximately half of fuel NO_x comes from stationary sources (power plants, primarily) and the other half is released by motor-vehicle emissions.

Nitrogen dioxide is approximately four times more toxic than nitric oxide, and it is a much more

serious air pollutant. At high levels, nitrogen dioxide is believed to contribute to heart, lung, liver, and kidney damage, and it is linked to the incidence of bronchitis and pneumonia. Moreover, because nitrogen dioxide occurs as a brownish haze, it reduces visibility. When nitrogen dioxide combines with water vapor, nitric acid (HNO_3), which is a corrosive substance, is formed. Oxides of nitrogen are also major precursors of smog.

Compounds of Sulfur

Sulfur enters the atmosphere in the form of *sulfur dioxide* (SO_2) from volcanic eruptions, as sulfate particles from sea spray, and as hydrogen sulfide (H_2S), which is produced when organic matter decays anaerobically. Those sulfur compounds are washed from the air by precipitation and taken up by soil, vegetation, and surface waters. In addition, many human activities release sulfur into the atmosphere, which contributes perhaps one-third the amount of sulfur that is emitted into the air by natural sources. Most of the sulfur dioxide that we contribute comes from the burning of sulfur-containing fuels (coal and oil) in electric generating plants (69 percent) and industrial boilers (8 percent). In addition, the smelting of sulfur-bearing ores—lead, zinc, and copper sulfides—is a source of sulfur dioxide (8 percent), as are petroleum refining (5 percent), motor vehicles (5 percent), and residential and commercial space heating (5 percent).

In the air, sulfur dioxide converts to *sulfur trioxide* (SO_3) and *sulfate* particles (SO_4). Sulfate particles restrict visibility and, in the presence of water, form *sulfuric acid* (H_2SO_4), which is a highly corrosive substance that also lowers visibility. Both sulfur dioxide and sulfur trioxide pose hazards to health by irritating respiratory passages and by aggravating such conditions as asthma, emphysema, and bronchitis. Also, many scientists believe that sulfate particles and sulfuric acid droplets increase our vulnerability to respiratory infection.

Certain industrial activities, including paper and pulp processing, emit *hydrogen sulfide* (H_2S) as well as a family of organic sulfur-containing gases, which are called *mercaptans*. Even in extremely small concentrations, those compounds are foul-smelling. Also, hydrogen sulfide tarnishes silverware and copper facings and discolors lead-based paints.

Photochemical Smog

When vehicular traffic is congested (for example, during morning and evening rush hours), *photochemical smog* is likely to develop. Smog is a noxious, hazy mixture of suspended particles and gases that is formed when oxides of nitrogen and hydrocarbons from the exhaust of automobiles react in the presence of sunlight. Although that reaction is most common in urban areas, winds occasionally transport automobile exhaust into suburban and rural areas, where the sun's rays trigger the development of smog. Consequently, smog can be a serious problem both within and downwind from a large metropolitan area.

Constituents of photochemical smog include PAN (peroxyacetylnitrate) and ozone (O_3). PAN damages vegetation and stings the eyes. Normal levels of ozone at the earth's surface are only approximately 0.02 ppm, but during very smoggy periods, the level of ozone may exceed 0.5 ppm. At those relatively high concentrations, ozone irritates the eyes and the mucous membranes of the nose and throat. It also degrades rubber and fabrics and damages some crops.

Suspended Particulates

So far, our survey of the major air pollutants has been focused primarily on gases. Now we turn our attention to the millions of metric tons of tiny solid particles and liquid droplets (for example, acid mist) that are suspended in the atmosphere, which, collectively, are called *suspended particulates* or *aerosols*. Approximately one-half of the atmosphere's total aerosol load of those substances is emitted by sea-salt spray, soil erosion, volcanic activity, woodburning stoves, and various industrial emissions. The other half is produced primarily by atmospheric reactions among various gases.

Perhaps the most common particulates are dust and soot. Most dust is produced when wind erodes soil, which is often accelerated by agricultural activity. *Soot*, which is composed of tiny solid particles of carbon, is emitted during the incomplete combustion of fossil fuels and refuse (Figure 10.3a). Particulates may be composed of a wide variety of materials, depending on the specific types of mining, milling, or manufacturing that occurs in a given area. For example, in urban–industrial air, particulates

(a)

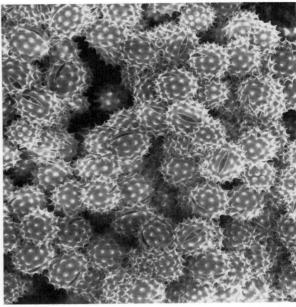

(b)

Figure 10.3 *Photomicrographs of (a) soot particles and (b) pollen grains. These examples show a vast variety of tiny particulates that are suspended in the atmosphere. (David Scharf, Peter Arnold, Inc.)*

usually include a diverse array of trace metals, such as lead, nickel, iron, zinc, copper, and cadmium. Those particulates pose significant health hazards, since their typically small size allows them to be inhaled deeply into the respiratory system. In addition, the air may contain such substances as asbestos fibers, pesticides, and fertilizer dust.

Air normally also contains fungal spores and pollen (Figure 10.3*b*). When the land is disturbed by farming and construction, the abundant growth of ragweed and other weeds are sources of pollen that evoke allergic reactions, such as hayfever, in roughly one out of every 20 persons.

Indoor Air Pollutants

Traditionally, air pollution has been associated with outdoor air or occupational exposure. In recent years, however, increasing concern has become focused on the quality of air that we breathe in our homes. After all, most of us spend at least 16 hours a day at home. A wide range of indoor air pollutant sources exist,

including building materials (especially those that contain formaldehyde resins or asbestos fibers), unvented cooking or heating appliances, smoking materials, cleaning compounds, and personal-care products. Table 10.2 summarizes indoor air pollutants, their sources, typical concentrations, and how the concentrations of indoor pollutants compare to outdoor pollutants.

Efforts that have been made to conserve energy have only compounded the problem of indoor air quality. Although tighter, better insulated homes reduce the amount of air that travels between indoors and outdoors, they allow the levels of indoor air pollutants to rise to potentially hazardous concentrations. Also, the increased use of wood, coal, and kerosene-burning heaters has caused greater exposure to toxic and carcinogenic gases and airborne particles.

For example, one health hazard that may develop in energy-efficient homes is caused by high radon levels. Like carbon monoxide, radon is a gas that defies direct human detection, because it is invisible, odorless, and tasteless. Radon is a radioactive decay product of radium-226, which is a natural trace component of soils and bedrock. It is released by building materials or enters homes through basement walls or well water. As radon decays, it forms highly radioactive elements that adhere to dust particles, which can produce lung cancer if they are inhaled.

Ordinarily, radon levels are too low to pose a health hazard, but with the restricted air exchange of energy-efficient homes and in regions where the bedrock is enriched in radium, radon levels (especially in the basements of homes) may climb to 200 times the background level. At that concentration, the risk of lung cancer is about the same as smoking

Table 10.2 Indoor air pollutants, their sources, typical concentrations, and a comparison with outdoor concentrations

POLLUTANT	MAJOR EMISSION SOURCES	TYPICAL INDOOR CONCENTRATIONS	INDOOR/OUTDOOR CONCENTRATION RATIO
	ORIGIN: PREDOMINANTLY OUTDOORS		
Sulfur Oxides (Gases, Particles)	Fuel Combustion, Smelters	0–15 $\mu g/m^3$	< 1
Ozone	Photochemical Reactions	0–10 ppb	≪ 1
Pollens	Trees, Grass, Weeds, Plants	L.V.*	< 1
Lead, Manganese	Automobiles	L.V.	< 1
Calcium, Chlorine, Silicon, Cadmium	Suspension of Soils, Industrial Emissions	N.A.†	< 1
Organic Substances	Petrochemical Solvents, Natural Sources, Vaporization of Unburned Fuels	N.A.	< 1
	ORIGIN: INDOORS OR OUTDOORS		
Nitric Oxide, Nitrogen Dioxide	Fuel Burning	10–120 $\mu g/m^3$‡ 200–700 $\mu g/m^3$§	≫ 1
Carbon Monoxide	Fuel Burning	5–50 ppm	≫ 1
Carbon Dioxide	Metabolic Activity, Combustion	2000–3000 ppm	≫ 1
Particles	Resuspension, Condensation of Vapors, Combustion Products	10–1000 $\mu g/m^3$	1
Water Vapor	Biological Activity, Combustion Evaporation	N.A.	> 1
Organic Substances	Volatilization, Combustion, Paint, Metabolic Action, Pesticides	N.A.	≫ 1
Spores	Fungi, Molds	N.A.	> 1
	ORIGIN: PREDOMINANTLY INDOORS		
Radon	Building Construction Materials (Concrete, Stone), Water	0.01–4 pCi/liter	≫ 1
Formaldehyde	Particleboard, Insulation, Furnishings, Tobacco Smoke	0.01–0.5 ppm	> 1
Asbestos, Mineral, and Synthetic Fibers	Fire Retardant Materials, Insulation	0–1 fiber/ml	1
Organic Substances	Adhesives, Solvents, Cooking, Cosmetics	L.V.	> 1
Ammonia	Metabolic Activity, Cleaning Products	N.A.	> 1
Polycyclic Hydrocarbons, Arsenic, Nicotine, Acrolein, and so forth	Tobacco Smoke	L.V.	≫ 1
Mercury	Fungicides, Paints, Spills in Dental-Care Facilities or Labs, Thermometer Breakage	L.V.	> 1
Aerosols	Consumer Products	N.A	≫ 1
Microorganisms	People, Animals, Plants	L.V.	> 1
Allergens	House Dust, Animal Dander, Insect Parts	L.V.	≫ 1

Source: From J. D. Spengler and K. Sexton, Indoor air pollution: A public health perspective. *Science* 221(4605):11, 1983. Copyright © 1983 by AAAS.
*L.V., limited and variable (limited measurements, high variation).
†N.A., not applicable.
‡Annual average.
§One-hour average in homes with gas stoves, during cooking.

2.5 to 10 packages of cigarettes a day. A solution to the problem is better ventilation, particularly of the radon-laden air in the basement.

Noise Pollution

So far we have presented air pollution as contamination of air by various gases, and suspended par-

---

BOX 10.1

THE GROWING THREAT OF NOISE POLLUTION

In trying to track a suspected Soviet submarine last month, the Swedish Navy had difficulty finding sailors who could hear well enough to operate the listening devices. The hearing of vast numbers of young people, a Navy captain said, apparently has been permanently damaged by years of listening to loud rock music.

Whether or not music is the culprit in Sweden, similar hearing losses have been noted among American high school and college students who are rock music afficionados or frequent discotheques, and hearing loss resulting from abusive noise has become a matter of pressing concern in this country, too.

For example, Dr. David Lipscomb, head of the noise laboratory at the University of Tennessee, recently found that more than 60 percent of 1410 college freshmen had significant hearing loss in the high-frequency range, a deficit he believes is increasing at an alarming rate. Just 1 year earlier he had found high-frequency hearing loss in 33 percent of the freshmen tested. He described the students as "two or three decades ahead of themselves in hearing deterioration."

While noisy work environments have long been the focus of research and regulatory efforts, in recent years avocational *noise has been attracting more attention. The explosive rise of noisy equipment in and out of American homes—ranging from snowmobiles, rock bands, and chain saws to hair dryers, food processors, and stereo headphones—has made nearly every American potentially vulnerable to noise damage.*

The Environmental Protection Agency estimated in 1978 that 10 million Americans are exposed to harmful levels of noise on the job. Other experts say this is a highly conservative figure. In addition to those who voluntarily expose themselves to high noise levels, such unwitting victims as premature infants in incubators, resi-

dents who live near airports, and students whose classrooms abut train tracks may suffer noise damage.

And the damage incurred may involve far more than hearing acuity. Though more and better research is needed to define precisely the nonacoustical harm caused by noise, studies thus far suggest that noise stress can result in high blood pressure, cardiovascular injury, ulcers, and possibly even susceptibility to infection and reproductive problems. Other studies have pointed to noise-related learning difficulties, irritability, fatigue, reduced work efficiency, increased accidents, and errors and socially undesirable behavior.

A number of findings have already been fairly well documented, including these:

Dr. [Sheldon] Cohen and his former colleagues at the University of Oregon showed that children whose schools were along the flight path of Los Angeles International Airport had higher blood pressure than similar children attending quiet schools. The noise-affected children also had more difficulty solving puzzles and math problems and were quicker to give up in frustration. Furthermore, as time passed, no improvement was seen in the noise-related effects on the children's abilities.

High levels of noise in the home, from television sets, radios, and other sources, were shown to disrupt the development of sensory and motor skills of children during the first 2 years of life. Babies living in noisy homes were slower to imitate adult actions and persisted in infantile habits longer than babies in quieter homes. Noise also delayed verbal development and exploratory behavior. The researcher, Dr. Theodore D. Wachs, a psychologist at Purdue University, believes that noise stress prompts babies to retreat into their own inner space.

A series of European studies indicated that workers exposed to noise were more likely to develop abnormal heart rhythms, balance disturbances, circulatory ailments, and ulcers. The work-

ticulates. This definition should be broadened to encompass unwanted sound or excessively high levels of sound. *Noise pollution* is not only an annoyance, but at sufficiently high levels, it may cause hearing loss. The problem of noise pollution is discussed in greater detail in Box 10.1.

---

ers complained more often of fatigue and irritability, and they reported more social conflicts on the job and at home. Studies in Britain and the United States suggested that people living in noisy areas, such as in airport flight paths, suffered more emotional disturbances and required medical treatment more often than those living in quieter areas.

But according to Dr. Cohen and others, all these studies suffer from methodological problems, primarily the failure to take into account other factors, such as age, socioeconomic status, and various on-the-job stresses, that could have influenced the effects attributed to noise.

Sound is measured in decibels, a scale that increases logarithmically. Zero decibels is the lowest level of sound that a young, healthy human ear can detect. A level of 140 decibels (a shotgun blast or jet takeoff) can be extremely painful. A rise of 10 decibels is perceived *by the human ear as a doubling of loudness. The most frequently used decibel scale, called dBA, measures* perceived *loudness by giving more weight to high-frequency sounds, which seem louder to people than the same intensity of low-frequency sounds.*

Thus the 100 dB sound of a power lawn mower or snowmobile is twice as loud as the 90 dB sound of a train roaring into a subway station. Dr. Maurice H. Miller, professor of audiology at New York University and chief audiological consultant at Lenox Hill Hospital and the New York City Department of Health, points out that noise well below the level of discomfort or pain can damage hearing.

Hearing loss will occur in 20 to 25 percent of workers exposed to the allowable limit of 90 decibels for 8 hours a day (the loudness of street traffic or a heavy-duty truck). Under the newest Federal regulations, employers must establish hearing conservation programs for all workers regularly exposed to 85 decibels or more.

Repeated exposure to loud noise destroys the delicate hair cells in the Organ of Corti, a part of the cochlea in the inner ear. These cells are responsible for picking up sound-induced pressure waves and transmitting them to nerve cells, which in turn carry them to the brain.

Sounds follow two paths into the brain. One path carries sound to the auditory center where it is perceived and interpreted. The other goes to an activating-regulating center in the brain, called the reticular formation, and then on to the brain centers that turn on the autonomic nervous system: the latter path is responsible for the wide range of nonaural effects of noise because it calls into play the classic fight-or-flight response to stress: arousal, increased heart rate and blood pressure, constriction of small blood vessels in the extremities, redirection of blood flow away from the skin and digestive organs and to the brain and muscles, muscular contraction, release of stress-related hormones from the adrenal glands, dry mouth, dilated pupils, and subjective feelings of tension, excitement, and anxiety.

This stress reaction to sound is believed to be an evolutionary holdover from preindustrial times when loud sounds usually meant trouble—a roaring lion, falling rock, or injured kinsman. The stress response enabled people to survive the danger by helping them either to run away or fight.

Noise researchers have found that most people get used to a sound that they hear often and know is not a cause for alarm, but their internal stress reaction continues unabated. Thus, if you live near train tracks, after a week or so you may no longer be awakened by passing trains. But internal reactions to the noise still occur and may eventually accumulate to cause bodily damage. Dr. Miller suggested that people who never become habituated to noise "may be better off because they tend to avoid noisy environments."

Excerpted from Jane E. Brody, "Noise Poses a Growing Threat, Affecting Hearing and Behavior." The New York Times, *November 16, 1982. Copyright © 1982 by the New York Times Company. Reprinted by permission.*

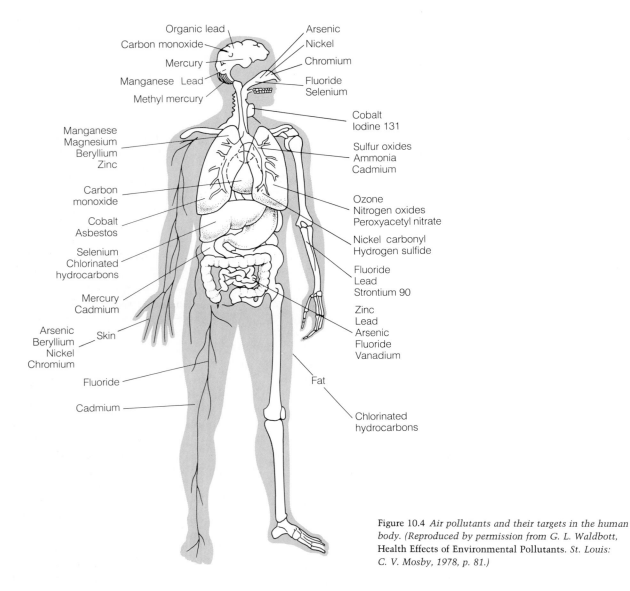

Figure 10.4 *Air pollutants and their targets in the human body. (Reproduced by permission from G. L. Waldbott,* Health Effects of Environmental Pollutants. *St. Louis: C. V. Mosby, 1978, p. 81.)*

Air Pollution and the Quality of Life

The impact of air pollution on human health is our primary reason for being concerned about it. Thus we need to know (1) whether a direct relationship exists between levels of air pollution and the incidence of respiratory and other diseases and (2) if air pollutants harm the animals and plants we depend on for food and fiber. In this section, we examine how air that is polluted by a variety of gases and aerosols affects living things.

Effects on Human Health

Many parts of the human body are targets of a wide variety of air pollutants (Figure 10.4), but the initial attack of pollutants occurs primarily through the respiratory system. Air that we inhale follows a long pathway through oral and nasal passages, the windpipe, and the bronchioles before it finally reaches the air sacs (alveolar sacs) of the lungs (Figure 10.5). Within the air sacs, the oxygen is removed from the air by red blood cells, which then carry it to all parts of the body. If a person's blood is deprived of its

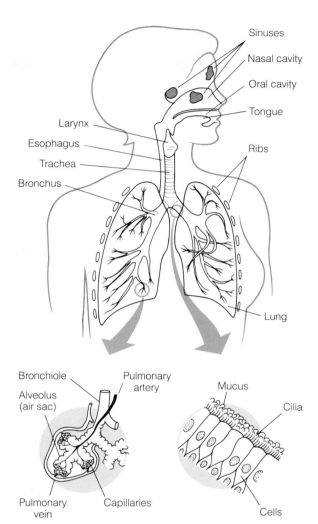

Figure 10.5 *The human respiratory system.*

bronchioles are covered with sticky mucus that captures small particles, including most dust and pollen, and absorbs some gaseous pollutants. Then mucus and the pollutants it captures are transported upward to the mouth by the continuous beating of millions of *cilia* (singular, *cilium*), which are tiny hairlike structures that line most of the respiratory tract (Figure 10.6). Once the pollutant-bearing mucus reaches the mouth, it is either swallowed or expelled.

Although the air sacs are neither lined with cilia nor produce mucus, they are equipped with formidable-looking cells, which are called *macrophages* (Figure 10.7). Those specialized, free-living cells engulf and digest foreign particles. In fact, each macrophage can consume up to 100 bacteria or pollen grains before it succumbs to the very poisons that it produces to kill the invaders. Many of those particle-laden macrophages reach the lower bronchioles, where they are caught up in the mucus and transported by the beating cilia to the mouth.

Some inhaled materials act as irritants and elicit reflex responses in the lining of the respiratory system. For example, coal dust triggers contractions of the muscle layers that surround the bronchi. Such a response narrows the interior of the passageway, which reduces further entry of inhaled coal dust. Many irritants also induce the reflex responses of sneezing and coughing. For example, sneezing ex-

Figure 10.6 *A photomicrograph of cilia within a rabbit lung. The coordinated flailing motion of these tiny hairlike structures transports pollutant-laden mucus out of the lungs into the mouth. (Margot Lundborg, The Karolinska Institute.)*

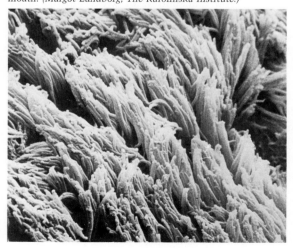

oxygen supply, the person will succumb to *asphyxiation*, or oxygen starvation. Since pollutants are inhaled into the respiratory tract along with air, they can follow the same route to the air sacs. In sufficiently high concentrations, some air pollutants, such as carbon monoxide, are *asphyxiating agents*. Other pollutants, for example, certain particulates, irritate the tissues of the respiratory tract itself.

Defense Mechanisms of the Body We are not entirely at the mercy of air pollutants, because the respiratory system is armed with defense mechanisms. For example, nasal hairs filter out large particles. Also, the linings of the nose, windpipe, and

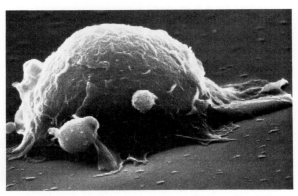

Figure 10.7 *A lung macrophage surrounded by particles of fly ash, viewed under a scanning electron microscope. Each of these free-living cells is capable of engulfing many particles of fly ash or other pollutants. (Courtesy of the Lawrence Berkeley Laboratory, University of California, Berkeley.)*

pels irritants by a rapid expulsion of air in the upper part of the respiratory tract, whereas coughing forces air out of the bronchi.

Gases   Our natural defenses cannot protect us from all attacks of air pollutants. For example, the quantity of gases that can dissolve in the mucus is quite limited. Unless a gas is degraded or removed by ciliary action, exposure for prolonged periods or to high concentrations may saturate the mucus, that is, exceed its capacity to dissolve any more. Then, if that occurs, the gases can penetrate deeper into the respiratory tract. If the gas is one that acts as an asphyxiating agent and enough of it reaches the air sacs, it can displace the oxygen on the hemoglobin molecules (the transporters of oxygen in red blood cells). Hence as increasing concentrations of those gases are inhaled, the quantity of life-sustaining oxygen, which the bloodstream transports from the lungs, decreases.

Asphyxiating agents also stress the heart. For example, persons who are afflicted with angina pectoris, which is a form of heart disease, appear to be particularly susceptible to asphyxiating agents. Lack of oxygen in the heart muscle causes angina victims (approximately 4.2 million in the United States) to suffer attacks of chest pain. In late 1977, the U.S. Council on Environmental Quality reported that the frequency of angina attacks increases as patients are exposed to higher levels of carbon monoxide.

Hydrogen sulfide, which is an anaerobic decomposition product of organic material, is also an asphyxiating gas, but, unlike carbon monoxide, it does not displace oxygen in the blood. Rather, in very high concentrations (which can develop in confined places, such as sewers and mines), hydrogen sulfide impairs the part of the brain that controls the chest movements that are essential for normal breathing, causing almost instantaneous death. However, hydrogen sulfide is relatively harmless in the low concentrations we normally encounter, although we may find its rotten-egg odor unpleasant.

Gases that act mainly as irritants of the respiratory tract include ozone, sulfur dioxide, and nitrogen dioxide. Although each one causes a different reaction, the effects of those irritants generally include the discomforting symptoms of persistent coughing and heavy secretion of mucus.

Particulates   Particulates that penetrate the natural defenses of the respiratory tract can cause illness and even death. The most severe problems result from lengthy exposure to relatively high concentration levels. For example, persons who experience long-term exposure to high levels of dust in occupations, such as mining, metal grinding, and the manufacturing of abrasives, can eventually develop lung disease, which is usually named for the type of particulates involved. Such lung diseases include *"black lung"* (from coal dust), *silicosis* (from quartz dust that is generated during mining), *asbestosis* (from asbestos fibers), *"brown lung,"* or *byssinosis* (from cotton dust). Depending on the type of particulate and the concentrations inhaled, the lungs may suffer irritation, allergic reactions, or scarring of tissue, which becomes a potential site for tumor development. Typically, victims experience coughing and shortness of breath and, in the long run, may develop pneumonia, chronic bronchitis, emphysema, or lung cancer.

Lead is a particularly widespread and troublesome particulate. It is emitted into the atmosphere primarily through the exhaust of vehicles that burn gasoline which contains lead as an antiknock additive. The danger of lead is exacerbated by the fact that it accumulates in the body more rapidly than it is excreted. We know little of the toxic threshold of lead in humans, but we are well aware that lead will poison children who eat lead-based paint chips. Lead poisoning attacks the blood-forming mechanism, the gastrointestinal tract, and, in severe cases, the central nervous system. It may also impair the functioning of the heart and the kidneys.

Only 5 to 10 percent of lead that is ingested is retained by the human body, but 30 to 50 percent of lead that is inhaled is retained. At one time, scientists thought that the primary sources of our daily intake of lead were food and water—not the air that we breathe. But in the mid-1970s, the EPA compared the levels of lead in children who lived in Los Angeles with those who lived in a rural area, which showed that the opposite is true. The higher levels of lead found in the Los Angeles group apparently resulted from the fact that those children breathed more lead.

As discussed already, once a toxic substance is inhaled, the capacity of the mucus to remove those gases and particulates from the inhaled air can become limited as a result of saturation but, sometimes, other respiratory defenses are impaired by various inhaled agents as well. For example, some particles interfere with the cilia-mucus transport system, thereby increasing the amount of time that pollutants will remain in the respiratory tract. Such impairment may increase the likelihood of bronchial cancer if carcinogenic agents are retained in the lungs. Likewise, damage to macrophages may play a role in chronic lung diseases, such as emphysema and silicosis.

It is sometimes argued that no certain link has been demonstrated between air pollution and respiratory diseases, such as emphysema and chronic bronchitis. But symptoms of bronchitis and emphysema are similar to those attributed to inhalation of air pollutants, which suggests that those diseases may be at least aggravated by polluted air. In addition, the incidence of respiratory diseases is greatest in cities that have the highest levels of air pollution.

Still, direct evidence that unequivocally links elevated atmospheric levels of a specific air pollutant to a particular respiratory disease is unusual. It is extremely difficult to prove the existence of such a connection, since the urban–industrial atmosphere contains numerous gases and particulates whose relative concentrations continually fluctuate and whose interactions may be complex. Also, age, level of physical activity, and general health as well as the amount and frequency of exposure affect a person's reaction to polluted air. See Box 10.2 for more information about the problem of identifying a link between specific pollutants and illnesses.

Impact on Other Species

Experiments that involve laboratory animals contribute somewhat to our understanding of the effects of air pollutants on human health although, as noted in Box 10.2, there are some serious limitations. Very little is known about the impact of polluted air on the well-being of other animals, most of which do not inhabit urban–industrial areas. In general, it appears that the two air pollutants that cause the most damage to other species are fluoride and lead. In both cases, organisms are affected most commonly as a result of ingesting contaminated vegetation rather than by inhaling the pollutants.

The processing of ceramics and phosphate rock releases fluorides into the atmosphere. Some plant species are damaged by hydrogen fluoride at a concentration of only 0.1 part per billion (ppb). Others, including alfalfa and orchard grass, detoxify fluorides by combining them with organic acids, thus rendering them harmless to the plants themselves. Fluorides can reach concentrations of 40 ppm in those forage plants. However, when livestock consume those plants, the organic compounds that contain fluorides break down, and the fluorides that are released can be lethal.

Dairy cattle are most sensitive to fluoride poisoning, which is called *fluorosis*. Fluorides reduce their milk production and attack their bones, producing lameness. Chronic fluorosis eventually leads to severe emaciation and death. For example, substantial losses of cattle in Florida were attributed to fluoride emissions from factories that process phosphate deposits for fertilizers. In fact, in the middle and late 1950s, fluoride poisoning claimed the lives of 30,000 cattle in two Florida counties.

Fewer documented cases exist of lead poisoning in animals than of fluorosis, but the symptoms of this condition are no less debilitating. Afflicted animals lose their appetites, develop dry coats and muscle spasms, and frequently suffer paralysis of the hindquarters. Some zoo animals (especially members of the cat family) have been poisoned by their instinctive self-cleaning activities after their coats have been contaminated by airborne lead.

Some of the most dramatic instances of air-pollution damage to vegetation have been caused by sulfur dioxide fumes from iron and copper smelters. Those emissions have virtually eliminated vegetation over large areas that are located adjacent to

---

BOX 10.2

ASSAYING THE HAZARDS OF POLLUTANTS

A hint of the difficulty of relating specific adverse effects on health to specific pollutants is provided by the example of cigarette-smoking. From the analyst's point of view, assaying the health hazards of smoking should be far easier than is the case for most contaminants. In smoking, the source of the pollutant is relatively uniform in character (although the number of potentially hazardous substances in tobacco smoke is large), and the intensity and duration of exposure are known with reasonable accuracy for a large number of individuals. Even with these advantages, it took many years to do the research required to demonstrate convincingly a close association between smoking and heart disease and lung cancer. (Whether the smoking public even today is convinced is arguable).

The difficulties facing toxicologists and epidemiologists in most other cases are much greater. We summarize some of the most important of these difficulties here. (Some apply to the smoking problem, too, but many do not.)

1. Pollutants are numerous and varied, and many of them are difficult to detect. Their concentrations vary geographically. In many areas, techniques for monitoring pollutants are highly inadequate, and long-term records are unavailable. Yet long periods of study are usually needed to reveal delayed and chronic effects.

2. It is often impossible to determine with precision the degree of exposure of a given individual to specific pollutants.

3. Which pollutant among many candidates is actually responsible for the suspected damage is often unknown. In many if not most cases, the agent doing the damage is a "secondary" substance produced by the chemical or radioactive transformation of the "primary" pollutant produced by human activity.

4. The preceding difficulty is multiplied many times over by the importance of synergisms, wherein the effect of two agents acting in concert exceeds the sum of the effects to be expected if the two acted individually. One or both agents in a synergism might be secondary substances produced from pollutants, one might be a naturally occurring agent, or there might be more than two factors involved.

---

smelters (Figure 10.8). Also, alfalfa, lettuce, barley, and white pine are particularly sensitive to sulfur dioxide. A white pine, showing the effects of sulfur dioxide poisoning, and a healthy tree of the same species are shown in Figure 10.9.

Ozone is another air pollutant that is hazardous to plant life, when it occurs in relatively high concentrations. The ill effects of ozone have been exhibited by grapes, sweet corn, lettuce, and pine. For example, windborne ozone that originates in the photochemical smog that is produced by vehicular traffic in Los Angeles is destroying hundreds of thousands of ponderosa pines in the San Bernardino and San Jacinto Mountains, which are located 125 kilometers (80 miles) from the city.

Most commonly, air pollution damages the leaves of plants. The waxy layer that covers leaves severely limits the amount of water that is lost through transpiration. However, the surfaces of leaves are also equipped with tiny pores, which allow carbon dioxide to enter the plant for photosynthesis. But the open pores also allow a considerable amount of water vapor to escape from the leaves. When transpiration losses exceed the absorption of water by roots, as often happens on a hot, dry, windy day, those pores may close, which greatly reduces both the loss of water vapor and the entry of carbon dioxide and other gases into the leaf. When enough water is available to the plant, the pores open to permit

Figure 10.8 *The Wawa, Ontario, iron-sintering plant, whose emissions have destroyed the forest downwind for 8 kilometers. Severe damage to vegetation extends at least an additional 8 kilometers downwind. (National Film Board of Canada. Photo by George Hunter.)*

5. Persons being studied for effects of pollutants may differ in many respects besides their degree of exposure to the suspected agent—for example, in socioeconomic class, sex, age, physiological susceptibility to certain contaminants, occupational exposure to pollutants, and smoking habits. Variation of many factors at once in the sample population makes it easy to be misled about what is causing what.

6. Official records of incidence of disease and causes of death are somewhat unreliable, for two main reasons: first, diagnostic procedures, nomenclature, and the knowledge of the average physician vary with time, and probably, to some extent, regionally as well; second, final causes of death are often different from the malady that brought death near, and the primary maladies are often not listed on the death certificate.

7. Controlled experiments with laboratory animals are of limited usefulness because of differences in physiology between the experimental animals and humans. Effects with a long latency period between insult and symptoms may not appear at all in animals whose natural lifespan is short. For these reasons, the extent to which animal data can be used to infer effects of pollutants on humans is often controversial. The animals that most resemble humans physiologically—the large primates—are so expensive (and, in some cases, endangered as species) that experiments involving substantial numbers are impracticable.

8. Very large sample populations (of animals or people) are needed to investigate effects at low doses of pollutants, where only a small percentage of those exposed will manifest the suspected effect even if the cause-effect relationship is genuine. It cannot be assumed that the dose-response relation is linear with dose, that is, that effects occur in the same direct proportion to dose from low doses to high ones. For some agents, low doses may produce more damage per unit of exposure than high doses; for others, there may be a threshold in dose below which no ill effect occurs. It has been customary in regulation of pollution to assume that evidence obtained at high doses can be extrapolated linearly to low ones; this will be overcautious in some cases and not cautious enough in others.

Excerpted from P. R. Ehrlich et al., Ecoscience, Population, Resources, Environment. New York: W. H. Freeman, 1977, p. 548.

(a)

(b)

Figure 10.9 *(a) A 10-year-old white pine that was severely dam-aged by relatively low concentrations of sulfur dioxide (possibly mixed with ozone). (b) A healthy white pine from the same area, grown in filtered air. (U.S. Department of Agriculture.)*

carbon dioxide to enter and water vapor to escape. Under those circumstances, gaseous air pollutants may enter the pores along with the carbon dioxide.

When gaseous pollutants, such as sulfur dioxide and ozone, enter leaves, they dissolve in the water that adheres to the cell-wall surfaces. At that point, a variety of complex, poorly understood chemical reactions take place. The first sign that a plant has suffered air pollution damage is the appearance of *chlorosis*, which is a yellowing of the leaves as a result of chlorophyll loss. That condition is illus-trated in Figure 10.10. The pattern of leaf damage can sometimes be used to identify the pollutant re-sponsible. For example, fluorides characteristically accumulate at the tips and edges of leaves, which turns the tips and edges yellow. When pollution damage is extreme, the plant tissues die and its leaves turn brown (which is called *necrosis*). When enough leaves die, photosynthesis and plant growth are re-tarded. When respiration exceeds photosynthesis, the plant's energy reserves are used up, and the plant

Figure 10.10 *Sulfur dioxide can destroy the chlorophyll-bearing cells of leaves. The affected leaves develop white areas in be-tween the veins. The leaf on the right is healthy. (U.S. Depart-ment of Agriculture).*

dies. Although gases can enter leaf pores readily, particulates are usually too large to gain access, so they have few adverse effects. However, in some cases, particulates may coat the surfaces of leaves, which reduces the amount of sunlight that is available to the plant for photosynthesis.

The tolerance of vegetation to air pollution varies considerably with the species of plants and the types of pollutants that they encounter. The susceptibilities of various plants to selected pollutants are shown in Table 10.3. Scientists do not know why plants vary in their susceptibility to gaseous pollutants but, in a few cases, they have suggested explanations. For example, they have found that in varieties of onions that are resistant to ozone pollution, the leaf pores are sensitive to ozone and close in its presence, thereby protecting the interior of the leaf.

Threats to the Ozone Shield

In general, residents of urban–industrial areas run the greatest risk of adverse health effects as a result of polluted air. But some pollutants threaten the well-being of all living things everywhere, because they threaten the delicate atmospheric ozone layer that

shields organisms from dangerous ultraviolet radiation.

Ironically, although ozone (O_3) is a pollutant in the lower atmosphere, its presence in minute concentrations (less than 12 ppm) in the stratosphere is essential for the continuation of life on earth. Most atmospheric ozone forms at altitudes that are between 15 and 35 kilometers (9 and 22 miles), when oxygen absorbs ultraviolet radiation (UV). Then the ozone also absorbs UV. As a result of the absorption of UV by both oxygen and ozone, organisms are protected from exposure to potentially lethal intensities of that radiation (hence the term *ozone shield*).

In the longer-wavelength region of the ultraviolet-radiation band, which is called UV-B, absorption by oxygen and ozone is only partial and is very sensitive to changes in ozone concentration. UV-B radiation is responsible for sunburn and either causes or contributes to skin cancer. Scientists distinguish among three forms of skin cancer: basal cell carcinoma, the more serious squamous-cell carcinoma, and, most virulent of all, malignant melanoma. Malignant melanoma afflicts about 15,000 Americans each year and eventually kills approximately one-third of its victims. There is a real possibility that the disruption of UV-absorption processes will de-

Table 10.3 Susceptibility of selected plants to air pollutants

PLANT	SULFUR DIOXIDE	OZONE	FLUORIDE	PEROXYACETYL NITRATE (PAN)	NITROGEN DIOXIDE
Alfalfa	S	I	R	I	
Cabbage	I	R	R		
Citrus			I		I
Corn (Sweet)	R	S	S	R	
Grape	I	S	I		
Lettuce	S	S	I	S	S
Onion	R	I	I	R	
Pine	S	S	S		
Potato	R	S	I		
Ragweed	S	R	R		
Rhubarb	S	I	R		
Rose	R		I		
Soybean	S		R	I	
Sunflower	S	R	I		S

Source: After A. C. Stern et al., *Fundamentals of Air Pollution*. New York: Academic Press, 1973.
S = sensitive; I = intermediate; R = resistant.

plete stratospheric ozone, thereby increasing the intensity of UV-B that reaches the ground, which will raise the incidence of skin cancer.

In the early 1970s, scientists were concerned that oxides of nitrogen, which would be released in the exhaust of a proposed fleet of supersonic aircraft, might react with and reduce stratospheric ozone levels. Around the same time, some concern was voiced about the extensive use of nitrogen fertilizer, which, it was supposed, could drift into the stratosphere and destroy ozone. However, subsequent research indicates that the net effect of aircraft exhaust and nitrogen fertilizers is negligible on stratospheric ozone concentrations.

Today, many scientists believe that chlorofluorocarbons (CFCs), which are commonly called *Freons*, comprise the most serious threat to the ozone shield. Most of us have used two CFCs, knowingly or unknowingly, in our daily lives. One was used until late 1978 as a propellant in common household aerosol sprays, such as deodorants and hairsprays; the other is still widely used as a coolant in refrigerators and air conditioners. Both those substances are relatively inert (chemically nonreactive) in the troposphere, where they have been accumulating for many years. But when CFCs enter the stratosphere, they break down chemically, releasing chlorine, which can react with and destroy ozone.

Reports on the potential health effects of ozone depletion have been ominous. In 1982, the National Research Council (NRC) of the National Academy of Sciences projected that if CFC emissions continued at the 1977 rate, they eventually could reduce stratospheric ozone concentrations by 5 to 9 percent. Also, for every 1-percent reduction in stratospheric ozone, the amount of UV that reaches the earth's surface increases by as much as 2 percent. Thus the NRC estimated that every 2-percent increase in UV will raise the incidence of basal-cell skin cancer by 2 to 5 percent and the incidence of squamous-cell skin cancer by 4 to 10 percent. Currently, the combined incidence of those diseases in the United States is 400,000 to 500,000 per year. Because the factors that are involved in the development of malignant melanoma are so complex, the NRC has been unable to project the effect of ozone depletion on the incidence of that most serious form of skin cancer.

In early 1984, after careful evaluation of available evidence, the NRC backed off from its earlier estimates and predicted that CFCs would reduce ozone levels by only 2 to 4 percent by the year 2100. Their report also stated that there was "no discernible" change in stratospheric ozone levels between 1970 and 1980. Further, atmospheric ozone levels could actually increase as a result of photochemical reactions involving motor-vehicle exhaust.

Clearly, more research and monitoring is needed to resolve questions concerning threats to the ozone shield. Although the potential health effects are our primary concern, the effects of more intense ultraviolet radiation on vegetation and climate also need to be addressed.

Air-Pollution Episodes

On the morning of October 26, 1948, a fog blanket that reeked of pungent sulfur-dioxide fumes spread over the town of Donora in Pennsylvania's Monongahela Valley. Before the fog lifted 5 days later, almost half of the area's 14,000 inhabitants had fallen ill, and 20 of them had died. That killer fog resulted from the combination of mountainous topography and stable weather conditions that trapped and concentrated deadly effluents from the community's steel mill, zinc smelter, and sulfuric acid plant.

Air pollutants are especially dangerous when atmospheric conditions cause them to become concentrated. When they are emitted into the atmosphere, their concentrations begin to decline at a rate that is partially determined by the extent to which they mix with cleaner air. The more thorough the mixing, the more rapid the *dilution*. When conditions in the atmosphere favor rapid dilution, the impact of pollutants is usually minor. However, on other occasions, which are called *air-pollution episodes*, conditions in the atmosphere minimize dilution, and the impact can be severe, particularly on human health. The two weather conditions that have the most influence on the rate of dilution of air pollution are wind speed and air stability.

Wind Speed

Intuitively, we know that air is likely to mix more vigorously on a windy day than on a calm one. As a general rule, a doubling of wind speed will reduce the concentration of air pollutants by half (see Figure 10.11). Since certain weather patterns favor light

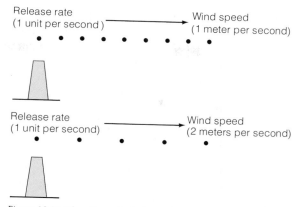

Figure 10.11 *The effect of wind speed on air-pollution concentrations. A doubling of wind speed from 1 meter per second to 2 meters per second increases the spacing between puffs of smoke by a factor of two, thereby reducing concentrations by one-half.*

winds, they inhibit the dispersal of contaminants. For example, within a high-pressure system, the wind is very light or calm near the center, so pollutants do not readily disperse.

Wind speed is also influenced by friction. For example, in a city, winds are slowed by the rough surface that is created by the canyonlike topography of tall buildings and narrow streets. In fact, average wind speeds may be 25 percent lower in a city than in the surrounding countryside. When light regional winds (less than approximately 15 kilometers, or 10 miles, per hour) prevail, the contrast is even more pronounced, which amounts to a wind-speed reduction of up to 30 percent. Hence the dilution of air pollutants by wind is particularly impeded in urban areas—the very places where most of the contaminants are generated.

Air Stability

Air stability determines how rapidly polluted air is dispersed vertically within the atmosphere. When the air is *unstable*, vertical (both up and down) motion is enhanced and polluted air is diluted. But when the air is *stable*, vertical motion is suppressed and polluted air maintains its concentration.

Thus, for convenience, to see the influence of air stability on pollutant concentrations, we distinguish between polluted air and the relatively clean air into which polluted air is emitted. We refer to

polluted air as *air parcels,*[*] which are discrete volume units, whereas we describe clean air simply as surrounding, or *ambient*, air in the following discussion.

Suppose that a parcel of air, which is polluted, is released, say, from a smokestack. If it is warmer than the ambient air, it rises. Such a parcel is buoyant with respect to the surrounding air. Air pressure on a rising air parcel decreases steadily, so that the parcel expands (the way a helium-filled balloon expands) as it ascends (Figure 10.12). But during the process of expansion, while it is rising, the parcel works against its surroundings. As a result of that work, heat energy is expended, which causes the parcel's temperature to drop. That process is known as *expansional cooling*. A parcel of air normally cools at a rate of about 9.8°C per 1000 meters (about 5.5°F per 1000 feet) of lifting (as long as condensation does not take place. See Chapter 6).

If the vertical temperature profile of the ambient air is such that the temperature at every altitude (within the ambient air) is lower than the temperature of the air parcel that is rising through it (see Figure 10.13), the air parcel will continue to rise.

---

[*] Meteorologists often use the term *air parcel* when referring to a unit mass of air, clean or polluted.

Figure 10.12 *A parcel of air expands and cools at the constant rate of 9.8°C per 1000 meters (5.5°F per 1000 feet) as it ascends in the atmosphere.*

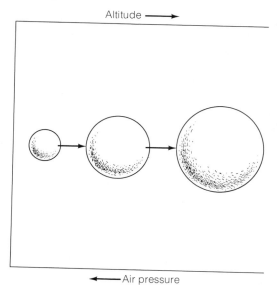

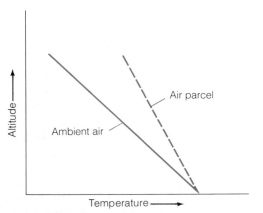

Figure 10.13 *Unstable air favors dilution of air pollutants. Air parcels cool at the rate indicated by the dashed line. The temperature profile of the ambient air is shown by the solid line. Air parcels are warmer (and hence, lighter) than the ambient air, so they are buoyed upward.*

In such a situation, the ambient air is said to be *unstable*. If, on the other hand, an air parcel enters ambient air whose temperature at every altitude is higher than the temperature of the air parcel (see Figure 10.14), the parcel will lose its buoyancy, and it will begin to sink. In that case, the ambient air is said to be *stable*. Thus a parcel of air undergoes more mixing when it is emitted into unstable air than when it is emitted into stable air. For this reason,

Figure 10.14 *Stable air inhibits dilution of air pollutants. Air parcels cool at the rate that is indicated by the dashed line. The temperature profile of the ambient air is shown by the solid line. Air parcels are cooler (and hence, heavier) than the ambient air, so upward motion is inhibited.*

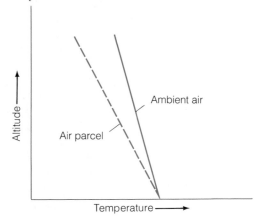

stable air inhibits the transport of air pollutants and may even act as a lid over the lower atmosphere to trap air pollutants. Continual emission of contaminants into stable air results in the continual accumulation and concentration of pollutants at lower altitudes.

The *mixing depth* of air is the vertical distance between the ground and the altitude to which convection and hence, mixing, extends. When mixing depths are great (thousands of meters), the relative abundance of clean air allows pollutants to mix and dilute rapidly. However, when mixing depths are shallow, air pollutants are restricted to a smaller volume of air and concentrations may approach hazardous levels. In addition, when the air is stable, convection is suppressed and mixing depths are low. But when the air is unstable, convection is enhanced and mixing depths increase. Therefore, since solar heating triggers convection, mixing depths tend to be greater during the afternoon than in the morning, during the day than at night, and during the summer than in the winter.

We can sometimes estimate the stability of air layers by observing the behavior of a plume of smoke that belches from a stack. If the smoke enters an unstable air layer, the plume undulates downwind, as shown in Figure 10.15. In general, this type of plume behavior indicates that polluted air is mixing readily with the surrounding cleaner air, thereby facilitating dilution. The net effect is improved air quality (except if and where the plume loops to the ground). On the other hand, a plume of smoke that flattens and spreads slowly downwind, as shown in Figure 10.16, indicates very stable conditions and minimal dilution.

Figure 10.15 *During unstable atmospheric conditions, a smoke plume undulates downwind as it readily mixes with the ambient air.*

Figure 10.16 *During stable atmospheric conditions, a smoke plume forms a thin ribbon that drifts downwind; mixing and dilution is minimal.*

Air stability, then, influences the rate at which polluted air and clean air mix. Stable air layers inhibit dilution; unstable air layers promote dilution.

Temperature Inversions

An air-pollution episode is most likely to occur when a persistent temperature inversion develops. In a *temperature inversion*, the air temperature increases with altitude, that is, warm, light air overlies cooler, denser air—an extremely stable stratification that strongly inhibits convective mixing and dilution (Figure 10.17). A temperature inversion can form by (1) subsidence of air, (2) extreme radiational cooling, or (3) flow of air masses. The resulting inversion may develop aloft or at the earth's surface.

A *subsidence temperature inversion* forms a lid over a wide area, often encompassing an area of sev-

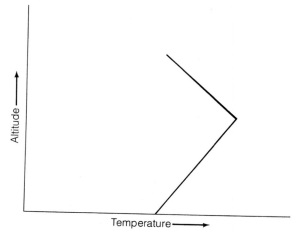

Figure 10.17 *A temperature inversion within a surface air layer, which indicates very stable conditions.*

eral states at one time. It develops during a period of fair weather when the hemispheric weather pattern causes a warm high-pressure system to stall. A high-pressure system is characterized by descending, or *subsiding*, and compressionally warmed air currents that spread out toward the earth's surface. However, the warm air is prevented from reaching the surface by the *mixing layer* in which the air is thoroughly mixed by convection (Figure 10.18). (The thickness of the mixing layer is the *mixing depth*, which was described earlier.) Air temperatures within the mixing layer decrease with altitude, but air that lies just above the mixing layer—having been warmed by compression—is significantly warmer than the air at the top of the mixing layer. Hence an elevated temperature inversion separates the mixing layer from the compressionally warmed air that lies above it. Under those conditions, as shown in Figure 10.18, pollutants are distributed throughout the mixing layer up to the altitude of the temperature inversion. This situation is sometimes referred to as *fumigation*.

Radiational temperature inversions are perhaps more common, and are often more localized, than subsidence temperature inversions. At night, under clear skies, the earth's warmth is rapidly lost to space through infrared radiation. That process (which was discussed in Chapter 9) is called *radiational cooling*, which causes surface air layers to become chilled by contact with the cooler ground. Then, because the air at the surface is coldest, a *surface temperature inversion* develops. Smoke that is emitted into such an air layer does not readily disperse, thereby reducing local air quality (Figure 10.19). Typically, by mid to late morning, as solar radiation is absorbed by the ground and the lower troposphere is heated, the inversion gradually disappears and a normal temperature profile is restored. During the winter, however, when snow covers the ground and the sun's rays are weak, a radiation inversion may persist for several days or even weeks at a time, inhibiting the dispersal of air pollutants.

Air-mass movement also can give rise to temperature inversions, which occur, for example, at the foot of the Rocky Mountains on occasion. As shown in Figure 10.20, a westerly air flow is compressionally warmed as it is drawn down the leeward side of the mountain range. But, at the foot of the range, northerly surface winds bring in cold air. As a result, an elevated temperature inversion occurs between the cold air at the surface and warmer air

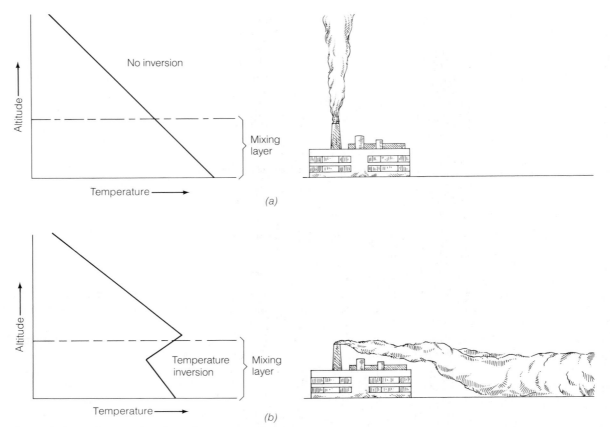

Figure 10.18 *Development of an elevated temperature inversion due to the subsidence of air over the mixing layer. (a) Temperature profile prior to subsidence. (b) Temperature profile during subsidence. The elevated temperature inversion acts as a lid over the lower atmosphere, trapping smoke in the mixing layer.*

aloft. Denver, situated just to the east of the Rocky Mountain Front Range is particularly susceptible to this type of temperature inversion and, consequently, the air quality in the city is adversely affected.

Potential for Air Pollution

Weather conditions that favor air stagnation, and hence air pollution, occur with varying frequency in different places and at different times of the year. Areas that have particularly high potential for air pollution include southern and coastal California, portions of the Rocky Mountain states, and the mountainous portions of the mid-Atlantic states. In general, in much of the West, the air quality is low-

est during the winter, while in the East, the air quality is lowest during autumn. In southern California, the potential for air pollution is highest during the summer. Those seasonal changes in air quality are caused primarily by normal seasonal shifts in atmospheric circulation patterns.

Many regions with high potential for air pollution tend to have great topographic relief, since hills and mountain ranges can block horizontal winds that could disperse polluted air. In addition, radiation temperature inversions that form in lowlands, such as river valleys, are often strengthened by an accumulation of cold, dense air that drains downward from nearby highlands. As illustrated in Figure 10.21, the result is a persistent stratification of mild, light air over cool, dense air.

Los Angeles is particularly susceptible to air-pollution episodes because of its topographic setting, high concentration of pollutant sources (more than 8 million cars), and frequent periods of air stability. Figure 10.22 shows the air circulation and

Figure 10.19 *The behavior of smoke emitted into the air that is characterized by a surface temperature inversion. (Bruce Roberts, Rapho/Photo Researchers.)*

Figure 10.20 *An elevated temperature inversion develops on the leeward (eastern) side of the Rocky Mountains when compressionally warmed air moves in over colder air that is flowing along the base of the mountain range.*

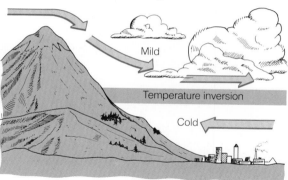

topographic features that influence the air quality in that city. The weather in Los Angeles, like that through much of California, is strongly influenced by the eastern edge of the semipermanent Pacific high-pressure system. That high-pressure system is responsible for California's famous fair weather as well as for descending air currents that are aloft, which generate a subsidence temperature inversion at approximately 700 meters (2300 feet) over Los Angeles. Consequently, low mixing depths occur on about two-thirds of the days throughout the year. In addition, the exceptionally high incidence of extremely stable atmospheric conditions in Los Angeles is aggravated by its topographic barriers. The city is situated on a plain that opens to the Pacific and is rimmed on three sides by mountains. Hence

347

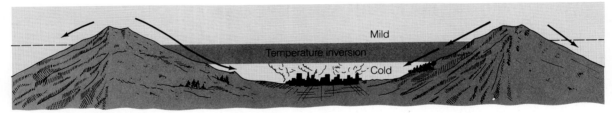

Figure 10.21 *At night, under the influence of gravity, cold air flows downhill, strengthening temperature inversions in valleys.*

cool breezes that sweep inland from the ocean are unable to flush the pollutants out of the city, and the mountains and an elevated temperature inversion encase the city in its own fumes. As a result, a complex photochemistry occurs to produce smog within this crucible.

Heat-Island Circulation

Large metropolitan areas typically exhibit a higher potential for air pollution than rural areas. For one thing, cities usually contain a higher concentration of pollution sources. For another, we have seen how the surface roughness of cities slows wind speeds, which impedes the dispersal of air pollutants. Large cities may have other effects on the circulation of air as well. For example, travelers who approach large cities often are greeted by an unsightly veil of dust, smoke, and haze, which can result from a convective circulation of air that concentrates pollutants over the city and is linked to a temperature contrast between the city and its surrounding rural areas.

The average annual temperature of a city is typically slightly higher than that of the surrounding countryside, but on some days it may be as much as 10°C (18°F) or more warmer. Consequently, snow melts faster and flowers bloom earlier in a city. That climatic effect is known as the *urban heat island*. Figure 10.23 illustrates the urban heat island of Washington, D.C., and Figure 10.24 is a satellite view of heat islands in southeastern New England. One factor that contributes to the development of an urban heat island is the relatively high concentration of heat sources (for example, people, cars, industry, air conditioners, and furnaces). Because all heat from every source eventually escapes to the atmosphere,

Figure 10.22 *The atmospheric circulation and topographic features that give Los Angeles an unusually high air-pollution potential.*

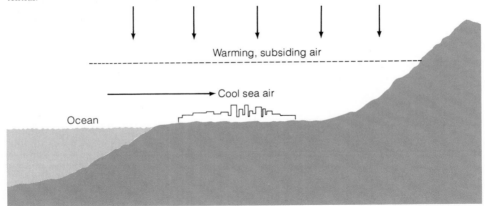

the air of a large city receives a considerable input of waste heat. In fact, on a very cold winter day in New York City, heat from urban sources may approach 100 watts per square meter, which is approximately 3 percent of the solar constant (Chapter 9).

The daytime absorption of solar energy in urban areas and the emission of heat into city air is facilitated by the thermal properties of urban building materials. For example, concrete, asphalt, and brick absorb radiant energy and conduct heat more readily than do the soil and vegetative cover of rural areas. Thus the heat loss at night by infrared radiation to the atmosphere and space is partially compensated for in cities by a release of heat from buildings and streets. The obstruction of the sky by tall buildings also blocks infrared cooling. The temperature contrast is further accentuated by a typically low rate of evapotranspiration. Urban drainage systems quickly and efficiently remove most runoff, so less of the available solar energy is used for vaporization of water (latent heating). Hence more solar energy is available to heat the ground and the air directly (sensible heating).

Figure 10.23 *Average winter low temperatures (in degrees Fahrenheit) in Washington, D.C., illustrating the urban heat-island effect. (From C. A. Woolum, Notes from a study of the microclimatology of the Washington, D.C. area for the winter and spring seasons. Weatherwise 17:6, 1964.)*

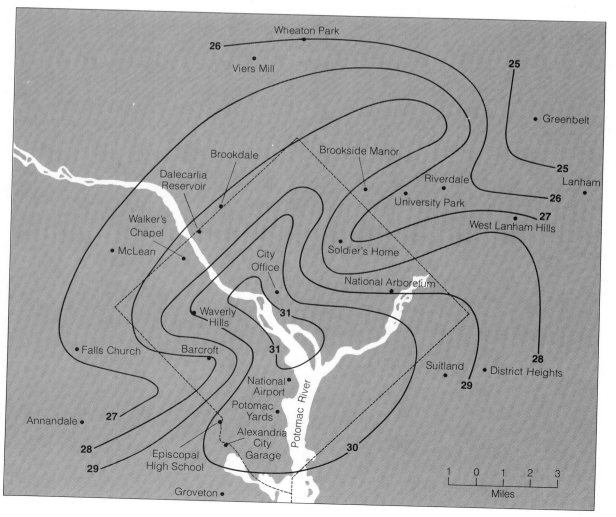

Figure 10.24 *A satellite image of 17 urban heat islands in south-eastern New England. Infrared sensors aboard the satellite detect areas of relative warmth that appear as light areas on this photograph. Warm areas on Cape Cod and the adjacent islands of Martha's Vineyard and Nantucket probably result from land-use patterns or soil type. (Courtesy of Michael Matson, Environmental Sciences Group, National Environmental Satellite Service, NOAA, Washington, D.C.)*

When regional winds are weak, the relative warmth of a large city, compared with its surroundings, can promote a convective circulation of air, which is shown in Figure 10.25. Warm air at the center of the city rises and is replaced by cooler, denser air that flows in from the countryside, and the rising columns of air gather particulates into a *dust dome* over the city. As a result, dust may be a thousand times more concentrated over urban–industrial areas than in the air of the open countryside. If regional winds strengthen to more than approximately 15 kilometers (10 miles) per hour, the dust domes elongates downwind in the form of a *dust plume*, which spreads the city's pollutants over the nearby countryside. For example, the Chicago dust plume is sometimes visible from the ground 240 kilometers (150 miles) from its source.

Natural Cleansing Processes

Conditions that favor the accumulation and concentration of pollutants in the air are countered to some extent by natural removal mechanisms. For example, some particulates are removed from the air when they strike and adhere to buildings and other structures by a process that is called *impaction*. Aerosols are also subject to *gravitational settling*, especially if they have radii that are greater than 1 tenth of a micrometer. Obviously, heavier (and larger) aerosols settle more rapidly than do smaller aerosols. Hence larger aerosols tend to settle out near their sources, whereas smaller aerosols may be carried for many kilometers and to great altitudes before they finally settle to the ground. Sometimes, impaction and

Figure 10.25 *An urban heat island may give rise to a convective circulation pattern that transports air pollutants aloft to create a dust dome over the city. (Modified after W.P. Lowry, "The Climate of Cities."* © *Scientific American 217:20, 1967. All rights reserved.)*

gravitational settling together are referred to as *dry deposition*.

The most effective natural pollution-removal mechanism is *scavenging* by rain and snow. In fact, in regions that experience moderate precipitation, as much as 90 percent of the aerosols are removed by scavenging. Although gaseous pollutants are somewhat less susceptible to scavenging than are aerosols, they do dissolve to some extent in raindrops and cloud droplets. Although air-pollutant scavenging enhances air quality, it degrades the quality of precipitation—sometimes to the point of polluting surface water so much that aquatic life is threatened. (That effect in the form of acid rain is discussed later in this chapter.)

The Impact of Air Pollution on Weather and Climate

Certain air pollutants that are usually found in urban air, including a variety of dust particles and acid droplets, can influence the development of clouds and precipitation within and downwind from the city. Those pollutants, many of which are hygroscopic, serve as nuclei for cloud droplets, thereby accelerating condensation (discussed in Chapter 6). The tendency for more frequent cloudiness and precipitation in and near urban–industrial areas than in rural areas is enhanced also by the rising of warm air over the urban heat island, which helps to lift and cool the air to the point of saturation.

The influence of urban air pollution on condensation and precipitation is illustrated by typical climatic contrasts between urban and rural regions. For example, winter fogs occur about twice as frequently in cities than they do in the surrounding countrysides, and downwind from cities, rainfall may be enhanced by 5 to 10 percent. Also, greater contrasts in the amount of rainfall tend to be exhibited on weekdays, when urban–industrial activity is at its peak, which suggests that increased precipitation is at least partially attributable to urban–industrial air pollutants.

Data from the Metropolitan Meteorological Experiment (*METROMEX*) indicate a significantly greater enhancement of precipitation downwind of St. Louis. During a 5-year intensive field study, METROMEX scientists analyzed weather observations and concluded that downwind of St. Louis,

Figure 10.26 *Contrails that are produced by the exhaust of jet aircraft add appreciably to the cloud cover over some regions of the United States. (Photo Researchers.)*

summer rainfall was up to 30 percent greater than upwind of the city. That rainfall anomaly was attributed to the combined effect of urban contributions of heat and "giant" cloud condensation nuclei that were released by various urban activities.

Because precipitation, fog, and cloudiness in urban areas often have adverse effects on both surface and air transportation, any artificial increase in them is potentially troublesome. For example, reduced visibility slows surface traffic, curtails air travel, and contributes to automobile accidents.

Also, jet-airplane traffic is significantly modifying the cloud cover, especially along heavily traveled air corridors between major cities. The visible jet contrails that etch the sky, such as those shown in Figure 10.26, are feathery cirrus clouds that can be traced to water vapor and condensation nuclei that are produced as combustion products in jet engines. Increased cloudiness, in turn, reduces sunshine penetration and may enhance local precipitation by serving as a source of ice-crystal nuclei.

Acid Rain

As we saw in Chapter 6, the atmospheric subcycle of the water cycle purifies water. As rain and snow fall from clouds to the ground, however, they wash pollutants from the air and, in the process, take up pollutants. Rainfall is normally slightly acidic because it dissolves some atmospheric carbon dioxide, which produces weak carbonic acid (H_2CO_3). But in regions where the air is polluted with oxides of sulfur and oxides of nitrogen, rainfall produces relatively strong sulfuric acid (H_2SO_4) and nitric acid (HNO_3). Thus precipitation that falls through such contaminated air may become 200 times more acidic than normal. (Dry deposition also delivers acidic materials to the earth's surface.)

The range of acidity and alkalinity, called the *pH scale*, is shown in Figure 10.27, which compares the normal acidity of rainwater with pH values of some other familiar substances. Note that the pH scale is *logarithmic*; that is, each unit increment corresponds to a tenfold change in acidity. Hence a drop on the pH scale from the normal value of rain—5.6 down to 3.6—indicates a 100-fold increase (10 times 10) in acidity or its equivalent, which is a 100-fold decrease in alkalinity.

When acid rain falls on soils or rock that cannot neutralize the acidity, regional surface waters become more acidic. For example, regions in North America in which the lakes are sensitive to acid precipitation are shown in Figure 10.28. The acids disrupt the reproductive cycles of fish, and acid rains leach heavy metals from soils, washing them into lakes and streams where they may harm fish, aquatic plants, and microorganisms. The fish populations in many lakes and streams in Norway, Sweden, eastern Canada, and the northeastern United States have declined or have been eliminated because acid rains have raised the acidity of their aquatic habitats. A

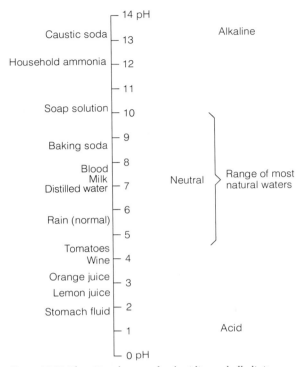

Figure 10.27 *The pH scale, or scale of acidity and alkalinity, and the pH of some common substances. It is a logarithmic scale, which means that an increase or decrease of 1 on the scale represents a tenfold increase or decrease in pH.*

Figure 10.28 *Areas in North America where lake waters are particularly susceptible to acidification. These areas lack natural buffers (neutralizers) in the soil or bedrock. (From J. N. Galloway and E. B. Cowling, The effects of precipitation on aquatic and terrestrial ecosystems: A proposed precipitation chemistry network. Reprinted with permission from the* Journal of the Air Pollution Control Association *28:233, 1978.)*

1975 Cornell University study of 214 high-altitude lakes in the Adirondack Mountains of New York found that more than one-third of the lakes contained no fish. The low pH values of their waters exceeded the tolerance limits of the fish. And Canada's Department of the Environment reports that salmon have been eliminated from nine Nova Scotia rivers in which pH values have fallen below 4.7. Furthermore, a 1979 study in the Catskill Mountains of New York reports that the pH of precipitation is as low as 2.9.

Recent studies have suggested that acid rains may be involved in the decline and dieback of conifer forests in West Germany and in upstate New York and northern New England. Another costly impact of acid rains is the accelerated weathering of building materials, especially limestone, marble, and concrete. Metals, too, corrode at faster than normal rates when they are exposed to acidic moisture.

Gene E. Likens of Cornell University and his associates reported an increase in rainfall acidity over the eastern United States between 1955 and 1973. Their findings were later confirmed and updated by measurements that were made by the National Atmospheric Deposition Program in the United States and the Canadian Network for Sampling Precipitation (Figure 10.29). During the summer of 1983, the National Research Council of the National Academy of Sciences issued a long-awaited report on the problem, which attributes 90 to 95 percent of acid rains in the Northeast to industrial effluents and motor-vehicle exhaust. Coal-burning electric power plants in the Midwest may be the chief emitters of acid-rain precursors.

Figure 10.29 *The acidity of precipitation over North America. Figures are average annual pH of precipitation for 1982. (From data developed by the National Atmospheric Deposition Program in the United States and the Canadian Network for Sampling of Precipitation.)*

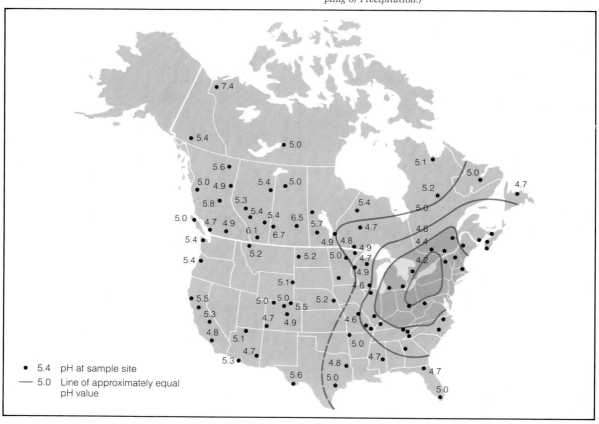

• 5.4 pH at sample site
— 5.0 Line of approximately equal pH value

Winds aloft can transport oxides of sulfur and nitrogen thousands of kilometers from tall stack sources and, as a result, acid rain is becoming a global problem. In fact, acid rains have been reported from such isolated regions as the Hawaiian Islands and the central Indian Ocean. The long-range transport of the precursors of acid rain has even strained the traditionally amiable relationship between the United States and Canada.

Canadian scientists are concerned about southerly winds that transport pollutants from the United States into Canada, which are causing acid rains in Canada, threatening its primary industries (lumber and fisheries). Perhaps half of the acidic rainfall in Canada originates from the emissions of industries and power plants in the Ohio Valley. Thus Canadian government officials are pressing their Washington counterparts to enact stricter controls on polluting industries and power plants to ease the problem, which some Canadians feel will become severe by the close of this decade. In response, in late 1981, five joint United States–Canadian research groups were appointed to study the problem, which is a first step in negotiating an air-quality treaty between the two nations.

Several possible solutions exist to alleviate the acid-rain problem. One possibility is to follow the example of Sweden, which spends $40 million annually to add ground limestone ($CaCO_3$) to acid-degraded lakes. The limestone neutralizes the acidic waters, allowing the lakes' fish populations to return to their former normal levels. But a more fundamental approach is to reduce the emissions of acid-rain precursors, which are principally oxides of sulfur. But that is no minor task, considering the tremendous costs that are involved and projections which show that the United States will triple its coal burning by 1995. And, for the world as a whole, the use of coal is expected to continue to climb—a trend that undoubtedly will increase sulfur oxide emissions and acidic precipitation worldwide.

Until recently, some researchers have argued that the reduction of sulfur emissions from industrial sources might not significantly alleviate the acid-rain problem. After all, they pointed out, human activities only account for about half of the total atmospheric sulfur. But, in 1983, the National Research Council reported substantial evidence that a reduction in industrial sulfur emissions would result in a proportional drop in acid rainfall.

Large-Scale Trends in Climate

In the late 1960s, the early days of the "environmental movement," many persons thought that air pollution was somehow responsible for the variations in hemispheric mean temperature that have been observed over the past 100 years (refer back to Figure 9.27). In fact, in one popular hypothesis, the warming trend of the first part of this century was attributed to rising levels of carbon dioxide, and the subsequent cooling episode was ascribed to a cooling effect of increasing atmospheric *turbidity*, or dustiness, which, by the late 1930s, supposedly had overwhelmed the carbon dioxide effect. Today, after more than a decade of extensive research on this issue, there is still no consensus of opinion about the effect—if any—of human activity on past variations in hemispheric-scale climate.

In terms of the future, many scientists fear that rising levels of atmospheric carbon dioxide eventually will trigger important changes in global climate. Some are also concerned that reductions in stratospheric ozone levels and increased atmospheric turbidity could, as time goes on, measurably alter the global climate.

Carbon Dioxide and Climate It is generally believed that higher concentrations of carbon dioxide (CO_2) intensify the "greenhouse effect," that is, increase absorption and reradiation of infrared radiation and, consequently, warm the lower atmosphere. Actually, that idea was proposed nearly 100 years ago by two scientists independently: Thomas C. Chamberlin, an American geologist, and Svante Arrhenius, a Swedish chemist.

The burning of fossil fuels (coal and oil, particularly) and wood and, to a much lesser extent, the clearing and burning of tropical forests (which releases the carbon that has been stored in trees) are responsible for the increased concentration of atmospheric carbon dioxide. Although rising levels of carbon dioxide were first reported in 1939 by G. S. Callendar, a British engineer, the increase apparently was initiated during the Industrial Revolution. Callendar suggested that the rising concentration of carbon dioxide was responsible for the global warming trend of the prior 60 years.

Systematic measurements of atmospheric levels of carbon dioxide began in 1958 at the Mauna Loa Observatory in Hawaii and at the South Pole

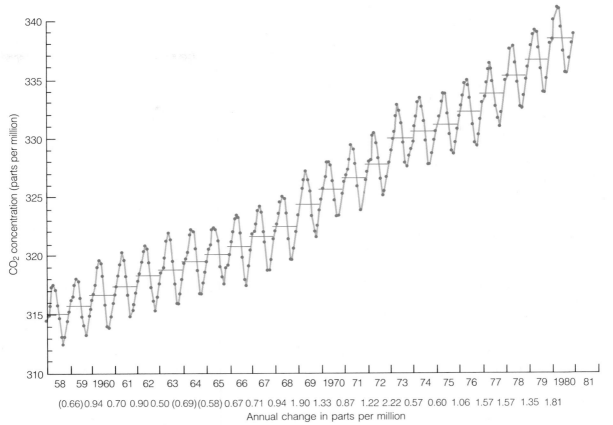

(0.66) 0.94 0.70 0.90 0.50 (0.69) (0.58) 0.67 0.71 0.94 1.90 1.33 0.87 1.22 2.22 0.57 0.60 1.06 1.57 1.57 1.35 1.81
Annual change in parts per million

Figure 10.30 *Recent upward trend in atmospheric levels of carbon dioxide as recorded at the Mauna Loa Observatory, Hawaii. The annual cycle of carbon dioxide, due to seasonal changes in vegetation, is superimposed on the steady upward trend. This trend is worldwide. (From C. D. Keeling et al., Scripps Institution of Oceanography as presented in M. C. MacCracken and H. Moses, The first detection of carbon dioxide effects: Workshop summary 8–10 June 1981, Harpers Ferry, W. Va.* Bulletin of The American Meteorological Society *63:1165, 1982.)*

station of the U.S. Antarctic Program; the Mauna Loa record is shown in Figure 10.30. At 3400 meters (10,200 feet) above sea level, the Mauna Loa Observatory is far enough removed from major industrial sources of air pollution that carbon dioxide levels, which are monitored in that area, are considered to be representative of the northern hemisphere as a whole. The South Pole station is also free from local contamination, and its record closely parallels that of Mauna Loa.

Both records show an annual carbon dioxide cycle that occurs as a result of seasonal changes in vege-

tation. Carbon dioxide levels fall during the growing season because of photosynthetic removal. Furthermore, that annual cycle is superimposed on an upward trend that is nearly exponential. In 1958, the rate of carbon dioxide increase was approximately 0.7 ppm (parts per million) per year; 23 years later, the yearly growth rate was 1.8 ppm, which translates into an 8-percent increase in total global carbon dioxide in the atmosphere over that 23-year period. Sketchy data suggest that during the past 100 years, the atmospheric concentration of carbon dioxide may have climbed by as much as 26 percent.

To determine what this exponential rise in atmospheric carbon dioxide could mean for the climatic future, researchers are experimenting with numerical models of the earth–atmosphere system that are programmed on high-speed electronic computers. In such experiments, the typical procedure is to test the sensitivity of the model by changing one of the variables—in this case, the concentration of carbon dioxide. Initially, the model is run to a state of equilibrium using current earth–atmosphere conditions. Then the concentration of carbon dioxide in the model atmosphere is raised and the numerical model is run to a new state of equilibrium. The difference between the initial and final states of equilibrium is assumed to be the result of an elevated concentration of carbon dioxide.

In a 1982 review of those modeling efforts, the National Research Council suggested that a doubling of atmospheric levels of carbon dioxide would elevate the global mean surface temperature by 1.5° to 4.5°C (2.7° to 8.1°F). Polar warming would be several times greater, while the stratosphere would cool. Because projected warming is greater in the lower troposphere than further aloft, the troposphere would be less stable than it is today. A reduction in stability would mean greater convection and perhaps more severe thunderstorms.

If levels of carbon dioxide were to double by late in the next century, as some studies predict, the consequent global climatic change could be greater than any experienced in the 10,000-year history of civilization. Virtually every sector of society would be affected to some degree, but agricultural efforts would probably experience the greatest disruption. Also, traditional farming practices would have to change. In many regions, warmer weather also would mean drier weather. Hence it is quite possible that drought would be more frequent, more lands would require irrigation, and world deserts would expand.

Perhaps the most ominous consequence of warming that is induced by carbon dioxide would be a substantial rise in sea level. The possibility of an accelerated global warming trend in polar latitudes has prompted much speculation on the fate of the ice sheets on Greenland and Antarctica. Some studies suggest that a doubling of global levels of carbon dioxide would trigger polar warming, which would be sufficient to melt enough ice to raise the ocean to a level that would inundate populated coastal areas. But before residents of Boston and New Orleans quit the city and head for high ground, it should be noted that the atmosphere is a complex and highly interactive system, which means that other processes may compensate for any warming that is induced by carbon dioxide. For example, the ocean has a high thermal inertia (resistance to temperature change), and that inertia may slow or delay a warming trend

that is induced by carbon dioxide. Also, warmer conditions at higher latitudes may result in more snowfall and, eventually, more—not less—glacial ice.

Some researchers point out that some benefits may result from warming that is induced by carbon dioxide. For example, warmer winters would reduce the fuel demand for space heating in middle and higher latitudes. F. Kenneth Hare of the University of Toronto reports that Canada could realize a 12 to 18 percent decrease in space-heating costs if carbon dioxide levels were to double. Moreover, warmer winters would lengthen the navigation season on lakes, rivers, and harbors, where ice cover is a problem.

Future trends in atmospheric carbon dioxide depend on the rate of growth of our energy demands and the portion of those demands that is met by fossil fuels. Energy demands that are met by means of solar-energy collection, hydroelectric-power generation, and other means that do not burn fuels do not produce carbon dioxide. But even conservative estimates on our net output of carbon dioxide point to climatically significant increases during the next century. For example, in the summer of 1981, researchers at NASA's Goddard Space Flight Center reported on the responses of numerical models of the atmosphere to elevated levels of carbon dioxide that are specified by various energy-use scenarios. They found that, according to their calculations, a slow growth in energy demand (about one-third of the current rate), with equal reliance on both fossil and nonfossil fuels, will elevate the global mean temperature by approximately 2.5°C (4.5°F) during the twenty-first century.

It is also possible that warming that is induced by carbon dioxide will be enhanced by the now-rising levels of certain trace atmospheric gases, whose concentrations typically are expressed in parts per billion (ppb). Those gases include methane (CH_4), whose atmospheric concentration is now at 1500 ppb, nitrous oxide (N_2O), which is now 300 ppb, and chlorofluorocarbons (CFCs), now lower than 1 ppb. The reason for the methane increase is not known; increases in the other two probably are linked to air pollution. The climatic importance of those trace gases lies in their strong absorption of outgoing infrared radiation in the atmospheric windows. (Recall from Chapter 9 that, within atmospheric windows, the absorption of infrared radiation by water vapor or carbon dioxide is minimal.) In fact, for those trace gases, absorption is directly proportional to concentration, which means that when the concentration is doubled, the absorption is also doubled. Thus the combined climatic impact of rising levels of those gases could equal that of carbon dioxide.

Turbidity and Climate Some atmospheric scientists believe that increased atmospheric turbidity has the opposite effect of that caused by elevated levels of carbon dioxide. They believe that, instead of warming the earth, an increase in turbidity adds to the reflectivity of the earth–atmosphere system, which reduces the amount of solar radiation that reaches the earth's surface, and thus cools the lower atmosphere. Other atmospheric scientists disagree, arguing that increased turbidity that results from human activity (for example, industry and agriculture) promotes warming at the earth's surface. They feel that aerosols tend to scatter solar radiation back toward the earth's surface, where it is absorbed, and the larger dust particles absorb and reemit infrared radiation, which enhances the "greenhouse effect."

To determine who is right, we need a better understanding of the net effect of aerosols on the radiation balance. It is known that aerosols absorb and reemit infrared radiation and absorb and scatter solar radiation. However, the percentage of insolation that is scattered downward toward the earth's surface versus the percentage that is scattered back into space varies with the optical properties of the aerosols. Furthermore, it may be that the albedo of the earth's surface which lies under an aerosol layer, helps to determine whether turbidity has a net heating or a net cooling effect. Although stratospheric aerosols of volcanic origin probably trigger cooling at the earth's surface, the net impact of aerosols that are produced by human activity—most of which do not reach the stratosphere—is not known.

Ozone and Climate Earlier in this chapter, we identified potential threats to the important stratospheric ozone shield. We were concerned primarily with the adverse health effects that would result from an increase in the intensity of ultraviolet radiation. But climatologists have raised the possibility that changes in stratospheric ozone levels will also affect global climate. Ozone plays a role in the global-radiation balance (by absorbing ultraviolet and infrared radiation). Therefore, variations in ozone concentration may influence air temperatures at the earth's surface. Moreover, changes in ozone levels might alter stratospheric temperatures, thereby disturbing stratospheric air circulation. Such a perturbation might be transmitted into the troposphere and somehow affect lower atmospheric circulation. Today, we simply do not know what the effect of changing ozone levels has on the climate. Our understanding of that phenomenon awaits a better understanding of both ozone chemistry and the linkages between stratospheric and tropospheric air circulations.

In this chapter, we have seen that many uncertainties surround the linkage between air pollution and future trends in hemispheric or global-scale climate. One thing is certain, though: the climatic future, like the climatic past, will be shaped by many interacting factors, including variations in solar output, volcanic eruptions, and air pollution that is caused by our own activities.

Conclusions

Although the case against air pollution is not yet complete, the weight of evidence strongly indicates that polluted air has detrimental effects on living organisms. Many questions remain, but continuing research will give us a more complete assessment of the total environmental impact of air pollution. In Chapter 11, we look at some of the more promising management techniques that are currently available to us for rendering air safer for all living things.

Summary Statements

Important air pollutants include carbon dioxide, carbon monoxide, hydrocarbons, nitric oxide, nitrogen dioxide, sulfur dioxide, sulfur trioxide, hydrogen sulfide, photochemical smog, and a wide variety of solid and liquid aerosols.

Most air pollutants are cycled naturally within the atmosphere, but their concentrations can reach levels that threaten human health or disrupt physical and biological processes.

In sufficiently high concentrations, some air pollutants are asphyxiating agents and others severely irritate the respiratory tract.

The human respiratory system is armed with mechanisms that defend against air pollutants. Those mechanisms include nasal hairs that filter air, glands that secrete mucus to trap pollutants, cilia that force pollutant-laden mucus up to the mouth, a cough reflex, and macrophages that engulf and digest foreign matter.

Much evidence indicates that air pollution threatens human health, although specific cause-and-effect relationships are difficult to isolate.

Little is known about the impact of polluted air on the well-being of nonlaboratory animals, although cases of lead and fluoride poisoning have been reported.

Pollutant damage to vegetation occurs primarily when certain gases enter leaf pores. The tolerance of vegetation to air pollution varies considerably with the plant species and type of pollutant.

Formation of ozone in the stratosphere protects life on earth by filtering out harmful intensities of ultraviolet radiation. That ozone shield is threatened by chlorofluorocarbons that accumulate in the atmosphere.

Strong winds and unstable air enhance the rate of dilution of air pollutants, while weak winds and stable air suppress dilution.

A temperature inversion consists of an extremely stable stratification of light, mild air over heavier, cooler air. An inversion may develop through subsidence of air, extreme radiational cooling, or movement of air masses.

Many regions with high potential for air pollution tend to have topographic relief, since hills and mountain ranges can block horizontal winds that disperse polluted air.

Human activities influence the climate of large cities by altering the local radiation balance. That occurrence gives rise to an urban heat island and, in some cases, a convective circulation pattern that concentrates pollutants in a dust dome over the city.

Conditions that favor accumulation and concentration of pollutants in the atmosphere are countered to some extent by natural cleansing processes, including gravitational settling and washout by rain and snow.

Urban air pollutants affect precipitation, cloudiness, and fog development within and downwind from large metropolitan areas.

In regions where air is polluted by oxides of sulfur or nitrogen, precipitation becomes strongly acidic. Acidic precipitation threatens aquatic life and corrodes structures.

Human activities modify weather and climate on a hemispheric scale, primarily through the release of carbon dioxide and dust into the atmosphere. The precise effects of those activities are subjects of controversy.

Questions and Projects

1. Explain why acid precipitation is a greater problem on the windward rather than leeward flanks of a mountain range.

2. How do taller smokestacks contribute to the acid-rain problem?

3. Comment on the notion that at least some air pollution is an inevitable consequence of our way of life.

4. Why are air-pollution levels so high in Los Angeles?

5. What factors contribute to the development of an urban heat island?

6. Describe how wind speed and air stability influence the dispersal of air pollutants.

7. Speculate on the sequence of weather events that would cause temperature inversions to develop both at the surface and aloft at the same time.

8. In view of your understanding of atmospheric stability, explain the following observation: During a spring day, convective clouds (cumulus) are observed to develop over land but not over lake waters in the Great Lakes region.

9. Speculate on steps that city planners might take to reduce the intensity of urban heat islands, thereby decreasing the likelihood of dust-dome development.

10. Evaluate the potential for air pollution in your community in terms of (a) frequency of stable atmospheric conditions, (b) topographical influences, and (c) locations of major sources of pollution.

11. The types of pollutants that foul the air of a particular area depends on the specific kinds of industrial, domestic, and agricultural activities that take place there. Prepare a list of sources and types of air pollutants for your community. Is information available on the amounts of air pollutants emitted by each source?

12. Based on your understanding of the global radiation balance (Chapter 9), speculate on how air pollution might have an impact on global climate.

13. Speculate on the major sources and types of air pollutants in nonindustrialized nations.

14. How does the size of particulates relate to their potential as a health threat?

15. Describe the mechanisms that protect the human respiratory system from the intrusion of particulates.

16. Why is it difficult for scientists to isolate specific cause-and-effect relationships between air-pollution levels and human illnesses?

17. Has your community ever experienced a severe air-pollution episode? If so, did an increased incidence of respiratory illnesses accompany the incident? You may wish to refer to past issues of local newspapers or consult with your local public health agency.

18. Collect a sample of snow, melt it, and filter the meltwater. Examine the residue on the filter paper under a microscope. Describe what you see and speculate on its origins. Compare the appearance of samples taken from different locations in your community.

19. Distinguish between primary and secondary air pollutants and provide examples of each.

20. Maintain a daily log, recording the appearance of local industrial smoke plumes. Determine whether the shape and behavior of the plumes indicates unstable or stable conditions. If a temperature inversion is evident, does it usually form at the surface or aloft? Taking daily photographs of plumes may aid your analysis.

21. In the course of a day's activities, each of us is responsible, directly or indirectly, for the emission of air pollutants. Enumerate your contributions as to types and sources.

22. Compile a list of air-pollutant sources within your home.

Suggested Readings

CHANGNON, S. A., JR.: More on the LaPorte anomaly: A review. *Bulletin of the American Meteorological Society* 61:702–711, 1980. This report concludes that precipitation anomalies at LaPorte, Indiana, from the late 1930s to the 1960s, was linked to urban–industrial air pollution.

DUTSCH, H. V.: Ozone research—Past-present-future. *Bulletin of the American Meteorological Society* 62:213–217, 1981. This report traces the historical roots of current research on stratospheric ozone. It also identifies avenues for future research.

HEIDORN, K. C.: A chronology of important events in the history of air pollution meteorology to 1970. *Bulletin of the American Meteorological Society* 59:1589–1597, 1978. Includes a useful bibliography.

LIKENS, G. E., ET AL.: "Acid Rain." *Scientific American* 241:43–51, 1979. A discussion of the trends and causes of acid precipitation in North America and Western Europe.

NATIONAL RESEARCH COUNCIL: *Causes and Effects of Stratospheric Ozone Reduction: An Update.* Washington, D.C.: National Academy Press, 1982. A summary report on what is currently understood about the causes and implications of stratospheric ozone depletion.

NATIONAL RESEARCH COUNCIL: *Carbon Dioxide and Climate: A Second Assessment.* Washington, D.C.: National Academy Press, 1982. A summary report, which concludes that a doubling of atmospheric carbon dioxide could raise global mean temperature by $3° \pm 1.5°C$.

POSTEL, S.: Air pollution, acid rain, and the future of forests. *Worldwatch Paper* 58:1–54, 1984. Reviews the actual and potential impact of acid deposition on world forest productivity.

REVELLE, R.: "Carbon Dioxide and World Climate." *Scientific American* 247:35–43, 1982. An excellent overview of the carbon dioxide problem and its potential implications.

WALDBOTT, G. L.: *Health Effects of Environmental Pollutants.* St. Louis: C. V. Mosby, 1978. Includes a lucid account of the impact of polluted air on the human body.

WISNIEWSKI, J., AND KINSMAN, J. D.: An overview of acid rain monitoring activities in North America. *Bulletin of the American Meteorological Society* 63:598–618, 1982. Includes a summary of precipitation chemistry monitoring in the United States, Canada, and Mexico.

Stacks at the open hearth furnaces of a steel works near Birmingham, Alabama, before and after an air pollution crisis forced operations to be cut back. (AP/World Wide Photos.)

CHAPTER 11

AIR-QUALITY MANAGEMENT

In Chapter 10, we saw how polluted air threatens the well-being of living things and how it may affect weather and climate. Some pollutants contribute to respiratory disorders; some threaten to destroy the delicate atmospheric ozone shield that protects organisms from high levels of ultraviolet radiation; and some are precursors of acid rains and snows. Furthermore, rising levels of atmospheric carbon dioxide and dustiness may influence climatic trends on a global scale.

All those adverse effects of air pollution have spurred a national effort to improve air quality. Today, in most areas of our nation, we breathe air that is significantly healthier for us than it was only a decade ago. Much of this improvement stems from strong pollution-control legislation and new air-quality control strategies (Figure 11.1). But advances in air quality are costly, and efforts to achieve them have met with resistance in some quarters. In this chapter, we discuss air-quality legislation, recent trends in air quality, air-pollution control technologies, the problems of implementation, and prospects for the future.

Air-Quality Legislation

After the Industrial Revolution, some steps were taken by state and local governments to control air quality, but generally those steps were limited to ordinances that regulated smoky nuisances that were caused by the burning of coal. By the late 1940s, however, smog-plagued Los Angeles began to make progress in controlling air pollutants that were being generated by motor vehicles. Since then, California has led the way in enacting strict legislation to control automobile emissions.

Federal law, more or less, has been modeled on the pioneering example of California. The first federal legislation that was directed at air-pollution control was enacted in 1955. That law provided funds for research on polluted air by the U.S. Public Health Service and decreed that states be supplied with the technical information that was compiled in that effort. In 1963, the Clean Air Act was passed, which expanded federally sponsored research on air pollution. The 1963 law also provided grants to the states to help them carry out their then meager control programs.

The first nationwide standards for automobile exhaust emissions were established in 1965, through amendments to the 1963 Clean Air Act. Again following the California lead, that legislation called for the reduction of carbon monoxide and hydrocarbon emissions in new automobiles by 1968. It also directed the Surgeon General to study the effects of automobile emissions on human health. Then air-pollution regulations were revised again in 1967, when the Air Quality Act was passed. That law set goals for air quality and called for state and federal cooperation in establishing standards, although enforcement rested with the individual states.

Figure 11.1 *This electric power generating plant in the Four Corners area of western United States had to be outfitted with air pollution control devices to ensure the quality of air. (© Michal Heron, Woodfin Camp & Associates.)*

Although the modifications of air-quality laws that were enacted during the 1960s were well-intentioned, they actually were weak and accomplished very little. But in the late 1960s, bolstered by a wave of public concern about environmental quality, Congress developed a new set of strict and comprehensive amendments to the Clean Air Act. Those amendments, signed into law by President Nixon on December 31, 1970, called on the Environmental Protection Agency (EPA) to develop uniform air-quality standards for the ambient air (outdoor air) in 247 air-quality control regions (AQCRs) nationwide. Each *air-quality control region* encompasses an area where at least two communities share common air-pollution problems. Meteorological conditions and the types of emission sources within each air-quality control region are relatively uniform.

The 1970 amendments also require the states to draw up state implementation plans (SIPs) and to enforce air-quality standards for stationary sources (such as incinerators, factories, and power plants).

All state implementation plans must be approved by the EPA. The standards for motor-vehicle emissions were specified by Congress, which assigned responsibility for their enforcement to the federal government. The distinctive feature of the 1970 amendments to the Clean Air Act is strictness. That landmark legislation set a schedule for compliance and was the first to mandate stiff fines for violators.

Since 1970, federal clean air legislation has undergone many revisions, but the basic intent and force of the law remains substantially unchanged—in spite of the 1980 campaign promises by President Reagan for sweeping reforms in air-quality regulations as part of his overall government deregulation plan. As of this writing, the Reagan administration has done little to weaken federal air-quality laws. The reasons for this turn of events are many, but strong public opinion and congressional support for continued strict regulation are major factors. Also, the National Commission on Air Quality (NCAQ) reported in 1981 that federal air-quality legislation was scientifically sound and necessary and required refinement rather than major overhauling. The National Commission on Air Quality was established by 1977 law, and it was charged with evaluating federal laws that are concerned with air quality.

Review and refinement of provisions of the Clean Air Act is ongoing. For example, in early 1984, the EPA reaffirmed the national ambient air-quality standard for nitrogen dioxide and announced plans to revise the national ambient air-quality standard for suspended particulates.

Ambient Air-Quality Standards

So far, national ambient air-quality standards have been established for six *criteria air pollutants* (Table 11.1). Two sets of ambient air standards were established by the 1970 amendments: primary and secondary. *Primary air-quality standards* are defined as the maximum exposure levels that can be tolerated by human beings without ill effects. *Secondary air-quality standards* are defined as the maximum levels of air pollutants that are allowable to minimize the impact on materials, crops, visibility, personal comfort, and climate. Actually, as shown in Table 11.1, the primary and secondary standards are identical in most cases. And, for particulates, the secondary standard is more stringent than the primary standard. Emission standards for stationary sources (such as power plants) and mobile sources (such as automobiles) are set to ensure, theoretically at least, that once pollutants are emitted, ambient air-quality standards are not exceeded.

Control standards for the six criteria air pollutants (suspended particulates, sulfur oxides, carbon monoxide, nitrogen dioxide, ozone, and lead) are derived from scientific studies which demonstrate that ill effects are likely only after the concentrations reach a specific threshold value. However, certain air pollutants pose a serious danger to human health even in extremely low concentrations. Thus the 1970 amendments also addressed those *hazardous air pollutants*, which enabled the EPA to identify particularly dangerous airborne substances. These amendments required the EPA to propose emission standards for hazardous air pollutants within 6 months of their formal listing, so that final emission standards would be issued within another 6 months.

Progress has been slow, however, because of economic pressures and scientific uncertainties. By 1984, the EPA had designated as hazardous only seven air pollutants: asbestos, beryllium, benzene, mercury, vinyl chloride, radionuclides, and inorganic arsenic. Of these, EPA has set emission standards for asbestos, beryllium, mercury, and vinyl chloride. Many of the potentially hazardous air pollutants are trace by-products of important industries, including steel making, petroleum refining, and chemical manufacturing. Thus strict adherence to the law would require those industries to eliminate *all* emissions at tremendous cost. Hence the EPA has opted to list as hazardous only those substances for which the scientific evidence is "almost irrefutable."

Table 11.1 National ambient air-quality standards for criteria pollutants (September, 1981 revision)

POLLUTANT	AVERAGING TIME*	PRIMARY STANDARD	SECONDARY STANDARD
Total Suspended Particulates (Micrograms per Cubic Meter)†	1 Year	75	60
	24 Hours	260‡	150
Sulfur Oxides (ppm)	1 Year	0.03	
	24 Hours	0.14‡	—
	3 Hours	—	0.5
Carbon Monoxide (ppm)	8 Hours	9‡	9
	1 Hour	35‡	35
Nitrogen Dioxide (ppm)	1 Year	0.053	0.053
Ozone (ppm)‡	1 Hour	0.12	0.12
Lead (Micrograms per Cubic Meter)†	3 Months	1.5	—

Source: U.S. Environmental Protection Agency.
*Averaging time is the time period over which concentrations are measured and averaged.
†A microgram is one-millionth of a gram.
‡Concentration not to be exceeded more than once (on separate days) per year.

Preventing Further Deterioration

The landmark 1970 legislation also called for limits on further deterioration of air quality in certain regions of the nation, including such pristine areas as national parks and wilderness areas. In effect, Congress sought to control air quality by regulating the sites of new industrial sources of pollution and any plans to expand existing sources. In accordance with that provision of the law—and in response to prodding by environmentalists—in 1974, the EPA drew up a classification scheme that puts constraints on future changes in air quality according to the air-quality control region (AQCR). But that program was not fully implemented until August 7, 1977, when President Carter signed into law the Clean Air Act Amendments of 1977.

Among the provisions of the 1977 law is one that classifies each national air-quality control region and certain public lands according to the extent to which air quality is allowed to deteriorate within its boundaries. For example, in air-quality control regions where ambient standards have not been attained, the 1977 law rigidly limits the expansion of industrial sources of SO_2 and particulates. Thus the EPA will not allow any new sources of these pollutants to be created unless it can be shown that they will be offset by reductions in the emissions from other sources within that air-quality control region.

Air Quality in the Workplace

In addition to regulating ambient air pollution, federal legislation also attempts to control air quality in the workplace (Figure 11.2). The importance of controls in the workplace is underscored by a survey of more than 4500 industrial plants, which shows that 25 percent of the nearly 1 million workers in those plants were exposed to substances that were capable of causing illness or death. Congress took steps to reduce occupational exposure to dangerous substances by enacting the Occupational Safety and Health Act in 1970. That law directed the Labor Department to set acceptable levels of exposure, to monitor causes of illness and injury in workers, and

Figure 11.2 *Examples of the kinds of safety precautions in the workplace mandated by the Occupational Safety and Health Administration (OSHA). (Kennecott Copper Corporation photograph by Don Green.)*

to establish training programs that promote on-the-job safety. Those responsibilities are now carried out by the Occupational Safety and Health Administration (OSHA), which receives technical guidance from the National Institute of Occupational Safety and Health (NIOSH). Among the dangerous substances that are under strict regulation are asbestos, vinyl chloride, benzene, and more than a dozen other carcinogens. In this regard, the efforts of the Occupational Safety and Health Administration are complemented by the Toxic Substances Control Act of 1976, which requires that chemicals be examined for potential negative effects on human health. However, this is an onerous and expensive task, given the enormous variety of industrial chemicals.

Although efforts by the Occupational Safety and Health Administration have been a step in the right direction, progress in controlling exposure to dangerous levels of chemicals in the workplace has been slow. The standards are generally less rigorous than EPA air-quality standards, which may be caused by political pressures, since most industries claim that the standards are too strict (while some labor unions claim that they are too lenient). However, it should be borne in mind that EPA standards are supposed to protect everyone (including the sick, elderly, and infants) for 24 hours a day. OSHA's standards are directed at healthy adults who are expected to face exposure no more than 8 hours a day.

Indoor Air Quality

Very few regulations seek to control nonoccupational exposure to indoor air pollutants, yet such exposure may pose significant health risks, as noted in Chapter 10. For example, smoking is prohibited in some public places and certain building materials, such as asbestos, are prohibited by local ordinances. So far, the federal government has no general policy on nonoccupational indoor air pollution and has tended to react to serious complaints on an individual basis. Also, the problem has failed to attract the attention of citizens' action lobbyists, with the exception of antismoking groups. However, all this may change as legislators and the public become aware of the health risks.

As indicated in Table 11.2, many approaches exist to control indoor air quality, depending on the specific pollutant, such as ventilation, source removal or substitution, source modification, air pu-

Table 11.2 Control techniques for indoor air quality

CONTROL MEASURE DESCRIPTION	POLLUTANT	EXAMPLE
Ventilation: Dilution of Indoor Air with Fresh Outdoor Air or Recirculated Filtered Air, Using Mechanical or Natural Methods to Promote Localized, Zonal, or General Ventilation	Radon and Radon Progeny; Combustion By-Products; Tobacco Smoke; Biological Agents (Particles)	Local Exhaust of Gas-Stove Emissions; Air-to-Air Heat Exchangers; Building Ventilation Codes
Source Removal or Substitution: Removal of Indoor Emission Sources or Substitution of Less Hazardous Materials or Products	Organic Substances; Asbestiform Minerals; Tobacco Smoke	Restrictions on Smoking in Public Places; Removal of Asbestos
Source Modification: Reduction of Emission Rates through Changes in Design or Processes; Containment of Emissions by Barriers or Sealants	Radon and Radon Progeny; Organic Substances; Asbestiform Minerals; Combustion By-Products	Plastic Barriers to Reduce Radon Levels; Containment of Asbestos; Design of Buildings without Basements to Avoid Radon; Catalytic Oxidation of CO to CO_2 in Kerosene Burners
Air Cleaning: Purification of Indoor Air by Gas Adsorbers, Air Filters, and Electrostatic Precipitators	Particulate Matter; Combustion By-Products; Biological Agents (Particles)	Residential Air Cleaners to Control Tobacco Smoke or Wood Smoke; Ultraviolet Irradiation to Decontaminate Ventilation Air; Formaldehyde Sorbant Filters
Behavioral Adjustment: Reduction in Human Exposure through Modification of Behavior Patterns; Facilitated by Consumer Education, Product Labeling, Building Design, Warning Devices, and Legal Liability	Organic Substances; Combustion By-Products; Tobacco Smoke	Smoke-free Zones; Architectural Design of Interior Space; Certification of Formaldehyde Concentrations for Home Purchases

Source: From J. D. Spengler and K. Sexton, Indoor air pollution: A public health perspective. *Science* 221(4605):13, 1983. Copyright © 1983 by the American Association for the Advancement of Science.

rification, and changes in behavior. However, indoor air-quality control is quite different in scope from ambient air-quality control. For example, if an industry is forced to clean up emissions for the common good, usually the costs are passed on to the consumers of that industry's products. On the other hand, a person who takes steps to regulate his or her household air quality reaps the entire benefit and assumes the entire cost.

Trends in Air Quality

The effectiveness of air-quality legislation and the response of industry in abating the nation's air-pollution problems is becoming apparent. Overall, much progress has been made in improving ambient air quality and reducing motor-vehicle emissions. In this section, we summarize recent trends in both ambient air quality and in industrial and motor-vehicle emissions.

Ambient Air Quality

One indication of improving ambient air quality is the recent trend in the *Pollutant Standards Index* (PSI). A PSI value is computed for some particular area using the *one* criteria air pollutant that exhibits the highest concentration when it is compared with its primary air-quality standard. If the concentration is at the primary standard, the PSI is assigned a value of 100; if it is twice the standard, the PSI value is 200, and so on. As shown in Table 11.3, PSI values are divided into five ranges, which are determined by the risk to human health. PSI values that are higher than 100 are considered to be unhealthy, and those that are higher than 300 are considered to be hazardous.

In 23 United States metropolitan areas for which PSI data are available, the average frequency of unhealthy air-quality days (in which the PSI is at or higher than 100) declined from 1974 through 1980 (Figure 11.3). Unfortunately, that trend does not hold for the frequency of hazardous days (in which the

Table 11.3 Pollutant Standards Index (PSI) of air quality

POLLUTANT STANDARDS INDEX VALUE STATEMENT	DESCRIPTION	EPISODE LEVEL	GENERAL HEALTH EFFECTS AND CAUTIONS
0–50	Good		
51–100	Moderate		
101–200	Unhealthful	Primary Standard	Mild Aggravation of Symptoms in Susceptible Persons, with Irritation Symptoms in the Healthy Population; Persons with Existing Heart or Respiratory Ailments Should Reduce Physical Exertion and Outdoor Activity
201–300	Very Unhealthful	Alert	Significant Aggravation of Symptoms and Decreased Exercise Tolerance in Persons with Heart or Lung Disease with Widespread Symptoms in the Healthy Populations; Elderly Persons and Those with Existing Heart or Lung Disease Should Stay Indoors and Reduce Physical Activity
301–400	Hazardous	Warning	Premature Onset of Certain Diseases in Addition to Significant Aggravation of Symptoms and Decreased Exercise Tolerance in Healthy Persons; Elderly Persons and Those with Existing Heart or Lung Diseases Should Stay Indoors and Avoid Physical Exertion; The General Population Should Avoid Outdoor Activity
401–Above	Hazardous	Emergency	Premature Death of Ill and Elderly; Healthy Persons Will Experience Adverse Symptoms that Affect Their Normal Activity; All Persons Should Remain Indoors, Keeping Windows and Doors Closed; All Persons Should Minimize Physical Exertion and Avoid Traffic.

Source: From the Wisconsin Department of Natural Resources, Bureau of Air Quality Management.

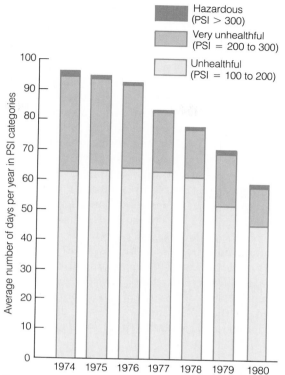

Figure 11.3 *The trend in the Pollutant Standards Index for 23 metropolitan areas in the United States, showing a decline in the frequency of "unhealthy" air-quality days from 1974 through 1980. (From the Council on Environmental Quality.)*

PSI is greater than 300). A particularly dramatic improvement was reported for Chicago, where the number of "unhealthful" days fell 80 percent—from 240 in 1974 to 48 in 1980. A substantial decline of 51 percent occurred during the same period of time in New York City. Not all cities were so fortunate, however. Los Angeles experienced only a slight improvement and Houston's air quality actually deteriorated.

An overall upswing in national air quality is also indicated by trends in levels of four of five of the criteria air pollutants.* The EPA estimates of changes in emission rates and ambient air levels are listed in Table 11.4. From 1970 through 1980, only the level of nitrogen dioxide increased in emissions and in ambient air levels. The picture becomes brighter when these trends are updated for the period from 1975 to 1982. As shown in Table 11.5, only nitrogen dioxide failed to register major improvement.

Before we proceed with a closer look at trends in criteria air pollutants, let us note the apparent discrepancy between the percent change in emissions and the percent change in the corresponding ambient air levels (Tables 11.4 and 11.5). A reduction in emissions is not necessarily accompanied by a comparable magnitude of improvement in ambient air quality. There are several reasons for this

* Sufficient data were not available to include lead prior to 1975.

Table 11.4 U.S. Environmental Protection Agency estimates of nationwide changes in emissions and ambient air concentrations of criteria air pollutants

POLLUTANT	EMISSIONS		AMBIENT AIR LEVELS	
	PERIOD	CHANGE (%)	PERIOD	CHANGE (%)
TSP*	1970–1980	−56	1960–1980	−31
SO_2	1970–1980	−15	1966–1971	−58
			1974–1980	−24
NO_2	1970–1979	+16.2	1970–1980	+20
	1975–1979	+9.7	1974–1980	+5.7
O_3†	1974–1980	−8.4	1974–1979	−9.9
CO	1970–1980	−23	1970–1980	−40.6

*Total suspended particulates.
†Emissions of volatile organic compounds (VOCs) that are ozone precursors.

Table 11.5 U.S. Environmental Protection Agency estimates of nationwide changes in emissions and ambient air concentration of six criteria air pollutants for 1975 to 1982

POLLUTANT	PERCENT CHANGE IN EMISSIONS	PERCENT CHANGE IN AMBIENT AIR LEVELS
TSP*	−27	−15
SO_2	−17	−33 (Annual Average)
NO_2	+5	0
O_3†	−13	−18
CO	−11	−31 (8-Hour Average)
Pb (Lead)	−69 (Gasoline)	−64

*Total suspended particulates.
†Emissions of volatile organic compounds (VOCs) that are ozone precursors.

Figure 11.4 *Carbon monoxide builds to unhealthy levels in areas where motor-vehicle exhaust is trapped in the canyonlike topography of congested cities. (U.S. Environmental Protection Agency.)*

occurrence. For example, the emissions values are estimates that apply only to point sources and do not account for contributions from area (nonpoint) sources, such as agricultural lands. (Recall from Chapter 10 that many natural sources of criteria air pollutants exist.) Also, the inherent variability of weather influences the quality of ambient air. For example, particulate emissions from a steel mill in April may be lower than they were in March, but if a temperature inversion overlies the area for 3 days in April, the PSI for those 3 days may be far higher than they were at any time in March. Then, too, instruments that are used to monitor air quality are not always sited at representative locations.

Of the criteria air pollutants other than lead, the total number of suspended particulates (TSP) has shown the most impressive improvement at the national level. In 1980, only approximately 7 percent of the nation's counties failed to attain the primary ambient air standard. Much of that improvement is linked to the installation of control devices on smokestacks and a reduction in coal and solid-waste

combustion. However, the statistics on the total number of suspended particulates may be unduly optimistic, because standard monitors fail to distinguish between sizes of particles, and extremely small particles pose the most serious health risk. Furthermore, the small particles are also the most difficult to control.

Only four major urban areas currently exceed the primary standard for sulfur dioxide: Chicago, Gary (Indiana), Indianapolis, and Pittsburgh. In addition, excessively high sulfur dioxide levels persist in areas that are adjacent to sulfide ore smelters in several western states. New control devices, taller stacks, siting of new facilities in rural areas, and burning of cleaner fuels are responsible for reductions in average sulfur dioxide levels nationally.

The downward trend in ambient carbon monoxide levels (Tables 11.4 and 11.5) is due to improved automobile emissions controls. Although average national levels have decreased substantially, an estimated 28 metropolitan areas still fail to meet the primary standard. Typically, those places are congested urban areas, where exhaust from heavy rush-hour traffic is trapped in the canyonlike topography of tall buildings and narrow streets (Figure 11.4).

Trends in ambient ozone levels (Tables 11.4 and 11.5) depend on trends in the ozone precursors: volatile organic compounds (hydrocarbons) and oxides of nitrogen. Although nationwide average levels are down, ozone is still a serious problem in the urban corridor that stretches from Philadelphia northeastward into southwestern Connecticut, in portions of southern California, and in Houston. Congested traffic is the problem in the Northeast. In southern California, bright sunshine, atmospheric stability, and topographic barriers (see Chapter 10) contribute to the ozone problem. And, in Houston, hydrocarbon emissions from the petrochemical industry are the culprit.

The upswing in national ambient levels of nitrogen dioxide during the 1970s (Table 11.4) was reversed in the early 1980s. This change was due to reductions in emission rates from both motor vehicles and electricity generating plants that are fired by fossil fuels. However, although automobile emissions have been reduced for every kilometer that is traveled, the total number of kilometers that are traveled has increased. And, for the electrical utilities, benefits of lowered emissions per unit of energy that is consumed have been offset to some extent by an increase in total electrical demand. Hence progress in reducing nitrogen dioxide levels has been slow.

Motor-Vehicle Emissions

Probably the chief reason for the recent upswing in the quality of ambient air is the control of motor-vehicle emissions. Recall that the 1970 amendments to the Clean Air Act stipulated that automobile manufacturers follow a strict schedule for reduction of hydrocarbons, carbon monoxide, and nitrogen oxide emissions. Although the automobile manufacturers requested and were granted numerous delays in complying with the schedule, they still achieved substantial reductions, especially in terms of carbon monoxide and hydrocarbons.

The effectiveness of control strategies on carbon monoxide emissions is indicated by Figure 11.5. Thanks to controls, the estimated emissions decreased substantially in the period between 1970 and 1979. If there had been no controls, the emissions would have increased sharply in response to greater motor vehicle travel.

Also, burning of unleaded gasoline has reduced substantially the amount of lead particles in the air.

Figure 11.5 *A comparison of motor-vehicle carbon monoxide emissions with and without controls from 1970 through 1979. (From the U.S. Environmental Protection Agency.)*

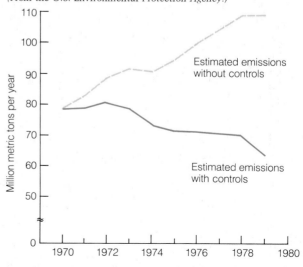

According to the National Center for Health Statistics, lead emissions from gasoline-powered vehicles were reduced approximately by half between 1976 and 1980. Researchers at the Center for Disease Control in Atlanta link that occurrence to a 37-percent reduction in the mean blood-lead level of United States' residents. And the trend toward lower ambient lead levels will probably continue in response to tighter regulations that were proposed recently.

Beginning in 1976, gasoline refiners were required to phase down the amount of lead that was being added to gasoline as an antiknock agent to a final standard of 0.5 grams per gallon by October, 1982. As of this writing, the EPA is expected to propose a further lowering of lead in gasoline to 0.1 gram per gallon by 1988. And there is even the possibility that lead will be banned totally as a gasoline additive. These proposals are spurred by clinical studies showing that adverse health effects are still possible under the present standard. Also, the expected phase-down of lead in gasoline reflects the EPA's concern about motorists who remove catalytic converters (an emissions control device) in an attempt to improve the performance of their vehicles or who use less expensive leaded gasoline that renders the catalytic converter ineffective in controlling pollutant emissions.

Air-Quality Control

Although efforts to control air and water quality are analogous in purpose, the abatement of air pollution is usually more challenging because air is more mobile than water. Hence although it is convenient and appropriate for us to consider water-quality control in the context of a watershed (Chapter 8), it is limiting and inappropriate to attempt to define an "airshed." Winds and the pollutants they transport are not restricted to geographical boundaries as are the tributaries of a river. For example, the air-pollution plume of Chicago may extend into Wisconsin on one day and into Indiana or Michigan on the next, depending on the wind direction. In short, we cannot define an "airshed" on which to focus our efforts. Thus the only feasible method of approaching the problem of air quality is to control emissions from individual point sources—that is, from exhaust pipes, chimneys, and smoke stacks.

An increasingly popular industrial strategy that is used to control air quality is to construct taller smokestacks. That strategy is one of the approaches that is taken by industry in Green Bay, Wisconsin, to combat the area's sulfur dioxide problem (Box 11.1).

Taller smokestacks reduce local concentrations of ground-level air pollutants. But taller stacks do not improve regional air quality, because they do not reduce the total quantity of pollutants that enter the atmosphere. Also, as we saw in Chapter 10, tall stacks may only exacerbate the acid-rain problem. Furthermore, local climatic or topographic conditions may negate any advantage of tall stacks for improving local air quality. For example, even the world's

Figure 11.6 *(a) With low smoke stacks, effluents can be trapped within the wake of nearby buildings or of the chimney itself. (b) With smokestacks that are constructed to the height of good engineering practice (2.5 times the height of the nearest obstacle), the effluents clear the wake, preventing downwash.*

tallest chimney† would not penetrate the elevated temperature inversion over the Los Angeles basin. In any event, tall stacks must be constructed to the so-called height of good engineering practice (GEP). The height of good engineering practice is generally taken to be 2.5 times the height of the nearest obstacle (Figure 11.6).

An alternative strategy that is used to control air quality is to locate new industrial sources of air pollution in rural sites rather than urban ones. But

locating industry outside an urban area may have serious economic and esthetic constraints. For example, displacing industry to rural areas will erode the tax base of cities and add to the industrial transportation costs, not to mention eliminating the potential of producing food on valuable agricultural land. Nor does this approach reduce the total amount of emissions.

Many strategies are, in fact, available today to reduce emissions from point sources. We can consider those techniques in relation to the major air pollutants that are emitted by industry and motor vehicles.

† The 380 meter (1246-foot) stack of the copper smelter at Sudbury, Ontario, Canada.

BOX 11.1

RESOLVING GREEN BAY'S SULFUR DIOXIDE PROBLEM

In 1980, the EPA announced that a portion of the city of Green Bay, Wisconsin, was not in compliance with the federal primary air quality standard for sulfur dioxide (SO₂). As a result, a plan was developed to meet the standard, which can be used as an illustration of how government, industry, and citizens' groups can interact in an effort to improve air quality without overburdening the local economy.

When Wisconsin's Department of Natural Resources (DNR) drew up its State Implementation Plan (SIP) in the early 1970s, no apparent problem with sulfur dioxide existed statewide. Hence Wisconsin's plan did not contain any provisions for limiting sulfur dioxide emissions. But between 1977, when the Department of Natural Resources began to monitor continuously Green Bay's air quality, and 1980, 13 violations of the 24-hour sulfur dioxide standard were recorded. (A violation of the 24-hour standard occurs when the standard is surpassed twice in a single year.) The highest measured value was 720 micrograms per cubic meter, which is well above the federal standard of 365 micrograms per cubic meter.

In Green Bay, the sulfur dioxide problem is linked to coal burning at an electric-power generating plant and pulp processing at six paper mills. Those sources are listed in Table 1 and are shown in Figure 1. The sulfur dioxide emissions from

those sources account for approximately 13 percent of Wisconsin's total sulfur dioxide emissions.

In 1980, in response to those violations of the sulfur dioxide standard, the EPA formally designated the area, which is outlined in Figure 1, as a nonattainment area. When an area is so designated, the Clean Air Act requires the state to develop a plan that will bring the area into compliance with federal standards. The plan must specify a target date for compliance, using "reasonably available control technology" (RACT)— technology that is currently available and economically feasible. Thus for sulfur dioxide control,

Table 1 Principal industrial emitters of sulfur dioxide in Green Bay, Wisconsin

SOURCES	TONS SO₂ PER YEAR (1981)	MILLIONS OF POUNDS SO₂ PER YEAR	PERCENT OF TOTAL
WPSC-Pulliam Plant	38,500	77.0	45.6
Proctor & Gamble			
Fox River Mill (Paper)	24,100	48.2	28.5
East River Mill (Paper)	700	1.4	0.8
Fort Howard Paper	18,900	37.8	22.4
Nicolet Paper	1,100	2.2	1.3
Green Bay Packaging	700	1.4	0.8
James River (Paper)	400	0.8	0.5
Other Industry	100	0.2	0.1
Total (1981)	84,500	169.0	100.0

Source: From the Wisconsin Department of Natural Resources.

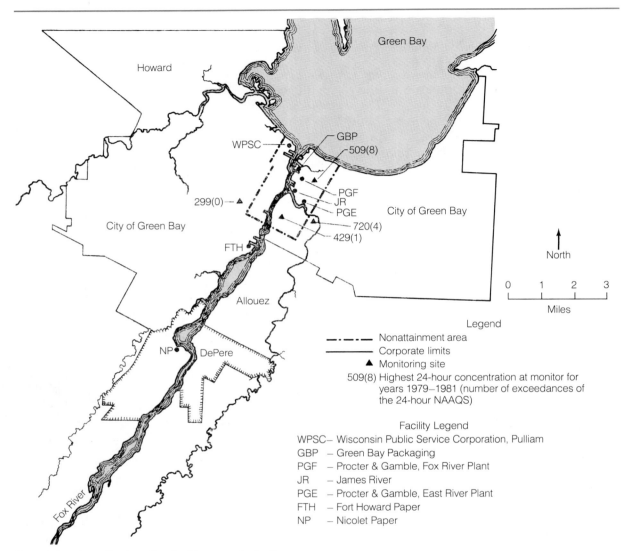

Figure 1 *Locations of major sulfur dioxide sources in the Green Bay, Wisconsin, area. The nonattainment area is enclosed by dashed lines. (Courtesy of the Wisconsin Department of Natural Resources.)*

"reasonable available control technology" includes taller stacks, the burning of low-sulfur fuels, and scrubbing. Each of those strategies is described elsewhere in this chapter.

To deal with the Green Bay problem, two alternative plans of action were drawn up. Alternative 1, developed by the Department of Natural Resources, was based in part on the experiences of other states with similar types of facilities.

Alternative 2 was designed by an industrial group that was composed of the companies which owned the seven polluting facilities, in cooperation with the Department of Natural Resources. Because the two plans were developed from different viewpoints, they differed in important ways. The two plans are contrasted in Table 2. Essentially, Alternative 2 was less stringent and less expensive than Alternative 1, and it was tailored to the emissions of each facility. Although Alternative 1 was more stringent than Alternative 2, both plans theoretically would have met the standard. (This was determined by the Department of

Table 2 A comparison of the features of the two alternative plans to bring Green Bay, Wisconsin, into compliance with federal air quality standard for sulfur dioxide

CRITERIA	ALTERNATIVE 1	ALTERNATIVE 2
New Stacks	5 GEP Stacks	4 GEP stacks§
Annual Allowable Emissions	104,700 Tons/Year	184,300 Tons/Year
Daily Allowable Emissions	287 Tons/Day	505 Tons/Day
Air-Quality Level (Standard = 365 $\mu g/m^3$)[*]	257 $\mu g/m^3$	355 $\mu g/m^3$
Scrubbers Required	1 Boiler Scrubber, P&G Fox 1 Process Scrubber, James R.	None Required
Sludge Disposal	Possibly Needed	No Sludge Produced
Costs[†] Capital	$39.3 Million New Stacks—58% Scrubbers—30% Other—12%	$23.9 Million New Stacks—93% Scrubber—1% Other—6%
Annual Operation and Maintenance Cost	$11.3 million Lower Sulfur Fuels—48% Scrubber Operation—52% (Including Sludge Disposal)	$2.75 Million Lower Sulfur Fuels—91% Scrubber Operation—6% Other—3%
Effect on Odors in Green Bay?	None Anticipated	None Anticipated
SO$_2$-Producing Industrial Growth Possible?[‡]	Everywhere, Including Downtown Area	Everywhere but in the Downtown Area
Acid-Rain Effects	Alternative 1 Minimizes Total SO$_2$ Emissions	Alternative 2 Does Not Minimize Total SO$_2$ Emissions

Source: Wisconsin Department of Natural Resources.
[*]The abbreviated $\mu g/m^3$ means micrograms per cubic meter, the way in which SO$_2$ concentrations are measured. The concentrations in the table are actually the "second-worst days" in a year, because the law allows the 365 $\mu g/m^3$ standard to be exceeded once per year. Thus under Alternative 1, in a given year the downtown area could experience one 24-hour average of 257 $\mu g/m^3$, and one other day in which the SO$_2$ concentration would be higher than 257. Under Alternative 2, the downtown area could experience one 24-hour average of 355 $\mu g/m^3$, and one other day in which the SO$_2$ concentration would be higher than 355.
[†]DNR cost estimates.
[‡]The computer model used to develop the Green Bay SO$_2$ strategy was also used to predict the effects of additional SO$_2$ emissions from new or expanded industry in the Green Bay area. Under Alternative 1, it would be possible to build an additional SO$_2$ source in the downtown area without causing violations of the standard. Under Alternative 2, a new source of SO$_2$ in the Green Bay area would have to be located outside the downtown area to prevent SO$_2$ standard violations there.
§GEP = good engineering practice.

Natural Resources with the help of a computerized model that predicts "worse-case" scenarios, in which weather conditions and emissions rates favor the maximum ground-level sulfur dioxide concentrations.)

The two alternative plans of action were brought before the citizens of Green Bay at a public hearing in the Spring of 1983. Local environmental groups argued for adoption of Alternative 1; industrial groups made their case for Alternative 2. Public concern focused on four major issues: potential economic impact, implications for the acid-rain problem, sludge disposal, and the health hazard.

Many citizens feared that the more costly Alternative 1 would put Green Bay's industry at a competitive disadvantage, which could displace local industry, inhibit industrial expansion, and contribute to unemployment. Environmentalists argued that provisions for taller stacks but no scrubbers in Alternative 2 would only exacerbate the acid-rain problem. Some persons feared that the scrubbers mandated by Alternative 1 would create a sludge disposal problem that would spawn other environmental problems. Finally, some persons stressed the adverse health effects of sulfur dioxide and called for the stricter emissions controls of Alternative 1.

In June, 1983, after a careful review of public response to the two alternative plans, the Department of Natural Resources selected a slightly modified version of Alternative 2, which limited the total annual sulfur dioxide emissions from each facility. The Department of Natural Resources

Figure 2 *A new tall stack under construction at Fort Howard Paper Company in Green Bay, Wisconsin. Construction of taller smoke stacks is one strategy that is employed by local industries in an effort to meet air-quality standards. (Courtesy of J. M. Moran.)*

opted for Alternative 2, because it minimized the cost of attaining and maintaining the public health standard and because it did not place Green Bay industry at a competitive disadvantage. In the absence of a federal policy on acid rains, this problem was not a consideration in the final decision.

A second round of public hearings was held in August, 1983, after which the plan went to the state legislature in Madison. Following review by committees in both houses of the legislature, the plan was made into law. The industries have until November, 1985, to implement control strategies and thereby attain the federal standard for sulfur dioxide emissions. Already, tall stacks are under construction (Figure 2). Note that the target date for compliance is a full 5 years after the EPA first designated downtown Green Bay as a nonattainment area for sulfur dioxide. It simply takes much time and effort to resolve such issues through the normal channels of public review and legislative action.

Industrial Control Technologies

Many industries have adopted a wide variety of technologies in an effort to lower their contributions to polluted air. Those technologies are tailored to the specific type of industry and its major emissions.

Particulates

Three methods of removing particulates from stack emissions are currently available: electrostatic precipitation, filtering, and gravitational settling. In *electrostatic precipitators*, effluent particulates are electrically charged and then collected on plates that have the opposite charge, which is a process that can remove up to 99 percent of the fly-ash emissions from the stacks of coal-fired power plants.

In *filtering*, a particle-laden airstream passes through a series of filters, most of which are composed of fiberglass. Those filters are porous enough to permit flue gases to pass through, while trapping particulates behind, and they may take the form of cylindrical bags, such as the one shown in Figure

Figure 11.7 *A typical industrial baghouse compartment for removal of particulates. Each bag is up to 11 meters (35 feet) in height and 300 millimeters (12 inches) in diameter. Accumulated dust is dislodged by some of the filtered gas that is channeled back into the baghouse ("reverse gas"). Dislodged particulates accumulate in the hopper. (EPRI Journal 8 (September):17, 1983.)*

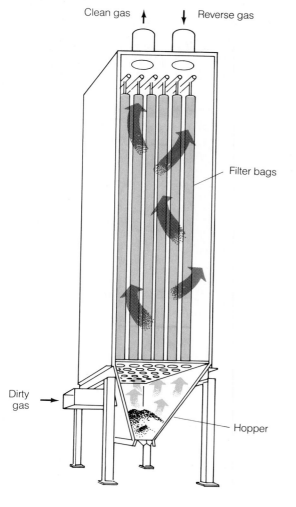

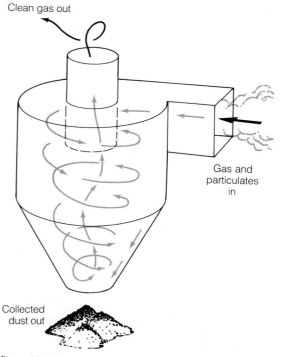

Figure 11.8 *A cut-away view of a cyclone particulate collector that removes dust from industrial effluents by inducing gravitational settling.*

11.7. For coal-fired electric-power generating facilities, bag houses are common alternatives to electrostatic precipitators. Bag houses match precipitators in efficiency, reliability, and overall cost, and they are more effective in collecting fine particulates, especially when low-sulfur coal is burned. Larger particulates can be separated from the effluent stream by a *cyclone collector* (Figure 11.8), which is a device that induces gravitational settling of heavier particles. By using these techniques, some industries have lowered their particulate emissions by as much as 90 percent.

Sulfur Dioxide

In recent years, some industries—especially coal-fired electric utilities—have made much progress in reducing sulfur dioxide emissions. They have done so in response to the EPA's stringent New Source Performance Standards (NSPS), which limit sulfur dioxide emissions from new facilities and from old facilities that are undergoing major modifications. (Recall that sulfur dioxide is a serious health threat and is the major precursor of acid rain.)

Many technologies that are designed to control sulfur dioxide emissions are either currently in use or in the testing and development phase. In Table 11.6, we summarize each technology in terms of its developmental status, advantages and disadvantages, economic feasibility, and waste by-products.

Today, *flue-gas desulfurization* (FGD), which is also called *scrubbing*, is the most commonly used method of controlling sulfur dioxide emissions.

Flue-gas desulfurization dates back to the early 1930s in Great Britain. At that time, a coal-fired power plant, which was operated by the London Power Company, used alkaline water that was drawn from the Thames River to wash (scrub) sulfur dioxide from flue gases. That system continued in operation until 1975, when the plant closed.

Table 11.6 Various technologies available to lower emissions of sulfur dioxide (SO_2)

TECHNOLOGY	DESCRIPTION	STATUS: CURRENT AND PROJECTED	ADVANTAGES
Physical Coal Cleaning	Removes Inorganic Sulfur by Using Difference in Specific Gravity of Iron Sulfide and Coal	Commercially Used Future Limited as Emission Standards Become more Stringent	Can Be Used in Conjunction with FGD to Reduce Costs Inexpensive Process Does Not Require Boiler Modifications (Normally)
Flue-Gas Desulfurization (FGD)	Remove SO_2 from Combustion Gases after Burning through Absorption in an Alkaline Solution	Commercially used Offers Most Efficient and Economical Method of SO_2 Control Now Available on a Widespread Basis	Can Be Retrofitted to Existing Boilers Efficient
Fluidized Bed Combustion (FBC)	Combustion of Pulverized Coal in a Bed of Limestone or Dolomite Fluidized by Air, SO_2 Absorbed by Absorbent (Limestone)	Emerging Technology in Pilot Plant Testing Commercial Availability for Atmospheric FBC in Mid-1980s; in Early 1990s for Pressurized FBC	SO_2 Removal: 80 to 90 Percent Lower Combustor Capital Costs Wide Fuel Flexibility Lower Operating Temperatures
Coal Liquefaction	Reaction with Hydrogen, Heat, and Pressure in Presence of Catalysts or Chemical Solvents to Produce Liquid Fuels	Four Processes in Pilot Stage: Direct Catalytic Hydrogenation, Solvent Extraction, Pyrolysis, and Liquid Hydrocarbon Synthesis Commercialization: 1990s	Wide Variety of Coals Produces Clean Burning Fuels SO_2 Removal: Greater than 95 Percent
Coal Gasification	Reaction with Steam or Hydrogen and Air/ Oxygen at High Temperature to Produce High or Low BTU Gas	Low BTU Commercialization: mid-1980s Current Low BTU Demonstration (Small-Scale Industrial) High BTU Commercialization: 1990	Can Be Used in Conventional Gas-Fired Power Plants SO_2 Removal: Greater than 95 Percent

Source: Modified slightly after the Environmental Protection Agency.

In one of the most effective flue-gas desulfurization systems that is currently in use, the effluents are channeled through a slurry of water and ground limestone ($CaCO_3$). Through that process, the calcium in the limestone combines chemically with the sulfur to produce calcium sulfate ($CaSO_4$), which subsequently is collected and disposed of. Up to 90 percent of the sulfur dioxide in flue gases can be removed this way.

Coal-fired electric power utilities can also control sulfur dioxide emissions by burning low-sulfur coal or by physically cleaning it. Most low-sulfur coal is mined in the West, and the cost of transporting it is so high that users in the Midwest and East find it to be unattractive economically. Those users find it cheaper to burn the high-sulfur coal that is mined in their own regions. However, from 40 to 90 percent of the inorganic sulfur (iron sulfide) that is contained in this high-sulfur coal can be removed by mechanical means before it is burned. Some re-

TECHNOLOGY	DISADVANTAGES	ECONOMIC FEASIBILITY	WASTE PRODUCT
Physical Coal Cleaning	Only 14 Percent of U.S. Coals Can Be Cleaned Enough to Meet NSPS Standards May Require Large Electrostatic Precipitator	Relatively Inexpensive but Limited in Application	Large Quantities of High Sulfur Refuse/Ash
Flue Gas Desulfurization (FGD)	Generates Large Quantities of Sulfur-Laden Sludge	Provides an Effective and Efficient Means of Meeting NSPS Emission Standards without Modifying or Replacing Existing Boilers	Sludge
Fluidized Bed Combustion (BFC)	Generates Large Quantities of Spent Absorbent Larger Precipitators Required	Appears to Have Economic Advantage over FGD Operating and Capital Costs	Dry Spent Sorbent
Coal Liquefaction	Unidentified Pollution Problems Potential for Toxic By-Products High Particulate Control Needed	High Overall Development Costs	
Coal Gasification	Same as Liquefaction	Same as Liquefaction	

sidual organic sulfur is chemically bound in the coal and cannot be removed by any standard technique prior to burning. Thus many power plants are able to meet the emission standards of sulfur dioxide economically by combining physical coal cleaning with partial flue-gas desulfurization.

In the future, fluidized bed combustion, coal liquefaction, and coal gasification (Table 11.6) should substantially reduce sulfur dioxide emissions from coal burning. The last two technologies produce cleaner burning fuel and permit a reduction in industrial consumption of oil and natural gas, thereby increasing the amounts of these fuels that are available for domestic consumption. We have more to say about this topic in Chapter 18.

Oxides of Nitrogen

Unlike emissions of the other major pollutants, industrial emissions of oxides of nitrogen are still rising. In fact, as the EPA points out, if current trends in emission rates were to continue, ambient air concentrations of NO_x (NO plus NO_2) would increase 50 percent by the year 2000. One of the principal reasons that NO_x emissions are increasing is explained by the slow development of appropriate industrial control technologies. The need for effective controls will become even more important as the nation's power plants burn more coal, because power plants that burn coal emit more NO_x than those that burn oil or natural gas.

However, several promising industrial control strategies for NO_x are currently under study. Those strategies address the two principal sources of NO_x: (1) high-temperature combustion, which causes atmospheric oxygen (O_2) and nitrogen (N_2) to react chemically, producing *thermal* NO_x; and (2) the chemical reaction of nitrogen that is contained in the fuel itself with oxygen in the air, producing *fuel* NO_x. Control strategies attempt either to prevent oxidation in the first place or to convert NO_x to harmless molecular nitrogen (N_2).

One industrial approach to NO_x control is *staged combustion*, which is a technology that can lower NO_x emissions by 30 to 50 percent. In that approach, fuel is burned in two stages at different fuel-to-air mixtures. In the first stage, so little oxygen exists that nitrogen fails to react with oxygen (oxi-dize) and is released as molecular nitrogen (N_2) along with carbon monoxide (CO) and carbon particles. In the second stage, sufficient oxygen is supplied to burn the carbon particles and convert the carbon monoxide (CO) to carbon dioxide (CO_2). However, a major problem with staged combustion is that it forms iron sulfide, which will corrode coal-fired boilers.

While the strategies to control NO_x emissions are promising, they are not likely to be used widely in the near future, because most of them are new technologies that are still in the small-scale testing phase of development.

Overall Problems

Although today's air-pollution abatement technologies can clean up industrial stack effluents to some extent, they are not a panacea for the problem of industrial emissions—many difficulties remain. For example, the particulates that cannot be readily removed from an effluent are often the very small ones (2.5 micrometers or less in diameter). But those particulates are the ones that pose the greatest health hazard, since they can evade the defenses of the respiratory tract and penetrate deeply into the lungs. Moreover, control devices operate at maximum efficiency only if they are properly maintained. And many devices lose efficiency quickly, because they come into contact with the corrosive substances they control, which damage or destroy them.

Ironically, adequate programs to control air quality create other environmental problems. After all, air contaminants that are removed from effluents do not disappear; they must be put to use or disposed of. In some cases, those products have an economic value and can be recycled (which is a subject that we deal with in Chapters 12 and 13). Sometimes, however, extracted air pollutants are improperly disposed of—they are dumped illegally into wetlands, for example, or onto sites where they penetrate the ground and contaminate the underlying groundwater. Thus an effective program to control air quality must require both adequate control strategies and reuse or proper disposition of collected air pollutants. For example, some roads are composed largely of processed scrubber sludge, which is a rare model of good recycling policy.

Control of Motor-Vehicle Emissions

In terms of the total quantity of pollutants that are emitted into air in the United States, the primary offender is the motor vehicle. Automobiles and trucks that are powered by internal combustion engines—which is to say, virtually all motor vehicles—emit several types of air pollutants (Figure 11.9). Carbon monoxide, oxides of nitrogen, and hydrocarbons are important constituents of exhaust. Hydrocarbons also escape from the crankcase and carburetor and evaporate from the fuel tank. And lead is emitted in exhaust when leaded gasoline is burned.

A number of measures are used to control the various pollutants that are produced by different components of the automobile. For example, the *positive crankcase ventilation* system (PCV) reduces hydrocarbon emissions from the crankcase by channeling crankcase blow-by gases back to the engine, where they are burned. Positive crankcase ventilation systems were first installed voluntarily by automobile manufacturers on all new cars beginning in 1963. The escape of hydrocarbons from the carburetor and fuel system was reduced by modifications that were introduced on 1971 models (1970 in California). Exhaust emissions, however, are more difficult to control, since concentrations of exhaust gases depend on many variables that relate to the performance of the engine. Thus in the late 1960s and early 1970s, automobile manufacturers attempted to control exhaust emissions primarily by making engine adjustments. For example, they increased the air-to-fuel ratio by making carburetor settings leaner. Although those alterations did re-

duce hydrocarbon and carbon monoxide exhaust, they also produced higher combustion temperatures and thereby increased the formation of nitrogen oxides (NO_x). Further refinements in engine adjustments have reduced exhaust emissions significantly, but at the expense of fuel economy.

Exhaust controls were improved further in approximately 70 percent of the automobiles that were manufactured in 1975, when *catalytic converters* were introduced. Catalytic converters, such as the one shown in Figure 11.10, chemically change the hydrocarbons and carbon monoxide in exhaust into carbon dioxide and water vapor. However, that device works only when unleaded gasoline is used; in fact, lead deactivates it.

Figure 11.10 *A catalytic converter chemically changes the hydrocarbons and carbon monoxide in automobile exhaust into carbon dioxide and water vapor. (*The New York Times, *March 15, 1984. Copyright © 1984 by The New York Times Company. Reprinted by permission.)*

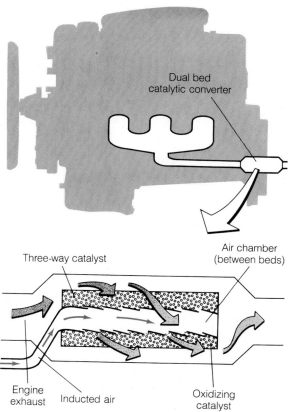

Figure 11.9 *Principal sources of air-pollutant emissions in the automobile.*

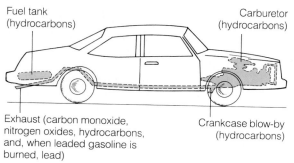

Fuel tank (hydrocarbons)

Carburetor (hydrocarbons)

Exhaust (carbon monoxide, nitrogen oxides, hydrocarbons, and, when leaded gasoline is burned, lead)

Crankcase blow-by (hydrocarbons)

A new generation of catalytic converters that is currently being developed promises to also lower the emissions of oxides of nitrogen. In the so-called three-way catalytic converter, the catalyst‡ causes nitric oxide (NO) to oxidize carbon monoxide (CO) and hydrocarbons, thereby releasing molecular nitrogen (N_2), carbon dioxide (CO_2), and water vapor (H_2O). The major difficulty with this new approach is that the proportions of carbon monoxide, hydrocarbons, and nitric oxide that enter the catalytic converter must be regulated precisely. That precision is accomplished by controlling the air-to-fuel ratio in the combustion chamber. Too much fuel increases the emissions of carbon monoxide and hydrocarbon; too much oxygen increases the emissions of nitrogen oxides. Thus motor vehicles that are equipped with three-way catalytic converters must be kept well tuned.

Motor vehicles are also minor sources of sulfur oxides, which are formed when gasoline is burned. Those emissions could increase substantially in the near future, however, should the number of diesel-powered vehicles increase. Although much more sulfur (about 2000 ppm) is contained in diesel fuel than in gasoline (350 to 400 ppm), the sulfur can be removed from diesel fuel during refining at a small cost to the consumer. The EPA estimates that desulfurization of diesel fuel (down to 200 ppm) would add only about 70 cents to the cost of an 80-liter (21-gallon) tank of diesel fuel.

Traditionally, the principal strategy for controlling automobile emissions has been to modify the internal combustion engine. Today, however, other options are under study. One suggestion is that alcohol—either mixed with or substituted for gaso-

‡ A catalyst is a substance that increases the rate of a chemical reaction without itself being altered by the reaction.

Figure 11.11 *Battery-powered vehicles, such as these postal service trucks, offer an advantage over vehicles that are powered by internal combustion engines in that they produce less air pollution. (Air pollution is a by-product of the electricity that is generated to recharge batteries.) However, they do not match traditional automobiles for performance. (Illustrators Stock Photos.)*

Figure 11.12 *Car pools and buses cut air-pollutant emissions and fuel consumption. This highway, in Honolulu, Hawaii, has a special lane for car-poolers and buses. (U.S. Department of Transportation.)*

line—be burned as fuel, which was a common practice in several countries during World Wars I and II and is used in Brazil today. Both ethanol (grain alcohol) and methanol (wood alcohol) burn cleaner than gasoline and both result in lower emissions of carbon monoxide and hydrocarbons.

Another option for reducing motor-vehicle emissions is the phasing out of the internal combustion engine entirely in favor of steam- or battery-powered engines (Figure 11.11). Air-pollution problems are not eliminated in such engines, but they are significantly reduced. However, nontraditional motor-vehicle engines require more refinement before they are likely to gain widespread acceptance. At present, they fail to match the internal combustion engine in terms of durability, performance, or cost.

By law, the automobile industry has a responsibility to reduce emissions from automobiles. However, we too, can contribute to improved air quality by voluntarily restricting our use of motor vehicles and keeping them well tuned. Municipalities that have extreme air-pollution problems might decide to enforce restrictions on motor-vehicle use. For example, by encouraging car pools, establishing bus lanes (Figure 11.12), restricting downtown parking, and expanding mass transit, we can help to alleviate air-pollution problems in metropolitan areas. Some regions that have failed to attain ambient air-quality standards, the so-called nonattainment areas, may enact mandatory vehicle inspection and maintenance programs.

Economics of Air-Quality Control

What does air pollution cost our society? The corrosion of buildings and the damage to crops are obvious effects whose cost in dollars can be estimated with relative ease. But much of the cost of air pollution is attributed to elusive, less obvious, and less direct effects. And while no doubt exists that esthetic losses result from dirty air, it would be virtually impossible to determine an estimate of their value that would be generally agreed-upon.

The indirect effects of air pollution are often linked inseparably with other factors that contribute to those effects—especially in the case of the effects of air pollution on health. For example, although high levels of air pollution may contribute to emphysema and other lung disorders, factors such as age, diet, and general health also contribute to the incidence and severity of those illnesses. Thus it is not possible to assign a dollar value to each contributing factor when we account for the costs of hospitalization and treatment. Such problems of cost assessment arise whenever the link between pollution and environmental damage is real but not measurable, which is a common circumstance. For more information on cost–benefit analysis, see Box 11.2.

Abatement Costs

The costs of federal air-quality standards are great and will continue to be so. A great deal of money will be required for the research and development of new control techniques, and industries will continue to spend enormous amounts on the purchase and maintenance of the devices themselves. According to the EPA, the cost of pollution abatement is particularly burdensome for electric power utilities. Between 1975 and 1985, those utilities spent an estimated $25 billion for pollution control, most of which was directed at air pollutants.

In early 1983, *U.S. News and World Report* projected that government and industry will spend $393 billion (in fixed 1982 dollars to discount inflation) on air-pollution abatement in the 1980s. That amount represents more than half of the total funds that are needed to combat all the environmental problems combined. On the other hand, at least through the mid-1980s, the White House Council on Environmental Quality has maintained that public and private funds that are spent on air-quality control are exceeded by the value of benefits that result from reduced-air pollution damage. However, some businesses and industrial leaders have expressed serious doubt concerning whether that positive cost–benefit ratio will continue through the remainder of this decade. For example, the economic burden of air-quality control is falling heaviest on the already depressed steel, rubber, and heavy manufacturing industries.

Some industrial spokespersons object to current air-quality standards. They argue that such standards are inappropriate, because no direct relationship between the adverse health effects and the concentration levels of specific pollutants has been established beyond doubt. In contrast, some environmentalists feel that existing standards are not strict enough and do not allow an adequate margin of safety. In response to both objections, the 1977 Clean Air Act Amendments called for the EPA to conduct a scientific review of the nation's ambient air-quality standards at 5-year intervals, beginning in 1980.

Partially because of economic considerations, in early 1979, the EPA relaxed its national ambient standard for ozone, which was considered to be an index of the photochemical smog level. Industry representatives had argued that the cost of compliance with the original standard was excessive. And their case was supported by medical studies which indicate that ground-level ozone is somewhat less hazardous to health than was previously believed. As a result, the maximum allowable 1-hour exposure concentration was raised from 0.08 ppm to 0.12 ppm.

Then, in early 1984, the EPA announced plans to rewrite the national standard for ambient air quality for suspended particulates, so that only particulates that are smaller than 10 microns in diameter will be regulated. This change was prompted by indications that serious health effects are associated only with the smallest particles that can penetrate the respiratory system's defense mechanisms.

Some industries claim that the costs of pollution control in general are so high that compliance would force them out of business. And, indeed, some industrial plants have had to close their doors because they were unable to afford control costs. In a now classic example, in the fall of 1977, Youngstown (Ohio) Sheet and Tube Company, a steel manufacturer, closed down–laying off 5000 employees in the process—because, they said, the cost of pollution controls plus declining market conditions made its continued operation unfeasible.

Another objection to enforcement is that appropriate technology is not yet available to meet the standards. Industries that take this position have gone to court to battle with environmentalists in an attempt to weaken or avoid regulations. Nevertheless, in spite of resistance in some quarters to full compliance with air-quality standards, as we have seen, ambient air quality has improved and emissions have been reduced considerably.

BOX 11.2

THE COSTS AND BENEFITS
OF COST–BENEFIT ANALYSIS

*How much would you be willing to pay to postpone
your death for 2 years? What is the value of being
able to look across the Grand Canyon without
having your view obscured by a haze of air pollution? Have the costs of the nation's clean air
program exceeded its benefits? These might seem
to be rhetorical questions, but for economists
using cost–benefit analysis, the answers are useful
in determining which policies, programs, or
projects governments should pursue. Those who
use cost–benefit analysis compare the estimated
costs and benefits of these activities to select
those that are economically efficient. Programs
are said to be efficient when their benefits exceed
their costs, and when they increase a society's
net economic output or well-being.*

*Advocates of cost–benefit analysis believe
that it has many desirable features. Not only is it
conceptually simple, but if used appropriately,
cost–benefit analysis can improve policymaking.
The method can help policymakers identify the
likely consequences of proposed actions and
encourage the selection of cost-effective policies or
projects. Applied retrospectively, cost–benefit
analysis can be used to evaluate public programs.
This form of analysis is especially useful when a
government has limited economic resources but
many competing claims on them. Under these
circumstances, advocates of cost–benefit analysis
assert, the approach can identify the policies
that produce the largest impacts for the smallest
expenditure as well as policies that cost more
than they produce in benefits. For example,
cost–benefit analysis can be used to compare the*
*relative desirability of public policies in different
areas, such as health, transportation, or environmental protection or to compare different ways of
achieving the same policy goal.*

*Consider the three hypothetical policy alter-
natives that are outlined in the table below. Using
cost–benefit analysis a policymaker would quickly
reject Policy C as a poor investment and as a
waste of limited resources. Policy A would have
the highest ratio of benefits to costs and is the
most efficient—it produces the highest value
of economic output per unit of investment. These
advantages do not necessarily ensure that policymakers will choose Policy A; they might choose
Policy B because it is cheaper than Policy A.*

*If we accept the attributes of cost–benefit
analysis, then we must also accept the assumptions
underlying its use. These assumptions are highly
controversial, and they explain why most persons
strongly oppose the use of cost–benefit analysis,
particularly in the field of environmental protection. The first assumption, as we have already
noted, is that economic efficiency should be the
sole evaluative criterion. Many persons find this
emphasis unacceptable because it gives no weight
to other values. Critics routinely note that the
method ignores concerns about equity, the identification of winners and losers, or the geographical
distribution of a policy's costs and benefits. In
terms of cost–benefit analysis, as long as a policy
is efficient, it is irrelevant that the residents of the
northeastern United States suffer most of the
unfavorable consequences of acid precipitation,
while Midwesterners benefit from the lower electricity prices that are associated with the use of
high-sulfur coal. Likewise, an air-pollution control
program that protects most citizens might be*

A hypothetical cost–benefit analysis

ALTERNATIVE POLICY	ESTIMATED BENEFITS (IN MILLIONS OF DOLLARS)	ESTIMATED COSTS	COST–BENEFIT RATIO
A	300	150	2
B	120	90	1.33
C	80	120	0.67

judged as being efficient. But this same program, if extended to protect those people who need it the most—namely, infants, the elderly, and people with asthma, emphysema, and heart disease— might be inefficient and, therefore, economically indefensible.

Another assumption of cost–benefit analysis is that a monetary value can be placed on all the consequences of a policy. Monetary values must be assigned because a cost–benefit analyst must identify both the short- and long-term costs and benefits of a policy or program and then compare them in common units of analysis, which are usually dollars. As a result, economists using cost–benefit analysis must predict the consequences of government activities 20 or more years into the future and then place a dollar value on such things as human life, the view across the Grand Canyon, the esthetic and recreational benefits of a day in a national park, and even the furbish lousewort, an endangered plant species. Since these items are not sold in the marketplace, and thus have no readily agreed upon economic value, a dollar value must be assigned to them. Economists' ability to do so is problematic and subject to claims that the assignment of monetary values to such things as human health and life is arbitrary at best.

The value assigned to human life provides a fascinating example of this process. Economists usually rely on one of two methods to value a human life. With the first, economists estimate the present value of a person's future earnings. This approach places high dollar values on persons in their mid to late twenties, who are just starting

their careers, and a lower value on retirees, children, and persons who are unemployed. Relying on this method, several cost–benefit analyses of government programs have placed the value of a single life at $200,000. In one study of air pollution conducted for the National Academy of Sciences in the 1970s, however, the health cost of a premature death was placed at only $30,000. This amount was selected because those most likely to die from air pollution were elderly, chronically ill, and unlikely to live much longer under any circumstances.

With the second approach, economists assign dollar values on the basis of a person's willingness to pay. For example, the Environmental Protection Agency used this technique when it assessed the economic benefits of eliminating lead from gasoline. Lead affects the mental development of children, so the agency posed this question: "How much would parents be willing to pay if they could buy a 2.2 point increase in the IQ of their child?" The answer, according to the Environmental Protection Agency, is slightly less than $30. The willingness-to-pay method can also be used to assess the value that persons would place on their ability to avoid certain undesirable costs or consequences, such as premature death. The major disadvantage of this approach is that a person's willingness to pay is often dependent on his or her income or wealth.

Critics of cost–benefit analysis are concerned not only with its assumptions, but also with its appropriate application. Since the advocates of a project or policy are usually responsible for conducting the analysis, a potential exists for bias

Bubbling and Banking

Bubbling and banking are two regulatory approaches that promise to ease the financial burdens of industries that are trying to comply with air pollutant emissions standards. Both policies were conceived during the Carter administration and given renewed emphasis during the Reagan administration.

With *bubbling*, an imaginary "bubble" is placed over all of the stacks of a single industrial plant or over a number of stacks of many industries within

a restricted area. A single emission limit is then designated; that is, all sources covered by the bubble are treated as a single aggregate source. This reduces abatement costs by eliminating the need for stack-by-stack monitoring and controls. However, bubbling cannot be applied to air-pollutant sources in nonattainment areas.

Banking enables industries within the same air-quality control region to accumulate, sell, or purchase emissions credits in the same way that they would negotiate for any other commodity. Thus an

and a conflict of interests. Unfortunately, evidence of such conflicts is readily found. In order to justify their projects, government agencies have often overestimated benefits, underestimated costs, and ignored some environmental consequences. For example, in its cost–benefit analysis of the Tennessee–Tombigbee Waterway through Alabama and Mississippi, the Army Corps of Engineers justified the project's benefits on the basis of estimated barge traffic that ranged from improbable to impossible. Although Congress had authorized a channel width of 170 feet, for purposes of computing benefits, the Corps used a channel width of 300 feet. When the costs of the project more than doubled, the Corps did not report the changes for fear that the cost–benefit ratio would be altered. Defenders of cost–benefit analysis respond to such examples by noting that the fault lies with the agencies conducting the analyses, not the method.

Many of these advocates also believe that cost–benefit analyses should be conducted more frequently. These advocates cite the federal government's clean air program as a prime opportunity for analysis. According to the Clean Air Act Amendments, the Environmental Protection Agency is not allowed to consider economic costs when it establishes national ambient air quality standards or standards for hazardous air pollutants. For these areas, the Environmental Protection Agency must be oblivious to all the costs of complying with these standards. In effect, Congress has made a judgment that whatever the ultimate costs, they are acceptable. Opponents of this position claim that without an effort to

balance costs and benefits, the public may be paying for far more pollution control than the benefits can justify. To support their case, opponents of the current policy point out that other sections of the Clean Air Act Amendments and other environmental laws, such as the Toxic Substances Control Act and the Clean Water Act, require consideration of costs. Several studies of the federal government's program to control motor vehicle emissions, including one by the Council on Environmental Quality, have found that the program's likely benefits are substantially smaller than the costs. The studies suggest that air quality standards for some mobile-source pollutants could be relaxed, with important cost savings, but with little or no reduction in benefits. In contrast, many cost–benefit studies of emissions from stationary sources of pollution have found that the benefits are much larger than the costs.

However we feel about cost–benefit analysis, we should be aware that it has gained increasing favor among government officials. Presidents Gerald Ford and Jimmy Carter both issued directives encouraging federal agencies to compare the anticipated costs and benefits of proposed new regulations. More recently, President Reagan initiated a program requiring all federal agencies, to the extent that laws allow them, to refrain from issuing new regulations unless their estimated benefits exceed their estimated costs and to choose regulations that maximize a society's net economic benefits while imposing the least net costs on society.

This box was written by Professor Richard Tobin.

industry can earn emissions credits for a specific pollutant when it reduces (or maintains) emissions of that pollutant to a rate that is lower than the maximum allowable level. Since the EPA requires that the emission credits be "surplus, enforceable, permanent, and quantifiable," the banking system allows a plant to meet abatement regulations by obtaining emissions credits that are earned by another industry. Many companies often find this option preferable because it is less costly than installing their own pollution control systems. For example,

in Louisville, Kentucky, the General Electric Company leased volatile organic carbon (VOC) credits that were "banked" by International Harvester at a cost of $60,000. In this way, General Electric realized a $1.5 million savings in control technology that they otherwise would have been required to purchase.

However, bubbling and banking have not made impressive gains. By 1984, only three air-quality control regions (San Francisco bay area, Puget Sound, and Louisville) had banks that were operating, and the EPA had sanctioned only 18 "bubbles" nationwide. Greater participation is likely, however, as the economic advantages of those two policies become more firmly established.

Monitoring Costs

Damage and abatement account for most of the economic impact of air pollution. But a significant sum of money also goes for monitoring air quality so that progress in cleanup can be documented. In 1980, state and local governments spent an estimated $215 million on air-pollution regulation and monitoring while the federal government spent an estimated $125 million. However, monitoring can also save money (as well as lives) indirectly by warning the public of potential episodes of air pollution. For example, city health departments issue air-pollution alerts when ozone levels approach hazardous concentrations, and they advise persons with respiratory problems to stay indoors and to avoid strenuous activity (Table 11.3). Thus in such cases, expensive monitoring procedures result in savings that are associated with human health.

What of the Future?

Some environmentalists assign full responsibility to industry and big business for the less-than-rapid pace of our progress toward improved air quality. But the situation is not that simple. Because ours is an industrial society, our nation's economic well-being is intimately tied to the well-being of its key industries. We depend on business and industry for employment and for goods and services that enhance the quality of life. However, if we opt for a growing economy over a steady-state economy, then our demands for more products will mean more resource utilization and more air pollution. Still, each of us is entitled to breathe air that will not impair our health or shorten our life expectancy. Thus our major problem is to reconcile our efforts to improve air

quality while maintaining a healthy and growing economy.

Typically, questions of policy must be resolved in the political arena, because the interests of the parties that are involved are often in fundamental conflict. Environmentalists may argue convincingly of the health benefits of cleaner air, but industries must also maintain their economic viability. For example, it is most unlikely that an industry would voluntarily install pollution-control equipment, because the cost of such an action would probably cause them to function at a competitive disadvantage, especially with foreign producers who may face only limited pollution-control requirements. Furthermore, consumers will not purchase a more expensive, but otherwise equal, product just because it was manufactured by a nonpolluting industry. Hence the courts and the legislature will continue to provide the arenas where industries and environmentalists seek resolution to their conflicts.

Conclusions

We are making real progress in our efforts to improve the quality of the air we breathe. Strict federal regulations are reducing pollution levels in the ambient air and the workplace. However, this progress has cost a great deal of money, it has generated an ongoing controversy, and the many problems that remain will continue to do both in the future. The dollar cost for abatement is a heavy burden for industries and consumers alike. The scientific validity of several existing standards has been challenged, and some of those challenges have brought about revisions of some standards (such as that of ozone). And we must overcome many major technical obstacles before we can control certain pollutants at all. In view of such difficulties, we may very well decide that we must learn to live with tolerable levels of air pollutants—that is, pollutant concentrations that pose some risk to public health rather than no risk at all. What is your opinion?

Summary Statements

The 1970 Clean Air Act Amendments set rigid standards for ambient air quality and for emissions from automobiles and stationary sources. Since 1970, that law has been refined, but it has not been changed substantially.

Primary standards for air quality are based on potential human health effects, while secondary standards are designed to minimize the impact of air pollution on crops, visibility, climate, materials, and personal comfort.

Control standards for criteria air pollutants are based on the assumption that ill effects are likely only after concentrations reach some threshold value. Hazardous air pollutants, on the other hand, pose a health risk even in extremely low concentrations.

The Clean Air Act Amendments of 1977 limit the amount of deterioration of air quality that is allowable in national air-quality control regions.

The Occupational Safety and Health Act of 1970 sets standards to reduce hazards that attend exposure to dangerous substances in the workplace.

An overall upswing in national ambient air quality is indicated by trends in (1) the Pollutant Standards Index and (2) levels of five of the criteria air pollutants.

Probably the chief reason for the recent upswing in the quality of ambient air is control of motor-vehicle emissions.

Two possible alternatives, building taller smokestacks and dispersing industry to rural sites, are ineffectual large-scale strategies to control air quality. The only effective alternative is to reduce point-source emissions.

Electrostatic precipitators, filters, and cyclone collectors can reduce particulate emissions by up to 90 percent. However, particulates that cannot be removed are often the very small ones that pose the greatest health hazard. Scrubbers effectively remove water-soluble gases, including sulfur oxides, from industrial effluents.

Industrial control technologies for NO_x have been relatively slow in developing. One of the most promising approaches to NO_x abatement is staged combustion.

An effective program to control air quality must not only require adequate control techniques, but also provide for proper disposal or reuse of collected air pollutants.

Control of automobile emissions so far primarily has involved modification of the internal combustion engine. But automobile makers have yet to comply with emissions standards set by the 1970 Clean Air Act Amendments. Because of technical constraints and the need to improve fuel economy, the government has moved the deadline for compliance forward many times.

Other options for reducing motor-vehicle emissions include using alcohol as fuel, phasing out the internal combustion engine, encouraging car pools, and establishing bus lanes.

The total monetary cost of air pollution is difficult to assess. Costs stem from pollution damage, monitoring, and cleanup. Those costs must be weighed against the benefits that accrue from cleaner air.

Bubbling and banking are two regulatory approaches that can reduce the financial burdens of industries that are trying to comply with emissions standards.

Questions and Projects

1. What are industries in your area doing to control emissions? Do special obstacles to adequate abatement exist in your area?

2. How and where is ambient air quality monitored in your community? Are summaries of pollution levels available? If they are, how does your community's air compare with national ambient standards for criteria pollutants, which are shown in Table 11.1? Have there been discernible trends in local air quality?

3. How do you rate your community's efforts to control air quality? What are the successes and the failures? Is the overall feeling one of complacency or is there evidence of a well-directed effort? What are you personally doing to improve your local air quality?

4. How might air-pollution control regulations influence land use and community population growth rates?

5. Large corporations may be more able to absorb the costs of emissions control than small companies. Can you think of ways in which this apparent inequity could be avoided?

6. Enumerate some of the "hidden" direct costs of air pollution. How do those costs compare with the obvious expenses that result from polluted air (damage, cleanup, and monitoring)?

7. Does current legislation to control air quality address the problem of interstate transport of air pollutants? Support your response.

8. One proposal for the control of air pollution is the siting of new industry in suburban and rural areas. Comment on the economic and social implications of that strategy.

9. Identify the factors that should be taken into account in designing a community air-pollution index.

10. Some communities have ordinances that ban cigarette smoking in such public places as restaurants and buses. Are such ordinances a form of air pollution control? Do they violate individual rights?

11. Distinguish between primary air-quality standards and secondary air-quality standards. Pollutant emission standards for stationary and mobile sources are based on what fundamental assumption?

12. Why is it inappropriate to design measures to control air quality in the context of an "airshed"?

13. What are the advantages and disadvantages of tall smokestacks as a strategy to control air quality?

14. Identify and briefly describe the various techniques and devices that are used by industry to control air quality.

15. How might effective air-pollution abatement technology actually aggravate water-pollution problems?

16. Identify some strategies that are being used to control automobile emissions other than modification of the internal combustion engine.

17. Why has full compliance with the strict air-quality standards that are specified by the 1970 Clean Air Act been delayed in numerous instances?

18. Why does progress in achieving ambient air-quality standards in urban areas depend on successful control of automobile-exhaust emissions?

19. Explain how efforts to cope with shortages in energy supply may conflict with efforts to improve national air quality.

20. In view of the fact that approximately 100 million buildings in the United States exist, speculate on the economic impact of strict regulation of indoor air quality.

Selected Readings

ENVIRONMENTAL PROTECTION AGENCY: *Research Summary, Controlling Nitrogen Oxides.* Cincinnati: EPA Center for Environmental Research Information, 1980. Brief discussions of the various strategies that are being researched by the EPA to control emissions of nitrogen oxides.

ENVIRONMENTAL PROTECTION AGENCY: *Research Summary, Controlling Sulfur Oxides.* Cincinnati: EPA Center for Environmental Research Information, 1980. Brief discussions of the various options that are available and being researched to control emissions of sulfur oxides.

HILEMAN, B.: Acid rain: A rapidly shifting scene. *Environmental Science and Technology* 17:401A–405A, 1983. The special problems of controlling acid rain.

JOSEPHSON, J.: Environmental quality standards development. *Environmental Science and Technology* 16:156A–159A, 1982. A consideration of peer-reviewed scientific data in the design of environmental quality standards.

RUBIN, E. S.: International pollution control costs of coal-fired power plants. *Environmental Science and Technology* 17:366A–377A, 1983. The economic aspects of controls on power-plant effluents.

THE CONSERVATION FOUNDATION: *State of the Environment 1982.* Washington, D.C.: The Conservation Foundation, 1982. Includes an update on trends in air quality nationwide.

A large open pit mine near Butte, Montana. (© Georg Gerster,
Photo Researchers.)

CHAPTER 12 RESOURCES OF THE EARTH'S CRUST

In June of 1859, one of the world's richest deposits of gold and silver—the Comstock Lode—was discovered near Virginia City, Nevada. Overnight, the town was transformed into a booming mining camp when prospectors, merchants, and other entrepreneurs who were in search of a quick fortune arrived from far and near. By the early 1870s, Virginia City's population had swelled to 30,000 and silver and gold production totaled hundreds of millions of dollars. But then the Comstock Lode met the fate of all mineral deposits: the gold and silver ran out. Miners, merchants, and others fled, the local economy collapsed, and by the 1880s, Virginia City had become a virtual ghost town, with fewer than 1000 residents. Today, the population is even smaller, and the town relies on tourists who come to view the artifacts of a more prosperous era (Figure 12.1).

Similarly, in Butte, Montana, which was once known as the nation's copper-mining capital, billions of dollars worth of copper and other metals were extracted during more than 100 years of mining. At one time, an estimated 20,000 persons were employed in the mines. But in January, 1983, the Anaconda Minerals Company announced its plans to suspend copper mining and milling operations in Butte. The company blamed unfavorable economic conditions for that move, which eliminated 700 jobs (approximately 5 percent of the local work force).

Although there is still plenty of copper in the ground, the cost of mining simply became too expensive as a result of the depressed market value of copper.

The boom-to-bust stories of mines in Virginia City and Butte have been told repeatedly throughout history as important mineral and fuel deposits have been discovered, exploited, and ultimately abandoned. In most cases, mining ceased when the deposit ran out or became so lean that continued operations were no longer feasible economically. In recent years, however, the pressures of environmental regulations have compounded matters. For example, the lead–zinc mines of southwestern Wisconsin were the longest continuously operating metallic mines in the nation until they closed in 1978. Mine operators attributed the shutdown to a combination of depressed zinc prices and their inability to meet strict water-quality standards for discharge waters.

In our nation's early days, new resource discoveries were more than adequate to meet the mineral and fuel needs of the growing population. Today, however, new discoveries are rarer, and as our demands for mineral and fuel resources continue to climb, we are continually reminded that the earth's mineral and fuel resources are finite. In this chapter, we survey the earth's crustal resources, the methods by which they are extracted, and the impact of their extraction on environmental quality. We also ex-

Figure 12.1 *The Comstock mines, Virginia City, Nevada. (U.S. Geological Survey photo by T. H. O'Sullivan.)*

A Historical Overview

When the Americas were discovered, they seemed to hold unlimited resources. Thus from colonial days into the early part of this century, mineral resources were extracted with little or no regard for conservation or environmental impact. For example, the Mining Law of 1872 essentially gave away the government's mineral rights to anyone who located a mineable deposit on federal lands, which reflects the spirit of the times in which that law was written.

During the first half of this century, the rapid growth of manufacturing industries created a strong dependency on a continuous supply of raw materials. However, during World War I, when the production of war materials rose sharply, industries began to feel the effects of materials shortages, and

amine the pressing problem of managing those resources in the face of growing demands and dwindling reserves.

some began to question the notion that limitless resources were available. After that war, the mining of nonmetallic resources on federal lands (such as coal and phosphate) was regulated to some extent by the Mineral Leasing Act of 1920. As a result of that law, some industries had to lease mineral rights instead of receiving them at no cost, but no provisions were made for environmental protection.

As consumer demand increased during the 1920s, the pace of resource exploitation accelerated until the "good times" gave way to the Great Depression in the 1930s. Then, with the onset of World War II, not only did the need for raw materials rise sharply, but the increasingly sophisticated weapons that were being developed demanded a greater variety of materials. By that time, the United States had begun to rely on imports from other countries for some of its mineral needs. Later, in 1952, when the growing dependency on foreign sources appeared to be threatening to national security, a presidential commission was formed to assess the status of our domestic mineral resources. However, the commission's recommendations failed to curtail our dependency on a variety of materials that are not always available from domestic sources. In fact, following the birth of space-age technology in the 1960s, more products (such as computers and spacecraft) were being produced than ever before, which required materials that had special properties. For example, tungsten is used widely in today's electronics industry and cadmium is used in the control rods and shielding within nuclear reactors.

By the late 1960s, the public's heightened environmental awareness and concern about the impact of mining spurred congressional action and in 1970, Congress passed the Resource Recovery Act. One provision of that law established a National Commission on Materials Policy (NCMP). The National Commission on Materials Policy recommended that mining procedures be modified to minimize waste and environmental disruption. The commission also encouraged the government to escalate conservation efforts and to reduce national dependency on foreign imports. But it was not until 7 years later that federal legislation was passed to reduce the environmental impact of surface mining. Then, in late 1980, President Carter signed into law the National Materials and Mineral Policy Research and Development Act. That law enabled the Bureau of Mines to (1) assess international mineral supplies, (2) increase research on the mining of critical and strategic minerals, and (3) make analyses of mineral data available for federal land use decision making.

Thus throughout this century, a gradual, although significant, shift in the United States government's regulation of resources has occurred. In the early days, regulation was minimal. The emphasis was on encouraging mineral development in the interest of settling frontier lands and spurring economic growth. In later years, the pendulum shifted toward stricter regulation at both the federal and state levels. And today's mining laws attempt to address both the need for continued economic development and national security as well as the growing demand for environmental protection.

Generation of Crustal Resources

Earth is a nearly spherical planet, which has a diameter of about 12,800 kilometers (8000 miles). From studies of earthquake vibrational waves, geologists have been able to deduce the structure and composition of the earth's interior. They have divided the

earth's interior into three major zones—core, mantle, and crust (Figure 12.2a)—according to changes in the earth's constitution. The *core* extends from a depth of approximately 2900 kilometers (1800 miles) down to the earth's center and is composed of iron, which is perhaps alloyed with some nickel. The core has a solid inner portion and a fluidlike outer por-

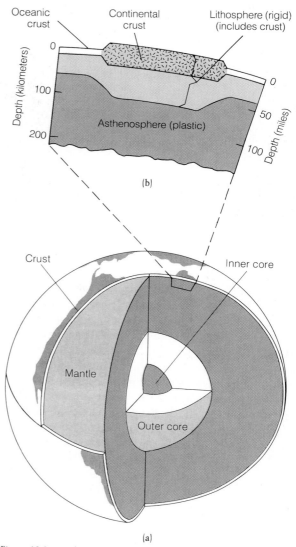

Figure 12.2 (a) The internal structure of the earth. All resources are extracted from the crust, which is the thin outermost layer. (b) The oceanic and continental crust plus the rigid upper portion of the mantle constitute the lithosphere. The lithosphere floats on the plastic asthenosphere. (From F. Press and R. Siever, Earth, 3rd ed. New York: W. H. Freeman, 1982, p. 17.)

tion. The *mantle* is sandwiched between the core and the crust. Its average thickness is almost 3500 kilometers (2200 miles). The mantle is composed of rock that is rich in iron and magnesium silicate,* which—although solid—is capable of slow flowage, much like an extremely viscous fluid. Above the mantle, forming the thin outer skin of the planet, is the *crust*, which is composed of a variety of rock types. Crustal thickness ranges from only 6 kilometers (3.8 miles) under the oceans to 70 kilometers (44.8 miles) in areas that have high mountains.

Although the earth's crust is insignificantly small when it is compared to the mass of the rest of the planet, it is the source of virtually all resources that are essential to our industrial way of life. Crustal resources include metals, nonmetallic minerals, rock, and fuels.

Fortunately, most commercially valuable resources are concentrated in specific areas within the upper crust of the earth. If those resources were disseminated uniformly throughout the crust, their concentrations would be so low that extraction would be neither technically nor economically feasible. As shown in Table 12.1, of the economically important elements, only four account for more than 1 percent of the total weight of the earth's crust: aluminum (8.00 percent), iron (5.80 percent), magnesium (2.77 percent), and potassium (1.68 percent). However, geologic processes, during millions of years, have selectively concentrated certain crustal elements into deposits that are commercially mineable today. For example, copper constitutes only 0.0058 percent of the earth's crust by weight. But in areas where that concentration has been enriched at least 60 to 100 times, copper can be mined profitably.

The same processes that concentrate resources in the earth's crust are also responsible for shaping the diverse features of the landscape. For example, landforms—mountains, canyons, plains, river valleys, and so on—develop in response to two types of processes that act on the earth's crust: internal crustal processes and surface crustal processes. *Internal crustal processes* include volcanic eruptions and the bending and fracturing of rock under great pressure—processes that are sustained by energy that is generated in the earth's interior. *Surface crustal processes* encompass interactions that occur be-

Table 12.1 The relative abundance of economically important elements in the earth's crust

NAME	CHEMICAL SYMBOL	CRUSTAL ABUNDANCE (PERCENT BY WEIGHT)
Aluminum	Al	8.00
Iron	Fe	5.80
Magnesium	Mg	2.77
Potassium	K	1.68
Titanium	Ti	0.86
Hydrogen	H	0.14
Phosphorus	P	0.101
Manganese	Mn	0.100
Fluorine	F	0.0460
Sulfur	S	0.030
Chlorine	Cl	0.019
Vanadium	V	0.017
Chromium	Cr	0.0096
Zinc	Zn	0.0082
Nickel	Ni	0.0072
Copper	Cu	0.0058
Cobalt	Co	0.0028
Lead	Pb	0.00010
Boron	B	0.0007
Beryllium	Be	0.00020
Arsenic	As	0.00020
Tin	Sn	0.00015
Molybdenum	Mo	0.00012
Uranium	U	0.00016
Tungsten	W	0.00010
Silver	Ag	0.000008
Mercury	Hg	0.000002
Platinum	Pt	0.0000005
Gold	Au	0.0000002

Source: From F. Press and R. Siever, *Earth*, 3rd ed. New York: W. H. Freeman, 1982, p. 553.

tween the earth's crust and the wind, running water, glaciers, and organisms—all of which are ultimately sustained by solar energy (Chapters 2, 6, and 9). In general, internal processes tend to build continents, while surface processes wear away (erode) continents. In surveying the processes that shape the earth's crust, we can appreciate the finite nature of the resources that they produce.

---

* A silicate is a compound of oxygen, silicon, and a metal.

Internal Crustal Processes

Much of the activity that results in crustal shaping is initiated by the movements of huge segments of the lithosphere, which are called *plates*. The uppermost portion of the mantle, which is rigid, is combined with the overlying crust to constitute the *lithosphere* (Figure 12.2*b*). The lithosphere is divided into about 10 large plates (Figure 12.3), each of which is approximately 100 kilometers (62 miles) thick. Those plates drift very slowly with respect to one another—typically less than 20 centimeters per year. The driving forces that cause the plates to move are convective currents that develop in the underlying plastic portion of the mantle, the *asthenosphere*, which has an average thickness of approximately 200 kilometers (120 miles) (Figure 12.2*b*). Heat

is generated in the earth's core and is released as a result of the decay of radioactive elements. That heat is then distributed within the mantle by convective currents, in which heated rock slowly flows upward and cool rock slowly flows downward. In places where convective currents reach the underside of the lithosphere, they push against and drag along the rigid plates.

Land masses are part of, and drift with, the plates. Geologic evidence indicates that 225 million years ago, Eurasia, Africa, and the Americas were combined as one continent, which is named *Pangaea*. Subsequently, Pangaea broke up as the result of the shifting of plates, and its constituent land masses, which are the continents that we know, slowly drifted apart and eventually reached their present locations (Figure 12.4).

Many large-scale landscape features, including mountain ranges, volcanoes, and deep oceanic trenches, are associated with geologic processes that

Figure 12.3 *The earth's lithosphere is divided into a number of shell-like plates that slowly drift across the face of the globe. (After W. M. Marsh and J. Dozier,* Landscape. *Reading, Mass.: Addison-Wesley, 1981, p. 368.)*

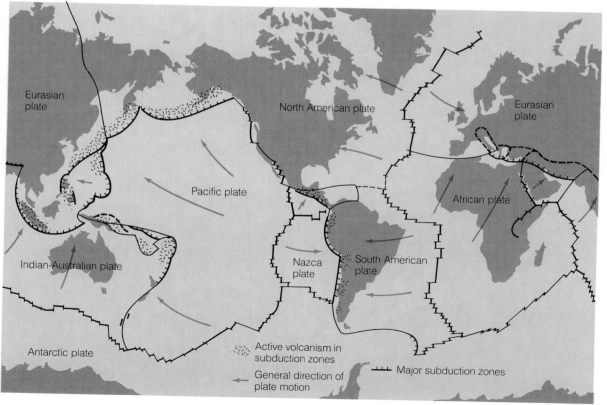

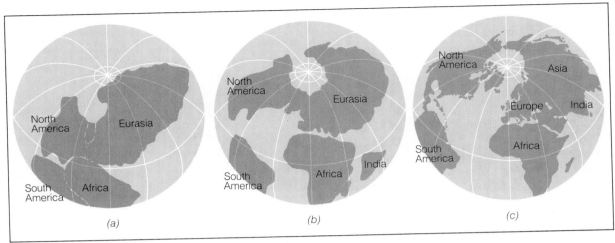

Figure 12.4 *The drifting continents: (a) 200 million years ago, (b) 65 million years ago, and (c) today. (From P. J. Wyllie,* The Way the Earth Works. *New York: Wiley, 1976, p. 2.)*

occur at the boundaries between plates. Those same processes concentrate elements that are important economically into mineable deposits. If we consider relative plate movement, we can define three types of boundaries: divergent, convergent, and parallel.

At *divergent plate boundaries*, adjacent plates drift apart and the melted mantle rock wells upward to fill the gap that is created (Figure 12.5). That molten rock, which is called *magma*, cools and solidifies to form new oceanic crust. For example, the mid-Atlantic ridge marks a plate boundary that has been

diverging for perhaps 200 million years. The divergence in that boundary split Pangaea apart and, in the process, opened the Atlantic Ocean (see Figure 12.4).

At *convergent plate boundaries*, one plate slides downward and under an adjacent plate, bending the crust into an oceanic trench (Figure 12.6). The downward-moving plate is forced into the *subduction zone*, which is at the convergent boundary, and enters the mantle, where it melts. Hence plate material forms at divergent boundaries and is destroyed at convergent boundaries (Figure 12.7). Meanwhile, the overriding plate is bent and fractured, which causes mountain ranges to form. The subduction zone is also the site at which volcanic eruptions and major earthquakes occur.

In some cases, plates move parallel to one another. Although crust is neither created nor destroyed in those instances, the slipping of plates horizontally past one another can trigger earthquakes. For example, the San Andreas fault of California,

Figure 12.5 *A divergent plate boundary. (a) As plates move apart, magma wells up from the mantle below and cools and solidifies as new oceanic crust. (b) The lateral spreading of the sea-floor opens the ocean basin. (From R. S. Dietz, "Geosynclines, Mountains, and Continent-Building."* Scientific American *226:37, 1972. Copyright 1972 by Scientific American, Inc. All rights reserved.)*

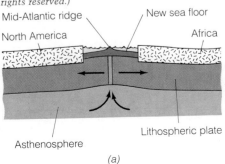

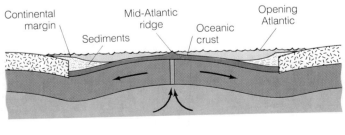

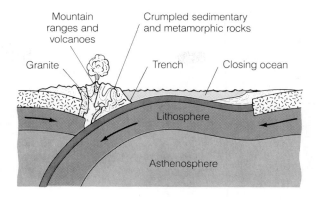

Figure 12.6 *A convergent plate boundary. In areas where plates collide, one plate plunges under the other and into the mantle. The overriding plate is deformed into mountain ranges. (From R. S. Dietz, "Geosynclines, Mountains, and Continent-Building."* Scientific American *226:37, 1972. Copyright 1972 by Scientific American, Inc. All rights reserved.)*

Figure 12.7 *Plates are formed at divergent boundaries and melted in subduction zones at convergent plate boundaries. (From P. R. Ehrlich et al.,* Ecoscience, Population, Resources, Environment. *New York: W. H. Freeman, 1977, p. 17, as modified after P. A. Rona, "Plate Tectonics and Mineral Resources,"* Scientific American *229 (July):95, 1973.)*

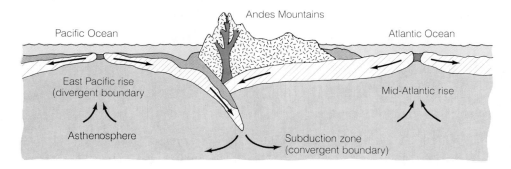

which was the site of the 1906 earthquake that destroyed San Francisco, marks such a boundary (Figure 12.8). There, the Pacific plate slides toward the Northwest, past the American plate.

Plate boundaries are zones of weakness that allow magma to invade the crust and, in some cases, to feed volcanoes. Magma that reaches the earth's surface is called *lava*. Lava may form huge volcanoes or spread over large areas. For example, the Hawaiian Islands consist of overlapping volcanoes that tower up to 10,700 meters (35,000 feet) above the Pacific Ocean floor; in the Pacific Northwest, the Columbia Plateau is covered by many layers of solidified lava, which, in some places, are more than 1400 meters (4500 feet) thick.

Intrusions of magma into rock and extrusions of lava on the earth's surface may be associated with commercially important mineral deposits, which are called *ores*. For example, along divergent plate boundaries, the magma that moves upward from the

mantle interacts with the seawater that enters fractures in the seafloor. Through a complex series of chemical rections, metallic sulfide† ore is deposited in those fractures, which are called *hydrothermal* (hot water) *vents*. Many such deposits were discovered recently along oceanic ridges, and that process is considered to be the origin of the metal-rich nodules that carpet the ocean floor.

Metallic deposits also may occur at convergent plate boundaries. The rising temperatures and pressures in those zones cause descending plates to melt partially, which liberates metallic elements. The metallic fluid then becomes concentrated in the magma, which ascends into the crust, cools, and solidifies as metal-rich rock.

In areas where magma invades solid rock, the heat, pressure, and chemically active fluids that are associated with magma trigger chemical alterations

---

† Metallic sulfide ore is composed of metals that are combined with sulfur.

399

Figure 12.8 *A LANDSAT satellite view of the San Andreas fault. The Pacific plate is to the left and the American plate is to the right. (EROS Data Center.)*

when the magma and host rock come in contact. That process is known as *contact metamorphism*, which is responsible for the formation of some important deposits of lead and silver. On a larger scale, the processes that occur within the subduction zone can elevate crustal temperatures and pressures over a wide area. That process, which is called *regional metamorphism*, has produced large ore deposits of asbestos, talc, and graphite.

Surface Crustal Processes

The processes that build the landscape are countered by surface crustal processes. Those processes include mechanical and chemical weathering as well as erosion. For example, when *bedrock* (solid rock)

is exposed to the atmosphere, it is subjected to *weathering*. Several physical processes, which are called *mechanical weathering*, fragment rocks: (1) the expansional pressure that accompanies the freezing of water that is trapped within cracks, (2) alternating extremes in air temperature, and (3) the wedging apart of rocks by plant roots. Chemical weathering occurs when rocks are decomposed as a result of their interaction with moisture or atmospheric oxygen and carbon dioxide. Chemical weathering produces solid alteration products, such as clay and some water-soluble products (for example, iron and calcium compounds).

The solid products of both mechanical and chemical weathering are rock fragments, which are called *sediments*. Some sediments are also composed of the remains (shells and skeletons, for example) of formerly living organisms. The ultimate product of weathering is *soil*, which is a substratum that supplies the nutrient and water needs of plants. Various agents of erosion remove, transport, and redeposit sediments and, in the process, expose fresh bedrock to weathering. As noted in Chapter 6, running water causes the most erosion on land, and rivers transport sediments from their drainage basins and deposit them along floodplains, in deltas, and in the sea. Wind causes erosion primarily in deserts and coastal areas, and ocean waves and currents redistribute sediments along shorelines. And in some high mountain regions, and in Greenland and Antarctica, glaciers provide the primary erosional force.

Weathering and erosion also concentrate resources in ore deposits either by adding desired mineral matter or by subtracting unwanted host rock or sediments. For example, gold, diamonds, and platinum are weathered out of their host rock and become mixed with sand and gravel in streambeds. Such sedimentary deposits of heavy minerals are called *placers.* Important sand and gravel deposits occur in areas where erosion by wind or water removes the finer sediments, thereby concentrating the larger sediment. Examples include beach and stream channel deposits. In some instances, dissolved minerals precipitate from groundwater (such as iron and copper compounds) or from seawater (such as manganese dioxide and table salt) to form mineable concentrations.

Rocks

Internal and surface processes that sculpt the landscape favor the formation of a multitude of rock types. Some of those rocks are economically important themselves without further processing; others are important because of valuable minerals that can be extracted from them. Rocks are classified into three groups—igneous, sedimentary, or metamorphic—according to the way in which they are formed.

Igneous rocks are created when magma cools and solidifies. That magma may remain within the earth's crust and solidify slowly to form coarse-grained rocks, or (as we noted earlier) it may spew forth as lava through volcanoes or fractures and solidify rapidly to form fine-grained rock or glassy rock. For example, granite is a particularly durable, coarse-grained igneous rock that is found in many mountain ranges. It is used extensively as monument stone and as a building material.

Sedimentary rocks are composed of compacted and cemented sediments that are the products of rock weathering, or consist of organic remains. For example, a common sedimentary rock-forming process occurs when waves break against a rocky coast and grind rock into fragments, which are sorted by size, carried away, and eventually deposited elsewhere as beach sand. As sand accumulates, its grains are packed by the weight of overlying sediments and, subsequently, are cemented together by fluids that migrate through the permeable sediments. As a result, the common sedimentary rock, which is called *sandstone,* is formed. Sandstone is sometimes used as a building material. Another widespread and economically important sedimentary rock is *limestone,* which is composed primarily of the calcium carbonate remains of marine organisms (for example, shellfish and coral). Limestone is used widely in making portland cement (which is used in concrete), as a building material, and as a roadbed for highways and railroads.

Like many sedimentary rocks, *metamorphic rocks* are derived from other rocks. A rock is metamorphosed, or changed in form, when it is exposed to high pressures, intense heat, and chemically active fluids that are located deep within mountain belts. However, by definition the metamorphic environment is never hot enough to melt the rock completely back to magma; rather, the rock changes form. Thus during metamorphism, the mineral components of rock sometimes become aligned. In some rocks, that process results in a banded structure that may facilitate its cleaving into platelike slabs. Slate, which is used as a roofing or flooring material, is a good example. In other rocks, the constituent particles recrystallize, becoming larger, and the overall quality of the rock is changed. For example, metamorphism of limestone produces the coarser rock, *marble,* which is an attractive material that is valued by sculptors and builders.

Although the earth's crust is composed of rocks that belong to all three rock families, igneous rocks are by far the most abundant. However, sedimentary rocks are the most conspicuous because they form a relatively thin veneer over nearly 75 percent of the earth's surface. In some areas, weathering and erosion have exposed coarse-grained igneous and metamorphic rocks that were formerly located deep in the earth's crust. In other regions, volcanic activity has covered, or is in the process of covering, the surface with new igneous rocks. In most land areas, bedrock is hidden underneath unconsolidated sediments, soil, and vegetation (Figure 12.9). Although bedrock is usually exposed in mountainous regions, in other regions, geologists often must search for scattered outcrops or drill through the surface soil and sediments to determine the composition of the local bedrock.

Figure 12.9 *Bedrock is often exposed where a road cuts through a hill. (Clyde H. Smith, Peter Arnold, Inc.)*

Minerals

Although rocks are sometimes valued for themselves, more often they are sought for the minerals they contain. A few rocks are composed of a single mineral. For example, rock salt consists entirely of the mineral *halite* (NaCl). However, most rocks are aggregates of minerals, typically consisting of a few dominant and several accessory minerals. *Minerals* are solids that occur naturally and are usually characterized by an orderly internal arrangement of atoms. This is another way of saying that most minerals are *crystalline*. All the atoms in a given mineral are arranged in a distinctive three-dimensional geometry, and that geometry, or structure, along with the mineral's elemental composition, determine its chemical and physical properties.

Frequently, an economically important mineral (ore) is only a minor component of a larger rock mass. Hence to obtain that mineral, miners must extract enormous quantities of rock from the ground and then separate the desired mineral from its host rock. For example, the average copper deposits that are mined today are composed of a grade that is approximately 0.6 percent copper. Thus every 1000 metric tons of copper-bearing rock that are mined will yield an average of only 6 metric tons of copper; the remaining 994 metric tons comprise waste rock. The mechanical processes, such as crushing and sorting, that separate the copper from the waste rock require large inputs of energy and water, which contributes substantially to the costs of copper mining.

Nonmetallic minerals (such as calcite, gypsum, and halite) typically are useful in their natural state, as are metals, such as gold and silver, which occur as elements. However, most metals are united chemically with other elements, such as oxygen or sulfur. For example, iron is mined from the minerals hematite and magnetite, which are chemical combinations of iron and oxygen, and zinc occurs in the mineral sphalerite, which is a combination of zinc and sulfur. To liberate the desired metallic elements, the mineral must be broken down chemically through smelting or refining processes. Those processes usually require an enormous input of energy and produce solid, liquid, or gaseous waste products that must be disposed of. Recall from Chapter 10 that smelting of metallic ore is a major source of air pollution, especially sulfur oxides and particulates. Mining operations also usually recover other substances that have value in addition to the target mineral, especially in the case of metals. Often, the extraction of small amounts of valuable accessory minerals, such as silver or gold, can make the difference between a marginal and a profitable enterprise.

Fossil Fuels

The fossil fuels—coal, oil, and natural gas—are the major sources of concentrated energy that are used by human beings. Coal is found in the earth's crust in the form of distinct layers, and oil and natural gas are trapped in permeable sedimentary rock that is sandwiched between less permeable layers of rock. As with all other crustal resources, a special set of circumstances was necessary to generate those invaluable energy sources.

When we burn coal, we are actually releasing solar energy that was stored in vegetation through photosynthesis tens or even hundreds of millions of years ago (see Chapter 2). At that time, luxuriant vegetation stood in swampy terrain, which probably resembled the modern Great Dismal Swamp of North Carolina and Virginia. Over many centuries, as giant tree ferns and other plants died, a thick vegetative mat accumulated in the swampy waters. High concentrations of carbon were gradually produced as a result of the partial decomposition of the matted plant remains by anaerobic bacteria. Eventually, the mass of highly carbonized, partially decomposed plant materials (which is called *peat*) was further compacted under the weight of more sediment and plant remains. Then the increasing heat and pressure transformed the peat into *lignite* (brown coal), and then into *bituminous coal* (soft coal). In regions where the heat and pressure were particularly intense, bituminous coal was metamorphosed to *anthracite coal* (hard coal). During that process, seams (layers) of coal are sandwiched between other sedimentary strata in various thicknesses. For example, west of the Mississippi, coal seams are very thick (3 to 60 meters, typically) and tend to lie relatively close to the surface. Eastern coal seams are thinner (1 to 3 meters) and many occur at great depths. As we see later, the depth of the coal seam dictates the method of mining that is needed to extract it.

The stages in the sequence from peat to anthracite are called *ranks of coal.* The higher the rank (from lignite to bituminous coal to anthracite coal), the greater is the concentration of carbon and the amount of heat that can be generated per unit of mass during combustion. Also, the higher the rank of coal, the smaller is the quantity of ash that is released during combustion. Thus anthracite yields the most calories per kilogram when burned and has the lowest potential for air pollution; it is thus a "clean" fuel. Anthracite, however, is in short supply. Also, the metallurgical industry uses most of the available anthracite. For these reasons, bituminous coal (which constitutes about half the world's fossil-fuel reserves) is the rank of coal that is used primarily in electric power plants and in some industrial processes.

Unfortunately, bituminous coal typically contains significant concentrations of sulfur and, during combustion, yields oxides of sulfur—a health hazard and a precursor of acid rain (see Chapter 10). Although western coal is relatively low in sulfur (0.6 percent is typical), the bulk of eastern coal contains 2 to 3.5 percent of sulfur. Thus pressure is mounting for further development of western coal resources, partly because western coal is cleaner to burn.

Like coal, the oil and natural gas that constitute petroleum are organic in origin. However, the process by which the original organic material (which included both plant and animal remains) was converted into petroleum is extremely complex and poorly understood. Oil is a mixture of thousands of different hydrocarbon molecules (molecules that contain only hydrogen and carbon) and other organic molecules. Natural gas is actually a mixture of gases that consists primarily of the hydrocarbon methane (up to 99 percent by volume), which is the most common fuel that is marketed commercially, and small quantities of the hydrocarbons ethane, propane, and butane.

Petroleum usually originates within layers of shale, but under pressure, the petroleum is squeezed into adjacent layers of permeable sandstone or limestone, which partially displaces the water that occupies the pore spaces. Then an overlying layer of less permeable rock traps the petroleum in the so-called reservoir rock. As illustrated in Figure 12.10, the reservoir rock may contain a stratification of natural gas over oil and oil over water. A cubic meter of reservoir rock may contain as much as 100 liters (26.4 gallons) of oil and 50 liters (13.2 gallons) of water.

Within the reservoir rock, oil is usually under pressure, which is caused by water pushing up from below or from a cap of natural gas. If the overlying rock is fractured, the oil may rise to the surface,

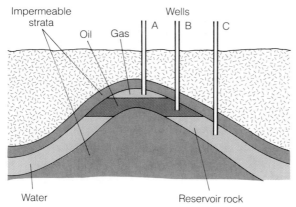

Figure 12.10 *Layers of rock, forming a trap for oil and natural gas. Well A yields gas, well B yields oil, and well C yields water. (After A. N. Strahler,* Principles of Earth Science. *New York: Harper & Row, 1976.)*

forming a natural oil seep. Although the pressure in the oil reservoir is normally advantageous for oil extraction, care must be taken during exploratory drilling to prevent oil from bursting to the surface as a gusher or to prevent a gas blowout. After the pressure is relieved, pumping is usually needed to extract the petroleum.

The air-pollution potential of burning or processing oil is much lower than that of coal. The sulfur content of oil varies, depending on the region in which the source is located, but it is usually less than 2 percent. (The sulfur content of Indonesian oil may be as low as 0.5 percent.) Since it contains only a trace amount of sulfur, which can be removed readily before combustion, natural gas is an even more desirable fuel than oil from the perspective of air quality.

Nuclear Fuels

Nuclear fuels, primarily uranium, are used to power nuclear reactors. (Virtually all large commercial reactors are used to generate electricity.) In the United States, the primary source of uranium is found in certain sandstone strata, which are located in the Colorado Plateau, Wyoming basins, and along the Texas coastal plain. Apparently, the uranium minerals, which weathered out of granite, precipitated from groundwater that circulated through those ancient sandstones. Hence uranium minerals primar-

ily occupy pore spaces within the sandstones. In other parts of the world, uranium ores occur as fillings within rock fractures and as placer deposits. In mineral form, uranium occurs in chemical combination either with oxygen or with oxygen and silicon. The most important domestic reserves occur in grades that are in a range of 0.08 to 0.30 percent uranium minerals.

The Rock Cycle

A rock in any of the three rock families (igneous, sedimentary, or metamorphic) may be altered as a result of geologic processes so that, eventually, it is transformed into rock of another family, as illustrated in Figure 12.11. That *rock cycle* implies a continual generation of rock, mineral, and fuel resources. For example, weathering reduces an igneous rock mass to sediments, and the sediments are subsequently eroded and redeposited elsewhere. Then the sediments that have accumulated are compacted to form sedimentary rock. As the sedimentary rock is carried deep into the crust, rising temperatures and pressure metamorphose the rock. Eventually, temperatures become so extreme that the rock melts to form magma. The magma may subsequently well upward, cool, and solidify to produce igneous rock, thereby completing the rock cycle.

The transformations that are involved in the rock cycle, however, are extremely slow. Typically, the regeneration of crustal resources takes millions of years. Thus from our point of view, the supply of crustal resources is essentially fixed, and for that reason they are often referred to as *nonrenewable resources*.

Extraction of Crustal Resources

Compared to other land uses, mining uses only a very small amount of land. The Bureau of Mines reports that during the half-century between 1930 and 1980, only 0.25 percent of our nation's land, some 2.3 million hectares (5.7 million acres), were utilized for resource extraction and processing. Approximately 70 percent of that land was devoted to actual excavating, while the rest was used for waste disposal. Almost half the land was used for the mining of bituminous coal (Figure 12.12).

Although small in terms of the area that is af-

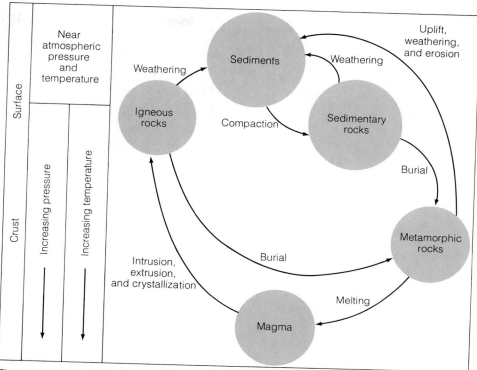

Figure 12.11 *Rocks of the three rock families are transformed from one family to another by processes that are involved in the rock cycle. The rock cycle implies a very slow generation of minerals, fuels, and rock resources.*

fected, mining is a very conspicuous form of land use because of its impact on the environment. For example, air and water are heavily polluted by refining and smelting processes. And lands that are being mined are stripped bare of vegetation and, subsequently, are eroded by wind and water (Figure 12.13). The severity of damage varies with the specific method of mining that is used, the physical and chemical properties of the resource and its waste by-products, and the site of the deposit. In this section, we examine the major techniques of surface and subsurface mining and consider their effects on the environment.

Surface Mining

Approximately 90 percent of the rock and mineral resources that are consumed in the United States and more than 60 percent of the nation's coal output

are extracted by methods of *surface mining*. Surface mining has a particularly disruptive effect on the landscape, because it requires the removal of the waste rock and soil (which is called *overburden*) and vegetation that lie over the deposit. But even in re-

Figure 12.12 *Land use for mining in the United States by resource from 1930 through 1980. (From the Bureau of Mines, U.S. Department of the Interior.)*

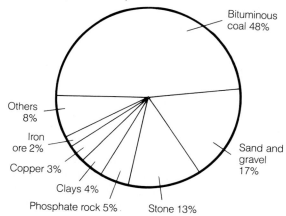

Bituminous coal 48%

Sand and gravel 17%

Stone 13%

Phosphate rock 5%

Clays 4%

Copper 3%

Iron ore 2%

Others 8%

Figure 12.13 *Severe soil erosion in an unreclaimed strip-mine area. (U.S. Department of Agriculture, Soil Conservation Service.)*

gions where underground mining techniques are feasible, surface mining is often favored by mine operators, because it permits more complete extraction and poses fewer safety hazards. The danger of collapse or explosion, which is considerable in underground excavations, is relatively low in surface mining. Furthermore, surface mining is less expensive than underground mining. For example, the surface mining of coal typically costs 40 percent less than subsurface mining.

Since several techniques of surface mining are practiced, the one that is used depends on the deposit. For example, sand and gravel are removed from *pits*, while limestone, granite, and marble are taken from *quarries*, such as that shown in Figure 12.14. In parts of the West and in the Mesabi Range near Lake Superior, metallic ores—copper and iron—are removed from *open pit mines*, which are huge, gaping excavations that are dug to great depths. For example, the Liberty open-pit copper mine in Ruth, Nevada, is 300 meters (900 feet) deep and more than a kilometer wide. Another surface mining technique is *dredging*, which is used to recover streambed sand

and gravel placer deposits. Dredging is performed with chain buckets and draglines.

Strip mining is the surface mining method that disrupts the landscape most extensively, primarily because it affects a much larger surface area for the resource that is recovered than the other surface techniques. When strip mining is done, huge power shovels or stripping wheels literally chew up the land in gulps of 10 to 100 cubic meters, as shown in Figure 12.15. Strip mining is used to extract coal from seams that average 2 meters thick, which occur within 30 meters of the surface. It is also used to recover phosphate rock and gypsum deposits.

Two basic types of strip mining exist: area and contour. *Area strip mining* is employed in flat or gently rolling terrain. For example, in Florida, it is used to extract phosphate. In Illinois and western Kentucky, it is used in coal fields. Area strip mining involves the stripping away of overburden to form a series of parallel trenches. Thus when coal extraction from one trench is completed, that trench is filled in with the overburden that is removed from the adjacent trench. When this technique is used, a rugged topography is left behind that resembles a washboard, as shown in Figure 12.16. In hilly or mountainous terrain, such as the Appalachian coal fields, near-surface coal seams are mined by the

Figure 12.14 *A small quarry that is used for mining of granite at Barre, Vermont. (Porterfield–Chickering, Photo Researchers.)*

Figure 12.15 *A huge power shovel at the ARCO coal mine, near Gillette, Wyoming. (John Blaustein, Woodfin Camp & Associates.)*

method of *contour strip mining.* Contour strip mining requires the cutting of a shelf, or bench, into the steep flanks of a mountain. One shelf is cut for each coal seam, and the overburden is then dumped downhill from each successively higher terrace onto the one below. Figure 12.17 shows a series of terraces that were created by contour strip mining. In areas where the overburden is too thick to be removed economically, the coal may be extracted by huge drills, which are called *augers,* that burrow up to 70 meters (200 feet) horizontally into a coal bed.

Most strip-mined land is in the Appalachian coal fields, where more than 1.6 million hectares (4 million acres) have been disturbed since mining began. In the future, however, western coal fields are likely to be the principal site of strip mining, because an estimated 75 percent of the nation's surface mineable coal is in the West. In 1981, approximately one-third of the nation's coal came from west of the Mississippi.

A major environmental problem that is associated with surface and subsurface mining is the disposal of mine waste: the overburden plus *tailings* (the residue of ore processing). Coal mining produces more waste than any other type of mining. Waste produced by the surface mining of metallic minerals, on the other hand, amounts to only about 5 percent of coal-mine spoils. Overall, in the United States, 1.7 billion metric tons of mine waste, or approximately 7.4 metric tons per person, are generated each year. Most of those spoils are left at the mine sites. Since the spoils lack nutrients and are excessively stony, those waste piles inhibit the growth of any type of stabilizing vegetation. Thus rapid erosion and dangerous sliding can occur. For example, in Wales, in 1966, a 125-meter (400-foot) pile of coal-mine wastes suddenly slid down into the town of Aberfan, smashing buildings, including a school house, and taking 144 lives, most of whom were children.

The seepage of rainwater through mine wastes that are rich in sulfur compounds produces sulfuric acid. More than 60 percent of the acid drainage in the United States comes from runoff in sulfur-rich coal-mine spoils. That acid runoff then drains into rivers, eliminates aquatic life, and contaminates water supplies. It has been estimated that in Appalachia, 16,000 kilometers (10,000 miles) of waterways have been polluted seriously by acid runoff from coal wastes. In some regions, mine drainage also contains

Figure 12.16 *Area strip mining for coal near Nucla, Colorado.* (EPA Documerica. Photo by W. Gillette.)

zinc, arsenic, lead, and other metals that are toxic to aquatic organisms even in minute amounts. In addition, dust that is lifted from mine dumps by winds adds to atmospheric pollution; sediments that are washed from mines make drainageways turbid; and the piles of mine wastes themselves foul wildlife and human habitats.

Subsurface Mining

When a mineral, rock, or fuel deposit lies at a depth that is so great that the cost of removing the overburden is prohibitive, the resource is extracted by subsurface mining. Obviously, in the case of fluid resources (oil or natural gas), subsurface methods are the only ones that are technically feasible. To extract deep ore and coal deposits, mine operators usually create a system of subsurface shafts, tunnels, and rooms by drilling and blasting the rock. In most cases, a portion of the deposit must be left behind to support the mine roofs. In the *room-and-pillar system* of subsurface mining, as much as half the coal is left in place to serve as supporting pillars, although the pillars are commonly "robbed" in the end.

Certain soluble minerals (such as potash and salt) are removed from the subsurface by *solution mining*. In that technique, water is pumped down an *injection well* to the deposit, where it dissolves the minerals. The solution is then brought back to the surface through *extraction wells*. One of the major problems with solution mining is its potential for contaminating groundwater.

Oil and natural gas are extracted by wells that are drilled into reservoir rock to depths as great as 9 kilometers (5.6 miles). The natural pressurized condition of oil and gas within reservoir rock forces a portion of those fluids toward the surface, but most must be pumped. Typically, less than 50 percent of the oil is recovered; the rest remains behind in pore spaces, clinging to grains of sediment. A second step, which is called *secondary recovery*, is necessary to remove a portion of the remaining oil. In that technique, water (or natural gas or both) is pumped down the well and into the reservoir rock to drive more oil toward the production wells. For example, secondary recovery in Alaska's Prudhoe Bay oil fields is expected to increase the oil yield by more than 1 billion barrels over the estimated 8 billion barrels that are recoverable with pumping and natural pressure (*primary recovery*). Although secondary recovery has been somewhat successful in boosting production, an estimated 300 billion barrels of domestic oil lie beyond the reach of traditional recovery techniques. (In comparison, the United States so far has withdrawn 121 billion barrels of its oil reserves.)

Figure 12.17 *Severely eroded strip-mine terraces on Bolt Mountain, West Virginia. (U.S. Department of Agriculture.)*

It is not surprising then that a considerable amount of the research that is being done today is directed at developing *enhanced oil recovery* (EOR) methods for both producing and abandoned wells. Several methods are being studied, because some important differences exist in the physical properties of oil reservoirs as well as in the chemical properties of oil. In the United States, the most widely used technique for enhanced oil recovery reduces the viscosity of thick, heavy crude oil. In that technique, steam is injected into the reservoir rock, which raises the temperature of the oil and makes it less viscous, so it flows more readily. In addition, the pressure that is exerted by the steam can be used to drive the oil toward extraction wells, as shown in Figure 12.18. By using the various techniques of enhanced oil recovery, we can increase yields by 10 to 15 percent over primary and secondary methods of recovery.

Although oil rigs usually have a minimal impact on the landscape, the danger of well blowouts and oil spills always exists. The hazard is particularly great in offshore drilling operations, where oil spills that threaten marine life and beaches are difficult to contain. Blowouts and oil spills are also particularly damaging in the frigid climate of the

Arctic tundra, where the breakdown of oil by microorganisms occurs very slowly. Still, such incidents are actually quite rare: only 11 major oil spills have occurred in the United States since 1953, and more than 18,000 wells were drilled during that period. Between 1973 and 1984, oil spills from offshore petroleum operations in the United States have totaled approximately one-tenth of 1 percent of the oil produced offshore during the same period. In comparison, natural oil seeps contributed almost 1000 times this amount to the world's oceans.

Although subsurface mining usually produces less waste than surface mining, the spoils are heaped on the ground at mine sites, where they can cause the same problems that attend surface mining spoils— acid runoff, landslides, and air pollution. Since materials are extracted without being replaced in both fluid and ore mining, mine collapse and ground subsidence over mines are constant dangers. Those developments, in turn, can have serious and costly effects on the landscape, such as damage to buildings, disruption of surface and subsurface drainage, and disturbance of wildlife habitats. In most cases, mine collapses occur in abandoned coal mines. More than 90,000 abandoned mines exist in the United States. Those mines collapse because the pillars of coal that are left to support the ceilings of subsurface caverns

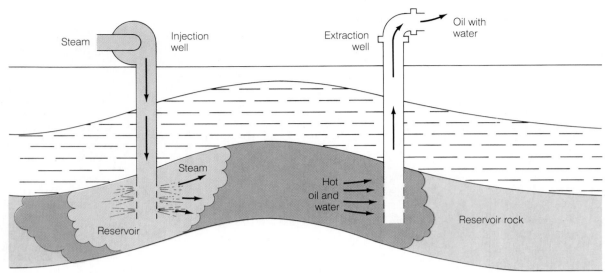

Figure 12.18 *Steam flooding, which is a technique of enhanced oil recovery, is used to reduce the viscosity of oil, which causes it to flow more readily within subsurface reservoirs. (Modified after Exxon Corporation.)*

are weak, and they sometimes fail. To prevent collapse, mine operators can inject a mud slurry into abandoned subsurface tunnels and rooms and then drain the water away, leaving behind a solid material that shores up mine tunnels.

Ground subsidence that results from the withdrawal of oil and natural gas from subsurface reservoirs can be especially costly when it occurs underneath urban areas. One particularly dramatic example occurred in the Wilmington and Signal Hill oil fields in the Los Angeles–Long Beach harbor area. Beginning in 1928, large-scale oil withdrawal triggered severe land subsidence, and in some places, the land over the well sites subsided as much as 9 meters (29 feet). Elaborate flood-control measures had to be taken, including the construction of levees and seawalls to hold back the Pacific Ocean. Subsidence was not halted until 1968, after the subsurface reservoirs had been recharged through the injection of seawater. But by then, powerplants, railroad terminals, docks, and most of the water and sewer systems of the city of Wilmington had to be rebuilt.

Resource Management

The rock cycle has provided us with limited quantities of rock, mineral, and fuel resources. When we extract those crustal resources, we also make considerable sacrifices in environmental quality. If we are to ensure an adequate supply of those resources and environmental quality to future generations, we must develop sound management practices. The principal objectives of such a management effort would be to (1) optimize the conservation of resources, (2) maximize the exploration for new reserves, and (3) minimize attendant environmental disruption. In this section, we discuss those aspects of crustal resource management, focusing primarily on nonfuel resources. The management of fossil and nuclear fuels is discussed in Chapters 18 and 19.

The Mineral Supply

Modern society relies on the properties of more than 90 metallic and nonmetallic minerals, some of which are listed in Table 12.2. In the past, we could not assess the quantities of the materials that were available to us because our terminology was imprecise. Recently, however, an attempt has been made to standardize the terms that are used to classify mineral resources in accordance with their availability now and in the future. Figure 12.19 depicts the classification system that is now being used by the U.S. Geological Survey and the U.S. Bureau of Mines.

According to that classification system, the quality *(grade)* and quantity of some resources, which

Table 12.2 Important crustal resources, their mineral source, geologic occurrence, and uses

MINERAL DEPOSIT	TYPICAL MINERALS	GEOLOGICAL OCCURRENCE	USES	MAJOR DEPOSITS REMARKS
METALS PRESENT IN MAJOR AMOUNTS IN EARTH'S CRUST				
Iron	Hematite, Fe_2O_3 Magnetite, Fe_3O_4 Limonite, $FeO(OH)$	Sedimentary Banded Iron Formation Contact Metamorphic Magmatic Segregation Sedimentary Bog Iron Ore	Manufactured Materials, Construction, Etc.	Mesabi, Minn.; Cornwall, Pa.; Kiruna, Sweden Resources Immense; Economics Determines Exploitation
Aluminum	Gibbsite, $Al(OH_3)$ Diaspore, $AlO(OH)$	Bauxite; Residual Soils Formed by Deep Chemical Weathering	Lightweight Manufactured Materials	Jamaica Resources Great, but Expensive to Smelt
Magnesium	Dolomite, $CaMg(CO_3)_2$ Magnesite, $MgCO_3$	Dissolved in Seawater Hydrothermal Veins, Limestones	Lightweight Alloy Metal, Insulators, Chemical Raw Material	Most Extracted from Sea Water; Unlimited Supply
Titanium	Ilmenite, $FeTiO_3$ Rutile, TiO_2	Magmatic Segregations Placers	High-Temperature Alloys; Paint Pigment	Allard Lake, Quebec; Kerala, India Reserves Large in Relation to Demand
Chromium	Chromite, $(Mg,Fe)_2CrO_4$	Magmatic Segregations of Mafic and Ultramafic Rocks	Steel Alloys	Bushveldt, S. Africa Extensive Reserves in a Number of Large Deposits
Manganese	Pyrolusite, MnO_2	Chemical Sedimentary Deposits, Residual Weathering Deposits, Seafloor Nodules	Essential to Steel Making	Ukraine, U.S.S.R. World's Land Resources Moderate, but Seafloor Deposits Immense
METALS PRESENT IN MINOR AMOUNTS IN EARTH'S CRUST				
Copper	Covellite, CuS Chalcocite, Cu_2S Digenite, Cu_9S_5 Chalcopyrite, $CuFeS_2$ Bornite, Cu_5FeS_4	Porphyry Copper Deposits Hydrothermal Veins Contact Metamorphic Sedimentary Deposits in Shales (Kupferschiefer Type)	Electrical Wire and Other Products	Bingham Canyon, Utah; Kuperschiefer, Germany; Poland
Lead	Galena, PbS	Hydrothermal (Replacement) Contact Metamorphic Sedimentary Deposits (Kupferschiefer Type)	Storage Batteries, Gasoline Additive (Tetraethyl Lead)	Mississippi Valley; Broken Hill, Australia Large Resources, Many Lower-Grade Deposits
Zinc	Sphalerite, ZnS	Same as Lead	Alloy Metal	Same as Lead
Nickel	Pentlandite, $(Ni, Fe)_9S_8$ Garnierite, $Ni_3Si_2O_5(OH)_4$	Magmatic Segregations Residual Weathering Deposits	Alloy Metal	Sudbury, Ontario High-Grade Ores Limited; Large Resources of Low-Grade Ores; Also in Seafloor Mn Nodules
Silver	Argentite, Ag_2S In Solid Solution in Copper, Lead, and Zinc Sulfides	Hydrothermal Veins with Lead, Zinc, and Copper	Photographic Chemicals; Electrical Equipment	Most Produced as By-Product of Copper, Lead, and Zinc Recovery
Mercury	Cinnabar, HgS	Hydrothermal Veins	Electrical Equipment, Pharmaceuticals	Almadén, Spain Few High-Grade Deposits with Limited Reserves
Platinum	Native Metal	Magmatic Segregations (Mafic Rocks) Placers	Chemical Industry; Electrical; Alloying Metal	Bushveldt, S. Africa Large Reserves in Relation to Demand
Gold	Native Metal	Hydrothermal Veins Placers	Coinage; Dentistry, Jewelry	Witwatersrand, S. Africa Reserves Concentrated in a Few Larger Deposits

Table 12.2 Important crustal resources, their mineral source, geologic occurrence, and uses (continued)

MINERAL DEPOSIT	TYPICAL MINERALS	GEOLOGICAL OCCURRENCE	USES	MAJOR DEPOSITS REMARKS
		NONMETALS		
Salt	Halite, NaCl	Evaporite Deposits Salt Domes	Food; Chemicals	Resources Unlimited; Economics Determines Exploitation
Phosphate Rock	Apatite, $Ca_5(PO_4)_3OH$	Marine Phosphatic Sedimentary Rock Residual Concentrations of Nodules	Fertilizer	Florida High-Grade Deposits Limited but Extensive Resources of Low-Grade Deposits
Sulfur	Native Sulfur Sulfide Ore Minerals	Caprock of Salt Domes (Main Source) Hydrothermal and Sedimentary Sulfides	Fertilizer Manufacture; Chemical Industry	Texas; Louisiana; Sicily Native Sulfur Reserves Limited but Immense Resources of Sulfides
Potassium	Sylvite, KCl Carnallite, $KCl \cdot MgCl_2 \cdot 6H_2O$	Evaporite Deposits	Fertilizer	Carlsbad, New Mexico Great Resources of Rich Deposits
Diamond	Diamond, C	Kimberlite Pipes Placers	Industrial Abrasives	Kimberly, S. Africa Synthetic Diamonds Now Commercially Available
Gypsum	Gypsum, $CaSO_4 \cdot 2H_2O$ Anhydrite, $CaSO_4$	Evaporite Deposits	Plaster	Immense Resources Widely Distributed
Limestone	Calcite, $CaCO_3$ Dolomite, $CaMg(CO_3)_2$	Sedimentary Carbonate Rocks	Building Stone; Agricultural Lime; Cement	Widely Distributed; Transportation a Major Cost
Clay	Kaolinite $Al_2Si_2O_5(OH)_4$ Smectite* Illite*	Residual Weathering Deposits; Sedimentary Clays and Shales	Ceramics: China, Electrical, Structural Tile	Many Large Pure Deposits; Immense Reserves of all Grades
Asbestos	Chrysotile, $Mg_3Si_2O_5(OH)_4$	Ultramafic Rocks Altered and Hydrated in Near-Surface Crustal Zones	Nonflammable Fibers and Products	Southeastern Quebec Limited High-Grade Reserves but Great Low-Grade Reserves

Source: From F. Press and R. Siever, *Earth*, 3rd ed. New York: W. H. Freeman, 1982, pp. 562–563.
*Formula highly variable; a hydrous aluminum silicate with other cations, such as Na^+, K^+, Ca^{2+}, Mg^{2+}.

Figure 12.19 *Classification of mineral resources. This is an attempt by the U.S. Geological Survey and the U.S. Bureau of Mines to standardize terminology. The leftmost column presents the economic status of the various crustal resources. Economic deposits can be mined immediately—subeconomic deposits cannot. With only a slight upswing in demand, marginally economic deposits can be mined. (From the U.S. Geological Survey Circular 831, 1980.)*

	IDENTIFIED	UNDISCOVERED	
		IN KNOWN DISTRICTS	IN UNDISCOVERED DISTRICTS OR FORMS
Economic	Reserves	Hypothetical resources	Speculative resources
Marginally economic	Marginal reserves		
Subeconomic	Subeconomic resources		

←————— Increasing degree of geologic assurance —————→

are called *identified* resources, are determined by actual field study. Following that step in the classification, an identified mineral or fuel resource is labeled a *reserve* when it can be extracted at the present time both legally and profitably. However, if an identified resource does not quite meet those criteria at the present time but would if relatively minor changes occurred in present economic, legal, or technical circumstances, then it is called a *marginal reserve*. Resources that are subeconomic, hypothetical, or speculative are called *nonreserve resources*. Thus when economic conditions do not warrant the extraction of an identified resource, it is described as being *subeconomic*. Resources that have not been discovered but are believed to exist in known mining districts are called *hypothetical resources*. And resources that have not been discovered but are believed to exist somewhere, based on geologic principles, are called *speculative resources*.

Our reserves and resources (both identified and undiscovered) of many rocks and minerals are so abundant that we are assured an adequate supply far into the future. Some examples include crushed stone, sand, and gravel. However, the reserves of certain minerals that are essential for the maintenance of our highly technological society are small and are declining under the pressures of accelerating demand. The domestic supply status of some important resources is summarized in Table 12.3.

Since the processes that are responsible for distributing ore deposits occurred sporadically around the globe through geologic time, no nation contains all the crustal resources that are important economically. Some nations are favored geologically more than others, but none is entirely self-sufficient in terms of metals, nonmetallic minerals, and fuels. Thus countries must engage in international trade to acquire their needed resources and, clearly, as the supply of those invaluable resources diminishes, the need for international cooperation becomes ever more pressing.

Although the production of 18 critical nonfuel minerals has soared worldwide (as Figure 12.20 indicates), production in the United States after the late 1930s did not keep pace with consumption, and the gap between production and consumption continues to grow. As a result, the United States has begun to import increasingly more minerals. In fact, we now depend on foreign nations for more than half of 24 of the 32 nonfuel resources that are most es-

Table 12.3 Projected availability of important domestic crustal resources

Group 1: Reserves in Quantities Adequate to Fulfill Projected Needs Well Beyond 25 Years*

Coal	Silicon
Construction Stone	Molybdenum
Sand and Gravel	Gypsum
Nitrogen	Bromine
Chlorine	Boron
Hydrogen	Argon
Titanium (Except Rutile)	Diatomite
Soda	Barite
Calcium	Lightweight Aggregates
Clays	Helium
Potash	Peat
Magnesium	Rare Earths
Oxygen	Lithium
Phosphorus	

Group 2: Identified Subeconomic Resources in Quantities Adequate to Fulfill Projected Needs Beyond 25 Years* and in Quantities Significantly or Slightly Greater than Estimated Undiscovered Resources

Aluminum	Vanadium
Nickel	Zircon
Uranium	Thorium
Manganese	

Group 3: Estimated Undiscovered (Hypothetical and Speculative) Resources in Quantities Adequate to Fulfill Projected Needs Beyond 25 Years* and in Quantities Significantly Greater than Identified Subeconomic Resources. Research Efforts for These Commodities Should Concentrate on Geologic Theory and Exploration Methods Aimed at Discovering New Resources

Iron	Platinum
Copper	Tungsten
Zinc	Beryllium
Gold	Cobalt
Lead	Cadmium
Sulfur	Bismuth
Silver	Selenium
Fluorine	Niobium

Group 4: Identified Subeconomic and Undiscovered Resources Together in Quantities Probably Not Adequate to Fulfill Projected Needs Beyond the End of the Century; Research on Possible New Exploration Targets, New Types of Deposits, and Substitutes is Necessary to Relieve Ultimate Dependence on Imports

Tin	Antimony
Asbestos	Mercury
Chromium	Tantalum

Source: From E. A. Keller, *Environmental Geology*, 3rd ed. Columbus, Ohio: C. E. Merrill, 1982, p. 343.
*For practical purposes, "beyond 25 years" may be a century or more.

sential for our nation's economic viability. Figure 12.21 shows the 25 countries that supply our strategic minerals. Although many of those nations are allies and politically stable, some (such as Zaire, Zambia, and Zimbabwe) are politically unstable developing nations.

Because of concern about the implications to our national security in terms of a growing dependency on foreign sources of critical minerals, many

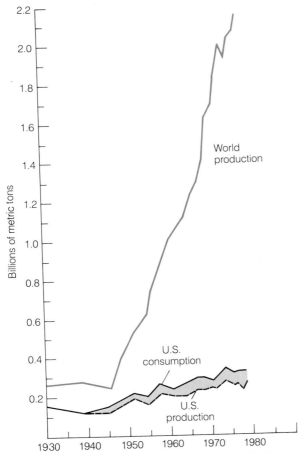

Figure 12.20 *World production and United States production and consumption of 18 critical minerals: iron ore, bauxite, copper, lead, zinc, tungsten, chromium, nickel, molybdenum, manganese, tin, vanadium, fluorspar, phosphate, cement, gypsum, potash, and sulfur. (After E. N. Cameron, ed., The Mineral Position of the United States, 1975–2000. Madison: University of Wisconsin Press, 1973. Copyright by the Regents of the University of Wisconsin, as updated by F. Press and R. Siever, Earth, 3rd ed. New York: W. H. Freeman, 1982, p. 561.)*

nations may form mineral cartels that are analogous to those formed by petroleum exporting nations (OPEC). After all, some nations contain within their borders large deposits of critical minerals that are in great demand by other nations. The purpose of a cartel is to regulate the supply of a material or product and thereby its market price, and, historically, oil cartels have enjoyed some success. However, attempts at establishing mineral cartels—with the possible exception of diamonds—have been dismal failures.

With respect to domestic reserves, some observers project that, at current rates of use and mine production, we have less than 50 years' supply of lead, zinc, copper, and iron. However, these estimates are conservative since they do not account for possible new discoveries, changes in the economic climate, and technological innovations that could make feasible the extraction of lower-grade deposits (that is, deposits that contain less of the desired material and more waste rock). For more information about the problem of assessing reserves, see Box 12.1.

The market value of each resource ultimately determines not only the grade of the deposit that can be mined economically, but also whether a deposit can be mined at all. For example, a worldwide slump in mineral prices occurred during the recession of the early 1980s, and a reduced demand for consumer goods, such as automobiles and appliances, lessened the demand for raw materials. Consequently, many mines closed, which disrupted local and regional economies. In Canada's Yukon Territory, an estimated 25 percent of the population left because mines closed. Then, by the spring of 1982, plummeting copper prices forced the shutdown of 15 of Arizona's 28 copper mines and the loss of 3200 jobs. Although the economic climate is expected to improve in the mid-1980s, which would revitalize mining in North America, some mines may never open again.

Although geologic processes ultimately determine the world's supply of minerals, economic conditions control the actual supply at any given moment. Thus as the demand for a mineral resource increases, its price rises, and it becomes more worthwhile to expend more energy and other resources to mine and process lower-grade deposits of that mineral. But the mining of low-grade minerals threatens greater environmental degradation, such as more air and water pollution and more waste heaps.

special-interest groups are trying to create a revitalization of the domestic mining and minerals processing industries. Their recommendations include opening more federal land to mineral exploration and mining, easing of environmental and health and safety laws, and more favorable tax laws. However, those efforts are countered by conservationists, who warn of the environmental degradation that is the likely consequence of accelerated resource development.

Shortages in domestic supplies of certain non-fuel resources has also led to speculation that some

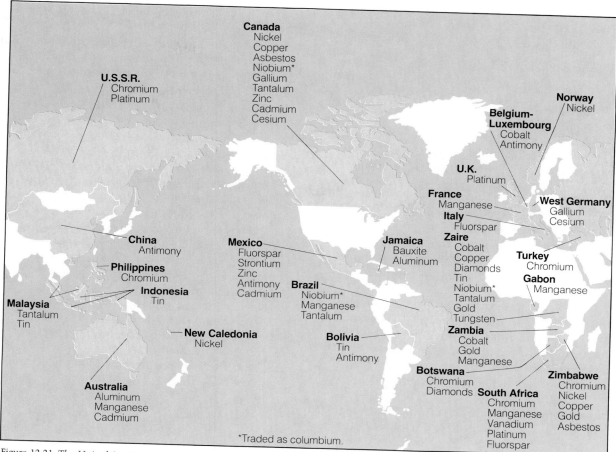

Figure 12.21 *The United States is dependent on 25 other nations for more than 50 percent of its supply of 24 strategic materials. (From Ember, L. R., "Many forces shaping strategic minerals policy." Reprinted with permission from* Chemical and Engineering News, *May 11, 1981, p. 21. Copyright 1981, American Chemical Society.)*

As a result, more resources must be employed to maintain environmental quality.

The principal obstacle to mining lower-grade deposits in the distant future may be the cost of energy. In the past, fossil fuels were abundant and inexpensive. Thus as advances in mining and processing techniques were made, miners could extract leaner and deeper deposits than ever before, since they had the fuel resources to do it. For example, in 1880, it was not feasible to mine copper deposits that contained less than 3 percent copper. That cutoff grade declined to 2 percent by 1910, 1 percent by 1950, 0.72 percent in 1960, and 0.59 percent in 1970. Thus throughout much of this century, the reserves of copper and other nonfuel minerals have risen steadily (although not fast enough to keep pace with the demand).

Heartened by past trends of ever-increasing reserves, some earth scientists dismiss the likelihood of a serious crisis in mineral supply. They point out that the quantity of critical minerals that is contained in the earth's crust in dilute concentrations is enormous. And they believe that, given the appropriate technological innovations, that supply will carry us thousands of years into the future. That optimism may be misplaced, however. Even though we have just cause to feel confident that we will devise a method for mining exceptionally low-grade deposits, time—and fuel—may run out before we can apply those innovations. Unless we develop new, extensive, and inexpensive energy sources, the mining of very low-grade mineral deposits will be impossible because of the high fuel demand of such efforts. Indeed, we may discover new technologies but be stymied by a lack of sufficient low-cost energy sources to employ them.

BOX 12.1

OUR RESOURCE FUTURE: A CASE FOR OPTIMISM

This chapter emphasizes the technological method of assaying future supplies of crustal resources— the method of geologists and engineers. A decidedly different approach, with different results, is used by some economists who base their predictions on historical price-trend data.

As we have seen, the rock cycle implies that, for practical purposes, the quantity of any resource within the earth's crust is fixed and finite. But we cannot actually measure the total quantity of any metal, mineral, or fuel resource. We can only estimate supplies by extrapolating from field samples.

Typically, geologists and engineers forecast resource supply in terms of "years of consumption left," that is, the time it will take for a given resource to be exhausted. This period is computed by dividing the estimated supply by the annual rate of consumption. Three different supply estimates are used: (1) known or proven reserves based on study of actual deposits, (2) an approximation of the total quantity of a resource in the

crust, and (3) an estimate of total recoverable resources, which the U.S. Geological Survey puts at 0.01 percent of the amount assumed to be in the upper 1 kilometer of the crust.

The table compares the "years of consumption left" for various resources, using the three supply estimates. The lengthy periods predicated on total crustal abundance are not very useful because we cannot say when, if ever, scientists and engineers will develop the technology needed to exploit very low crustal concentrations. The other two estimates are notoriously variable. Known global reserves of most important resources have increased with time. For example, between 1950 and 1970, proven reserves of iron increased 1221 percent; bauxite, 279 percent, and copper, 179 percent. Ultimately recoverable estimates have also trended upward with time.

The economist Julian L. Simon of the University of Illinois is a leading critic of these technological forecasting methods. Writing in the June, 1981, issue of The Atlantic Monthly, Simon notes:

One problem with all the technological methods is that they confuse abundance in the

Number of years of consumption potential for various elements

	KNOWN RESERVES/ ANNUAL CONSUMPTION	U.S. GEOLOGICAL SURVEY'S ESTIMATES OF "ULTIMATE RECOVERABLE RESOURCES"/ ANNUAL CONSUMPTION*	AMOUNT ESTIMATED IN CRUST/ANNUAL CONSUMPTION
Copper	45	340	242,000,000
Iron	117	2,657	1,815,000,000
Phosphorus	481	1,601	870,000,000
Molybdenum	65	630	422,000,000
Lead	10	162	85,000,000
Zinc	21	618	409,000,000
Sulphur	30	6,897	NA
Uranium	50	8,455	1,855,000,000
Aluminum	23	68,066	38,500,000,000
Gold	9	102	57,000,000

Source: William D. Nordhaus in American Economic Review, May, 1974, p. 23, as presented in J. L. Simon, "The Scarcity of Raw Materials." The Atlantic Monthly, June, 1981, p. 35, U.S. Geological Survey data.
*"Ultimate recoverable resources" equal 0.01 percent of materials in the top kilometer of the earth's crust.

earth with economic availability. Platinum, gold, and silver are far and away the least abundant minerals. But they have never been plentiful, and so even though it has been progressively harder to find these minerals, few people worry about a "shortage" of them.

A second difficulty with technological forecasts stems from this important property of natural-resource extraction: A small change in the price of a mineral generally makes a very big difference in the potential supplies that are economically available—that is, profitable to extract. . . .

Third, the technological inventory of the earth's "contents" is at present quite incomplete, for the very good reason that it has never been worthwhile to go to the trouble of making detailed surveys. Why should one count the stones in Montana when there are enough to serve as paperweights right in one's own back yard?

. . . A fourth difficulty with the technological forecasts is that the approaches that do go beyond the known-reserves concept are speculative, necessarily making strong assumptions about discoveries of unknown lodes and about technologies that have yet to be developed. Technological forecasting depends heavily upon how well the forecaster can imagine the methods of extraction that will be developed in the future. Making the conservative assumption that future technology will be the same as present technology would be like making a forecast of twentieth-century copper production on the basis of eighteenth-century pick-and-shovel technology.

Simon employs an economic-based approach to forecasting the supply of resources, and his outlook is optimistic. He argues that the past is key to the future—that long-term economic trends of the past will persist into the future. And what does the past tell us about resource supplies? Historically, the cost of extracting resources has

steadily declined, because technological innovations have made it economically worthwhile to mine progressively lower-grade deposits. These technological innovations, in turn, have been spurred by increasing demand for resources and the services and products they provide. According to Simon, there is no reason not to expect a continuation of this favorable relationship among rising resource demand, technological advances, and declining costs.

The historic decline in the costs of resource extraction becomes clear in the context of personal income. Simon points out, for example, that:

> *In the 1970s, the average hourly wage for factory work would buy approximately fifteen times more coal than the equivalent wage in 1800.*
>
> *In the 1970s, the average hourly factory wage would buy around twenty times more oil than it would around 1870, the earliest period for which prices are available. Relative to the prices of other consumer goods, oil was about four times cheaper in 1970 than in 1870.*
>
> *Since the 1920s, the earliest period for which data are available, the average hourly factory wage has come to buy about eight times more electrical power. Relative to other prices, the price of electricity, like the price of oil, was more than twice as cheap in the 1970s as in the 1920s.*

Simon is confident that human ingenuity will continue to devise the technological means to tap increasingly lean and remote deposits, thus assuring a more than adequate supply far into the future. He cautions against trusting known or proven reserve data, because—as he points out—the known reserve is inversely proportional to the cost of mining. Thus, as costs decline, the portion of a resource that can be mined and marketed (i.e., known reserves) increases.

Based on Simon, J. L. "The Scarcity of Raw Materials." The Atlantic Monthly, June, 1981, pp. 33–41. Excerpts from p. 34–35 and p. 37–38.

Exploration

Until new energy sources are developed, the prudent course appears to be to maximize our search for ore deposits. Exploration, however, has always been a financially risky venture that requires a heavy capital investment and, traditionally, a free-wheeling spirit. Typically, of 10,000 sites where a deposit might exist according to what is understood about the local geology, only 1000 merit detailed scientific study, only 100 warrant costly drilling, trenching, or tunneling, and, typically, only one of those excavations will prove to be a productive mine.

Exploration efforts have been aided in recent years by the development of new instrumentation and the evolution of new geologic concepts. For example, remote sensing by orbiting satellites (such as LANDSAT) has proved to be invaluable for selecting likely exploration sites for possible deposits. Figure 12.8 is an example of the sort of satellite view that aids geologists in analyzing bedrock structures for possible exploration sites. Also, the concept of *plate tectonics* has revolutionized geologists' ideas about how metallic ores are formed. As we noted earlier in this chapter, metallic deposits are generated at submarine vents along divergent plate boundaries. Also, the hypothesis that ores may be concentrated in areas where crustal plates have converged has led to recent discoveries of important deposits of copper, molybdenum, and mercury.

However, efforts to expand exploration have been hindered by environmentalists who are concerned about the environmental degradation that is caused by resource development. For example, in response to mounting pressure from those groups, the federal government has increasingly prohibited exploration and mining activities on federal lands. For example, the Department of the Interior imposed a moratorium on the leasing of federal coal lands that lasted from May, 1971, to January, 1981. (Up to 70 percent of western coal deposits occur on public lands.) By the late 1970s, miners had been excluded from 40 percent of public lands—national parks, wildlife refuges, and native lands. If those trends continue, that figure could increase in the 1980s to encompass nearly one-third of the nation's total land area. We say more about conflicts in land use in Chapter 15.

Seabed Resources

Some nations are looking to the seabed as a source of minerals to offset their present and future shortages and to reduce their dependency on foreign sources. Because continental shelves are actually submerged extensions of the continents, it makes sense that they probably contain many of the same mineral (and fuel) deposits that are found on the land. However, an unfavorable economic climate has so far discouraged extensive mining of the shelf. Today, some dredging of phosphorite rock (the source of phosphorus for fertilizers) is taking place, but, although deposits on the continental shelf are enormous, they are of lower grade than those on the land. Therefore, terrestrial mining of phosphorite rock is still favored. In some shelf regions, wave action and ocean currents have concentrated heavy minerals in sediments, and those marine placer deposits are profitably mined on a limited scale.

Offshore petroleum deposits are being extracted in the Middle East, the North Sea, and the Gulf of Mexico. Oil exploration is active (although unproductive so far) on the continental shelf in Georges Bank (off New England) and in the Baltimore Canyon (off the mid-Atlantic coast). It is possible that petroleum is trapped within the thick sedimentary accumulations that are found on the North American continental slope and rise as well as in the continental shelf.

Seabed *manganese nodules* have received considerable attention in recent years because of the metallic elements that they contain. Typically, nodules contain 30 to 40 percent manganese (which is used in the production of certain steel alloys) and, more importantly, small amounts of copper, cobalt, and nickel. The manganese nodules (Figure 12.22)

Figure 12.22 *Manganese nodules on the floor of the Pacific Ocean: the nodules are an important future source of manganese, copper, cobalt, and nickel. (Kennecott Copper Corporation.)*

vary in size from tiny grains to masses that weigh hundreds of kilograms; most, however, are the size of potatoes. They carpet about one-quarter of the deep ocean bottom—mostly in international waters. The technology for harvesting those billions of tons of nodules basically consists of a device that is similar to a vacuum cleaner, which pulls nodules from the muds of the ocean bottom and delivers them aboard a ship. That process has been likened to sucking grains of sand through a straw to the top of a high-rise building.

Until recently, the principal obstacle to the recovery of manganese nodules has been the absence of an international policy concerning the exploitation of seabed resources. Since 1958, an ongoing effort has been made under the auspices of the United Nations to formulate such a policy, and the negotiations have involved more than 150 nations. However, slow progress toward agreement has stemmed from conflicting interests between the developed and developing nations. Developing nations view ocean resources as the common heritage of all mankind, but they fear that the world's richest and most technologically sophisticated nations will reap the bulk of the harvest. The United States and many of the other developed nations, on the other hand, fear that too much power would be vested in the governments of developing nations if they were granted significant power in terms of ocean resource development.

In December, 1982, representatives of 100 nations signed the United Nations' Law of the Sea Treaty. Among other provisions, that pact would place deep-sea mining under international regulation. The United States, Italy, Belgium, West Germany, and Great Britain refused to sign the treaty. And, as of this writing, it has been rough sailing for advocates of the Law of the Sea. Only nine of the required signatory nations have ratified the treaty and progress has bogged down as a result of bureaucratic procedural difficulties and quarrels among participating nations. Meanwhile, consortia of developed Western nations, including the United States, have forged ahead independently with plans to begin mining manganese nodules off the coast of Hawaii.

Conservation Strategies

Prospects for the immediate recovery of substantial quantities of high-grade minerals from new discoveries on land and at sea are not very promising. This reality makes resource conservation crucially important. And the most fundamental approach to crustal resource conservation is a reduction of consumption. For example, although the United States' population represents only 5 percent of the globe's population and inhabits only 6 percent of the globe's land area, the United States consumes 23 percent of the globe's nonfuel rock and mineral resources. Also, a great deal of time is necessary to develop the technology and energy sources that are required for exploration and extraction of new reserves. Therefore, it is essential that responsible government representatives are elected who will work for sound resource conservation policies.

In the meantime, many efforts are being made to use the existing crustal resources more efficiently and to reduce the present rates of depletion. In the past, mine operators often bypassed low-grade deposits in the quest for more valuable grades. In fact, mines were frequently abandoned because it was not economically feasible to extract low-grade deposits. Now, however, miners are being encouraged to remove both high- and low-grade deposits, stockpiling the low-grade materials until demand or technology makes processing feasible. In fact, today, some metals (such as copper) are being recovered from old mine dumps at a significant energy savings (for copper, approximately 20 percent).

Another conservation strategy is the substitution of renewable resources for essentially nonrenewable mineral resources, such as using wood instead of metal in construction. That practice also exhibits a significant energy savings. For example, a study that was prepared under the auspices of the National Academy of Sciences' National Research Council compared the total input of energy that is used in the production of various wood and nonwood structural materials. The study showed that the wood materials were far more energy-efficient. As proof, the following figures, among others, were cited: (1) it takes five times as much energy to make a piece of aluminum siding as it does to produce the same amount of wood siding; (2) eight times more energy is used to produce a steel interior-wall stud than is used to produce a wood stud; (3) it requires 25 times more energy to produce brick siding than to produce plywood siding; and (4) it takes 50 times more energy to make a steel floor joist than it takes to create its wood equivalent!

In many applications, one depletable resource may be replaced by another that is in greater supply, such as when aluminum is substituted for copper; glass, for tin-plated cans; or stone, for copper (as a building facade). In addition, research is under way to increase the life span of metals by developing more corrosion-resistant alloys.

Recycling of metals is an essential component of a national resource conservation program. Although metals are not destroyed in manufacturing processes, they are widely dispersed over the earth's surface in the form of components of a multitude of goods (from airplanes to zippers). When those items lose their value, their metal components are lost to us unless efforts are made to recover them. Hence recycling initially requires the recovery and collection of metallic components.

Iron and aluminum, which account for 94 percent of all the metals that are used today, are recycled in great quantities. In the early 1980s, scrap iron was recovered at a rate that equalled 35 percent of the iron that is consumed in the United States. (This figure excludes scrap that is recirculated within steel mills.) The figure for aluminum was 32 percent. (In fact, our nation recovers more aluminum from recycled aluminum cans than Africa produces from all sources.) American industry also recycles zinc, copper, and lead in significant amounts, but for other strategic materials, the rate of recycling is minor. Typically, those materials, such as tungsten, are so widely dispersed that the energy required to collect and separate them virtually prohibits their recycling.

Some scrap metal is now favored over "new" metal that is produced from ores, because it is less expensive to recycle it, partly because of the high cost of energy. For example, it requires 20 times more electricity to produce a ton of aluminum from bauxite, the source ore, than it does to produce a ton of aluminum from scrap metal. Studies that were initiated just after the energy shortages of the mid-1970s also projected substantial energy savings for using scrap iron in place of iron ore, but recent studies that employ sophisticated economic models indicate that such projections may be unduly optimistic. For example, a 1982 study of the United States steel industry found that if all steel was manufactured from scrap, the total energy savings would be 6 percent, which is equivalent to approximately 0.6 percent of the nation's total energy use.

One major problem with recycling is its sensitivity to fluctuations in the economy. After a metal recycling boom in the early 1970s, recycling suffered serious setbacks as a result of the economic recession. Many small recycling centers went out of business and, in 1975, the use of recycled metals declined precipitously. For example, the use of recycled copper decreased by 30 percent; zinc, 27 percent; stainless steel, 43 percent; and aluminum, 7 percent. However, by early 1976, with a recovering economic climate, the downward trend reversed and the demand for scrap metals increased. Since then, the demand for recycled metals has generally followed the major fluctuations of the economy.

In many cases, recycling is a marginal operation that is not profitable unless it is operated at maximum efficiency, utilizing an uninterrupted flow of large quantities of recyclable materials. Hence recycling has the best chance of success in large urban–industrial areas, which have both a reliable source of appropriate waste and a market for recycled products. In such areas, a steady stream of recyclable waste of a specific composition can be assured on a contractual basis. But for recycled materials to compete favorably with raw materials, a large-scale recycling operation that is equipped with centralized facilities for separation, cleaning, and shredding is needed. This economic fact of life seriously limits the value of small-scale recycling efforts.

Land Reclamation

An integral component of resource management is the rehabilitation, or *reclamation*, of surface-mined lands. Reclamation encompasses those activities that foster ecological succession on land that is disturbed by mining. Generally, those activities include contouring (shaping) mine spoils and overburden to minimize erosion, applying topsoil and fertilizer, and planting and maintaining vegetation to match the species assemblage that is native to the area. The specific types of procedures that are used depend on the physical and chemical properties of the spoils. For example, the type of vegetation that is planted on acidic waste heaps may not be appropriate on spoils that have a high salt content.

Reclamation activities are most effective and economical when they are employed as an integral component of the mining operation rather than as a separate process after the mining is completed. For example, contour strip mining is not nearly so destructive of the landscape when miners employ what is known as the *haul-back method.* With that technique, trucks haul spoils back along the contour terrace to fill in and restore the original slope of the land rather than dumping the spoils downhill. That procedure virtually eliminates the danger of landslides and allows considerably more land to remain undisturbed than does conventional contour strip mining. It also lowers the cost of other reclamation techniques.

Reclamation was given a boost nationwide in August, 1977, with the long-awaited passage of the Surface Mining Control and Reclamation Act. First formulated in 1971, that law was passed in both houses of Congress, but it was vetoed twice by President Ford, first in 1974 and again in 1975. In explaining his actions, President Ford cited the prospect of reduced coal production and increased unemployment if strip mining were regulated.

The 1977 federal law is the first to regulate strip mining nationwide, and it sets standards for leasing, mining, and reclamation. It requires that land be restored to "a condition capable of supporting the uses it was capable of supporting prior to any mining." As a result, mine operators must demonstrate that they can reclaim the land before they are granted a permit to mine. To minimize erosion and contamination, they must store A and B soil horizons, and for prime farmland, they must store the C soil horizon as well. After mining, the land must be regraded to approximately its original profile and the topsoil must be replaced and reseeded. In areas where the mined land was originally agricultural, the restored land must be as productive as it was before the mining occurred. The law also requires that landowners give written consent for mining in those places where mineral rights are held by someone other than the landowner. Also, drainage waters must be impounded and treated. Finally, the law encourages miners to use the haul-back method of contour strip mining.

In a 1982 assessment of the reclamation situation, the Bureau of Mines concluded that 75 percent of the land in the United States that has been utilized for bituminous coal mining from 1930 through 1980 was reclaimed. And overall, for 1930 through 1980, 47 percent of the land that was used for all mineral extraction and processing also was reclaimed. However, only 8 percent of the land that was mined for metals was reclaimed, and only 27 percent of the land that was mined for nonmetallic

minerals was reclaimed. The slow progress in reclaiming land that is mined for nonfuel resources is attributed to several factors. Over the years, there has been much greater public pressure for the regulation of strip mining for coal. Hence no federal laws exist that specifically address reclamation of nonfuel mined lands and state laws are weak. Typically, large-scale nonfuel mining takes place in remote and sparsely populated areas, so the environmental impact is not as visible. ("Out of sight, out of mind.") Furthermore, open pit mines and large quarries disturb relatively small surface areas, considering the enormous quantity of ore that is removed.

In the coal fields of the eastern and central United States, reclamation has been underway for more than four decades. However, in the western coal fields, reclamation has been more difficult, primarily because the region gets less rainfall and rigorous research on revegetation has been underway only in recent years. Rainfall is particularly marginal in rangeland, making the reestablishment of seedlings quite difficult. Also, the rehabilitation of mines in the mountainous West is a great challenge, because revegetation proceeds very slowly in mountainous ecosystems: temperatures are low, growing seasons are short, soils are shallow, and few plant species are adapted to such habitat conditions.

Furthermore, compliance with regulations for surface mining is costly. Depending on the specific geologic structure of the deposit, the cost that is added to surface-mined coal can vary from $1 to as much as $5 per metric ton. Thus some mine operators view the provisions of the 1977 law as being too restrictive and the regulatory costs as being too burdensome. As a result, several challenges to the law are pending in the courts.

Conclusions

According to the National Research Council, efforts to explore and develop the earth's crustal resources will probably continue to conflict with efforts to protect environmental quality. Since the demand for these resources will more than double over the next two decades and the prospects for the immediate recovery of substantial quantities of high-grade minerals from new discoveries are not very promising, we will be forced to mine lower grades of minerals and thereby disrupt the environment ever more severely. The mining and processing of lower grades of minerals will also consume larger supplies of fossil fuels. Thus to reduce the conflict between resource development and environmental preservation, we must develop new techniques of exploration, recovery, and processing of crustal resources that will reduce environmental degradation. Of equal importance is our need to develop new inexpensive and abundant energy sources. But implementing any new technology takes time. Therefore, the conservation of crustal resources now should have a higher priority than ever before.

Summary Statements

Industrial nations are heavily dependent on a multitude of rock, mineral, and fuel resources that are extracted from the earth's crust.

A variety of processes have acted over geologic time to selectively concentrate crustal resources. Those processes include plate tectonics, weathering, and erosion.

Rocks are classified, according to their mode of origin, as igneous, sedimentary, or metamorphic. Igneous rocks are the most abundant, but sedimentary rocks are the most conspicuous.

A rock is an aggregate of minerals that consists of a few dominant and several accessory minerals. Some rocks and minerals are economically important.

Metallic or nonmetallic minerals may occur as minor components of large rock masses. Hence energy and water are needed for physical separation, smelting, and refining of these minerals. Solid, liquid, and gaseous waste products are generated by those processes.

Coal deposits were developed by the anaerobic decomposition of plant matter in swampy terrain millions of years ago. With increasing heat and pressure, peat layers changed to lignite, then to bituminous coal, and, in some regions, to anthracite coal.

Oil is organic in origin and occupies pore spaces in certain permeable sedimentary rocks.

Natural gas is derived from oil and is found either mixed with or on top of oil as well as in separate reservoirs.

In the United States, uranium deposits are mined primarily from certain sandstone strata in the West.

Rocks in any of the three rock families may be altered by geologic processes into other rock types. The typically slow pace of those processes, however, essentially fixes our supply of crustal resources.

Surface mining removes vegetation, soil, sediments, and rock that overlie mineral deposits. Surface-mining techniques include quarrying, open-pit mining, dredging, and strip mining.

Rain seeping into mine wastes triggers erosion, dangerous sliding, and the pollution of drainageways.

Deep ore and coal deposits are extracted from subsurface shafts and tunnels, which are blasted and drilled into the rock. Mine collapse and costly ground subsidence are potential risks.

Although the danger of oil spills and well blowouts is real in offshore drilling operations, they are actually rare.

For many rocks and minerals, reserves and resources are so abundant that we are assured an adequate supply far into the future. But for some materials that are necessary for our highly technological society, the situation is less favorable.

The principal obstacle to mining lower-grade deposits in the future is the cost of energy. Until new energy sources are developed, exploration and conservation efforts must be maximized.

Conservation strategies include reducing depletion rates, substituting renewable resources for essentially nonrenewable ones, and recycling.

Recycling saves energy, but it is very sensitive to fluctuations in the economy.

Reclamation encompasses those activities that foster ecological succession on land that is disturbed by mining.

Questions and Projects

1. What resources are mined in your area? Describe the mining techniques that are employed. Do local or state laws require restoration of the landscape after mining is completed? If so, are those laws enforced?

2. Suggest some uses for an abandoned quarry or sand and gravel pit that would not threaten groundwater quality.

3. Identify the geologic processes that are primarily responsible for the appearance of the landscapes in your area.

4. How does the unequal distribution of crustal resources influence world trade and our national security?

5. Dollars and energy are used to redistribute mineral resources across the face of the earth, and dollars and energy must be expended to recover those materials for recycling. Suggest measures that might be taken to reduce that double expenditure.

6. Speculate on the basic reasons for the marked contrasts between lunar and earth landscapes.

7. How might more strip mining in the Midwest influence the price of food?

8. Modern industrial societies rely on numerous minerals that have specific properties and, in some cases, other minerals exhibit the same or similar properties. Does this observation suggest a strategy for resource conservation?

9. Compare and contrast the advantages and disadvantages of surface mining and subsurface mining.

10. How does an understanding of geologic processes aid our search for mineral and fuel resources? Give some examples.

11. The classification of rocks as igneous, sedimentary, or metamorphic may be viewed as an *environmental* classification. Explain this statement.

12. Explain why deposits of rock, mineral, and fuel resources do not occur uniformly over the earth's surface.

13. Describe the implications of the rock cycle for our supply of rock, mineral, and fuel resources.

14. Explain how subsurface mining might threaten water quality.

15. Identify some of the environmental problems that are caused by wastes from surface and subsurface mining. Also, describe measures to reduce the impact of those wastes.

16. Explain the differences between reserves and resources.

17. What is the primary obstacle to mining very low-grade mineral deposits?

18. What are the prospects for recovering high-grade minerals from the seabed and new discoveries on land?

19. List some conservation strategies that make more efficient use of resources and reduce the present rates of depletion.

20. How does the grade of a mineral deposit influence the environmental impact of mining?

Selected Readings

ABELSON, D. H., AND HAMMOND, A. L. (EDS.): *Materials: Renewable and Nonrenewable Resources.* Washington, D.C.: American Association for the Advancement of Science, 1976. A compendium of articles that first appeared in *Science* magazine between November, 1973, and February, 1976. It is somewhat dated but, nevertheless, it contains much important information concerning the supply and use of materials.

BORGESE, E. M.: "The Law of the Sea." *Scientific American* 248(3):42–49, 1983. A review of the international convention on the Law of the Sea, which was signed in December, 1982.

CHANDLER, W. U.: "Materials Recycling: The Virtue of Necessity." *Worldwatch Paper*, No. 56, 1983. An update on the economics of recycling paper, aluminum, and iron.

EDMOND, J. M., AND VONDAMM, K.: "Hot Springs on the Ocean Floor." *Scientific American* 248(4):78–93, 1983. Describes a link between hydrothermal processes and the generation of metallic resources on the seafloor.

JEANLOZ, R.: "The Earth's Core." *Scientific American* 249(3):56–65, 1983. Describes evidence for the composition and structure of the earth's core.

MCKENZIE, D. P.: "The Earth's Mantle." *Scientific American* 249(3):66–78, 1983. Includes an interesting discussion of convective currents within the earth's mantle.

PRESS, F., AND SIEVER, R.: *Earth*, 3rd ed. New York: W. H. Freeman, 1982. A well-illustrated introductory textbook on the principles of physical geology.

SIEVER, R.: "The Dynamic Earth." *Scientific American* 249(3):46–55, 1983. A comprehensive description of the processes that shaped the earth through the millions of years of geologic time.

TANK, R. W.: *Focus on Environmental Geology: Text and Readings.* New York: Oxford University Press, 1983. Includes selected readings on mineral resources and the environmental impact of mining in the United States.

WYLLIE, P. J.: *The Way the Earth Works: An Introduction to the New Global Geology and Its Revolutionary Development.* New York: Wiley, 1976. An unusually lucid and classic account of geologic processes, developed in the context of continental drift.

Thoughtless disposal of waste, hazardous to health and threat-ening to environmental quality. (U.S. Department of Health and Urban Development.)

CHAPTER **13**

WASTE MANAGEMENT

In 1983, investigations revealed that approximately 100 sites in Missouri had been sprayed with waste oil that was contaminated with the highly toxic chemical dioxin. The contaminated oil was spread over gravel roads to control dust. Dioxin, which is a chlorinated hydrocarbon, has been shown in animal tests to be one of the most toxic chemicals in existence. The cost of cleaning up such sites is very high in terms of both dollars and human anxiety. In the case of one of the contaminated areas, Times Beach, Missouri, the federal government finally purchased the city for $32 million, so the residents of the city could afford to move elsewhere.

In 1972, the Stringfellow acid pits (a dump site for hazardous wastes located in Riverside, California) was closed because acids, pesticides, organic solvents, and heavy metals were leaking from the site. Ineffective and meager efforts by all levels of government since 1972 have failed to control the problem. Those efforts were concentrated on the removal of contaminated surface and groundwater rather than on the prevention of further groundwater contamination. However, in 1984, the contaminated plume of groundwater was found to be moving rapidly toward the Chino Basin aquifer—the source of water for 500,000 persons and scores of farms and industries. Scientists predicted that the contaminated groundwater would reach the aquifer in 1 to 2 years. Because the dump site has underground springs and fractured bedrock beneath it, there appears to be no means of containing the leaking chemicals. In 1984, the ultimate cost of removal and

treatment of the wastes at the site was estimated at $60 million, which was in sharp contrast to the $3.4 million estimate in 1977.

We are no longer content to live in a polluted environment. For both personal and legal reasons, we no longer tolerate smoldering dumps, smoky incinerators, leaky landfills, and factories that disgorge toxic wastes into our waterways and atmosphere. But we continue to generate enormous amounts of wastes. Thus we must determine where we should dispose of our garbage, trash, hazardous chemicals, and radioactive wastes. Also, we must resolve the problems created by old, abandoned, and faulty disposal sites.

In this chapter, we examine some of the requirements for the safe disposal of municipal wastes, hazardous chemicals, and radioactive wastes, and we examine some of the factors that cause disposal problems to remain unresolved. We also examine some of the new management practices that are being used to control and to keep track of wastes. And we consider the potential for recovering valuable materials and energy from those wastes. Alternatives for sewage sludge are also discussed in Chapter 8.

A Historical Overview

Throughout much of our nation's history, we have freely tossed garbage and other refuse into our backyards, streets, fields, marshes, and rivers. Back in colonial days, hogs were allowed to wander freely through city streets, rooting through garbage and leaving their excrement and stench behind. Then, in post-Civil War days, the cities swelled with new industries and workers, and the problem of waste disposal became acute. For example, Chicago's population soared from only 5000 in 1840 to almost 1 million by 1890. The meager sanitation efforts that were made at that time could not keep pace with the amount of wastes that were being generated, and city streets became clogged with trash. And in the late nineteenth century, refuse accumulated in such large heaps along some streets in New York City that it impeded pedestrian and vehicular traffic (Figure 13.1).

Public concern over the growing health hazard eventually forced some reforms. For example, hogs were banned from city streets, and garbage and other refuse was hauled to hog farms, to open dumps that were adjacent to cities, and to nearby rivers or it was barged out to sea. But the waste problem was not solved by those actions; it was merely displaced. Not until recently have health and environmental quality regulations been implemented to control waste disposal.

Although hauling wastes from cities improved conditions in urban areas, it worsened environmental conditions around the disposal sites. For example, the open, uncompacted wastes made ideal breeding grounds for disease-carrying pests, such as flies and rats. And over the years, rainwater and melting snow seeped into dump sites, producing contaminated leachate, which percolated into groundwater reservoirs. (*Leachate* is defined as the contaminated liquid that results as water seeps through wastes.) Many of those dumps received hazardous chemical wastes, which are still migrating from those sites as part of the leachate. Also, smoldering fires in those dumps polluted the air, which caused a continuing source of irritation for nearby residents. And exposed hazardous wastes have caught

Figure 13.1 *Heaps of refuse clogging the streets of New York City in the 1880s. (The Bettmann Archive, Inc.)*

fire at times, sending toxic fumes downwind, which has forced many persons to evacuate their homes, sometimes for days.

During the late 1970s, the leaking dump site at Love Canal in New York called widespread public attention to the problem of sloppy practices in the disposal of hazardous waste. When persons who lived nearby learned about the danger of being exposed to unknown levels of unknown pollutants in their neighborhoods, most grew fearful and many experienced panic. Some persons attributed their childrens' illnesses and birth defects to exposure to chemicals that were oozing from the dump site.

However, it is difficult to correct problems that were caused by the thoughtless dumping practices of the past. In many states, dump sites have not been regulated, and, as a result, the exact locations and contents of many dumps are not known. And small (1 or 2 hectare) sites can pose as great a threat to nearby residents as larger ones. Furthermore, once an old site is located, it is an expensive task to determine whether it is leaking, what specific chemicals are present, and how to arrange financing to clean it up, which is usually a legal nightmare.

A few states, notably Massachusetts and California, have had standards for the collection and disposal of solid waste for many years. Typically, however, until recently most states exercised meager control over land disposal sites. The first federal legislation on waste disposal was not enacted until 1965. The Solid Waste Disposal Act of that year called for research on the solid-waste problem and provided funds to states and municipalities for planning and developing waste-disposal programs. The 1965 law was amended with the passage of the 1970 Resource Recovery Act, which provided funds to states for constructing waste-disposal facilities. Also, the 1970 law was the first federal legislation to encourage the recycling of solid waste.

The strict regulation of solid-waste disposal is a major objective of the Resource Conservation and Recovery Act of 1976. This law required the states to close all open dumps by 1983 and assists them in developing waste-reduction programs. The 1976 act also calls on the Environmental Protection Agency (EPA) to draw up guidelines for the siting of sanitary landfills to minimize the environmental impact of solid-waste disposal. Provisions also seek to strictly regulate hazardous wastes, from the point of generation to ultimate disposal. The law requires the states to identify hazardous wastes and sets national standards for the generation, transport, treatment, storage, and disposal of hazardous wastes.

Hazardous materials became subject to further regulation in 1976 with passage of the Toxic Substances Control Act, which requires premarket screening of new chemicals (excluding pesticides) for potential toxicity. Chemicals that pose an unreasonable risk in their manufacture, distribution, use, or disposal can be regulated through the EPA. Although the intent of this law is well placed, its provisions have generated considerable controversy that is likely to delay full compliance until later in this decade. Disagreements have developed over the degree of risk that results from exposure to different chemicals. Also, chemical companies fear that the law will force public disclosure of trade secrets and aid their competitors. Those problems are compounded by the great number of chemicals that are currently in commercial production (approximately 70,000), which ensures that the screening process will be both expensive and time consuming.

One of the major milestones in hazardous-waste management legislation occurred in the early 1980s with the passage of the Superfund Act. That law established a $1.6 billion fund for the cleanup of abandoned hazardous-waste sites. In addition, the law requires the discharger to compensate victims for illness and injury as well as for property damage. The Superfund is financed primarily (87 percent) by taxes on the chemical and petroleum industries, with the remainder (13 percent) coming from federal revenues. But in 1984, it became apparent that the initial allotment of money would not be sufficient to meet the needs to finance the cleanup that is necessary across the country. A late 1984 EPA study estimated that an additional $8 to $23 billion would be needed to clean up the 1800 to 2200 high-priority dump sites that have been identified.

Properties of Waste

Proper management of waste hinges on an evaluation of certain key properties of the waste. Those properties include the degree of hazard and the persistence of the waste in the environment. The degree of hazard that a waste poses depends on its corrosivity, reactivity, ignitability, or toxicity, and those catagories are specifically defined by the Resource

Conservation and Recovery Act. Hazardous wastes, for example, include acids, explosives, combustible solvents, and toxic chemicals. Radioactive wastes are also hazardous, but because of their special properties, they are managed separately by the Nuclear Regulatory Commission.

Some hazardous wastes are particularly troublesome because they tend to persist in the environment. Usually, those wastes are persistent because the physical, chemical, and biological processes operating in the environment are ineffectual in breaking them down into harmless products. (Heavy metals cannot be broken down because they are elements.) Environmental persistence of hazardous materials causes problems for organisms if those materials enter food webs. They bioaccumulate as they pass from one trophic level to the next higher trophic level and they may eventually be ingested by humans.

Wastes can also be classified as being *biodegradable* (that is, subject to decomposition by microorganisms) or *nonbiodegradable*. In many instances, biodegradable substances (such as paper, lawn clippings, and wood from demolition) can be used directly as fuels or can be processed to produce fuel products. That distinction is crucially important when we assess the potential of recovering fuel materials from wastes. Also, less concern exists about the potential pollution problems that occur as a result of biodegradable substances, because they can be degraded by the action of aerobic or anaerobic decomposers. Nonbiodegradable materials, on the other hand, are not broken down, and some of them (for example, lead, mercury, cadmium, and other heavy metals) pose a long-term threat. Agricultural wastes consist primarily of biodegradable materials, whereas mining and mineral wastes consist of nonbiodegradable materials. Municipal and industrial wastes are usually complex mixtures of biodegradable and nonbiodegradable components.

Sources of Waste

Virtually all of our activities generate waste. We create waste when we extract and process raw materials, fuels, and foods and when we use services. When products wear out, become outmoded, or otherwise outlive their utility, we merely throw them out. Fundamentally, air and water pollution are prob-

lems of waste disposal. Thus a wide variety of materials are referred to as being waste. For convenience, these wastes can be classified according to source: municipal, industrial, mining-mineral, and agricultural.

The constituents of municipal waste are as diverse as those of our surroundings: they range from bricks to tree branches, aluminum cans to newspaper, explosives to sludge, and cinders to food. The typical components of refuse collected by a municipality are listed in Figure 13.2. In the United States in 1980, each of us threw away an average of 1.7 kilograms (3.8 pounds) of solid waste each day, which totals 150 million metric tons (160 million tons) each year. This figure is expected to grow to 200 million metric tons (220 million tons) by 1990.

Yet, as Figure 13.3 shows, those amounts are small when they are compared with the amounts of waste that are produced by mining, industrial, and agricultural activities. Mining and mineral wastes include mill tailings, slag, fly ash, and various mine wastes exclusive of overburden (see Chapter 12). Those wastes are generated at a rate that is approximately 11 times greater than the rates at which municipal wastes are produced. Because mining and mineral processing usually take place in remote areas, those wastes are not as evident as municipal wastes— but their quantities are enormous. For example, in the southwestern United States, when 1 metric ton of copper is produced by open-pit mining (which in-

Figure 13.2 *The constituents of typical municipal trash. (The relative proportions of trash materials have changed little since this study was done.) (After J. G. Abert et al., The economics of resource recovery from municipal solid waste. Science, 183:1052, 1974. Copyright © 1974 by the American Association for the Advancement of Science.)*

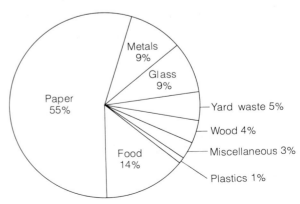

cludes extracting, grinding, separating, smelting, and refining), more than 500 metric tons of solid waste are generated. Air pollutants that are generated in the process are not included in this total.

Tremendous quantities of wastes are also produced by the burning of coal at electricity-generating plants. The noncombustible fraction of coal that remains after coal is burned, which is called *fly*

Figure 13.3 *Each year, some 4 billion metric tons of nonradioactive wastes are generated in the United States. The absolute (a) and relative (b) contributions of their several sources are shown here.*

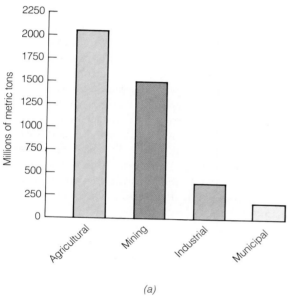

(a)

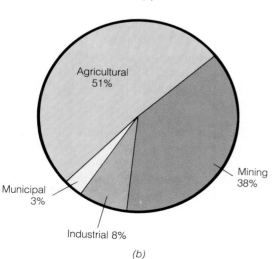

(b)

ash, typically makes up 10 percent of the total weight of the coal that is burned. Thus a 1000-megawatt plant (the approximate size of most plants being built today) disposes of 250,000 metric tons (280,000 tons) of fly ash each year. During their 40-year lifetime, such plants typically require 2.5 square kilometers (1 square mile) of land to dispose of that fly ash.

Agricultural wastes are approximately 15 times greater than municipal wastes. Most of the 2.1 billion metric tons of agricultural wastes (approximately 75 percent) that are produced each year consist of animal manure, crop residue, and various by-products of food production. However, most of the animal wastes are disposed of by plowing or injecting them into the soil, thereby returning unused nutrients to the earth and enhancing the soil's capacity to retain moisture.

Of the total amount of wastes that are generated in the United States each year, approximately 260 million metric tons, or approximately 1 metric ton per person, is considered to be hazardous if it is improperly managed. Hazardous wastes are produced in the manufacture of inorganic and organic chemicals (including plastics), in the smelting and refining of ores, in electroplating, and in petroleum refining. Nearly two-thirds of the hazardous wastes are generated by the chemical industry and its associated industries. Paper and glass product industries and metal production each account for between 3 and 10 percent of the nation's hazardous-waste output.

Also, for more than 40 years, our nuclear weapons systems and nuclear power plants have been producing radioactive wastes that have been "temporarily" stored, pending the development—and public acceptance—of permanent storage techniques and sites. The wastes that are produced by those weapons systems and power plants emit radiation, which is extremely hazardous to humans as well as other organisms. And those wastes continue to accumulate at military sites and nuclear power plants across the country. To date, the United States government has not decided on a specific way to dispose of those wastes safely.

Management of Nonhazardous Wastes

Most of the garbage and rubbish that is collected from our community's residences, commercial establishments, and industries probably ends up in a sanitary landfill, although some may be incinerated. For coastal communities, sewage sludge and some nonhazardous industrial wastes may be dumped at sea.

There are, however, other ways of managing municipal waste besides disposal. Indeed, provisions of the 1976 Resource Conservation and Recovery Act recognize two ways of alleviating problems that are associated with solid waste: recovering resources from solid waste (that is, recycling) and reducing the waste potential of products before those products reach the consumer. Both strategies ultimately reduce the number of facilities that are needed for disposal. Recycling, in effect, converts a community liability into a community asset by adding to the supply of available materials and energy. Waste reduction lowers collection and disposal costs and lessens demands on raw materials and energy.

Disposal Techniques

Because municipal waste is a complex mixture of materials and is worth very little, it is usually too expensive for communities, especially the smaller ones, to recover and recycle any of its constituents. In such instances, the least expensive alternative often is to bury the wastes in a landfill. However, before a community can use that technique to dispose of its wastes, a landfill site must be chosen that meets rigorous engineering and environmental standards that are set by the EPA or an appropriate state regulatory agency. Landfills are now run in a manner that contrasts sharply with the casual operation of the open dumps of earlier decades, where disease-carrying insects and rodents were allowed to breed and smoldering fires fouled the air continuously with acrid smoke.

Sanitary Landfills A *sanitary landfill* eliminates most of the problems that are encountered at an open dump site by sealing wastes between successive layers of clean earth each day. No open burning is allowed at sanitary landfills; rodents and insects cannot thrive in them; and odor is seldom a problem. Furthermore, the landfills can be located in areas that have been mined and can be designed to restore scarred landscapes. Indeed, many communities are creating new parks and other recreational sites by revegetating areas that have been filled with wastes in recent years.

Three types of sanitary landfills exist: trench landfills, area landfills, and mounded landfills. *Trench landfills*, such as the one shown in Figure 13.4, are only appropriate in regions where the soils are deep and have a high clay content. The clay greatly reduces the rate at which leachate seeps through the soil. Also, the water table must lie well below the zone of fill. To make a trench landfill, a trench is dug and solid waste is spread in it and then compacted by heavy machinery. Then it is sealed each day with a layer of dirt that was excavated previously, when the trench was dug. That process is repeated until the trench is filled.

Area landfills are made from natural valleys or canyons as well as abandoned pits and quarries. In that method, the site is first lined with a layer of clay that is 2 meters thick if the natural soils on the site will not retain leachate. Then the wastes are placed in layers on top of the bottom liner, compacted, and covered with soil each day until the site is filled. Sites for area landfills are usually easier to locate, because an existing site can be lined with clay, thus making it unnecessary to search for a site with exactly the right soil conditions. Although area landfills are usually more expensive to operate because of the additional excavation that is needed, overall, the cost of transporting wastes is the dominant economic consideration. As a result, area landfills are usually chosen over trench landfills if a site can be found that is closer to the source of waste.

In regions where the water table is close to the surface, a *mounded landfill* is often necessary to prevent groundwater contamination. Alternating layers of compacted wastes and dirt are mounded over a clay base on the surface to form a hill. After the mounded landfill is completed, the hill is contoured and vegetated to control erosion.

Although sanitary landfills are far more desirable than open dumps, they do have some drawbacks. Groundwater contamination is always a potential problem. In most areas, surface water that seeps through wastes (leachate) must be pumped out and then either piped or hauled to a sewage treatment plant. As an additional precaution, groundwater that exists in the vicinity of the landfill must be checked frequently to determine whether it is being contaminated by unexpected leaks in the liners of landfills.

Another drawback of sanitary landfills is the possible migration of gases that are formed during the anaerobic decomposition of waste. Methane and hydrogen sulfide (which are gaseous products of an-

(a)

(b)

Figure 13.4 *(a) Refuse is spread and compacted in a trench land-fill. (b) At the end of the day, the compacted wastes are covered with a layer of clean fill. (Deere and Co.)*

aerobic decomposition) can seep through landfill liners into neighboring buildings. Since methane can explode, care must be taken to ensure that it does not seep into enclosed spaces, such as basements of nearby houses. Such explosions have destroyed homes. In addition, hydrogen sulfide can act as an asphyxiating agent. Therefore, if gas migration from landfills become a problem, vent pipes and vent trenches must be installed to collect the gases so they can be flared (burned) or allowed to escape into the open air.

However, methane can be recovered from landfills and used as an energy resource. (Natural gas is composed primarily of methane.) The first methane recovery facility went into operation in 1975 at Rolling Hills Estates in California. Since then, additional landfill operations have been established that collect methane by using a relatively simple technology. In that process, vertical holes are drilled into the waste, and then perforated collecting pipes are inserted into the holes. A compressor is connected to the pipes, which is used to draw the methane from the site. All moisture and other gases (such as toxic hydrogen sulfide) are removed before the methane is piped to a nearby facility (such as an industrial boiler), where it is burned as a fuel source. The Fresh Kills landfill site on Staten Island in New York City (the world's largest landfill that collects methane) produces enough methane to heat 10,000 homes through a New York winter.

Once a landfill operation has been completed, the site must be inspected periodically for uneven settling of the ground, which can expose wastes and increase gully erosion. Settling always occurs, primarily because of the anaerobic decay of the wastes, which causes further compaction of the refuse. Approximately 90 percent of the total settlement usually occurs within 5 years. Thus after a substantial amount of settling has occurred, more filling and grading may be necessary to make the site smooth and to reduce erosion.

Finally, a significant nuisance problem can occur during daily operation, when the wind scatters litter and raises dust. Maintenance personnel can

433

minimize those problems by installing litter fences around the site and regularly applying water from sprinkler trucks. Still, the staff will have to collect litter by hand to maintain a neat site.

Although a completed landfill does not make a suitable building site, it is suited for recreational activities (such as parks, golf courses, and playing fields), botanical gardens, or wildlife areas. However, even though landfills offer many advantages over open dumps for solid-waste disposal, a major obstacle can be public opposition. Most persons equate a sanitary landfill with a dump and simply refuse to accept it in or near their neighborhood, even though it may eventually be turned into a park, which would benefit the entire community. This type of land use is a classic example of the sometimes intractable problem of reconciling individual preferences with the public good.

Incineration *Municipal incinerators* burn combustible solid waste and melt certain noncombustible materials. Incineration is more advantageous than open dumping from a health perspective, since the high temperatures destroy pathogens and their vectors. However, because installation, maintenance, and operational expenses are higher, the average cost of incinerating solid waste is generally somewhat higher than operating a sanitary landfill. This difference does not apply in localities where appropriate sites for sanitary landfills are unavailable and where land prices are high. Ironically, those very circumstances usually prevail in and around urban areas, where most household, commercial, and industrial wastes are generated. The solution in such situations may be a combination of incineration and sanitary landfills. Incineration can reduce the volume of solid waste by 80 to 90 percent. Hence disposing of the material that remains after incineration requires considerably less land than would the nonincinerated waste.

Incineration has additional advantages over landfills of any kind. Incinerators do not directly endanger groundwater quality. And, if they are equipped with adequate air-pollution control devices and attractively housed and landscaped, incinerators are more likely to gain public acceptance than are landfills. Unfortunately, most of the incinerators operating in the United States during the late 1960s were constructed before air-quality control devices were required. When the Clean Air Act amendments were enacted in 1970, the incinerators were forced to comply with air-quality standards, and the added cost of installing the pollution-control equipment forced most of them to shut down. Another drawback of conventional incinerators is their inability to handle certain types of wastes, for example, explosives and potentially smoky wastes, such as tires.

Considerable progress was made in the 1970s and early 1980s in incineration technology and in air-pollution control devices for incinerators. Hence, in some large metropolitan areas, new incineration techniques are a central component in resource recovery facilities. These facilities are described later in this chapter.

Ocean Dumping More than 80 percent of the waste material that is barged out to sea and dumped consists of spoils that are generated by the dredging of ship canals in harbors and rivers. Typically, one-third or more of those spoils are contaminated by industrial and municipal effluents and runoff from farmlands. Most of the remaining waste that is dumped at sea is composed of industrial waste and sewage sludge.

Sludge from sewage treatment facilities poses a special hazard to ocean life, because it usually contains heavy metals that can concentrate in food webs (see Chapters 2, 7, and 8). Also, sewage sludge usually contains pathogenic microorganisms. The industrial wastes dumped in the oceans encompass a wide variety of substances, many of which may be toxic to sea life (for example, arsenic compounds, strong acid and alkaline solutions, and PCBs). Often those substances are dumped at sea because water-quality regulations prohibit them from being dumped into rivers or sanitary sewers and because no appropriate land disposal site is available. Concern about the dumping of such wastes into the ocean or onto the seabed focuses not only on toxicity to organisms, but also on the possibility of bioaccumulation in food webs that humans utilize. Also, because mixing processes in the oceans are poorly understood, there is a possibility that ocean dumping that seems safe enough at the present time will cause unsuspected problems in the future.

With the enactment of the Marine Protection, Research, and Sanctuaries Act of 1972—commonly referred to as the Ocean Dumping Act—industries were required to obtain an EPA permit to transport wastes and dump them in oceans. That law also empowered the EPA (and the U.S. Army Corps of Engineers, in the case of dredge spoils) to locate proper ocean disposal sites (beyond the continental shelf, where possible) and to regulate the types and amounts of substances that could be disposed of. Thus toxic industrial wastes must be dumped several hundred kilometers offshore. In the past, most ocean dumping sites were situated within 40 kilometers (25 miles) of shore. At those locations, waste could be transported back to shore by ocean currents, where it could adversely affect commercial shellfish beds and force the closing of beaches. The long-term objective of the 1972 law is the eventual phasing out of all ocean dumping.

In 1977, the EPA called for an end to ocean dumping of wastes that threaten to "unreasonably degrade" the marine environment. Amendments to the 1972 Ocean Dumping Act, which were enacted in 1977 and 1980,-prohibited disposal of all sewage sludge and industrial wastes in the Atlantic Ocean (the site of 90 percent of the United States' ocean dumping) after December 31, 1981. These regulations would force coastal municipalities to seek on-land alternatives. However, federal court action taken by New York City against the EPA in November of 1981 delayed compliance to these regulations and forced the EPA to examine the environmental impacts of sludge disposal on land and to reevaluate its ocean dumping policy.

As a consequence of more rigorous ocean dumping laws, disposal of industrial wastes at sea declined 68 percent between 1975 and 1982. However, because of improved levels of sewage treatment, dumpers of sewage sludge were allowed to increase the volume of sludge they dumped by 54 percent during the same period. It is likely that in the future large coastal cities will increasingly look to the sea as a disposal site for their sewage sludge. This occurrence is partially because of the economic advantage of ocean dumping over on-land disposal and partially because of the expected increase in sludge generation. A 1982 estimate by the National Oceanic and Atmospheric Administration (NOAA) projects a 130 percent increase in the generation of sewage sludge by coastal cities by the year 2000. It is also reasonable to assume that tightening controls for land disposal may put more pressure on industry to seek permits to dump their wastes at sea.

In regulating ocean dumping, the EPA must compare the environmental consequences of dumping at sea with the potential environmental impact of land-based alternatives. Although ocean dumping may be economically attractive over the short term, unquantifiable future risks remain. Limiting dump sites, where feasible, to deeper waters off the continental shelf is important for reducing potential future problems. But in order to determine whether or not ocean dumping is a viable long-term waste-disposal alternative, there must be more research on transport and cycling of pollutants and their ultimate fate in the oceans.

Reuse Techniques

Resource recovery is a broad term that is used for the retrieval of valuable materials or energy from a waste stream. Hence resource recovery is classified into two basic categories: that which recovers materials and that which recovers energy.

Reclamation is the separating out of materials, such as rubber, glass, paper, and scrap metal, from refuse and the act of reprocessing them for reuse. In the United States, waste reclamation is a multibillion-dollar-a-year business, based primarily on the recovery of scrap metal from the more than 8 million motor vehicles that are junked each year. Also, a significant fraction of the aluminum, lead, copper, and zinc used by industry is derived from the recycling of those metals. For example, in industries that machine metal parts, metal turnings and scraps are kept separate from other plant refuse so they can be sold for recycling. Automobile repair shops keep worn-out batteries because the lead they contain can be recycled. Today, many persons recycle their aluminum cans to receive cash for the aluminum at recycling centers (Figure 13.5). And wastes that are generated in office buildings, which contain approximately 50-percent high-grade paper, are separated by more than 600 companies and 100 government facilities and are sold to paper recyclers.

Some communities have separation systems that include the collection of all wastes, but they depend on their citizens to separate the valuable materials from their rubbish or to bring separated wastes to recycling centers. More than 200 communities in the United States have collection systems that use some sort of separation scheme. Approximately 80 percent of those communities collect newspapers; the rest collect metals, glass, or plastics as well. Some

Figure 13.5 *Aluminum can recycling centers such as the one shown operate 24 hours a day at some shopping centers. They operate as reverse vending machines: customers deposit cans in a receiving bin (left side) and receive cash in return. (Golden Recycle Co., a subsidiary of the Adolph Coors Co.)*

communities even collect waste oils and greases. Then the separated materials are kept segregated during collection and are sold. Although source separation methods depend on voluntary compliance, increasing numbers of communities are requiring compliance to increase the amount of material that is collected and to reduce the cost of recovering the materials.

Magnetic separation is utilized by community recycling systems to remove iron-containing, or ferrous materials, which can be picked out of a waste stream by magnets. In most magnetic separation systems, the wastes are first shredded into small pieces and then are passed through a magnetic separator.

Composting is an age-old process, wherein the organic fraction of wastes is heaped into piles and allowed to decompose. After about 6 months, a humuslike substance remains, which makes an excellent soil conditioner. In the composting of municipal wastes, the organic fraction (such as paper, leaves, and lawn clippings) of the waste stream first must be separated out. That fraction is then piled into windrows that are aerated approximately once a month by turning them over mechanically. The final humus product is then sold. Only a few communities have tried composting as a method of recycling wastes, and most of those efforts have been discontinued, because the market for compost is small. The humuslike material cannot compete with chemical fertilizers, because it is relatively low in nutrients. However, it can be used as a soil conditioner.

Hydropulping is a method that is used to recycle paper products. In that process, wastes are mixed with water to prepare a slurry of paper fibers. Then, a centrifuge (a device that is similar to a clothes washer in its spin cycle) is used to separate the paper fibers from the heavier materials, such as metals, glass, and rocks. The lighter paper fibers are left in suspension and are drawn off to produce recycled paper products by conventional paper-making techniques.

Reclamation is not limited to the recycling of materials for reuse as new products. Some refuse is reclaimed directly as a fuel, which is called *refuse-derived fuel* (RDF). When municipal waste is shredded or milled, combustibles are separated from noncombustibles, and the combustible fraction is then burned with coal, using a mixture that contains 5 to 25 percent refuse, to generate steam. Currently, only a dozen small-scale facilities employ refuse-derived fuels in the United States. The major obstacle to using refuse-derived fuels is that communities are reluctant to invest the capital that is necessary for recovery facilities. In addition, reliable markets for refuse-derived fuels are often difficult to identify.

Another method of recycling is *conversion*, that is, the chemical or biochemical transformation of waste material into useful products, such as fuels, and a variety of industrial chemicals. In one such process, *pyrolysis*, organic wastes are broken down by exposure to intense heat in the absence of oxygen. The yield of this conversion technique depends on the composition of the refuse, but normally it includes char (which is similar to charcoal), alcohols, light oils, and combustible gases. All these substances are potential fuels. This process can convert nonbiodegradable refuse (plastics and rubber) into fuels and emits no air pollutants. Commercial pyrolysis facilities have experienced technical problems and are still considered to be in the developmental stage.

Composting (which was discussed earlier) is actually a bioconversion technique. A variety of other bioconversion techniques involve the aerobic or anaerobic treatment of organic waste. When these techniques are applied to agricultural wastes, the organic fraction of municipal refuse or sewage sludge, they yield such fuels as alcohol and methane (Figure 13.6). Although those methods are at present still in the small-scale testing phase, they are slowly gaining favor. For example, the European countries, especially Belgium, have become quite proficient at producing methane from animal wastes by anaerobic digestion (treatment).

Figure 13.6 *This facility at Pompano Beach, Florida, is an effort to prove the concept of the anaerobic digestion of a mixture of solid waste and sewage sludge, to develop design parameters, to establish economics, and to collect data on the quantity and quality of the methane gas that is produced. Urban waste is shredded and separated into light and heavy fractions. The light fraction is combined with up to 10 percent sewage sludge and is digested. (Department of Energy.)*

An ideal approach to solid-waste recycling is the development of centralized community *resource recovery facilities* that combine the techniques of materials reclamation and conversion processes (Figure 13.7). In these facilities, wastes are separated first according to size. The larger organic fractions (such as newspaper, cardboard, plastic packaging, and branches) are separated from the noncombustible inorganic fractions (such as cans, bottles, and rocks) and are shredded to produce waste that is smaller and more uniform in size. Then, the smaller organic materials that are mixed in with the inorganic wastes are shredded and a stream of air separates the wastes by blowing the lighter organic material from the waste stream. Then, both organic fractions are combined. At that point, they can be incinerated to produce steam; they can be processed further to produce refuse-derived fuels; or they can be pyrolized.

The inorganic wastes, which are the denser, or heavier components of the waste stream, are separated further. First, the iron-containing metals are separated out by electromagnets. Then, aluminum and glass usually are removed by passing them through a series of screens of varying size. The remaining residue (which is mostly dirt and grit) has no value, so it is hauled off to a landfill.

The number of resource recovery facilities is continuing to grow. In the early 1980s, some 70 plants were in operation and nearly that many more were in the planning stage. Those plants can process approximately 12 million metric tons annually, or slightly less than 1 percent of the municipal waste stream that is generated in the United States. Although the energy that is recovered from wastes probably never will contribute more than 1 percent of the total energy demand in the United States, it can be an important local source and can save money for a community.

Most communities are opting for energy recovery systems rather than for materials recovery systems because the market for energy is more dependable. For example, changes over time in the price of electricity are small when they are compared to the wide swings in prices that a recycling facility can charge for the iron, aluminum, or paper it recovers.

Although great strides have been made in resource recovery technology, further development can be stalled if the public opposes it or if lending institutions are unwilling to underwrite recovery facilities because of past failures. For example, some persons vehemently oppose the siting of resource recovery facilities in their immediate neighborhoods, despite the fact that such plants can be designed to blend unobtrusively into their surroundings. And in some cases, representatives of industry refuse to even consider wastes as a possible source of fuel because of the additional costs that are associated with the planning that is required to implement the use of fuels that are recovered. Therefore, industries that do cooperate in recovery endeavors often are recognized for their corporate community spirit.

Waste Reduction

One important waste-reduction strategy is to encourage the manufacture of reusable products rather than disposable ones. That approach is exemplified by the returnable beverage container. Following Oregon's lead in 1972, eight other states—Vermont, Maine, Michigan, Connecticut, Iowa, New York, Massachusetts, and Delaware—have passed "bottle laws" that impose a deposit on beverage containers. Deposits are refunded when the containers are returned for subsequent reuse. Such legislation is aimed at stemming the proliferation of containers that are used only once and encouraging the circulation of recyclable ones. Oregon's law, in an attempt to address related issues that are concerned with unnecessary gimmicks in packaging and unsightly litter, also prohibits the distribution of flip-top or pull-tab cans and bottles. One of the big benefits of the returnable beverage container has been the reduction of litter along highways.

Another waste-reduction strategy is to redesign products so that less material is required for each unit. With respect to automobiles, that approach has a double advantage. Because cars that require less metal in production are smaller and lighter, they also use less fuel in their operation. Spurred by foreign competition and by federal requirements to improve fuel economy, United States' automobile manufactures have trimmed the weight and size of most models that are now on the market.

Current trends in the packaging of products in the United States results in the consumption of prodigious amounts of raw materials and the creation of tremendous quantities of waste. Packaging of consumer goods uses 75 percent of the glass, 40 percent of the paper, 29 percent of the plastic, 14 percent of the aluminum, and 8 percent of the steel that is consumed each year in the United States. Typically, 30 to 40 percent of a city's solid waste consists of discarded packaging materials. One strategy for reducing this type of waste is to develop packaging methods that conserve raw materials.

Although planned obsolescence directly results in major waste-disposal problems, planned obsolescence has governed the operations of many important industries, particularly since World War II. Advertizers encourage consumers to purchase new products (and dispose of old products) by changing styles every year. We could counter the effects of that practice to some extent by developing more durable products that are more easily and economically repaired. But probably the most fundamental waste-reduction strategy is for each of us to eliminate wasteful consumption and strive to maximize our efficiency in use.

Management of Hazardous Wastes

Hazardous-waste disposal is recognized by the EPA as the primary environmental problem of the 1980s, and we are just now beginning to deal with it. In the

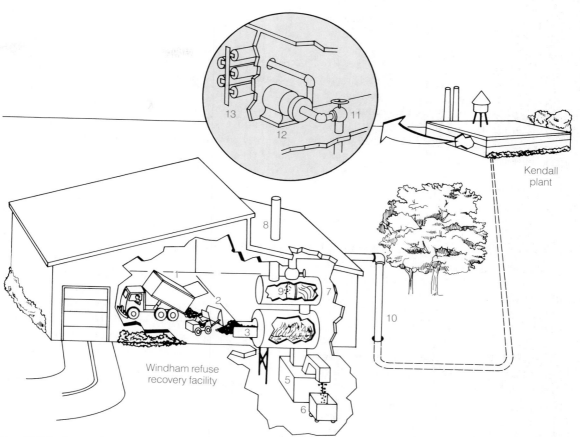

Figure 13.7 *In the Windham, Connecticut, refuse-to-energy facility, 13 steps are followed: (1) The truck dumps the refuse on the floor. (2) The loader operator scoops up the refuse and dumps it into the refuse loading hopper. (3) The refuse is moved into the combustion unit at a carefully controlled rate to produce a flammable gas. (4) Moving rams, which are located under the fire, slowly push the ash from burnt refuse toward the rear of the combustion unit, where it drops into a water tank. (5) The ash is cooled in a water tank, reducing its volume by 90 percent. (6) Ash is put into containers for removal to a landfill. (7) Hot gases that are generated in the combustion unit flow into a secondary chamber, where they are mixed and burned. (8) Cooled stack gases—the combustion product of the secondary chamber—are vented to the atmosphere; they must meet EPA standards. (9) Heat from the secondary combustion chamber produces steam in the boiler tubes. (10) The steam is piped to Colgate-Palmolive's Kendall plant, (11) which manufactures nonwoven fibers. (12) The steam passes through and powers a turbine that generates electricity. (13) After leaving the turbine, the steam is used to heat fabric dryers. (Solid Wastes Management 24:11, 1981.)*

United States, an estimated 22,000 to 25,000 hazardous dump sites exist (see the photo at the beginning of this chapter). Those sites received chemical wastes before any regulatory guidelines were established. And many of the owners of those sites are no longer in business or cannot afford to clean up their leaky sites. Furthermore, most of the problem sites are located in the heavily industrialized parts of the East and Midwest and are associated with chemical and petrochemical industries. In 1984, the EPA had a list of 538 priority dump sites (dump sites where existing problems pose immediate threats to humans or the environment) for immediate cleanup. Ultimately, the list is expected to reach 1800 to 2200 sites.

When legal responsibility can be established, the cost for cleanup must be paid by the owners of a hazardous-waste site. But records of who dumped what, when, where, and in what quantity are usually nonexistent, which makes the establishment of financial responsibility very difficult. For example, at one site it was determined that more than 100 companies had dumped many different kinds and amounts of wastes. Thus in an effort to clean up those sites where responsibility cannot be established fully, Congress passed the Comprehensive Environmental Response and Compensation Act of 1980. That act, which is more commonly referred to as the Superfund, collected $1.6 billion by January of 1985 to clean up the hundreds of sites where legal responsibility could not be established or where the owners could not afford to clean up their wastes. The initial funding for the Superfund has proved to be totally inadequate and in 1985, Congress was wrestling with legislation to finance the estimated $8 to 23 billion required to complete the task.

Legislative Mandate

Under the Resource Conservation and Recovery Act (RCRA) of 1976 and subsequent amendments, the EPA has been given the legal responsibility of defining hazardous wastes and protecting the public's health and the environment from them. The law directs the EPA to "develop and promulgate criteria for identifying the characteristics of hazardous (chemical) wastes, . . . taking into account toxicity (including carcinogenicity), persistence, and degradability in nature, potential for accumulation in tissues, and other related factors, such as flammability, corrosiveness, and other hazardous characteristics." In an attempt to meet those mandates, EPA officials first had to develop policies that would control the estimated 86,000 different chemical formulations that are produced by some 67,000 companies that generate hazardous wastes. Today, the EPA regulates several hundred commercial chemical products and the residues that are produced from certain manufacturing processes. Although an estimated 1000 chemicals are considered to be hazardous, the EPA, at present, does not have specific regulations for many of them.

Industries that generate more than 1000 kilograms (2200 pounds) of hazardous wastes each month at any one site must report the nature of the material and the quantity that is generated to the EPA or to their state regulatory agency. In the case of a few extremely toxic substances, the generation of as little as 1 kilogram (2.2 pounds) of wastes per month must be reported. The disposal of hazardous wastes from those sites is tracked by a *manifest* (cargo-invoice) *system*, which follows the waste from its point of generation to its ultimate disposal site. As a re-

sult, each group or organization in the disposal sequence is responsible for that waste while it is in its possession. Then, copies of the manifest go to the appropriate regulating agency to ensure that the waste was delivered to a designated disposal site and that it was disposed of properly.

A major weakness in present EPA regulations is that industries which produce less than 1000 kilograms of hazardous wastes each month are not regulated. For example, approximately 760,000 industries in the United States generate hazardous wastes, but approximately 91 percent of them produce less than 1000 kilograms each month. As a result, approximately 1 percent (2.6 million metric tons) of hazardous wastes that are generated are not regulated. Nevertheless, most of the hazardous wastes that are generated (approximately 99 percent) are regulated. Furthermore, those industries that generate small quantities of hazardous wastes are not free to dispose of them in an irresponsible manner; they are, in fact, legally responsible for any complications that might arise from improper handling or disposal. But since more than 690,000 industries generate an estimated 220,000 metric tons of unregulated hazardous wastes each month, that policy could result in many future problems if small industries do not act responsibly. Thus in 1984, federal legislation was proposed that would regulate all industries that produce 100 kilograms or more of hazardous waste each month.

Waste Classification

When disposing of industrial wastes, an industry must first classify the waste as being either hazardous or nonhazardous. Although the procedures that are used to classify wastes are defined by the EPA, it is not always clear how to classify a particular type of waste. And some of the tests that are used to classify wastes are controversial because the results are difficult to reproduce. If a type of waste is classified as nonhazardous, disposal costs are low, because nearby sanitary landfill sites can be used.

However, if a waste is classified as hazardous, the particular hazardous substance, or substances, that it contains must be identified so the proper method of disposal can be selected. For example, heavy metals (such as mercury, chromium, and zinc) are not destroyed by incineration, so some type of landfill must be used to dispose of them. Organic

Table 13.1 Federal legislation governing hazardous-waste disposal

Comprehensive Environmental Response, Compensation, and Liability Act (1980)	Creates a Multibillion Dollar "Superfund" for Cleanup or Containment of Abandoned Hazardous Waste Dumps that Are a Threat to Public Health or the Environment
Resource Conservation and Recovery Act (1976)	Controls Hazardous Wastes Disposal, Present and Future, from the Point of Generation to Ultimate Disposal.
Safe Drinking Water Act (1974)	Controls Disposal into Underground Disposal Sites
Toxic Substances Control Act (1976)	Regulates the Manufacture, Distribution, and Disposal of Primarily Newly-Developed Chemicals

wastes (such as pesticides and solvents) might be incinerated. Other methods of hazardous-waste disposal include chemical and biological detoxification, deep-mine disposal, and deep-well injection. In some instances, valuable products may be recovered from the waste stream, although that method is usually the most costly alternative. No matter which method of disposal is chosen, it is controlled by one of four federal legislative acts, which are summarized in Table 13.1.

Predisposal Treatment

Many procedures are available to reduce or eliminate the hazards or toxicity of some wastes so they can be managed in more traditional ways. For example, the toxicity of some hazardous wastes is virtually eliminated when they are mixed with various solidifying agents, such as cement, bitumen, and glass. Also, those materials have been shown to be effective for sealing wastes, so they do not become exposed to the leaching action of water. Another alternative is to encapsulate wastes in plastic containers which then can be disposed of in sanitary landfills. Those landfills charge significantly lower disposal fees than those levied at hazardous-waste sites.

Also, some industries, especially the petroleum industry, have found that soil organisms can be used to detoxify certain types of wastes. For example, refineries produce small amounts of unusable toxic wastes that can be incorporated into the soil, where bacteria decompose and, therefore, detoxify them. Furthermore, since tilling aerates the soil, the de-

composition rate can be accelerated. The EPA estimates that more than half of the 275 refineries in the United States are using this so-called land-farming technique to dispose of wastes.

Many chemical industries recognize that the best way to avoid handling hazardous wastes is to avoid generating them in the first place. Thus new plants are incorporating technologies that minimize the generation of hazardous wastes. Unfortunately, it is often too costly to retrofit older plants with that technology.

Resource Recovery

In some cases, the hazardous waste stream can be reduced by recovering and selling discarded chemicals. For example, *waste exchanges* have been established in some areas, where chemicals that cannot be used by one industry can be exchanged with another. Thus waste exchanges act as brokerage houses for used chemicals. (Industries submit a list of available waste chemicals to the waste exchange, and prospective customers examine that list and buy the chemicals they can use.) For example, at one time the Monsanto Company landfilled a toxic acid by-product of a herbicide that they manufacture. Because the chemical is useful for removing sulfur emissions from coal-fired electric power plants, the company now sells that acid to Cities Utility of Springfield, Missouri, for use in its scrubbers. Both companies benefit, waste-disposal costs are eliminated, and the cost for chemicals to remove sulfur from the stack gases is reduced.

Also, some industries have found that they can recover valuable products from their own wastes. For example, when chlorinated hazardous wastes are destroyed by incineration, hydrochloric acid can be recovered. In fact, enough is recovered each year to provide approximately 90 percent of the hydrochloric acid that is used in the United States. Usually, those facilities also recover much of the heat energy that is liberated in the incineration process. Other examples of chemical recovery are discussed in Box 13.1.

Disposal Techniques

The three principal methods of disposal of hazardous wastes are: secure landfills, incineration, and deep-well injection.

BOX 13.1

THE RECYCLING OF
CHEMICAL WASTE

In place of landscaping, huge and unsightly ponds surround the Allied Corporation's chemical plant in Metropolis, Ill., a rural town on the Ohio River. The ponds are filled with a dull white liquid, thick as oatmeal, that would begin to burn the skin off a hapless wader within minutes.

The substance is calcium fluoride, an unwanted by-product of chemical manufacturing that has accumulated at the Metropolis plant for 7 years as hazardous waste. To keep it from human contact indefinitely—as federal law requires—a network of five ponds, each as large as a football field and 10 feet deep, was built across the nearby fields and woodlands and surrounded by fence.

But a year ago, the pond-building stopped and Allied took a new tack. It invested $4.3 million in a "recovery plant" that recycles the waste into a safe raw material for the production of commercially valuable fluorine-based chemicals that are a specialty of Allied's Metropolis plant. The result: Allied saves $1 million annually on new storage ponds and also because it no longer needs to purchase as much raw material.

Allied's strategy at Metropolis is being matched throughout the chemical industry. Increasingly, the hazardous waste problem is being dealt with by recycling the waste into raw material—or by investment in new equipment that produces less waste. The technologies involved are not new, but until recently they were considered too costly to use.

That has changed as Federal regulations governing hazardous waste disposal have become more stringent—and more expensive. The proliferating rules have pushed up storage costs for hazardous chemical waste from about $24 a barrel in the late 1970s to more than $100 a barrel today, according to James Gutensohn, commissioner of the Massachusetts Department of Environmental Management.

The industry has also been pushed into recycling by the public outcry over inadequate or careless storage of chemical waste. Most recently, for example, the Federal Environmental Protection Agency charged that trichloroethylene—a suspected carcinogen—might be seeping into the Minneapolis water supply from a landfill site in Fridley, Minn., where containers of chemical waste once were buried.

That sort of publicity has reinforced the industry's growing belief that investment in recycling plants will turn out to be less costly than settling future lawsuits filed by people claiming they had developed cancer and other ailments from contact with stored chemical waste. "The bottom line is the basic unpredictability in the quality and endurance of chemical storage," said Robert D. Stephens, chief of the Hazardous Materials Laboratory of the California Health Department. "That's something nobody wants to bet on."

Theoretically, recycling could eliminate all

Secure Landfills Most hazardous wastes are disposed of at special landfill sites that are now regulated by the EPA or state regulatory agencies. Places that are designated for the burial of hazardous wastes are called *secure landfills* (Figure 13.8), which have very strict requirements for operation. For example, the site must be located higher than the 100-year floodplain (see Chapter 6) and away from fault zones. Impermeable liners must be installed, and a network of pipes must be installed (beneath the landfill and just above the liner) to collect leachate. Then, any leachate that is collected must be pumped out and treated. Also, monitoring wells are required to check the quality of groundwater in the area. The Resource Conservation and Recovery Act also requires that specific procedures be followed to operate and to monitor secure landfills after they are filled.

Incineration Incineration is the method that is used to dispose of combustible liquid and solid hazardous wastes, such as solvents, pesticides, and petroleum-refinery wastes. Although that method greatly reduces the volume of those wastes, more importantly, it destroys organic chemicals by converting

the millions of tons of hazardous waste that the nation's industries produce each year, solving a major environmental issue. But in fact, decades will be needed even to approach this goal. For one thing, most of the nation's factories were not designed to recycle their by-products. Converting those factories is costly. In addition, there is no standard recycling procedure.

At its Metropolis, Ill., plant, Allied has managed to recycle all of the calcium fluoride thrown off as a hazardous by-product of the fluorine-based chemicals made there to produce gases for refrigerators and air-conditioners. In fact, the company says that over the next decade, the network of pools in the nearby fields will be drained to feed the recycling plant and the land possibly restored. The Metropolis innovation earned Allied a National Environmental Industry Award last November, and the huge chemical company is also making changes at other plants.

Dow Chemical of Midland, Mich., also is recycling some hazardous wastes into useful products. Stacy Daniels, research leader in the company's Environmental Sciences Research Laboratory, talks of recycling "stillbottoms" from the production of caustic soda and chlorine into the useful raw material hydrochloric acid. But he says that Dow nevertheless relies on incineration to destroy most of its hazardous wastes.

Incineration is a major industry method of reducing the volume of hazardous wastes that must be stored indefinitely, with its attendant risk that seepage will contaminate the environment, creating future liability. But some experts say that incineration merely trades one type of pollution hazard for another. John Ehrenfeld, consultant in the hazardous-waste management group at Arthur D. Little, maintains that incineration "converts a small probability of high risk into low-level but continuous exposures" from air-pollution emissions.

As for burial of waste in sealed containers, state legislation may eventually limit or end this alternative. California, for exaple, began last year to enforce a law that bans hazardous waste from landfills, and similar bans may soon appear at the national level. The Federal Resource Conservation and Recovery Act of 1976 is currently up for renewal in Congress, and experts predict a new law will most likely emerge that specifically prohibits the burial of some hazardous wastes.

Even in its present version, the Resource Conservation and Recovery Act has been a catalyst for change. The reduction of hazardous waste through recycling or incineration, said Mr. Stephens of the California Health Department, "is a very significant result of regulation."

Thus "some of the more forward-looking companies," Mr. Stephens said, "are sounding like environmentalists. People were calling us fuzzyheads a few years ago, but now companies like Getty, Dow, and 3M, who know how chemicals behave and what to do with them, are speaking as we did."

them to carbon dioxide, water, and other gases that are removed by scrubbers. Many incinerators also recover some of the energy that is released in the combustion process. But before an incinerator can be used to destroy a particular type of hazardous waste, it must be approved for burning that type of waste by the EPA. That requirement is especially important when the chemical substances to be destroyed are very stable at high temperatures, such as PCBs (Chapter 7). Low-temperature incinerators (650 to 850°C) do not completely destroy those substances, so high-temperature incinerators (around 1200°C) must be used. To be approved, incinerators must demonstrate that they can destroy 99.99 percent of the combustible toxic wastes, and they must be equipped with air-pollution control devices to remove acidic by-products and noncombustible materials.

Since 1969, various industries in Europe and the United States have used incinerators aboard ships, such as the one shown in Figure 13.9, to burn hazardous wastes at sea. The first such incinerator ship, the Dutch-owned *Vulcanus I*, a small, converted merchant vessel, was used to destroy large quan-

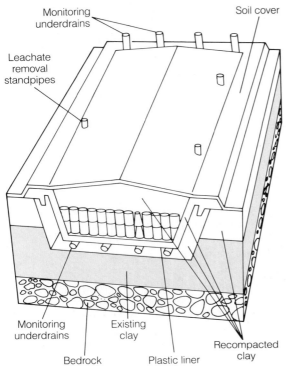

Monitoring underdrains

Soil cover

Leachate removal standpipes

Monitoring underdrains

Existing clay

Bedrock

Plastic liner

Recompacted clay

Figure 13.8 *A cross section of a secure chemical landfill.*

Figure 13.9 *The incinerator ship* Vulcanus II, *which is owned by Chemical Waste Management, Inc., is used to incinerate hazardous wastes at sea to avoid burning them near humans. The ship is equipped with high-temperature incinerators and air-pollution control devices. (Chemical Waste Management, Inc.)*

tities of Agent Orange (the highly toxic defoliant that was used in the Vietnam War). PCBs and other toxic wastes have been successfully destroyed on this ship as well. Because the *Vulcanus I* has been so successful, and because a single ship can destroy many thousands of tons of waste each year, several other ocean-going incinerator ships have been constructed. But persons who live in port cities do not want hazardous wastes hauled through and temporarily stored in their areas for loading on those vessels. Furthermore, many persons are concerned about the effect that a shipwreck or spill would have on marine life.

Deep-well Injection *Deep-well injection* is a waste-disposal technique that was developed by the petroleum industry in the late 1800s because oil exploration and extraction often brings brines to the surface. The harmful effects of those solutions can be avoided by pumping them back into the subsurface rock formation from which they were drawn. Although deep-well injection is still used by the petroleum industry, it also is used to dispose of other types of hazardous liquid wastes.

Only certain kinds of geologic formations can be used for disposal by deep-well injection. The formations must be deep, porous enough to provide storage space, and sandwiched between impermeable layers of rock. Moreover, the formation should not contain any valuable mineral or fuel resources. Since the installation of a deep injection well requires drilling and encasing the shaft down to a minimum of 1000 meters (3000 feet), as shown in Figure 13.10), such wells can easily cost $1 million or more. However, the cost of operating those wells is very low (except when the casings corrode). Some authorities oppose deep-well injection, primarily because of the possible occurrence of blowouts and ground tremors.

Economic Considerations

The cost to dispose of hazardous wastes is much higher than that for nonhazardous wastes. Those costs are summarized in Table 13.2. (Note that those costs do not include transportation, which can increase the total cost by 20 to 80 percent.) Unfortunately, those high costs have provided an incentive for "midnight dumpers" to dispose of hazardous wastes illegally. For example, companies that generate haz-

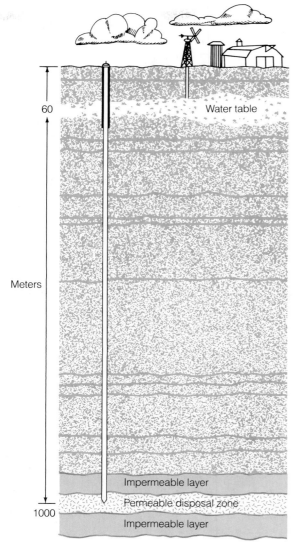

Figure 13.10 *Deep-injection wells used to dispose of various types of hazardous wastes require specific geologic conditions. (Environmental Protection Agency.)*

Table 13.2 Reported prices for hazardous waste management services, 1980

WASTE MANAGEMENT METHOD	DOLLARS PER METRIC TON
Landfill	
Wastes that Are Not Acutely Hazardous, Including Sludges	20–90
Highly Toxic, Explosive, or Reactive Wastes	100–400
Deep-well Injection	
Oily Wastewaters	15–40
Dilute Toxic Rinse Waters	50–100
Resource Recovery	50–200
Land Treatment	5–25
Chemical Treatment	
Acids, Alkalines	15–80
Cyanides, Heavy Metals, Highly Toxic Wastes	100–500
Incineration	
Wastes with High BTU Value, Presenting No Acute Hazard	50–300
Highly Toxic, Heavy Metal Wastes	300–1000

Source: *State of the Environment, 1982.* Washington, D.C.: The Conservation Foundation, 1982, p. 172.
Prices are exclusive of transportation costs, which may represent 20 to 80 percent of the total costs of off-site disposal, depending on the method of hauling and distance shipped.

Illegal disposal of hazardous waste continues and no one knows just how much waste is illegally disposed. A 1985 report prepared by the General Accounting Office (GAO), however, indicated that the EPA received more than 240 allegations of illegal disposal between 1982 and 1984—more allegations than it can check up on. Furthermore, by 1985 the EPA had allocated $500,000 from the Superfund to clean up illegal sites.

Some of those "midnight dumpers" have been linked to organized crime, and the FBI and state police have become involved in efforts to stop them. In fact, it has been proposed that EPA officials be given the power to make arrests and carry weapons. As a result, the EPA, for the first time in its history, has included criminal investigators on its staff. Of the 36 disposal cases involving enforcement proceedings (included in the 1985 GAO report) 27 involved charges against generators of hazardous waste and 9 involved charges against transporters. Thirty-four of the 36 cases were found as the result of a tip from a private citizen, employee, or local government employee. EPA inspectors found the other two cases.

ardous wastes frequently subcontract the disposal of those wastes to firms that offer hazardous-waste disposal services, and they pay the going rate for disposal of their particular hazardous wastes. Then, rather than pay the high fees charged by a secure landfill or incinerator, the hazardous-waste disposal firms collect the wastes and simply dump them into a storm sewer, along an isolated roadway, or into a nonapproved dump site.

Management of Radioactive Wastes

Since humankind learned to manipulate the nucleus of the atom a half-century ago, our society has had to contend with radioactive wastes. Nuclear weapons production and nuclear power plants generate most of the radioactive wastes that are produced today. However, many persons are even more concerned about the spectre of radioactive fallout from the detonation of a nuclear bomb or from a catastrophic accident in a nuclear power plant. Because radioactive wastes differ from other hazardous wastes in several significant ways, we need to describe those unique properties so that we can understand the extraordinary precautions that are required to safely handle and store radioactive materials, including wastes.

Unique Characteristics

Radioactive substances owe their radioactivity, that is, the emission of energy and particles (which are collectively called *high-energy radiation*), to the fact that their nuclei are unstable. When those nuclei convert to a more stable configuration by changing their structure within the nucleus, highly energetic particles, rays, or both are emitted from them. Three types of high-energy radiation are emitted from radioactive nuclei: alpha particles, beta particles, and gamma rays. (Recall from Chapter 9 that gamma rays are a form of electromagnetic radiation.) All three types of radiation possess enough energy to break chemical bonds in any medium through which they pass.

Herein lies the reason for their adverse effects on health. When those forms of high-energy radiation penetrate living cells, they collide with and break apart molecules in the cells. The health effects of this molecular damage can be categorized as either somatic or genetic. *Somatic effects* are direct impacts on the person and often include the conversion of normal cells to cancer cells and other changes that lead to a decline in life expectancy. For example, large doses of radiation can lead to radiation sickness (vomiting, diarrhea, and nausea) and possible death. *Genetic effects* are changes in the genetic makeup of sex cells (egg and sperm) that are transferred to the offspring. Scientists only have limited knowledge regarding the genetic effects caused by exposure to radiation, and currently, they can only be detected as an increased incidence of some health problem in the offspring of persons who have been exposed to radiation compared to the incidence of the same health problem in the offspring of persons who have not been exposed to radiation.

Radioactive materials differ in the types of radiation they emit. Some are alpha emitters, whereas others are beta and gamma emitters. This difference is important because all three types of radiation—alpha, beta, and gamma rays—differ in terms of their ability to penetrate materials (Figure 13.11). For example, the fact that a block of concrete that is 1 to 2 meters thick is required to stop gamma rays attests to their highly energetic nature. Thus gamma rays readily penetrate entire organisms, which results in what is known as *whole-body exposure.*

Beta particles, which have less penetrating power than gamma rays, are stopped more readily, even by body tissues. Thus, although the skin and tissue directly beneath the skin can be damaged by an external source of beta emitters, internal organs are generally protected. But if material that emits beta particles becomes concentrated in one part of the human body, that area will receive a higher dose of radiation, which may result in extensive localized radiation damage. For example, because strontium has chemical properties that are similar to those of calcium, strontium tends to become concentrated in bone. Strontium-90 is a radioactive isotope* that is a beta emitter. Therefore, if a person ingested food (for example, milk) that was contaminated with strontium-90, the strontium-90 would be taken up by the blood and become concentrated in the bone. Since bone marrow inside the bone is the site of red blood cell production, one of the possible results of an accumulation of strontium-90 in the bone is leukemia. To protect oneself from beta particles from an external source, a person would have to place several centimeters of solid material (this book, for example) between himself and the source.

Alpha particles have the least penetrating power of the three types of nuclear radiation. Even the most energetic alpha particles cannot penetrate beyond the outer layer of skin. Thus exposure to sources of alpha particles can be prevented by a thin protective covering, such as a sheet of aluminum foil, that is

---

* Isotopes of an element are different forms of an element, for example, strontium-89 and strontium-90, which are attributable to differences in atomic mass. Isotopes of an element vary in their nuclear properties but behave similarly with respect to their chemical properties.

Exposure to external radiation source

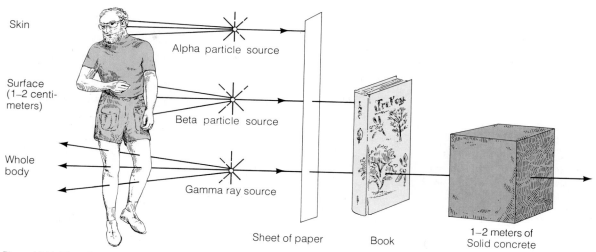

Figure 13.11 *The relative penetrating power of particles and rays that are emitted by radioactive materials.*

placed over the source. However, if an alpha emitter enters the body through breathing or ingestion, it can cause extensive localized damage. For example, when a person inhales air that contains radon gas from sources, such as uranium mine tailings, their lungs can suffer damage from exposure to the alpha particles that are emitted. Thus protection from gaseous alpha emitters, such as radon, requires that they be stored in closed gas containers or diluted (for example, through better ventilation).

As each radioactive isotope spontaneously emits its radiation, it gradually becomes transformed into a stable form that is no longer radioactive. That shift toward stability is called *radioactive decay*, and scientists indicate the rate of decay of a particular isotope by measuring its *half-life*—the time that it takes half the nuclei of a radioactive isotope to decay to a more stable form. Figure 13.12 illustrates the decay of the radioactive isotope phosphorus-32 (^{32}P), which has a half-life of 14 days. Phosphorus-32 decays to form sulfur-32 (^{32}S), which is a stable, and thus a nonradioactive, isotope. If we start with 1 gram of phosphorus-32, after 14 days, 0.5 grams of phosphorus-32 remain. The other 0.5 grams has converted to sulfur-32. After another 14 days have passed, 0.25 grams of phosphorus-32 remain, and so on.

The half-lives of radioactive wastes that are produced in nuclear reactors vary widely—from 0.96 seconds for xenon-143 to as long as 160 million years for iodine-129 (Table 13.3). Usually at least 10 half-lives must elapse before a radioactive isotope is considered to have decayed to the point at which it no longer constitutes a serious radiation threat. Thus the half-life of a radioactive isotope is important in determining how long it must be kept isolated from human exposure (and, ideally, all other living things).

Table 13.3 Half-lives of typical radioactive-waste components

RADIOISOTOPE	HALF-LIFE (YEARS)
Americium-241	460
Americium-242	150
Cesium-135	2,000,000
Cesium-137	30
Curium-242	0.45
Curium-243	32
Curium-244	18
Iodine-129	160,000,000
Iodine-131	0.022
Neptunium-237	2,100,000
Plutonium-239	24,000
Plutonium-241	13
Radon-226	1600
Strontium-90	28
Technetium-99	200,000
Thorium-230	76,000
Tritium	13

447

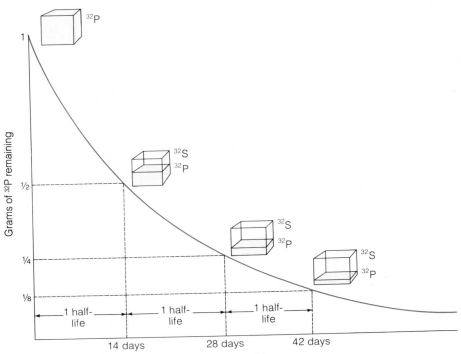

Figure 13.12 *The amount of radioactivity that is emitted by a radioactive element decreases by one-half for each half-life that passes. For example, phosphorus-32 loses one-half of its radioactivity every 14 days, which is its half-life. The phosphorus-32 decays to form nonradioactive sulfur-32.*

Because of the long half-life of many radioactive wastes, such wastes can be considered to be persistent. Thus once radioactive wastes enter ecosystems, there is a strong possibility that some isotopes will accumulate in food webs. One of the lessons learned from the above-ground testing of nuclear weapons is that bioaccumulation of some radioisotopes does occur. For example, during bomb testing in the 1950s, strontium-90 was spread over grasslands through fallout. When the grasses were eaten by cows, they also ingested the strontium-90. Since strontium has properties similar to calcium, much of the strontium found its way into cows' milk. When persons drank the contaminated milk, they also accumulated small amounts of strontium-90. This transfer of strontium-90 through food webs to humans was one of the reasons that above-ground testing of nuclear weapons was halted.

Sources of Radioactive Wastes

In the United States, radioactive wastes are generated by two principal sources: nuclear weapons pro-

duction sites and some 87 nuclear electric power plants. Those sources produce wastes that are classified as high-level wastes because they emit intense levels of radiation. Low-level wastes pose much less of a threat for exposure and are more easily managed. The estimated 16,000 sources of low-level wastes include hospitals, research laboratories, universities, and some industries.

Radioactive wastes, including low-level wastes, are generated at about one-sixtieth the rate of hazardous chemical wastes. The specific rates at which nonmilitary radioactive wastes are generated in the United States are listed in Table 13.4. Although the volumes of radioactive wastes are relatively small, the *specific hazard* (danger per unit of weight) is extremely high, especially for high-level and transuranic wastes. *Transuranic wastes* are defined as radioactive alpha-emitting elements that are heavier than uranium. Those wastes are isolated during the reprocessing of nuclear fuel and nuclear weapons production.

Nuclear power plants produce high-level radioactive wastes in the form of spent *fuel-rod assemblies*, such as those shown in Figure 13.13. Those assemblies are temporarily stored in water-filled basins at nuclear power plants. Each year, during annual refueling operations, a typical nuclear (1000

Figure 13.13 *Radiation emanating from spent fuel-rod assemblies under water. The brightest assemblies (top and center) were just removed from a reactor. In front of them are assemblies that were removed a month earlier. At the extreme lower left are assemblies that have cooled for 3½ months. (E.I. du Pont De Nemours & Company.)*

megawatts) power plant adds approximately 30 metric tons of spent fuel rods to its storage facility. In 1980, approximately 7300 metric tons of such wastes were stored at the 76 licensed plants around the United States. That quantity of waste is expected to grow to 72,000 metric tons by the year 2000!

Table 13.4 Nonmilitary nuclear-waste generation in the United States in 1980

	METRIC TONS		
High-Level Wastes			
Spent-Fuel Assemblies from Nuclear Reactors	1,580		
Nuclear Fuel-Production Wastes (Primarily from Reprocessing Nuclear Fuels)	450	Subtotal:	2,030
Transuranic Wastes	1,500	Subtotal:	1,500
Low-Level Wastes			
Nuclear Fuel Cycle Wastes	490,000		
Hospital Research Laboratories, Medical Laboratories	440,000	Subtotal:	930,000
Total			933,530

Source: *Cato Journal* 2(2): 571 (Fall), 1982.

Figure 13.14 *Storage tanks for high-level liquid radioactive wastes under construction at the U.S. Department of Energy's Hanford site near Richland, Washington. The double-walled steel tanks are shown here prior to encasement in concrete. Each tank holds 4 million liters (1 million gallons). (Department of Energy.)*

The annual addition of spent-fuel assemblies to "temporary" storage facilities is already causing a space-storage problem at many nuclear power plants. But until the federal government approves a method and a site for permanently storing radioactive wastes, such wastes will continue to accumulate. During the next few years, the storage problems will probably be dealt with by expanding the size of "temporary" storage facilities and by storing the spent-fuel assemblies closer together.

High-level military wastes represent about the same amount of radioactivity as those from nuclear power plants, but they exist in a more dilute, liquid form. In 1980, those wastes, which totaled about 280 million liters (74 million gallons), were stored in large double-lined tanks (such as those shown in Figure 13.14). These liquid wastes will remain in those storage facilities until they can be concentrated, solidified, and permanently stored underground at a nuclear waste repository. By the year 2000, military radioactive wastes are expected to nearly double in quantity.

Unfortunately, the lack of permanent storage facilities for military-related liquid wastes has led to several major spills. Nearly 2 million liters (500,000 gallons) of those wastes have leaked from storage tanks at sites in Hanford, Washington, and on the Savannah River in South Carolina. The worst spill occurred in 1973, when 450,000 liters (115,000 gallons) leaked into the ground from a tank at the Hanford site. Fortunately, no human casualties (as far as we know) have resulted from any of those spills. Most of those wastes have been retained in the soil at the site, although at the Hanford site, a small amount is migrating through the soil into the Columbia River.

Permanent Storage

Recall that the decay of radioactive materials is spontaneous, so it cannot be prevented. Therefore, if all forms of life are to be protected from exposure to radiation, those wastes must be isolated so that they have virtually no chance of reaching the earth's surface and cycling through the environment. In addition, because 10 half-lives must elapse before the material is considered to be safe, and because some radioisotopes have half-lives of decades or more, those radioactive wastes must remain isolated for centuries.

For nuclear wastes from power plants, the first several hundred years of storage are the most critical, because those wastes emit the highest levels of radiation during that period. The lethal levels of spent nuclear fuel and nuclear reprocessing wastes change over time and are compared to those of some nonradioactive chemicals in Table 13.5. The comparison is based on the assumption that the materials are ingested, which is the most likely route of exposure. The table indicates that, initially, nuclear wastes are extremely toxic, but the toxicity of spent fuel decreases by a factor of 400 between 10 and 500

Table 13.5 Lethal quantities of hazardous materials that are ingested

MATERIAL	COMPOUNDS	AVERAGE LETHAL DOSE* (mg/kg)	EXPERIMENTAL ANIMAL	EXTRAPOLATED† LETHAL DOSE TO MAN (g)
Selenium	Na_2SeO_3	5	Rabbit, Mouse Rat, Guinea Pig	0.35
Cyanide	KCN	10	Rat	0.7
Mercury	$HgCl_2$	23	Rat, Mouse	1.6
Arsenic	As_2O_3	45	Mouse, Rat	3
Barium	$BaCl_2$, $Ba(NO_3)_2$	250	Rat	18
Copper	CuO, $CuCl_2$	300	Rat	21
Nickel	$Ni(NO_3)_2$	1620	Rat	110
Aluminum	$AlCl_3$, $Al_2(SO_4)_3$	4000	Rat, Mouse	280
High-Level Reprocessing Wastes After 10 yr				0.03
After 500 yr				170
Spent Fuel‡ After 10 yr				0.15
After 500 yr				57

Source: C. M. Koplik, M. I. Kaplan, and B. Ross, The safety of repositories for highly radioactive wastes. *Review of Modern Physics* 54:274, 1982.
*The quantity of material such that half the affected people die.
†The extrapolation to man from the test animal data is scaled by weight to a 70 kg (150 pound) man.
‡Spent fuel was not included in the original table by Cohen (1977). The values shown here were computed using the same procedures as were used by Cohen for high-level waste.

years of storage. The toxicity of reprocessed nuclear fuel wastes decreases by a factor of 5500 during that period of time. And after 500 years, the toxicity of nuclear wastes is lower than that of some of the hazardous chemicals that are used in industry. Nevertheless, some isotopes in nuclear waste have long half-lives, so they will continue to emit low levels of radioactivity for thousands of years.

The EPA considers a span of 10,000 years to be a sufficient time for storage of spent-fuel rods because it takes that amount of time for the total hazard (effective radiation dose) of those rods to reach the radioactive decay rate that existed in the uranium ore naturally when it was mined. In 1983, a panel from the National Academy of Sciences evaluated the technology for containment of radioactive wastes and reconfirmed the 10,000-year storage criteria set by the EPA. They defined effective disposal as procedures that would limit the release of radioactive materials to levels that would be no greater than 10 percent higher than the radiation dose that persons receive from natural radiation (see Table 13.6). The panel concluded that only a small fraction of the radioactivity in storage facilities could ultimately reach the environment during the 10,000 years proposed by the EPA as an adequate design criteria.

Furthermore, they concluded that a disposal system "is not necessarily capable of continuing to protect people and the environment beyond 10,000 years,"

Table 13.6 Typical average annual doses of radiation in the United States

RADIATION TYPE	SAMPLE SOURCE	AVERAGE ANNUAL DOSE TO REPRODUCTIVE ORGANS (mrem/year)*
Alpha (and Gamma)	Natural Radioactivity (Uranium) in Rocks and Minerals	30.0
Beta	Natural Radioactivity (Potassium-40) in Rocks and Minerals	20.0
	Natural Radioactivity in the Atmosphere (Tritium)	2.0
	Television, 1 hr/day	0.5
Gamma	Medical X-Rays	20.0
	Cosmic Radiation at Sea Level	40.0

Source: P. A. Vesilind and J. J. Peirce, *Environmental Pollution and Control*, 2d ed. Boston: Butterworth, 1983, p. 218.
*mrem = millirems, a measure of the effective dose of radiation in humans.

because small amounts of radiation could be released slowly for hundreds of thousands of years.

Several methods have been proposed for the permanent storage of nuclear waste. These methods include burial in deep-sea trenches and burial in deep underground repositories. At present, deep underground repositories are being pursued more actively, because scientists believe that the wide variety of environmental questions that are associated with underground disposal can be answered more quickly.

However, to bury those wastes in deep underground repositories, a storage facility must be built that is at least 600 meters (2000 feet) below the surface in a stable geologic formation (Figure 13.15). Locations of some possible sites for repositories that have been (or will be) studied, and the various types of rock formations that exist in those areas, are shown in Figure 13.16. Those sites must be evaluated with respect to the same factors that are used for the siting of storage facilities for hazardous chemical wastes—groundwater movement, rock type, rock strength, earthquake activity, human intrusion, surface characteristics, the economic impact on nearby communities, and the impact on the environment.

For some time, investigations have been underway to evaluate four different rock types as possible sites—salt, basalt, volcanic tuffs, and crystalline rocks. In December of 1984, the Department of Energy narrowed down to three the number of rock types that would be evaluated extensively by selecting three locations for detailed study. These are: (1) salt deposits in Deaf Smith County, Texas, 50 kilometers (30 miles) west of Amarillo, Texas, (2) volcanic basalt formations on the United States' Hanford nuclear reservation in southeastern Washington, and (3) compacted volcanic ash, called volcanic tuff, 160 kilometers (100 miles) northwest of Las Vegas, Nevada. Salt deposits are being considered because of their long-term stability, strength, dryness, and capacity to dissipate heat from decaying wastes. Basalt is strong and has good heat-dissipating characteristics. And volcanic tuffs have excellent sorptive properties. (*Sorption* is the capacity of a material to remove soluble chemicals from water.)

Once a site is picked, it will require approximately 160 hectares (400 acres) of land, with an additional 1200 hectares (3000 acres) as a protective perimeter. Other drilling and mining activities in the area would be strictly forbidden.

The isolation of radioactive wastes in an underground repository will be achieved through a system of natural and engineered barriers. First, the wastes will be solidified into a durable, leach-resistant material, such as vitrified glass, which is similar to a ceramic glaze. Next, those solidified wastes will be sealed in corrosion-resistant cannisters, which will be placed in holes that are lined with corrosion-resistant sleeves, in the floor of the

Figure 13.15 *A plan for permanent storage of high-level radioactive wastes in deep underground salt deposits.* [B.L. Cohen, "The Disposal of Radioactive Wastes from Fission Reactors." Scientific American 236:25(June), 1977. Copyright © 1977 by Scientific American, Inc. All rights reserved.]

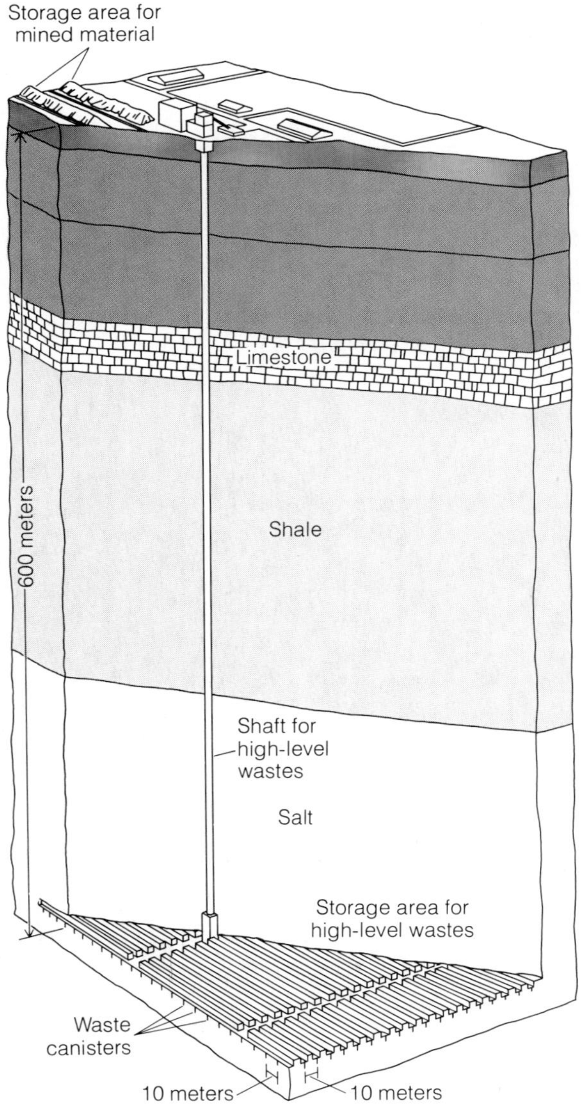

repository. And then each sleeved hole will be surrounded by materials that have high sorptive characteristics, so the migration of chemicals will be prevented if water should somehow leak into the repository. Initially, the waste cannisters will be monitored and remain retrievable in the event that problems arise. However, if monitoring indicates that future problems are unlikely, the holes will be sealed, and the entire excavation will be filled with sorptive materials. Following that process, if a repository were to experience a problem, only very small releases would seem likely. At present, it is not possible to predict how releases could occur.

Many scientists feel that such sites would be safe, permanent storage sites as a result of evidence that has been gathered at a natural nuclear reactor site, called Oklo, in the Gabon Republic in equatorial Africa. At Oklo, 2 billion years ago, a rich uranium deposit developed a natural nuclear chain reaction which produced radioactive wastes that were identical to those produced in nuclear reactors today. Although the site is still under investigation, initial results show that the long-lived hazardous by-products have remained adsorbed to the rock and soil materials despite the fact that they have been in contact with water. Sites such as Oklo are extremely important to scientists, because they provide unique opportunities to study what happens to radioactive chemicals under natural conditions.

To hasten the disposal schedule, the Nuclear Waste Policy Act was signed by President Reagan in early 1983. That law authorized the Department of Energy to find two sites for nuclear waste repositories. The President is required to choose the first repository site by 1991, and the site is supposed to accept its first wastes in 1998.

However, one big worry that remains is whether any state will allow the disposal of radioactive wastes within its boundaries once a site has been chosen. Once a qualified repository site has been chosen and a plan has been submitted to Congress by the President, the Governor, state legislature, or Indian tribe that is affected has 60 days to submit a notice of

Figure 13.16 *The various types of rock formations that are being studied as possible sites for the permanent storage repository for high-level solid radioactive wastes. The first repository will be at one of the following three sites: The Hanford site (in Washington), the Nevada test site, or the Permian Basin (in Texas). (U.S. Department of Energy, Office of Nuclear Waste Isolation, 1982.)*

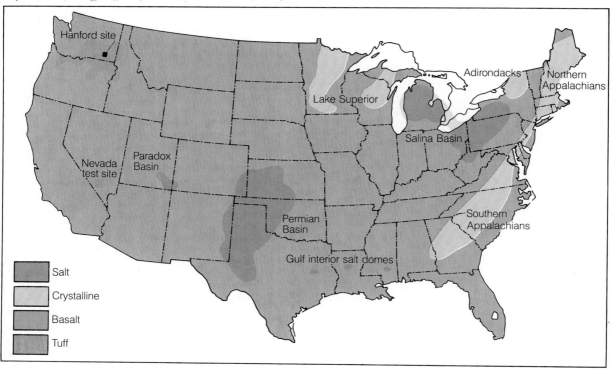

disapproval to Congress. If that happens, the site will remain unapproved unless Congress, within 90 days, overrides the state or tribal disapproval. Quick negative reaction by state and local politicians in the three areas selected in late 1984 for potential repository sites well illustrates the difficulty that will be encountered in gaining political approval for sites that are geologically sound repositories.

Conclusions

Public concern about human health risks and environmental quality have resulted in laws that require municipalities and industry to adopt better (and more costly) methods of waste disposal. Strict federal legislation regulates where and how municipal wastes, hazardous chemical wastes, and radioactive wastes can be disposed of, stored, or both. The key

to safe disposal is proper isolation of wastes. And eliminating or minimizing migration of water through disposal sites is the most important factor to control. Although new disposal technology should reduce future problems, the enormous task of cleaning up an estimated 22,000 to 25,000 old, leaky, environmentally unsound disposal sites still remains for the 1980s and beyond.

Reclamation of energy, materials, or both from wastes is considered to be a reasonable alternative for handling wastes. As a result of the costs of traditional disposal methods and increased income from recovered energy and materials, waste-reclamation facilities are becoming more attractive to many industries and communities. Reclamation of valuable materials from radioactive fuel-rod assemblies, however, is unlikely at present, because materials that could possibly be used for nuclear weapons construction would also be isolated (see Chapter 18).

Summary Statements

Our activities generate a wide variety of municipal, industrial, mining, and agricultural wastes. Toxic chemical wastes and radioactive wastes require more technologically sophisticated methods of disposal.

Disposal alternatives for municipal wastes include sanitary land-filling, incineration, and ocean dumping. But unless properly regulated, those measures may also prove to be hazardous to human and environmental quality. All three methods are regulated by the EPA.

Trench landfills are appropriate in regions where the soil is thick and the water table is low. Area landfills are constructed in canyons or mined areas and are provided with clay liners. Mounded landfills are used where the water table is near the surface. Groundwater contamination is the primary environmental concern with landfills.

Incineration offers many advantages over sanitary landfilling, especially in areas where land prices are soaring and where heat from the incineration process is recovered.

Ocean dumping is used primarily for the disposal of dredge spoils and sewage sludge. The disposal of those wastes is regulated by the U.S. Army Corps of Engineers and the EPA, respectively.

Resource recovery methods reduce the amount of waste that must be disposed of and add to energy and material resources. At present, energy recovery facilities are the most economically feasible alternative. Those facilities may include materials recovery, especially of metals. Magnetic separation, composting, hydropulping, producing refuse-derived fuels, pyrolysis, and aerobic and anaerobic digestion are other methods of producing useable products from nonhazardous wastes. To be economically successful all methods of resource recovery require that the user of recovered products be located nearby.

Hazardous waste-disposal methods include secure landfills, incineration, and deep-well injection. Waste exchanges and chemical recovery are techniques used to reduce the quantity of hazardous wastes. Some hazardous wastes can be treated or encapsulated so that they can be disposed of as nonhazardous materials. Disposal costs for hazardous wastes are up to 100 times more expensive than those for municipal wastes. Illegal dumping of hazardous wastes enables unscrupulous disposal firms to realize large profits and presents a societal nightmare.

High-level radioactive wastes must be isolated to protect organisms from radiation and to prevent radioactive substances from cycling through the environment. The technology to dispose of radioactive wastes has been developed and is being tested. At present, deep underground storage is the preferred method. A permanent repository is to be chosen by 1991, with disposal scheduled to begin in 1998. Undoubtedly, opposition will be voiced concerning any repository site that is chosen.

Questions and Projects

1. In your own words, write a definition for each of the terms that are italicized in this chapter. Compare your definitions with those in the text.

2. Develop a history of the disposal methods that have been used in your community during the past 50 years. Comment on why one method was abandoned in favor of another.

3. How and where are the hazardous wastes that are generated in your state disposed of?

4. Debate the pros and cons of beverage-container deposit ("bottle bill") laws.

5. Why are methods of solid-waste disposal so slow in developing?

6. What sort of waste reclamation efforts are underway in your community? What other materials might be recovered?

7. Find the location of some old dump sites that need cleaning up in your state. State the nature of the problem at each site. Are Superfund monies being used to renovate those sites?

8. How does a sanitary landfill differ from a secure landfill?

9. Should a state have to provide disposal facilities for all the different types of wastes that are generated within its boundaries? List some pros and cons.

10. Summarize the various methods for disposing of hazardous wastes. Arrange the alternatives in order, from the least costly methods to the most costly ones.

11. Have conflicts occurred in your community regarding the siting of landfills or resource-recovery centers? If so, what factors contributed to those conflicts?

12. How would you assess your community's strategies for solid-waste management? Is progress evident or do conflicts prevail that are

interfering with progress? What can be done to help resolve the difficulties?

13. Should metals be given priority over glass and paper in recycling efforts? Support your opinion.

14. Suggest several reuses in the home for each of the following common household items: (a) newspapers, (b) metal cans, and (c) nonreturnable glass bottles.

15. How does the proper management of solid waste reduce our air-pollution problem? Identify some incentives that would reduce the per-capita generation of wastes.

16. List the advantages and disadvantages of sanitary landfilling.

17. Distinguish between trench, area, and mounded landfills. Under what conditions is each landfill method appropriate?

18. What precautions must be taken during deep-well injection to protect the quality of groundwater?

19. Describe several techniques of waste conversion. What is the principal obstacle to their widespread use?

20. Why have most commercial composting facilities closed down in the past decade? What factors could alter that situation in the future?

21. What happens to radioactive wastes generated at nuclear power plants? How are they to be disposed of ultimately?

Selected Readings

AMERICAN CHEMICAL SOCIETY: *Cleaning Our Environment—A Chemical Perspective*, 2d ed. Washington, D.C.: American Chemical Society, 1978. Discusses the problems of, and alternatives to, solid-waste and radioactive-waste disposal in detail. Ocean dumping is also discussed. An excellent overview of problems.

BEEBE, G. W.: Ionizing radiation and health. *American Scientist* 70:35, 1982. Discusses the uncertainties that are involved in, and the risk that is associated with, exposure to low levels of radiation.

BELFIGLIO, J., LIPPE, T., AND FRANKLIN, S.: *Hazardous Wastes Disposal Sites: A Handbook for Public Input and Review.* Stanford, Calif.: Stanford Environmental Law Society, 1981. Deals with the problems of siting hazardous-waste facilities and encourages public participation.

CARTER, L. J.: WIPP goes ahead, amid controversy. *Science* 222:1104, 1983. Discusses technology and politics of using only repositories for nuclear wastes (military) in the United States.

CARTER, L. J.: The radwaste paradox. *Science* 219:33, 1983. Discusses problems of evaluating the geology of prospective repository sites.

COHEN, B. L.: *Before It's Too Late.* New York: Plenum Press, 1983. A scientist's case for nuclear energy, which includes a discussion of the hazards of nuclear waste.

COUNCIL ON ENVIRONMENTAL QUALITY: *Annual Reports, 1977–1984.* Washington, D.C.: U.S. Government Printing Office. Includes a yearly summary of progress and major issues that involve solid-waste management.

DEPARTMENT OF ENERGY: Background information on the National Waste Terminal Storage program and related issues. Pamphlets available from the Office of Nuclear Isolation Library, 505 King Avenue, Columbus, Ohio 43201.

EPSTEIN, S. S., BROWN, L. O., AND POPE, C.: *Hazardous Waste in America.* San Francisco: Sierra Club Books, 1982. Overview of the problems that are associated with the generation and disposal of hazardous wastes. Written by authors who are intimately familiar with the problem.

HAYES, D.: *Repairs, Reuse, Recycling—First Steps Toward a Sustainable Society.* Washington, D.C.: Worldwatch Institute, 1978. Identifies the major issues that are involved in resource recovery and conservation.

HILL, D., PIERCE, B. L., METZ, W. C., ROWE, M. D., HAEFELE, E. T., BRYANT, F. C., AND TUTHILL, E. J.: Management of high-level waste repository siting. *Science* 218:859, 1982. Survey of social-choice problems that are associated with the siting of a high-level radioactive waste repository.

HINGA, K. R., HEATH, G. R., ANDERSON, D. R., AND HOLLISTER, C. D.: Disposal of high-level radioactive wastes by burial in the sea floor. *Environmental Science and Technology* 28A:16, 1982.

KOPLIK, C. M., KAPLAN, M. F., AND ROSS, B.: The safety of repositories for highly radioactive wastes. *Review of Modern Physics* 54:269, 1982.

LEAVITT, J. W.: *The Healthiest City.* Princeton, N. J.: Princeton University Press, 1982. A historical analysis, which demonstrates that improvements in public health depends on community leaders and politicians as much as scientific and medical knowledge.

MAUGH, T., II: Just how hazardous are dumps? *Science* 215:490, 1982. Describes the state of the art of assessing health effects from exposure to hazardous wastes.

MELOSI, M. V.: *Garbage in the Cities.* College Station, Texas: Texas A&M University Press, 1982. A well-researched book that covers how wastes were handled between 1880 and 1980.

NORMAN, C.: High-level politics over low-level wastes. *Science* 223:258, 1984. Discusses problems that states face in disposing of hospital and laboratory radioactive wastes after 1985.

SMITH, R. J.: The risks of living near Love Canal. *Science* 217:808, 1982.

TCHOBANOGLOUS, G., THEISEN, H., AND ELIASSEN, R.: *Solid Wastes.* New York: McGraw-Hill, 1977. A textbook that deals with engineering principles and management issues.

THE CONSERVATION FOUNDATION: *State of the Environment.* Washington, D.C.: The Conservation Foundation, 1982. Chapter 4 surveys the hazardous waste problem in the United States.

VESILIND, P. A., AND PEIRCE, J. J.: *Environmental Pollution and Control,* 2d ed. Boston: Butterworth, 1983. Six chapters in this textbook are devoted to the topics discussed in this chapter.

WIGHT, G. D.: "Onsite Landfill Gas Projects." *Waste Age* 13:50, 1982.

The bald eagle, our national symbol and a major symbol in the battle to save endangered wildlife. (Gordon S. Smith, National Audubon Society, Photo Researchers.)

CHAPTER **14** MANAGING ENDANGERED SPECIES

The enormous flocks of passenger pigeons—sometimes more than 2 billion strong—that once darkened the skies are gone forever. The last passenger pigeon, an aging female that was named Martha, died in the Cincinnati zoo in 1914. It was the savory flesh and gregarious nature of those birds that led to their demise; they were good to eat and easy to kill. With the advent of railroads, which could transport harvested birds to major markets before they spoiled, thoughtless commercial harvests of the pigeons began. In the 1840s, several thousand persons made their living by hunting those birds. They shot them (Figure 14.1), clubbed them to death while they were in their nests, or trapped them. In Michigan alone, more than 1 billion passenger pigeons were killed in a single year. Thus within only a few decades, the highly prized species had become extinct.

Today, another bird that also once numbered in the millions is close to extinction—the short-tailed albatross (Figure 14.2). That seabird was so abundant in Alaskan waters at one time that it was a common source of food for the native Aleut Indians. But because their plumage was prized, those birds were plundered by hunters in their primary breeding area— the island of Torishima off the coast of Japan. As a result, the species was nearly extinct by the time World War II began. Since then, Torishima has been declared a nature reserve and a national monument by the Japanese government, and that rare species is now making a slow comeback in that region.

Endangered species are defined as those species of organisms that are considered to be in immediate danger of extinction. Today, many species of plants and animals throughout the world are considered to be endangered. In 1985, the U.S. Fish and Wildlife Service listed 461 foreign species of animals as being endangered, including the leopard, the giant panda, the white rhinoceros, the Japanese crane, the St. Vincent parrot, and the Galapagos tortoise (Figure 14.3). Furthermore, it listed 271 domestic species of mammals, birds, reptiles, amphibians, fish, snails, shellfish, and insects. (This number includes 47 species that are found in both the United States and other countries.) Some representatives of those animal groups are listed in Table 14.1. The population levels for all those endangered animals are precariously low. For example, in 1984 only 18 California condors remained in the wild, and only 10 to 20 breeding female American crocodiles remained in Florida's Everglades National Park.

The U.S. Fish and Wildlife Service also lists *threatened species*, which include 51 domestic species, 37 foreign species, and 10 species that reside in both the United States and other countries. Those listings include the southern sea otter, the pigmy chimpanzee, the African elephant, and the desert tortoise. Although members of those species are still abundant in some parts of their territorial ranges, they are classified as being threatened because their numbers have declined significantly in other areas.

Figure 14.1 *Passenger pigeons being hunted for market. (American Museum of Natural History.)*

Figure 14.2 *Unlike the passenger pigeon, the short-tailed albatross may yet be saved from extinction. (Gilbert S. Grant, Photo Researchers.)*

(a)

Figure 14.3 *Examples of animals on the endangered species list: (a) leopard (© Terence O. Mathews, 1973, Photo Researchers); (b) white rhinoceros (© Terence O. Mathews, 1973, Photo Researchers); (c) giant panda (© Stan Wayman, 1972, Photo Researchers); (d) Japanese Crane (© Tom McHugh, Photo Researchers); (e) St. Vincent parrot (© Tom McHugh, Photo Researchers); (f) Galapagos tortoise (© Miguel Castro, 1981, Photo Researchers).*

(b)

(d)

(c)

(e)

(f)

Table 14.1 Some endangered animals of the United States

MAMMALS	BIRDS
Ozark Big-Eared Bat	Masked Bobwhite (Quail)
Eastern Cougar	Hawaiian Coot
Key Deer	Mississippi Sandhill Crane
San Joaquin Kit Fox	Hawaii Creeper
Florida Manatee	Hawaiian Duck
Utah Prairie Dog	Everglade Kite
Sonoran Pronghorn	California Clapper Rail
Gray Wolf	Red-Cockaded Woodpecker

REPTILES	AMPHIBIANS
American Alligator	Desert Slender Salamander
American Crocodile	Texas Blind Salamander
Blunt-Nosed Leopard Lizard	Houston Toad
San Francisco Garter Snake	Santa Cruz Long-Toed
Green Sea Turtle	Salamander

FISHES	INSECTS
Humpback Chub	El Segundo Blue Butterfly
Mohave Chub	Mission Blue Butterfly
Maryland Darter	Oregon Silverspot Butterfly
Pecos Gambusia	San Bruno Elfin Butterfly
Comanche Springs Pupfish	Smith's Blue Butterfly
Colorado River Squawfish	
Gila Trout	

(a)

Figure 14.4 *Examples of plants on the endangered species list: (a) San Clemente Island larkspur (T. Oberbauer); (b) spineless hedgehog cactus (T. Oberbauer); (c) Tennessee purple coneflower (Paul Somers).*

For example, although the grizzly bear remains relatively secure in Alaska and western Canada, its numbers have fallen dramatically in the Rocky Mountain states. In the Yellowstone National Park area, the population of grizzly bears has plummeted 40 percent in the past decade. Box 14.1 discusses the uncertain future of the grizzly bear in the lower 48 states.

The public has shown much concern over the plight of many endangered animal species, but the fact that many plant species are also endangered is less well known. Thus in 1977, the U.S. Fish and Wildlife Service suggested that over 1800 plant species in the United States should be listed as being either endangered or threatened. However, because well-documented scientific evidence is needed and detailed procedural requirements must be fulfilled before a species can be listed as endangered or threatened, only 72 of those plant species had been listed officially as being endangered, and only 11 had been listed as being threatened by 1985. As a result, only those species were given federal protection. With 21 species of cacti on the list, the cactus family is particularly imperiled. Three endangered plant species are shown in Figure 14.4. No worldwide listing is currently available, but approximately 10 percent

(b)

(c)

of the world's estimated 220,000 species of flowering plants are considered to be endangered.

Although considerable efforts have been made in recent years to protect endangered species, many scientists, resource managers, legislators, and concerned citizens are pessimistic about the continued existence of many species. A growing human population, with its needs for more food, water, energy, space, and other resources continually threatens the survival of our fellow organisms on earth. In this chapter, we explore the many reasons why species are endangered and evaluate our efforts to save them.

463

BOX 14.1

THE GREAT GRIZZLY BEAR CONTROVERSY

The grizzly bear is rapidly disappearing in the lower 48 states. Numbering more than 100,000 when the first European settlers arrived, the population of grizzly bears has dwindled to less than 1000. Although it is still prominently displayed on the California state flag, the last grizzly has long disappeared from the wilds of California. In the San Juan Range of Colorado in 1979, a hunting guide killed a grizzly when the bear attacked him, but since then, there has been no further evidence of grizzlies in Colorado.

Altogether, six remnant grizzly populations exist in the United States outside of Alaska. But only two populations contain a significant number of grizzlies. An estimated 450 to 650 grizzly bears reside along the northern continental divide in an area that encompasses Glacier National Park, Bob Marshall wilderness, and parts of several national forests. A second enclave, which contains probably fewer than 200 bears, is centered in Yellowstone National Park and includes parts of five adjacent national forests. The population in the northern divide appears to be relatively stable in number, but the population of grizzly bears in Yellowstone is dropping at a perilous rate— perhaps a decline of 40 percent since the early 1970s. Simply put, the mortality rate of grizzly bears is much higher than their birth rate.

The grizzly has much going against it in its fight for survival. Grizzly bears are strong K-strategists—they start reproduction at a late age; they produce small litters, and a long interval occurs between their litters. Hence a grizzly population usually cannot make up for the relatively high losses that accrue from human-related mortality.

Each grizzly also requires a lot of undisturbed territory, because it is a large animal, which has a great demand for food. Moreover, individual members usually live in isolation rather than in a group. For example, many grizzlies in the Yellowstone population travel 65 to 90 kilometers (40 to 55 miles) each year between their dens and a variety of feeding areas. Hence a large region (thousands of square kilometers) is needed to accommodate a population of even a few hundred bears.

Today, the last sanctuaries of grizzly bears are being invaded increasingly by loggers, housing developers, road builders, coal miners, oil drillers, sheepherders, and hikers. To stem that tide, government agencies are evaluating grizzly habitats and are formulating management plans to preserve them. For example, in the Yellowstone region, all of Yellowstone National Park is considered to be a favorable grizzly habitat and, consequently, it has been categorized in Management Situation 1. In that category, all management decisions will favor the bear over competing land use. Areas that make poorer habitats for grizzly bears are placed in different management categories. The presence of bears will not be encouraged in habitats that are in those categories. Of course, management decisions (such as deciding where to draw the boundaries between management zones and which land uses favor the bears and which do not) have raised many unresolved conflicts.

Not only is the grizzly bear losing areas within its habitat, but encounters between bears and humans are becoming much more frequent. On very rare occasions, such encounters lead to a human fatality. For example, in early summer of 1983, a grizzly dragged a sleeping man from his tent and killed him in the Gallatin National Forest, which is adjacent to Yellowstone National

The grizzly bear, Ursus arctos horribilis. (© 1978 by Tom Mc-Hugh, Photo Researchers.)

Park. Since the early 1900s, four persons in Yellowstone National Park and three persons in Glacier National Park have been killed by grizzlies. During the same period of time, well over 100 persons have died in those two parks from other accidents, including drownings, falls, and being hit by falling rocks and trees. But the prospect of being mauled by a grizzly bear is particularly frightening to most persons.

To reduce the number of encounters between humans and grizzlies, the Forest Service and the Park Service temporarily close campsites and trails in areas where grizzlies are present; they relocate or destory bears when it is necessary to protect human life; they close garbage dump sites,

and they establish camping rules, such as cooking away from campsites, locking food in cars, and allowing no food in tents.

As a rule, encounters between humans and grizzlies much more frequently result in a dead grizzly than in a dead person. Since 1976, an average of 11 grizzlies have been killed each year in the Yellowstone area. Because they are listed as a threatened species, a grizzly bear can be killed only when it is necessary to protect human life. However, illegal hunting contributes significantly to the high mortality rate for grizzly bears. And sheepherders who graze their sheep in the national forests have traditionally shot any grizzly bears that have appeared within their sights. Furthermore, poachers can sell a bear's claws and pelt for far more than they would be fined if caught, prosecuted, and convicted. Unscrupulous hunters and outfitters care little about protecting endangered and threatened species.

While most attention centers on the Yellowstone grizzlies, the future of the grizzly bears that live in the northern divide is also in jeopardy. Industries are rushing into oil and gas exploration in that area, and proposals are being submitted to improve existing roads as well as to build new roads. That development will further penetrate and dissect the prime habitat of grizzly bears. Also, increasingly more persons are visiting that area, and more are building summer residences there.

Should token populations of grizzly bears be protected in the lower 48 states? (No doubt the real future of the grizzly lies in Alaska and western Canada, where between 18,000 and 38,000 bears are still found.) Should federal agencies control human activities in an attempt to save the bears? Or should the grizzlies be extirpated to allow human access to those regions without fear from grizzly attacks? What do you think?

Why Preserve Them?

Some persons might ask why we should care about the loss of a few species or what good they do for us. After all, we seem to be getting along fine without the Carolina parakeet, the California grizzly bear, and the eastern elk—all of which are now extinct. Although most persons—including many government policymakers—are, at best, lukewarm toward saving endangered species, many scientists, policymakers, and interested citizens argue that species extinction is one of the most serious problems that humankind must face in the coming decades. Thus we wonder why opinions are so divided.

Value as a Genetic Reservoir

One major reason for preserving as much diversity of life as possible is the indisputable fact that plants and animals possess undiscovered or undeveloped traits that cannot be replaced—traits that may have an inestimable value to the survival of our own species. As we discussed in Chapter 3, every living thing is a unique combination of genetic characteristics, which enables it to adapt to a variety of environmental conditions. Thus when all the genes of all the individual members in a given population are added together, a *gene pool* is created, which is representative of that species. Many biologists and other scientists believe that it is vitally important to preserve those gene pools, which might prove to be useful to us in the future. For example, 45 percent of the medical prescriptions that are written in the United States contain at least one product of natural origin. The lifesaving antibiotic penicillin is produced by a fungus *(Penicillium notatum)*, and certain species of bacteria produce the important antibiotics tetracycline and streptomycin. Thanks to those antibiotics and more than a 1000 others, scourges, such as typhoid fever, scarlet fever, bubonic plague, diphtheria, syphilis, and gonorrhea can be treated more effectively. Certain flowering plants also produce medicinal compounds. For example, quinine is used to treat malaria (from the cinchona tree); digitalis is used to treat chronic heart trouble (from the foxglove plant); and morphine and cocaine are used to reduce pain (from the opium poppy plant and the coca shrub, respectively). Drugs that have been extracted from plants are used to treat leuke-

mia, several forms of tumorous cancer, various heart ailments, and hypertension (high blood pressure). Therefore, the discovery of other lifesaving drugs depends on the survival of microorganisms, plants, and animals that some persons might consider to be insignificant. Furthermore, fewer than 5000 of the earth's 220,000 species of flowering plants have been analyzed by biochemists for the presence of valuable drugs.

The medicinal value of a seemingly insignificant species is illustrated by the role of the nine-banded armadillo in the work that is being done to eventually control leprosy. That disease has been a curse to humankind since ancient times, and a cure has been difficult to find, because the bacterium that causes the disease grows in humans but not in laboratory animals. Hence it has been almost impossible to investigate the nature of the infection and to develop an effective vaccine for it. However, in 1971, biochemists discovered that the bacterium flourishes in the nine-banded armadillo (Figure 14.5). Fortunately, because that species is still with us, scientists now have a good opportunity to study and, perhaps someday, to conquer leprosy.

The maintenance of a large gene pool is also of great interest to agriculturalists. All domestic crops and livestock originated from native plants and animals. And those native species are still needed to provide the new genetic characteristics that we need to help solve our present and future food-production problems. For example, the new varieties of wheat and rice, which have significantly increased food production in many countries (see the discussion of the Green Revolution in Chapter 17), are products of breeding experiments that utilized thousands of native and domesticated varieties of rice and wheat.

Because pests, disease-causing organisms, and technology evolve over time, many challenges that we cannot predict are certain to present themselves—challenges that will require the genetic resources of other varieties and breeds. Such an event occurred in the early 1970s because of a fungus that causes southern corn-leaf blight. Although the fungus had been present in the southern states for many years, it had only reduced the yield of corn by less than 1 percent. But in 1970, a genetic change (a mutation) produced a new strain, or variety, of the blight fungus, which was more lethal than the earlier strain. Borne by the wind, the new strain rapidly spread

Figure 14.5 *The nine-banded armadillo, an animal of considerable importance to research into a cure for leprosy. (Jen and Des Bartlett, Photo Researchers.)*

from the South to the central Midwest. To compound the problem, in 1970, most of the corn that was being grown in the United States (80 percent) was of a single type. Although that variety of corn gave a high yield, it was very susceptible to corn-leaf blight. Thus as a result of the new strain of blight and the susceptibility of much of the nation's corn, an estimated 15 percent of the genetically vulnerable corn crop was destroyed. Fortunately, during the winter of 1970–1971, hybrid seed companies planted blight-resistant varieties of corn in Hawaii and South America. Thus farmers had adequate amounts of seed corn of blight-resistant varieties to plant the following spring, and blight damage was much less severe.

Despite those reasons to preserve genetic resources, agricultural gene pools are being swept away in the wake of human activities. For example, 50 years ago, 80 percent of the wheat that was grown in Greece consisted of native strains that were well adapted to the local environment and resistant to local pests. Today, only 5 percent of those wheat varieties still exist.

Industries also utilize a wide variety of plant and animal products and benefit from a rich diversity of species. As many as 2000 plant species throughout the world have economic importance. For example, American building, furniture, and paper-making industries utilize more than 100 different species of trees, and cotton, flax, hemp, jute, and agave are used for fibers. Also, a multitude of plant species are utilized for such products as rubber, medicinals, dyes, tanning agents, perfumes, lotions (such as aloe), resins, gums, insecticides (such as pyrethrum and rotenone), oils (such as oils in paints and varnishes), and so on.

What if the wild ancestors of rice, wheat, corn (the three grains that account for more than one-half of the world's food production), cattle, and horses had become extinct before they were domesticated? And what if humankind never discovered that many medicines and industrial products could be produced from natural substances before the organisms that contained them became extinct? Truly, those gene pools have been the very foundation of human civilization.

Although those arguments for species preservation appear to be quite compelling, they do have some shortcomings. For example, scientists know practically nothing about most of the 1.6 million identified species on this planet. Furthermore, 62 species have become extinct in the United States alone since the 1600s without causing any well-documented impact on our society. Thus it is difficult to convince citizens, legislators, and persons in business that they should be concerned about some obscure species (that has no currently known value) that may become extinct if, for example, a dam or highway is built. It is easy to ignore the possible consequences of species extinction, particularly if those consequences to society are nebulous. Many persons do not even worry about the increased risk of developing cancer or emphysema from smoking, even though there is ample evidence that a relationship exists.

Value in Maintaining Ecosystem Stability

Besides serving as a valuable genetic reservoir, each species interacts with other species and plays a role in the transfer of energy and materials within and between ecosystems; hence each one, in its own way, contributes to the stability of ecosystems. Some species, of course, appear to be more important than others in this regard. For example, the alligator is a dominant member of subtropical wetland ecosystems, such as the Florida Everglades. Depressions that are excavated in the ground by alligators ("gator holes") hold fresh water during the relatively dry winters, which serve as reservoirs that preserve aquatic life as well as provide fresh water and food for birds and mammals. If the alligators were eliminated, "gator holes" would fill quickly with sediment and aquatic plants, and wildlife would not be able to depend on them for survival during the dry periods. Alligators also affect the stability of the Everglades' ecosystem as a result of their feeding habits. They eat large numbers of gar, which is a major predatory fish, and thereby help to maintain the populations of game fish, such as bass and bream.

Although few species appear to be as important to an ecosystem as the alligator, ecologists know too little about most species to say that one particular species is not important. Moreover, we can safely conclude that as more species disappear, the diversity diminishes, and the number of checks and balances on plant and animal populations decreases. As species are lost, the stabilizing influences of predation, parasitism, and competition are disrupted, and an ecosystem thus becomes more vulnerable to disturbances that, in some cases, threaten to destroy it (see Chapter 4).

Although we tend to mourn the passing of an animal species much more than the loss of a plant species, the extinction of a plant species is often more critical to ecosystem stability. Because plants occupy the base of food webs, a single disappearing plant species may cause 10 to 20 species of animals to become extinct, because they were dependent on that species for food or shelter at some time during their life spans.

Some persons have suggested that if a particular population or species becomes extinct, another one can be introduced into the ecosystem to take its place. But many species, particularly climax species, are intricately adapted to particular habitat conditions. For example, many insect species that are herbivores have made such delicate adjustments to both host plants (coevolution) and the local climate that even members of one population may not be able to substitute for another population of the same species. For example, when all the British populations of the large copper butterfly became extinct, several attempts were made to introduce butterflies from Dutch populations of the copper butterfly. All attempts failed. Similarly, many plant popultions are so intricately adapted to such factors as the local photoperiod and the soil type (see Chapter 5) that

attempts to transplant them from one region to another are often unsuccessful. Thus even though a location may appear to have habitat conditions that would support a new species or another population of a species that formerly existed there, such introductions often fail. Thus it is clear that the genetically based tolerance limits of the organisms that are being introduced are often not sufficiently suited to the new habitat.

These arguments for species preservation suffer from the same shortcomings that were described for the arguments based on maintaining genetic diversity. Ecologists know very little about the specific functions (and their relative importance) of most species in ecosystems. As far as ecologists can ascertain, few species play all the roles that are critical to maintain an ecosystem, and the value of many species in the functioning of ecosystems may be relatively minor. Thus in the view of many persons—including many policymakers—these nebulous values often are not considered to be as important as jobs and industrial development when the issue of species preservation conflicts with resource development.

Economic Value

Economic value can be a major factor in the arguments that favor species preservation. A vast multitude of natural products are used by industry, and those products, along with agricultural products, add trillions of dollars to the world's economy. In addition, billions of dollars are generated by tourism and recreation—and, generally, the most popular tourist attractions are not human-made marvels, but national and state parks and forests. Sport hunters and fishermen, as well as their commercial suppliers, spend hundreds of millions of dollars every year in the pursuit of their quarry, and they have a deeply vested interest in the preservation of their target species. As a result of those economic considerations, many local conservation groups have worked toward habitat and species preservation, but in essentially all instances, their efforts have been restricted to only a few species of game mammals, upland game birds, water fowl, and fish. Nongame animals have been virtually ignored.

Since most species have no obvious economic value, it might seem that the economic considerations of species preservation is flawed. However, although the economic value of a given species may not be apparent, we cannot assert that a species has no economic value. Indeed, the very fact that ecologists know so little about the importance of so many species should caution us against questioning a species' ultimate value. For example, what if it turns out that some inconspicuous species plays a vital role in, say, an aquatic food web that supports a commercially important fish species? In such a case, that obscure species would actually be as important economically as the commercial species that it supports.

Esthetic Value

Some proponents of species preservation are guided by esthetic values rather than economic considerations. For example, the thrill of watching an eagle soar silently overhead, an afternoon spent listening to the capricious chatter of chipmunks, the taste of wild berries, the refreshing fragrance of wildflowers, and the softness of a bed of moss have no monetary value, because they have no equal. Still, those who base their arguments on esthetics must face the fact that most plants and animals hold little esthetic value for most people. Although the great blue whale or the bald eagle may be esthetically valuable to many persons, few of us are attracted to the California slender salamander, which is shown in Figure 14.6, or the orange-footed pearly clam. Furthermore, when persons are hungry, poor, and unemployed, esthetic concerns usually disappear from their list of priorities.

Inherent Value

The arguments for species preservation that have been presented so far have one common element—they are all based on what we want; they are person-centered. But some persons feel that the inherent value of a species cannot be measured properly by the extent to which human beings can get along

Figure 14.6 *The California slender salamander—endangered, but not impressive enough to elicit much human concern. (Jack Wilburn, © Animals Animals.)*

without it. They argue that each species has its own rights. According to that view, if a species exists, then it has a fundamental right to continue to exist without being driven to extinction by human activities. Moreover, such noted thinkers as the microbiologist René Dubos, the ecologist Aldo Leopold, and the naturalist John Muir as well as many great religious writers argue that we have a responsibility to be faithful stewards of the earth and its resources, including other organisms with whom we share the planet.

Others, such as Gardner Moment of Goucher College, have a differing view. Although he states that persons should have a strong sense of stewardship, he also contends that we should not succumb to a mindless absolution and claim that every species should be preserved. And, therefore, he feels that we should not abdicate the use of human reason, which is our most powerful and unique tool. He concludes that no one is obligated to save a species without regard for the human cost in money, health, or lives. Many persons would agree that we have the right to discriminate when the human cost of preserving a species is perceived as being too great.

Thus we might ask, Should humans attempt to preserve all species or just some of them? If we try to save only some species, what criteria should be used to select those that will be protected? And who will participate in making those decisions? These are difficult questions to answer, but if we do not answer them soon and take intelligent action, untold numbers of species will soon disappear from this planet.

A Historical Overview

Extinction is not a new phenomenon. Organisms have been disappearing from the face of the earth since life began, billions of years ago. Furthermore, extinction is the rule rather than the exception. For example, 20,000 species of vertebrates (animals with backbones) were alive at the end of the Paleozoic era, approximately 230 million years ago (see Appendix II), but only approximately 24 of those species have any living descendants today. Interestingly, however, approximately 50,000 new species have descended from those 24 ancestral species.

Some extinctions have occurred relatively slowly and can be explained by gradual, large-scale climatic changes. Other extinctions have occurred relatively quickly and have involved many species. The largest mass extinction of all time occurred 225 million years ago (in the Late Permian period) when the world's continents coalesced into the supercontinent, Pangaea (see Chapter 12). The resulting sea level and geographical changes that occurred shrank the area that was occupied by shallow seas by nearly 70 percent, and the types of distinct marine habitats were reduced by nearly 50 percent. As a result of those habitat losses, most of the marine species (up to 95 percent) became extinct.

A more recent mass extinction occurred 65 million years ago, when the dinosaurs disappeared. Although many ideas have been suggested to explain the loss of those fascinating creatures, perhaps the best-accepted theory today is that a large asteroid struck the earth at that time. As a result of that catastrophic impact, an enormous cloud of dust was raised, which largely obscured the sun for several years. With sunlight greatly reduced the vegetation died, which, in turn, led to the starvation of all the large herbivores and their predators. Scientists estimate that the largest surviving animals only weighed 25 to 30 kilograms (55 to 65 pounds). The only vertebrate species (animals with backbones) that survived were those that could exist on a diet of insects, carrion, and other detrital materials.

A recent mass extinction took place in North America at the end of the last Ice Age. That Pleistocene extinction also involved many large herbivores and their associated predators—horses, camels, mammoths, mastodons, ground sloths, giant beavers, jaguars, and saber-toothed tigers. Figure 14.7 illustrates some of the animals that became extinct a little more than 11,000 years ago. Only 30 percent of the species of big game animals that were in existence at that time, including moose and caribou, survived that period.

What caused that mass extinction? Nobody knows. The climate certainly changed dramatically during that period, when the continental ice sheets receded. As the North American ice sheet melted back, a corridor opened along the MacKenzie River between the Canadian Rockies and the retreating ice front, which allowed large masses of extremely cold air to flow out of the Arctic and onto the Great Plains

Figure 14.7 *A reconstruction of a scene from the Pleistocene era at Rancho La Brea Tar Pit, near present-day Los Angeles. Illustrated are vultures, saber-toothed tigers, mammoths, giant ground sloths, and dire wolves. These species became extinct about 11,000 years ago. (Charles R. Knight, American Museum of Natural History.)*

(Figure 14.8). As a result, the cold temperatures may have killed the animals or the vegetation that served as their source of food. However, a significant criticism of this explanation is that major climatic changes occurred many other times, during previous advances and retreats of Ice Age glaciers, without any notable increase in extinctions. Thus we do not know how the last glacial retreat was so different climatically that it resulted in the mass extinction of large animals.

Professor Paul S. Martin of the University of Arizona points out that the most recent mass extinctions occurred at the same time that human beings arrived in North America. He believes that those early immigrants crossed the Bering land bridge from Siberia into Alaska and then moved south and east across North America, where their hunting activities obliterated most of the large animals. The major objection to that explanation is that the human population was not large enough to cause such a massive extinction over so great an area as North America. However, Martin counters that objection by suggesting that the immigrants experienced a population explosion when they reached North America with its abundant wildlife. Moreover, Mar-

tin suggests that they advanced across the continent as a wave, keeping the greatest number of hunters at the front, as illustrated in Figure 14.9. He further suggests that those primitive hunters were very efficient at killing animals in large numbers (such as by setting fires to drive large herds over cliffs or into rivers, where they drowned). Hence large grazing animals could have succumbed in great numbers to the concentrated onslaught of the hunters. Then, Martin believes that when the hunters moved on to a

Figure 14.8 *The corridor through the continental and mountain glaciers that allowed humans from Siberia to enter the southern portion of North America some 11,000 years ago. The darker area is glaciated and the lighter area is unglaciated.*

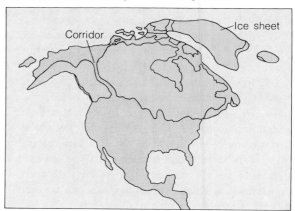

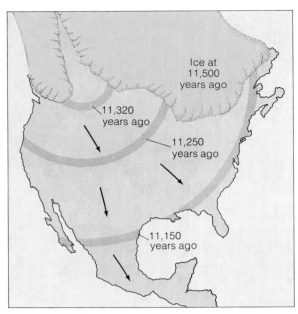

Figure 14.9 *The postulated sweep of migrating hunters across North America (represented by the dark bands). As local extinction of large mammals occurred, hunters moved on. (After P. Martin, "The Discovery of America," Science 179:969–974. Copyright © 1973 by the American Association for the Advancement of Science.)*

new area, large carnivores eliminated the few large prey that were left behind. As a result, the population of large carnivores may have become greater than usual for awhile, because of the unusual availability of food (in the form of carrion) that was left from kills by humans. But, eventually, as human hunters moved on and the large herbivores became extinct, the large, specialized carnivores, such as the saber-toothed tiger, could have become extinct because they lacked the ability to capture other animals, which are particularly wary and fleet (such as deer) or that have group defenses (such as musk oxen, which form a protective circle) or both.

Which factor—climate or humans—was more important in eliminating so many species of animals? We will probably never know for sure. Perhaps one or more other as yet unknown factors may have played a role. Quite likely, however, both the changes in the physical environment and the influx of human beings exerted a stress on wild animal populations, but human hunting pressure was probably the factor that tipped the balance.

How Our Species Endangers Other Species

Although rates of extinction are very difficult to estimate, probably less than 1 percent of the plant and animal species became extinct during the thousands of years between the Ice Age extinction and the year 1600. But since the beginning of the seventeenth century, the number of species that has been disappearing has risen dramatically, as indicated in Figure 14.10. Roughly 1 percent of the species of birds and mammals that existed in 1600 have become extinct in just the last 385 years. Moreover, the rate of extinction is accelerating at an alarming rate. The total number of species that becomes extinct in the twentieth century may be 10 times greater than it was in the seventeenth century. That recent sharp upturn is directly linked to the unprecedented growth of the human population and the increasing stresses that we are creating by exploiting the earth's resources. Humankind has brought about extinction in several ways: by destroying natural habitats, by hunting for food and commercial products (such as furs, hides, and tusks), and by introducing alien species. The relative significance of each type of stress varies among species. Today, however, the alteration of natural habitats has by far the greatest adverse impact on native wildlife populations.

Destruction of Natural Habitats

Human beings have grossly modified many ecosystems. For example, complex forest ecosystems have been transformed into farms, where corn, soybeans, hogs, and cattle are the dominant species; marshes have been filled in for the construction of housing tracts, shopping centers, and industrial parks; and stream channelization, damming, strip mining, and highway construction continue to raise havoc with the remaining natural habitats. Each year, 4750 square kilometers (1900 square miles) of land in the United States are developed intensively. And natural habitats in tropical nations are disappearing at an even faster rate.

However, those massive changes in habitats do not occur without some consequences. Recall from Chapter 3 that every organism has specific, genetically based tolerance limits. And, frequently, human beings modify a habitat so drastically that the tol-

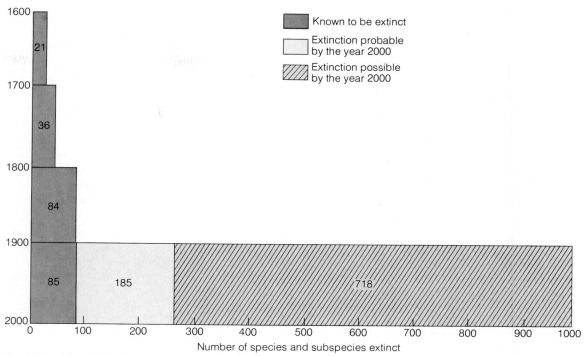

Figure 14.10 *The number of extinct species and subspecies of vertebrate animals (animals with backbones) by century, with projections for the twentieth century. (After G. Uetz and D. Johnson, "Breaking the Web."* Environment 16(10):31–39, 1974.)

erance limits of most, if not all, of its inhabitants are exceeded. Then, when the critical factors of a habitat become limiting, localized extinction occurs. Thus when natural habitats for a particular species are destroyed, its members have fewer places to live and the very existence of the species is threatened.

Although many human activities, such as lumbering, grazing, and clearing, actually create areas with favorable conditions for successional species (such as rabbits, deer, elk, and grouse), they also destroy areas for animals that require climax conditions (such as caribou, wolverines, and prairie chickens). The climax ecosystems that have the greatest potential for species extinction are the tropical rain forests. Scientists estimate that 3 million species of plants and animals reside in the tropics as compared with some 1.5 million species in temperate regions. (Only approximately 500,000 of the tropical species have been identified and cataloged.) Recall from

Chapter 5 that tropical rain forests have already been reduced to approximately one-half of their former extent, and much of the remaining forests will be gone by the turn of the century.

Also, damming, draining, and filling have destroyed many wetland areas, which, in turn, have resulted in wildlife and plant losses. Box 14.2 describes how human intervention has effected the flow of water into southern Florida. And although drainage can destroy wetlands, irrigation can disrupt dryland habitats. For example, the endangered San Joaquin kit fox population now numbers less than 3000. In 1960, the kit fox roamed 1.2 million hectares (3 million acres) of the dry San Joaquin Valley. Today, as a result of irrigation, the valley's desert land has been reduced to less than 400,000 hectares (1 million acres), and the number of kit foxes is critically low.

Pollution of natural habitats has also taken its toll. Industrial wastes have had a severe impact on aquatic habitats (see Chapter 7). For example, although the impact of acid rain is far from being fully understood, it may be a slow, subtle agent that is driving some species in the lakes and streams of

BOX 14.2

THE SHRINKING EVERGLADES: WHAT'S HAPPENING TO ITS ENDANGERED WILDLIFE?

It is a land of contradictions, a semitropical wilderness dotted by fields of corn and tomatoes and slashed by superhighways like Alligator Alley and the Tamiami Trail. Wedged into the southern tip of the Florida paw, the Everglades are a curious marriage of wet and dry, salt water and fresh water, temperate and tropic. More than 5 feet of rain falls here annually, twice the national average—yet the area has been ravaged by drought and fire. The landscape is deceptive: it looks montonously flat—yet a change of a mere inch or two in elevation creates swampy marshes next to sawgrass prairies and pine forests. Says Bob Doren, a National Park Service biologist, "There are nothing but extremes in the Everglades."

Those extremes were once entirely natural; now man has exacerbated them. The question of water—too much or too little—is paramount in the Everglades. Efforts to control it have produced floods in many years, drought in others—clear evidence of man's failure to master nature. Encroaching agricultural and urban development has cut the Everglades wetlands in half, destroying animal and plant habitats and allowing foreign species with few natural enemies to intrude. The wilderness that settlers once called the Forever Glades now has all too apparent limits.

Before man intervened, Lake Okeechobee overflowed annually, sending a flood of water south through the Everglades, nourishing its wetlands. Today, the natural inundations from Okeechobee have slowed to a trickle. Sugar cane grows in what were once the headwaters of the Everglades, and water is shunted eastward for cities and farms. On one occasion, the man-made drainage system of South Florida caused much of the Everglades to dry up; water flow to Everglades National Park stopped completely during the severe droughts of the early 1960s, allowing massive brush fires to devour thousands of parched acres.

Paradoxically, what now plagues the park is too much water. Since 1981 the South Florida Water Management District has dumped excess water from neighboring water conservation areas into the park to help keep surrounding agricultural and urban land dry, and to relieve flooding during the rainy summer months. To deal with the unusually heavy rainfall of the past 2 years, the Army Corps of Engineers opened floodgates to the north, creating a rush of water into the park that destroyed almost all the alligator nests. Even

New England and Canada toward extinction (see Chapter 10). Furthermore, during the 1950s and 1960s, insecticides, particularly chlorinated hydrocarbons (such as DDT), reduced the population levels of several large birds of prey, including the bald eagle, the peregrine falcon, and the brown pelican, to such low levels that they were officially listed as endangered species.

Interestingly, the decline of those species was not caused by direct poisoning and death of the adult birds but, rather, to a complex disruption of successful hatching. Recall from Chapter 2 that DDT bioaccumulates and, consequently, those organisms at the upper trophic levels (such as birds of prey) may have significantly higher concentrations of DDT in their tissues. Research has shown that some of that DDT is changed by means of metabolic processes to another chlorinated hydrocarbon, called DDE. That compound interferes with the transport of calcium in the female bird, which makes calcium less available for the development of eggshells. Much evidence exists to confirm the fact that eggs which were laid prior to the widespred use of DDT had thicker shells than those eggs that were laid thereafter. Since the thinner shells led to greater breakage of eggs during normal nesting activities, the number of hatchlings fell dramatically. As a result, the population of those birds of prey declined drastically during the 1950s and 1960s.

Habitat destruction is a by-product of many human activities. So, if the accelerating trend toward species extinction is to be slowed, we must give

the porous, honeycombed limestone that under-girds much of the Everglades has been inadequate to sop up the excess. Says Joel Wagner, a National Park Service hydrologist, "We've got unnaturally timed amounts of water coming into a wetland area that is only half the size it used to be. The water simply has no place to go."

Last year, park officials called for emergency measures that would restore a more natural water flow by plugging up portions of canals within the Everglades, removing some levees, and pumping excess water east. The plan has been challenged by farmers and others who would get the overflow. Says one angry tomato farmer, "We don't want a single drop of their damn water."

Supporters of the plan are most concerned about the dozens of animal and plant species that depend on a regular cycle of wet and dry seasons. For some of these species, the consequences of flood-control efforts have already been devastating. Nearly half of South Florida's deer herd starved to death last year because of flooding. The best known resident of the Everglades, the American alligator, has also suffered: according to National Park Service biologist James Kushlan, alligator eggs have failed to hatch in 3 of the past 5 years. The park has not had a normal bird-nesting season for 10 years. The wading bird population, now only a tenth of what it was before 1930, is still dwindling. Some other birds, especially the osprey and the brown pelican, have also been affected.

The fate of the wood stork worries biologists the most, because they consider it a living barometer of the health of the Everglades. This majestic wading bird depends on the right mix of water, food, and nesting habitat—and it has nested successfully only once since 1975. It relies on pools of water to concentrate enough fish to feed its young. Those pools never lasted long enough to sustain the bird through its most recent nesting season last winter.

The Florida panther, one of North America's rarest mammals, has fared better. Only 8 years ago, biologists thought it had disappeared from South Florida, a victim not of flooding but of poachers and urban growth. Now some 20 panthers are known to live in Big Cypress Swamp. Oil exploration in the area, however, threatens the panther's only foothold: new roads between drilling sites have opened its habitat to an increasing number of swamp buggies.

Excerpted from Sana Siwolop, "The Shrinking Everglades: Water, Water, Everywhere." © Discover Magazine 5:39 (April), 1984, Time Inc.

greater priority to species and ecosystem preservation when we make decisions regarding land use and development. Chapter 15 has more on land use conflicts.

Commercial Hunting

Within the last century or so, several species have become extinct because they were hunted out of existence as a source of food. Those species include the passenger pigeon, the great auk (which was a large, penguinlike, flightless bird that once inhabited the North Atlantic from Newfoundland to Scandinavia) and the heath hen (a bird that was similar to the prairie chicken and was once common from Massachusetts to the Potomac River).

Today, many whales (such as blue whales, gray whales, sperm whales, and several others) are on the endangered species list, in part because they have been overharvested for food. Other species that are endangered today because of their food value include the giant tortoise (for its flesh and liver), which is found on the Galapagos and Seychelle Islands, and the green sea turtle (for its flesh and eggs), which is found around the globe in tropical and temperate seas.

Also, many animals are endangered today because they are overexploited for their furs, hides, tusks, horns, or feathers. For example, big cats—tigers, leopards, jaguars, cheetahs, and ocelots—have been particularly hard hit by hunters, who sell their skins to furriers to make fashionable coats and capes.

Commerce in wild-animal products is big business. Recently, in Hong Kong, police seized 319 cheetah skins that arrived illegally from Switzerland. The skins were valued at $160,000 and represented 5 to 10 percent of the world's cheetah population.

Several species of rhinoceros are facing extinction because of the human demand for their horns. For example, 60 percent of the illegal world trade in rhinoceros horns takes place in the Far East, where the powdered horn is used as medicine to reduce fevers. In Hong Kong, the black market value of rhinoceros horns (which are composed of a mass of compact hair) is $20 a gram ($600 an ounce), and an entire horn may weigh 1.2 kilograms (40 ounces). The remaining market for rhinoceros horns occurs in North Yemen, where it is a prized material for dagger handles. In a similar manner, the future of elephants in the wild is in doubt because of the high commercial value of their ivory tusks.

The pet trade also consumes an astonishing number of wild animals. Each year, millions of live mammals, birds, reptiles, amphibians, and fish are imported into the United States for commercial sale. And that number represents only a small percentage of the animals that are sacrificed, since many more animals die when they are captured or during transport than make it alive to the United States. Those losses are examples of frivolous waste, because most of these animals make very poor pets. And when they lose their novelty, many are flushed away by their owners or released to fend for themselves in a foreign habitat.

Species that are hunted for commercial purposes are frequently caught in a vicious circle. The more they are hunted, the rarer they become; and the rarer they become, the more highly they are prized, and thus the more they are hunted (Figure 14.11).

Figure 14.11 *"Had to bag one, Harry—in case this damned Conservation thing doesn't work and they become extinct."* (Copyright © 1970 Punch/Rothco.)

Introduction of Alien Species

As long as human beings have traveled around the world, they have carried with them (accidentally or intentionally) many species of plants and animals, which they have introduced to new geographical areas. In some instances, an opening has existed in the new environment, and the alien has been able to establish itself without seriously interfering with the well-being of native species. But, in other instances, the alien has been a superior predator, parasite, or competitor, and has brought about the extinction or near extinction of native species. For example, the starling and house sparrow, which were introduced intentionally into the United States from Europe in the mid-1800s, have displaced many native songbirds from their preferred habitats. Also, the Melaleuca tree was intentionally introduced into southern Florida to vegetate "useless swamps" with a wood-producing plant. Although few efforts have been made to develop commercial uses for its timber, the Melaleuca tree has invaded native plant communities with alarming speed and vigor, turn-

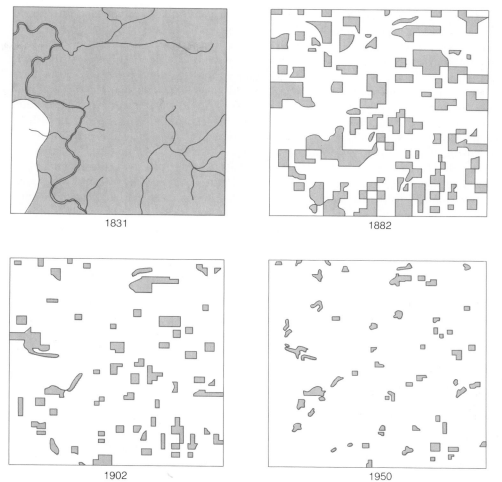

1831

1882

1902

1950

Figure 14.12 *The reduction and fragmentation of woodland (shaded areas) in Cadiz Township, Wisconsin (6 miles by 6 miles) between 1831 and 1950. The fragmentation created small, isolated habitat islands. (After J. Curtis, The Modification of Midlatitude Grasslands and Forests by Man. In W. L. Thomas (ed.),* Man's Role in Changing the Face of the Earth. *Chicago: University of Chicago Press, 1955.)*

ing many areas into Melaleuca monocultures and causing the localized extinction of native plants.

The introduction of alien species is most devastating on islands, where endemic species are often inadequately equipped to deal with invasions of humans and their domesticated animals. For example, the dodo bird lived only on Mauritius, which is a small island in the Indian Ocean. The dodo bird possessed two characteristics that were its eventual un-

doing: it had no fear of people and, therefore, could be easily clubbed to death; and it was flightless, so it had to lay its eggs on the ground. The dodo bird became extinct by 1681, after pigs that were introduced to the island devoured its eggs. Today, more than half of the nearly 50 species of endangered birds in the United States are endemic to the Hawaiian Islands.

On an island, both the variety of habitats, and the area that comprises each habitat are physically limited. As a result of such constraints, often the native animals' ability to adapt to new predators and competitors is reduced. For example, the habitat may not be large enough or diverse enough for a prey species to avoid new predators. And the diversity of resources may not be adequate to allow an existing species to coexist with a new competitive species. Hence, in some instances, competitive exclusion leads to the extinction of the native species.

Competitive exclusion can be illustrated by the demise of the tortoises on one of the Galapagos islands after fishermen released some goats on the island in 1957. When a research team from the Charles Darwin Station arrived in 1962 to investigate the tortoise population on the island, they found only empty tortoise shells, many of which were wedged among the rocks that covered the upland slopes. Investigation of the lowlands indicated that all the vegetation that had previously been within reach of the tortoises had been consumed by the flourishing goat herd. As a result, the hungry tortoises apparently were forced to obtain food on the rocky slopes, where they subsequently either became entrapped between rocks or fell over precipices and starved to death.

Island floras, too, have been devastated by the introduction of alien species. For example, in the Hawaiian Islands, no large herbivorous animals existed until the 1700s, when Europeans introduced domestic livestock. Because no prior selection pressures had existed to favor them, thorns or poisons to ward off herbivores were not characteristic of native plants. Moreover, the limited size of habitats on the islands favored overgrazing. Thus species that had not been threatened previously were devastated by grazing animals. As a result, nearly one-half of the native vegetation in the Hawaiian Islands is thought to be in danger of extinction.

An island, in the usual sense of the word, is a land mass that is surrounded by water. Ecologically, however, a *habitat island* is defined as any restricted area of habitat that is surrounded by dissimilar habitats. Hence a lake without outflow is considered to be an aquatic island. Also, when natural habitats are bisected by highways and converted into farms and cities, natural and seminatural areas become increasingly smaller and more isolated, as illustrated in Figure 14.12. To save what little remains, some of those isolated areas have been set aside as parks, conservancies, forest preserves, or wildlife preserves. But such preserves are usually widely scat-

Figure 14.13 *The California condor, which is a relic of the past. (Carl B. Koford, National Audubon Society Photo Researchers.)*

tered, and the intervening highways, farms, and cities serve as barriers that prevent the migration of many species of plants and animals into or out of them. Hence those natural and seminatural habitats have become islands as well, and the animals and plants that live in them are subject to the same pressures as those living on islands that are surrounded by water. Furthermore, the smaller the habitat island, the greater is the pressure on its inhabitants. Thus we should consider those factors when we plan how to use land rather than indiscriminately invade and carve up natural habitats into increasingly smaller pieces.

Characteristics of Endangered Species

Some species are more vulnerable to human activities than others. The characteristics that are often associated with endangered or threatened animals

are listed in Table 14.2. For example, many endangered species are specialists. They have narrow habitat tolerances and, consequently, have a restricted geographical distribution—a condition that is exacerbated by habitat destruction. Thus the populations that exist in habitat islands are particularly vulnerable to extinction.

As we discussed in Chapter 4, many endangered species are extreme *K-strategists:* that is, their reproduction begins at a late age, they produce few young, and long intervals occur between the births of offspring. For example, the endangered California condor, which is shown in Figure 14.13, lays just one egg per clutch. Once the egg is hatched, the chick is usually dependent on its parents for as long as a year. A mother condor who has a dependent chick will not lay another egg until the chick is 2 years old. However, if they are disturbed, condors may abandon their nest or chick, and, occasionally, condor parents squabble over "incubation rights" and,

Table 14.2 Some characteristics that increase, and some that lessen, the chances of extinction

ENDANGERED	EXAMPLE	SAFE	EXAMPLE
Individuals of Large Size	Cougar	Individuals of Small Size	Wildcat
Predator	Hawk	Grazer, Scavenger, Insectivore, Etc.	Vulture
Narrow Habitat Tolerance	Orangutan	Wide Habitat Tolerance	Chimpanzee
Valuable Fur, Hide, Oil, Etc.	Chinchilla	Limited Source of Natural Products and Not Exploited for Research Purposes	Gray Squirrel
Hunted for the Market or Hunted for Sport Where There is no Effective Game Management	Passenger Pigeon (Extinct)	Commonly Hunted for Sport in Game Management Areas	Mourning Dove
Has a Restricted Distribution: Island, Desert Watercourse, Bog, Etc.	Bahamas Parrot	Has Broad Distribution	Yellow-Headed Parrot
Lives Largely in International Waters, or Migrates Across International Boundaries	Green Sea Turtle	Has Populations that Remain Largely Within the Territory(ies)	Loggerhead Sea Turtle
Intolerant of the Presence of Man	Grizzly Bear	Tolerant of Man	Black Bear
Species Reproduction in One or Two Vast Aggregates	West Indian Flamingo	Reproduction by Pairs or in Many Small or Medium Sized Aggregates	Bitterns
Long Gestation Period: One or Two Young per Litter, and/or Maternal Care	Giant Panda	Short Gestation Period: More than Two Young per Litter, and/or Young Become Independent Early and Mature Quickly	Raccoon
Has Behavioral Idiosyncracies that Are Nonadaptive Today	Red-Headed Woodpecker: Flies in Front of Cars	Has Behavior Patterns that are Particularly Adaptive Today	Burrowing Owl: Highly Tolerant of Noise and Low-Flying Aircraft; Lives Near the Runways of Airports.

Source: From David W. Ehrenfeld, *Biological Conservation*, p. 129. Copyright © 1970 by Holt, Rinehart and Winston, Inc. Reprinted by permission of Holt, Rinehart and Winston, Inc.

in the process, may destroy an egg. (Fortunately, if that occurs, the female condor may lay another egg that year.) Reproduction in condors is further reduced because young birds require 6 to 7 years to reach mating age. As a result, today, few California condors exist. Surveys that were done in 1984 showed that only approximately 20 California condors remained in their only habitat—the mountainous areas that are north of Los Angeles. Therefore, if not enough young reach reproductive maturity to offset mortality, the species will be doomed to extinction.

Attempts to save endangered species frequently raise the question, What is the species' *minimum viable population*? That is, What is the smallest number of individual members that is needed to ensure the survival of an isolated population? As we noted in Chapter 4, a population that becomes too small is subject to many stresses—a tendency toward inbreeding; an increased vulnerability to temporal changes in the density and activities of predators, competitors, or diseases; and an increased susceptibility to natural catastrophes, such as fires, droughts, or severe storms.

The extinction of the heath hen well illustrates the difficulties that a small population experiences in terms of coping with those stresses. Once common from New England to Virginia, the numbers of heath hen began to decline as European settlement progressed. By 1900, fewer than 100 survivors existed—all on Martha's Vineyard, which is an island off Cape Cod, Massachusetts. To save the remaining birds, a predator control program was instituted and, in 1907, a portion of the island was set aside as a refuge. In response, the population of heath hens rose to 800 birds by 1916. However, during that summer, a fire destroyed most of their nests and habitat, and that setback was followed by unusually heavy predation from a high concentration of goshawks during the winter. Then, by the following spring, the birds numbered between 100 and 150. In 1920, the population increased again to about 200 members, but disease took its toll and the population fell below 100 again. Although the species endured a while longer, the last survivor died before 1932. In the last stages of the population's decline, the proportions of males to females increased, and an increasing number of birds appeared to be sterile, which was perhaps a result of inbreeding.

In some cases, a population can survive such events if members of the same species immigrate from other colonies. Thus if more than one population of the species exists, localized extinction may eventually be followed by recolonization. But, as mentioned previously, as habitat islands become more isolated, barriers that prevent dispersal become more restrictive to migration.

The unique genetic makeup of each species, with its attendant tolerance limits, and the great temporal and spatial variability that is inherent in natural stressors (including fires, hostile weather, and the presence of competitors, predators, and parasites), make it very difficult to determine the minimum viable population size for any species. Furthermore, observations suggest that the minimum viable population varies considerably among species. For example, although several thousands of passenger pigeons remained after intensive hunting of them ceased in the early 1880s, the population continued to decline until the last bird died in 1914. In contrast, at one time the eastern cougar was assumed to be extinct in eastern North America. However, it is apparently recovering as a result of just a few breeding members that have survived and now live in the forests of eastern New Brunswick and western Ontario. Thus the cougar's minimum viable population size is demonstrably much smaller than that of the passenger pigeon.

Efforts to Save Endangered Species

Habitat Preservation

Habitat destruction is the major factor in the extinction of plants and animals. Thus the most important strategy that we can use to prevent extinction is to preserve a representative cross section of the world's ecosystems. In the United States, federal and state agencies as well as many private groups have set aside millions of hectares of land, many of which house endangered and threatened species. And the United States is not the only nation that is taking such steps toward species preservation. In tropical and subtropical nations, where the rich diversity of plants and animals is being seriously threatened by human activities, some sizable natural preserves have been established. Nations that now support such preserves include Thailand, India, Peru, Costa Rica, Columbia, Kenya, and Tanzania.

Preserves are beneficial in several ways. For example, in most instances, more than one endangered species can be protected in the same area. Furthermore, a naturally functioning ecosystem may not need much management to continue to serve as an effective preserve, particularly when it is a climax ecosystem. Hence maintenance costs and consumption of resources are usually minimal.

Today's preserves, however, do not comprise a complete solution to the problem. As mentioned previously, ecologists do not have enough information to determine, with reasonable certainty, the size and number of preserves that are needed to maintain a minimum viable population of even one species on today's endangered species list. Moreover, it should be remembered that many species that occupy a relatively large geographical area actually exist as a set of ecotypes. (*Ecotypes* are defined as populations of a species that are particularly adapted to specific conditions which are only found in a part of the total range of the species.) For example, although many species of prairie plants exist in areas that extend from Texas to Saskatchewan, transplanting experiments have shown that for some species, a population from Saskatchewan will not flower and produce seeds in Texas, and vice versa. Each plant ecotype requires a different photoperiod to flower and perpetuate itself. Also, as we saw earlier in this chapter, the Dutch ecotype of the copper butterfly would not survive in England, where the English ecotype of the copper butterfly once flourished. Hence any attempts to save a species should include the establishment of a set of preserves that include all the habitat conditions in which the species does exist. Unfortunately, however, the degree of ecotypic differentiation in most endangered species remains uncertain.

Furthermore, the establishment of preserves does not always ensure adequate protection of endangered species. For example, poaching of animals and plants on preserves by hunters, collectors, and traders is a serious problem worldwide. In an effort to curb poaching, the Convention on International Trade in Endangered Species of Wild Fauna and Flora (CITES) was negotiated in 1973. As a result, CITES regulates the export and import of wild specimens and product derivatives of plants and animals that are listed. (All the species that are listed have been chosen as a result of votes by member nations.) Although the list is not identical with the U. S. Fish and Wildlife Service's list of endangered species, considerable overlap exists. Thus in accord with CITES, it is illegal to import living specimens of endangered species (such as the orangutan) for sale in pet shops in the United States. Likewise, it is illegal to import tiger-skin coats for sale or to export products that are made from American crocodiles. Although more than 70 nations are now active participants in the convention, many nations (some of which are major importers or exporters of live animals or their product derivatives) still have not signed the agreement. Moreover, as with all laws, the convention does have some loopholes, and it is impossible to prevent totally the very lucrative, illegal trafficking in endangered species. Nevertheless, the treaty has significantly diminished the amount of endangered species and their products that are traded in international markets.

Captive Breeding

Some species have been reduced to such small numbers that habitat preservation, by itself, will not save them, so more intensive management of those species is needed. One strategy that is used in those instances is to take eggs from the nests of endangered birds and hatch the eggs in captivity. Such attempts have been successful with the California condor and several species of cranes. Captive breeding in zoos, animal breeding parks, and research centers has also been attempted with some success. For example, the Arabian oryx became extinct in the wild by the mid-1960s, and the remaining members of that small, almost pure-white antelope species existed in only a few zoos in the United States and the Middle East. Thus captive breeding was attempted and, fortunately, it was successful enough so that a few animals have been released in reserves in the Middle East. But although some species (such as lions) breed so prolifically in zoos that they must be sterilized or put on birth control pills, many endangered species appear to have exacting and usually unknown requirements for mating. For example, cheetahs, pandas, bats, cranes, and penguins are not inclined to breed in captivity. Although this problem has been circumvented successfully in whooping cranes by artificial insemination, that technique has not been successful with pandas.

A major goal of captive breeding programs is the eventual release of the species back into the wild. But those individuals that are released are especially vulnerable. They must face all the dangers that occur in small populations; in addition, since they have been raised in captivity they frequently fail to develop vital behavior patterns, such as those necessary for finding food. Hence when left to fend for themselves in the wild, many animals die soon after their release. For example, over 200 pen-raised peregrine falcons have been released at several sites in the eastern United States, but only a few (less than 10 percent) have survived to breeding age.

The biggest drawback to captive breeding is that only a very few of the hundreds of endangered species can be given enough time and resources to address their needs. Hence some conservationists argue that the time and money that is devoted to saving individual species would be much better spent on habitat preservation. They believe that habitat preservation would help many more species in return for the same amount of effort. Still, because the natural habitats of many endangered animals are quickly disappearing, zoos and animal breeding parks may be the last refuges for some wildlife species.

Plant Reestablishment

Similar efforts are under way to save specific endangered plant species. In many areas, special arboretums and gardens are being established, where the habitat requirements of endangered plants can be studied, so that, eventually, those plants might be reestablished in the wild. For example, at the Waimea Arboretum in Hawaii, biologists are studying tropical and subtropical plants in the hope that those plants can be preserved for future generations. In addition, the International Board for Plant Genetic Resources, which is funded by several agencies of the United Nations is establishing regional seed banks to collect and maintain samples of traditional crop strains, which will be made available to crop breeders.

However, these efforts are restricted by the same types of problems that are encountered in captive animal breeding. For the vast majority of species, botanists possess too little information to know how long and under what conditions of temperature and humidity the seeds can be stored without losing their viability. And accidents in storage facilities (such as power failure, fire, or unwitting disposal of seeds) are inevitable. Moreover, the storage of seeds curtails the natural evolution of the plant species. In the meantime, in the species' native habitat, the physical conditions continue to change and animal and other plant species continue to evolve. Hence, with time, plants grown from stored seeds become

less fit for their native habitat. Hence the reasons in favor of promoting habitat preservation seem to keep reappearing.

The Endangered Species Act of 1973

Probably the single most important relevant legislation is the Endangered Species Act of 1973. In addition to providing guidelines for identifying endangered species, the act also seeks to identify threatened species early, so that preventive action can be taken before such a species reaches its minimum viable population size. Another section of that act implements CITES. (The United States was the first country to ratify this treaty.) And a third section outlines procedures for designing recovery plans for endangered species. As a result, by 1985, 164 recovery plans had been developed and approved for 195 species. Those plans include a variety of procedures, including (1) monitoring population status and trends, (2) reducing losses caused by poaching, road-kills (collisions with cars and trucks), and other human-caused mortality, (3) preserving or improving habitats or both, (4) establishing additional population(s) outside the current population range, and (5) maintaining captive breeding programs.

Probably the most significant part of the act is Section 7, which prohibits any federal agency from funding or carrying out a project that threatens the existence of—or alters the habitat that is critical to—the survival of an endangered species. That clause has been the most controversial. For example, it set the stage for the battle to save the endangered snail darter by stopping the construction of the Tellico Dam on the Little Tennessee River, in Tennessee, in 1977. As a result of that confrontation, in 1978, Congress amended the act to create an exemption process to Section 7, which included the establishment of a cabinet-level committee to review requests for exemption. The committee has reviewed only two cases since 1978. In the Tellico Dam controversy, it ruled that the dam should not be exempted from the act. (Subsequently, Congress passed a law that gave the dam special exemption.)

In the second case, the proposed construction of the Graylocks Dam and reservoir threatened the habitat of the whooping crane on the Laramie River in Wyoming. In that case, the committee granted a special exemption for the construction of the dam in accordance with a simultaneous court settlement that required the establishment of a fund to support a habitat for whopping cranes on the Platte River in Nebraska. Although the Tellico Dam controversy received considerable attention nationwide, virtually all conflicts between endangered species and public projects have been resolved quietly as a result of modifications to the project or the development of reasonable alternatives.

Other Federal Actions

In addition to the Endangered Species Act, other federal actions have helped endangered species. During the 1970s, the widespread use of several chlorinated hydrocarbon pesticides (DDT, aldrin, dieldrin, heptachor, and chlordane) was banned. As a result, birds of prey (such as the bald eagle, brown pelican, and osprey) are staging a comeback. For example, by 1981, the population of bald eagles in northwestern Ontario had nearly tripled its numbers from its lowest ebb in 1974. During that same period of time, residues of DDE found in eagle eggs had declined significantly. Similarly, increased reproduction and decreased DDE contamination has been reported for the populations of ospreys on Long Island and in Connecticut.

In 1977, President Carter signed an executive order that strengthened the existing controls on the release, escape, or establishment of foreign species into natural ecosystems.

What of the Future?

A few species have taken a step back from the brink of extinction. For example, since the time that the American alligator was given protection from hunting, it has rebounded so well that it once again can be legally hunted in several parts of its range. The whooping crane population which numbered 15 in 1941, now has more than 100 members in the wild and over 35 in captivity. And birds of prey are now appearing in increasing numbers and in the same areas from which they had disappeared just a few years ago.

Yet the numbers of many species are still so precariously low that their long-term survival remains in serious doubt. Moreover, at this time, researchers know too little about most endangered species to confidently implement recovery plans. Most recovery plans are on a "learn-as-you-go" basis.

Human attitudes remain the biggest threat to the survival of many endangered species. The continued efforts to weaken or repeal the Endangered Species Act of 1973 provide ample evidence that some persons do not give a high priority to species preservation. Because Section 7 of that act can jeopardize any development project that threatens endangered species, special-interest groups (such as for utilities, mining, construction, and land development) have lobbied strongly to change the law. Those groups want to ensure that their projects will not be delayed or prohibited by concern over the extinction of an endangered species. Although the act remains fairly well intact, the 1978 and 1983 amendments have reduced its effectiveness in some areas. Perhaps the greatest impact has occurred because the listing of new species has slowed down considerably. The amendments have added various procedural requirements and have reduced the time limit for listing endangered species to 1 year. As a result, once a listing is proposed, the Department of the Interior has only 12 months to accomplish its required procedures and reach a decision. Although such requirements prevent administrators from dragging their heels to reach a decision, they also make it difficult for the Fish and Wildlife Service to obtain the information that it needs to fulfill its own procedural requirements within a short period of time. A more subtle political impact is reduced funding, which results in an understaffed agency that does not have enough resources to carry out its legal mandates.

There is no doubt that the development of land will continue to put pressure on plants and wildlife. Therefore, if habitat preservation is accomplished, economic development will be prevented on some lands. But, by and large, affluent countries could give species preservation a high priority without imposing serious economic hardships on its citizens, assuming that they can muster the political will to do so. Unfortunately, the same is not true for the poorer countries.

On a worldwide basis, the forecast for endangered species is, indeed, bleak. For example, in tropical and subtropical nations of Africa, South America, and Asia, burgeoning human populations are putting tremendous pressures on wildlife habitats—and the worst is yet to come. Already short of food, land, water, and firewood, many of those impoverished nations face spectacular human population growth in coming years (see Chapter 16). In East Africa, the home of such famous wildlife areas as the Serengeti Plains and the Ngorongoro Crater, the human population is expected to double from 145 million to 290 million within less than 25 years. Hence many conservationists urge that action be taken quickly if the wholesale loss of species is to be avoided.

Governments of developing nations face many difficulties in their efforts to save endangered species. Many do not possess the monetary resources or enough properly trained persons to set aside and manage large preserves for wildlife. Furthermore, those countries often lack the means to enforce ex-

---

BOX 14.3

"THE FACT REMAINS THAT WE HAVE CONSERVED THE WILDLIFE": PROGRESS AND PROBLEMS IN TANZANIA

African wildlife is in dire trouble. An estimated 10 million large mammals remain on the continent's great savannahs and in its broad river valleys. But human populations, which have almost doubled since the early 1960s, are likely to double again to almost 850 million by the turn of the century. Already Africa contains 21 of the world's 45 poorest nations, and by the year 2000, Africans are expected to be able to provide just 65 percent of their own food. Pressed to feed themselves, these impoverished people are pushing wild creatures and wildland to the limit. For both animals and people, it is a desperate struggle for survival.

Nowhere are the issues as clearly crystallized as in Tanzania, and no country better represents the dilemma facing Africa. It is the country which has placed the largest wager on conservation and it is the country with the most to lose. Perhaps a quarter of Africa's large mammals are found within Tanzania's borders, particularly in the 300 miles stretching from Lake Victoria, the continent's largest lake, to Mt. Kilimanjaro, at 19,340 feet, Africa's highest peak. Included in this area are the legendary Serengeti Plain, Ngorongoro Crater, Olduvai Gorge and the Great Rift escarpment beside Lake Manyara. The total mammal population of Serengeti alone exceeds 4 million, including 30 species of hoofed animals and 15 major species of predators.

Since 1961, the first year of Tanzania's independence, President Julius Nyerere has emphasized conservation. "The survival of wildlife," he proclaimed in a document known as the Arusha Manifesto, "is a matter of grave concern to all of us in Africa. These wild creatures amid the wild places they inhabit are not only important as a source of wonder and inspiration, but are an integral part of our natural resources and of our future livelihood and well-being. In accepting the trusteeship of our wildlife, we solemnly declare that we will do everything in our power to make sure that our children's grandchildren will be able to enjoy this rich and precious inheritance."

These commendable words were soon followed by commensurate deeds. Before independence, Tanzania had only one national park. Since then 10 more have been established along with 18 game reserves and 49 game-controlled areas—all in all, almost 30 percent of the nation's land area. Moreover, Tanzania spends approximately 4 percent of its annual budget on conservation. On a per capita basis, Tanzanian conservation funding is eight times that of the United States.

But noble efforts of the past may not prove enough to secure the future. The condition of Tanzania's agriculture has immediate implications for the nation's wildlife areas. Declining agricultural productivity, especially since the resettlement in 1975 of half the country's citizens into collective villages, assures recurrent food shortages and occasional famines. Almost 90 percent of the population is engaged in hand-tool subsistence agriculture on less than 20 percent of the land. Moreover, the arable areas—concentrated around Lake Victoria, across the northern tier toward Mt. Kilimanjaro and along the Indian Ocean coast—include some of the nation's most spectacular national parks and game reserves.

For the moment, demand for more farm and pasture land is being held in check, but it may not last. "There is enormous pressure from people for more land," says Solomon ole Saibull, 47,

---

isting poaching laws or to prevent illegal commercial traffic in products that are derived from endangered species.

Undoubtedly, the most difficult problem is the establishment of priorities. For example, few persons are interested in saving wildlife if they are hungry and do not have enough fuel to cook the little food they do have. Hence, food and wood production must be enhanced outside wildlife preserves. In addition, the wildlife and vegetation that reside in the preserves but are not endangered also will have to be made more accessible to humans as sources of food and fuel. Otherwise, we will have no chance of reducing the loss of endangered species. Box 14.3 presents the progress and problems of saving wildlife in Tanzania—the home of the Serengeti Plains.

---

the conservator of the Ngorongoro Conservation Area, "and the pressure is genuine."

In the long run, says Saibull, the greatest threat to East African wildlife will be a man with a hoe, not a man with a gun. For this reason, he warns, "If we want to save our wildlife, agricultural productivity is the only answer. Their future will ultimately be decided by the effectiveness of the Ministry of Agriculture, not the Ministry of Natural Resources."

Given [the] variety and complexity of difficulties confronting Tanzanian wildlife, it is not surprising that the country's wildlife officials hold a diversity of points of view on the important issues. In the main, their individual attitudes reflect, as in the cast of their counterparts in more developed countries, the goals of the institutions which they serve.

In Tanzania, six agencies, which report directly or indirectly to the Ministry of Natural Resources and Tourism, have an explicit stake in the future of the country's large mammals: Tanzania National Parks (Tanapa), the Ngorongoro Conservation Area Authority, the Division of Tourism, Mweka (the College of African Wildlife Management), the Division of Wildlife, and the Tanzania Wildlife Corporation (Tawico).

The first two, Tanapa and the Ngorongoro Authority, are, like the U.S. National Park Service, decidedly "preservationist" in outlook. Their representatives are clearly embattled. "The parks should not be an economic factor," says Tanapa's David Babu. "But how do you convince the politicians of the needs of wildlife? They'll look at you as a crazy man." Ngorongoro's Solomon ole Saibull adds: "My opponents like to say that I love animals more than people."

Tourism officials typically side with the preservationists, for the parks and reserves are recognized as one of Tanzania's principal tourist attractions.

Representatives of Mweka and the Wildlife Division, like many U.S. academics and the U.S. Forest Service, generally talk in terms of "multiple use," tending to give more weight to "utilizationist" arguments. "It's not fair to tax people to protect animals, which have no direct benefit to them" says Mweka's Hermann Mwageni.

Tawico, the wildlife corporation, anchors the camp of "utilizationists." Its spokesmen propose making wildlife pay for itself. In addition to overseeing the licensed sport hunting of more than 1500 game animals a year, by which Tawico earns an average of $13,000 per hunting client, it is authorized to conduct its own "control safaris"—hunts of game for meat, skins, and live sales to zoos—which earn almost $1.5 million a year. At present, Tawico crops approximately 2000 zebras a year, each worth $450 per hide and $150 in meat.

So where does all this leave the horizonless herds of Serengeti wildebeest, the ponderous rhinos of Ngorongoro Crater, the lions that doze in the trees on the shores of Lake Manyara?

Feroz Kurji, 33, a Tanzanian graduate student in land-use policy, speaks with informed passion about his country's wildlife efforts: "Never forget that conservation is ultimately a question of political will, not economics. And I believe that Tanzania's political will has been amply demonstrated. It may look stumbling and bumbling, but there is a coherence to it. Perhaps we have made mistakes in the last 20 years. Our performance may have been poor by Western standards. But the fact remains that we have conserved the wildlife."

Excerpted from David Abrahamson, "What Africans Think About Wildlife," International Wildlife (July/August), 1983. © 1983 David Abrahamson. All rights reserved.

As a final consideration, persons in developed nations should realize that they also contribute to the loss of habitats in developing nations. For example, consumer demands in the United States create a market for such products as wood veneer and plywood, which are supplied from tropical forests. And during the last 25 years, more than one-quarter of the tropical rain forests in Central America have been cleared to provide grazing land for cattle. Almost all the beef produced on that land was imported by U.S. fast-food restaurant chains.

Conclusions

Through its diverse and pervasive activities, our species is probably the single most important force in the process of natural selection today. We are deter- mining the course of evolution on the basis of short-term considerations, and we are proceeding in spite of our ignorance about the full ramifications of our activities. Hence even without knowing it, the human species may well be in the process of selecting against itself. It is true that few persons care about the survival of plants and animals until their own needs for survival—adequate food, clothing, shelter, and health care—are met. But even when basic needs are met and persons are relatively prosperous, most of us are still more concerned about improving our material standard of living than we are about saving endangered species. Also, policies that are based on short-term economic goals will probably result in serious ecological losses, which, in turn, could seriously diminish the quality of human life. The successful resolution of those conflicting interests should become a high priority for humankind.

Summary Statements

Hundreds of species of animals and plants throughout the world are in danger of extinction.

Each species possesses a unique set of adaptive characteristics that might prove to be useful someday in medicine, agriculture, or industry. Each species also contributes to ecosystem stability.

Other arguments for species preservation include the esthetic value of each species and human responsibility to be faithful stewards of the earth and its resources. In view of short-term economic considerations, however, including the need for jobs, housing, and food, species preservation often ranks low on most persons' lists of priorities.

Millions of species have become extinct since life began on earth because of their inability to adapt to a changing environment. More than 11,000 years ago, many large mammals became extinct in North America as a result of a changing climate and predation by a wave of human hunters that immigrated from Siberia.

During the past two centuries, human activities have caused a sharp rise in the number of species that have become extinct. Humans have caused extinction in several ways: destruction of natural habitats, commercial hunting, and introduction of alien species.

Habitat destruction has made a tremendous impact on populations. Most severely affected are plants and animals that occupy climax forests and prairies and unique habitats, such as bogs and swamps., The pollution of habitats, particularly with insecticides and herbicides, has also taken a toll.

Many species are extinct and others are endangered because they have been overexploited for their furs or hides or because they have been hunted as food, pets, or collectors' items.

An introduced species is sometimes a superior predator or competitor and, therefore, can bring about the extinction or near extinction of a native species. Introduced species have been especially devastating to native plants and animals on islands.

Endangered species often possess one or more of the following characteristics: they are specialists that require particular types of food and other habitat conditions; they are extreme K-strategists; or they have a relatively large minimum viable population.

The best way to prevent extinction of plants and animals is to preserve their natural habitats. As human pressures increase, zoos, animal breeding parks, and arboretums may prove to be the only safe havens for many endangered species.

The outlook for endangered species is mixed. Some species, such as the American alligator, whooping crane, and peregrine falcon, appear to be making a comeback. In general, affluent countries can afford to save the habitats of endangered species, without imposing serious economic hardships, if the citizens have the will to do so. On a worldwide basis, the forecast is, indeed, bleak. Already short of food, land, water, and firewood, a continually growing human population in many less-developed nations has little interest in species preservation.

Questions and Projects

1. In your own words, write a definition for each of the terms that are italicized in this chapter. Compare your definitions with those in the text.

2. What are the differences between endangered species and threatened species?

3. List at least five ways in which a plant or animal species may contribute to the well-being of humankind.

4. Describe how human beings may have contributed to the extinction of large animals in North America 11,000 years ago. How have human activities changed since that time? What is the significance of those changes for the survival of endangered species?

5. Consider the following argument: Conservationists are very naive; they are just not living in the real world. When it's a question of, "What's more important, me or an eagle?" nobody is going to choose the eagle. Do you agree or disagree? Defend your answer.

6. Consider this argument: Innumerable species of plants and animals have become extinct since life began. Humans are no different from any other factor in the environment that has acted as a natural selection factor. We are just selecting for those plants and animals that can survive in the environment that we have helped to create. Do you agree or disagree? Defend your answer.

7. Describe why habitat destruction is the most important factor today in species extinction.

8. Some persons believe that if a species becomes extinct, another species can be introduced as a substitute and the ecosystem will continue to function as it did prior to the extinction. Comment on the ecological validity of this statement.

9. Go to your city or regional planning office and ask for a current land-use map as well as a map that is 30 to 50 years old. Compare the two maps with respect to residential areas and regions that are used for commercial development, industry, and recreation. How much wildlife habitat has been lost to development? Can you detect any habitat islands? If so, are they being managed to preserve valuable habitats? Are they large enough to support the forms of wildlife that are native to the region? What are your community's land-use planning policies regarding habitat preservation?

10. Human-caused pollution is a relatively new threat. Describe several ways in which pollutants can endanger plants and animals.

11. Explain how a small park in the middle of a large metroplitan area is analogous to a small piece of land that is surrounded by water. What is the importance of habitat islands for the successful management of endangered species?

12. Why are species that live only on islands particularly vulnerable to extinction from the introduction of alien species?

13. Should the United States develop a policy to prevent the importation of all alien species? Discuss the relative merits of importing animals and plants for (a) pet shops, (b) zoos, (c) use in scientific and medical research, and (d) use as a predator or parasite in the control of pests that are already in the United States.

14. Why do ecologists generally believe that extreme K-strategists will become extinct relatively soon, even without human-related pressures?

15. Why is the minimum viable population size an important consideration in attempts to save endangered species?

16. With reference to the characteristics listed in Table 14.1, comment on the potential for survival of each of the following hypothetical animals: (a) a mouse-sized omnivore that gives birth to 30 young each year, tolerates the presence of humans well, and inhabits weedy fields and lots throughout the temperate portion of the northern hemisphere; (b) a predatory bird that is the size of an eagle, which raises only two young every other year, and is found only on a few remote islands in the South Pacific; and (c) a weasel-sized predator that has valuable fur, raises five young every spring, can tolerate the presence of humans, and resides in logged forests of the Pacific Northwest.

17. What method or methods would you advocate as being the best to save endangered species in North America? Defend your choice.

18. Design a realistic multistep approach to preserve endangered species in a heavily populated, impoverished nation. What steps would receive the highest priority?

19. How can the enormous amounts of time and money that are spent on preserving species in zoos and arboretums be justified?

20. What plants or animals in your area are on the endangered species list? Check with your state Department of Natural Resources or

Department of Conservation. What efforts are being made to preserve those species? How can you contribute to those efforts?

Selected Readings

DOHERTY, J.: "Refuges on the Rocks." *Audubon* 85:74–116 (July), 1983. An extensive examination of the many problems that the National Wildlife Refuge System faces in its efforts to save the wildlife legacy of the United States.

EHRLICH, P. R., AND EHRLICH, A. H.: *Extinction.* New York: Random House, 1981. An interesting look at the causes and consequences of the disappearance of species.

EHRLICH, P. R., AND MOONEY, H. A.: Extinction, substitution and ecosystem services. *Bioscience* 33:248–254, 1983. An examination of attempts to substitute one species for another to preserve ecosystems.

ENDANGERED SPECIES TECHNICAL BULLETIN: Washington, D.C.: U. S. Fish and Wildlife Service. A monthly bulletin that describes the many international, national, and regional efforts to save endangered species.

LENARD, L.: "The Frozen Zoo." *Science Digest* 85:77–79(June), 1981. An interesting look at the use of sperm banks to save endangered species.

MARANTO, G.: "Saving South Florida.' *Discover* 5:36–40 (April), 1984. A look at how flood-control projects have destroyed wetland habitats and drastically reduced wildlife populations in south Florida.

MARTIN, P. S., AND KLEIN, R. G. (EDS.): *Quaternary Extinctions: A Prehistoric Revolution.* Tucson, Ariz.: University of Arizona Press, 1984. A lively debate over the role of humans in the extinction of many large land animals and birds within the past 15,000 years.

MC NAMEE, T.: "Breath-holding in Grizzly Country." *Audubon* 84:68–83 (November), 1982. An account of 1 year in the life of a grizzly bear in the Yellowstone area and of the controversies over management plans for the grizzly bear.

SHAFFER, M. L.: Minimum population sizes for species conservation. *Bioscience* 31:131–134, 1981. An analysis of the concept of minimum viable population and approaches to determining its size.

SIMBERLOFF, D.: "Big Advantages of Small Refuges." *Natural History* 91:6–14 (April) 1982. An examination of the debate concerning how to design refuges for saving endangered species.

SOULÉ, M. E., AND WILCOX, B. A.: *Conservation Biology.* Sunderland, Mass.: Sinauer, 1980. An examination of many topics that are related to species preservation, including the consequences of insularization, captive propagation, and habitat management.

STANLEY, S. M.: "Mass Extinctions in the Ocean." *Scientific American* 250:64–72 (June), 1984. An interesting examination of the factors that may have caused mass extinctions of marine plants and animals.

TANGLEY, L.: Protecting the "insignificant." *Bioscience* 34:406–409, 1984. A look at the difficulties of getting endangered species of invertebrates on the endangered species list.

A slope of once valuable agricultural land near Concord, California, that has just been contoured by tractors to prepare for construction of a housing development. (1978, Barrie Rokeach.)

CHAPTER 15

LAND USE: PATTERN AND CONFLICT

During the late 1970s, environmentalists and resource developers became embroiled in heated arguments concerning land-use priorities in Alaska, because the bedrock is believed to contain vast reserves of oil, natural gas, and metallic minerals. But Alaska also contains the nation's last frontier of pristine wilderness. And unbridled exploitation of those crustal resources probably would cause irreparable damage to Alaska's wilderness and wildlife. Thus conservationists wanted to protect and preserve as much of Alaska's wilderness as possible for future generations, while groups that were interested in mining and petroleum cited our growing needs for crustal resources and argued for freedom of exploration. Also, many Alaskan citizens argued that the federal government should not "lock up" their state's resources and meddle in private-sector development of their land.

After much Congressional debate, in December, 1980, a compromise of sorts was struck when the Alaska National Interest Lands Conservation Act (ANILCA) was passed. As a result of that law almost 43 million hectares (105 million acres) of land were placed under federal protection as national parks, wildlife refuges, and units of the national wilderness preservation system. That compromise left nearly 9 million hectares (22 million acres) less land protected than was originally proposed by environmentalists. Furthermore, ANILCA opened the coastal plain of the Arctic Wildlife Range to oil and gas exploration. All this has left the majority of Alaskan citizens feeling less than satisfied. They generally view ANILCA as being unduly restrictive of their state's options for land use, which will probably trigger further debate on the issue in the future.

However, the Alaskan problem is just one of many conflicts that arise as a result of our multiple demands for the use of land. We are constantly changing the land by modifying it and reshaping it for housing sites and industrial and commercial purposes. We terrace hillsides, drain and fill in wetlands, cover the ground with concrete, asphalt, and water that is impounded by dams, and whittle out campsites and playgrounds. We grow our food and fiber on land, and we build our great cities on it. Furthermore, we tear up the land, extract its rock, mineral, and fuel resources, and we use it as a dumping ground for our wastes.

We have carried out those activities with increasing fervor over the years because our population and its demands have continued to grow. But the land and its resources are finite. Already, many conflicts have occurred in terms of the ways in which we use the land, which, in many cases, compete with one another. For example, prime farmland is being gobbled up by urban sprawl and flooded by dams, miners are vying for wilderness preserves and sheep and cattle are grazing on range land and competing with indigenous wildlife. Meanwhile, our search for living space has prompted many persons to build in regions that are prone to flooding, earthquakes, and other natural hazards.

In this chapter, we survey our nation's land-use patterns and consider the current efforts that are being made to manage the land. We also examine natural hazards, their influence on land use, and possible ways of alleviating their impact. Although this chapter is focused primarily on domestic land-use problems, many international land-use conflicts are discussed throughout the book. For example, the impact of land use on the preservation of wildlife in Tanzania is discussed in Chapter 14, and desertification in the Sahel of North Africa is discussed in Chapter 17.

Land-Use Conflicts and Controls

Figure 15.1 portrays the extent of development in the United States, which includes 0.92 billion hectares (2.3 billion acres) of land, including inland water. The pattern of development—which varies from highly developed urban areas to essentially undisturbed wilderness—reflects the intensity of demands that we place on the land. The major demands, or uses, are enumerated in Table 15.1. In many cases, those demands conflict, such as when a freeway is routed across farmland. In this section, we focus primarily on problems that involve public lands, the coastal zone, and community growth.

Public Lands

Several federal agencies regulate the activities on approximately one-third of the nation's total land area (national forests, parks, wildlife refuges, and other *public lands*). However, that regulatory approach did not always prevail. As we noted in earlier chapters, during the first 100 years of our nation's history, our natural resources were exploited without any restrictions at all. At that time, the government handed over enormous parcels of land to anyone who was willing to exploit its minerals, timber, or water. As a result, rapid economic development occurred as well as extensive environmental damage (Figure 15.2). Then, late in the nineteenth century, it became apparent that relentless development of the land was despoiling the environment, and the wisdom of practicing resource conservation was voiced. Many persons felt that the federal government should ex-

Table 15.1 A comparison of land-use activities in the United States in 1980*

ACTIVITY	MILLION ACRES	MILLION HECTARES
Agriculture (1977 Data)		
Cropland	413.0	167.2
Grassland, Pasture, and Range	985.7	399.1
Forest Land Grazed	179.4	72.6
Farmsteads, Farm Roads	10.9	4.4
Total	1589.0	643.3
Urban and Built-Up Areas†	68.7	27.8
National Park System‡	77.0	31.2
Wildlife Refuge System‡	88.7	35.9
Forest Service Wilderness‡	25.1	10.2
Highways (1978)	21.5	8.7
Railroads (1978)	3.0	1.2
Airports (1978)	4.0	1.6
Mining§	5.7	2.3
Other	388.6	157.3
Total	2271.3	919.6

*Modified slightly after the Bureau of Mines Information Circular, No. 8862, 1982. Estimates based primarily on reports and records of the Bureau of Census, the Department of Agriculture, and other federal and state agencies.
†Includes residential, farm, industrial and recreational sites, and highways, railroads, and other transportation facilities within urban and built-up areas.
‡Includes areas designated under the Alaska National Interests Lands Conservation Act, December 2, 1980, Public Law 96-487.
§For the period from 1930 through 1980; includes 2.7 million acres reclaimed.

ercise more prudent stewardship over public lands for the common good. As a result, the first step in that direction occurred when Yellowstone National Park was founded in 1872. Later, during the same year, California took steps to protect Yosemite Valley. Then, 20 years later, the first federal forest reserves were set aside, which were later called national forests, and six national parks were established by 1900.

In the early part of this century, the conservation movement was championed by President Theodore Roosevelt, who was concerned about the cost of widespread mismanagement of the nation's natural resources. In 1908, Roosevelt called a White House Conference on National Resources, and one outcome of that meeting was the appointment of a 50-member National Conservation Commission. Gifford Pinchot (who had been appointed Chief Forester of the Department of Agriculture in 1898) along with Roosevelt, persuaded Congress to enlarge the nation's forest reserve holdings nearly fivefold. That major goal was accomplished in spite of vigorous

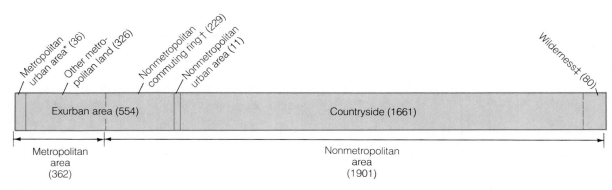

Total U.S. land area is 2,264 million acres

*Metropolitan urban area land includes 12.1 million acres in central cities and 24.2 million acres in suburban areas.

†Nonmetropolitan counties with 10% or more of resident workers commuting to metropolitan areas.

‡Wilderness includes land in the National Wilderness Preservation System only.

Figure 15.1 *The intensity of development of lands in the United States (in millions of acres) as of 1980. (From* The Conservation Foundation, State of the Environment, *1982, p. 293.)*

objections by ranchers, miners, and others, who were accustomed to having free rein over public lands.

Pinchot's conservation philosophy was the guiding principle for federal management of public lands throughout much of this century. He favored resource development over preservation, and he advocated development of the nation's resources for the public good. However, he believed in minimizing the environmental impact wherever it was possible. Until the 1960s, preservation of wilderness areas (unexploited lands) was left to the efforts of small, private organizations, such as the Sierra Club. Then, prodded by public concern over the nation's diminishing reserves of wilderness lands, Congress acted to safeguard them.

As a result, throughout the 1960s and 1970s, increasing amounts of land were set aside for the preservation of wildlife habitats and natural amenities. That movement led to the Alaska National

Interest Lands Conservation Act, which Congress passed in 1980. As noted earlier, ANILCA more than doubled the acreage of federally protected lands (Figure 15.3). But in the early 1980s, under the leadership of President Reagan and his first Secretary of the Interior, James Watt, federal government policy began to shift away from resource preservation and toward resource development. James Watt opened more federal lands to exploration of fossil fuels and other resources, curbed acquisition of new lands, and even proposed to sell public lands to private special-interest groups.

Today, public lands under federal management total almost 325 million hectares (800 million acres)—approximately 1.2 hectares (3 acres) per citizen. More than half of Idaho, Oregon, Nevada, and Utah and more than 75 percent of Alaska are comprised of public land. Based on use, public land falls into two very general categories: (1) regions that are set aside for the primary purpose of preservation, and (2) areas that are intended for regulated multiple use. Areas that are categorized as wilderness areas, wild and

Figure 15.2 *In the past, timber was harvested with little regard for the scars left behind. This land, unprotected by a vegetative cover, is subject to severe soil erosion. (U.S. Department of Agriculture, Soil Conservation Service.)*

scenic rivers, national parks, and wildlife refuges are designed primarily to protect and preserve invaluable natural resources. In national forests and in those regions that are called *national resource lands*, or *Bureau of Land Management* (BLM) *lands*, various activities are permitted. Those areas usually are subject to regulations that are intended to prevent multiple uses from interfering with one another and to ensure that the uses of those areas are compatible with environmental quality. Additionally, some federal lands (15 percent) are used for military installations or Indian reservations.

National Forests

National forests were established primarily to foster both a sustained yield of timber and watershed protection as a result of careful management. National forests are also used for the grazing of domestic cattle and sheep, recreation, and mining as well as for preserving fish and wildlife. The management of those activities was first mandated in the public interest

Figure 15.3 *In 1980, 105 million acres (43 million hectares) of Alaskan lands were placed under federal protection. Those additions more than doubled the total area of the nation's parks, wildlife areas, and wilderness. (U.S. Department of the Interior, National Park Service, and U.S. Department of Agriculture, Soil Conservation Service.)*

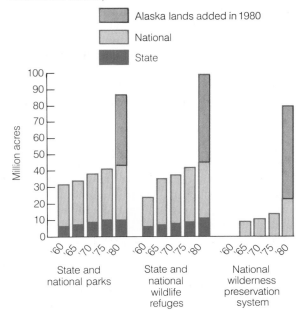

when the Multiple Use–Sustained Yield Act of 1960 was passed. Then, more stringent guidelines for management became law in 1976, when the National Forest Management Act was passed. That law requires that measures be taken to maintain plant and animal diversity in national forests, to minimize soil erosion, to protect water flow and quality, and to achieve a balance among competing uses.

Although the goal of the U.S. Forest Service is to see to it that the various activities that are allowed proceed harmoniously, it is seldom possible to realize that goal fully. For example, if too many people tramp or drive through national forests, the natural amenities that drew them there in the first place can be destroyed. If livestock are allowed, they can outcompete deer, elk, and other wildlife for available forage. And mining can undo all the gains that are accomplished in soil conservation as a result of sound management.

Another challenge that faces our national forests is the rising world demand for wood products. In 1970, domestic timber was growing faster than it was being removed. But since then, the gap between supply and demand has been closing, and some forestry experts have predicted that by the year 2000, the domestic demand will exceed the supply. However, it now appears that such predictions are not justified. Analysts in both the private and government sectors underestimated how much the supply would benefit as a result of both sound timber management and advances in wood technologies, which permit greater use of waste wood and hardwoods. Also, analysts overestimated the future domestic needs for wood, especially for new home construction.

Encouraged by that turn of events, in 1983, the Congressional Office of Technology Assessment (OTA) proposed that the United States play a greater role in meeting the rapidly growing world demand for wood products. Although the United States has been a net importer of wood products during the past 30 years, the Office of Technology Assessment projects that the United States could become a net exporter by 1990 if current trade barriers are decreased.

If the policy in the United States moves in that direction, the U.S. Forest Service will probably be pressured to adopt more intensive timber management practices, so that forests will yield more wood per hectare. Half of all the softwood (evergreen) forests in the United States are on public lands and, typically, those national forests yield less wood per hectare than private commercial forests. Production could be increased in those forests by applying more pesticides and fertilizers and maintaining monoculture forests by using controlled burning to remove competing vegetation. However, such practices could also jeopardize the wildlife, whose protection is a major objective of the U.S. Forest Service. Increased soil erosion, reduced water quality, and decreased recreational opportunities are also probable results of establishing such simplified ecosystems. Furthermore, monoculture forests run a greater risk of being decimated by insects or virulent pathogens.

Although the concept of multiple use has considerable appeal, we have seen that there are many practical management problems. Managing for high timber yield may conflict with wildlife management aims. Moreover, as national needs for one particular resource (such as timber or coal) rise, the demand to exploit that resource further will make it more difficult to successfully manage the other resources that exist on public lands. Such conflicts are inevitable in the future and require that citizens keep well-informed on the resource-management priorities of public lands.

National Resource Lands

Regulations of land use are not always as strict on the nation's vast *national resource lands*, which comprise 40 percent of the total public lands, as they are in national forests. Those lands, which are scattered throughout the nation, have been used mainly for grazing and mining under the relaxed management of the Bureau of Land Management (BLM). However, many of those lands have deteriorated considerably because of overgrazing and, especially in the West, damage to ground cover by motorcycles and dune buggies. In an effort to halt the steady degradation of land that is controlled by the Bureau of Land Management, Congress passed the Federal Land Policy and Management Act of 1976 (also known as the BLM Organic Act), which strengthens the Bureau's authority in protecting and managing national resource lands for multiple use. Thus in 1978, the Bureau of Land Management issued strict new regulations on grazing. And, with similar intent, in 1977, President Carter signed an executive order that curbs the use of off-road vehicles on public lands.

National Parks

National parks are regulated to preserve a specific natural feature, or set of features, and to ensure that the educational and recreational needs of the public are met. The National Park System also includes historic monuments, national lakeshores, and national seashores. Today, the National Park System encompasses almost 32 million hectares (80 million acres) and includes 48 national parks, 78 national monuments, and 88 national historic sites and parks. Delaware is the only state that does not contain an area in the National Park System.

More than 20 years ago, the National Park Service began a vigorous campaign to attract visitors to national parks, and it built new camping facilities, trails, and access roads. The public's response was overwhelming—persons visited the parks in droves, campgrounds became overcrowded, traffic was congested, and the noise all but drowned out nature's tranquility. Altogether, 243.6 million persons visited national parks in 1983, an increase of 300 percent in 25 years.

Today, however, the pendulum is swinging the other way, and the original preservation function of our national parks is being reemphasized. Although visitors are still being encouraged, their numbers are being controlled more closely. For example, reservations may be required for campsites, and visitor centers and campgrounds are centralized (to minimize recreational sprawl). In many parks, the traffic problem has been eased by free shuttle buses, as illustrated in Figure 15.4. In Yosemite Valley, the National Park Service has reduced automobile traffic and, eventually, plans to eliminate it altogether.

Wilderness Areas

In 1964, Congress established the National Wilderness Preservation System and, by 1982, almost 32 million hectares (80 million acres) of our nation's public land was designated as wilderness. Approximately 70 percent of that land is situated in Alaska under the provisions of ANILCA. *Wilderness areas* are defined as portions of national forests, parks, and wildlife refuges that are preserved in their original, primitive, undeveloped state. Thus timbering, most commercial activity, motor vehicles, and human-made structures are prohibited in those areas. Then, in 1968, scenic or wild rivers also became protected

Figure 15.4 *A shuttle bus in Yosemite National Park. Yosemite receives 2.5 million visitors each year, and its buses reduce traffic congestion on the park's roads. (U.S. Department of the Interior, National Park Service.)*

by law. The National Wild and Scenic Rivers System aims to block development on, or along, selected rivers that either have particular esthetic or recreational value or still exist in a natural, free-flowing state. By 1981, portions of 54 rivers were included in the system, covering approximately 11,000 kilometers (7000 miles).

Currently, our public lands also include *National Wildlife Refuges*, which protect the habitats of threatened or endangered species. In 1983, there were 769 areas covering 35 million hectares (89 million acres), including areas of the U.S. Fish and Wildlife Service.

Certain human activities are permitted in wilderness areas as long as they do not infringe on the fundamental objectives of preservation. For example, ranchers may allow their herds to graze in some wildlife refuges and wilderness areas, and public access is provided for recreation. But use of wilderness areas is rising steadily and, in some regions, overuse by human beings is threatening to destroy the fragile vegetation, seriously disturb wildlife, and make solitude impossible. Unfortunately, traditional methods of controlling use—for example, by paving trails to direct foot traffic and cut soil erosion, constructing fences, and planting vegetation that resists trampling—are inappropriate in wilderness areas. Thus in some wilderness areas, the number of overnight campers is being limited and reservations are required.

The management of land is a particularly vexing problem in wilderness regions. Although most of us probably would agree that we have a right to enjoy nature's pristine wilds—its serenity and natural beauty—our presence, even in small numbers, disturbs and even threatens to destroy those qualities that lie at the very essence of wilderness areas. And although preservation of those lands is a serious responsibility, we may have to accept some deterioration of them as an inevitable consequence of our presence—in the same way that we accept some air and water pollution as an inevitable by-product of our industrial way of life.

Although human activities are stringently controlled in wilderness areas and on other public lands, mining is a glaring exception. In essence, under current mining laws (which are discussed in Chapter 12), the federal government exercises little control over mining sites that are located on most public lands. However, although mining is still allowed in wilderness areas, after 1984, the filing of new claims was no longer permitted.

Until the 1984 cut-off date, a person only needed to file a claim in the local county courthouse to mine nonfuel minerals (such as iron, copper, gold, or uranium) that were discovered on public lands. Then, the miner could maintain the right to that claim indefinitely by only demonstrating that at least $100 worth of work was accomplished in working the claim each year. Furthermore, the government does not exact royalties on such claims. However, government control is much stricter for fuels, such as oil, gas, and oil shale, in which case the federal government oversees the development by leasing land to developers and sharing royalties with the states. Despite such regulation, any mining activities can radically interfere with the other purposes of public lands.

Our public lands are coming under increasing scrutiny by both conservationists and commercial resource developers. Conservation groups are pressing for expansion of protected areas and the tightening of use regulations within areas that are already under federal management, whereas developers, pointing to our needs for fuels, timber, minerals, and water resources, are pressing for the easing of restrictions. Public lands hold perhaps half of the nation's undiscovered oil and gas; approximately 10 percent of domestic coal production is from public lands and this figure is expected to climb to 25 percent by 1990; and approximately 54 percent of the nation's softwood timber inventory is on public lands—mostly in the Pacific Northwest. Since both conservationists and resource developers address real needs, new public land use policies will be required. This, like other public issues, ultimately will be resolved in legislatures and the courts.

The Coastal Zone

Today, increasing numbers of persons are beginning to realize that coastal zones—estuaries, marshes, and other wetland habitats—sustain aquatic life, which is irreplaceable and all too fragile. But the coastline was not always a focus of public concern. Until recently, coastline areas have been relentlessly and irreversibly modified by uncontrolled and poorly directed development. More than half our population lives along the 160,000 kilometers (100,000 miles) of coastline that border the Atlantic and Pacific Oceans and the Great Lakes, and the coastal population is growing at three times the national rate. The sheer density of the population alone has stressed the coastal environment in countless ways—for waste disposal, recreation, the siting of industry, homes, and power plants, and the exploitation of mineral, natural gas, and oil resources. As a result of such activities, in just 9 years—between 1955 and 1964—California lost 67 percent of its estuarine habitats. And in a recent 2-year period, so many requests for permission to modify the shoreline of the Chesapeake Bay were filed with the U.S. Fish and Wildlife Service that, if they had been granted, 190 kilometers (120 miles) of shoreline would have been altered seriously. As of 1983, the nation's wetlands were disappearing at a rate of more than 180,000 hectares (450,000 acres) per year.

A particularly vexing problem in many coastal areas is the rate of shoreline erosion, which has been accelerated by shortsighted human disturbances of the natural processes of beach formation. A 1983 estimate of annual shoreline erosion rates is 1 meter along the Atlantic coast, 2 meters along the Gulf coast, and up to 3 meters along the Pacific coast.

Sand is supplied to beaches by the combined action of waves and rivers that empty into the sea. Normally, as rivers empty into the sea, they deposit enormous quantities of sediment. Although a portion of that sediment may build up a delta, another portion is transported along the shoreline by *long-shore currents*, which result from coastal wave action. Long-shore currents are produced by sea waves that approach the shore at an angle, which produce a current that flows parallel to the shore. That current transports sand (delivered to the beach by rivers) along the shore and ensures a continual supply of sand for coastal beaches (Figure 15.5).

However, the long-shore flow of sand can be interrupted by artificial structures. For example, when a river is dammed upstream, it eliminates the source of sand for the beaches downstream. Then, long-shore currents transport the sand away from beaches until nothing but rocky rubble is left behind. Also, in some coastal areas, rock barriers or jetties are constructed so that boaters can enjoy calm waters for docking; these structures reduce the force of the waves and decrease long-shore sand transport. As a result of the reduced wave action, the amount of sediments increases in those areas, and the harbors become choked with sediments, which must be dredged. Also, because most of the sediments are deposited in the harbor, they become unavailable for deposition on down-current beaches. Thus in regions where breakwaters have been constructed (such as in Santa Barbara, California), it has become necessary to pump sand from the upstream to the downstream sides of the harbor. In effect, energy-consuming pumping must be used to transport the sand, which was formerly accomplished by natural long-shore currents.

Powerful ocean storms compound the hazards of living along coasts, because the driving force of waves and water undermine seaside roads, homes, and businesses. However, the greatest danger exists for the residents who live on the nearly 300 *barrier islands* that fringe portions of the nation's Atlantic and Gulf coasts. Barrier islands protect coastal beaches, wetlands, and estuaries by dissipating the energy of approaching storm waves. Thus sea waves dissipate their energy by shifting the sands that compose barrier islands and, in the process, change the shapes of the islands. However, when barrier islands are developed for human habitation, the sands are stabilized and are less able to absorb (or dissipate) the energy that is produced by storm waves. For decades, barrier islands have undergone rapid development and urbanization; and some coastal cities, including Miami Beach, are built entirely on barrier islands. Federal, state and private agencies have preserved only approximately one-third of the total land area of the nation's barrier islands from development.

Population growth on barrier islands worries meteorologists, who note that such exposed locations are particularly vulnerable to the ravages of tropical storms and hurricanes (Figure 15.6). Much of that growth occurred during the 1960s and 1970s, which was a relatively quiet period in terms of storm activity. However, weather forecasters now fear that residents could not receive adequate warning to evacuate the islands and avert disaster if a hurricane strike occurred. This issue is examined in greater detail in Box 15.1.

The need to balance the use and preservation of coastal zones has spurred the states that are located on the shoreline to develop guidelines for coastal management. Probably the most stringent of those guidelines is Delaware's regulation, which prohibits the siting of all new industry within 3 kilometers (2 miles) of its shoreline. The federal government is

(a)

Figure 15.5 *Longshore currents transport sand along a beach. (a) Along Westhampton Beach, Long Island, a series of rock jetties (called groins) retain beach sand. (U.S. Army Engineer Waterways Experiment Station.) (b) The sand builds up on one side of a jetty, functioning as a breakwater, and is eroded back on the other side.*

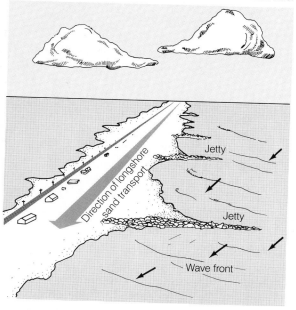

(b)

helping the coastal states to manage their resources under the auspices of the 1972 Coastal Zone Management Act. As a result of that law, all 30 eligible states are now receiving federal funds to help them inventory their coastal resources and set priorities for the orderly use of coastal land and water. And development on barrier islands has been discouraged as a result of the 1983 Coastal Barrier Resources Act, which limits federal funds for bridge, road, and sewer construction and eliminates federal flood insurance for barrier islands.

Another approach to preserving the natural resources of coastal zones includes the designation of *marine sanctuaries*. For example, Point Lobos, California, (which is near Monterey) was founded in 1960 and became the nation's first underwater sanctuary. Since then, California has set aside 10 other sites for sanctuaries and has been studying five more possible sites.

BOX 15.1

BARRIER ISLANDS:
A FOOL'S PARADISE?

At the entrance to the U.S. National Hurricane Center [in Coral Gables, Florida] is posted a weather bulletin for the city of Galveston, Texas, dated September 8, 1900. It advised: "Rain, brisk to high northerly winds becoming variable."

That night, however, a mighty hurricane descended on Galveston, sending a wave of water 20 feet high and 50 miles wide across the city. At daybreak, Galveston was a crescent-shaped pile of rubble with 6000 dead.

Hoping to prevent such a disaster from happening again, scientists at the hurricane center here are stationed within a glowing circle of computer screens 24 hours a day. On some screens, they can spot a cold front moving up the East Coast or a splotchy disturbance in the Caribbean. On another, a satellite spies on storm breeding grounds near the African coast. With more hurricanes striking the United States in September and October than in all other months combined, research planes are poised at a nearby airport to fly into the eye of any big storm.

But for all their scientific hardware, meteorologists at the hurricane center concede that predicting the direction of the big storms is difficult. For instance, when a hurricane is 72 hours offshore, the weather service has only a 10 percent chance of predicting accurately where it will finally come ashore—if it comes ashore at all. Even when a storm is just 24 hours off the coast, scientists say, their predictions are wrong more than half the time.

What's more, they say that the predictions are only slightly better today than they were in the 1950s, that they are no better than they were a decade ago and that they probably won't get any better in the foreseeable future.

Imperfect as it is, the forecasting system worked well in recent years. But while the ability to predict the course of hurricanes has plateaued, the population growth along coastal areas hasn't. From Brownsville, Texas, along more than 3000 sandy miles to Eastport, Maine, the population of coastal counties has increased 34 percent since 1960 to some 40 million people. An even greater rate of growth has occurred on the 1.6 million acres of barrier islands that lie offshore, places where vacation homes and tourist towns sit shoulder to shoulder.

So herein lies the rub: With such enormous growth in the coastal lands, and no real improvement in hurricane forecasting, scientists have concluded that they can no longer predict the landfall of the big storms in time to guarantee that everyone can get out alive.

Hurricane scientists are obviously more concerned with a possible hurricane catastrophe than the public is. Along the nation's barrier islands, the summer homes and seaside towns extend across 300,000 acres today, compared with about 90,000 acres in 1950. Construction continues at 6000 acres a year; development is four times as dense as on the mainland.

The public has a short memory for the darker side of the islands' history, Mr. [Neil] Frank [of the National Hurricane Center] says. He cites these examples:

At Westhampton Beach on Long Island in 1938, a hurricane destroyed all but 26 of 179 homes. Today there are more than 900 homes there.

On Long Beach Island south of Wilmington, N.C., in 1954, Hurricane Hazel destroyed all but five of 357 homes. Now there are more than 1000 homes on the island.

The barrier islands off Georgia and South Carolina were swept in 1893 by a 20-foot-high

Although the federal government has moved more slowly than California, federal legislation holds much promise in this endeavor. For example, provisions of the 1972 Marine Protection, Research, and Sanctuaries Act authorize the President to designate marine sanctuaries in coastal waters of the continental shelf and in the Great Lakes. Any of a number of characteristics may allow a region to be qualified as a marine sanctuary (Table 15.2). However, basically, they are usually regions that have special bio-

wave that killed 2000 people. The islands now are the site of resorts that attract thousands of vacationers.

On most barrier islands, evacuation can be completed quickly. It is in the more populated areas that scientists see far greater risks. Consider, for instance, the Florida Keys, a 150-mile-long archipelago of coral rock rising at most 5 feet out of the sea. The home of 70,000 people, the keys are connected to the Florida mainland only by 42 narrow bridges, the longest of which spans 7 miles of ocean like a piece of string. A Miami newspaper columnist noted of the long bridge:

"Already the smallest emergency puts it in a fearsome snarl. As a hurricane route, the prospects are scary."

The U.S. Army Corps of Engineers recently calculated that it would take 31 hours to evacuate the keys—48 hours counting the time needed to get the process under way. At 48 hours before landfall, a forecast that the storm would actually hit a stretch of the keys would have well under a 20 percent chance of being right, yet this would be the last time a decision could be made to evacuate the people with the assurance that everyone would get off the keys alive.

This is also assuming that there really are 48 hours left. Hurricanes are erratic storms, sometimes surprising the forecasters by arriving several hours early, sometimes looping back on themselves, and often changing intensity.

In the densely populated Miami area, where one hurricane can be expected to hit every 7 years, a hurricane of much lesser intensity could send a 10-foot surge across Biscayne Bay, sweeping across the islands and into downtown Miami. With nearly half a million people in South Florida vulnerable to a big storm, evacuation time itself is estimated at 21 hours, with 6 to 12 hours beyond that required to get the process working.

But Florida isn't the only place vulnerable. Galveston, despite its 15-foot seawall, still requires about 36 hours to evacuate. Studies aren't available for most other regions, but the New Orleans area and the New Jersey shore are just two of many that probably need long evacuation times.

Until this year the hurricane center sent out a hurricane "watch" at 36 hours and hurricane "warnings" at 24 hours and 12 hours. But as evidence mounted that many areas needed 36 hours to clear out, forecasters decided to issue the odds that a storm would hit a particular city.

Under the new system, when a storm is estimated to be 72 hours from the coast, cities and towns in a 700-mile stretch get a one-in-10 chance that the eye of the storm will pass within 65 miles of any one of them. As the storm gets closer, the 700-mile zone is progressively narrowed—to 500 miles at 48 hours, 400 miles at 36 hours, 300 miles at 24 hours, and 100 miles at 12 hours. And the chances that the storm will hit any one city or town within the zone are progressively increased— to one in eight at 48 hours, one in six at 36 hours, one in three at 24 hours, and an almost even chance at 12 hours.

The system of probabilities raises almost as many questions as it answers. Should the decision maker in a town order an evacuation if there is only a 15 percent to 20 percent chance that the town will be hit? Will coastal residents evacuate knowing that the odds are good that they don't really have to?

Mr. Frank is aware that credibility is an issue, but he answers: "In the fairy tale, the boy cried wolf and there was no wolf. But here there is a wolf. It's just a question of whether he'll appear at your door or your neighbor's."

logical, esthetic, or historical significance. Thus the objective of federal legislation is to preserve and protect those areas by managing the multiple demands that are placed on them. Hence those sanctuaries are not, in the strictest sense, places of refuge and

protection for marine life. Indeed, some harvesting of aquatic resources is permitted in them.

The National Oceanic and Atmospheric Administration (NOAA) currently is evaluating 29 potential sanctuary sites. Among those proposed is

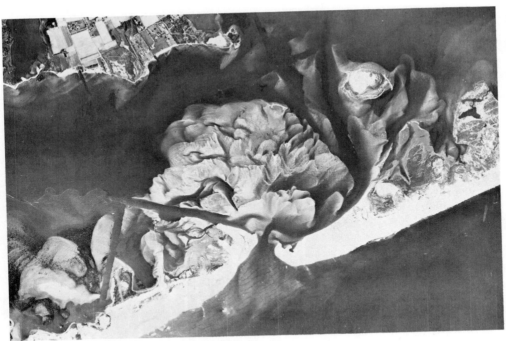

Figure 15.6 *A barrier island (white) that was breached by energetic waves, which were driven by hurricane force winds in September, 1947. (Courtesy of M. M. Nichols, U.S. Department of Agriculture.)*

a humpback whale wintering ground off the coast of Hawaii. An example of an existing federal marine sanctuary is Key Largo in Florida, which consists of coral reefs that are home to more than 500 species of fish.

Table 15.2 Types of marine environments that may qualify for sanctuary status under the 1972 Federal Marine Protection, Research, and Sanctuaries Act

Special Habitats	Essential Marine or Estuarine Communities, Such as Seaweed Forests
Special Areas	Spawning, Breeding, or Nursery Grounds
Research Sites	Locales of Archeological or Historical Importance
Recreational/Esthetic Areas	Areas that Are Pristine
Unique Areas	Areas Featuring Unusual or Spectacular Geology or Oceanographic Characteristics. Examples Include Submarine Canyons; Locales of Extraordinary Tides

Community Growth

A third category of land-use problems stems from community population growth and urban expansion. Perhaps foremost among those problems is the loss of prime agricultural land to development for industry, housing, highways, reservoirs, or other urban-related land use. Of the 500,000 hectares of cropland that is removed from production each year, more than one-half is used for urban and suburban development. That trend conflicts with the future food needs of a growing population, and although food productivity per hectare could be increased, such a course threatens environmental quality, as we will see in Chapter 17.

We may view community growth from two different perspectives—the rate at which land is converted to urban uses and the differential rates at which the human population is growing in urban and rural areas.

Between 1960 and 1980, the portion of the nation's land that was classified as being urban increased by 84 percent. In the 1970s alone, 4.9 million hectares (12 million acres) were given over to urban development—up from 3.7 million hectares (9.2 million acres) during the previous decade. Today, approximately 2 percent of the nation's land is covered by cities, suburbs, and small towns.

Although the land area of urban settlement continues to expand rapidly, the urban population is increasing at a slower pace. For example, from 1960 through 1980, urban population increased by one-third, which implies that urban population density declined. In fact, on a national average, urban population density declined from 1235 persons per square kilometer in 1960 to 890 persons per square kilometer by 1980. That trend is due to a remarkable change in the rates of urban and rural population growth.

Analysis of 1980 national census statistics reveals that the long-standing urbanization (in terms of population) of the United States is declining. In addition, during the 1970s, for perhaps the first time since the nation was founded, the populations in rural areas grew as fast as populations in urban areas. As a result, the percentage of the population who live in cities is no longer increasing.

The U.S. Census Bureau defines an *urbanized area* as a city that has a population of 50,000 persons or more, plus contiguous areas that have at least 386 persons per square kilometer. On that basis, the portion of the United States population in urbanized areas stabilized at 58 percent during the 1970s (Figure 15.7). If we broaden the definition of an urbanized area to encompass small cities and towns (with populations of 2500 to 50,000), the percentage of population in urbanized areas actually declined slightly during the 1970s.

The growth that occurred in rural populations was primarily in the open countryside rather than in small cities and towns, and is generally attributed to economic factors. For example, new industrial plants increasingly have been attracted to more rural sites and, simply put, where the jobs are, the population follows. However, there has been no reversal in the long-standing decline in the number of persons who are engaged in farming—and the increase in rural population places even more stress on our ever-dwindling agricultural lands.

Furthermore, the decline in our rate of urbanization has not curbed the continuing development of supercities. (A *supercity* is defined as a coalescence of adjacent cities and towns, with a combined population of at least 1 million.) The number of supercities in the United States rose from 24 in 1970 to 29 by 1980 (Table 15.3). A southern California

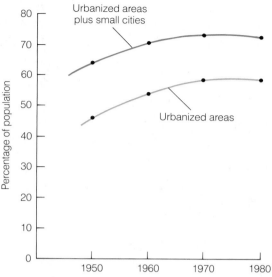

Figure 15.7 *Changes since 1950 in the percentages of the United States' population living in urbanized areas and in urbanized areas plus small cities (From L. Long and D. DeAre, "The Slowing of Urbanization in the U.S."* © Scientific American 249(1):33–41, 1983. All rights reserved.)

supercity is mapped in Figure 15.8. And the construction of expressways that facilitate commuting between suburbs and the inner city has contributed significantly to urban expansion. Those highways have allowed industries to locate on the fringes of metropolitan areas, and they have spurred the growth of suburbs. The original goal of the National System of Interstate and Defense Highways, which was begun during the Eisenhower administration, was to construct a network of highways that would link 90 percent of the cities which had populations that were greater than 50,000. Today, with the system nearly complete, the 70,000 kilometers (43,000 miles) of interstate highways consume 0.73 million hectares (1.8 million acres) of land. Although states and cities enthusiastically participated in the extensive interstate highway construction during the 1950s and 1960s, today, many areas are reevaluating the merits of more new freeways. Those highways crisscross cities (as shown in Figure 15.9) breaking up old, established neighborhoods, threatening the economic viability of central business districts, encouraging the relocation of industries and population to formerly rural areas, and forcing an increased dependency on motor vehicles. Highway construction also

Table 15.3 The 29 supercities as determined in the 1980 census

SUPERCITY	1980 POPULATION	LAND AREA (SQUARE MILES)	POPULATION DENSITY (PEOPLE PER SQUARE MILE)
New York City/Northeastern New Jersey/Southern Connecticut	17,606,680	3,656	4,816
Los Angeles/Long Beach/San Bernardino/Riverside	10,184,611	2,187	4,657
Chicago/Northwest Indiana/Aurora/Elgin/Joliet	7,212,778	1,675	4,306
Philadelphia/Wilmington/Trenton	4,779,796	1,270	3,764
San Francisco/San Jose	4,434,650	1,122	3,952
Detroit	3,809,327	1,044	3,649
Boston/Brockton/Lowell/Lawrence/Haverhill	3,225,386	1,084	2,975
Miami/Fort Lauderdale/Hollywood/West Palm Beach	3,103,729	816	3,804
Washington, D.C.	2,763,105	807	3,424
Houston/Texas City/Lamarque	2,521,857	1,169	2,157
Cleveland/Akron/Lorain/Elyria	2,493,475	989	2,521
Dallas/Fort Worth	2,451,390	1,280	1,915
St. Louis	1,848,590	597	3,096
Pittsburgh	1,810,038	713	2,539
Seattle/Everett/Tacoma	1,793,612	672	2,669
Minneapolis/St. Paul	1,787,564	980	1,824
Baltimore	1,755,477	523	3,357
San Diego	1,704,352	611	2,789
Atlanta	1,613,357	905	1,783
Phoenix	1,409,279	641	2,199
Tampa/St. Petersburg	1,354,249	527	2,570
Denver	1,352,070	439	3,080
Cincinnati/Hamilton	1,228,438	465	2,642
Milwaukee	1,207,008	496	2,433
Newport News/Hampton/Norfolk/Portsmouth	1,099,360	614	1,790
Kansas City	1,097,793	589	1,864
New Orleans	1,078,299	230	4,688
Portland	1,026,144	349	2,940
Buffalo	1,002,285	266	3,768

Source: From L. Long and D. DeAre, "The Slowing of Urbanization in the U.S." © *Scientific American* 249(1):39, 1983. All rights reserved.

has had an adverse effect on the landscape. It has destroyed wildlife habitats, disrupted drainage patterns, and distorted scenic panoramas.

In response to those problems, Congress passed the Federal Aid Highway Acts of 1973 and 1976, which provide cities with an alternative to building more highways. That law allows some federal funds that are earmarked for highway construction to be used, instead, for upgrading urban mass transit systems. However, the federal government covers 90 percent of the cost of highway construction but only 80 percent of the cost of mass transit systems. Hence, from a financial perspective, new highways are still favored.

Nonetheless, in Los Angeles, which has long been a model for the love affair between persons and their automobiles, a movement is underway to construct a high-speed, modern mass transit system to relieve traffic congestion. With an urban population that continues to soar, the freeway system is becoming so overcrowded that freeway speeds decline steadily each year. The proposed initial phase of the mass transit project is a 30-kilometer (18.6-mile) subway that will link the San Fernando Valley to downtown Los Angeles. Proponents see this as the first step in the development of a commuter rail network that would be comparable, at least in size, to the one in Chicago or New York City. Critics of

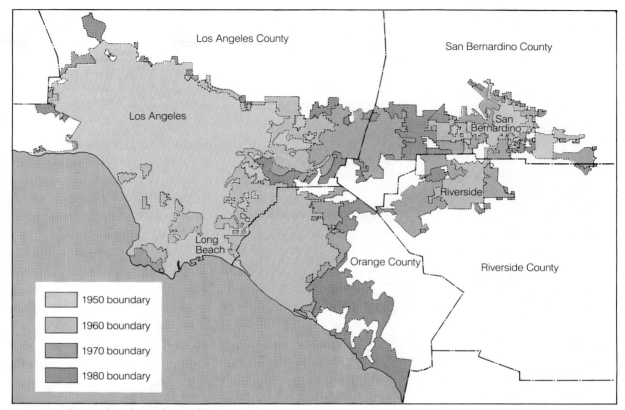

Figure 15.8 *The growth of the southern California supercity, formed by the merging of adjacent urban areas. (1950 boundary—lightest shading; 1960 boundary—light shading; 1970 boundary—dark shading; 1980 boundary—darkest shading.) (From L. Long, and D. DeAre, "The Slowing of Urbanization in the U.S."* © *Scientific American 249(1):35, 1983. All rights reserved.)*

this movement cite the problems—particularly financial—that plague existing urban commuter lines. For example, many of them, such as in New York City and Boston, run deeply into debt each year and are onerous burdens for taxpayers. New equipment generally has been less than satisfactory, with costly breakdowns and delays. And in some cities, subway cars are a target for vandals and thieves, and those threats to the safety of passengers usually necessitate a costly security system.

Growth-Control Strategies

Increasingly, states are turning to their legislatures to solve the conflicts in land use. For example, by the mid-1970s, more than half the states had enacted general land-use programs. Dozens of states now have laws that regulate the siting of power plants and control strip mining. Many states have passed laws to protect wetlands and to identify and protect "critical areas"—regions that are either particularly sensitive to disturbance or have historical importance. And some states even provide tax incentives to encourage landowners to retain their land for agriculture or to preserve it for its esthetic or recreational value.

Furthermore, many towns and suburbs have long employed zoning ordinances to control residential growth. Usually, those regulations specify minimum lot sizes for each home; they regulate the size of structures; and they specify how far structures must be set back from the road. And some communities have growth-control ordinances that limit the number of new dwellings. For example, after its population increased by 25 percent in only 2 years,

Figure 15.9 *Freeways, such as this one in Los Angeles, cut a wide swath through a city and physically separate neighborhoods. (© Georg Gerster, Photo Researchers.)*

Petaluma, California, passed an ordinance that allows the construction of no more than 500 new subdivision homes each year.

A recent growth-control strategy is the *land trust*, in which private citizens donate undeveloped land to a nonprofit conservation agency for a specified period of time, with the stipulation that the land will not be developed. In return, the donor is granted a property tax break. Several communities on Cape Cod, Massachusetts, have adopted this approach as a way of curtailing the rapid rate of development. Cape Cod is a 113-kilometer (70-mile) long arm of sand that reaches into the Atlantic Ocean, and it is the fastest growing section of New England. Between 1970 and 1980, the year-round population of Cape Cod climbed 53 percent, to 147,900. And each summer, that population is swelled by hundreds of thousands of tourists. As a result, shops, fast-food restaurants, motels, cottages, and condominiums are increasingly dotting the Cape.

However, growth-control strategies have been criticized by persons who argue that the regulations create economic barriers to open housing and restrict free enterprise unjustly. Some persons view zoning ordinances to be unwarranted intrusions on property rights and personal freedom. Those disputes are ultimately resolved in the courts.

Urban Planning and New Towns

Virtually every city has a planning commission that typically hears challenges to zoning regulations and decides on the location of highways and other public service facilities. Unfortunately, policy is more often determined by political pressures applied by special interests than by carefully reasoned and visionary planning. And so we see, in most of our cities, costly demolition of houses and other structures before their time to make way for expressways or water and sewer lines.

Many city planning agencies have come under severe criticism for urban renewal projects that favor the well-to-do over the low-income segments of the population. Critics point out that too often slums are cleared only to be replaced by luxury high-rise apartment complexes. Meanwhile, housing opportunities for low-income persons decline, pushing up demand and rents for existing low-income housing. Critics also complain that urban renewal projects have produced sterile, bland, and uninviting buildings, walkways, and public parks.

Although they are generally politically independent, a city and its suburbs are highly interdependent for such public services as police protection, water supply and treatment, transportation, and garbage collection. Many suburbs, in fact, are nothing more than bedroom communities, whose residents are heavily dependent on city services but provide the city with little or no tax support. Hence the most logical approach to planning is for the various component communities of a metropolitan area (or supercity) to coordinate planning through a single authority. Yet very little of this regional scale of planning currently occurs in the United States.

Theoretically at least, some of the problems of existing cities can be relieved by planning and constructing new communities virtually from scratch. Such *new towns* would be self-sufficient, efficient in operation, and protective of the environment. New towns would relieve population pressures on urban areas and permit a more orderly and deliberate approach to the problems of rising housing costs and social integration. Actually, the concept is not new and dates back to the turn of the century, although popular interest in new towns was spurred by the housing needs of post-World War II years.

Among the best known and more successful of the new town projects is that authorized by the Greater London Plan of 1944. New towns were intended to curb population growth and reduce congestion in London. The 1944 plan set limits on population density within London (250 persons per hectare) and established a permanent greenbelt to surround the city. Citizens and industry were encouraged to locate in new planned communities that were situated well beyond the greenbelt (at least 40 kilometers). In subsequent decades, many new towns were planned and constructed (Figure 15.10). Although a unique plan was drawn up for each town, they generally featured green areas (parks and conservancies), efficient traffic patterns and mass transit

Figure 15.10 *Harlow New Town in Essex, Great Britain. (The British Tourist Authority.)*

systems, and placement of homes, businesses, and industry to optimize esthetics and environmental quality. Although the new town approach did reduce congestion in London to some extent, population density increased dramatically in areas just beyond the greenbelt. That occurrence led to a revision of the original Plan so that now all new towns must be located at least 100 kilometers beyond London.

About 5 years after the Greater London Plan was adopted, plans were drawn for building a number of satellite communities near Stockholm, Sweden. Those new towns were also intended to relieve urban congestion, but unlike their English counterparts, they are not self-sufficient. Satellite communities were built in conjunction with construction of new subway lines, thereby encouraging commuting to the city. Typically, each new town consists of several neighborhoods that are clustered within walking distance of a subway station, and each town is located within a 30-minute transit ride to downtown Stockholm. Most residents are apartment dwellers and, in the first new towns, buildings were designed to blend with the natural landscape; subsequent plans, however, exhibit more regular geometric and denser layouts of buildings.

New towns in the United States are quite different from those in Europe in that they typically feature lower population densities, greater emphasis on leisure facilities and automobile transport, and primarily white-collar employment opportunities. Also, median incomes of residents tend to be relatively high. Some new towns were planned near research or educational facilities (such as Irvine, California) or regional shopping centers (such as Columbia, Maryland); other new towns have very specialized purposes (such as retirement communities).

Until the early 1960s, new town development in the United States seldom adhered to a master regional plan. That practice changed with the planning and construction of Reston, Virginia, and Columbia, Maryland. Reston, which is 23 kilometers (15 miles) west of Washington, D.C., is a leisure-oriented community that features much open space and green areas; employment is limited to offices and research facilities. Columbia, which is located midway between Baltimore and Washington, D.C., was originally developed around a large shopping center, but then it attracted substantial industrial employment opportunities.

Although Reston, Columbia, and a few other new towns begun in the 1960s were relatively successful financially, this has not been true for most recent projects. The federal New Communities Program, authorized in 1968 and greatly expanded in 1970, provided grants and loan guarantees with the intent of spurring new town construction that would be sufficient to meet the nation's anticipated population growth of 75 million between 1971 and 2000. However, the program was scrapped in 1978 when it was realized that its original objective was far too ambitious; new towns require an enormous initial investment of funds for planning and construction and most new towns in the United States are experiencing financial difficulties. Also, critics of new towns claimed that they lack the character and spontaneity of older established cities and do little to improve opportunities for low-income people.

Land Use and Natural Hazards

A primary objective of federal, state, and local land-use control strategies is to provide for our resource and recreational needs now and in the future, while minimizing environmental disturbances. However, that goal requires that the land be assessed in terms of what it can and cannot be used for. As we have seen, some lands can be used for multiple purposes as, for example, when a national forest is used for recreation, grazing, and lumbering. However, other lands are severely limited in their use by natural hazards, such as floods, earthquakes, and landslides. Thus comprehensive land-use policies also regulate the habitation of those areas to minimize the risk to life, limb, and property. Such policies must be determined with an understanding of the geologic characteristics of the area in question and the potential hazards that are created by those characteristics.

Flood Hazard

Floods have caused more damage in our country than any other type of natural hazard, and they account for the majority of events that are officially declared disasters by presidents (whereby residents become eligible for federal assistance). For example, during the first 5 months of 1983, President Reagan issued nine declarations of federal disasters—all but one

because of flooding. Floodwater drowns humans and livestock, flattens buildings, erodes valuable topsoil, disrupts municipal water and sewage systems, and interferes with communications, transportation, and commerce. Moreover, as floodwaters recede, they leave thick layers of silt and mounds of debris behind. It is not difficult to imagine the extent of the damage that was sustained from the flood shown in Figure 15.11. The National Weather Service estimated that in 1983 the dollar loss due to floods in the United States was more than $4.2 billion. The average annual loss over the prior 40 years, in contrast, was $2.0 billion. Floods also take their toll in personal injuries and deaths, and that toll is mounting. During the 1970s, flash floods took an average of 200 lives per year, which is twice the number of flood fatalities that occurred during the 1960s and three times those that occurred during the 1940s. In 1983, 204 persons died in floods.

The damages and fatalities that are attributed to flooding are increasing, because increasing numbers of persons are living on floodplains. The fertile soils, plentiful water for irrigation, and the potential for inexpensive transportation has always been attractive to some persons. However, with the shift from an agrarian to an industrial-based society, the attraction has become even stronger. The flat terrains of floodplains are economically attractive as sites for highways, railroads, industrial plants, and homes. And with the steady growth of cities, the populations on floodplains has soared. Today, many cities in the United States (such as New Orleans) are built almost entirely on floodplains. Approximately 15 percent of the nation's total urban land and almost 10 percent of the total agricultural land are comprised of floodplains and, therefore, are vulnerable to flooding whenever a river or stream overflows its channel.

Another factor that has contributed to increased flood fatalities in recent years is the fact that remote areas (where flood warnings are not readily communicated) have become more accessible to campers and other visitors. Thus increasing numbers of persons are driving their cars, campers, and mobile homes into flash flood-prone mountainous areas. Those vacationers set up camps and build cabins in narrow canyons, which are subject to unexpected and rapid rises in stream levels. For example, on July 31, 1976, more than 130 lives (mostly campers) were lost in the Big Thompson Canyon in Colorado, when

Figure 15.11 *Floods such as this one of the Mississippi River in Muscatine, Iowa, can take lives, destroy homes and businesses, and disrupt utilities. (Myron Wood, Photo Researchers.)*

heavy rains caused the water level to rise unexpectedly, causing a flash flood.

To determine whether floods can be prevented or whether flood damage can at least be reduced, we need to understand the causes of flooding. A flood occurs whenever runoff exceeds the discharge capacity of a river channel, flows over the river's banks, and spreads out over the floodplain. That excessive runoff, which is sufficient to cause flooding, is caused by natural events, human activities, or some combination of the two. Hurricanes and intense thunderstorms produce heavy rainfall, which exceeds the infiltration capacity of soil and causes flooding. For example, in the summer of 1972, torrential rains from Hurricane Agnes triggered flooding that claimed 122 lives and caused property damage that exceeded $2.1 billion in the northeastern states. In that same year, 237 lives were lost in Rapid City, South Da-

kota, when exceptionally heavy rains caused flash flooding. Rapid spring snow melts also cause a sharp rise in river levels, which is sometimes compounded by the breaking up of river ice. When that happens, huge slabs of ice pile up at bridges or narrow parts of channels, which cause the water to back up and flood.

Furthermore, as cities grow and expand, new roads and buildings render increasing amounts of land impervious to water, and sewer systems in many cities are unable to accommodate the enormous volumes of water that can accompany, for example, a summer downpour. As a result, basements, viaducts, and other low-lying areas are subject to rapid inundation (Figure 15.12). Another human factor that contributes to flooding is the removal of the land's vegetative cover. Vegetation slows the flow of runoff and thereby reduces the threat of floods. But logging, overgrazing, and mining are notorious for removing vegetative cover and contributing to the hazard of floods. In addition, the failure of dams, levees, and

Figure 15.12 *Automobiles trapped by flash flooding under a viaduct. Urban storm sewer systems may be unable to carry off the waters of a torrential rain. Water accumulates in low-lying areas such as this and may trap motorists. (AP/World Wide Photos.)*

other structures that are designed, ironically, to prevent flooding, is a major cause of calamities that are caused by floods.

Until recently, our nation's flood-control efforts consisted almost entirely of engineering projects that were designed to keep floodwaters out of populated areas. In that approach, structural flood-control measures are matched to the specific characteristics of a flood-prone area. Thus in areas where it is possible, the land along river valleys is shaped into terraces, benches, or levees to confine floodwaters. And, in some cases, the discharge capacity of a river is enlarged by dredging a channel or by channeling and straightening a stream. In other cases, earthen or concrete dams, dikes, or floodwalls are built to detain or divert floodwaters. Those flood-protection measures are enhanced by erosion control, which protects the soil and vegetative cover of the watershed.

One of the more controversial flood-control approaches is *stream channelization*, which is the straightening and ditching of meandering stream channels. Channelization alleviates the danger of upstream flooding by causing water to flow downstream rapidly. However, one of the major problems with stream channelization is that it merely displaces the flood problem downstream. And sometimes, it triggers a chain of environmental problems. In some areas, stream channelization has disrupted fish spawning grounds and destabilized stream banks, which has led to accelerated erosion. Also, it has increased the sediment load of streams, thus reducing aquatic life, and, in some rivers, the potential for recreational fishing has been virtually destroyed. Channelized streams also drain water from adjacent wetlands, thereby destroying wildlife habitats. For example, the negative effects of channelization on the Kissimmee River in Florida have caused the public to exert pressure to reverse its channelization. In 1962, the Kissimmee River, which connects Lake Okeechobee and Lake Kissimmee in south central Florida, was a gently meandering stream that was approximately 160 kilometers (100 miles) long. By 1971, it had been drastically transformed by stream channelization into a straight, swiftly flowing canal that is only 93 kilometers (58 miles) long (Figure 15.13). Studies have shown that channelization of that river was not an effective strategy to control flooding. Also, more than 75 percent of the river's original wetlands have been drained as a result of the channelization, which has caused serious declines in animal populations, including bald eagles (down 74 percent since channelization), ducks, and coots.

Although wildlife biologists and sportsmen have opposed the channelization since the project was first proposed, it was not until 1983 that enough political clout was mustered to begin to correct the problem.

Figure 15.13 *Looking south along the channelized Kissimmee River in south-central Florida. Meandering segments of the river's former course prior to channelization are visible on either side. (James Balog, Black Star.)*

Then, in August of that year, Governor Graham of Florida announced that a program was underway to halt the destruction of the wetlands from Lake Kissimmee to the Everglades. As a part of that effort, engineers reestablished the Kissimmee's original course along a 16-kilometer (10-mile) stretch during a 1984 testing phase. As of this writing, plans for complete restoration of the river are progressing, although it is expected to be very costly. The test project alone cost $1.5 million.

In spite of more than $15 billion that has been spent over the years on structural flood control, flood-related deaths, injuries, and property damages continue to rise. The presence of any flood-protection device often engenders a false sense of security among local residents and encourages land development and home construction. However, since most flood-control structures are actually designed to accommodate only moderate flooding, the stage is set for tragedy when a great flood occurs.

Several disastrous dam failures in the 1970s demonstrated the vulnerability of those flood-control structures. For example, in June, 1976, the Teton Dam in eastern Idaho failed while its reservoir was being filled for the first time. (The gap in the dam is shown in Figure 15.14.) Fourteen lives were lost and property damage approached $1 billion. Subsequent investigations revealed that a combination of geologic factors (the most significant being a highly permeable rock foundation) and inadequate design was responsible for the dam's failure. And on November 6, 1977, an earthen dam failed during heavy rains above the campus of Toccoa Falls Bible College in Georgia, which resulted in 39 deaths.

Congress responded to the catastrophic floods in 1972 by enacting the Federal Flood Disaster Protection Act of 1973, which reflects a new emphasis on floodplain management as opposed to structural flood-control measures. That law requires that local governments adopt floodplain development regulations before they can become eligible for federal flood insurance. Also, any construction projects that are planned within regions that are identified as flood hazard areas (which do not qualify for flood insurance) are denied federal funding altogether. Floodplain development regulations require that buildings be elevated or flood-proofed and that provisions be made for a floodway, which will allow floodwaters to pass through a community without causing severe damage.

Figure 15.14 *Failure of the Teton Dam in eastern Idaho in 1976. This downstream view shows a huge gap in the dam. (U.S. Department of the Interior, Bureau of Reclamation.)*

The goal of the 1973 law is to foster the wise use of floodplains and other flood-prone areas and thereby to reduce the toll of floods. That law encourages communities to use floodplains in ways that are compatible with periodic flooding (such as for agriculture, forestry, some recreational activities, parking lots, and wildlife refuges) rather than as construction sites for homes and industries. For example, in the 1960s, the U.S. Army Corps of Engineers proposed construction of a concrete drainage channel through the center of Scottsdale, Arizona, which would carry off floodwaters from the Indian Bend Wash. The proposed channel would be 11-kilometers (7-miles) long and 13-meters (40-feet) deep. However, the local civic leaders, fearing that the channel would divide the community both physically and psychologically, searched for an alternative. The result was the Scottsdale greenbelt, which was completed in 1983 (Figure 15.15). The Scottsdale greenbelt is an 11-kilometer grass-lined drainageway that consists of five parks, featuring tennis courts, golf courses, bicycle paths, and other recreational amenities. Although the main purpose of the greenbelt is to control floods, it also serves other community needs. Other approaches to floodplain management are described in Box 15.2.

Earthquake Hazard

In late July, 1976, one of the greatest natural disasters of recorded history struck T'ang-shan in northeast China. Perhaps 240,000 persons perished when severe earthquakes devastated the homes and businesses that were located in that congested city. T'ang-shan's buildings had not been constructed to withstand earthquake tremors, and the city was situated on unstable river sediments that shifted during the violent quakes.

However, violent shaking of the earth is not the only cause of earthquake damage. Earthquakes can trigger landslides and avalanches of rock, mud, and snow; they can disrupt the flow of rivers and groundwater, causing rivers to flood and wells to run dry; and, in some coastal areas, major earthquakes can generate the very dangerous *tsunamis*, or seismic sea waves.

At sea, the energy in a tsunami wave is dispersed within a large volume of water, so the wave poses no threat to ships and, in fact, is not even noticeable. It travels along at tremendous velocities (perhaps more than 800 kilometers per hour) at an inconspicuous height that is usually less than 2 meters. However, when a tsunami approaches certain shorelines (particularly around the Pacific coast), its energy gradually becomes focused into a wall of water that can be taller than a three-story building. For

Figure 15.15 *The Scottsdale (Arizona) greenbelt, completed in 1983, was built as an alternative to more traditional flood-control structures. (© Tom Johnson, Scottsdale Scenes.)*

example, during the Alaskan earthquake of March 27, 1964, a tsunami that was more than 10 meters (33 feet) above the high-tide level crashed into Kodiak harbor, causing extensive damage to ships and harbor buildings. The loss of life that accompanies a tsunami can be staggering. In August, 1976, a very strong earthquake that was centered off the island of Mindanao in the Philippines created a series of tsunamis that claimed 5000 lives—even though a tsunami warning system has operated in the Pacific since 1948.

Millions upon millions of earthquakes have occurred through the course of human history—some catastrophic (Table 15.4) and most others so slight as to be inconsequential. In fact, more than 1 million tremors occur each year (about one every half-minute), and nearly 80 percent of those are barely strong enough to rattle dishes.

Almost all major earthquakes occur along boundaries of the gigantic lithospheric plates that are described in Chapter 12. In those instances, adjacent plates grind against one another along convergent or other fault boundary zones and, occasionally, they lock together, which causes stress and strain to build up and accumulate within the rock. Then, abruptly, when the plate breaks free, the trapped energy is released in the form of vibrations (waves), which travel within the earth or along the earth's surface. In other instances, the plates creep slowly past one another which produces frequent microearthquakes.

The West Coast is the primary site of earthquake activity in the United States. And it was determined in a study by the Federal Emergency Management Agency in 1980 that a 50-percent probability exists (a 1 out of 2 chance) that a catastrophic earthquake will occur in California within the next 30 years. Furthermore, the report identified seven fault systems in the state which could trigger earthquakes

---

BOX 15.2

FLOODPLAIN MANAGEMENT: SOME EXAMPLES

Downtown Soldier's Grove, Wisconsin, is no longer in Soldier's Grove, Wisconsin. It's 3 miles away.

Soldier's Grove still exists, of course. The residents, all 616 of them, just wanted to keep their feet dry.

Soldier's Grove is one of several towns and cities all across the country that are taking new approaches to flood control.

Located on the banks of the Mississippi River, the town has always been subject to severe flooding. But according to Tom Hirsch; flood-control consultant to Soldier's Grove, the $200,000-plus proposal to build levees in the town "would not have been justifiable." A more reasonable approach was to relocate the whole downtown area from the river's banks.

Mr. Hirsch says that the project was "a common-sense, grass-roots movement," and that to finance the project, they "bullied their way into a remarkable package of government funding." As a result, the town is now high and dry and closer to interstate commerce routes. The new buildings even include passive solar-energy construction.

In years past, flood control has been synonymous with monumental construction projects—building dams, canals, and reservoirs—many under the auspices of the U.S. Army Corps of Engineers. But the public outcry they have caused and the need to be more cost efficient have led many communities to rethink their flood-control plans.

One town took its case to Congress. Littleton, Colorado, lies on the South Platte river. The Corps proposed building a channel around the city to handle floodwater. But according to Jon Payne, town planning director, this "would have destroyed the natural river valley." Town residents argued that the valley itself was adequate to handle floodwater if some parts of it which had been developed could simply be returned to their natural state. They proposed buying up 640 acres in the river's floodplain.

Because the Corps' proposal was part of a larger federal dam project, the town had to petition Congress to alter the plan. The town's plan, now almost completed, has been successful so far, says Mr. Payne. Besides acting as a buffer against flooding for the town, the floodplain teems with wildlife, such as beaver, red fox, and moose.

As a result of the Littleton project, Payne says, congressional legislation was eventually enacted which requires the Corps to consider "nonstructural" alternatives to flood control. It

---

that could devastate major population centers. Foremost among those faults is the segment of the San Andreas fault that is near Los Angeles. That fault was produced by the sliding of the Pacific plate past the North American plate (Figure 12.8)—a 430-kilometer (270-mile) fracture zone—and it has been the locus of many severe earthquakes. Eight or more major earthquakes have occurred near Los Angeles during the past 1200 years—the most recent one occurred in 1857.

Although the likelihood of earthquake activity is greatest in the geologically active Pacific coast states, earthquakes have been reported far from plate boundaries—some of them severe. For example, the most violent earthquakes in the nation's history

struck New Madrid, Missouri, in the winter of 1811–1812; and they changed the course of the Mississippi River. Also, Charleston, South Carolina, experienced a major earthquake in 1886, during which 60 persons were killed. Figure 15.16 shows how the risk for earthquakes varies across the United States.

Thus we might ask, can earthquakes be anticipated? In recent years, with the growth of population in seismically active regions, the prediction of earthquakes has become a high-priority objective for research. And as a result, several forecasting schemes are currently being studied. One technique relies on the computation of recurrence rates (that is, the determination of earthquake frequency in a specific area from past records). However, that method is not

also places budget restrictions on Corps projects, according to a Corps spokesman.

Joseph Ignazio, chief of planning in New England for the Corps, says "we now have more tools than ever before to deal with the problem of flooding." These will be used whenever possible because "the federal government's financial participation [in flood control] is diminishing."

One of the Corps' most successful projects is nearing completion in Massachusetts, where the answer was a matter of working with nature, instead of trying to change it.

Periodic flooding on the Charles River posed a threat to heavily developed areas of downtown Boston and Cambridge. A study by the Corps found that upstream marshes and meadows acted like sponges, absorbing excess rainwater, and releasing it slowly.

But the upper Charles region faced development in the late 1960s. Pavement and drainage systems would have greatly decreased the land's storage capacity, threatening cities downstream. The Corps proposed buying up 9000 acres of wetlands which offered the greatest storage capacity. Landowners who wished to keep their property were allowed to by promising to leave it undeveloped.

The land bought by the Corps is being preserved in its natural state and is administered by the Massachusetts Fisheries and Wildlife

Division. The project, 90 percent complete, has successfully prevented flooding so far.

Such foresight wasn't exercised in Warwick, Rhode Island, however. Construction was allowed on the watershed of the Pawtuxet River, and one low-lying neighborhood has been subject to repeated flooding.

In Warwick, the Corps proposes buying up and demolishing more than 100 houses and relocating the residents. Barbara Sokoloff, Director of Planning in Warwick, says that although the project is still in the early stages, they "expect no opposition to the plan."

Because the Corps requires that all the residents be relocated, there is opposition to similar plans in other areas. In Prairie du Chien, Wisconsin, about 20 percent of the home and business owners in an area scheduled for relocation are opposed to the idea. Karen Troxell, Assistant Administrator for Community Development, says while "there have been some conflicts with the Corps administration, the relationship between the Corps and city has been very good." She says she feels a solution will be found for the residents.

satisfactory, since, historically, earthquakes have been random events. Also, several violent earthquakes have occurred in locations that have no record of significant earthquake activity.

A more promising technique that is used to predict earthquakes is based on the identification of *seismic gaps*, which are regions that are relatively inactive within a seismically active belt. Although the existence of seismic gaps may indicate that energy is gradually accumulating in advance of a major earthquake, it is not possible to predict the timing of an earthquake using this method alone. Seismic gaps may only pinpoint areas that are not appropriate for further development of housing.

The primary objective of most research that is

being done to predict earthquakes today is the identification of *premonitory events*, which are signals that precede potentially destructive earthquakes. Scientists use extremely sensitive instruments to monitor those events by measuring the buildup of strain (deformation) in rocks along fault zones. They use other instruments to detect slight changes in land elevation or changes in very small earthquake waves that may occur before a major earthquake. Changes in the behavior of animals prior to an earthquake are also being studied because some animals apparently are sensitive to premonitory events.

If and when the methodology to predict earthquakes has been perfected, precautions can be taken to minimize the damages and injuries that are sus-

Table 15.4 Some of the world's most disastrous earthquakes in terms of lives lost

YEAR	PLACE	DEATHS (EST.)	YEAR	PLACE	DEATHS (EST.)
856	Corinth, Greece	45,000	1915	Avezzano, Italy	30,000
1038	Shansi, China	23,000	1920	Kansu, China	180,000
1057	Chihli, China	25,000	1923	Tokyo, Japan	99,000
1170	Sicily	15,000	1930	Apennine Mountains, Italy	1,500
1268	Silicia, Asia Minor	60,000	1932	Kansu, China	70,000
1290	Chihli, China	100,000	1935	Quetta, Baluchistan	60,000
1293	Kamakura, Japan	30,000	1939	Chile	30,000
1456	Naples, Italy	60,000	1939	Erzincan, Turkey	40,000
1531	Lisbon, Portugal	30,000	1948	Fukui, Japan	5,000
1556	Shen-shu, China	830,000	1949	Ecuador	6,000
1667	Shemaka, Caucasia	80,000	1949	Khait, U.S.S.R.	12,000
1693	Catania, Italy	60,000	1950	Assam, India	1,500
1693	Naples, Italy	93,000	1954	Northern Algeria	1,600
1731	Peking, China	100,000	1956	Kabul, Afghanistan	2,000
1737	Calcutta, India	300,000	1957	Northern Iran	2,500
1755	Northern Persia	40,000	1960	Southern Chile	5,700
1755	Lisbon, Portugal	30,000–60,000	1960	Agadir, Morocco	12,000
1783	Calabria, Italy	50,000	1962	Northwestern Iran	12,000
1797	Quito, Ecuador	41,000	1963	Skopie, Yugoslavia	1,000
1822	Aleppo, Asia Minor	22,000	1968	Dasht-e Bayaz, Iran	11,600
1828	Echigo (Honshu), Japan	30,000	1970	Peru	20,000
1847	Zenkoji, Japan	34,000	1972	Managua, Nicaragua	10,000
1868	Peru and Ecuador	25,000	1976	Guatemala	23,000
1875	Venezuela and Columbia	16,000	1976	T'ang-shan, China	240,000
1896	Sanriku, Japan	27,000	1976	Philippines	3,100
1897	Assam, India	1,500	1976	New Guinea	9,000
1898	Japan	22,000	1976	Iran	5,000
1906	Valparaiso, Chile	1,550	1977	Rumania	1,500
1906	San Francisco	500	1978	Iran	15,000
1907	Kingston, Jamaica	1,400	1980	Algeria	3,500
1908	Messina, Italy	160,000	1980	Italy	4,000

Source: From F. Press and R. Siever, *Earth*, 3rd ed. New York: W. H. Freeman, 1982, p. 405.

tained as a result of major earthquakes. But even before those methods are perfected, we can institute certain safety procedures in earthquake-prone regions. For example, we can write building codes and zoning regulations that limit the height of buildings and otherwise minimize earthquake damage. And we can curb building on terrains that are prone to sliding during earthquakes or in coastal areas where the threat of tsunamis exists.

Volcano Hazard

On May 18, 1980, a massive explosion blew the top off Mount St. Helens in Washington State, reducing its summit from 2950 meters to 2560 meters above

sea level. Until that catastrophic eruption occurred (Figure 15.17), little concern was expressed over the hazard of volcanoes in the United States. Active Alaskan volcanoes are isolated and too remote to threaten population centers, and the Hawaiian eruptions generally consist of relatively quiescent flows of lava. But the Mount St. Helens eruption occurred near major population centers in the Pacific Northwest. Moreover, many persons fear that the eruption of Mount St. Helens may herald a reawakening of volcanic activity in the Cascades and in other portions of western United States.

In the 90 years before the Mount St. Helens eruption, only two relatively minor eruptions occurred among Cascade volcanoes: an ash eruption

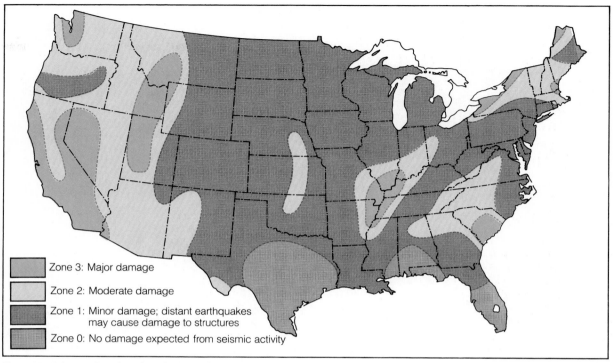

Zone 3: Major damage

Zone 2: Moderate damage

Zone 1: Minor damage; distant earthquakes
       may cause damage to structures

Zone 0: No damage expected from seismic activity

Figure 15.16 *Zones of earthquake risk across the nation. (From D. N. Cargo and B. F. Mallory,* Man and His Geologic Environment. *Reading, Mass.: Addison-Wesley, 1977; originally developed by the U.S. Coast and Geodetic Survey,* ESSA Rel. ES-1, *January 14, 1969.)*

at Mount Hood in 1906 and a series of small eruptions of Lassen Peak, between 1914 and 1917. Mount St. Helens itself had been dormant since 1857. However, in the mid-nineteenth century, northern Cascade volcanoes exhibited considerable activity. Thus it is possible that the eruption of Mount St. Helens marks a renewal of that activity. Such speculation prompted the U.S. Geological Survey to assess the likelihood of future volcanic activity throughout the western states. And based on historical and reconstructed records of past volcanic activity, many volcanoes were singled out as being potentially hazardous (Figure 15.18).

A *volcano* is defined as a landform that consists of eruptive material, which accumulates around a vent in the earth's crust. And the type of products that are extruded depends on the type of activity that is exhibited by the volcano. Although eruptions of some volcanoes are relatively peaceful, others are explosive and violent. And some volcanoes have an

eruptive history that begins explosively and then becomes quiet.

Volcanoes whose ejections consist mainly of lava flows are the most peaceful and the most massive. Their gently sloping flanks are built up as a result of numerous flows of viscous lava that form layers on top of one another. The Hawaiian Islands are composed of several such overlapping volcanoes. Nevertheless, lava flows can cause considerable damage, setting forests and other vegetation ablaze, inundating and destroying dwellings, and covering the land with a blanket of rock rubble. However, as a general rule, lava flows so slowly that the threat to human life is not great. In some cases, lava flows can even be diverted away from populated areas by structural means (such as by diversion walls).

The most hazardous volcanoes are those which are prone to explosive eruptions. Apparently, eruptive intensity depends on the temperature and the composition of the magma that feeds the volcano. Magmas that have the greatest potential for explosive activity have relatively low temperatures and are rich in silicon. Such magmas are very viscous and retain large quantities of gases. And if those gases cannot escape readily, the pressure may build to ex-

517

Figure 15.17 *Eruption of Mount St. Helens on May 18, 1980.
(© Roger Werth/Longview Daily News, Woodfin Camp & Associates.)*

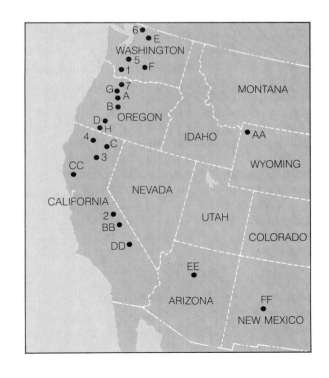

Figure 15.18 *The volcanic hazard in the western United States.
Volcanoes that have erupted on an average of every 200 years or
less, during the past 300 years, or both include the following: (1)
Mount St. Helens, (2) Mono-Inyo Craters, (3) Lassen Peak, (4)
Mt. Shasta, (5) Mt. Rainier, (6) Mt. Baker, and (7) Mt. Hood. Vol-
canoes that erupt less frequently than every 1000 years and have
not erupted in more than 1000 years include the following: (A)
Three Sisters, (B) Newberry Volcano, (C) Medicine Lake Volcano, (D) Crater Lake, (E) Glacier Peak, (F) Mt. Adams, (G) Mt.
Jefferson, and (H) Mt. McLoughlin. Volcanoes that last erupted
more than 10,000 years ago include: (AA) Yellowstone, (BB)
Long Valley, (CC) Clear Lake Volcanoes, (DD) Coso Volcanoes,
(EE) San Francisco Peak, and (FF) Socorro. (From R. A. Kerr, Volcanoes to keep an eye on. Science 221:634–635, 1983. Copyright
1983 by the AAAS as adapted from USGS Open-File Report 83-
400.)*

plosive proportions. For example, such a situation could arise if the vent through which the magma is moving becomes blocked.

An explosive volcanic eruption can send fine ash particles as well as steam and other gases into the stratosphere, and huge rock slabs and partially congealed masses of lava can be catapulted down its flanks. In some cases, an incandescent cloud of ash that is mixed with steam and other gases can roll down the flanks of a volcano at express-train speed, incinerating all the vegetation in its path. If the volcanic summit was originally covered with snow and ice, the heat of the eruption can cause rapid melting which can trigger avalanches of mud and rock.

Occasionally, a volcanic eruption can be predicted but, at present, such predictions are more an art than a science. In making such predictions, geologists rely primarily on premonitory events. For example, as magma flows in the subsurface, it sets off minor seismic activity. Also, prior to an eruption, intruding magma may cause a volcano to swell, thereby causing slight changes in the tilt and elevation of the land. Those events can be detected by sensitive instruments, but none of those events are unequivocable indicators of a pending volcanic eruption. In fact, 2 years before Mount St. Helens erupted, the U.S. Geological Survey scientists, D. R. Crandell and D. R. Mullineaux, after analyzing a study of premonitory events in the volcano's area, warned that an eruption was likely sometime before the end of this century.

Landslide Hazard

Landslides are the downslope movements of masses of rock, soil, mud, or a mixture of those materials. Like floods, earthquakes, and volcanic eruptions, landsliding is a natural phenomenon that contributes to the evolution of the landscape. Although some landslides are gradual, others are abrupt and sometimes catastrophic. In remote regions, landslides may pass unnoticed, but in populated areas, they wreck homes, destroy sections of highways, and, in some cases, bury entire towns.

Landslides occur when the land slips along inclined fracture planes between rock layers or when a sloped mass of rock or soil becomes saturated with water. The actual slide is usually triggered by an earthquake or the infiltration of great quantities of water during periods of prolonged rainfall. Heavy rainfall may infiltrate the ground to the extent that the soil becomes saturated with water. The mass of soil and water then flows rapidly in the form of a mud slurry. But human activities can also set the stage for landsliding. For example, the terracing of mountains and hillslopes for home sites or highways can create dangerously steep slopes that fail under heavy rains. Furthermore, irrigation and other activities that alter the flow of groundwater may also saturate sloped land and trigger sliding.

The West Coast of the United States is particularly prone to landsliding, as Figure 15.19 indicates. That region is hilly or mountainous, seismically active, and subject to intense rainfall—all characteristics of landslide-prone areas. And, unfortunately, the hazard to humans in that region is increasing because, as the demand for living space increases, more new housing subdivisions are built on slide-prone hillsides and mountain slopes. Figure 15.20 shows the result of landsliding on a construction site.

Although we are able to map areas of landslide susceptibility, our ability to control sliding is limited. For example, we can stabilize hillslopes that are denuded of vegetation by revegetation, construct retaining walls at the base of hillslopes, and install elaborate drainage systems to divert surface and subsurface water away from steep hillslopes. But those strategies are not always successful, because many variables contribute to landsliding. The most effective strategy at the present time is to discourage construction in landslide-prone regions.

Conclusions

As the human population grows, the competition for our finite lands also increases. Conflicts over land use are bound to become more frequent, and, in the near future, as we step up our efforts to find new sources of fossil fuels and other resources, they will doubtless grow more severe. Our public lands will become the target of increasing pressures from resource developers; and as our cities expand, more persons may live in regions that are prone to floods, earthquakes, and other natural hazards. Assessment of land capability is a logical first step in wise land-use planning. Ideally, that process will enable us to balance multiple uses of land and to permit both use and preservation.

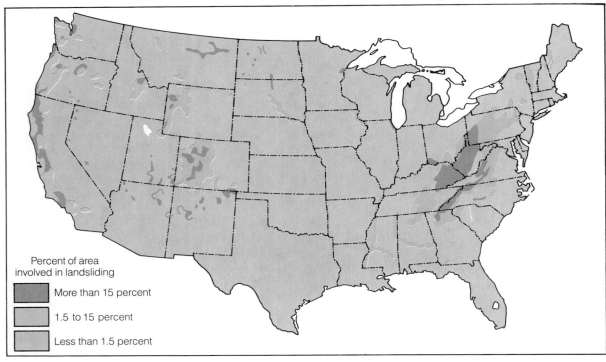

Percent of area
involved in landsliding

More than 15 percent

1.5 to 15 percent

Less than 1.5 percent

Figure 15.19 *Landslide potential across the United States. (After
the U.S. Geological Survey,* Annual Report, *1976.)*

Figure 15.20 *Landslide damage to homes in the Stone Canyon
Road area of Los Angeles, California. Heavy rains soaked soil,
which slid, wrecking two homes and sending one into the backyard of a third home. (Tom McHugh, Photo Researchers.)*

Summary Statements

Agencies of the federal government regulate activities on one-third of the nation's lands through management of national forests, parks, wildlife refuges, and other public lands.

Some public land is set aside for the prime purpose of preservation, and other lands are intended for regulated multiple use.

Mining is the one activity that is not subject to stringent controls in wilderness areas and in other public lands.

Conservationists are increasingly pressing for expansion of protected areas, while resource developers argue for the easing of land-use restrictions.

Disturbance of beach-forming processes has accelerated shoreline erosion. And the building of homes on barrier islands threatens to disrupt their basic protective function.

The 1980 national census reveals a slowing in the long-standing urbanization of the nation's population. Nevertheless, the number of supercities is on the rise.

Communities control development through zoning ordinances and, recently, by land trusts. Regional planning is not commonly practiced in the United States.

Land-use control strategies must account for the fact that some lands are vulnerable to natural hazards.

Damage that can be attributed to flooding is on the rise, because floodplains are more densely inhabited than ever and remote flood-prone areas are more accessible to increasing numbers of persons.

Floods may be caused by natural events (such as excessive rains and rapid spring snowmelts) and by human activities that alter the land's vegetative cover.

In the past, our nation's flood-control efforts were focused almost exclusively on structural alternatives (such as dams and levees). Today, however, the emphasis is on floodplain management as a strategy to avoid floods.

An earthquake is a violent shaking of the land that can trigger landslides, avalanches, floods, and tsunamis.

Earthquake forecasting research focuses primarily on recurrence rates, seismic gaps, and premonitory events.

The Mount St. Helens eruption in 1980 may be signaling a resurgence of volcanic activity in the western United States.

The terracing of mountains or hillslopes for home sites or highways may create dangerously steep slopes of rock and soil that might slide under heavy and prolonged rainfall.

The future is likely to see increasing competition for finite lands as our resource demands continue to grow.

Questions and Projects

1. How is land use regulated in your community? Your local planning agency can provide you with valuable information.

2. Is your region prone to floods, earthquakes, or landslides? What precautions are taken to minimize the impact of those hazards? Are approaches merely structural, or do they include management efforts?

3. Some critics argue that community growth-control ordinances are discriminatory. What is your view? Support your opinion.

4. Strong arguments support our need for more mining and the preservation of wildlife habitats. However, most mining activities are not compatible with the preservation of sensitive wildlife habitats. Devise a plan to resolve this conflict in land use.

5. Someday, scientists may be able to accurately pinpoint locations where future earthquake or volcanic activity is likely to occur months in advance. Suppose such a prediction were made for your community. How would you and your fellow residents react? What would be the economic impact?

6. List activities that are appropriate along a river floodplain. Order those activities in terms of their importance. What criteria do you use to set priorities?

7. Identify means other than zoning and growth-control ordinances that a community could use to curb its growth.

8. Comment on the statement that some conflict in land use is an inevitable consequence of our way of life.

9. Prepare a list of the demands that each person places on the land. List those demands in order of importance.

10. Enumerate those environmental factors that must be considered in selecting a suitable site for a new (a) shopping center, (b) copper smelter, (c) hospital, and (d) recreational park.

11. What is the primary purpose of national forests and parks?

12. How strictly are mining activities regulated on public lands?

13. Summarize the strategies that are employed to regulate development of barrier islands. Why is such regulation important?

14. Outline the advantages and disadvantages of urban freeway systems.

15. Why is the toll of fatalities and property damage due to floods on the rise?

16. Explain the recent shift away from structural control of floods and toward floodplain management.

17. What human activities contribute to the threat of floods and the hazards of landslides?

18. What measures can be taken in earthquake-prone areas to reduce the toll of deaths, injuries, and property damage?

19. Enumerate the methods of predicting earthquakes that are currently being studied.

20. Explain how an evaluation of land capability might contribute to the resolution of land-use problems.

Selected Readings

CONSERVATION FOUNDATION: *State of the Environment 1982.* Washington, D.C.: The Conservation Foundation, 1982. Includes a chapter on national perspectives and trends in land use.

COUNCIL ON ENVIRONMENTAL QUALITY: *Eleventh Annual Report.* Washington, D.C.: U.S. Government Printing Office, 1980, pp. 294–337. Includes a summary of land-use trends and critical issues that are related to the nation's land-use policies.

KELLER, E. A.: *Environmental Geology,* 3rd ed. Columbus, Ohio: C. E. Merrill, 1982. Presents the geologic principles that underlie natural hazards.

LONG, L., AND DEARE, D.: "The Slowing of Urbanization in the U.S." *Scientific American* 249(1):33–41, 1983. Examines an important demographic shift, which is revealed by the nation's 1980 census.

SOIL CONSERVATION SOCIETY OF AMERICA: Managing floodplains to reduce the flood hazard. *Journal of Soil and Water Conservation* 31 (March/April): 44–62, 1976. A collection of papers on floodplain dynamics, the nation's increasing vulnerability to floods, and approaches to floodplain management.

TANK, R. W.: *Focus on Environmental Geology: Text and Readings.* New York: Oxford University Press, 1983. Includes case histories and selected readings on natural hazards.

UNITED STATES GEOLOGICAL SURVEY: "Nature To Be Commanded . . ." In *Earth-Science Maps Applied to Land and Water Management.* U.S. Geological Survey Professional Paper 950. Washington, D.C.: U.S. Government Printing Office, 1978. An exceptionally well-illustrated demonstration of how the geologic sciences provide valuable information for land-use analyses.

WESSON, R. L., AND WALLACE, R. E.: "Predicting the Next Great Earthquake in California." *Scientific American* 252(2):35–43, 1985. Reviews the techniques and status of earthquake prediction.

PART III

FUNDAMENTAL PROBLEMS: POPULATION, FOOD, AND ENERGY

In previous chapters, we studied the ways in which the environment works and the ways in which human activities have disrupted the environment's normal functions. We also considered human attempts to rectify past mistakes in resource management, and we evaluated strategies for improving environmental quality and for utilizing natural resources more efficiently. Throughout these discussions, we have repeatedly pointed out that the success of any strategy is affected by two critical variables: future population growth and future supplies of energy. Furthermore, we have seen that agriculture has a significant impact on environmental quality and that it is a central issue in many land use controversies. Moreover, because we need a continual supply of food to stay alive, the quality of our lives in the future will depend greatly on world food supplies.

In this section, we consider the three variables that represent the fundamental problems facing humankind: the size of our own population, the sources of our energy supply, and our patterns of food production. In Chapter 16, we first examine the impact of continued population growth on environmental quality and resource management, and then we consider the principles governing human population growth. In the context of those principles, we evaluate strategies to slow human population growth. In Chapter 17, we evaluate the promises and environmental limitations of various efforts to enhance food production. In Chapters 18 and 19, we consider present and future energy supplies and energy consumption in the various sectors of our society. Chapter 18 deals mainly with the role of conventional energy sources (coal, oil, natural gas, and uranium) in meeting our energy needs; and in Chapter 19, we examine the promise of alternative sources of energy and conservation of energy. We hope that in examining those critical problems, each reader will consider the potential consequences in global as well as personal terms and begin adopting ways that lead to solutions.

Inadequate housing, unsanitary conditions, and overcrowding (as shown in this street scene in Manila, Philippines) are often associated with too many persons competing for a finite number of resources and opportunities. (Charles Steiner, Sygma.)

CHAPTER 16

HUMAN POPULATION: GROWTH AND ISSUES

Famine, crowding, inflation, unemployment, pollution, and dwindling resources are all tied to the fact that too many persons are competing for a finite number of resources and opportunities. Today, more than 4.8 billion persons inhabit this planet. Tomorrow, 225,000 more, and next year, 82 million more persons will be added to the world's population. Although the population in the United States is growing more slowly, it is expected to increase by nearly another 30 percent—to 305 million persons—within the next 50 years.

The pressures of population growth have already forced us to make changes in our way of life. For example, in the United States, communities that are faced with dwindling resources are formulating zoning ordinances to limit their growth. In the preceding chapters, we have identified the kinds of conflicts that arise when too many persons compete for the same resources (such as whether water should be used for municipalities or irrigation, whether forest lands should be used for timber or recreation, and whether land should be used for housing tracts or agriculture). Our high level of affluence largely determines the nature and frequency of those conflicts and has implications not only for ourselves but also for other persons around the world.

Although Americans comprise less than 5 percent of the world's population, we consume nearly 30 percent of the resources that are utilized in the world each year. Hence each American actually has a much greater overall impact on the environment than does the average citizen of a less-developed nation, such as India or Kenya.

In turn, we are influenced in many ways by population growth in other countries. For example, the lack of gainful employment is a growing problem worldwide. Each year, 3 to 6 million illegal immigrants enter this country seeking work (perhaps two-thirds are citizens of Mexico) (Figure 16.1), and another 400,000 to 500,000 persons enter legally in the hope of finding a better life. Box 16.1 describes the problems that are associated with unemployment in Central American countries.

Because the world population is growing, the demand for goods is also rising. Unfortunately, it has become increasingly difficult to enhance the supply of available resources (such as water, minerals, timber, and food) to meet those demands. For example, a growing population means a growing demand for food. And attempts to meet that increased demand have met with only limited success. Because the worldwide production of grain has not kept ahead of population growth in recent years (as it did in the 1960s and early 1970s), the per-capita production of foods that are made from grain has leveled off. Moreover, those attempts to increase food production have accelerated soil erosion, which has led to a decline in soil fertility and water quality. Furthermore, the consumption of petroleum has increased in the process of cultivating crops, which has increased the level of carbon dioxide in the atmosphere. Hence attempts to feed an ever-increasing human popula-

BOX 16.1

CENTRAL AMERICA'S "LOOMING NIGHTMARE"

At the airport, they elbow each other to carry your bag. Outside hotels and office buildings, they clamor to shine your shoes. When you park your car, a dozen of them appear from nowhere, promising to watch over the vehicle while you are gone.

They dart among the rush-hour traffic selling newspapers. They run errands, help truck drivers unload deliveries, walk the streets selling combs or chewing gum. If they are lucky, an uncle or cousin may need occasional help on a carpentering or painting job.

They are Central America's looming nightmare—the constantly swelling ranks of teenage males who can't find work and who, unless an unexpected economic miracle occurs, won't find it any time soon. Even if the area's current military conflicts are resolved, the teens' accumulating numbers add up to a monster problem certain to trouble governments of right, left, or center.

All through the five countries of Central America—Costa Rica, El Salvador, Guatemala, Honduras, and Nicaragua—vaccination, chlorination, and other health and sanitation advances have been sharply reducing infant- and child-mortality rates since the mid-1960s; those rates have been dropping far faster than birth and fertility rates. As a result, each year larger and larger numbers of Guatemalans, Salvadorans, and other Central Americans come of working age and enter the job market, the product not so much of a U.S.-style baby boom but of a death dearth. In societies already plagued by widespread poverty and economic and social inequality, these expanding cohorts of unemployed youths are the all-too-available raw material for future instability and unrest.

"Here in Guatemala, the labor force is growing much faster than the jobs," says Roberto Santiso, the head of Aprofam, the local family-planning organization. "The 14- and 15-year olds

are leaving the farm and coming to the cities, and there isn't work for those already here." Declares Dr. Ricardo Lopez-Urzua of the Ministry of Health: "It is a dangerous situation that too many people refuse to see."

Given Central America's inability to provide jobs for the growing work force, "social discontent, political violence, and open warfare are simply to be expected," Loy Bilderback, a longtime Central America specialist now at California State University in Fresno, told the Kissinger commission on Central America last fall. "The remarkable thing isn't that Central America is coming apart at the seams. The real mystery is what has held it together this long."

In addition to their potential for local explosion, these unemployed teenagers and young men raise another specter for the U.S. Illegal migration to the U.S. is an obvious escape valve for many of them; and there is ample evidence that just such a movement is under way, adding to (and often mingling with) the far larger flood of illegal Mexican immigrants.

Most horror stories about a "population bomb" deal with the mammoth numbers certain to overwhelm the Third World 40 or 50 years from now if fertility rates don't soon begin to fall dramatically. But the problems posed by these young unemployed Central Americans don't depend on uncertainties about future fertility. Here the working age starts at 10 or 11, and most of the young people who will be looking for jobs from now until the end of the century are already born, growing up—and hoping for a better future that may not be there.

"Young people are going to demand a job, a farm, a house, other things," says Jorge Arias de Blois, Guatemala's leading demographer. "Larger and larger numbers are coming each year, and they'll have bigger and bigger frustrations."

Central America's population has almost doubled in the past 23 years, from 11.2 million in 1960 to 21.9 million in 1983. Assuming only a modest drop in fertility rates over the next 15 years—by far the likeliest assumption—the numbers will increase sharply again, to 37 million, by the year 2000.

Even more worrisome than the total population figures, though, are the figures on the young age groups, those who not only will be needing jobs now and in the years ahead but also will soon be forming families and having still more children. Nearly half the region's population is under 15, almost two-thirds under 25. In 1950, there were 886,000 Central Americans in the politically volatile 15-to-19-year-old age group. By 1980, the figure had almost tripled, to 2.3 million, and by 2000 it probably will be close to 3.8 million.

"Much more of U.S. and other foreign-aid money is going into lowering mortality rates than lowering fertility," says Marshall Green, a former top-ranking U.S. diplomat who now works with the Population Crisis Committee in Washington. "Out of the goodness of our hearts, we are aggravating the population explosion. It's fine to improve health conditions and bring down death rates, but then it's all the more important to work on the family-planning side of the ledger."

When death rates go down, demographers say, the drop almost always starts with babies and very young children; and the drop in infant- and child-mortality rates that began here some three decades ago but really picked up steam in the late 1960s and early 1970s is now pouring the youngsters into the work force.

The Futures Group, a Washington research organization that works on population projections for the Agency for International Development, estimates that close to 2.1 million new jobs will have to be created between 1980 and 2000 here in Guatemala just to absorb the new workers, without making a dent in previous unemployment and underemployment rates. Considerably more than that amount will be needed for the other four Central American countries together.

"This big a gap between labor force and employment has never happened before anywhere,"

claims Robert Fox of the Inter-American Development Bank in Washington.

That is where the shoeshine boys, paper vendors, and errand runners come in. Many young people drop out of school to help eke out family income however they can; over one-third of Guatemala's primary-school age population doesn't go to school.

As in other Third World areas, the army and police absorb many of the young men. Crime provides another outlet; figures are impossible to come by, but there is universal agreement that burglaries, prostitution, car thefts, and other crime have been rising sharply here.

Wider use of birth-control measures would certainly reduce future population pressures; but even with substantial assistance from U.S. foreign aid and International Planned Parenthood groups, family-planning programs here and in other parts of Central America make only slow headway against strong resistance.

The Catholic Church, extremely powerful in a country that is more than 90 percent Catholic, has intensified its opposition since John Paul II became pope, Guatemalans say. Many right-wing nationalists believe that more people make for a stronger country. Much of the left denounces birth-control programs as another Yankee imperialist trick to keep the Third World weak. "It is one issue on which the far right and far left see eye to eye," says a U.S. businessman here.

Even if fertility rates began to drop more rapidly than anyone expects right now, there would still be bewildering population growth for decades because of the high proportion of young men and women now in, or coming into, their prime reproductive age. Declares a U.S. embassy official here: "I don't know what the answer is, but someone better come up with it damned soon."

Excerpted from Alan L. Otten, "Population Explosion Is a Threat to Stability of Central America." The Wall Street Journal, February 17, 1984. Reprinted by permission of The Wall Street Journal © Dow Jones and Company, Inc. 1983. All rights reserved.

Figure 16.1 *Each year the Border Patrol arrests nearly 1 million persons who try to enter the United States from Mexico illegally. (J. P. Laffont, Sygma.)*

Table 16.1 U.S. imports of some leading commodities, 1983

COMMODITY	VALUE (IN MILLIONS OF DOLLARS)
Petroleum and Related Materials	52,325
Metals and Metal Products	18,717
Food and Live Animals	15,412
Chemicals and Related Products	10,779
Natural Gas	5,530
Ores and Metal Scrap	2,498
Lumber	2,717
Wood Pulp	1,470

Source: U.S. Department of Commerce

tion could also contribute to climatic change.

Today, our well-being in the United States depends on a continual supply of many resources from other countries (Table 16.1), many of which are becoming limited as a result of increasing demands on them. Thus political upheaval in any of those countries could cut off our access to essential resources.

At this time, many persons in the world are unaware of, or indifferent to, the problems that are associated with population pressures. But population growth affects us all. For example, it is an unfortunate fact that each new baby places additional pressure on the earth's food, energy, water, mineral, and space resources. And the more affluent the parents, the greater will be the pressures that the child will place on those resources. Already, even the wealthiest persons feel at least some of those effects, such as pollution, which touches everyone; no ecological islands exist where they can go to escape it.

However, population growth is not the only factor that is responsible for our problems. Rising affluence, inappropriate technologies, and mismanagement of resources also contribute directly or indirectly to our difficulties. But unless population growth is slowed significantly, few countries will be able to meet the needs of future generations, let alone provide adequate food, water, shelter, health care, and education to those persons who are already living. In this chapter, we explore some of the causes of the recent human population explosion, we examine more fully the differences that exist in population growth between developed and less-developed nations, and we evaluate strategies for slowing the growth of the human population.

A Historical Overview

Throughout most of history, the human population has been quite small. It has grown relatively slowly, and it has even experienced occasional declines. (Figure 16.2 shows the general trend of population growth over the last 10,000 years.) For example, our early ancestors were vulnerable to a hostile environment. Food often was scarce, so famine (a time of extreme scarcity of food) was a common visitor; pro-

tection against bad weather and harsh climatic conditions was limited, because clothing and shelter were primitive; and outbreaks of disease took heavy tolls. Thus population growth rates were very low because death rates were very high and, therefore, nearly matched the birth rates. Then, about 10,000 years ago, humankind underwent a transition from a hunting-and-gathering existence to an agricultural one. As a result, crops were cultivated and animals were domesticated, which provided a somewhat more reliable food supply, and populations began to grow slightly faster. Nevertheless, famines continued, and both human health and human lives continued to be at risk (Figure 16.3).

Figure 16.3 *A child in East Africa suffering from a form of protein-calorie malnutrition called kwashiorkor.* (FAO Photo.)

Figure 16.2 *The growth of the human population from 10,000 years ago to the present.*

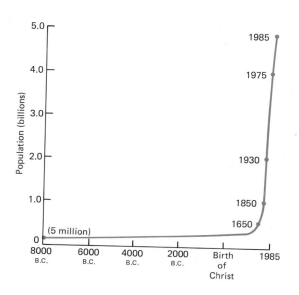

Historically, epidemics of disease have also severely limited the growth of populations. For example, it is believed that during the fourteenth century, the bubonic plague (which is caused by a bacterium) killed more than half the population of Europe and Asia. The impact of the plague on the population of Europe from 1000 to 1700 A.D. is shown in Figure 16.4. Those epidemics were frequently accompanied by severe outbreaks of influenza and typhus. As a result, entire populations were terror-stricken by those outbreaks, since no one knew the causes of the diseases, and they could not prevent the epidemics.

However, by the early 1800s, medical science had begun to control many of those diseases by means that we now take for granted. For example, one of the first great advances occurred in 1796, when Edward Jenner demonstrated that smallpox could be prevented by inoculating human beings with a substance that was extracted from cowpox lesions. Today, smallpox, which once killed hundreds of thousands of persons each year, has been virtually eliminated from the face of the earth. Other diseases were shown to be caused by pathogenic bacteria by Louis Pasteur and Robert Koch, and those findings also helped to prevent the spread of disease.

As our understanding increased through research, ancient beliefs that demons and noxious vapors caused illness were slowly dispelled. And the helplessness that many persons felt as a result of disease was gradually reduced by improved sanitation and medical care. For example, when the germ theory of disease was accepted, measures to protect the public (such as the chlorination of drinking water and the development of sewage treatment systems) became possible. Also, antiseptic techniques were developed that could be used in doctors' offices and hospitals. (Surprising as it may seem today, as recently as 100 years ago, enlightened doctors who advised their colleagues to wash their hands between examinations of patients were often ridiculed and banished from their profession.) And personal hygiene was promoted. For example, bathing became more desirable and washable clothing became popular.

Another major breakthrough was Walter Reed's research, which proved conclusively that a species of mosquito, *Aedes aegypti*, is the transmitter of yellow fever. As a result of that discovery, it became possible to limit some diseases by controlling their vectors. (*Vectors* are defined as organisms that transmit parasites from one host to another.) And Alexander Fleming's accidental discovery—that a culture of mold, which is called *Penicillium notatum*, prevented the growth of many species of pathogenic bacteria—represented still another breakthrough. Since that discovery, many other antibiotics (such as streptomycin and tetracycline) have been isolated from microorganisms.

Today, medical advances continue to reduce the number of deaths and to lengthen human life spans. For example, in the United States, heart disease and strokes are the major causes of death. Yet in the past decade, a combination of better medical treatment and improved life-styles (including regular exercise, less smoking, and a better diet) has reduced the number of deaths as a result of heart disease by nearly 30 percent. Medical advances also have increased the rate of survival following the treatment of certain types of cancer. Thus we can be confident that future medical advances and improvements in life-styles will further reduce the annual death rate.

The combination of medical advances and improved sanitation have led to victories over many diseases, such as smallpox, cholera, tuberculosis, and polio. However, the resulting worldwide reduction in deaths has, in turn, contributed significantly to increases in human population. And another factor

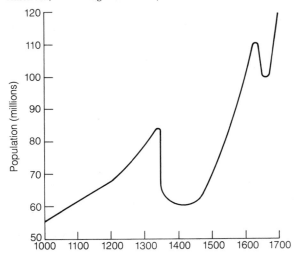

Figure 16.4 *The impact of recurrent plagues on the population of Europe. (After W. Langer, "The Black Death."* Scientific American *210:14–19 (February), 1964. Copyright © 1964 by Scientific American, Inc. All rights reserved.)*

that has contributed to population growth has been the generally greater availability of food. For example, the continual improvement of agricultural technology has reduced (but has not eliminated) the chances of crop failure. Furthermore, improvements have been made in the storage and transportation of food. And the advances that have been made in communications have softened the blow of local famines by providing access to distant emergency food supplies. Unfortunately, however, political and logistical considerations sometimes thwart the sending or receiving of emergency food supplies, which can result in starvation for some persons.

The fact that this period of explosive growth is unique may be exhibited by considering some major mileposts in terms of population growth. For example, it took several hundred thousand years (approximately 99 percent of human history) for the human population to reach 1 billion persons, which occurred sometime around 1850 (Figure 16.2). In marked contrast, humankind doubled its population to 2 billion persons in only 80 years more and redoubled it to 4 billion persons in only 45 years. Hence we are living in a period of population growth that is unequaled during the span of human existence. Thus we turn now to a consideration of the principles of human population growth, and then we examine the implications of this period of unprecedented growth for the future.

Characteristics of Human Population Growth

Human populations grow in much the same way as populations of any other type of organism. Thus the principles of population growth and regulation that are described in Chapter 4 also apply to humans. For example, if we disregard the effects of migration, the population growth of a country depends on the difference between its *crude birth* and *death rates*. (Those rates are called "crude" because they provide only an approximation of the actual growth of a population in one given nation. Crude death rates, in particular, can be affected by a population's age structure, which will be described further in the section on age structure.)

The crude birth and death rates can be used to calculate the *percent annual growth rate*, which is determined by using the following formula:

Percent annual growth rate

$$= (\text{crude birth rate} - \text{crude death rate})$$
$$\times 100 \text{ percent}$$

In the United States, in 1984, the crude birth rate was 16 per 1000 persons (0.016) and the crude death rate was 9 per 1000 persons (0.009). Thus by using these values for crude birth and death rates in the United States, we have

Percent annual growth rate

$$= (0.016 - 0.009) \times 100 \text{ percent}$$
$$= 0.7 \text{ percent}$$

The significance of a percent annual growth rate is easier to visualize when we consider its corresponding *doubling time*, that is, the length of time that is needed for a population to double at a fixed rate of growth (see Table 16.2). Doubling time can be calculated by using the following formula:

$$\text{Doubling time} = \frac{70}{X} \text{ , where}$$

$$X = \text{percent growth rate}$$

For example, if the population of the United States continues to grow at the current 0.7 percent annual growth rate, the population will double in approximately 100 years. In contrast, if the population of Syria continues to grow at the current 3.9 percent annual growth rate, Syria's population will double in 18 years.

Table 16.2 Doubling times associated with some population growth rates

ANNUAL POPULATION GROWTH RATE (IN PERCENT)	NUMBER OF YEARS IN WHICH POPULATION WILL DOUBLE	EXAMPLE
0.5	140	Greece
1.0	70	Poland
1.5	47	Taiwan
2.0	35	Ethiopia
2.5	28	Peru
3.0	24	Iran
3.5	20	Guatemala
4.0	18	Kenya

Source: 1985 World Population Data Sheet, Population Reference Bureau, Washington, D.C.

Exponential Growth

Percent annual growth rates demonstrate the exponential nature of human population growth. Recall from Chapter 4 that *exponential*, or *geometric*, growth occurs when a factor increases by a constant percentage of the whole during a constant period of time. Also, exponential growth has two very important characteristics. First, exponential growth can generate enormous numbers in a short period. For example, the world's population is currently doubling (experiencing a 100-percent increase) approximately every 40 years. If we assume for illustrative purposes that this doubling time will continue, the world's population in 2090 will be 32 billion persons. That is, shortly after the third centennial of our country, the world will need to support nearly eight times more persons than it does today. At that rate, in 40 years more, in 2130, the world's population will be 16 times greater than it is today.

Second, exponential growth is explosive and can reach a fixed upper limit within a surprisingly short period of time. Recall from Chapter 4 how an animal population can overshoot the carrying capacity of its habitat by growing exponentially. Human populations can also overshoot their nations' carrying capacities in terms of certain resources. For example, the recent history of production and consumption of petroleum in the United States shows how consumption can outstrip the production of a resource. During the 1950s and 1960s, the increased production of petroleum in the United States kept up fairly well with the increased consumption that is associated with a growing population and economy. As a result, imports of crude oil increased only approximately 25 percent—from 372 million barrels in 1960 to 483 million barrels in 1970. But in the early 1970s, production did not increase much, while consumption continued to grow at a relatively high rate. And to make up the deficit, more petroleum was imported. Thus from 1970 to 1977, the amount of crude oil that was imported increased by 500 percent. The most troublesome impact was economic. The monetary costs of petroleum imports rose from 2.7 billion dollars in 1970 to 44.5 billion dollars in 1977—a 16-fold increase. Although conservation measures and an economic recession have reduced our needs for petroleum imports since 1977, domestic production has not increased significantly since the late 1960s. Chapter 18 discusses this problem further.

Also, it is important to realize that seemingly small differences in exponential growth can produce surprisingly great changes in population size within just a few generations. For example, Figure 16.5 illustrates the dramatic differences that can occur in population growth between two nations—Poland, which has a 1-percent annual growth rate, and Kenya, which has a 4-percent annual growth rate. For purposes of comparison, we have set the population of both nations as being equal to 1000. Although Poland's population would more than double within 100 years if it maintained its 1-percent growth rate, such population growth would pale in comparison with Kenya's, which would increase more than 32-fold within 100 years if its growth rate held at 4 percent.

Actually, Poland, today, has a population that is greater than 36 million, which is essentially double that of Kenya. But if each nation were to continue to grow at its current rate, Kenya's population would overtake that of Poland's within approximately 20 years. More importantly, Kenya's population would be nearly 15 times greater than Poland's population within 100 years. Thus even small reductions in exponential growth rates are important contributions to efforts to maintain the human population within the earth's carrying capacity.

Age Structure

To predict future population growth more precisely, we need to know more than crude birth and death rates, because those rates do not tell us anything

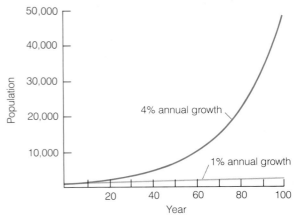

Figure 16.5 *A population that grows at an annual rate of 4 percent will be some 15 times greater in number at the end of 100 years than a country that grows at an annual rate of 1 percent.*

about the *age structure*, or age profile, within a particular population. Age structure is important because not all individual members who make up a particular population are equally likely to have children or to die. For example, only those women who have reached an intermediate age can give birth (the reproductive age of most women is between 15 and 44 years). Also, persons in the intermediate age groups are less likely to die than those persons who are very young or old. Hence a country's population can be subdivided into three major age groups: *prereproductive* (0 to 14 years old), *reproductive* (15 to 44 years old), and *postreproductive* (45 years old and older). Thus by classifying the persons in a country in one of those three categories, we can generate an age-structure diagram for that nation. Then, we can refine the diagram further by subdividing it into age groups, or *cohorts*, of 5-year intervals (such as 0 through 4 years, 5 through 9 years, and so on). Furthermore, we can differentiate between males and females by placing them on opposite sides of the diagram.

Age-structure diagrams can help us to determine the current growth status of a population and predict what might happen in the future. For example, an age-structure diagram that has the shape of a broad-based pyramid describes a population that has a large percentage of persons who are quite young, which indicates a recent history of high birth rates. Such a diagram describes the population of Mexico, which totaled nearly 8.0 million persons in 1985 (Figure 16.6). A broad base also indicates that a large number of persons soon will be entering the reproductive age. As a result, a significant increase in the total number of births can be expected. In contrast, the upper portion of the pyramid is narrow, which indicates that a small percentage of the population is at, or is approaching, old age. Hence the total number of deaths over the short term can be expected to remain relatively small. Thus when the number of births increases and the number of deaths remains stable, the population can be expected to grow rapidly in the years to come.

Figure 16.6 *The age structure of (a) Mexico, which has a recent history of high fertility, compared with that of (b) Sweden, which has a recent history of low fertility. (United Nations,* Demographic Indicators of Countries: Estimates and Projections as Assessed in 1980, *New York, 1982.)*

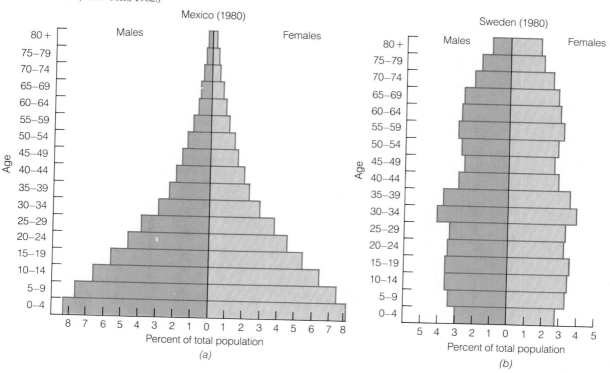

A contrasting age-structure diagram is illustrated for Sweden (also shown in Figure 16.6). In Sweden, the population was 8.3 million persons in 1985, but birth rates have been low since the mid-1950s. Thus the prereproductive group is approximately equal in size to the reproductive group, and the postreproductive group is the smallest of the three. Because the prereproductive and the reproductive groups are approximately equal in size, we can expect little change in the number of births as the prereproductive group reaches reproductive age (assuming that the preferred family size of this generation remains the same as that of its parent's generation). And as the present reproductive group reaches old age, the number of deaths will increase. Hence the age-structure diagram of Sweden describes a population that is not expected to grow much in the years to come.

Thus although crude birth and death rates are quoted most frequently for describing the growth of a population, we must be careful to interpret those rates in terms of age structure. Furthermore, another important factor—the *fertility rate* (which is the number of births per woman)—also has an impact on the growth of populations. For example, in the United States, the fertility rate has become so low that women in the United States are now averaging only 1.9 births during their childbearing years, which is below the *replacement rate* of 2.1. In other words, parents are not having enough children to replace themselves. Furthermore, because some children die before they can raise a family, and because some women are either physically unable or unwilling to bear children, the average family size in the United States must be 2.1 children, rather than 2.0, to ensure replacement. In nations that experience higher infant mortality and shorter life expectancies, the replacement rate might be as high as 2.5 children per family.

However, the fact that fertility in the United States has declined below the replacement rate does not mean that our nation's population will not continue to grow. In fact, our population age structure, which is illustrated in Figure 16.7, indicates that our population will continue to grow. For example, the baby boom that occurred during the two decades that followed World War II produced a significant number of persons in the United States who are now 20 to 30 years old—the prime childbearing age span. In contrast, the number of persons who are older

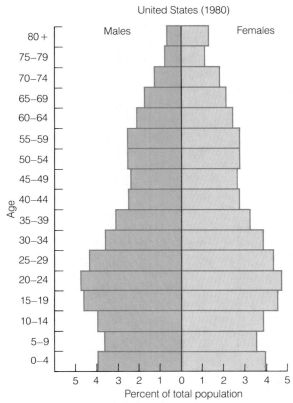

United States (1980)

Figure 16.7 *The age structure of the United States, which shows that the members of the baby-boom generation are in their prime reproductive years (ages 20 to 30 years).* (United Nations, Demographic Indicators of Countries: Estimates and Projections as Assessed in 1980, *New York, 1982.*)

than 60 years is smaller than the number of persons who are 15 to 30 years old. Even if those young persons average only 1.9 children per family, our population will continue to grow for another 50 years, because so many new families are forming. Thus the total number of births will continue to be greater than the total number of deaths until 2035. At that time, the sizes of all age groups should be much more similar—that is, the shape of our age profile will be much like that of Sweden's (Figure 16.6)—and the population will be nearly *stationary*. Thus disregarding migration, the size of the population will not be changing. As a result, the U.S. Census Bureau projects that by 2035, the population in the United States will be 305 million persons—nearly 70 million persons more than today. When the growth of a population essentially ceases, it is often referred to as *zero population growth* (ZPG). However, the projection that the United States will reach zero population growth by the year 2035 is based on the assumption that during the coming decades, the number of children that are born to each family will remain essentially the same as it is today. If, however, the fertility rate increases again at any time during the next 50 years, the population would continue to grow beyond current projections.

Population Momentum

An analysis of the age structure in the United States (Figure 16.7) also illustrates the concept of *population momentum*, which also was introduced in Chapter 4. According to this concept, once a population begins to grow, it usually includes a large percentage of persons who have reached or soon will reach reproductive age, so the population will continue to grow for some time, even if each couple (on the average) only replaces itself. Thus since the United States experienced a baby boom that peaked in the late 1950s, the country's population will continue to grow, at least until 2035.

Population momentum is particularly crucial for the many nations in South and Central America, Africa, and Asia, which are in the midst of unprecedented population growth rates. In those countries, nearly 40 percent of the persons are younger than 15 years old and will soon be raising families. Hence it will be many years before a stationary population is attained in those countries, even if the birth and death rates approach each other soon. For example, if we assumed that the replacement rate could be achieved immediately and then maintained (which will not happen), Mexico's population would still double within 50 years; indeed, it would take more than a century for Mexico to achieve a stationary population.

Variations among Nations

The overall pattern of population growth that is illustrated in Figure 16.2 is not representative of all the countries in the world. In fact, population growth rates vary considerably among nations. Although some important exceptions exist, most countries can be classified roughly into one of two groups according to their rates of growth. The first group comprises the *developed countries*, or *more-developed countries*, which includes the nations of Europe and North America, the Soviet Union, Japan, Australia, and New Zealand. The second group comprises the *developing countries*, or *less-developed countries*, which includes the nations of Central and South America, Africa, and Asia.

The developed countries usually have a low rate of population growth—less than 1 percent per year. For example, the populations of several European nations, including Sweden, East Germany, Luxembourg, Austria, and Belgium are essentially not growing at all. And the populations of West Germany, Denmark, and Hungary are actually declining. In contrast, the developing countries have an average annual growth rate that is higher than 2 percent, which means that their populations double every 35 years or less (refer back to Table 16.2). Many African nations (such as Kenya, Ghana, and Uganda) and Middle East nations (such as Syria, Jordan, and Kuwait) have growth rates that are higher than 3 percent. Thus their populations are doubling about every 24 years.

What accounts for those dramatic differences in the rates of population growth among nations? As we saw earlier, death and birth rates have been nearly equal throughout much of human history, so rates of population growth have remained small. However, in western Europe, death rates began to fall in the seventeenth and eighteenth centuries. And since high birth rates continued, population growth rates rose. However, within a few decades, the birth rates also began to decline, and today birth and death rates are nearly equal again in Europe. That decline in death rates, followed by a decline in birth rates, is called a *demographic transition*. A similar transition has also occurred in the other more-developed countries. As a result, the more-developed nations are now growing at a rate of 0.6 percent per year— a doubling time of approximately 118 years. (Although this growth rate may appear to be quite low, it is actually quite high in historical terms. For most of human history, the doubling time of the global population has been on the order of 30,000 years.)

While a demographic transition was occurring in Europe and North America, most of the rest of the world was still suffering high mortality rates. For example, many of the less-developed countries were not introduced to improvements in health care and sanitation until after World War II, at which time their death rates declined precipitously. But birth rates remained high—and they are still high—in most less-developed nations, so the populations in those nations continue to swell. The less-developed nations (excluding the People's Republic of China) are now growing at a rate of 2.4 percent per year—a doubling time of 29 years.

During the past two decades, more than 10 less-developed countries, including Singapore, Taiwan, South Korea, Cuba, and Costa Rica, have experienced a 50 percent, or greater, decline in fertility. In addition, several other nations, such as Colombia, Thailand, and Indonesia, have made impressive headway in terms of reducing their fertility. During the last decade, the rate of population increase has declined faster in the People's Republic of China (which accounts for 25 percent of the world population) than in any other large less-developed nation. But in many nations, particularly in Africa, Southwest Asia, and parts of Latin America, little decline has occurred in fertility rates from their historically high rates. Therefore, population experts are concerned that the demographic transition in many less-developed nations will not occur soon enough or fast enough to prevent grave economic and ecological consequences. Thus in the following sections we discuss some of the ways in which the demographic transition can be facilitated.

Slowing Population Growth

Motivational Factors

A growing body of evidence, as well as common sense, indicates that birth rates decline when parents find that it is desirable to have smaller families. Thus if a country wants to curb its population growth, its citizens must want to have fewer children. But in many countries, most persons still want to raise large families.

In the latter half of the 1970s, the World Fertility Survey, which was directed by the International Statistical Institute in London, interviewed more than 200,000 women in 29 less-developed countries. The survey showed that the mean preferred family size was between 4 and 5 children for most parents, but that the desire to have 6 to 7 children was not uncommon. (Recall that the replacement rate for less-developed nations is in a range of 2.2 to 2.5 children

per family.) Thus, obviously, many women in those countries still desire to have families that exceed their replacement rate. What, then, motivates persons to have small families?

The answer to this question is not at all clear and remains the subject of considerable debate. Some population experts believe that many of the differences between developed and less-developed nations are responsible for the differences in preferred family size. Some of those factors are listed in Table 16.3. For example, infant mortality rates, on the average, are five times higher in less-developed nations than in more-developed nations. Thus some population experts have proposed that if infant mortality rates are reduced, a decline in birth rates will follow. Those experts believe that in less-developed countries, children are a form of social security for their parents' old age. Thus in nations where the infant mortality rates are high, large families are necessary to ensure that enough children will survive to take care of their aging parents. For example, in India, a couple must have six children to ensure that one male offspring will survive until the father is 65 years old. Hence, it is argued, high infant mortality rates motivate parents to have large families.

Some population experts have also proposed that

Table 16.3 Major characteristics of less-developed and more-developed nations

LESS-DEVELOPED NATIONS*	MORE-DEVELOPED NATIONS
Moderate to High Infant Mortality (50 to 200 Deaths of Infants Under 1 Year Old per 1000 Live Births, Average 101)	Low Infant Mortality (5 to 25 Deaths of Infants Under 1 Year Old per 1000 Live Births, Average 18)
Low to Moderate Average Life Expectancy, (40 to 65 Years, Average 56)	High Average Life Expectancy (69 to 75 Years, Average 73)
Largely Rural Population (Average 66 Percent)	Largely Urban Population (Average 72 Percent)
High Illiteracy Rate (25 to 75 Percent)	Low Illiteracy Rate (Less than 5 Percent)
Moderate Daily Food Consumption (Average 2180 Calories per Person)	High Daily Food Consumption (Average 3315 Calories per Person)
Low to Moderate Average Per-Capita Income ($200 to $3000, Average $880)	High Average Per-Capita Income ($3000 to $14,000, Average $9380)

Source: Annual Population Data Sheet, 1985, Population Reference Bureau, Washington, D.C., and Agriculture: Toward 2000, Food and Agricultural Organization of the United Nations, 1981.
*Excludes People's Republic of China.

illiteracy has a great impact on birth rates. They reason that illiterate persons cannot understand the economic and social advantages of having small families, so they are not receptive to family-planning programs.

The conclusion that is drawn from those inferences is that socioeconomic development is the key to motivating parents in less-developed nations to reduce the size of their families. To support this position, some population experts claim that a basic education for all children, basic health care, and an equitable distribution of income are factors in some of the recent reductions in birth rates. For example, recent, dramatic declines in the birth rates in Taiwan, South Korea, Malaysia, and Singapore have been attributed to socioeconomic development. But other population experts point out many exceptions. For example, Thailand and Indonesia have experienced significant declines in their birth rates, yet their socioeconomic development lags far behind the other Asian nations just mentioned. And a particularly dramatic example of the disparity between birth rates and economic development is Kuwait. That nation has one of the highest per-capita incomes in the world, and the distribution of income, goods, and services among the citizens is so equitable that little poverty exists in the country. However, Kuwait's population growth rate (3.2 percent per year) and crude birth rate (35 per 1000 per year) are among the highest in the world. Indeed, the population of that tiny Middle Eastern country almost doubled between 1970 (774,000) and 1980 (1,353,000). Interestingly, the mean preferred family size for nations in the Middle East (such as Kuwait) includes six or more children. So, obviously, socioeconomic development is not the predominating factor, at least in some nations, in motivating parents to have fewer children.

Further insights into the factors that motivate persons to have smaller families have been gained from a recent reexamination (published in 1980) of the demographic transition in Europe. That study concludes that a marked decline in birth rates occurred simultaneously in most European nations. But, contrary to popular belief, large differences existed among those nations in terms of such socioeconomic conditions as industrialization, urbanization, and literacy. The study concluded that no easily definable threshold existed for social and economic progress that could account for the decline in family size.

To explain why the reduction in family size occurred in Europe during the demographic transition, the authors of the study suggest several emerging cultural changes. New attitudes began to emerge. For example, parents may have begun to reject the societal and religious traditions of having large families in favor of their own individual interests. Also, society may have begun to accept the notion of birth control rather than leaving conception to chance. The authors also suggest that changes in attitude may have occurred as a result of a change in the status of women. The position of women in society, particularly in the family, had begun to change: Women were becoming more active in socioeconomic activities that occurred outside their homes rather than remaining isolated in their homes (Figure 16.8).

However, such changes in attitudes and women's status are, of course, quite difficult to measure. Moreover, it is even more difficult to document a direct cause-and-effect relationship between a change in women's status and a decline in family size. But many sources of circumstantial evidence show that such a causal relationship does exist. For example, expanding employment opportunities for women in most developed nations are associated with declining family size. And in many less-developed nations, we also find a correlation between women who work outside the home and a smaller family size. In fact, a recent summary of studies that were conducted in 20 less-developed nations found that women who work outside the home (and who also have a primary-level education and live in a city), have an average of 4.2 children. In contrast, women who do not work outside the home (and who also have less than a primary-level education and live in a rural area) have an average of 6.9 children. In the Islamic countries in the Middle East, the association between birth rates and the subordinate position of women is clear. Most of those countries have annual birth rates of 30 to 48 per thousand persons as opposed to the average of 15 births per 1000 persons in developed countries.

Common sense tells us that the motivational factors that influence family size are probably interrelated and that the most important motivational factors will probably vary from one nation to another. For example, although one measure of a woman's status may be that she works outside the home, the opportunity to work depends on the economic

Figure 16.8 *Employment opportunities for women outside the home, such as this garment manufacturing venture near Cairo, Egypt, are associated with declining family size. (World Bank Photo by Kay Chernush cIFC/IRBD 1980.)*

conditions of the country she lives in. Furthermore, a woman's job status and level of income are, in turn, determined largely by her level of education. Thus it is counterproductive to argue which influence (social, economic, or cultural) has the greatest influence on family size. Rather, policymakers should look for the most effective blend of policies that would be compatible with the particular social, economic, and cultural characteristics of each nation.

Birth-Control Methods

Once a person or a couple is motivated to avoid having children, either temporarily or permanently, a variety of methods of birth control exists to choose from. The various alternatives are listed in Table 16.4. Because cultural and social conditions vary among different societies and persons, and even during different phases of a couple's life span, no single birth-control method ever has been, or probably ever will be, accepted universally. However, we can enumerate the key characteristics of an ideal method: it should be safe, effective, inexpensive, convenient, free of side effects, and compatible with local cultural, religious, and sexual attitudes.

Safety The safety of birth-control methods is of utmost importance.* Figure 16.9 shows the annual number of deaths that are associated with several methods of birth control, along with the age of the

---

* Because each person is unique genetically, the selection of a birth-control procedure should be done in consultation with a physician to reduce the chances of unforeseen complications.

women who use them. (These data are based on women in the United States and Great Britain.) However, to properly assess the safety of birth-control methods, we must keep in mind that even pregnancy presents risks to the mother. Complications during pregnancy and childbirth result in approximately 23 deaths per 100,000 expectant mothers each year in the United States and Great Britain. And in less-developed nations, the death rate is much higher—60 deaths per 100,000 expectant mothers each year.

Much concern has been voiced over the health risks that are associated with *the pill*, which is essentially a combination of female hormones that controls the reproductive cycle. Specifically, questions have been raised about a possible increased incidence in heart attacks, strokes, and cancer among women who take oral contraceptives. As a result, Johns Hopkins University initiated a study, which included data for more than 20 years of oral contraceptive use. In 1982, they submitted a report, which summarizes the risks and benefits of using the pill. According to that study, use of the pill does increase the chances of heart attack and stroke, but the increased risk is confined almost entirely to women who are older than 35 years and who smoke. However, it has not been demonstrated that an increase in breast cancer—the most common cancer among women in developed countries—occurs as a result of using the pill. Interestingly, accumulating evidence indicates that women who take the pill run a lower risk of contracting uterine cancer (66 percent

Table 16.4 Methods of birth control

Methods that Prevent Entry of Sperm
  Abstention
  Coitus Interruptus (Withdrawal)
  Condom
  Spermicides
  Diaphragm
  Sterilization

Methods that Avoid or Suppress the Release of the Egg
  Rhythm
  Oral Contraceptive (the Pill)

Methods that Prevent Implantation
  Intrauterine Device (IUD)

Methods that Prevent Birth in Case of Pregnancy
  Abortion

Source: After S. J. Segal and O. S. Nordberg, Fertility Regulation Technology: Status and Prospectus. *Population Bulletin 31*. Washington, D.C.: Population Reference Bureau, 1977.

lower risk) and ovarian cancer (33 percent lower risk) than nonusers. Thus at this time, the evidence seems clear that older women who smoke should use some other form of birth control. Younger women and nonsmokers run a far lower health risk as a result of taking the pill than from becoming pregnant.

Few serious health risks are associated with the use of the *intrauterine device* (IUD), which is a small plastic or metal object that is placed inside the uterus and, presumably, works by interfering with the implantation of a fertilized and developing egg. Although some bleeding and discomfort usually occur for a short period of time after an IUD is inserted, if those symptoms continue or are excessive, it should be removed. However, no evidence is available that shows any relationship between the IUD and cancer. Also, the *diaphragm* (which is a rubber cap that covers the cervix and prevents entry of the sperm into the uterus) is nearly free from side effects and complications. As a matter of fact, the death rates that are associated with the use of diaphragms and condoms, as illustrated in Figure 16.9, actually relate to

the deaths of women in childbirth as a result of pregnancies that are caused by contraceptive failure.

Following the discovery (in Europe, in the early 1960s) of the catastrophic effects of the sedative thalidomide on fetal development, birth-control methods have been regulated and monitored with increasing strictness. As a result, the regulatory requirements for contraceptive drugs are now more stringent than for most other classes of medication. Also, new federal legislation requires more safety regulations for nondrug methods of birth control as well, such as IUDs. And more tests for the effectiveness and safety of all new birth-control procedures are required, including tests for side effects, reversibility of the procedure, and potential carcinogenic effects. As a result, as many as 15 years of research, development, and testing may now ensue before the federal government approves a new contraceptive procedure. Despite all those precautions, however, the long-term risks remain unknown.

Effectiveness The effectiveness of birth-control devices varies considerably, as Table 16.5 indicates. *Abortion* (the surgical or medical termination of pregnancy) and *sterilization* (which is caused by surgical alteration of either the female or male repro-

Figure 16.9 *Annual number of deaths in women of different ages as a result of using various methods of birth control or no birth-control method. (After C. Tietz, "New Estimates of Mortality Associated with Fertility Control."* Family Planning Perspectives *9:74–76, 1977.)*

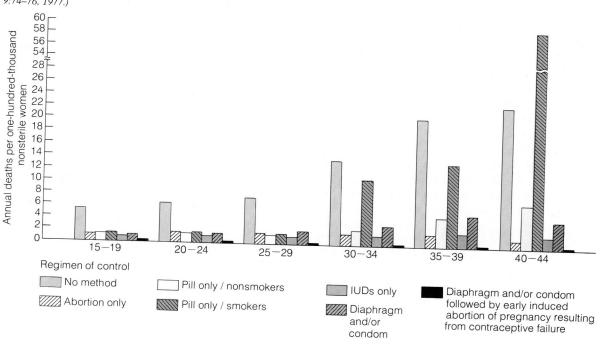

Table 16.5 Percent of women in the United States who experience contraceptive failure within the first year of use of contraception

METHOD	PERCENT FAILING
Pill	2.4
IUD	4.6
Condom	9.6
Spermicides	17.9
Diaphragm	18.6
Rhythm	23.7
Others	11.9

Source: After A. L. Schirm, J. Trussell, J. Menken, and W. R. Grady, "Contraceptive Failure in the United States: The Impact of Social, Economic, and Demographic Factors." Reprinted with permission from *Family Planning Perspectives* 14:68–75, 1982.

ductive system to prevent pregnancy) are essentially 100-percent effective. *The pill* is also nearly 100-percent effective, but a woman must remember to take it every day, because each pill that is forgotten increases the chances of pregnancy. The *rhythm method* (the timing of intercourse to avoid the most fertile time) generally is less successful for two reasons. First, the period of time during which a woman is capable of conception is often difficult to determine, particularly in women who have irregular menstrual cycles. Therefore, to allow an adequate margin of safety, the woman should abstain from coitus for a considerable fraction of the month. And second, since hormonal changes occur when a woman is ovulating, which make some women particularly inclined to have sexual relations, a high degree of self-restraint is required by the rhythm method.

Moral and Ethical Considerations Cultural, religious, and sexual attitudes greatly influence persons' choices of birth-control measures. Undoubtedly, the most controversial method of preventing birth is abortion. But, actually, the primary goal of those who support legalized abortion is not to further birth control. Rather, those persons want to ensure that women can exercise their fundamental right to control their own fertility. Moreover, proponents of abortion contend that the abolition of legal abortions would result in more unwanted births and recourse to potentially dangerous illegal abortions. For example, before the 1973 Supreme Court decision to legalize abortion, an estimated 250 women died each year in the United States as a result of infec-

tions and internal hemorrhages that were caused by self-induced abortions or unsterile techniques used by untrained persons. Today, fewer than 40 women die each year in this country from legal or spontaneous abortions. In fact, the risk of dying from abortions that are performed by trained personnel, under sanitary conditions, during the first 15 weeks of pregnancy is only one-seventh the risk of dying from pregnancy and childbirth.

Opponents of legalized abortion champion the rights of the unborn child. They believe that from the moment of conception, every unborn baby is a human being with all the rights of a human being. However, from the time that an egg and a sperm unite until the time at which a fetus has developed that is capable of living on its own, pregnancy is a very complex process, which consists of a sequence of stages that have no clear demarcations between them. Although scientists can describe those events, human personhood, itself, is not a state that can be evaluated by scientific experiments or observed under a microscope. There is nothing that a scientist can measure to determine whether the entity that is developing in the uterus has become a person who has the right to live. Therefore, the question of personhood in the uterus is not a scientific question. Rather, the controversy about abortion is a very complex philosophical and ethical question that is beyond the scope of this textbook. (For a presentation of the various philosophical viewpoints on abortion, the reader is referred to the collection of papers edited by Joel Feinberg, which is listed in the Selected Readings section at the end of this chapter.)

Approximately two-thirds of the countries in the world generally permit medically induced abortion. However, in Latin America, Africa, and the Middle East, abortion is severely restricted or illegal. Worldwide, approximately 40 million abortions are performed each year. The highest abortion rates may be in Italy and Portugal, where approximately 1000 abortions are performed for each 1000 live births. In the United States, the abortion rate is approximately 425 for each 1000 live births. And in most of Europe, the rates are slightly higher.

Abortion patterns vary markedly among nations. For example, in the United States, approximately 70 percent of the women who obtain abortions are unmarried, approximately 50 percent are childless, and approximately 30 percent are younger than 20 years old. In Mexico, more than 90 percent of the women who seek abortions are married and have more than three children.

In 1982, on a worldwide basis, voluntary sterilization was the most popular form of birth control—an estimated 100 million couples used this method. Voluntary sterilization is expected to become more popular, because surgical procedures have been simplified and persons now realize that sterilization does not hinder their sexual activities. Another factor that would contribute to the increased acceptance of sterilization is the successful completion of recent efforts to develop surgical procedures that can be reversed at some later time. With 54 million persons taking the pill, that method is the second most preferred contraceptive. And the *condom*, which is one of the oldest and simplest means of contraception, is now used by an estimated 30 million couples.

Nevertheless, research is continuing to find better means of birth control. Table 16.6 lists several new contraceptive procedures that will probably be available within the next decade. The priorities that have been established in the investigation of new methods of birth control include improved safety and the development of long-acting forms of contraception, which would make the daily chore of taking

a pill or frequent checks for expulsion of an IUD unnecessary. For example, injections of female hormones, which are given at intervals of up to 3 months, are now being tested. Also, trials are being conducted on contraceptive implants that may last as long as 5 years. Those implants are shaped in the form of tubes, which are made of rubberlike compounds, that slowly release synthetic hormones to prevent ovulation. They are implanted under the skin, usually in the arm or the buttock.

Until the early 1970s, little research had been done on male contraceptive techniques. However, today, a major strategy that is being studied involves using synthetic male hormones to modify the normal functioning of the male reproductive system. The focus of most research in that area has been to suppress the formation of sperm. But progress is slow, since many combinations of hormones have caused side effects, such as impotency and incompatibility with alcohol consumption. Furthermore, a major hazard that must be overcome is the danger that a partially damaged sperm could fertilize an egg and lead to defects in a fetus. Thus procedures to suppress the formation of sperm probably will not become available for another 15 to 20 years, if then.

Public Policy

To speed up the demographic transition, many policymakers believe that national governments must make a stronger commitment to fertility reduction. We have noticed the success that several less-developed countries have experienced in reducing their fertility rates. And those declines took only a few decades, which is a considerably more rapid reduction than has occurred in developed countries. Although their approaches to fertility reduction were different for each of those less-developed nations, virtually all involved the following factors: (1) widespread availability of family-planning services, (2) educational campaigns concerning the impact of population growth, and (3) some combination of social and economic rewards, penalties, or both, to encourage small families. Thus we now examine those approaches to reducing fertility rates.

Family-Planning Services *Family planning* is a voluntary program of fertility control used by a couple to achieve the family size of their choice. Family-planning services provide information on birth con-

Table 16.6 Improved and new contraceptive procedures that likely will be available to the public during the next decade

1. Safer Oral Contraceptives—with Lower Doses, Lower Drug Levels in the Bloodstream, or Drugs More Specifically Targeted at a Particular Part of the Ovarian Cycle

2. Improved IUDs—Improved Copper-Releasing IUDs that Require Replacement only Every 5 to 10 Years; Postpartum IUDs that Can Be Safely Inserted After Delivery

3. Improved Barrier Methods for Women—Including One-Size-Fits-All Diaphragms, Cervical Caps, and More Efficient Spermicides

4. Improved Steroid Injections—Enabling Extremely Effective Contraceptive Injections that Last 1 to 6 Months

5. Improved Ovulation-Detecting Methods for Use with Periodic Abstinence—Including a Simple Test for Reliably Telling when Ovulation Has Occurred. (A Kit for Testing Urine is Under Development.) Predicting Ovulation Remains a More Difficult Goal

6. Steroid Implants—Steroid Capsules Implanted Under the Skin that Could Provide a Contraceptive Effect for as Long as 5 Years

7. Steroid Vaginal Rings—to Be Inserted by the User for 3 Weeks or More and Provide Protection Similar to Oral Contraceptives

Source: *World Population and Fertility Planning Technologies: The Next 20 Years.* Washington, D.C.: U.S. Congress Office of Technology Assessment, 1982.

trol. Many also provide contraceptive devices and procedures. To be effective, family-planning agencies must make special efforts to reach the poor and persons who live in rural areas. In addition, successful programs provide a wide array of contraceptive procedures, so that persons with differing needs and different sociocultural backgrounds can be served.

Educational Programs Successful educational programs have linked fertility reduction with future benefits to individual families as well as to society. By emphasizing future relationships between population and resources, governments can shift the focus of childbearing from the well-being of parents (such as having children as old-age security) to the well-being of children. For example, when policy planners in the People's Republic of China began to project the future trends concerning their population and resources, they noted that each Chinese person currently subsists on barely one-tenth of a hectare of land. Then they projected that if each family had an average of two children, another 300 to 400 million persons would be added. Thus they were able to show that each member of the next generation would have to exist on 40 percent less land. And if the People's Republic of China grows beyond 1.2 billion persons (its current population is about 1 billion persons), it probably will not be able to feed itself and maintain its current standard of living. As a result, government officials used such population and resource projections to build support for their one-child family program. The nature of that program is described further in Box 16.2.

Rewards and Penalties Governments have used various types of incentives and disincentives in their attempts to reduce family size. One of the oldest family-planning incentives is to make a one-time payment to persons who become sterilized or use contraceptives. For example, in India, a person who is sterilized receives about $15, which is approximately 2 weeks' pay for an agricultural worker. And in countries that have stronger economies, more extensive rewards are offered. Although each person in South Korea who is sterilized only receives approximately $15, couples who have two children and are sterilized gain priority in purchasing subsidized housing and in qualifying for housing and business loans. Moreover, their children can receive free medical treatment at nearby clinics until they are 5 years old.

In a few cases, incentives that improve the welfare of the community have been offered. For example, in a northern province of Thailand, piglets were given to women who agreed not to become pregnant while the animals were being fattened for market. The local family-planning group then marketed the pigs and shared the profits with the women who raised them. As a result, during one 3-year period, no woman who contracted to raise a pig became pregnant.

The alternative is to penalize persons who have large families. For example, Singapore is currently the only nation with a comprehensive disincentive program that is specifically designed to promote two-child families. Features of that program include limiting paid maternity leave to the first two births, and only allowing the full income-tax deduction of $750 for the first two children. The tax deduction for a third child decreases to $500. Also, fees in government maternity hospitals are lowest for the first two children, and they accelerate rapidly with additional births.

Given our previous discussion of motivational factors, future family-planning programs should be aimed more directly at women. Those programs can include educating young women, creating more jobs for women, increasing women's earning power, and improving health care for mothers and their babies.

Moreover, public policy could encourage women to return to breast-feeding, which acts as a natural contraceptive, because it delays ovulation as long as the child is being nursed. Mothers in several nations, including Thailand, Indonesia, and Bangladesh, commonly breast-feed their infants for more than 20 months after their birth. If a child is nursed for 2 years, the next child could not be born much sooner than 3 years after the birth of the first child. Thus breast-feeding is an effective, safe procedure to space children. However, women in developed nations have all but abandoned breast-feeding and, ironically, as young women in less-developed countries become better educated, they also prefer bottle-feeding.

Another effective way to reduce fertility is to delay marriage and, subsequently, to delay the birth of the first child. To encourage such practices, incentive payments could be made to young women. In most instances, such payments would be much less than the cost of maternity and child care.

Although programs such as those just described definitely have benefits, they also have their shortcomings. For example, disincentive programs can punish innocent persons by withholding resources and services from the "extra children." However, if the programs are designed and administered properly, rewards and penalties can be effective ways to reduce family size.

Public Policy and Common Sense Common sense should prevail in all public policies to reduce family size. For example, government policies to reduce fertility should nudge—not push—parents to have fewer children. And most persons would accept these policies more readily if they were directed at future births and did not penalize families that are already large. Furthermore, stronger policies are more likely to be accepted if they are preceded by milder policies. Similarly, success is more likely if the policies are phased in gradually and include widespread advance notice.

To be successful, small-family policies must be tailored to the cultural, political, social, and economic setting of the country. For example, China's stringent one-child policy, described in Box 16.2, would not work in most other countries. Yet there are so many different policies available (and more can be developed) that governments do not need to be reluctant to confront the difficulties of influencing family size. Certainly, countries should not wait to slow population growth until compulsory measures seem to be the only answer. Compulsory measures not only infringe on human rights, but they also run a high risk of failure as a result of noncompliance or open rebellion.

Public Policy in the United States Before we become critical of governments of less-developed nations who are slow to devise and vigorously implement small-family policies, we must remember that the United States does not itself have a population policy. We can better appreciate the reluctance of other governments to accept policies that influence childbearing by recalling the public controversies in our own country concerning much simpler issues, such as sex education in the schools and the provision of contraceptive services to unmarried teenagers. Although our federal government usually avoids controversial issues, it does provide funds for contraception research and limited financial assistance for family-planning programs in this nation and others.

Perhaps it can be argued that our population and

resource balance is such that more stringent population policies are not needed in the United States. Certainly, as Table 16.3 implies, the socioeconomic conditions of the United States are much better than those of most less-developed countries. But we must realize that significant increases in fertility rates in the United States probably would lead to the need for stringent policies.

Transition to a Stationary Population in the United States: Problems and Prospects

During the late 1970s, the population of the United States reached a major milestone—the fertility rate of American women declined below the natural replacement rate of 2.1 children per woman. Since then, the fertility rate has remained lower than the replacement level, and population growth continues to decline. In fact, projections by the U.S. Census Bureau in 1982 showed that by 2035, total deaths will equal total births and, disregarding migration, the nation will arrive at zero population growth. Figure 16.10 illustrates the approximate age structure that will exist for the United States in 2035. Because that transition to a stationary population has begun to affect our social and economic systems already, let us examine those effects.

To better appreciate the consequences of moving toward zero population growth, we must first realize that the current low fertility in the United States was preceded by the extraordinary baby-boom generation of the 1950s and 1960s, when men and women married earlier and had children sooner than couples do today. The fertility rate also increased to a peak of 3.7 births per woman in 1957. Between 1955 and 1964, nearly 42 million babies were born in the United States—more than during any other 10-year period, before or after, in the nation's history.

That unprecedented population growth forced a rapid expansion of elementary schools, secondary schools, and colleges in the 1960s. As a result, during the 1970s, more than 17 million young persons entered the job market, which placed a great amount of stress on the economy's ability to absorb them.

In the meantime, the fertility rate declined precipitously, and the baby-boom generation was followed by the baby-bust generation. And as fertility rates fell to 1.9 births per woman, the painful task

BOX 16.2

POPULATION CONTROL IN THE PEOPLE'S REPUBLIC OF CHINA

There have been many changes in the People's Republic of China since the death of Mao [Zedong in 1976] but none of them has been more dramatic than the reversal in population policy. Mao believed that the Chinese people, no matter how numerous, could construct the material basis for prosperity; it was said that "revolution plus production" would solve all problems. In the regime of Deng Xiaopeng, however, population growth is perceived as the main obstacle to improving the standard of living. The emphasis given to birth control by the current regime is reflected in many official statements and actions. For example, in December, 1982, the National People's Congress adopted a new national constitution that includes a provision making family planning the duty of all married couples.

Between 1949 [when the Communist Party took power] and 1980 the birth rate fluctuated sharply with changes in economic organization and social policy. In 1949 the rate was 35 births per 1000 people per year, a level that had prevailed for many decades. As a result of the elimination of private ownership of agricultural land and early birth-control campaigns the rate fell to 25 per 1000 in the late 1950s. During the Cultural Revolution and other political upheavals of the 1960s birth planning was abandoned as an official goal, whereupon the birth rate rose rapidly to a maximum of 44 per 1000 in 1 year.

It was in the 1970s that the full effect of organized population control began to be felt. The Third Birth Planning Campaign, with the slogan of "Later, Longer, Fewer," advocated later marriage, a longer interval between births, and fewer children. It was succeeded in 1979 by an even more energetic program: the One-Child Campaign. By 1980 the birth rate had decreased to 18 per 1000. Official reports may have underestimated the number of births; there has also been a recent increase in fertility. Hence the current birth rate could be about 20 per 1000.

The magnitude of the decrease that has been achieved in fertility is impressive. The One-Child Campaign appears to be taking hold, at least in some parts of the country. In Shanghai, Beijing, and 5 of the 26 provinces and autonomous regions between 80 and 90 percent of all births are births of first children.

The effectiveness of the recent birth-control campaigns results chiefly from the political structure of Chinese society. The structure is the legacy of the 20-year struggle for power waged by Mao and his associates. In the struggle cadres were carefully selected and trained in secret party schools; each cadre then became a link in the unified command structure. Many cadres were recruited from the peasantry, and their job was to transform the rural populace from the unprotesting, taxed people that Asian peasants have tended to be into conscious agents of change.

When the Communist Party took power, the increased political consciousness and organization that had been developed in war were turned to civilian purposes, including family planning. In part because the political structure originated in an armed conflict, it has a hierarchical form much like the form of a military organization. The hierarchy has six main levels. At the top is the nation as a whole and at the bottom are the production teams, each with from 250 to 800 members. Every administrative body, regardless of its level, includes an agency of the executive branch, a Communist Party unit, and a unit responsible for the delivery of health services. Birth-control services are integrated into the health-care system, and in each administrative body there are personnel whose job is family-planning education. In small discussion groups guided by party cadres the concept of population control is passed down from the highest executive agency to married couples.

In addition to arguments for population control, birth-control devices and rewards and punishments are delivered through the administrative apparatus. Contraceptive pills, intrauterine devices (IUDs), and sterilization operations are available free to married couples, as are premarital physical examinations that include explanations

A billboard on a main street in Peking encourages the one-child family. (Owen Franken, Sygma.)

of contraception. Unlike most other nations, the Chinese have favored the IUD over all other contraceptive devices.

The means of persuading married couples to limit their fertility is left to the local authorities, but in all localities there are rewards for family planning and punishments for having too many children. Newly married couples are asked to sign a statement committing them to having only one child. By the middle of 1981 more than 11 million couples had signed the agreement, which entitles them to free hospital delivery of their child, free medical care and education for the child, special consideration for better housing, and an extra month's salary each year. Single children are given preferential treatment in school and in obtaining jobs, particularly if the child is a girl.

The penalties for unsanctioned pregnancies can be substantial. The birth of a second child to a couple that has signed the one-child document results in an official reprimand. The birth of a third child results in a pay cut. Many women who are pregnant without permission are persuaded to have an abortion. Having an abortion entitles a woman to a paid vacation, as does a contraceptive sterilization operation. The decentralized execution of political decisions, with zealous local party cadres competing to make a good showing, can lead to coercion in both abortion and sterilization.

The combination of free contraceptives, the system of rewards and punishments, and the well-organized apparatus of social persuasion has led to one result that is quite unusual. In most countries where the demographic transition (from high birth rates and death rates to low birth rates and death rates) has begun the reduction in fertility has been greatest in the areas with the highest level of urbanization and income and the lowest death rate. Work done by H. Yuan Tien of Ohio State University shows that until recently there was a correlation in China between urbanization, income, and low mortality on the one hand and reductions in fertility on the other. By 1980, however, the relation was much weaker and indeed was no longer significant in the statistical sense. The family-planning campaign has apparently now reached the poor rural areas without the health, wealth, and urbanization that have been associated with the demographic transition in Europe and many developing countries.

The regime has considerable support from the Chinese people and its guidelines elicit a degree of spontaneous compliance. Nevertheless, newspapers and other periodicals in China disclose and criticize many elements of compulsion. Accounts are published of forced abortions, and there are cases of infanticide when the firstborn turns out to be a girl. Protests are heard when a female only child and her parents are given exceptional privileges. One can expect this form of resistance and others to increase.

Excerpted from Nathan Keyfitz, "The Population of China." © Scientific American 250:38–47 (February), 1984. All rights reserved.

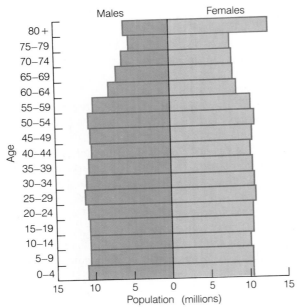

Males   Females

Figure 16.10 *The approximate age structure of the United States if it achieves zero population growth by the year 2035. (After L. F. Bouvier, "America's Baby Boom Generation: The Fateful Bulge." Population Bulletin 35 (April), 1980.)*

of rapid retrenchment followed. For example, schools closed and teachers were laid off. Moreover, the demand for goods and services for children and young adults declined. As a result, advertising campaigns began to focus on older persons.

As a result of that trend, we can expect some improvements in social conditions in the future. Because high crime rates and unemployment are closely associated with young, unskilled persons, those problems are expected to decline. Moreover, as fewer unskilled persons enter the work force and workers of the baby-boom generation continue to improve their skills, overall job productivity should increase.

But problems also lurk in the future. One problem that we can anticipate is that the total number of births is very likely to fluctuate rather than remain constant. For example, during the late 1970s, the total number of births per year began to increase despite an unchanging fertility rate. The reasons for that increase were twofold. First, most of the women of the baby-boom generation had reached their childbearing years. And second, some of the many women who had delayed childbearing to start their careers decided to have their first child. Because a very large percentage of women are now 20 to 29 years old and are still childless, and because surveys

of those women indicate that many of them still expect to have two children, the total number of births should remain high during the next decade. As a result, almost as many children may be in elementary schools in 1995 as were enrolled during the peak year of 1970, when the current decline began. Hence communities that are laying off teachers and closing schools today will need more schools and many more teachers within the next decade. However, most of those new schools will be needed in the sunbelt and western states, where birth rates are higher than in the Northeast and Midwest. Unfortunately, these alternating periods of boom and bust will make it difficult to maintain a high-quality educational system. Furthermore, they will force cyclic adjustments to be made in all the other facets of our social, economic, and political systems as well.

Some of the greatest difficulties will arise when the baby-boom generation reaches retirement age. As Figure 16.11 illustrates, twice as many persons will be at least 65 years old by 2020 than there were in 1980. Thus concern has been voiced about the burden that will be placed on the social security system and private pension plans. For example, today, approximately 3.2 workers exist for each beneficiary in the social security system. However, in 2030, only 2 workers will probably exist for each beneficiary. Moreover, just as the baby-boom generation put tremendous pressure on the educational system in the 1960s and early 1970s, a similar demand will occur for medical care, housing, and recreational facilities for the elderly in the future.

Although it is not clear at this time how those needs will be met, some optimism is warranted. For example, if birth rates remain low, the total number of dependent persons relative to working persons will be approximately the same in the 2010s as it is today. That is, the increased proportion of elderly persons will be offset by a reduced proportion of children who are under the age of 18. Thus the overall number of dependents relative to the number of persons who are working (persons who are 18 to 64 years old) will remain virtually unchanged. As a result, the major problem that may confront society may be a need to reallocate resources from the young to the elderly. To accomplish that transition, a major reorganization of our society will be needed. Our society has clung tightly to its youth culture, and Americans have generally abhorred the idea of growing old. But the baby-boom generation has forced

American society to restructure in the past and, no doubt, its members will continue to change society as they approach retirement age.

Changes in the age structure of the United States will occur gradually, and we have the time to plan ahead for a smooth transition. Although some persons may fear that transition, it is inevitable, and, clearly it is better to provide for an orderly transition now.

World Population: What of the Future?

Some of our insights concerning the future give us hope; others cause us to have deep concern. For example, population experts can point to many positive changes that have taken place in recent decades. The world's annual growth rate reached a peak of 2 percent in the early 1960s and, since that time, it has slowly declined to approximately 1.7 percent. In fact, the population growth rate is not increasing

Figure 16.11 *A doubling of the number of Americans who are 65 years old or older will occur within the next four decades. But their increased numbers will be offset by the reduction in the size of the population that is younger than 18 years old (assuming that birth rates remain low). This shift in population structure will necessitate a major reallocation of resources from the young to the elderly. (After L. F. Bouvier, "American's Baby Boom Generation: The Fateful Bulge." Population Bulletin 35 (April), 1980.)*

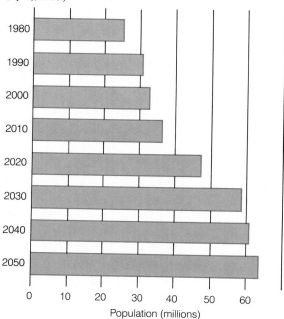

significantly in any region of the world at this time. Most of the developed nations are approaching zero population growth, and growth rates have declined significantly in some less-developed nations, particularly in Asia. And some indications show that the demographic transition will occur more rapidly in the less-developed countries, once it begins, than it has in the developed nations. Four out of five persons in the less-developed world today live in countries that, at least, have an official policy to slow the growth of population. Perhaps most encouraging are results of a recent survey of 27 less-developed nations, which showed that nearly half of the women who were surveyed claimed that they desired no more children. Because of these developments, population experts are predicting a continued decline in the world's population growth rate. And a few believe it may decline to as low as 1.3 percent by 2000.

But optimism for the future must be tempered by several factors, including the consequences of exponential growth. Although the percent annual growth rate may be declining, the actual number of persons that are added to the world's population each year will actually increase for some time. For example, in 1985, the world's population of 4.8 billion persons was growing at an annual rate of 1.7 percent. Hence 82 million new persons were added to the world's population that year. In 2000, the world's population is expected to be nearly 6.1 billion persons. Thus if we assume a conservative decline in growth rate, to 1.5 percent, nearly 92 million persons would be added that year—over 10 percent more than were added in 1984, despite the lower percent growth rate.

Another reality is population momentum—35 percent of the world's population today is younger than 15 years of age, while only 6 percent is older than 64 years of age. Hence six times more persons will be soon reaching reproductive age than will be fulfilling their life expectancy.

Some nations, particularly many in Africa, are not experiencing any decline in their growth rates. Thus population doubling times of 25 years or less are common in those nations. Moreover, the preferred family size in many less-developed countries, including countries outside the African continent, is still much higher than the replacement rate. Hence if growth rates are to be lowered to replacement levels in those nations, considerable efforts are still needed to create social, economic, cultural, and political conditions that will motivate persons to have smaller families.

Once persons are motivated, family-planning programs must be available. But even if that happens, much work must be done. For example, even though many women do not want any more children and most nations have an official policy to slow population growth, at least half of the world's couples do not actually have access to family-planning services. And access to those services is particularly inadequate in rural areas. Even in the United States, public policy restricts the availability of family-planning services for the poor. Hence socioeconomic developments, cultural developments, and the use of contraceptives must be improved considerably in less-developed nations before they can attain the low population growth rates that are now common in developed nations.

Therefore, we can conclude that the world's population will continue to grow for many years to come. As we have mentioned previously, the world's population is expected to grow to approximately 6.1 billion by 2000—nearly one-third more than it is today. Assuming that policymakers continue to create an environment that motivates persons to limit the size of their families, that all persons have access to family-planning programs, and that parents decide to limit the size of their families to the replacement level, the United Nations projects that the world's population will reach 10.5 billion before it stabilizes in approximately 100 years. The World Bank foresees a somewhat lower level (9.8 billion). The all-important question is: Can food, water, and energy resources keep pace with the additional demands from a growing population?

We have evidence that the race is already being lost. Table 16.7 indicates that, on a worldwide basis, production of food and fuel is no longer keeping ahead of population growth. And a related concern is the growing disparity between developed and less-developed nations. As Figure 16.12 indicates, almost all future population growth will occur in the less-developed nations, yet many of them do not have the resources to meet the needs of their current populations, let alone the additional millions that will be born during the next 15 to 20 years. For example, already, impoverished nations, such as Bangladesh and the Sahel-zone nations of Africa, are experiencing rising death rates, primarily as a result of hunger and nutritional stress. And droughts and floods have

Table 16.7 World per-capita production of basic commodities, 1950–1983, with peak year underlined

YEAR	GRAIN (KILOGRAMS)	BEEF (KILOGRAMS)	FISH (KILOGRAMS)	WOOD (CUBIC METERS)	OIL (BARRELS)
1950	251	—	8.4	—	1.5
1955	264	—	10.5	—	2.0
1960	285	9.3	13.2	—	2.5
1965	275	9.9	16.0	0.66	3.3
1970	299	10.6	_19.2_	0.72	4.5
1971	318	10.4	19.0	0.71	4.7
1972	304	10.6	17.5	0.71	4.9
1973	326	10.5	17.4	0.72	5.3
1974	309	11.0	17.4	0.71	5.0
1975	310	11.3	16.5	0.69	4.8
1976	331	_11.6_	17.0	0.70	5.2
1977	319	11.5	16.5	0.70	5.3
1978	_343_	11.4	16.5	0.71	5.4
1979	327	10.6	16.4	0.71	_5.5_
1980	326	10.2	16.4	_0.72_	5.2
1981	332	10.1	16.2	0.70	4.9
1982	337	9.6	16.7	—	4.7
1983	310	9.5	—	—	4.4

Source: Food and Agricultural Organization, U.S. Departments of Agriculture and Energy, and U.N. Demographic Division. Compiled by Worldwatch Institute, April, 1985.

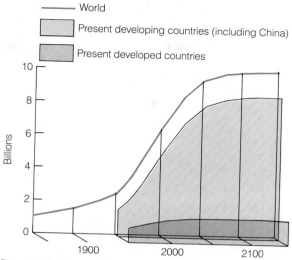

Figure 16.12 *Almost all future population growth will occur in the less-developed nations. (Based on estimates by World Bank, published in* Agriculture: Toward 2000. Rome, Italy: FAO, 1981.)

reduced the food supply in those countries, which have forced a decline in food stocks and sharp rises in food prices. Furthermore, food that is available is often insufficient and, in many cases, persons are too impoverished to buy enough food to sustain themselves even when it is available. Millions of persons—most of whom are children or elderly—die from starvation and famine-induced diseases every year. And, for each person who dies, hundreds more suffer. The millions of children who somehow survive semistarvation inherit the grim legacy of irreparable brain damage.

But even this is not the whole story. Stressed by overpopulation, some regions are evolving into ecological disaster areas. Deforestation and overgrazing, and their consequence—severe soil erosion—are reducing the land's productivity. Thus many subsistence farmers and nomads are losing their means of earning a living and producing an adequate food supply. But the developed countries also face similar problems. For example, accelerated soil erosion in the United States is reducing the fertility of the world's richest agricultural land. Tragically, the land's carrying capacity for humans is diminishing while the human population continues to grow.

Conclusions

In the global perspective, we are approaching the earth's limited capacity to sustain us. To further complicate matters, the persons of the world are divided by political boundaries. Few, if any, countries are self-sufficient; all rely on resources from other nations. Thus when the demands of a country's population outweigh its capacity to produce or trade for needed resources, that country becomes overpopulated. Many countries have already exceeded their carrying capacity, and others are rapidly approaching it. A nation that exceeds its carrying capacity is faced with three alternatives: it can (1) rely on the good will of other nations to provide the needed resources, (2) raid its neighbors for needed resources, or (3) allow death to reduce its population lower than its carrying capacity. In any event, international conflicts are the inevitable result of intensified competition for finite resources and opportunities.

In response to these problems, the human population must cease to grow. Nature dictates that our numbers (and those of all other species) must reach a balance with the supply of needed resources. However, the question remains whether we will achieve that balance by rational and orderly efforts to lower the birth rate or reach it through chaos and a grim rise in the death rate.

Summary Statements

Famine, unemployment, inflation, crowding, dwindling resources and pollution are all related to the fact that too many persons are competing for a finite number of resources and opportunities. Population pressures are felt both in this country and abroad. The more affluent the population, the greater is its impact.

For thousands of years, because of the vulnerability of human beings to a hostile environment, the human population remained quite small and grew very slowly. The development of agriculture, advances in disease prevention and cures, and improved sanitation were major factors that contributed to the recent unequaled population growth.

A population's growth rate is a function of its birth rate, death rate, and age structure. If a large percentage of a population is in the prereproductive age group, then the population will probably continue to grow. And even a population that has reached its replacement rate will grow if its reproductive and prereproductive groups are large. A stationary population will have nearly equal numbers in the prereproductive, reproductive, and postreproductive age groups.

Developed nations have gone through a demographic transition, in which a decline in death rates has been followed by a decline in birth rates. Hence their population growth rates are low. Most less-developed nations have experienced a recent decline in death rates, but birth rates remain high. Hence those nations have a high population growth rate.

Reducing birth rates is the key to slowing population growth. Several interacting factors, including the socioeconomic status of women, basic education for all children, basic health care for all citizens, and equitable job opportunities appear to play a role in motivating parents to have smaller families.

A variety of means exist for limiting family size. Although no perfect method is available, key characteristics include safety, effectiveness, low cost, convenience, and compatibility with cultural, religious, and sexual attitudes.

Although much concern has been raised over the health effects of the pill, younger women and nonsmokers run far lower risks from taking the pill than from becoming pregnant. Women who are older than 35 years and who smoke should use some other form of birth control. Few serious health risks are associated with the use of IUDs and diaphragms. Abortion and sterilization are essentially 100-percent effective. The rhythm method is much less successful. Abortion is the most controversial method of controlling the birth rate in terms of cultural and religious attitudes.

Research on improved birth-control methods continues. But extensive testing is required, and any new advances in birth-control technology will be slow in reaching the public.

To speed up the demographic transition in less-developed countries, national governments need to make stronger commitments in terms of allocating adequate resources for fertility reduction. Successful public policies include widespread availability of family-planning services, educational programs that link population growth with future resource availability, and a combination of social and economic incentives, disincentives, or both.

As the population of the United States becomes stationary, it will contain more older persons and fewer younger persons. That shift in age structure carries many social, political, and economic ramifications for our society, but because the shift will be gradual, we have the time to plan ahead for a smoother transition.

Some of our insights into the future give us hope; others are the cause of deep concern. The world's population growth rate has begun to decline, the demographic transition is occurring relatively rapidly in some less-developed nations, and increasingly more persons desire to have smaller families. Hence population growth rates are expected to continue to decline. On the other hand, the continued high population growth rates in many African nations, the limited availability of family-planning services (even in the United States), and the fact that nearly 35 percent of the world's population today is less than 15 years old portend for serious population pressures in the future. Already, in a few impoverished nations, the death rate appears to be rising because of starvation and famine-induced diseases. And ample evidence exists to suggest that we are approaching the earth's capacity to sustain us. Just how the human population will reach a balance with the supply of needed resources remains unclear.

Questions and Projects

1. In your own words, write a definition for each of the terms that are italicized in this chapter. Compare your definitions with those in the text.

2. How is the well-being of the United States related to fertility reduction policies of other nations?

3. List five factors that contributed to the growth of the human population during the last 10,000 years.

4. Describe why age structure must be considered along with birth rates and death rates in predicting the population growth of a nation. In view of the most recent trend—a small decline in the world's birth rates—what is the significance of the interactions of these factors for resource management during the next 50 years?

5. Why is it necessary to understand the concept of population momentum before one can realistically predict when a country will achieve a stationary population?

6. Describe the process that is known as demographic transition.

7. A desire for greater personal freedom has contributed to the decision of many couples in developed nations to have few or no children. But some persons are concerned that those couples will take on less responsibility in their communities; that is, couples with few or no children may not be concerned about the quality of schools, youth opportunities, or the kind of world that will be left for the next generation. How would you respond to someone who voiced this concern?

8. What factors could motivate persons to have smaller families? What factors do you think are most important?

9. What are your personal plans concerning marriage and family size? What factors most influence your decisions, and why?

10. What are the characteristics of an ideal birth-control method? In your opinion, which currently available method best meets those criteria? Defend your choice.

11. Describe how improperly designed and implemented programs of incentives and disincentives for fertility reduction can violate hu-

man rights and create a backlash. How can those problems be minimized?

12. The transition in the United States to a stationary population will require many adjustments. Describe some of the expected social, environmental, and economic changes. How will your life be influenced by those adjustments?

13. What factors would tend to raise the American birth rate? What factors would lower it? Which set of influences do you think will be more important during the next decade?

14. Comment on the following statement: Because the entire population of the United States could stand on an area that is roughly the size of New York's Manhattan Island, we really do not have a population problem.

15. When is a country considered to be overpopulated? What countries meet those criteria? What are those countries doing to solve their population problems?

16. Although we strongly encourage other countries to develop and implement population policies, the United States does not have an official policy on population. Suggest some probable reasons for this apparent paradox.

17. Evaluate the following statement: Population control is necessary, but, by itself, it is not sufficient to solve many of society's problems.

18. Visit a Planned Parenthood Office in your community. What are the major problems that the agency is attempting to solve? What are the prevailing attitudes of the agency's clients about their desired family size? Also visit a Birthright Office. How do the attitudes of the Birthright employees differ from the Planned Parenthood employees? How are they similar? Could those two groups work together toward a common goal?

19. Design a survey to assess attitudes concerning family size. Administer the survey to several diverse groups, such as your class, a religious organization, and your neighbors. Compare the responses and identify the reasons for similarities and significant differences among the several groups.

Selected Readings

BOUVIER, L. F.: Planet Earth 1984–2034: A Demographic Vision. *Population Bulletin 39* (Feburary). Washington, D.C.: Population Reference Bureau, 1984. An examination of the present demographic situation of the world and an intriguing speculation of what it might be 50 years from now.

BOUVIER, L. F.: America's Baby Boom Generation: The Fateful Bulge. *Population Bulletin 35* (April). Washington, D.C.: Population Reference Bureau, 1980. An examination on the causes of the baby boom and the impact that this oversized generation will continue to have on American society.

CATES, W.: Legal abortion: The public health record. *Science* 215:1586–1590, 1982. Evaluates the safety of legal abortions as well as some potential adverse outcomes on future desired pregnancies.

COALE, A. J.: Recent trends in fertility in less-developed nations. *Science* 221:828–832, 1983. An examination of the wide variation among less-developed countries in their fertility rates and some of the causes of such variations.

FEINBERG, J.: *The Problem of Abortion*, 2d ed. Belmont, Calif.: Wadsworth, 1984. A collection of papers that examines the issue of abortion from a wide range of philosophical and ethical viewpoints.

GOLIBER, T. J.: Sub-Saharan Africa: Population Pressures on Development. *Population Bulletin 40* (February). Washington, D.C.: Population Reference Bureau, 1985. An in-depth study of the population characteristics, pressures, and policies that will have a major influence on the future of Africa's 42 sub-Saharan nations.

GWATKIN, D. R., AND BRANDEL, S. K.: "Life Expectancy and Population Growth in the Third World." *Scientific American* 246:57–65 (May), 1982. A study of the influence of a declining death rate on population growth in developing nations.

JACOBSON, J.: *Promoting Population Stabilization: Incentives for Small Families.* Washington, D.C.: Worldwatch Institute, 1983. An objective evaluation of a variety of incentives and disincentives that are available to policymakers for encouraging small families.

KEELY, C. B.: "Illegal Migration." *Scientific American* 246:41–47 (March), 1982. An examination of the impact of foreign workers on the economic and legal systems of the United States.

POPULATION REFERENCE BUREAU: *Annual Population Data Sheet.* Washington, D.C.: Population Reference Bureau. A concise summary of population data, listed for more than 125 nations, that is updated yearly.

SAI, F. T.: The population factor in Africa's development dilemma. *Science* 226:801–805, 1984. An excellent overview of the relationships between population growth and social and economic development in sub-Saharan Africa.

SHORT, R. V.: "Breast Feeding." *Scientific American* 250:35–41 (April), 1984. An examination of how breast-feeding acts as a contraceptive and how a trend toward bottle-feeding in developing nations is contributing to rising fertility rates.

VAN DE WALLE, E., AND KNODEL, J.: Europe's Fertility Transition: New Evidence and Lessons for Today's Developing World. *Population Bulletin 34* (February). Washington, D.C.: Population Reference Bureau, 1980. A reexamination of the causes of the demographic transition in Europe and the implications for slowing population growth in less-developed nations.

WULF, D.: "Low Fertility in Europe." *Family Planning Perspectives* 14:264–270, 1982. An evaluation of the impact of subreplacement fertility and economic and social systems of Europe and the varying responses of European nations.

The two extremes of the world food situation—starvation in the Sahel and surplus wheat in the United States. (FAO photo; Carl Davaz/Topeka Capital-Journal.)

CHAPTER 17

FOOD RESOURCES AND HUNGER

In 1983 and 1984, one of the worst famines in history struck the African continent as severe drought held nearly 30 African nations in a deadly grip. One belt of drought swept across the continent along the southern edge of the Sahara Desert from the Atlantic Ocean eastward to Ethiopia. This sub-Saharan region, called the *Sahel zone*, encompasses 6 nations—Mauritania, Senegal, Mali, Burkina Faso (Upper Volta), Niger, and Chad. A second area of drought crossed the southern portion of the continent from Angola to Mozambique and then northward along the Indian Ocean to Somalia.

As the drought progressed in those regions, much of the livestock died and many nomadic persons lost their means of earning a living. As the drought further intensified, many persons (with no food and no means of support left) were forced to walk many kilometers to relief stations in search of food. Many persons arrived at those camps in such a weakened condition that they soon died. Others never arrived at the camps, having died en route.

Although there had been numerous warnings from U.N. officials that a widespread famine was imminent, the world failed to notice the desperate plight of those persons until November of 1984. As a result of television broadcasts that vividly depicted the tragedy of starving children, several developed nations rushed food to a few African nations that appeared to be suffering most from the famine—Ethiopia, Mozambique, and Chad. In the meantime, famine had killed as many as 500,000 persons and

no one knew how many more would die before the tragedy would end.

Famines are not new to Africa. The last big African famine occurred only 15 years earlier, when another period of drought in the Sahel zone resulted in the death of 100,000 to 250,000 persons. The tragedy that occurred in the Sahel in the early 1970s and was repeated soon after in 1984 is representative of a potentially worldwide problem. Thus in our attempts to explain what happened in the Sahel zone, we can illustrate some of the complexities that are involved in feeding the inhabitants of a single, isolated area as well as the rapidly growing world population.

The Sahel zone is located in the savanna transition zone, between the hot, steamy equatorial region to the south and the arid Sahara Desert to the north. Normally, the Sahel zone experiences a rainy summer and a dry winter. But beginning in the late 1960s and generally continuing through the early 1980s, the monsoon winds that normally bring rain to the region arrived later than usual. And when the monsoons finally did appear, they lasted for only a short period of time and produced less rain than usual. The average annual precipitation for that region in the late 1960s declined by 45 percent and 1983 and 1984 were the driest years of the century (Figure 17.1)—a particularly disastrous turn of events for a region that has only marginal rainfall in the best of times. As a result, the soil dried up and blew away, livestock perished, and persons starved.

However, causes of that tragedy included more than just the climate. During the 1950s (when the rainfall was particularly great), the Sahel zone received aid from developed nations and its social conditions improved—increases in medical services lowered death rates, and imported technology spurred food production. As a result, the human and livestock populations experienced a burst of growth. Therefore, even though prolonged droughts had occurred previously in the Sahel zone, when the drought returned in the late 1960s, its effects were intensified because of the increased populations of humans and livestock.

Also, restricted migration compounds the plight of the Sahelians during periods of drought. In earlier times, the nomadic herdsmen had been able to flee when drought struck, and they traveled southward to regions where water and food were more plentiful. However, by the late 1960s, rigid political boundaries existed between those countries, and that option was no longer available. And even if those persons had been able to move southward, they may not have fared much better. The nations that border the Sahel zone, such as Nigeria and Ghana, were already having difficulty providing adequate resources for their own populations. And an influx of an additional 10 to 20 million persons from the Sahelian countries would have overwhelmed the resources of their neighboring countries.

Although the monsoon rains returned briefly in 1975, the legacy of overpopulation remained. For example, during the drought, the large herds of livestock had overgrazed the land, stripping it bare of vegetation, and the dry, thin layer of topsoil had vanished with the wind, leaving only rock and sand behind. As a result of those conditions, the Sahara Desert continued to expand slowly and unrelentingly southward, engulfing land in the Sahel zone that once supported the nomads and their herds.

As a result, the Sahel zone was even more vulnerable to disruption when drought returned and further aggravated the ecological damage that had been caused by overpopulation in an area that is too fragile to sustain the demands for resources. This relentless process of land degradation (which has been caused by overgrazing and poor soil management coupled with drought) threatens to reduce further the capability of Sahelian nations to feed their rapidly growing populations.

The basic problems that were experienced in the Sahel zone are the same as those that are asso-

Figure 17.1 *The sub-Saharan region of Africa experienced unusually wet weather in the 1950s. But that time of plenty has turned into an extended period of extreme drought. (Prepared by Peter Lamb, in R. A. Kerr, "Fifteen Years of African Drought."* Science 227:1453–1454, 1985. Copyright 1985 by the AAAS.)

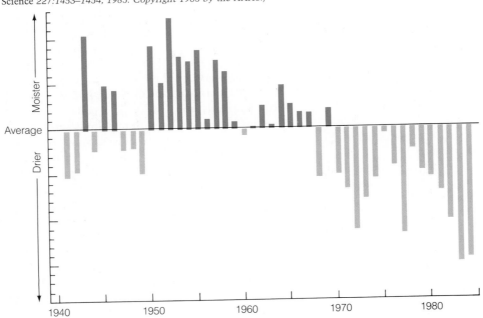

ciated with the world's food supply, and they can be summed up succinctly. Never before have so many humans existed on earth. Furthermore, current efforts to feed the world's growing population are degrading the earth's ecosystems to an unprecedented degree. During the past decade, food production has barely kept pace with population growth. In fact, as Figure 17.2 illustrates, the citizens of more than 20 nations actually receive less than 90 percent of the calories that they need, while the citizens of another 30 nations receive just enough calories to meet their needs. And of the nearly 150 nations on this planet, only 10 are net food exporters. (Of those nations, the United States is the largest net food exporter.) Therefore, quite simply, the human population may be approaching the limits of the earth's capacity to sustain it. Nevertheless, as we saw in Chapter 16, despite that possibility, the human population is continuing to double approximately every 40 years.

Figure 17.2 *The geography of hunger, 1978 through 1980. Many developing nations barely receive an adequate supply of calories.* (1983 World Population Data Sheet. *Washington, D.C.: Population Reference Bureau, Inc., 1983. Based on the U.N. Food and Agriculture Organization's estimates, provided by the World Bank.)*

In this chapter, we explore the prospects for feeding the growing human population. We also survey the constraints that are placed on food production, which all the nations of the world have to contend with. And, finally, we analyze some contrasting projections that have been made concerning how long we can realistically expect to feed the human population at its current rate of growth.

Reducing Food Losses

Approximately half of the world's food production is lost to humankind each year as a result of the damage caused by an amazing variety of organisms. For example, in the United States alone, nearly 19,000 species of agricultural pests exist, and approximately 1000 of those pests are considered to be a major problem. Some insects and rodents compete directly with us for food. And other insects, bacteria, and fungi weaken or kill crop plants or domestic animals, whereas still others induce rotting of crops either in the fields before harvesting or in storage facilities after harvesting (Figure 17.3). Also, weeds

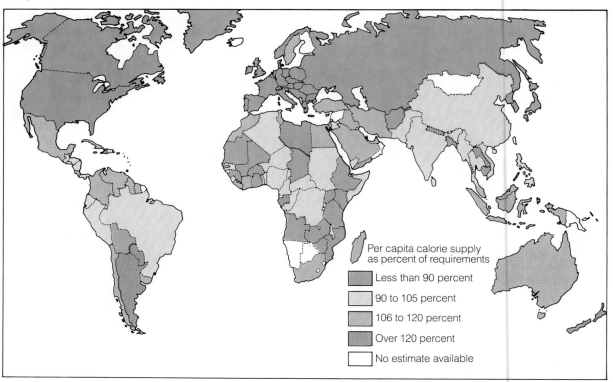

Per capita calorie supply as percent of requirements

- Less than 90 percent
- 90 to 105 percent
- 106 to 120 percent
- Over 120 percent
- No estimate available

Figure 17.3 *The tumorlike galls of corn smut (a fungus) will destroy this ear of corn. (Michel Viard, © Peter Arnold, Inc.)*

reduce food production by competing with crop plants for light, mineral nutrients, and water. Therefore, reducing the amount of food that is lost to those pests is one obvious way to feed more people.

Pesticides

Reducing food losses to pests can be partly accomplished by the use of pesticides. For example, in the United States, approximately one-third less food is lost to pests than is lost on the average throughout the world. Because of our widespread use of pesticides and well-maintained storage facilities, food prices in the United States are 30 to 50 percent lower than they would be otherwise.

Other benefits also have accrued from the use of pesticides. Because pesticides have helped to increase food production per hectare of land, we have not had to expand the amount of land that is being cultivated. Hence more land is available for forests, recreation, and wildlife preserves. Moreover, because much of the land that would have been used for agricultural expansion is hilly and subject to excessive erosion, the use of pesticides has reduced soil erosion indirectly and, consequently, has aided in maintaining water quality. However, unfortunately, some pesticides also have significant costs and risks.

Ironically, pesticides diminish food availability in some instances. For example, most pesticides are formulated to kill specific types of pest species. *Insecticides* are directed at insects; *rodenticides*, at

rodents (such as mice and rats); *fungicides*, at fungi (such as blight and rust); and *herbicides*, at weeds (that is, any plants that grow where they are not wanted). But, in addition to killing pests, some pesticides also kill species that benefit food production, such as bees. For example, when honey bees forage for pollen and nectar, they are frequently exposed to pesticides, which may kill them. In fact, pesticide poisoning contributed to a 20-percent decline in the honey bee population during the past decade in the United States. And that decline has serious ramifications, since nearly 100 crops depend on honey bees for pollination. Without insect pollination, there would be no apples, cherries, cucumbers, melons, or strawberries to eat, and no seeds to grow such forage crops as alfalfa and clover.

One potential problem that results from killing nonpest species accidentally is that it may lead to a disruption of the natural pest-control mechanisms, such as parasitism and predation (which are discussed in Chapter 4). For example, such a disruption occurred when efforts were being made to control insect pests on cotton that was growing in the Rio Grande Valley. Initially, the boll weevil, pink bollworm, and the cotton fleahopper were the major pests that had to be controlled every year to achieve a profitable yield of cotton. Two potential pests that threatened crops of cotton—the tobacco budworm and the bollworm—were usually kept in check by their natural enemies. However, when the new chlorinated hydrocarbon pesticides, such as DDT, became available shortly after World War II, farmers commonly applied these insecticides 10 to 20 times during each growing season. (Pesticides, such as DDT, that kill a wide variety of pests are frequently called *broad-spectrum pesticides*.) As a result, cotton yields spiraled upward as the major insect pests were destroyed. But within a few years, the tobacco budworm and the bollworm started to reduce the cotton yields. Further research revealed that the insecticides that were applied to control the three major pest species had also devastated the natural enemies of the tobacco budworm and bollworm. And without their natural population-control agents, those two minor pests became major pests. Thus in their attempt to control three pests, the cotton growers had produced two additional pests.

A further look at the use of insecticides in the Rio Grande cotton fields illustrates another frequent problem that occurs when pesticides are used—*pes-*

ticide resistance. For example, by 1960, the boll-worm and tobacco budworm were becoming increasingly resistant to DDT. Thus the growers increased the dosage and frequency of applications, which was a common practice. However, by 1965, those pests could no longer be controlled by DDT or any other chlorinated hydrocarbon pesticides. As a result, the growers then switched to the organophosphate pesticide, methyl parathion, but by 1968, the tobacco budworm had developed a resistance to that pesticide as well. Subsequently, severe damage to cotton yields occurred even though many fields were treated 15 to 20 times each season with all conceivable combinations of pesticides.

Although the development of pesticide resistance by insects in the Rio Grande Valley is an extreme case, increased resistance to pesticides is becoming more common. For example, the U.S. Department of Agriculture claims that genetic strains of more than 450 species exist that are resistant to one or more pesticides. To determine how such resistance develops, we must note that (1) thousands of members of a single species are commonly present in a hectare of land and (2) within such a large population, considerable genetic variability exists. Therefore, some members of that varied population may be naturally resistant to a particular insecticide, so they will not be killed when they are exposed to it. However, after several applications of the insecticide, most nonresistant members are selected against (that is, eliminated), and a small resistant population is left. In the absence of intraspecific competition (between resistant and nonresistant members), the numbers of resistant insects will explode. A schematic representation of the selection for a pesticide-resistant population is shown in Figure 17.4.

Thus when growers apply pesticides to their crops, they are, in effect, taking natural selection into their own hands. However, inadvertently, they are selecting for those members in the pest population that are more resistant to the pesticide. As a result, instead of eliminating the pest population, they are contributing to the growth of a pest population that is composed primarily of resistant members. And when growers start increasing dosages, applying more frequent applications, and using other pesticides, they are usually continuing the selection process for pests that become increasingly more difficult to control.

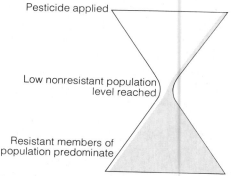

Figure 17.4 *A schematic representation of the selection for a pesticide-resistant population. The shaded area represents pesticide-resistant members of the population; the unshaded area represents pesticide-sensitive members. The passage of time goes from the top to the bottom of the figure.*

Another problem with some pesticides, particularly the chlorinated hydrocarbons, is that they break down slowly and, therefore, persist in the environment for many years. An obvious disadvantage of persistent pesticides is that the longer they remain in the environment, the greater the chances are that they will be transported into adjacent ecosystems. For example, aquatic ecosystems are easily contaminated by pesticide runoff. And since those chemicals are particularly soluble in the fatty tissues of organisms, they bioaccumulate in food webs. Also, the bioaccumulation of pesticides in aquatic food webs reduces the natural rate of reproduction in some fish populations and, as we saw in Chapter 7, renders their flesh unfit for human consumption. The accumulation of pesticides has also contributed to population declines of several endangered species (such as the bald eagle and the brown pelican), which occupy the higher trophic levels in aquatic food webs (see Chapter 14).

Thus we see that pesticides possess both benefits and costs in terms of both food production and environmental quality. And our assessment is further complicated by the enormous variety of pesticides that are on the market. For example, more than 1000 different chemicals are used as pesticides, and those chemicals occur in more than 30,000 different formulations. Furthermore, each of those formulations has different properties, and each one behaves differently in the environment.

During the 1970s, as the public became more

aware of the problems that are associated with pesticides, those chemicals came under much closer scrutiny. As a result, the chlorinated hydrocarbons were found to be the most hazardous to the environment, so the EPA banned the widespread use of several insecticides in that group, including DDT, aldrin, dieldrin, and heptachlor. However, the agency may permit their use in an emergency, such as during an epidemic of a disease that is transmitted by an insect vector. Moreover, those insecticides are still being manufactured by companies in the United States and by companies in other nations for sale throughout the world. Although most European nations have also banned one or more pesticides that contain chlorinated hydrocarbons, virtually all nations in Central and South America, Africa, and Asia still use large quantities of those pesticides.

When some of the chlorinated hydrocarbons were removed from general use, other insecticides—for example, the organophosphates—became more popular. Organophosphates are more compatible with the environment than chlorinated hydrocarbons, because they are generally less persistent and less likely to accumulate in food webs. However, organophosphates are more toxic to humans than are chlorinated hydrocarbons and thus they present a greater direct hazard to humans, especially to those persons who apply the pesticides. Persons have been fatally poisoned as a result of improper handling of organophosphates.

During the last decade, major changes in pesticide research have taken place. And chemical companies are now focusing their research efforts on developing new pesticides that are less persistent and more selective. In addition, more precise methods of application are being studied, because most pesticides (nearly 65 percent) are applied by aircraft, and nearly half of them miss their target areas.

Other Pest-Control Methods

The best way to avoid the harmful side effects of pesticides is to use other methods of pest control. Therefore, in recent years, pest control research has undergone a major shift in direction. Today, nearly 70 percent of the budget for pest-control research that is being funded by the U.S. Department of Agriculture is applied to nonpesticide projects. Ideally, the alternative methods of pest control should affect only the target pests; they should be able to control

pests permanently by reducing their populations to harmless densities; and they should not select for resistant pests. We would expect nature to be a major source of such alternatives. For example, we can reason that natural means of pest control are functioning all the time, since much of the plant material that is produced each year survives the onslaught of parasites, rodents, and herbivorous insects. Thus we must determine how potential pests are normally controlled in nature.

Biological Control   We have already described the role of predators and parasites in population regulation. In Chapter 4, we described how natural predators and parasites can be used to control pest populations by using a procedure that is known as *biological control*. That procedure has several advantages. It is nontoxic and, usually, host-specific. Furthermore, biological control is often self-perpetuating; that is, once a population of predators or parasites is established, it does not have to be reintroduced. Also, the possibility that the pest might develop a resistance to the control agent is minimal as compared with pesticides, largely because both the pest and its control agents often coevolve to maintain a stable interaction. Worldwide, approximately 70 pests have been successfully controlled by the introduction of their natural enemies. For example, success stories include control of the spruce sawfly in Canada, the coffee mealy bug in Kenya, the prickly pear cactus in Australia, and the spotted alfalfa aphid in the United States.

However, biological control does have some limitations. For example, following its introduction, the control population needs time to reduce the pest population effectively, but to avoid economic loss, farmers usually want to kill pests immediately. In addition, since considerable research is required to understand how a pest interacts with potential enemies, 15 to 20 years of research usually are needed to determine the best control agent. And even after a control agent has been identified, there is always a chance that it will also become a pest. Thus in some instances, it appears that biological control just might not work. For example, nearly 100 natural enemies have been introduced to control the gypsy moth in the United States, but none have significantly controlled that major pest, which defoliates a great variety of trees, including pines, birches, and oaks.

Natural Pesticides In the coevolution that occurs between plants and their pests, plants have developed many chemicals that are injurious to their enemies. In fact, some researchers suggest that chemicals that are produced by plants may be more important than any other factor that occurs in nature to control insects. For example, citrus plants produce antifeedants, which are chemical agents that prevent certain insects from feeding by destroying their sense of taste. Some plants produce chemicals that interfere with the production of *chitin*, which is the hard, protective outer covering of insects. And other plants produce toxic chemicals, such as pyrethrum. Pyrethrum is a natural pesticide, which is produced by the flowers of a daisylike plant (shown in Figure 17.5). Several pesticide companies extract pyrethrum from those flowers and incorporate it into formulations that are available in the United States to control insects in gardens.

A fantastic array of plant defenses are just being discovered by researchers. Thus the task of screening thousands of plant species for chemicals that are harmful to their natural enemies is just beginning. And researchers in university, government, and commercial laboratories face a monumental challenge.

Disease-Resistant Crops The development of disease-resistant crops is a major advancement in the struggle to reduce pest damage. For example, at one time, the Hessian fly caused damage to wheat that cost several hundred million dollars annually. However, by breeding strains of wheat that were resistant to that pest, crop geneticists have relegated the Hessian fly to a minor status. Other pests, such as the corn borer, aphid, and loopworm, have also been controlled as a result of the development of resistant crop varieties. But in plant breeding, tradeoffs often occur. It is difficult to develop a hybrid that will have all the characteristics that we want. Thus to gain a greater resistance to insects, we may have to grow varieties that produce lower yields. Further-

Figure 17.5 *Harvesting pyrethrum flowers in Kenya. The flower heads will be ground into a powder and used as a commercial insecticide in the United States.* (FAO photo.)

more, pest populations eventually overcome the defense mechanisms of existing crop varieties through natural selection, so geneticists must continue to develop new crop strains. For example, in the wheat-growing regions of the northwestern United States, a new variety of wheat can be expected to maintain its resistance for only approximately 5 years before evolving pests will begin to reduce its yield.

Control by Cultivation   Another natural approach to pest control is to use cultivation procedures that discourage or inhibit the growth of pest populations. Some pests feed only on one particular crop, so the rotation of crops can prevent pest populations from building up in the soil and in the crop residue from year to year. For example, by alternating corn with wheat or soybeans from one year to another, farmers can inhibit the growth of corn rootworm populations in the soil. Also, the use of fence rows and a diversity of crops may prevent the migration of pests and create a refuge for the natural enemies of those pests. Unfortunately, however, economic conditions today force most farmers to plant the particular crop each year that they believe will likely produce the greatest monetary return for their efforts. As a result, crop-rotation practices typically are given little weight in the decision. In addition, few fence rows remain today, because they would stand in the way of the large machinery that farmers use today to provide a better economy of scale.

Integrated Pest Management   Beginning in the early 1960s, the concept of integrated pest management began to receive considerable attention. *Integrated pest management* is defined as the coordination of all the suitable procedures and techniques that can be used (in as environmentally compatible a manner as possible) to maintain a pest population at levels that are low enough so that no economic damage is incurred. For example, in the early 1970s, when cotton farmers no longer were able to control the insect pests with pesticides in the Rio Grande Valley, a new system of pest management was developed. That system included three basic components: (1) cotton stalks were shredded and plowed under by mid-September, which reduced the number of weevils; (2) a rapid-fruiting, short-season cotton variety was cultivated, which could be harvested before the weevils would normally attack the cotton bolls; and (3) a

limited application of insecticides was carefully timed in the spring to kill overwintering adult weevils and to minimize the impact on the insect enemies of the bollworm and tobacco budworm. When those procedures were followed carefully, there was usually little need for additional insecticide applications. As a result of this pest-control program, production costs have declined in the Rio Grande Valley and yields have more than doubled since the late 1960s.

Integrated pest-management programs have also been successful with several other crops, including alfalfa, apples, and peaches. Thus many researchers believe that integrated pest management holds great promise for economic and environmentally sound pest control. However, many difficulties still must be overcome before that practice is widely accepted. Integrated pest-management programs require a great deal of research and testing, and the resulting procedures usually need to be developed to fit each crop and geographical region. For example, the program that was described to control cotton pests in the Rio Grande Valley is not effective in the San Joaquin Valley of California, because many differences exist between the two regions. The insect pests and their behavior in terms of how and when they attack cotton plants are different; the pests' enemies and their behaviors are different; and the climate and soil conditions are different.

Thus we might wonder what the future holds for pest management. Because the human population is continuing to grow, we must continue the battle to reduce food losses. In this battle, pesticides remain highly favored by agriculturalists because of their uniform and rapid effectiveness, ease of application and shipment, and relatively long shelf-life. Also, pesticides represent less than 5 percent of total food-production costs in the United States. Hence many farmers view pesticides as a relatively inexpensive way to protect their large investments. In contrast, many farmers do not perceive crop rotation and maintenance of fence rows as being economically feasible in terms of controlling pests. Moreover, many of the new pest-control methods require much more research before their effectiveness and economic feasibility can be demonstrated. Hence the use of pesticides probably will continue to be the major method of pest control in developed nations.

Problems that are associated with pests are particularly severe in less-developed countries, where persons can least afford to lose food. In those areas, farmers often work at the subsistence level and cannot afford pesticides even when they are available. Hence those impoverished, technologically unsophisticated farmers are in great need of inexpensive, simple-to-use alternatives. Thus pest-control programs in those areas should focus on the development of pest-resistant crop varieties and the improvement of local cultivation practices. Also, crops must be protected from pests after harvesting, since approximately one-third of the world's food that is lost to pests occurs after the crops are harvested. In particular, rat-proof structures are needed, which can be fumigated with nonpersistent pesticides to eliminate insects and fungi. In many areas today, even bumper crops do little to alleviate hunger, since surplus crops must be stored outside, where they either rot or are devoured by rats and birds because storage and transportation facilities are inadequate.

Reducing Food-Chain Losses

Reducing food-chain losses could also increase the amount of available food. For example, in Chapter 2, we saw that less than 20 percent of the food energy that is present at one trophic level in agricultural ecosystems is transferred to the next. Hence it has been suggested that more persons could be fed if everyone gave up eating meat and animal products and became vegetarians. But although the grains—wheat, rice, and corn (which account for most of the world's diet)—are relatively rich in calories, they are relatively poor in protein. And humans need both protein and calories.

The United Nations estimates that as many as 500 million persons (approximately 1 out of every 9 persons) may be suffering from protein malnutrition today. In the future, when several billion more persons inhabit the globe, the extent of protein deficiency could be much greater. Thus one of our tasks is to determine where that additional protein will come from.

Although many food calories are lost in livestock production, foods that are derived from animals have a greater protein value than those that are derived from plants. But raising livestock means losing calories through food-chain inefficiencies. Box 17.1 assesses the tradeoffs that are involved when we obtain protein from animal sources as opposed to plant sources.

Cultivating New Lands

From the time that agriculture began, some 10,000 years ago, until approximately 1950, the major strategy for increasing the world's food production has been the cultivation of new lands. Today, however, no readily available reserves of arable land are left in the world. Furthermore, little suitable space is left to expand croplands in the temperate regions of North America, Asia, and Europe. And the very limited potential for expansion that does exist lies primarily in humid tropical regions. Thus in this section, we examine the possibilities of, and limitations on, expanding agriculture into those areas.

The Amazon Basin of South America and the rain forests of Africa are the only remaining large tracts of arable land that receive enough rainfall to accommodate intensive agriculture. The lush vegetation of the tropical rain forests seems to suggest high soil fertility and an excellent potential for crop growing but in this case, appearances are deceiving. Recall from Chapter 5 that the soil in most tropical rain forests is actually quite low in nutrients and humus content. Essentially all the mineral nutrients are contained in the vegetation, not in the soil.

Farmers in tropical rain forests adapt to this mode of nutrient distribution by practicing *slash-and-burn agriculture*. In that practice, farmers cut down and burn vegetation on small plots, which are usually less than 1 hectare (2.5 acres) in area (see Figure 17.6). And the ashes of the burned vegetation, which contain some of the nutrients that were contained in the plants, are used as fertilizer. However, within 3 to 5 years, the harvesting of crops (along with their incorporated nutrients), coupled with the runoff and percolation that is associated with heavy rainfall, depletes the soil of most of its nutrients. As a result, the soil becomes too infertile to produce adequate crops, and the farmers move on to other areas. Then, after 20 to 25 years, the first plots become overgrown again with vegetation that has accumulated more nutrients from the soil and air, so the farmers can return to those areas and start the cycle again. Although slash-and-burn agriculture is a successful and effective adaptation to tropical soils, the yield from using that technique is small.

However, during the past 15 years, research has demonstrated that high crop yields can be sustained in the Amazon Basin of Peru if adequate fertilizers are applied. In fact, three crops can be grown each

---

BOX 17.1

ASSESSING PROTEIN SOURCES

Protein is essential in our diet for normal physical and mental development. If deprived of sufficient protein, a baby or young child may experience stunted growth and may become mentally retarded. Tragically, those effects are not reversible; no amount of remedial education can correct the mental damage that is sustained as a result of early protein deficiency.

Our bodies build proteins by assembling 20 component substances, amino acids, *in innumerable combinations. Of those 20 amino acids, 8 are not synthesized by the body and must be supplied by food—the* essential amino acids. *Thus we assess the nutritional value of protein by determining its content of essential amino acids. The higher this amount, the greater is the nutritional value of the protein.*

Foods that are derived from animals have greater nutritional value than those that are derived from plants. Animal protein contains the greatest quantities of all eight essential amino acids. And protein from eggs has the best mix of amino acids for humans. The most valuable protein sources other than eggs are fish, milk, cheese, and meat. Those animal products also supply valuable minerals and vitamins.

Plant protein is poorer in quality than animal protein, since it is usually low in one or two of the essential amino acids. Proteins that are pro-vided by cereals (rice, wheat, and corn) are particularly low in the essential amino acid lysine. And soybeans, while high in lysine, are low in another essential amino acid, methionine.

Presently, the world's population obtains approximately 70 percent of its dietary protein from cereals, vegetables, and legumes. Thus a significant way to improve the protein content of the human diet would be to educate people to eat more high-protein plant materials and to develop means of incorporating those materials into their regular diets. Soybeans are the best source of plant protein that is available today. After the oil is extracted from soybean seeds, the residual soybean meal contains approximately 50 percent protein. The protein in soybean meal has the highest nutritive value of any plant protein.

Much soybean meal is fed to livestock as a protein supplement but, recently, human beings have begun to eat more of that material in various forms. For example, soy flour that is made from soybean meal can be added to wheat flour, and the bread and pastries that are made from the mixture (even though their texture and appearance remain relatively unchanged) have a significantly higher protein value than those that are made from pure wheat flour. Soybean protein is also used as a meat extender.

Food that is made with cereal grains (such as bread and breakfast cereals) can be fortified by adding essential amino acids, such as lysine. Those amino acids can be synthesized or taken

---

year on the same plot, which yield several times more food than can be produced by slash-and-burn agriculture. Also, it has been found that a rotation of rice, corn, and soybeans will produce the most successful yields. In contrast, continuous planting of the same crop resulted in much lower yields because of a pest buildup. Therefore, if that form of agriculture can be applied more widely in the Amazon Basin, less land will be needed to support more persons, and many hectares of tropical forest can be spared from clearing to meet food demands.

Although that discovery in the Amazon Basin is encouraging, this new agricultural technology faces several limitations. For example, its success is based largely on the availability of fertilizers. Without fertilization, yields decline to zero after the third consecutive crop, which again demonstrates the inherent infertility of those soils. And without terracing, soil erosion becomes excessive in hilly regions. Furthermore, although weeds and pests generally have been controlled during the past 15 years by crop rotation, selection of resistant varieties, and a limited use of pesticides, researchers expect that pest problems will become more severe over time as a result of the land remaining in continuous cultivation. Therefore, to obtain continual high yields in new

from natural high-protein sources (such as soy flour). In the United States today, white bread is fortified with replacements for the nutrients that are lost in the processing of grain.

The meat from cattle, sheep, goats, and buffalo accounts for another 25 percent of the world's protein supply. Hence the world's protein supply could be significantly enhanced if livestock management were improved, particularly in less-developed nations, where livestock grow slowly and suffer high mortality rates. For example, approximately 67 percent of the world's milk cows reside in less-developed countries, but those cows produce only 20 percent of the world's milk supply. Much could be done to improve meat and milk production without adding to the land area that is used to support livestock.

Some persons, citing the inefficiencies that are involved in raising livestock, believe that human beings should eat plants only. Because approximately 7 kilograms of grain are needed to produce 1 kilogram of meat, they argue, many more people could be fed if we all became vegetarians. Proponents of that approach also point out that the production of animal protein puts stress on reserves of fossil fuels. As the table indicates, producing animal protein generally requires five to ten times more fossil-fuel energy than producing an equivalent amount of plant protein.

Nevertheless, a worldwide shift to vegetarianism could actually increase protein deficiency

Fossil-fuel input per protein yield for various foods

FOOD PRODUCT	FOSSIL-FUEL INPUT PER PROTEIN YIELD
Soybeans	2.06
Wheat	3.44
Corn	3.63
Rice	10.01
Beef (Rangeland)	10.10
Eggs	13.10
Pork	35.40
Milk	35.90
Beef (Feedlot)	77.70

Source: After D. Pimentel et al., Energy and land constraints in food protein production. *Science* 190:754–761, 1975. Copyright 1975 by the AAAS.

rather than reduce it. In order to design a well-balanced diet that contains the necessary vitamins and minerals as well as adequate protein, vegetarians must have a sophisticated understanding of the principles of nutrition. Most persons are unwilling to make the effort that is required to gain such knowledge. Besides, global vegetarianism would waste the forage plants that grow on grazing lands. Those grasses and shrubs are eaten by livestock and are converted by them into animal protein but they are of little value to humans directly. Thus in the interests of using our fuel and food resources in the most efficient ways possible, it appears that most of us will continue in our traditional role as omnivores.

areas, soil conditions must be carefully tested and monitored over time. Also, to adapt to local conditions, such modifications as the use of different crop varieties, rotations, and fertilizer rates probably will be needed. Hence successful local adaptation of that new technology will require skilled personnel, receptive farmers, and time.

However, the time that is necessary to adapt that new technology to local regions and to transmit the knowledge that is necessary to maintain it to local farmers may be running out. Because of the increasing population pressures in those areas, more land each year is being subjected to slash-and-burn

agriculture. Clearings are being made larger and closer together, and the time that is allowed for a clearing to regenerate is being shortened. As a result, soil erosion is becoming quite severe in many regions. More than 2 percent of the Amazon Basin's forests are cleared every year for agriculture and forest products. And at that rate, the great Amazonian rain forest will essentially vanish in about 40 years. Some persons may view this destruction as being necessary for providing food and forest products to impoverished people. However, others are greatly concerned that the loss of the vast genetic reservoir that lies within the disappearing forest will have serious

Figure 17.6 *A site of slash-and-burn agriculture in the Amazon rainforest. (Earth Scenes.)*

long-term ramifications, as described in Chapter 14.

In summary, most of the arable land in the world is already being cultivated. And the successful application of new and appropriate agricultural technologies in the tropics will be costly and slow. Reports by the U.S. Department of Agriculture (in 1978) and by the United Nations (in 1981) both projected only a 6-percent increase in the amount of land that will be cultivated worldwide by the year 2000. In the meantime, the world's population will have grown more than 30 percent. Clearly, the food that will be required by a growing world population cannot be supplied primarily by expanding the land that is being cultivated.

Increasing Production on Cultivated Land

Since 1950, more than 70 percent of the increased production of grain throughout the world has stemmed from improved yields on land that is already being cultivated. Higher yields have been achieved not only in nations with advanced agricultural technologies, but also in many of the less-developed countries, which have imported some American and European agricultural technologies to improve their food production.

Initial attempts at increasing yields in less-developed nations did not meet with much success. For example, varieties of corn and wheat from the United States often fared poorly in foreign soil and climatic conditions, and they were often particularly susceptible to native pests. Also, when fertilizers were applied to varieties of wheat and rice that had been grown traditionally in those countries, the grain heads often became too heavy for the tall, spindly stalks, which caused the plants to fall over. Thus those grains were difficult to harvest and vulnerable to spoilage and pests. To overcome those difficulties, crop geneticists spent many years developing new, high-yielding varieties of grain—mainly short-stemmed rice and wheat—which were genetically engineered to make more efficient use of fertilizers. As a result, the new varieties produce shorter, thicker stems, which can support the heavier heads of grain (see Figure 17.7), and many of those improved varieties also mature in a shorter period of time than native strains. Hence if sufficient water is available, more than one crop can be raised during a single growing season. Also, some of the early varieties were highly susceptible to pests, but strains that are currently being used are more pest-resistant, at least temporarily.

In the mid-1960s, when those high-yielding varieties were introduced and the use of fertilizers was increased, crop yields improved significantly in many nations in South America, Asia, and North Africa. In fact, the rate of food production exceeded the rate of population growth in many countries. Those developments, together with increased food production in developed nations, contributed to a nearly 15-percent increase in the world's per-capita production of cereals. As a result, the term *Green Revolution* was coined to refer to those remarkable events. Some scientists believed that the Green Revolution would solve most of the world's food problems.

However, the Green Revolution has not been a total success. Since the 1960s, many countries, particularly those in Africa, have actually experienced a significant decline in their per-capita food supply, and food production in most other less-developed countries has barely stayed ahead of the demands of their growing populations (Figure 17.8). Overall, per-capita grain production has been erratic throughout

Figure 17.7 *A new dwarf strain of high-yielding wheat (the wheat on the left with the shorter stalks) compared with a native strain. The development of this strain contributed significantly to increased wheat production in many nations. (FAO photo.)*

the world, and the growth that occurred in the late 1960s has not continued.

Many factors contribute to successful agriculture. As we saw in Chapter 5, plant productivity—whether it is in the form of natural vegetation or agricultural crops—is influenced primarily by the weather (mainly moisture and temperature) and soil fertility. Other significant factors that affect the yield of crops are pests and the genetic potential of the crop varieties that are available. Furthermore, strong economic incentives are needed to encourage farmers to increase their yields. In this chapter, we have already discussed the limitations on crop yields that are imposed by pests. Thus we now explore several other factors that influence the production of food on land—climate variability, soil erosion, water supply, fertilizer supply, and energy supply.

Adapting to Climatic Variability

Climate is the primary factor in controlling the production of food on land. Climate regulates soil development, the length of the growing season, sufficient heat during the growing season, the amount of water and sunshine that is available for crops, and

suitable conditions for animal husbandry. The climate also influences the growing conditions for agricultural pests. Thus the climate governs the types of crops and animals that thrive in particular regions and determines both crop yields and animal production. Even in this age of advanced technology, our food supply remains largely at the mercy of the weather.

This vulnerability was especially evident in the United States during three periods of time in just the last 12 years. For example, during the 1974 growing season in the United States, heavy spring rains, a midsummer drought, and an early autumn freeze combined to reduce the corn harvest by 11 percent and the soybean harvest by 16 percent from levels that were attained in 1973. Then, during the growing season of 1980, the weather dealt agriculture an even greater setback. A severe summer-long drought, which was accompanied by searing heat, ravaged much of the eastern half of the country. Corn and soybean production plummeted 23 percent and 18

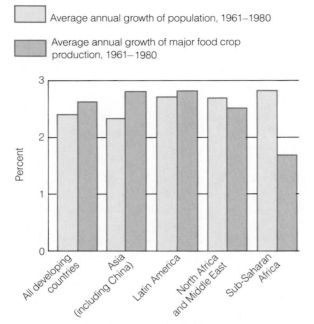

□ Average annual growth of population, 1961–1980

▨ Average annual growth of major food crop production, 1961–1980

Figure 17.8 *Food production in most developing nations is barely keeping pace with population growth. In many African nations, population growth is outstripping food production. (FAO and the International Food Policy Research Institute. After J. W. Mellor and R. H. Adams, Jr., "Feeding the Underdeveloped World."* Chemical and Engineering News *62(17):32–39, 1984.)*

percent, respectively, from levels that had been attained in 1979. In contrast, very favorable weather conditions prevailed in 1981 and 1982, which led to record harvests in the United States. Then, in 1983, parts of the central Midwest suffered perhaps the worst drought in 50 years. Corn production declined by 25 percent, and the soybean yield declined by 17 percent. (Those reductions do not include the impact of a large voluntary program that was sponsored by the federal government to reduce the amount of land under cultivation.)

The crop that has been hit the hardest in recent years by inclement weather is Florida's citrus fruit (oranges and grapefruit). Five freezes during the past decade have been so severe that citrus growers are either going out of business or they are relocating further south. Box 17.2 describes the impact of this southward migration of Florida's citrus belt.

Portions of the Soviet Union also experienced several years of poor weather in the early 1980s, when poor wheat harvests forced them to import large

quantities of grain. Earlier in this chapter, we described the devastation that was caused by prolonged drought during the 1970s and early 1980s in the Sahel region. In addition, in the early 1980s, the drought spread to eastern and southern Africa. South Africa, which is the agricultural powerhouse of Africa, was so devastated by this drought that it was forced to become a net importer of food after having been a net exporter for many years. Meanwhile, extreme drought in Australia led to a $2 billion loss in crops and the destruction of millions of sheep and cattle. Interestingly, in comparison, South America suffered relatively little from droughts during this time.

Many climatologists believe that such spatial and temporal variations in the weather probably will continue, which may pose the greatest threat to efforts that are being made to provide a reliable food supply for a growing world population. At the same time, it appears that the world's agricultural system is becoming increasingly susceptible to the vagaries of weather. Given (1) the expansion of agriculture into climatically marginal land, (2) the increasing reliance on new high-yielding hybrids that are more vulnerable to inclement weather than traditional native varieties, (3) the greater use of agricultural technology, and (4) shrinking water resources, world agriculture is developing a production base that will generate excellent harvests during favorable weather but sharply lower harvests during unfavorable weather.

To assess the impact of climatic variability on food production, we also need to consider the role of agroclimatic compensation. *Agroclimatic compensation* is defined as the effect that results when good growing weather in one area offsets poor growing weather in another area. Recall from Chapter 9 that the magnitude and direction of climatic anomalies vary with geographical location. Hence for a nation that is as large as the United States, we would not expect all regions to suffer significant crop losses during the same year if poor weather occurred. In fact, there could even be an abundant harvest of a particular crop, depending on the region and its season of growth. For example, by the time the heat wave and drought struck the Midwest in 1980, much of the wheat crop had already matured. Thus although yields of the later-maturing corn and soybean crops declined significantly, the 1980 wheat harvest set a record high for total production.

BOX 17.2

THE FLORIDA CITRUS INDUSTRY: A VICTIM OF WEATHER VARIABILITY

Florida's worst crop freeze of the century has damaged citrus groves throughout the state and cast a pall of uncertainty over the future of the industry in the north central part of the state.

Low temperatures were an average of two to three degrees lower across the state than the record-breaking cold of Christmas 1983. That damaged or destroyed 250,000 acres of citrus in north central Florida and cut the state's orange juice production by 40 percent.

This year's freeze is believed to have finished off many of the trees that survived last year's.

The latest industry setback has raised concerns about the ability of growers in north central Florida—the area near Orlando—to stay in business. A few years ago, that area represented the heart of the state citrus industry.

The industry has been battered by five freezes over the last decade. Even before the damage caused by the latest freeze, Florida's orange crop was estimated at only 119 million 90-pound boxes, down from 206.7 million harvested in the 1979-80 season.

Lenders and other observers of the citrus industry said they expect this year's freeze to accelerate two trends set in motion by the four earlier cold snaps. Growers, they said, are either moving south or moving out of the industry. As a consequence, land prices are dropping to as low as $3000 a live acre at the northern end of the citrus belt, and rising to as high as $15,000 an acre in the south.

Rather than relocate, some growers have been persuaded by the erratic weather of the last few years to abandon the industry altogether. "For the last year, I've gotten pathetic little notes saying, 'I'm out of the business, I won't resubscribe,'" said Nancy Hardy, editor of Florida Citrus Reporter, an industry newsletter.

Lake County, an agricultural region just west of Orlando, was particularly hard hit by both last year's freeze and this year's. One of the two top citrus-producing counties in Florida as recently

as the 1982-83 season, Lake County is "hurting incredibly badly," said Ms. Hardy of the Florida Citrus Reporter.

She said she expects a "very small" percentage of the county's growers to replace trees killed by the frost.

The question now facing many growers—and their creditors—is what to do with land populated by dead orange trees. "We're sitting here wondering what a totally frozen grove is worth," said James C. Beck, president of the branch of the Federal Land Bank Association that serves 13 central Florida counties.

"From a lending standpoint, we're going to have to look at alternative uses for collateral," said A. G. "Buddy" Johnson, senior commercial lending officer of Florida National Bank of Orlando. He said he expects bankers to encourage many growers to consider planting other crops, such as vegetables, or to sell out to developers.

Citrus-industry observers and land brokers said they didn't expect panic selling of land to be widespread, but do foresee further drops in land values at the northern end of the citrus belt. Ronald Stevens, a Lake County realtor and orange grower, said that even before the most recent freeze, "There was more acreage for sale than at any time in recent memory."

But housing, shopping centers and hotels are possibilities mainly in the southern and eastern parts of Lake County, which are neighbors of the city of Orlando and Walt Disney World.

Elsewhere in the region, the options are more limited. Asked what might replace the orange business, John DelVecchio, an official of the state Department of Commerce's Division of Economic Development, responded with a question of his own: "What we're doing is sitting down with these folks and saying, 'What kind of industry can you support, besides citrus?'"

Excerpted from Ed Bean and Thomas E. Ricks, "Florida's Worst Freeze in Century Casts Doubt on Future of North Central Groves," The Wall Street Journal, January 23, 1985. Reprinted by permission of The Wall Street Journal, © Dow Jones and Company, Inc. 1985. All rights reserved.

However, agroclimatic compensation also has limitations. For example, favorable weather conditions in a region that has low soil fertility (which already limits crop yields) probably would not compensate for a decline in yields that is caused by unfavorable weather in a region that has high soil fertility. In that situation, the total food production would decline. Another limitation on agroclimatic compensation is that intensive cultivation of most crops generally remains confined to specific regions. For example, in March and April of 1983, cold winds and heavy rainfall delayed the planting and harvesting of vegetables in California, which significantly reduced the nation's supply of those vital foodstuffs (Figure 17.9). Nearly half of the country's vegetables, which are marketed fresh and used for food processing, are grown in California. Thus while agroclimatic compensation may provide some flexibility, it also has serious limitations.

From the discussion in Chapter 8, we can conclude that we cannot control the climate to ensure that the world can produce a more dependable food supply. For example, weather-modification projects, such as rainmaking, have met with very limited success. And because scientists still have a very limited understanding of the processes of precipitation, dependable weather-modification technology is still at least several decades away. Moreover, during a drought, clouds are not always amenable to cloud seeding. Hence our only practical alternative is to learn to live with the inherent variability of climate.

However, one strategy that we can use is to modify our agricultural practices to fit current weather conditions. For example, the response of growers to the drought that occurred in California from 1976 to 1977 illustrates the kinds of modifications that are possible. In the spring of 1977, California's reservoirs contained less than 20 percent of their capacity. And when California farmers learned that they would obtain only a small fraction of their normal allocation of surface irrigation water, they turned to groundwater supplies. Many farmers drilled new wells or reconditioned old wells that had not

Figure 17.9 *Flooding in the Salinas Valley in March of 1983 destroyed vegetables that were ready for harvest (such as these artichokes) and prevented the planting of new crops. (AP 1983, Clay Peterson.)*

been in use. They substituted crops that require less water (such as oats and barley) for rice, which has a high water requirement. And some farmers stopped using their flood irrigation systems and installed more costly trickle irrigation systems, which cut their water use by 30 to 50 percent (see Figure 17.10). Many farmers even installed pumps to recirculate the runoff from irrigation water back through their fields. The combined effect of those procedures was successful. Although the drought was severe, California's crop production in 1977 was about equal to that of former years, when the weather was good. Nevertheless, the costs of maintaining those crop yields were high. Although gross farm receipts nearly equaled those of earlier years, the California Department of Food and Agriculture estimated that the state's net farm income declined by 17 percent that year as a result of the added expenses of installing trickle irrigation equipment and new pumps and drilling new wells.

Although California crop growers were able to adapt to the drought, cattle producers were not so fortunate. Because supplies of grass and water were inadequate during the drought, ranchers had to truck in hay and water. But those measures were very expensive, and many ranchers were forced to sell all their cattle except for those which they kept for breeding. As a result, beef prices declined and California ranchers suffered losses of approximately $500 million during 1977.

Then, the California drought ended as suddenly as it had begun, which further illustrates the variability of weather. Heavy rainfall began in mid-December of 1977 and continued through March of 1978. In just 3 months, river valleys were flooded and reservoirs were nearly filled, and more than enough water was available for irrigation.

Nevertheless, the relative success that California's growers experienced in adapting to climatic variability during that drought (at least in the short run) could not be repeated in many less-developed nations. California farmers are highly skilled and have access to the financial resources, modern machinery, energy, and the information that is necessary to adapt quickly to change. However, in most less-developed nations, such resources are not available, and farmers must rely on their own meager resources to cope with adversity. Hence inclement weather in those regions often leads to an extreme scarcity of food and starvation.

Another strategy that is available to us is continued research to develop crop varieties that are more tolerant of weather variability. But, as we mentioned earlier, it is difficult to develop a hybrid that will have all the characteristics that we want. Thus to gain a greater tolerance to inclement weather, we may have to elect to grow varieties that produce lower yields, have less resistance to pests, or both. As a further countermeasure, the United Nations has proposed that a world food bank be established to store grain in the good years to provide food for the lean years. However, given the inevitable international disagreements concerning where the reserves would be stored and who would control and pay for them, a world food bank is a long way from becoming a reality.

Controlling Soil Erosion

Although fertile soil is absolutely essential for crop and livestock production, that priceless resource is continuing to be lost at an alarming rate. Although soil erosion occurs naturally (see Chapter 2), human activities have greatly accelerated the process. Each year, nearly 4 billion metric tons of sediments are carried into waterways in the United States, and three-quarters of that soil originates in farmlands. Also, winds blow away another 1 billion metric tons of soil each year. A 1981 report by the U.S. Department of Agriculture shows that the inherent productivity of one-third (56 million hectares, or 140 million acres) of the farmland in the United States is declining as a result of soil erosion. For example, a Midwestern farm field loses 2.5 centimeters (1 inch) of topsoil every 10 to 20 years, but it takes 200 to 300 years to reform 2.5 centimeters (1 inch) of lost soil. Hence soil from American farmland is being lost up to 20 times faster than it is being added by natural soil-forming processes. Recall from Chapter 7 the impact that soil erosion has on aquatic ecosystems and the longevity of reservoirs. No wonder many scientists see soil erosion as one of our greatest environmental problems.

Furthermore, soil erosion in the less-developed countries is at least twice as severe as it is in the United States—and it continues to grow worse as the demand for food increases. Already, persons are moving into marginal lands where cultivation induces severe soil erosion. And as a result, hillsides

(a)

(b)

Figure 17.10 *Flood irrigation of a mature almond orchard (a),
and trickle irrigation of a young almond orchard (b). In trickle
irrigation, small amounts of water are metered out next to the
plants from buried pipes. Both orchards are in California's Central Valley. (U.S. Department of the Interior, Bureau of Reclamation.)*

are washing away and deserts are expanding in both developed and less-developed countries.

Recall from Chapter 8 that many methods of controlling soil erosion are available to farmers, but few of them are used. For example, when farmers plow furrows up and down rolling hills, runoff flows rapidly downhill, which accelerates soil erosion (see Figure 17.11). However, if the farmers plowed those furrows parallel to land contours (*contour farming*) and flattened local slopes into terraces, soil erosion would be reduced significantly. Although such procedures require an increase (albeit small) in farming time and fuel use, the savings in valuable soil would make up for the added cost.

Also, crops differ in terms of their ability to anchor the soil against erosion. As a result, the planting technique that is called *strip cropping* (Figure 17.12) is frequently used. In strip cropping, a crop that is planted in widely spaced rows, such as corn or soybeans, is alternated with a crop, such as alfalfa, which forms a complete cover and thus reduces soil erosion. Furthermore, during the winter months, such crops as annual rye or clover can be sown to protect the land. However, most farmers do not favor the latter procedure, because they prefer to plow the land in the fall to allow for earlier planting in the spring.

(Delays in spring planting lead to shorter growing seasons and reduced harvests.)

Leaving crop residues on the soil surface significantly reduces soil erosion. As a result, recent research has focused on *no-till* techniques. In that strategy, loosening of the surface soil, planting, and weed control are combined into one operation through the use of specialized machinery that plants seeds and applies herbicides and fertilizers to unplowed soil. By combining those activities, the disturbance to the soil surface is minimized, and the potential for soil erosion is greatly reduced. But there is a tradeoff—reducing tillage increases weed and insect populations, which necessitates the use of more pesticides.

Finally, the planting of trees to form *shelterbelts* will reduce soil erosion that is caused by the wind (see Figure 17.13). Following the Dust Bowl days of the 1930s, many shelterbelts were established in the Upper Great Plains.

Soil erosion is a multifaceted problem. And even though many methods of controlling soil erosion are available, relatively few farmers in the United States employ them. Since the land is a farmer's principal economic asset, we would expect most farmers to protect it. However, recent economic studies have shown that the short-term costs of implementing soil-conservation measures often exceed the short-

Figure 17.11 *Severe soil erosion on a hillside, resulting from improper cultivation practices. (U.S. Department of Agriculture, Soil Conservation Service.)*

Figure 17.12 *Strip cropping and contour cultivation help to reduce soil erosion on slopes near Lewistown, Idaho. (Georg Gerster, Photo Researchers.)*

term benefits. (Earlier in this chapter we mentioned the reluctance of many American farmers to employ crop rotation, because it limits their profits during any given year to a level that is lower than they would earn by planting the most profitable crops.) Thus although the long-term benefits of conserving valuable topsoil are obvious, few farmers can afford to consider those benefits if their immediate economic survival is at stake.

Another major difficulty in dealing successfully with soil erosion is that farmland is owned privately. A long-standing American tradition is that landowners are free to manage (or mismanage) their land as they please. Thus voluntary government programs, such as those administered by the U.S. Soil Conservation Service, have had only limited success. In contrast, government programs that manage the quality of air and water have been more successful, because those resources belong to everyone, so government agencies can more easily establish and enforce cleanup policies.

Better Management of Irrigation Systems

Inadequate soil moisture is a common limiting factor for crop production. Thus in areas where rainfall is commonly too unreliable to grow crops profitably, water must be brought in by means of irrigation projects. The United Nations estimates that in the less-developed nations, more than 105 million hectares (260 million acres) of land are now being irrigated for crops. In the United States, 20 million hectares (50 million acres) of land are irrigated. Rice, which is the food staple for more than half of the world's population, is almost always grown as an irrigated crop. And without irrigation, some fruit and vegetables would be scarce. In the United States, most of the nation's lettuce, carrots, celery, cantaloupe, and radishes (more than 70 percent) are grown on irrigated land. Thus irrigation is a significant factor in world food production.

Figure 17.13 *Acting as windbreakers, these shelterbelts in North Dakota reduce soil erosion. (U.S. Department of the Interior, Bureau of Reclamation.)*

But irrigation also involves some tradeoffs. For example one significant drawback is that only 1 out of every 4 liters of water that are drawn for irrigation is actually taken up by plants. (Figure 17.14 shows the sources and losses of water in an irrigation project.) And seepage from irrigation canals can be so substantial that it can raise the water table and lead to a water-logged soil that is unsuitable for cultivation. Although methods do exist to prevent waterlogging and to rehabilitate water-logged soils, they are often sophisticated and expensive.

A more common management problem that is associated with irrigation is the buildup of salts at the soil surface as a result of the evaporation of water (Figure 17.15). Crop yields decline when the salinity of the soil increases. And the process that is used to flush away the excess salts requires large quantities of additional water that, otherwise, could be used directly to increase crop production.

The United Nations estimates that 20 percent of the world's irrigated land is adversely affected by salt buildup and waterlogging. And the U.S. Office of Technology Assessment reported in 1982 that in the western United States, salinity has already reduced production on 25 to 35 percent of the irrigated land. Furthermore, salination can be costly. For example, in the San Joaquin Valley, salt-contaminated aquifers that underlie 162,000 hectares (400,000 acres) of land are costing $32 million annually as a result of reduced yields. Figure 17.16 shows the effect of salt buildup on cultivated land.

Thus a high agricultural priority should be the rehabilitation of irrigation systems. In many cases, siltation of canals, waterlogging, and salinity could be controlled with proper management. However, management is costly: irrigation systems must be continually maintained, and drainage, groundwater, and salinity must be monitored closely. Nevertheless, if we allow faulty systems to continue functioning, millions of hectares of marshes and salt flats will be produced.

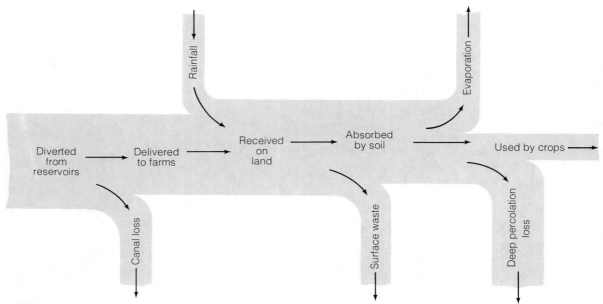

Figure 17.14 *Water is gained and lost at several points in irrigation operations. Gains and losses at various points are roughly proportional to the widths of the shaded areas.*

Management is not the only cost that is involved in maintaining existing irrigation systems. Because irrigation is energy-intensive, crops that involve irrigation require three times more energy than is needed for normal rain-fed crops. Thus as the cost of energy has continued to climb, the rising cost of irrigation has forced some farmers in the Southwest to abandon low-value crops (such as alfalfa) for higher-value crops (such as cotton and strawberries). Furthermore, the use of water in food production' increasingly conflicts with other uses for water. And in some instances, irrigation may be seen as a poor investment of water as compared with other enterprises. For example, the economic returns from the industrial use of water are approximately 0 times greater than those that can be obtained from irrigation. Furthermore, since irrigation is a consumptive use of water and since it accounts for up to 80 percent of the water that is used in the water-poor western states, it may well be sacrificed to other uses (in which water is recoverable) as the competition for water grows more intense. For example, in 1980, Arizona passed legislation that is aimed directly at limiting water for irrigation in favor of supplying water to more industries and city dwellers.

As we noted in Chapter 8, conflicts between agricultural and urban–industrial uses are likely to intensify as future demands for the fixed water supply increase.

Figure 17.15 *Irrigation water contains soluble salts. As the water evaporates, it leaves the salts behind at the soil surface. As the soil's salinity increases, crop yields decrease.*

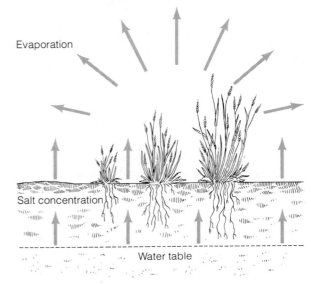

Figure 17.16 *Salt damage to a carrot crop in the Coachella Valley of California. This scene is common where arid lands are irrigated, but in the Coachella Valley land usually can be reclaimed. (U.S. Department of the Interior, Bureau of Reclamation.)*

Increasing the Fertilizer Supply

Today's high crop yields are dependent on fertilizers, which contain the major plant nutrients—nitrogen, phosphorus, and potassium. In fact, the commercial use of fertilizers throughout the world has increased by more than 650 percent in the last 30 years—from 18 million metric tons in the early 1950s to more than 120 million metric tons in 1982–1983. Furthermore, the United Nations estimates that if growing food demands are to be met, the use of fertilizers must increase another 200 percent by the end of this century. As a result, a question arises concerning where the nutrients will come from to meet this increased demand for fertilizers.

In the past, the application of *organic fertilizers* in the form of crop and animal residues was a major way to improve soil fertility. Also, many farmers helped to restore their soil's fertility by growing leguminous crops—peas, beans, alfalfa, or clover—on a rotational basis with other crops, such as corn. As we noted in Chapter 2, bacteria that live in nodules on the roots of those plants increase the nitrogen (ammonia) content of the soil.

Although organic fertilizers and leguminous plants continue to be important ways to improve soil fertility, major new supplies of plant nutrients have been tapped during recent decades as a result of the development of new technologies. For example, phosphorus, in the form of phosphates, and potassium, in the form of potash, are extracted from rich deposits in the earth. And those inorganic forms of phosphorus and potassium are granulated for commercial sale. Also, the Haber process, which is described in Chapter 2, can be used to synthesize ammonia from atmospheric nitrogen. Thus because those *inorganic fertilizers* are easy to handle, relatively inexpensive, and produce high yields, they have been widely adopted. They also allow farmers to obtain higher crop yields without having to raise livestock to produce manure.

If the United Nations projections of future needs for fertilizers are accurate, worldwide supplies of commercial inorganic fertilizers will be adequate for some time. But a major limitation to fertilizer production—and thus to improving the world's food production—is the unequal geographical distribution of potash deposits, phosphate deposits, and natural gas reserves (which are needed in the Haber process for ammonia production). For example, the largest reserves of phosphate are in Morocco, the People's Republic of China, the United States, the Soviet Union, Jordan, and Mexico. Large potash deposits exist in Germany, Canada, the Soviet Union, and the United States. And Mexico and the Middle East nations, which have large petroleum reserves, have the greatest potential for manufacturing ammonia. In contrast, phosphate deposits, potash deposits, and natural gas reserves are quite limited in most less-developed countries. Furthermore, those nations usually lack the financial resources to develop the deposits that they do have or to import sufficient quantities of fertilizers.

As a result of those regional differences, the use of fertilizers contrasts markedly among nations. For example, fertilizers are used most heavily in Europe, where applications average more than 200 kilograms per hectare (180 pounds per acre) of arable land. In contrast, applications of fertilizers average only approximately 15 kilograms per hectare (13 pounds per acre) of arable land in Africa, and the application rates of other regions fall somewhere in between. On the whole, fertilizer use per hectare of arable land in less-developed nations is only 17 percent of the total amount that is used in the developed nations. Such a disparity is tragic, because increased use of fertilizers would be more beneficial in less-developed countries than in developed countries. When additional fertilizers are applied to land that is already heavily fertilized, relatively small improvements occur in yields. However, when fertilizers are applied to nutrient-depleted soil, such as in the less-developed nations, dramatic increases in yield result. But before the use of fertilizers in impoverished nations can be increased significantly, many difficult problems concerning international aid and trade will have to be solved.

The unequal distribution of reserves is not the only limitation on inorganic fertilizer production. Considerable quantities of fossil fuels are required to extract and process inorganic fertilizers. For example, in the United States, the manufacture of fertilizers consumes more energy than any other sector of farming operations—nearly 28 percent of the total. Hence alternatives to using commercial inorganic fertilizers would not only allow impoverished farmers to improve their soil fertility, but they would permit farmers in developed countries to lower their accelerating production costs.

At first glance, favoring animal manure over commercial inorganic fertilizers might seem to be a real alternative. However, 90 percent of the total amount of manure that is produced annually in the United States is already being returned to the land. And the small amount of unused manure would not be economically competitive with commercial inorganic fertilizers, except in areas where crops and livestock are raised very close to each other. Because of the costs of labor and energy that are sustained in loading, hauling, and spreading manure, the costs exceed those of using inorganic fertilizers if the manure has to be transported more than 2 kilometers (1.2 miles). And in some less-developed countries, manure is the only fuel that is available for cooking and space heating—uses that are given higher priority than fertilizing the soil (see Figure 17.17).

Figure 17.17 *Animal manure to be used for cooking because firewood is not available. (Agency for International Development.)*

Crop rotation using leguminous plants (such as soybeans and alfalfa) is beneficial in two ways: (1) the nitrogen-fixing bacteria in the nodules of those plants increase soil fertility for future crops, and (2) the soybeans and alfalfa themselves require less nitrogen to grow than other crops, such as corn. Because cereal grains provide most of the world's food, and because those grains (especially corn) require large quantities of nitrogen fertilizer, major research efforts are currently underway to develop strains of nitrogen-fixing bacteria that will live in nodules on the roots of grain plants (as they do on the roots of legumes). But success in that area is not likely to occur in the near future, if ever. Two other ongoing research efforts have better chances of success: (1) improving the nitrogen-fixing capabilities of the legumes that are currently being grown, and (2) inoculating the soil with species of free-living, nitrogen-fixing, blue-green algae.

Unfortunately, increased use of fertilizers in less-developed areas, such as tropical rain forests, will not occur until significant social, political, and economic changes take place. Those changes would include educating farmers about the benefits of using fertilizers, ensuring that fertilizer supplies reach the farmers, and ensuring that farmers have access to credit at reasonable interest rates. Moreover, national policies should ensure that fertilizer does not become too expensive to be considered.

Managing the Fossil-Fuel Supply

The significance of fossil-fuel energy in raising crop yields is illustrated in Table 17.1, in which the use of energy in rice production is compared for modern, transitional, and traditional production methods. In traditional rice cultivation, fossil fuels are used only for the manufacture of tools. Humans and animals supply the power for sowing and harvesting, and manure serves as the fertilizer. In areas where the Green Revolution has gained a foothold—that is, where farming methods are transitional, between traditional and modern—additional fossil fuels are required to manufacture and operate small machinery and to produce and apply small amounts of fertilizers and pesticides. And in modern rice production, much larger quantities of pesticides and fertilizers are applied and larger farm equipment is employed. Also, in modern methods of rice farming, fossil fuels are used to pump irrigation water and to run the grain-drying process. Thus although rice yields that result from using modern methods are two times greater than the yields that result from transitional rice farming methods, fossil-fuel consumption is nine times greater. For other crops, the energy expenditure that is necessary to double crop yields from those that are raised by transitional farming methods is not as great.

Modern agriculture consumes fossil fuels in many ways. As Table 17.2 illustrates, the relative amount of energy that is devoted to each purpose varies considerably among crops. For example, the difference in energy that is required to provide nitrogen fertilizer for growing corn and soybeans well illustrates the energy benefits of planting legumes. It is considerably more economical for plants to make their own nitrogen (ammonia) than it is for humans

Table 17.1 Fossil-fuel energy that is required for rice production by modern, transitional, and traditional methods (thousands of kilocalories per hectare)

INPUT	MODERN (UNITED STATES)	TRANSITIONAL (PHILIPPINES)	TRADITIONAL (PHILIPPINES)
Machinery	4,184	335	172
Fuel	8,983	1,602	—
Fertilizer	11,071	2,447	—
Seeds	3,410	1,724	—
Irrigation	27,330	—	—
Pesticides	1,138	142	—
Handling and Transportation	8,531	29	—
Total Energy Consumption	64,647	6,279	172
Crop Yield (kg/ha)	5,800	2,700	1,250

Source: After F.A.O., *Monthly Bulletin of Agricultural Economics and Statistics* 25:3 (February), 1976.

Table 17.2 Fossil-fuel energy required for United States production of corn, wheat, and soybeans (kilocalories per hectare)

USE	CORN	WHEAT	SOYBEANS
Labor	5,580	3,255	4,650
Machinery	558,000	360,000	360,000
Diesel	1,278,368	604,942	79,898
Gasoline			50,545
Nitrogen	1,881,600	715,000	58,800
Phosphorus	216,000	78,000	54,000
Potassium	128,000	48,000	75,200
Limestone	31,500	11,025	110,250
Seeds	525,000	699,600	480,000
Insecticides	86,910		
Herbicides	199,820	49,955	499,550
Drying	426,341		
Electricity	380,000	200,000	28,630
Transportation	34,952	45,489	25,700

Source: From D. Pimentel, and M. Pimentel, *Food, Energy, and Society*. London: Edward Arnold Ltd., 1979.

to manufacture and apply nitrogen fertilizer. Also, the energy that is used to produce and apply herbicides varies among crops, but the amounts that are needed are usually smaller than those that are consumed by machinery and fertilizers.

Barring future oil embargos, fossil-fuel supplies for agriculture should be adequate for several decades, but farmers' energy expenses are likely to increase faster than their other costs. And during recent years, the low prices of farm products have forced many farmers to cut back on their applications of fertilizers and pesticides. Therefore, farmers in developed nations can save even more fossil-fuel energy by modifying their crop-production practices, such as substituting manure for commercial fertilizers (where manure is available nearby), using no-till techniques to reduce fuel that is used in cultivation, and rotating crops of grains with legumes to decrease the need for pesticides and commercial fertilizers. But the effectiveness of those alternatives is limited by significant constraints. As we have seen, unused manure is scarce. The no-till techniques require more pesticides than standard cultivation techniques, and the energy costs to produce the pesticides can approach the energy that is saved by reducing cultivation. And systematic crop rotation restricts the choice of crops that can be planted, whereas farmers prefer to plant the crops that are likely to

bring them the best monetary return in any given year. Thus economically hard-pressed farmers are unlikely to choose energy-efficient practices that would actually cost them more money than the fuel that is required in energy-intensive procedures.

For less-developed nations, the United Nations sees the quadrupling of fossil-fuel use as an essential step in providing an adequate amount of food by the year 2000. Fertilizers account for nearly 60 percent of that increase, and machinery accounts for the next largest increase. However, the resulting annual consumption would still be 40 percent lower than the projected amount of use in developed nations. Because agricultural production is only a modest consumer of energy—only 4.5 percent of the total amount of commercial energy that is used in less-developed nations—the United Nations is optimistic that the fourfold increase can be attained.

Of course, difficulties will arise. For example, less-developed nations must establish national energy policies which ensure that fossil fuels will be available at prices that will encourage farmers to continually improve their yields. And if choices must be made in terms of allocating scarce energy supplies, agriculture must be a top priority. To alleviate energy shortages, energy policies should encourage the search for, and use of, alternative energy resources, as described in Chapter 19.

Prospects for Increased Production

Clearly, food production on cultivated land must be increased to feed the additional 82 million persons who are added to the world's population each year. However, as we have seen in this section, many obstacles stand in the way of that goal. For example, although the largest gains can be realized in less-developed countries, food production in most of those countries is limited by inadequate resources. Moreover, achieving significant gains depends, to a large extent, on government policies that encourage agricultural research and ensure the availability of needed resources at prices that farmers can afford. Also, governments must ensure that farmers will receive a fair price for their crops and livestock. To date, however, most governments in less-developed countries have focused heavily on industrial and commercial development. And pricing policies have strongly favored urban consumers at the expense of farmers. Such public policies must change if the less-developed nations are to have any chance of feeding their burgeoning numbers.

Although farmers in developed countries are often able to employ advanced agricultural technologies and to gain access to needed resources, they too face difficulties in enhancing their future yields. Because of rising costs and declining net incomes, farmers are finding it increasingly more difficult to purchase the resources that they need for production. And the alternative, switching to less-expensive agricultural practices, would probably result in a diminished yield. Hence some agriculturalists believe that by using their current crop-production strategies, the farmers in developed nations are approaching the upper limits of their yields per hectare of arable land.

John Boyer, a plant physiologist at the University of Illinois, has proposed an alternative means of increasing crop production. As we have seen, the primary strategy that has been used has focused on improving the growth environment (such as adding fertilizers and water, preventing the growth of weeds and other pests by cultivation and pesticides, and selecting crop varieties that produce high yields under highly favorable conditions). But adverse conditions are common phenomena around the globe. For example, weather is rarely optimal throughout a growing season, and essentially all soils require fertilizers to achieve maximum yields. Hence Boyer suggests the development of plant varieties that will produce reasonably favorable yields under growing conditions that are less than optimal. Such a strategy would decrease the need for—and thus would conserve—resources (such as fertilizers, water, and fossil fuels). And because those agricultural inputs are becoming more costly, a strategy that reduces them would be readily accepted by farmers if, on balance, they came out ahead economically. Obviously, the traditional approaches of increasing crop production will continue, but Boyer's alternative strategy should gain favor with time, both in developed and less-developed nations.

Harvesting the Oceans

So far, in this chapter, we have focused on terrestrial food production. We turn our attention now to the oceans, which are also a source of food. Although the world's fish harvesting contributes only approximately 6 percent of the total annual supply of protein, that figure belies the importance of fish protein in many countries. The United Nations lists more than 30 countries in which fish protein represents 40 percent of the total supply of animal protein. In many of those nations, persons rely so heavily on diets of rice or starchy roots that, without fish protein, many persons would suffer from severe protein deficiencies. Refer back to Box 17.1 for a description of the vital significance of protein in the human diet.

However, at this time, the future yield of the world's fisheries is in question. One reason for that uncertainty is that fish harvests have not improved significantly since the early 1970s. During the 1950s and 1960s, harvests increased from 21 million metric tons to 70 million metric tons annually, which represented a 5-percent annual growth rate. The world's per-capita fish supply was thereby enhanced, and the supply of much-needed protein was increased. But since 1970, the annual fish harvest has leveled off, and per-capita production of fish has declined more than 10 percent.

Despite the lack of growth in fishing harvests over the last 15 years, some fishery experts believe that a yield of 100 million metric tons of fish per year can be attained. However, to achieve that goal, the worldwide harvest would have to be increased by one-third. As a result, new approaches will have to be taken to increase our harvests of fish. For example, one possibility is to step up fishing efforts in areas that are considered to be underfished, such as the Indian Ocean and the southwest Atlantic. Also, it is possible that a number of species could support a heavier harvest, such as squid in the eastern Pacific, capelin in the northwest Atlantic, and anchovies off the coast of West Africa. Also, some observers expect that technological developments (such as satellites) will aid fishermen in finding and harvesting large schools of fish.

However, all those possible methods of increasing world fish harvests are still merely sophisticated means of hunting and gathering. Thus another approach to improving the fish harvest is to treat the sea as a vast resource, which can be cultivated through the technology of *aquaculture* (which is analogous to the technology of agriculture on land). When aquaculture is practiced in marine environments, it is often called *mariculture*. Although current annual production from aquaculture is approximately 4 million metric tons, United Nations experts believe that the yield could be tripled if aquaculture efforts were intensified.

To date, the most successful form of aquaculture has been freshwater fish farming. And to achieve high yields, several management practices have evolved that are based on ecological principles (described earlier). Thus to improve the yield of fish without increasing competition, ponds are stocked simultaneously with several species that require different types of food. For example, fishes that are commonly used in such polycultures include Chinese, Indian, and European carp, various *Tilapia* species, and channel catfish. And to increase the production of organisms that are consumed by the fish, nutrients are added in the form of inorganic fertilizers, agricultural wastes, or human sewage. Use of such methods has produced quite high harvests in such diverse locations as India, Israel, Thailand (Figure 17.18), and the United States.

Marine fish farming (mariculture) has focused on the raising of shellfish and salmonid fishes (salmon and trout). Yields of shellfish (such as oysters and mussels) are enhanced by using racks, rafts, or trays. By growing on those submerged structures rather than on the bottom, shellfish have access to more food, and they are protected from bottom-dwelling carnivores and burial by sediments. In addition, many shellfish can be mass-produced rather easily in hatcheries and then released to enhance old fisheries or to establish new ones. For example, programs of selective breeding of trout and salmon have created stocks that grow larger and reach maturity more quickly than their ancestors. And some of those species do particularly well in floating marine cages.

In addition, more complex forms of aquaculture are currently under study. For example, two possible techniques include (1) the fencing off of an entire bay for fish raising and (2) the growing of shrimp or turtles in large tanks. Although such efforts could yield valuable protein, they are still in developmental stages. We still have much to learn about

Figure 17.18 Tilapia *fish being harvested in the canals that irrigate rice paddies in Thailand. (FAO Photo.)*

the nutritional and water-quality needs of the species that we want to farm, and many technological problems need to be overcome, such as waste removal and disease prevention. Furthermore, the huge capital investment that is required by such enterprises will doubtless prove to be an obstacle as well. Although some successes will be gained with this type of aquaculture, the products, clearly, will be too expensive for those persons who have the greatest need for fish protein.

In direct contrast to scientists who believe that we can increase the world's fish harvest, others are convinced that we are already at or near the sustainable level of harvest. To justify this outlook, they cite continuous overfishing and the ongoing pollution and destruction of aquatic habitats, which already limit fish production.

To ensure a sustained catch in future years, fishermen must leave enough fish behind to maintain an adequate reproduction rate. But *overfishing*—the harvesting of more fish than produced in a given period of time—is currently resulting in lowered harvests of the East Asian sardine, California sardine, Northwest Pacific salmon, Atlantic salmon, Atlantic herring, Atlantic cod, and Atlantic haddock. Furthermore, annual harvests of such species as the bluefin tuna, yellowfin tuna, cod, ocean perch, and yellowtail flounder have failed to grow despite increased efforts.

The repercussions of overfishing will continue to intensify unless fishermen begin practicing conservation. However, fishery conservation is limited in several ways. For example, because little is known about the population dynamics of marine fish, appropriate quotas for fish harvests (which are limits on allowable takes) are usually established on a trial-and-error basis. And if a quota is set too high, the fishery will fail. As a result, commercial fishermen will suffer either way: if the quota is too low, their harvest will be limited; and if it is too high, their stocks will be depleted.

A more severe limitation on conservation has been the lack of cooperation among nations in establishing and enforcing quotas. Many fish species migrate over large areas of oceans, and, until recently, many fishing areas were located beyond national territorial limits. As a result, no one has been in charge to be sure that overfishing does not occur. Therefore, to provide for the orderly use of ocean resources, the United Nations sponsored a series of Law of Sea Conferences. After 15 years of negotiations, nearly 120 nations finally signed the Law of the Sea in December of 1982. That law establishes a 320-kilometer (200-mile) economic zone off the coast of each nation that borders the sea, and coastal nations have the sovereign right to establish and enforce fishing quotas within that zone. But many problems remain. For example, several leading sea powers, including the United States, the United Kingdom, and the Soviet Union withheld their signatures from the law. Moreover, only nine of the signatory nations have ratified the treaty, and progress toward ratification in other nations has been bogged down in bureaucratic procedural difficulties. Thus the future of the law remains clouded. Even if the law is ultimately ratified, it will mean little unless all the major sea powers support it.

The other threat to sustained yield from the oceans is the pollution and destruction of habitats that are vital for fish production. That danger results from the fact that the most productive areas of the oceans—estuaries, reefs, and regions of upwelling—are adjacent to continents. And these zones are becoming seriously polluted by eroded sediments, dredge spoils, pesticides, and domestic and industrial wastes (Chapter 7). For example, in the United States today, some 30 percent of the shellfish beds are now closed because of the risk of possible contamination by human pathogens that are in domestic wastewater. Furthermore, commercial, industrial, and recreational development is threatening to eliminate fish and shellfish habitats altogether. And, finally, coastal waters are being exploited with increasing frequency for petroleum and mineral deposits. Hence if coastal zones are not managed properly, fish production throughout the world might actually decline.

Although many experts have focused their attention on enhancing fish harvests, the surest way to increase the amount of available fish is to minimize the number of harvested fish that spoil or are wasted. For example, a United Nations study suggests that as much as 40 percent of the catch in some less-developed nations is lost or spoiled before it can be consumed. Hence if adequate refrigeration and processing facilities were provided to preserve the fish that have already been caught, protein malnutrition would be significantly reduced.

If recent history is a reliable indicator, the world's harvest of fish may have already reached its maximum. Moreover, even the current harvest level may not be sustained unless better management is implemented in fisheries and coastal zones around the world. In addition, the contribution of aquaculture is limited by technological and economic constraints, and quite probably, it will never produce anywhere near the quantity of fish that is produced naturally in the oceans.

Meanwhile, the human population is growing rapidly in less-developed nations—many of which are heavily dependent on fish protein. As a result, the United Nations projects that the demand for fish will double by the year 2000. Therefore, although fish protein will continue to be important in the human diet, it is painfully obvious that, relative to the needs of the human population, the seas are too small to meet the growing demands.

Conclusions

In our attempt to predict whether the world's food supplies will be adequate to meet future needs, we can be certain of one thing: barring a holocaust, such as nuclear war, more than 6 billion persons will inhabit the planet by the year 2000. Projecting further, the United Nations and the World Bank foresee that the world's population will continue to grow for at least a century before it levels off at approximately 10 billion persons. Therefore, if all those persons are to be fed adequately, more than twice as much food must be produced than is available today.

It is much more difficult to determine the amount of food that will be available in years to come than it is to evaluate the needs. Efforts to reduce food losses and increase yields depend on both the availability of resources (such as seeds, fertilizers, pesticides, water, energy, and storage facilities) and environmental factors (such as favorable weather conditions and the control of soil erosion, salination, and pollution). However, even if we can overcome those limitations, economic, social, and political factors will also have a great influence on food losses and production. For example, effective land-use policies, economic incentives for increased production, improved food storage and preservation, and international cooperation are all essential for an improved global availability of food. Furthermore, the greatest reduction in food losses and the greatest increase in yields (a fivefold increase by United Nation's estimates) must occur in the less-developed nations. Yet those very same nations face the most severe constraints in terms of resources and institutional factors. Because of the complex nature of those variables, estimates of future food supplies vary widely.

Some predictions are optimistic. Some forecasters, relying primarily on the continued development of agricultural technology, suggest that the earth's resources are sufficient to raise food production by at least tenfold. Such estimates, however, rest on two assumptions: (1) that the resources that are necessary to implement various forms of agricultural technology will be available in sufficient quantities in essentially all nations; and (2) that environmental degradation will not limit food production. However, those assumptions appear to be quite tenuous.

In view of the limitations, such as those described in this chapter, some researchers suggest that a more realistic expectation regarding potential food production is one more doubling. However, even meeting that limited goal will require considerable effort and involve many environmental risks. For example, Norman Borlaug, who won the Nobel Peace Prize in 1970 for his work on improving wheat varieties, estimates that we managed to gain 30 years through the Green Revolution in the race against mass starvation. Thus before this 30-year period ends, we must bring the demand in line with the supply by slowing population growth significantly.

Undoubtedly, we will continue to increase our food yield for some time. But history warns that we should expect to experience years in which harvests will be poor and production will decline. In the long run, whether or not the population will level off before we exceed the earth's sustained capacity to feed us is still an open question.

Summary Statements

About half of the world's food production is consumed by pests—insects, rodents, fungi, and weeds. In developed countries, the use of pesticides has reduced crop loss significantly.

Some pesticides kill beneficial animals and disrupt natural pest-control mechanisms. As a result of selection, populations of more than 450 species of pests have become resistant to at least one pesticide. And because some pesticides are persistent, they may accumulate in food webs in toxic concentrations.

Each pesticide must be evaluated on its own merits. The pesticides that are most dangerous to the environment are the persistent chlorinated hydrocarbons.

Alternatives to pesticides include biological control, naturally occurring pesticides, breeding of pest-resistant crop varieties, and crop rotation. All these alternatives have both advantages and disadvantages.

Integrated pest management is based on a combination of several pest-control procedures that are selected to provide the best control of a pest under certain environmental conditions.

No decline in the use of pesticides is expected until alternative pest-management techniques, which are economically acceptable to farmers, are perfected.

A potential exists for expanding agriculture into humid tropical regions. But many tropical soils presently limit agricultural expansion because of their low fertility. And because of rising costs and limited resources, areas that are being cultivated worldwide are unlikely to be increased much more.

The development of dwarf varieties of wheat and rice—the Green Revolution—reduced the threat of starvation in the late 1960s. During the 1970s, however, grain production became erratic and the world's total production of grain leveled off.

Climate is one of the primary controls of food productivity. For example, crop failures that have occurred during recent years in both developed and less-developed countries illustrate the vulnerability of our food supply to weather. Moreover, world agriculture continues to develop excellent harvests during favorable weather, but it suffers sharply lower harvests during unfavorable weather.

Although they are essential for food production, current agricultural practices are allowing the soil to erode at a rapid rate. Many procedures are available to reduce soil erosion, but because of economic hardship, few farmers around the globe are employing those techniques.

Irrigation is necessary in those regions where rainfall is commonly unreliable for growing crops. However, the supply of water for irrigation is becoming increasingly limited. Associated irrigation problems include waterlogging and salination.

Today's high yields depend on the improvement of soil fertility as a result of fertilizer application. Although supplies of inorganic fertilizers seem to be adequate for the near future, the unequal distribution of fertilizer resources around the world could severely limit the growth of the world's food production.

Modern agricultural practices consume large amounts of fossil fuels. The United Nations estimates that the use of fossil fuels will need to quadruple in the less-developed nations by the year 2000 if they are to provide an adequate amount of food. Much of that increased energy expenditure will be allocated to fertilizer production.

Fish are an important source of protein for persons in many countries. However, the future yield of the world's fisheries is unclear. Some experts predict that fish harvests will be significantly improved as a result of advances in harvesting technology, greater utilization of less-desirable fish, and developments in aquaculture. Other experts believe that continued overfishing and destruction of aquatic habitats will preclude significant increases in ocean harvests.

Because of the complex nature of growing, processing, and distributing food, estimates of future supplies of food throughout the world vary greatly. Given the expected limitations on resources and economics, one more doubling of the world's total food production is a realistic estimate. Meanwhile, the world's population is continuing to grow and climatic and economic uncertainties cloud the future.

Questions and Projects

1. In your own words, write a definition for each of the terms that are italicized in this chapter. Compare your definitions with those in the text.

2. Describe how an insect population develops resistance to an insecticide.

3. List the characteristics of an environmentally sound insecticide.

4. List the advantages and disadvantages of biological control.

5. List and discuss the factors that you would consider in deciding how to control a pest (insect) in your garden. Would you base your decision on other considerations if you owned a large farm?

6. Persons do not like to find insects in their food or to eat blemished fruit. But it has been estimated that 10 to 20 percent more insecticides (than would otherwise be required) are used to reduce the incidence of insects in food and to improve the appearance of fruits and vegetables. Considering the energy costs and the health and environmental problems that are associated with insecticides, is this additional use justified? What factors must be taken into consideration?

7. Although the forest vegetation in tropical areas is usually very lush, the soils are often quite infertile. Explain this apparent paradox.

8. What is the Green Revolution? Why has it not lived up to earlier expectations as a panacea for feeding the world?

9. Why is climate a critical influence on crop production? In recent years, has severe weather reduced crop production in your area? If so, how have the reductions affected your community's economy?

10. How do strip cropping and no-till techniques help to reduce soil erosion?

11. Discuss the pros and cons of irrigating cropland. If possible, visit an irrigated farm. Is the irrigation system well managed?

12. Vegetables are important sources of vitamins and minerals, but vegetable farming is energy-intensive and often requires irrigation. As a society, can we afford to continue to grow vegetables this way? Do alternative methods of growing vegetables exist?

13. Compare the level of energy that is consumed by agriculture in the United States with that of less-developed countries. Speculate on how agriculture in different countries might be affected in the near future if prices of fossil fuels were to rise again substantially.

14. Describe the factors that currently limit crop production on land that is already being cultivated. How might those limitations be overcome? Might attempts to overcome those limitations result in adverse consequences? If so, what are they?

15. Why did world fish harvests level off in the 1970s? Discuss the potential of the oceans to continue to serve as an important source of protein.

16. What is the most important thing that you could do as an individual to help solve the world's food problem?

17. The United States is one of a handful of countries that regularly has extra food to export. What should we do with that food? Store it for the lean years? Sell it to the highest bidder? Give it away to the needy? Cut back production to meet our own needs alone, thereby reducing the deterioration of American farmland? Name some other alternatives. How would your answers differ if you were a farmer, a conservationist, a stockholder of a grain-exporting company, or a person who lives in a large city?

18. Have land-use conflicts arisen between agriculture and other interests in your community? What is your community doing to resolve those conflicts?

Selected Readings

ADKISSON, P. L., NILES, G. A., WALKER, J. K., BIRD, L. S., AND SCOTT, H. B.: Controlling cotton's insect pests: A new system. *Science* 216:19–22, 1982. A description of an integrated pest-management procedure that is used to control pests of cotton in the Rio Grande Valley.

BATIE, S. S., AND HEALY, R. G.: "The Future of American Agriculture." *Scientific American* 248:45–53 (February), 1983. A comprehensive outline of the future impact of land, water, climate, and demand for exports on agricultural production in the United States.

BATRA, S. W. T.: Biological control in agroecosystems. *Science* 215:134–139, 1982. A description of biological-control methods and their integration with other agricultural practices to control pests.

BORGESE, E. M.: "The Law of the Sea." *Scientific American* 248:42–49 (March), 1983. An account of the often difficult negotiations that led to the signing of the Convention on the Law of the Sea and the potential significance of this new international treaty.

BOYER, J. S.: Plant productivity and environment. *Science* 218:443–448, 1982. A proposal to improve agricultural productivity by selecting from crop varieties that yield well under adverse growing conditions.

BREMAN, H., AND DE WIT, C. T.: Rangeland productivity and exploitation in the Sahel. *Science* 221:1341–1347, 1983. An analysis of consequences of low soil fertility and low rainfall on the success of new management schemes for increasing livestock production in the Sahel.

BREWER, M.: The Changing U.S. Farmland Scene. *Population Bulletin* 36 (December). Washington, D.C.: Population Reference Bureau, 1981. Examines the adequacy of cropland in the United States to meet future domestic and international demands for food and fiber.

BROWN, L. R., AND SHAW, P.: *Six Steps to a Sustainable Society.* Washington, D.C.: Worldwatch Institute, 1982. An examination of the worldwide deterioration of the resource base, followed by policy proposals to reach a sustainable society, including the protection of cropland.

BROWN, L. R., AND WOLF, E. C.: *Soil Erosion: Quiet Crisis in the World Economy.* Washington, D.C.: Worldwatch Institute, 1984. An examination of the causes of soil erosion, its impact on food production, and the economics of conserving soil.

DONALDSON, L. R., AND JOYNER, T.: "The Salmonid Fishes as a Natural Livestock." *Scientific American* 249:51–58 (July), 1983. A fascinating look at recent successes of breeding programs to develop new genetic strains of salmon and trout, which make it possible to adapt fish to ranching operations.

FOOD AND AGRICULTURAL ORGANIZATION: *Agriculture: Toward 2000.* Rome, Italy: Food and Agricultural Organization, 1981. A synopsis of the challenging efforts that are needed to meet the food needs of less-developed nations.

KERR, R. A.: Fifteen years of African drought. *Science* 227:1453–1454, 1985. An account of the extended drought in sub-Saharan Africa and its causes.

MELLOR, J. W., AND ADAMS, R. H., JR.: "Feeding the Underdeveloped World." *Chemical and Engineering News* 16:32–39 (April 23), 1984. An assessment of the agricultural technology that developing countries will need if they are to adequately feed their rapidly growing populations.

NICHOLAIDES, J. J., BANDY, D. E., SANCHEZ, P. A., BENITES, J. R., VILLACHICA, J. H., COUTU, A. J., AND VALVERDE, C. S.: Agricultural alternatives for the Amazon Basin. *Bioscience* 35:279–285, 1985. A description of an agricultural technology that permits continuous production of crops on the infertile soils of the Amazon Basin.

PIMENTEL, D., AND EDWARDS, C. A.: Pesticides and ecosystems. *Bioscience* 32:595–600, 1982. An analysis of how pesticides affect ecosystem functioning by changing patterns of energy flow and nutrient cycling and by reducing species diversity.

RHYTHER, J. H.: Mariculture, ocean ranching, and other culture-based fisheries. *Bioscience* 31:223–230, 1981. Discusses the potential of various types of aquaculture to enhancing food production.

ROTHSCHILD, B. J.: More food from the sea? *Bioscience* 31:216–222, 1981. Discusses the biological and economic limitations on future harvests of marine fish.

SHELDON, R. P.: "Phosphate Rock." *Scientific American* 246:45–51 (June), 1982. A description of the distribution, mining, and world trade of this vital plant nutrient.

SLATER, L. E., AND LEVIN, S. K., EDS.: *Climate's Impact on Food Supplies.* Boulder: Westview Press, 1981. A detailed analysis of strategies and technologies to cope with the influence of climate on crop production.

Power lines stretch across rice farms in the fertile Sacramento Delta area north of Davis, California. (© 1978 Barrie Rokeach.)

CHAPTER 18

CONTEMPORARY ENERGY ISSUES

A Historical Overview

For centuries, people have been fascinated by devices that save human labor, provide entertainment, make life more comfortable, and amplify human muscle power in an innocent desire to improve the quality of life. In modern times, that trend has led to the development of a tremendous variety of energy-consuming machines and devices, which we depend on so heavily now that life is nearly unimaginable without them. For example, less than a century ago, horsepower still meant just that—horsepower—to most persons. Stagecoaches and surreys were pulled by horses, and teams of horses augmented farmers' sweat and muscle power.

The development of the steam engine marked the beginning of our dependence on fuel. Initially, steam engines were used to pump water from mines. But by 1817, steamboats provided regular service on the Ohio and Mississippi Rivers, and after 1825, steam locomotives—iron horses—pulled loads over rapidly expanding rail networks. By the 1850s, clumsy steam engines provided enough power to drive factory machinery and farm equipment, such as threshing machines. Although the first steam engines burned wood, coal soon replaced wood, because it provided more heat per kilogram and was easier to handle.

In 1859, the first commercial oil well gushed forth in Titusville, Pennsylvania (Figure 18.1), and by the 1870s, small engines that were fueled by natural gas or gasoline had been developed. And those inven-tions were soon followed by larger, more powerful petroleum-burning engines (Figure 18.2). For example, the first horseless carriages, which were hand-built in backyard garages, appeared near the turn of the century and were the harbingers of America's most voracious energy-consuming machines.

During the following years, one invention after another came into common use. For example, in 1882, the first centralized electricity-generating systems in the United States began operating in Appleton, Wisconsin, and New York City. Initially, electricity was used only for lighting; electric appliances did not begin to appear until near the turn of the century. Then, the electric stove was introduced in 1896, and the electric vacuum cleaner and clothes washer were introduced in 1907.

In succeeding years, as the cities swelled with newcomers, the demand for energy increased. As a result, around 1930, Chicago was linked to Texas natural gas fields by a 60-centimeter (24-inch) diameter pipeline. And the electrical energy supply for Los Angeles received a boost in 1936, when the city was connected to hydroelectric facilities at Hoover Dam by the first long-distance, high-voltage transmission lines. Elsewhere, electricity became more readily available as larger, more efficient coal-fired electricity-generating plants replaced antiquated power plants that had been built in the 1890s. Those new plants operated at higher steam pressures and were ten times more efficient than the old ones.

Between 1935 and 1955, high-compression die-

Figure 18.1 *The Drake oil well, near Titusville, Pennsylvania, was the first oil well in the United States. (Drake Well Museum.)*

Figure 18.2 *Henry Ford set up this experimental piston–cylinder assembly in his kitchen to prove to himself that gasoline engines could be used as a source of power. He later helped to usher in the era of the horseless carriage. (From the Collections of Henry Ford Museum and Greenfield Village.)*

sel engines (which could operate on less-expensive fuel) replaced gasoline and steam engines as industry's work horses. During that same period of time, electric motors came into widespread use. Furthermore, after World War II, new pipeline networks increased the use of natural gas in stoves, clothes dryers, refrigerators, and other appliances. And the appearance of petroleum-based plastics (polystyrene in 1937, Nylon in 1938, Teflon in 1944, and Orlon in 1948) ushered in an era of synthetic fibers and inexpensive manufacturing materials. In the years that followed, a seemingly endless series of energy-consuming devices appeared, such as air conditioners, electric clothes dryers, color television sets, video-cassette recorders, and a myriad of power tools—even elec-

tric carving knives and electric shoe buffers. Fuel is required not only to operate those appliances, but also to produce, distribute, and market them.

The seemingly endless supply of inexpensive energy spurred this great surge of inventions, and one important result of that occurrence was a great increase in human productivity. For example, assembly line workers were able to accomplish more in a day because the energy that ran their machines, in effect, supplemented their own muscle power. Teamsters shifted from horse-drawn wagons to 18-wheel trucks. Construction crews and miners stopped using picks and shovels and started driving huge power shovels and bulldozers that were literally capable of moving mountains. And, as illustrated in Figure 18.3, farmers stopped guiding horsedrawn, single-bottom plows that could turn only a single furrow and began to employ multibottomed plows (which are capable of turning a dozen or more

(a)

(b)

Figure 18.3 *The effect of fossil fuels on agriculture during the past 75 years. (a) A farmer plowing in Montana in 1908 (U.S. Department of Agriculture.) (b) A modern multibottomed farm plow. (Deere & Company.)*

furrows) that are drawn by four-wheel-drive, 225-horsepower air-conditioned tractors.

Cheap, plentiful fuel engendered complex changes in life-styles and social structure, too. For example, millions of persons were able to flee from rural America and move to cities and suburbs in search of what they hoped would be a better life. And improved communications resulted, which changed the very fabric of our society.

However, throughout this dramatic period of transition, few persons concerned themselves with the possibility that we might run low on fuel. The first clear signal that energy supplies were finite came to the American public in the fall of 1973, when the Arab oil countries shut off all shipments of oil to the United States to press for higher prices. Although only a small amount of oil was withheld (only 2.7 million barrels per day), the impact on most citizens caused a great deal of personal inconvenience and discomfort. For example, persons had to wait for hours in line for gasoline at local service stations, live in cooler homes during the winter, and adjust to a speed limit of 88-kilometers (55-miles) per hour. But the long-term significance of the Arab oil embargo was that it made us aware of our dependence on an uninterrupted and copious flow of energy. And that new consciousness inspired us to reevaluate the state of our domestic fuel reserves and our energy-consumption practices.

Since our first energy problems erupted in 1973, the supply–demand picture for energy in the United States has improved significantly. But the easing of problems that are associated with energy resources does not mean that our energy problems have been resolved. Rather, reduced consumption has given us a little more time to evaluate future schemes to enhance our energy resources. The roots of a future crisis remain. Today, arguments abound concerning how conservation efforts should be financed and which new energy sources should be given special attention through research and tax credits. And many persons do not believe that the energy problems are real because energy prices fluctuate so widely. Furthermore, that attitude is compounded by the public's now prevalent distrust of persons in authority, whether they are government officials, scientists, or oil-company executives. Yet, in truth, the government's decisions concerning our energy policies will influence the quality of our environment, economy, national security, and international relations for decades to come.

Different Viewpoints on Energy

The role of energy in our society is viewed differently from one person or group to another. For example, some persons view energy resources as a group of *commodities*—as something to be bought and sold. Others view energy as a *natural resource*, which, when used, has implications (especially ecological) that range far beyond the buyer and seller. And some persons view access to energy supplies as being a *social necessity*, because any curtailment in supply would affect their life-styles and health. Also, government officials who are responsible for national security view energy supplies as a *strategic resource* in the national security system. Thus it is clear that no singly shared concept prevails concerning the role of energy resources in our society.

When energy is viewed solely as a commodity, discussion focuses on policies to develop and market gas, oil, coal, and electricity. However, even though the commodity concept has dominated energy policies in the United States for most of this century, its critics are growing in number. Many persons feel that the commodity view (in which supply and demand determine prices) is too narrow. They point out that persons or groups who embrace this viewpoint are too insensitive because they ignore important long-term considerations, such as future access to dwindling supplies, dependence on foreign supplies, and poor accountability for pricing by regulated monopolies (utility companies).

When energy is viewed as a natural resource, the specific effects of utilization of energy on ecosystem components—water, air, soil, climate, and plant and animal communities—are considered. Persons or groups who take this viewpoint consider the long-term implications of pollutants and resource availability and classify energy resources as being polluting or nonpolluting, renewable or nonrenewable, and exhaustible or inexhaustible. And although they recognize that the final use, or energy service, that is provided by energy resources is the main concern of most persons, they also recognize how an energy service is provided. For example, all other things being equal, persons who have this viewpoint might choose a combination of weatherization and solar methods to meet their space-heating needs—the energy service—rather than natural gas or oil. Thus a nonpolluting energy resource is preferred over a polluting resource, and a renewable resource is favored over a nonrenewable energy resource. In this view, substitutes for fossil fuels, such as solar technologies and fuel conservation, are very important.

During most of this century, few persons have challenged the view that energy is anything more than a commodity. After all, energy was relatively inexpensive, energy resources appeared to be large in comparison to the rates at which they were being depleted, and pollution that was caused by the use of certain types of energy resources was largely ignored. However, during the 1970s, energy shortages and a growing concern about environmental pollution brought the natural–resource viewpoint to the public's attention. And the rapidly rising energy prices that occurred during this period, as well as the hardships they inflicted on the poor, helped to focus attention on the social necessity of energy. The energy shortages during 1973 and 1974 also pointed out the strategic importance of conserving energy resources in industry (because the most energy-efficient companies are the most likely to survive) and for national security. Undoubtedly, the relative importance of these competing viewpoints on energy will change, depending on circumstances. For example, another oil embargo would strengthen the natural–resource viewpoint. Nevertheless, widespread recognition of these viewpoints should focus the debate on energy policies and facilitate the policymaking process.

Unfortunately, low energy prices during earlier decades—up to 1973—left us using many energy-inefficient technologies, which is illustrated by the many energy-inefficient homes that could be operated at much lower costs. For example, the cost of energy in Sweden declined by 50 percent between the mid-1950s and the early 1970s. Then, prices tripled during the next decade. Other developed countries experienced similar changes in the relative cost of energy during that period. And as long as energy was inexpensive, businesses had few economic incentives to manufacture and build more expensive, but also more energy-efficient appliances and homes, and consumers had few incentives to buy them. Now, however, because energy prices are much higher, we find ourselves concerned about our energy-inefficient homes and devices. And it will take a great deal of education, effort, and capital to change our energy-consuming technologies so they will become energy-efficient and cost-effective. Furthermore, the wide swings in energy prices make it impossible to design technology that will be cost-effective for a long period of time. For this reason alone, strong incentives exist to stabilize energy prices, so that over the long term, newly installed technology will not become outdated as a result of poor efficiency if energy prices rise substantially.

Stable energy prices are also important to industries that are developing new energy technologies, such as coal gasification and the production of fuels from oil shale. For example, a decline in the price of petroleum, which occurred in the early to mid-1980s, has caused serious problems for companies (and regions) that have made enormous investments in new types of technology which were justified only on the basis of stable or rising prices for their products.

In this chapter, we examine some of the problems that must be resolved during the next decade and beyond as we attempt to change the basic technology that provides us with energy services. Those problems are complex and solutions to them must consider not only energy sources and supplies but, also, protection of the environment, public health, economic growth, equity among geographical regions and economic classes, and national security. Thus we begin our examination of the problems we are having concerning energy by identifying current trends in the use of energy and assessing the prospects for future energy supplies.

Energy Use by Economic Sector

During most of the twentieth century, the development and expansion of our highly energy-dependent society has been a major endeavor, which has succeeded remarkably well. Because domestic production of fossil fuels could not keep pace with rising demands, many countries, including our own, are now importing energy—primarily oil, from politically unstable mid-Eastern countries. Therefore, to be less vulnerable to interruptions of their energy supply, these countries are beginning to revamp their strategies to maintain their energy services. However, such changes in the delivery of energy services may well affect the environment and the quality of life.

To gain insight into both energy supply and demand, we need quantitative information. For these discussions, we use the nonmetric *British thermal unit* (BTU) as our basic measure of the energy content of fuels, because it remains in widespread use, especially in literature that is produced by the United States government agencies. One BTU is defined as the amount of heat that is required to raise 1 pound of water 1 degree Fahrenheit. Thus the BTU is a small unit. For example, a ton of bituminous coal

yields 25 million BTUs; and a 42-gallon barrel of oil contains 5.8 million BTUs. A much larger energy unit, the *Quad*, is generally used in discussions of supplies and demands in the United States. One Quad is equal to 1 quadrillion, or 1.0×10^{15}, BTUs. Thus if approximately 172 million barrels of oil were burned, 1 Quad of energy would be released. (The United States consumed 73.7 Quads of energy resources in 1984.) By using the BTU and the Quad to calculate amounts of energy, we can avoid making conversions between the traditional measures that are used to buy and sell energy resources. For example, natural gas is sold by the cubic foot (1020 BTUs) or therm (100,000 BTUs); oil is sold by the barrel (42 gallons = 5.8 million BTUs); coal is sold by the ton (approximately 22 million BTUs); and electricity is sold by the kilowatt-hour (3412 BTUs). Students who desire to become more familiar with energy-resource management must become familiar with these units and with the laws of thermodynamics (see Box 2.1) that explain energy flow in machines as well as the environment. (See Appendix I for conversion factors for measurements of energy.)

Today, the United States consumes approximately 15 times more energy than it did a century

Figure 18.4 *Energy use in the United States during the last century. Approximately 2.5 Quads of biofuel energy are not accounted for in recent data. The amount of each type of energy used is the distance between the curve for that fuel and the curve immediately below it. Thus nuclear and hydroelectric energy each provided approximately 4 Quads of energy in 1984, whereas oil provided 31 Quads. The sharp downturns in 1979 were due to a combination of economic recession and stepped-up conservation. (Data for 1880–1976: Federal Energy Administration.* Energy in Focus—Basic Data. *Washington, D.C., 1977; data for 1977–1984: U.S. Department of Energy,* Monthly Energy Review.)

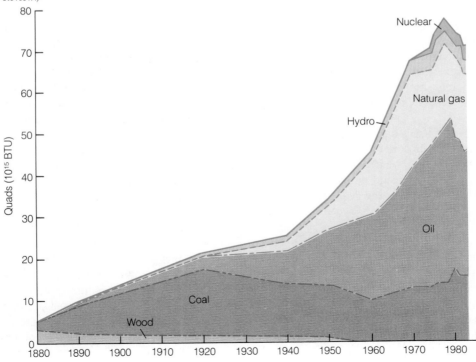

ago (Figure 18.4). That dramatic rise in consumption is attributable, in roughly equal parts, to the growth in population and the soaring energy demand per capita. However, the demand per capita has eased somewhat since the Arab oil embargo. Hence, as Figure 18.5 suggests, all sectors of our economy—residential, commercial, industrial, and transportation—have shown little change in use of energy since 1973. Just how much energy is necessary to keep an average citizen in the United States supplied with energy services is revealed by the statistics of average annual use. For example, 24 barrels of oil, 81,000 cubic feet (2300 cubic meters) of natural gas, and 2.9 metric tons of coal are used by each person every year. Furthermore, we can gain perspective on the vast amount of energy that is used every year by examining the trends in consumption within each of our society's economic sectors.

Energy is used in both homes and commercial buildings for basically the same purposes: heating, air conditioning, lighting, ventilation, and the powering of small electrical appliances. Thus natural gas and oil are the two most important fuels that are used directly (see Figure 18.6). Between 1950 and 1972, the rate at which energy was used in the residential sector increased by 4 percent per year, which was about double the rate of population growth. But since 1973, the total amount of energy consumption in the residential–commercial sector has increased approximately 0.5 percent each year on the average. In 1984, the residential–commercial sector accounted for 35.3 percent of the total amount of energy that was used in the United States. And resi-

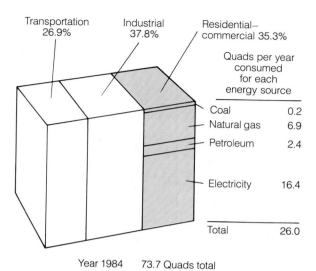

Figure 18.6 *Sources of energy for the residential–commercial economic sectors of the United States. The total amount of energy that was consumed by all sectors in 1984 was 73.7 Quads.* (Monthly Energy Review, *December 1984. Energy Information Administration, Washington, D.C.*)

Figure 18.5 *Energy consumption by the various economic sectors of the United States. The amount of energy that is consumed by each sector is the amount shown between the curve for that sector and the curve immediately below it. (*Monthly Energy Review, *December, 1984. Energy Information Administration, Washington, D.C.)*

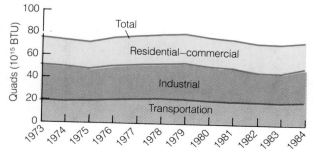

dential and commercial demands for energy were about equally divided: homes consumed approximately 19 percent and business establishments consumed approximately 16 percent of the total.

Space heating is the single most important use of energy in homes and businesses. And air conditioning has become a major contributing factor to the rise in energy consumption during the past several decades. For example, in 1950, fewer than 1 percent of all homes in the United States had any type of air conditioning but, today, approximately 50 percent are so equipped. Other appliances that are becoming increasingly popular in homes are clothes dryers, self-defrosting refrigerators, color television sets, water beds, video-cassette recorders, and improved, but often excessive, lighting. Thus given the fact that our population is increasing, it is clear that our total energy consumption will not decline in the residential–commercial sector unless we learn to use energy more wisely.

In 1984, the transportation sector of the United States economy accounted for 26.9 percent of the total amount of energy that was consumed. And transportation depends almost entirely on petroleum products that are refined from oil (Figure 18.7). In fact, approximately 45 percent of the oil that is used in the United States fuels our cars, trucks, buses,

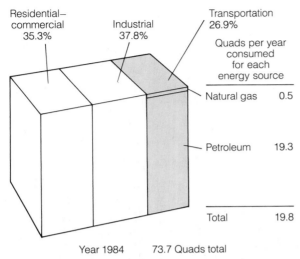

Figure 18.7 *Sources of energy for the transportation economic sector of the United States. The total amount of energy that was consumed by all sectors in 1984 was 73.7 Quads.* (Monthly Energy Review, *December, 1984. Energy Information Administration, Washington, D.C.)*

trains, planes, ships, and barges. In total, the fleet of vehicles in the United States burns up the equivalent of 6.4 million barrels of oil each day, and 74 percent of that oil is consumed by cars and light trucks. Thus, an important goal is to improve the fuel economy of U.S. automobiles.

The industrial sector of the United States economy consumes the largest amount of energy, which accounts for 37.8 percent of the 1984 total (Figure 18.8). And most of that energy is used to fire boilers. For example, the chemical industry is the largest single consumer within the industrial sector, since it not only requires energy for manufacturing, but it uses petroleum as the raw material for many of its products (such as fibers, pharmaceuticals, plastics, and pesticides). Also, the steel-producing, petroleum-refining, paper, and aluminum industries consume enormous amounts of energy. However, to a large degree, industries can switch from one type of fuel to another if they have an economic incentive. For industries, the most important fuel characteristic is assured, uninterrupted supply—and in the foreseeable future, coal is the fuel whose supply seems to be most assured. However, some industries may not be able to convert to coal if they are located in a region that does not meet federally mandated air-pollution standards (see Chapter 11).

Energy Resources

The sight of an oil-filled supertanker, a railroad yard that is lined with coal cars, or a mountain of coal that is located near a giant electric utility gives us an inkling of the tremendous quantities of fossil fuels that are consumed by our energy-hungry technology (Figure 18.9). And we might wonder how long we can expect those resources to last.

To determine how long our resources will last, we must consider two factors: (1) our rate of consumption and (2) the amount of resources that exist which are economically recoverable (see Chapter 12). Once we have that information, we can calculate the expected lifetime for a resource. For example, a small automobile usually consumes approximately 8 liters (2.1 gallons) of fuel per hour and has a fuel capacity of 40 liters (10.6 gallons). Thus if a car were driven continuously for 5 hours, it would consume a tank of fuel (40 liters divided by 8 liters per hour). Similar calculations could be made for fossil-fuel resources in the United States and worldwide, but the answer would not be as exact as the calculation shown for automobiles for the following reasons.

Although our present consumption rates of fossil fuels are known with a fair degree of accuracy,

Figure 18.8 *Sources of energy for the industrial economic sector of the United States. The total amount of energy that was consumed by all sectors in 1984 was 73.7 Quads.* (Monthly Energy Review, *December, 1984. Energy Information Administration, Washington, D.C.)*

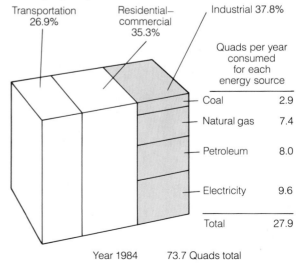

Figure 18.9 *Unit trains consisting of 100 rail cars filled with low-sulfur coal are used to transport coal from western states to power plants that are located in the Midwest.*

the total amount of recoverable fossil fuels that exist is much more speculative. Therefore, calculations to determine how long our fossil-fuel reserves will exist are only rough approximations (some would say guesses). Furthermore, future patterns of consumption will probably change in response to varying prices and availability, which will introduce more error into the projections. Nevertheless, these calculations can tell us which resources will be depleted first, and they can indicate where efforts should be made to reduce rates of consumption. This type of analysis also gives us some idea concerning which alternative sources of fuel are best to provide our desired energy services. Thus we now examine the supply and demand of our fossil-fuel resources—oil, natural gas, and coal.

Oil

During the twentieth century, crude oil has become one of the primary raw materials for our technolog-

ical society. For example, petroleum products power our transportation system, fuel our farms and factories, and heat many of our homes and businesses. A petroleum-refining residue, which is called asphalt is used to pave roads. And oil products are used in more than 3000 commercial products (lubricants and paints) as well as being the raw materials that are used to synthesize another 3000 petrochemicals (plastics and pesticides). In total, the production of goods and services for an average family of four requires approximately 73 liters (0.45 barrels) of oil-derived products each week.

In 1984, the United States produced 18.3 Quads of energy in the form of oil and consumed 31.0 Quads of energy as oil. We consume a total of 73.7 Quads of energy from different sources each year, and oil supplied 42.1 percent of our total energy demand in 1984. Yet, of all the fossil fuels that exist, oil is in shortest supply. Furthermore, as the statistics for 1984 production and consumption show, we import approximately 30 percent of the oil that we need, because domestic sources cannot meet our demands. Thus if the demand for oil increases, we will need to import more oil. Some persons find this reliance on foreign oil objectionable, because it renders our economy—and our life-styles—dangerously dependent on decisions that are made in other countries (as we discovered during the 1973 Arab oil embargo).

Thus we might wonder how much longer oil can continue to be our primary fuel resource. Figure 18.10 shows the actual rate of oil production in the United States since we started to use it and the predicted rates for future domestic oil production. The calculations of future supplies are based on both the number of known oil reserves and petroleum geologists' estimates of how much oil remains to be economically discovered and recovered.

Figure 18.10 illustrates the fact that although we did not start to use oil until the beginning of the twentieth century, today, our known and estimated domestic oil resources have been depleted by approximately 62 percent. Thus our fuel resources are lower than one-half of their original level, and we are still using oil at very near the historical maximum rate of consumption. However, some persons argue that the projected production curve will be much longer than the one that is shown in Figure 18.10, which would require an increase in the projected discovery rates on the right-hand side of the figure. But that occurrence would require that

oil be discovered at a rate that is greater than the actual rate of current discoveries, and several lines of evidence suggest that this is unlikely. For example, the average amount of oil that has been discovered as a result of drilling wildcat wells in the United States has declined from 23 million barrels in the early 1950s to only 10 million barrels in the early 1980s.

Not only is each drilling effort less successful on the average, but the total amount of oil that is discovered each year in the United States has declined from approximately 38 billion barrels per year in the early 1960s to approximately 10 billion barrels per year in the 1980s. Even worldwide, the rate of discovery has declined during the past 15 years, despite increased exploration activity and improved methods of prospecting. Furthermore, only one possible new giant oil field has been discovered—off Alaska's shores—since 1965, when Alaska's North Slope oil field was discovered. (*Giant oil fields* are defined as fields that contain more than 100 million barrels of oil). Giant oil fields, such as those that were discovered in the 1930s in East Texas and in Wilmington, California, still dominate current production in the United States. And petroleum geologists do not believe that there is much possibility that many new giant oil fields will be discovered. This is particularly true in the United States, where extensive drilling has already been done. Furthermore, even if a new giant oil field were discovered, it would extend our oil supplies for a few weeks or, at most, a few years if our current rate of consumption continued and if it were the sole source of production. For example, the output of a giant oil field, which might produce 100 million barrels of oil, would be used up in the United States in a week at the current rate of consumption, which is 14.3 million barrels of oil per day.

Late in 1983, a U.S. Geological Survey report assessed the oil resources throughout the world as well as in the United States. The report estimated that worldwide oil reserves, which are ultimately recoverable by conventional means, comprise 10,000 Quads (1700 billion barrels) of energy, and 3200 Quads (550 billion barrels) in that category have not been discovered yet. If those figures are correct, then at the world's current rate of consumption, our current oil supply will last for 60 years. The report estimates that proven reserves and undiscovered reserves in the United States constitute 580 Quads of energy

Figure 18.10 *Actual and projected production rates from oil fields in the United States. Projected values made by J. Steinhart are essentially identical to the 1983 estimates made by the U.S. Geological Survey. (L. C. Ruedisili and M. W. Firebaugh,* Perspectives on Energy, *3rd ed. New York: Oxford University Press, 1982.)*

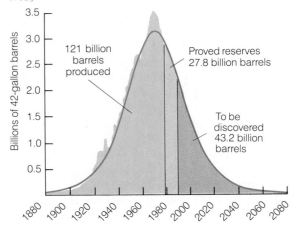

and would last for 36 years at our current rate of consumption—and only 19 years if we stop importing oil.

But these estimates of the lifetime of United States' oil reserves hide the fact that existing reserves in the United States will not keep pace with demand. In fact, in 1984, Joseph Riva, an oil resource expert at the Library of Congress's Congressional Research Service, projected a 17 percent decline in production in the United States by the year 2000. This figure of declining production is his most optimistic estimate; and if explorations for oil in such areas as the Atlantic Coast, Oregon-Washington, and the Rocky Mountain, Northern Great Plains area do not meet expectations, he projects a 29 percent decline in production.

Thus the issue of oil supply is not one that is likely to fade away. Rather, many experts feel that competition for oil will intensify in the future. They cite both increasing worldwide competition for oil from less developed nations and decreasing worldwide production within the next decade. Indeed, if every person in the world used as much oil as citizens in the United States, the total amount of estimated oil resources throughout the world would be used up in 8 years!

It is important to note that the shape of the curve in Figure 18.10 is typical of the rate-of-use cycle for a finite resource. It shows the curve that is formed when a resource is being used up, which illustrates the fact that use tends to tail off rather than experience a sharp cutoff. Moreover, as supplies dwindle, petroleum will be used only for those purposes that can justify the premium price that it will command.

Thus our examination of past consumption and future oil supplies points out the unique "oil era" in which we live. It is a period in which *relative energy costs* (the proportion of income that we spend for energy) has been small, because access to oil and products that are derived from it has been relatively easy and, therefore, relatively inexpensive. However, to maintain our current living standards, we will need to make a slow but determined transition to other sources of energy and use more effective methods of energy conservation.

Natural Gas

Natural gas is often found in the same locations as oil, but it may also be found in regions where no oil exists. And although more places on earth contain natural gas than oil, both resources are associated with sedimentary rock environments (Chapter 12).

Natural gas is an ideal fuel because it is clean-burning, producing only carbon dioxide and water as end products. If the dirtier fuels—oil, coal, or wood—were used exclusively in home heating systems in large urban areas, the resultant air-pollution problem would necessitate air-pollution control devices on individual systems, pollutant removal from fuels, or both. Also, natural gas is transported easily by pipelines directly to where it is needed and where it can be used more efficiently than if it were first converted to electricity. On a per-energy unit basis, natural gas is also the safest fossil fuel in terms of the number of accidents and deaths that are caused during its extraction and use.

Natural gas has been available to consumers at exceptionally low prices, relative to the cost of other forms of energy (see Figure 18.11), because its price has been controlled by government regulations. However, in 1978, the Natural Gas Policy Act (NGPA) was passed to allow gas prices to rise, which, in turn, will increase exploration to uncover new gas supplies. The Natural Gas Policy Act provides for the phased deregulation of the price of "new" natural gas—gas wells that have been developed since 1978—so that the price per energy unit of new gas will rise until it equals the price that is commanded by oil. (In some parts of the United States, the cost of natural gas might exceed that of oil.)

Gas resources that were known to exist prior to the enactment of the Natural Gas Policy Act are classified as being "old" gas and, therefore, its price is controlled. The regulated price for gas depends on the geology, depth, year of discovery, and existing contractual arrangements for the gas. The act also established a category for new gas in which the price ceiling is allowed to increase faster than inflation. For example, pipeline companies that cannot acquire enough gas to meet the demand in their distribution area can make bids for the unregulated new gas. Then, consumers pay the average price that the pipeline pays for the gas, plus a tariff that is controlled by the Federal Energy Regulatory Commission. Thus when the pipeline companies buy more unregulated gas, the price to consumers increases. Under present law, by 1985, approximately half of domestic gas production will be free of price controls. In the absence of new regulations, the average price of gas is expected to increase markedly in 1985, when price controls on new gas are removed. And consumers will pay even higher prices (Figure 18.12).

The Natural Gas Policy Act is very complex and controversial. Continuation of gas price controls is favored by most consumers, especially older citizens who have fixed incomes, but price controls slow exploration and encourage hiding (not reporting) of newly discovered gas in anticipation of a future higher price structure. Also, by maintaining prices at artificially low levels, conservation measures are also discouraged. Thus low prices will hurt the public over the long term, because natural gas is a finite resource, which must be used efficiently. Although amendments to the 1978 law are expected to be passed before full decontrol occurs, the nature of those amendments is impossible to predict because of the intense lobbying that is being carried out by a large number of special-interest groups—consumers (residential and the chemical industry) and producers of gas that fall under the various regulations of the Natural Gas Policy Act.

Supplies of natural gas in the United States now come from the lower 48 states, but financing for a huge pipeline to transport natural gas from Alaska's North Slope region to the upper Midwest is pending.

The United States imports only 5 percent of its natural gas, mostly from Canada.

Natural gas plays an important role in the United States' economy and accounts for 52 percent of the total amount of energy that is used in residences for space heating—a use that could not be converted easily to another energy source without great cost and increased air pollution. And industry consumes even more gas than the residential sector. However, it would be possible for industries to convert to other energy sources, because their larger sizes can justify the financial investment in situations where air-pollution control devices are needed.

Between 1967 and 1980, we used natural gas reserves faster than new sources of gas were being discovered. Then, during the early 1980s, the short-term supply of natural gas improved somewhat. But still, only in 1981 did the rate of discovery match our rate of consumption. Moreover, recent exploratory drilling for natural gas in the Cordilleran overthrust belt in the western United States has indicated large potential sources of both gas and oil. On the other hand, exploratory drilling in what once were thought to be very promising areas, such as in the Gulf of Alaska, southern California borderland, and south Atlantic continental shelf regions, have indicated much less potential for yielding oil and natural gas than was previously thought.

Further complicating the issue of future availability of natural gas is the presence of methane in geopressurized brines in deep (approximately 6000 meters, or 20,000 feet) sedimentary layers. In those layers, *geopressurized methane* exists in three

Figure 18.11 *Cost of fuels to consumers in the United States, 1973–1983, in constant 1972 dollars. The rising slope of the lines indicates that fuel costs have risen faster than costs for other goods and services. Because of inflation, the actual costs in 1983 were 2.38 times higher than the values shown on the graph.* (Monthly Energy Review, *December, 1984. Energy Information Administration, Washington, D.C.)*

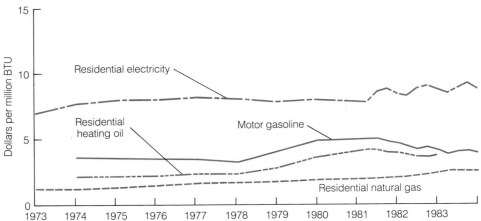

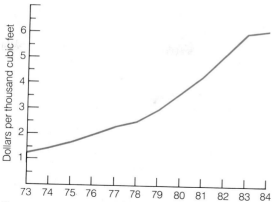

Figure 18.12 *The change in price of natural gas to residential consumers. (Monthly Energy Review, 1984. Energy Information Administration, Washington, D.C.)*

forms—as free gas, as gas that is dissolved in brine, and as immobile gas that is trapped in sediments. Some experts speculate that those poorly defined resources are immense. But before we can count on those resources, we need to determine the cost of recovery and the amount of methane that can be extracted from them. The cost of developing those resources undoubtedly will be higher, because they are located at great depths and the extraction of methane from them requires more expensive handling and cleanup procedures. However, if it becomes economically feasible to develop geopressurized methane, supplies of natural gas will be much larger than present estimates indicate.

We cannot determine how long natural gas supplies will last in the United States, because changes in regulations and conservation efforts change the rate at which natural gas is used. Furthermore, the size of undiscovered natural gas resources is even more difficult to predict than the size of oil resources. The U.S. Geological Survey estimated in 1981 that discovered reserves contained 380 Quads of energy and undiscovered reserves contained 600 Quads of energy. Thus if we use the 1984 natural-gas consumption rate of 18 Quads per year, the discovered and undiscovered resources would last 21 and 33 years, respectively, which would give us a 54-year supply. Considering various uncertainties, the range of estimates for the length of time that our natural gas resources will last is from 46 to 62 years, providing that we do not change our rate of consumption.

In late 1983, the Office of Technology Assessment, Congress's technical information arm, also summarized the natural-gas situation. Their report, which reflected a mixed forecast for future supplies of natural gas, stated that there is "no convincing basis for the argument that the lower 48 states have become so well understood, that a consensus can be reached about the size of the gas resources." And their range of estimates for recoverable natural gas resources is from 410 to 920 Quads of energy, which is more optimistic than the U.S. Geological Survey's estimate. The report concludes that "This range varies from a level that would seriously constrain gas production by the year 2000 to a level that might allow production to continue at current levels for the remainder of this century."

It would be possible to extend our natural gas supplies by importing natural gas in its liquefied state from the Middle East countries, which flare (burn) off natural gas as a waste product of oil production because they have no local market for it. *Liquefied natural gas* (LNG) can be shipped in tankers in which the natural gas is liquefied by chilling it to −162°C (−259°F), and it can be kept in that state by maintaining those low temperatures. Then, liquefied natural gas tankers are unloaded by regasifying the liquefied gas at its destination. In the liquefied state, natural gas occupies only approximately 1/600 of its normal volume. Hence an enormous amount of liquefied natural gas can be transported on each tanker. But liquefied gas is extremely hazardous to handle, and elaborate precautions must be taken to prevent gas leaks and possible explosions during loading, transport, and unloading. In addition, liquefied natural gas is considerably more expensive than natural gas that is supplied by pipelines.

It is also possible to produce synthetic gas from coal, oil shale, agricultural wastes, municipal wastes, and some industrial wastes (see Chapter 13). We examine how synthetic gases can be produced from coal and oil shale later in this chapter.

Coal

In 1984, the United States derived 23.3 percent of its energy from coal, compared with 93 percent at the turn of the century. For example, in 1950, coal heated 35 percent of the homes in the United States, a value which has declined to less than 1 percent, and 57 million metric tons of coal were burned in steam locomotives, a value which has declined to less than 0.05 million metric tons. Today, coal is used primarily in boilers to produce steam that generates electricity. Indeed, the increased demand for electricity has been responsible for the resurgence in the use of coal since 1970. For example, in 1983, 570 million metric tons of coal were used to generate electricity. Also, coal is an important chemical in metallurgy, which is the production of metals from their native ores.

Also, coal can be used directly as a substitute for other fuels, except in motor vehicles and aircraft. But in terms of economics, environmental considerations, and convenience, it is not advantageous in rail transportation, residential–commercial, or small industrial uses. However, coal has a big competitive advantage in the generation of electricity, because it can be delivered to utility plants at a very low cost per unit of energy. For example, in 1983, coal was one-half as expensive as natural gas and one-third as expensive as oil for generating electricity. Even though capital costs for equipment to burn coal are higher than the costs of oil- or gas-burning equipment, most utilities find that their total costs are lowest when they use coal as fuel.

Because the United States has immense reserves of coal and much smaller reserves of oil and natural gas, the policy decisions that are made regarding how and where coal will be extracted and how it will be used will have a great impact on our economy and environment in the future. Thus we now examine some of the current uses of coal resources in the United States and some of the factors that will influence those decisions.

The coal reserves in the United States are the largest in the world. They are estimated to represent 25 percent of the world's known recoverable reserves. The latest estimates, which were made in 1978, showed that our coal reserves include 10,600 Quads (430 billion metric tons) of energy, of which 52 percent is considered to be minable by conventional techniques. Thus the known coal reserves are approximately nine times larger than the combined total amount of recoverable oil and gas reserves throughout the United States, which is enough energy to supply our needs for three centuries at our current rate of consumption.

Coal deposits are scattered across the lower 48 states (Figure 18.13). Three-quarters of the surface minable coal lies west of the Mississippi, and vir-

tually all of those deposits are composed of subbituminous coal and lignite (Chapter 12). Approximately equal quantities of subsurface minable coal are found east and west of the Mississippi River. The East also has small reserves of anthracite coal. The characteristics of eastern and western coals are summarized in Table 18.1.

The geographical location of coal production shifted markedly during the past decade. In 1970, 93 percent of our coal was extracted from eastern mines, mostly from the Appalachian region. But, today, 33 percent of our coal is extracted from western mines. Two major factors caused that shift. First, western coal is low in sulfur content and is bought by many utilities and some large industries to reduce sulfur dioxide emissions, so they can comply with air-pollution standards. Second, because western coal is primarily "surface" coal, it can be mined for approximately one-third the cost of underground, eastern coal. However, those savings are partly offset by the higher costs of transporting western coal to eastern consumers. In fact, the cost of transporting coal from Wyoming to Illinois accounts for 73 percent of the price per delivered metric ton (which was approximately $43 in 1984). However, the trend toward the use of more western coal is expected to continue. And the United States is expected to export more coal as other nations attempt to decrease their oil consumption.

Also, changes in the methods of transporting coal have a significant impact on the environment. For example, because of the increased rail transport of western coal, many towns have requested that railroads be rerouted around them to avoid the noise and the interruption of local traffic patterns. Coal trains are about a mile long and take several minutes to pass a highway crossing. And in the eastern mid-Atlantic states, a number of canals and locks must be enlarged to handle the larger barges that are required to haul the amount of coal that is needed to meet the increased demand. Furthermore, truck transportation, which is used primarily in the East, is the most costly means of transporting coal, and it increases dust, traffic, and noise.

Coal-slurry pipelines comprise a relatively new method of transporting coal, which alleviates some of the negative impacts of the other alternatives. Those pipelines pump a 1:1 mixture of ground-up coal and water from coal mines to power plants or ports. The major incentive for building those pipe-

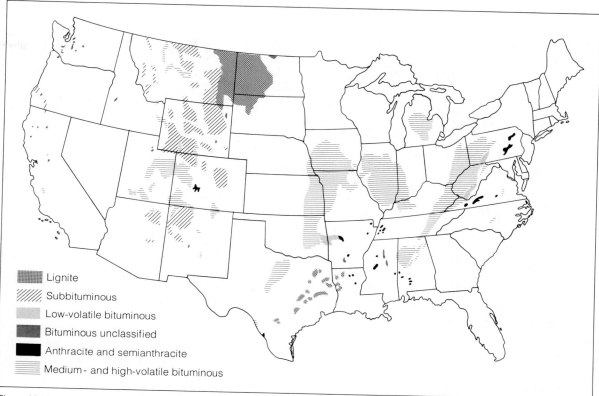

Lignite

Subbituminous

Low-volatile bituminous

Bituminous unclassified

Anthracite and semianthracite

Medium- and high-volatile bituminous

Figure 18.13 *Locations and types of coal deposits in the lower 48 states. Alaska also has a considerable amount of coal. (Illinois Geological Survey, Circular 499,* Trace Elements in Coal, *1977.)*

lines is that they can transport coal more cost effectively in regions where rail lines need to be expanded or upgraded, or where the terrain limits access to rail lines. For example, a 1982 study by the Virginia Electric Power Company found that it could transport coal from West Virginia to its plants in the east for 25 to 40 percent of the cost of rail transport.

Also, fewer traffic deaths result when coal is transported by pipeline rather than by rail, which is another incentive to build pipelines. The two major impediments to coal-slurry pipelines are rights-of-way for building pipelines and water availability. For 20 years, the coal industry has tried to gain the right of *eminent domain* (the right to condemn land and purchase it for public benefit) so it could build coal-slurry pipelines, but it failed in its most recent (1983) effort to get such legislation enacted. (Railroads are

Table 18.1 Characteristics of eastern and western coal

	HEATING VALUE (BTU/POUND)	SULFUR CONTENT (PERCENT)	SEAM THICKNESS (METERS)
Western Coal			
Lignite (Brown Coal)	6,000–7,500	0.6	
Subbituminous Coal	9,000–11,000	0.6	3–60
Eastern Coal			
Bituminous Coal	11,000–14,000	0.6–6.0	1–2.5
		(2.0–3.5 ave.)	
Anthracite	13,500–15,000	0.5	—

one of the major industries that lobby against coal-slurry pipelines.) In addition, although water shortages are not expected to be a problem in eastern states, complex laws concerning water rights may complicate water use in the water-short western states. Nevertheless, a 440-kilometer (275-mile) coal-slurry pipeline has been transporting 5 million metric tons of coal per year from Black Mesa, Arizona, to Mohave, Nevada, since 1970.

The greatest environmental impact of coal slurry is likely to be its effect on water quality. The water that is removed from the coal slurry at its destination must be treated chemically to remove contaminants before it can be discharged into surface waters. Also, the use of coal can have another significant effect on the environment. As we have already discussed in Chapter 10, the combustion of coal can contribute significant quantities of particulates and sulfur dioxide to the atmosphere. As a result, utilities and other large industrial operations that burn coal are required to meet environmental regulations for air quality, water quality, and solid-waste disposal, which can add as much as 10 percent to the cost of using coal. Those costs are much lower when oil or natural gas are used as fuel.

The use of coal to generate electricity became an even more complex issue in the early 1980s, when sulfur dioxide was identified as one of the major causes of acid rain. Furthermore, the insistence of the Canadian government that the United States do something to control its emissions of sulfur dioxide has made the issue an international one. As a result, proposed amendments to the Clean Air Act would seek sharp reductions in sulfur dioxide emissions from sources in all states east of the Mississippi and in five states that are west of it. Estimates made by the Department of Energy indicate that $200 billion to $300 billion would be added to the cost of generating electricity from coal during the next 30 years if that legislation is enacted. And that increase would amount to $30 to $45 per person every year in additional costs for electricity. Furthermore, the utilities would generate more than twice as much solid wastes, since plants that use flue-gas desulfurization scrubbers produce a larger quantity of wastes than plants that do not use them. Also, holding those wastes in ponds, which is the current method, is viewed as being an unacceptable long-term method of disposal. For a different and promising approach to the problems of "cleaning up" coal, see Box 18.1.

Thus although coal can be substituted for other fuels in many applications, it has major drawbacks. The costs in terms of environmental degradation of land, air, and water must be considered when planning is initiated to mine, transport, and use coal. Also, coal is the most dangerous fuel in terms of the number of deaths, illnesses, and injuries that result from handling it when it is compared to other fuels, including nuclear fuels, on an energy-equivalent basis. (The risks of the nuclear and coal fuel cycles are compared in more detail later in this chapter.)

Synthetic Fuels (Synfuels)

Fuel substitutes for oil and natural gas, which are called *synthetic fuels*, or *synfuels*, can be produced from coal or oil shale, extracted from tar sands, or they can be obtained by mining from oil formations that have already been drained by oil wells. (A typical oil well recovers only 32 percent of the oil that is present in the ground.) By mining tar sands and oil deposits, oil that is attached to sand grains or other rock materials can be recovered. And by mining oil shale, a petroleum-like liquid can be recovered by heating rock deposits that contain organic material. In addition, coal can be processed to produce gaseous fuels, petroleum liquid substitutes, or specific chemicals, such as methanol (wood alcohol). Furthermore, in more advanced and expensive processes, gasoline can be manufactured from coal.

Although synfuels are not economical to produce today, they offer some noneconomic benefits to the general public, such as increasing the dependability of supply, reducing the amount of oil that is imported, and providing stronger leverage in terms of foreign policy. For these reasons and in the wake of the Arab oil embargo, the Iranian crisis, and sky-rocketing fuel prices, legislators during the late 1970s felt that the government should explore commercial-scale synfuel plants to determine their economic and environmental feasibility for the future.

To develop the technology that is necessary for the commercialization of synfuel production, President Carter signed the Energy Security Act of 1980 and set up the U.S. Synthetic Fuels Corporation (SFC). That act committed $88 billion over a 12-year period—through 1992—for loans, price guarantees (for products that cost more to manufacture than they could be sold for), and joint-venture participation for projects that could not obtain private funding, in which the Synthetic Fuels Corporation would be a part owner. The Synthetic Fuels Corporation began with one early big award to the Exxon Corporation and the Tosco Corporation for an oil-shale project in Colorado. However, when those companies saw that project costs were doubling, they backed out in May of 1982 and returned the money to the government. Then, other synfuel projects died in 1982 and 1983. For example, a coal-to-gas project in Wyoming, a coal-to-gasoline project in Wyoming, and a coal liquefaction project in Kentucky were some of the casualties. Thus the massive effort by the government to develop a profitable synfuels industry has been scrapped, and only limited funds for synfuel development are available. Box 18.2 describes the complex political wranglings that have paralyzed the federal synfuels program.

One of the reasons for the very limited success of the Synthetic Fuels Corporation resulted from the increased petroleum supplies and the subsequent lowering of energy prices that occurred in the early and mid-1980s. Companies that were working with the Synthetic Fuels Corporation were counting on continued escalation of oil prices. Thus when sup-

plies increased and prices fell, most industries lost interest and Congress deserted the synfuels effort by shifting away from price supports and tax breaks. And the Reagan Administration changed the political climate by convincing the nation that the oil problems could be solved by drilling more holes in the ground—a prediction that has not come true. After 2 years of unconstrained drilling, reserves were still shrinking and production remained unchanged.

However, a few companies, such as American Natural Resources (coal gasification) and Union Oil of California (oil shale), continue to develop synfuel technology. Union Oil of California started its first oil-shale plant near Parachute, Colorado, in the fall of 1983, with a goal of producing 10,000 barrels of synfuel per day (see Figure 18.14). The company plans to increases its production to 90,000 barrels per day by the mid-1990s.

We now take a brief look at some of the techniques and limitations in the production of synfuels.

Coal Gasification and Liquefaction *Coal gasification* is not a new technology. It has been practiced on and off during the past two centuries. As late as 1932, gas for the residential market in the eastern United States was derived mainly from coal. However, less-expensive natural gas became more popular when pipelines were built into the area. Nevertheless, since coal is the largest fossil-fuel resource in the United States today, renewed interest in the process is being expressed.

Figure 18.14 *The first commercial-scale oil-shale project in the United States. Located in Parachute Creek, Colorado, it is run by Union Oil of California. The retort complex is located on a 5-acre bench carved out of the mountain side at the mouth of the mine 1000 feet above the valley floor. Raw oil shale is extracted from crushed ore in the retort and then piped to another facility for further processing. (Union Oil of California.)*

BOX 18.1

FLUIDIZED BED COMBUSTION: THE KEY TO CLEANER-BURNING COAL?

The first "oil crisis," in 1973, caused a major re-evaluation of our country's fuel supplies, fuel-pricing policies, energy consumption, and methods for generating power. Discussions of arcane technologies, such as coal gasification, coal liquefaction, fluidized bed combustion, oil-shale recovery, biomass conversion, magnetohydrodynamics, and geothermal energy moved from technical journals to the front pages of newspapers. The United States rediscovered its vast coal reserves, and coal was suddenly a glamorous commodity. Substantial political and economic pressures developed for finding new and better ways to use coal.

Concurrently with this resurgent interest in coal, the environmental movement was gaining momentum. The 1970s marked the formulation, for the first time, of many nationwide regulations to control pollution. If we were going to use coal, it was clear that it would have to be done cleanly and with minimal damage to the environment.

In the years following the oil crisis, the federal government spent hundreds of millions of dollars on the design, planning, and in some cases the initial construction stages of giant synthetic fuel plants that would test new or rediscovered technologies for coal gasification, coal liquefaction, and oil-shale refining. In terms of funding for research and development, advanced combustion techniques prospered too. Among these advanced techniques was fluidized bed combustion; in the period 1975–79, about 20 different coal-fired pilot plant or demonstration plant projects involving fluidized bed combustion were initiated under various government-sponsored research and development programs.

Times change. As the global recession spread, oil prices dropped, the Reagan administration deemphasized the Department of Energy (DOE) and deregulated fuel prices, and OPEC's solidarity began to crack. The artificially created synfuels market dried up. The only coal combustion technology from the Energy Era of the 1970s that still seems destined for commercial success in the 1980s is fluidized bed combustion.

In fluidized bed combustion, a bed of solid particles, such as ash or sand, rests on a grate at the bottom of a boiler, and a fan is used to force air up through the grate at a velocity high enough to lift the particles and overcome their settling velocity. At the same time, crushed solid fuel is fed into the boiler. The result is a dense mixture of fuel and sand particles suspended in air.

The use of a fluidized mass as a combustion medium offers intriguing advantages for boiler design and also for controlling pollution. Thus fluidized bed combustion is sometimes thought of as a "new" technology, since it addresses many of the new energy and environment concerns. In fact, the concept has been known for decades.

Two unique features distinguish this technology from other methods of burning solid fuels. First, the particles of fuel, ash, or sand suspended in a burning "fluid" conduct heat with high efficiency. Direct contact between burning fuel and the other particles, and between hot particles and the components of a boiler, can yield a rate of heat transfer three to four times more efficient than the rate of transfer achieved through convection or radiation alone. The second advantage is that fluidized bed combustion offers unique opportunities for controlling pollution at its source—within the boiler itself.

One of the most attractive features of fluidized bed combustion is the capability of suppressing sulfur dioxide at the time of combustion rather than removing it from flue gases later with expensive and sometimes difficult-to-operate "scrubbing" devices. The primary source of SO_2 is the sulfur contained in solid fuels as an impurity. Coal typically contains anywhere from 1 percent to 4 percent sulfur, and about 95 percent of this sulfur is converted to SO_2 during combustion. In fluidized bed combustion, the key to keeping it out of emission gases lies in using limestone as part of the bed material.

Limestone captures SO_2 by means of the following reactions:

$$CaCO_3 \text{ (limestone)} \rightarrow CaO + CO_2$$
$$2SO_2 + 2\,CaO + O_2 \rightarrow 2CaSO_4$$

Operating fluidized beds at lower temperatures has an extra attraction in that it helps to reduce nitrogen oxide emissions as well. In combustion, nitrogen oxide is formed in two ways:

$$N_2 + O_2 \rightarrow 2NO$$
$$\text{Fuel-N} + O_2 \rightarrow NO$$

The first reaction is just the oxidation of atmospheric nitrogen. The second reaction is the oxidation of nitrogen contained in the fuel. Coal, for example, contains anywhere from 0.2 percent to 1.2 percent nitrogen. Molecular nitrogen begins to oxidize at about 2200°F; hence, fluidized bed combustion, with its low temperatures (1500–1600°F), suppresses this contribution to NO_x formation. Significant amounts of nitrogen oxides can still be produced through the second reaction, but with care in the selection of operating parameters, NO_x levels can be reduced below those produced by conventional combustion.

From the data now available, it appears that fluidized bed combustion will meet the NO_x emission limits which currently prevail. A modest redesign to incorporate staged combustion may be required, however, for reductions in NO_x emissions below 0.5 lbs/10^6 BTU.

Fluidized bed wastes have high concentrations of calcium, aluminum, iron, silica, and sulfur; these are always present at percent levels. Present at a factor of ten or so lower are elements, such as barium, magnesium, sodium, potassium, strontium, titanium, boron, and chlorine. Other elements, in particular the heavy metals, such as lead, zinc, cadmium, nickel, vanadium, and mercury, are present at 100 parts per million or fewer. These wastes are not much different in composition from ash produced in conventional coal combustion, except for high concentrations of calcium oxide and sulfate because of the limestone used in the beds.

In disposing of such waste material, it is important to verify the extent to which chemicals leach into the ground. In a recent study, all the leachates from fluidized bed wastes remained well below current EPA concentration limits.

What are the prospects for fluidized bed combustion? The advantages over conventional combustion are clear, when studied one by one, but there is not yet a reliable comparison across the board of total operating costs or pollution control capabilities. Questions continue to surface, because of the diffuse data base, about the true long-term operating costs of fluidized bed boilers.

Nevertheless, for a prospective buyer of a coal-fired power plant, fluidized bed combustion is one of the few alternatives to a coal-fired boiler coupled with a flue-gas SO_2 scrubber. The most recent and extensive analyses indicate that the costs of operating fluidized beds overlap with the costs of operating conventional boilers (including SO_2 scrubbers) within the error limits of the analyses.

Several areas of uncertainty still hold back full-scale commercial development. First, the corrosive nature of the fluidized particles—which, in effect, sandblast the inside of a boiler—place high demands on construction materials. Only long-term operation can expose the points where improved materials must be substituted. Second, a sudden change in air pollution regulations could make most first-generation designs obsolete. Second-generation designs might cope, but substantially more development could be required. Indeed, the current concern about acid rain will most likely force stricter emission controls. Finally, there are questions in some quarters that the inherently variable qualities of coal and limestone may be too much for a fluidized bed system to handle within currently accepted guidelines for capital expenditure and operating costs.

According to a recent review, the United States has more than 16 fluidized bed boilers in commercial operation using coal as the primary fuel. About 25 additional boilers use wood or wood waste as the primary fuel. Worldwide, more than 100 fluidized bed facilities are burning either coal or wood waste, and their output ranges from 4 to 350,000 lb/hr of steam. An additional figure comes from the Chinese, who say they are using 2000 small fluidized bed boilers to provide electricity and heat to villages all over China. Some of these units burn peat, and others burn a gravelly material containing as much as 70 percent ash. In other applications, wood chips, pulp, and municipal wastes containing up to 50 percent water have been used as fuel. It is precisely this adaptability—to inexpensive fuels, to more compact designs, to the retrofitting of old heating plants, and to versatile means of controlling pollution—that has pushed fluidized bed combustion to the verge of success.

Excerpted from Paul F. Fennelly, Fluidized bed combustion. American Scientist 72:254–261, 1984.

BOX 18.2

SYNTHETIC FUELS: A FADING DREAM

WASHINGTON—A group of tired, uncomfortable-looking congressmen sat through a sweltering hearing recently, each one doggedly waiting his turn to criticize the government's synthetic-fuels effort.

"The people in charge are obviously incompetent and incapable of running the program," asserted Illinois Democrat Sidney Yates, the usually mild-mannered chairman of the House appropriations subcommittee that controls the Synthetic Fuels Corp.'s budget. "The management is appalling," chimed in Joseph McDade of Pennsylvania, the panel's ranking Republican member. "We're going to fire more of them, if I have my way."

The entire session seemed to be scripted by foes of multibillion-dollar federal subsidies for synthetic-fuels projects. But it was just the opposite. The harsh words, in fact, came from some of the most outspoken and persistent advocates of synfuels development still left on Capitol Hill.

Today, even supporters of the government-backed Synthetic Fuels Corp. are bitter over a dream turned sour: The 4-year-old effort to reduce U.S. reliance on imported oil is dead in the water. Although opponents contend that costly alternative-fuel projects don't make sense in the face of today's oil surplus, the problems that have bedeviled the Synthetic Fuels Corp. have been primarily political, not economic. The corporation has been crippled by months of political bickering, allegations of financial scandal, an exodus of senior officials and general industry apathy.

As a result, dozens of once-promising synfuels projects have been scrapped or delayed by confusion. Last week the House voted to cut the corporation's spending authority to $8.25 billion from $13.25 billion, but the White House wants the Senate to reduce spending by an additional $4 billion, leaving synfuels with $4.25 billion. Few in government or business stand ready to defend the record of the corporation, which seems likely to perish without new leadership or policy changes.

Edward Noble, the corporation's chairman, says he sometimes feels that he is sitting on a "ticking bomb." The corporation is at the mercy of "an industry that's not ready and half the Congress that doesn't want me here," complains Mr. Noble, who is a conservative Oklahoma businessman and a Reagan appointee.

With only two members, the board he heads lacks the quorum needed to conduct routine business. Nominations for the other five places are stalled because of internal administration disputes and arguments between the White House and lawmakers over the program's future. "I'm as anxious as I can be to go home," the exasperated chairman told another House panel in June. "Maybe more so than people are to send me home."

The critics, meanwhile, are stepping up their attacks and steadily gaining the upper hand. Veteran Republican Congressman James Broyhill of North Carolina, for example, used to be an important—although grudging—supporter of the program. But after analyzing the assistance earmarked for faltering projects, he now accuses Mr. Noble of running "the biggest program of waste and abuse I've ever seen," and adds: "I think it's high time to bring it to an end." A majority of the House is cosponsoring a bill to keep the corporation in limbo until Congress reasserts tighter management controls and decides how much aid should be doled out.

Such proposals are a far cry from the bipartisan groundswell that created the synfuels program in 1980. That was in the wake of the revolution in Iran and the accompanying fear of widespread fuel shortages in the U.S. At the time, lawmakers optimistically predicted the corporation would become a unique "investment bank," able to cut through red tape to quickly help produce hundreds of thousands of barrels of synthetic fuel daily from coal, tar sands, oil shale, and other unconventional sources.

The idea was praised as a model of cooperation between the public and private sectors; others hailed it as "America's ace in the hole" for achieving energy independence. Today those hopes have been dashed and the corporation, in Energy Secretary Donald Hodel's words, is "absolutely moribund." Hobbled by bureaucratic delay, it has provided only a total of $740 million in aid to two relatively small projects. The board hasn't met since the end of April, and nobody takes the ambitious production goals seriously any more.

Some of the difficulties reflect radical market changes outside the government's control. The unexpected combination of stable world oil prices and excess crude-production capacity has made most synfuel projects uneconomical. Deficit-conscious legislators running for reelection don't look kindly on subsidizing plants to turn out synthetic products at prices as high as $67 a barrel when crude oil costs $29 a barrel.

At the same time, the program's credibility has been undermined by persistent and much-publicized personnel problems. With the blessing of the White House, Mr. Noble brought in a group of former business associates and longtime friends to help run the program. But some soon were in hot water because of allegations of ethical violations or a lack of political savvy and management skill.

Even corporate giants that have the commitment and financial wherewithal to stick it out until the political storm dies down are bound to "cut back on planning dollars" and postpone detailed engineering work, concedes Robert Hanfling, a former Energy Department official who helped set up the federal synfuels under President Carter. The result, critics contend, is that some of the top-priority projects in line for subsidies haven't obtained all the necessary state and federal permits to begin actual construction.

Although some of those projects "will never get off the ground," the corporation "continues to carry them as viable projects," Congressman Broyhill says.

Proponents assert that cutting off funding now violates good-faith agreements with sponsors. In addition, they argue that the current oil glut is only temporary and that the U.S. should be working to build a synthetic-fuels industry for the 1990s and beyond. In the face of current problems, "you don't kill the program, you fire the managers," says Rep. McDade.

"It would be tragic to see it all stop now," Mr. Noble says, arguing that projects could move forward quickly with congressional support.

Many synfuels experts, even those working for Mr. Noble, contend that his argument is much too optimistic. The synfuels corporation, for example, earlier this year signed a letter of intent to award $2.7 billion to the second phase of a controversial Colorado oil-shale project sponsored by Unocal Corp. But the plant's much-smaller first phase is about 8 months behind its construction schedule. There have been a variety of unexpected technical problems with the equipment. And even the sponsors agree that additional federal subsidies shouldn't be granted until the bugs are worked out and the government is convinced the technology works properly.

Congressman Wolpe asserts that the Unocal project shows how "premature commercialization of ill-advised" technologies can hurt taxpayers. At the very least, the lawmaker argues, Congress should make the corporation more accountable by subjecting it to the routine budget and bureaucratic controls other agencies operate under.

For Mr. Noble and the White House, that may be an easy pill to swallow. "As one who came uninitiated to" the ways of the capital, the chastened chairman told a House oversight panel at the end of June, "I have learned that perceptions are as important as facts." In retrospect, Mr. Noble concluded, "I probably should have been more sensitive" to the criticism.

Coal is gasified by reacting crushed coal with steam and either air or pure oxygen. Under those conditions, hydrogen atoms from the steam and oxygen from the air combine to form a mixture of carbon monoxide, hydrogen, and varying amounts of methane. If air is used, the gas mixture is called *low BTU gas*, because it has a heating value of approximately 100 BTUs per cubic foot, which is roughly one-tenth the heating value of natural gas. If oxygen is injected into the process, *medium BTU gas* is produced, which has a heating value of approximately 300 BTUs per cubic foot. Neither of those gas mixtures is economical to pipe long distances, but they are both ideally suited for industrial purposes in nearby boilers or as raw materials for local chemical industries.

The only synfuel that can be transported economically over long distances is high BTU gas *(substitute natural gas)*. But the process for manufacturing substitute natural gas requires two additional steps in the conversion process. Although substitute natural gas has a heat content of 950 to 1000 BTUs per cubic foot, which is the same as natural gas, it costs 20 to 40 percent more to produce.

While the production of substitute natural gas involves the removal of many of the pollutants (sulfur in particular) in coal, the process that converts coal into substitute natural gas produces both gaseous emissions and solid wastes that must be dealt with. For example, the air-pollution problems that face substitute natural gas plants are similar to those of an oil refinery because of the tars and aromatic hydrocarbons that are produced. Although some of those hydrocarbons are known carcinogens, emissions of those substances can be controlled with existing technology. However, no regulations have been set regarding the level of emissions that are allowable from substitute natural gas plants.

Coal gasification can also be accomplished *in situ* (in place). In this method, part of a coal seam is burned by supplying oxygen through an injection well (Figure 18.15). After the coal is ignited, water is injected into the seam, which converts the coal to a mixture of carbon monoxide, hydrogen, and methane. Then, the product gases, which constitute a medium BTU gas, are removed. It is also possible to manufacture substitute natural gas by using this method, and initial estimates of cost are $4.35 per million BTUs for the medium BTU gas and $6.25 per million BTUs for substitute natural gas. If those estimates are correct, the substitute natural gas that

is produced by this method will compete with natural gas. The 1985 price for natural gas is expected to be in the range of $5 to $7 per million BTUs.

Underground gasification technology has special appeal, because it can withdraw energy from coal deposits that cannot be extracted by conventional techniques. The amount of coal that is estimated to be available by underground gasification constitutes 35,000 Quads (1.6 trillion metric tons of coal)—nearly five times the amount of coal that is minable by conventional means. The technique also avoids some of the environmental problems that are associated with the use of coal, such as sulfur dioxide emissions and fly-ash disposal. However, a few problems do occur, namely, occasional ground subsidence and the possible linking of adjacent aquifers, which might lead to contamination or a change in the direction of groundwater flow.

In addition to coal gasification, the manufacture of petroleum-like liquids from coal *(coal liquefaction)* has received much research attention, because liquid fuels are better suited for use in motor vehicles. However, coal liquefaction is more complicated and costly than coal gasification. For example, it took the Exxon Company 16 years to develop their Exxon Donor Solvent process. The process decreases the size of coal molecules, increases their hydrogen content (coal has a very low hydrogen content compared to oil and natural gas), and separates the resulting liquids from the solid mineral materials. In

Figure 18.15 *An advanced coal gasification method. In this method, heat for the gasification process is initially provided with a separate initial connecting well. After the burning is started, the point of combustion is controlled by the position of the air-injection pipe, which is cut off as the burning progresses. Gas that is produced is removed through a second pipe.* [Reprinted with permission from Chemical and Engineering News, 45:15 (July 18), copyright 1983 ACS.]

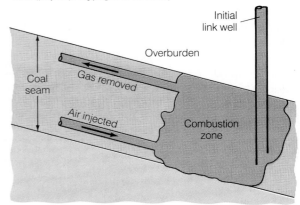

1982, Exxon announced that it had gained the technical experience to build a coal liquefaction plant that would be capable of processing 18,000 to 27,000 metric tons per day when it became commercially justifiable. All told, approximately 35 other major research projects across the country have been focused on liquefying, or partially liquefying, coal.

Oil Shale *Oil shale* is neither oil nor shale, but a fine-grained rock that is called *marlstone*, which contains varying amounts of gray-to-brown organic material that is called *kerogen*. When oil shale is heated to approximately 480°C (900°F), it decomposes, forming combustible gases and petroleum-like liquids. Vast high-grade deposits of oil shale exist in the Colorado–Wyoming–Utah area, which is known as the Green River formation. And those deposits yield at least 8.5 percent, or 105 liters (28 gallons) of premium-grade oil per metric ton. Eventually, recoverable high-grade oil shale deposits in the United States may yield as much as 3500 to 4060 Quads (600 to 700 billion barrels) of energy—roughly a 100-year supply of petroleum at present rates of consumption.

Lower-grade oil-shale deposits also exist; and they are found in a roughly triangular region that extends from Michigan to western Pennsylvania to Mississippi. However, those deposits contain less than 62 liters (17 gallons) of oil per metric ton. Furthermore, they are not economically extractable by current technology, but they may contain as much as 17,000 Quads (3 trillion barrels) of energy, which makes oil shale second to coal in terms of the amount of fossil-fuel resources that are available in the United States. *In situ* gasification is considered to be the most applicable technology for lower-grade oil shale.

Although the technology for processing oil shale is available (Figure 18.16), the present economic climate is unfavorable for large-scale exploitation. Imported oil is less expensive at this time. In addition, serious environmental considerations inhibit exploitation. For example, for every liter of oil that is produced, 2 to 4 liters of water are required. And ironically, essentially all high-grade oil-shale deposits are located in the semiarid states. Thus water shortages in those areas would probably limit oil-shale production to approximately 170 million liters (1 million barrels) per day, which is about 7 percent of the nation's 1984 rate of oil consumption.

A second and almost overwhelming obstacle to oil-shale development is disposal of waste from the sheer volume of material that must be processed. For example, for each 170 liters (1 barrel) of oil that is produced, 1.4 metric tons of shale must be processed. Thus to realize the probable maximum production of 170 million liters (1 million barrels) of oil per day, 1.4 million metric tons of spent marlstone (shale) must be processed, resulting in 1.3 million metric tons of waste. Also, that residue is highly alkaline, poor in nutrients, and, especially in semiarid climates, difficult to revegetate.

Because most high-grade oil-shale deposits are located under federal lands (80 percent of the total), oil companies must obtain leases to exploit those sources. Six small tracts of land in the Central Rockies were recently leased for the development of some small oil-shale operations under the authority of the Department of the Interior. Those efforts involve conventional mining techniques prior to processing.

Given the problems associated with adequate water resources and waste disposal, conventional oil-shale mining operations will remain small. Because *in situ* mining reduces many environmental problems, research has been conducted on this technology. However, *in situ* mining decreases the percentage of oil recovered from the shale. Thus large-scale, *in situ* oil shale production will probably have to wait for a substantial increase in world oil prices or perhaps a change in political climate—a change that probably would require another oil crisis, such as the one that occurred in 1973—which would support subsidies for the development of oil shale. Over the short term, oil shale will only make a minor contribution to our domestic oil supply, but over the long term, its potential is tremendous.

Tar Sands Tar sands are another potential source for petroleum liquids. The U.S. Department of Energy estimates that some 30 billion barrels of oil (a 5- to 6-year supply at the 1984 rate of consumption) exist in deposits that are scattered across 21 states, but Utah has the largest deposits. Although several companies are pursuing ways to extract oil from tar sands, the most promising technique involves injecting steam into them to reduce the viscosity of oil and then collecting it at recovery wells—a technique similar to the one used for enhanced oil recovery (see Chapter 12). Most deposits of tar sands in the United States are covered by a thick overburden, which requires recovery by *in situ* techniques, whereas in Alberta, Canada (as shown in Figure 18.17), surface deposits can be mined. Canada has much

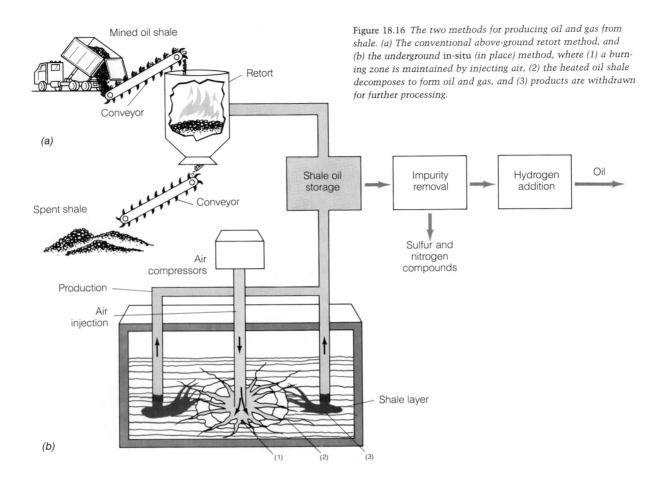

(a)

(b)

Figure 18.16 *The two methods for producing oil and gas from shale. (a) The conventional above-ground retort method, and (b) the underground in-situ (in place) method, where (1) a burning zone is maintained by injecting air, (2) the heated oil shale decomposes to form oil and gas, and (3) products are withdrawn for further processing.*

more extensive deposits of tar sands. Its two major plants now produce a total of 155,000 barrels of oil per day. By the year 2000, Canada expects to obtain 40 percent of its oil from tar sands.

Other candidate methods for enhancing the future supply of petroleum liquids include peat gasification, enhanced oil recovery, and mining oil from depleted fields, which may have up to 90 percent of the original amount of oil still remaining in the ground. The amounts of petroleum-type fuels that will be recovered by those various techniques have been predicted by the U.S. Department of Energy. Their forecast is detailed in Table 18.2.

Electricity

Our society has become totally reliant on electricity. We fully expect to have it available whenever we flick a switch. As a result, if it were not for backup

Figure 18.17 *Canada has extensive regions of tar sands, such as the one being mined here in Alberta, Canada, near Fort Mc-Murray. A bucket-wheel excavator is used to pick up tar sands, which are then transported by conveyor to a processing plant. (Hans Blohm, NFB Phototheque ONF ©.)*

Table 18.2 Oil supply in the United States (Quads)

SOURCE	1979 ACTUAL	1990 ESTIMATE
Lower 48 States	15.5	10.2
Alaska	2.5	3.2
Enhanced Oil-Recovery Techniques	0.2	1.5
Oil Shale	—	0.6
Natural Gas Liquids	3.6	2.3
Synthetic Coal Liquids	—	0.6
Refinery Efficiency Gains	1.1	1.1
Total Domestic Supply	22.9	19.5
Net U.S. Oil Imports	17.0	13.8
Total Consumption	39.9	33.3

Source: *Reducing U.S. Oil Vulnerability.* Washington, D.C.: U.S. Department of Energy, November, 1980.

diesel-power electric generators, we would have serious problems. For example, life-saving equipment in hospitals would stop; frozen foods in homes, restaurants, and warehouses could spoil; stalled pumps could cause basements, tunnels, and sewage-treatment plants to flood; and telephone communications could cease. Major cities come to a near standstill during power blackouts; traffic lights go out, elevators stall, and computerized trains stop. In fact, life without electricity is almost unimaginable to most of us.

However, to generate electricity, we need to use other sources of energy, such as: coal, oil, natural gas, nuclear fuels, falling water, and wind. Fossil fuels or nuclear fuels provide heat, which converts water to pressurized steam, and the steam drives turbines, which turn electric generators. In addition, moving water or air (wind) directly turns the blades of turbines and windmills. (Water and wind are considered to be renewable sources of energy and are discussed in Chapter 19.)

In 1983, the production of electricity accounted for 35.2 percent (25 Quads) of the total amount of energy that was consumed in the United States, which included both the residential–commercial (63.7 percent) and industrial (37.3 percent) sectors of the economy. And coal was the most important source of energy that was used to produce it. The amounts of the various fuels that are used to generate electricity are shown in Figure 18.18. Furthermore, the Department of Energy made some midrange fore-

casts[*] in 1982, which project that by the year 2000, the United States will be using 39.2 Quads more of electrical energy (57 percent) than was used in 1983.

Thus we might ask how future electrical demands will be met and what environmental, space, health, and safety problems will be created as a result. Here we examine the environmental and health impacts of coal and nuclear fuel cycles and their relative economic costs. Natural gas and oil are not considered to be viable fuels for the generation of electricity. They are in relatively short supply and are considered more valuable as fuels for other purposes, such as home heating and transportation. In Chapter 19, we examine the impact of alternative sources of energy. Valid comparisons among the various strategies for obtaining energy can be made only by comparing the entire *fuel cycle* of each source.

The Nuclear Fuel Cycle

Every fuel cycle has unique characteristics. The steps in the nuclear and coal fuel cycles are compared in Figures 18.19 and 18.20. The nuclear fuel cycle begins with either underground or open-pit uranium mines. On those sites, radioactive radon gas poses the major health hazard to uranium miners. For example, if radon gas is inhaled in sufficient concentrations over extended periods of time, it will induce lung cancer in some workers. However, proper ventilation of the mines greatly reduces that hazard. Also, the radon that is contained in mining wastes endangers the health of persons who live downwind from the wastes or who spend much time in homes, workplaces, or schools (for example, in Grand Junction, Colorado) that are built of materials (cement block; concrete) which contain those wastes. Old uranium mining tailings are still a problem in Salt Lake City, where a pile of uranium mine tailings releases radon within the city limits.

Once uranium ores have been mined, they are converted into fuel elements for nuclear reactors by means of a complex process, which includes enrichment of the usable (fissionable) form of uranium (uranium of mass 235) in the fuel. After the uranium-235 has been enriched from its natural ore concentration (0.7 to 3.0 percent), it is formed into

---

[*] Typically, three types of energy-use forecasts are made: low, mid, and high range. Assumptions made for low energy-use forecasts include slow economic growth and high energy prices, whereas high energy-use forecasts assume rapid economic growth and low prices for energy.

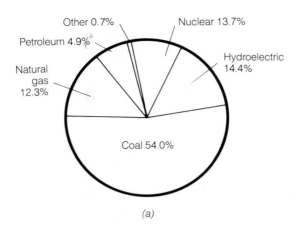

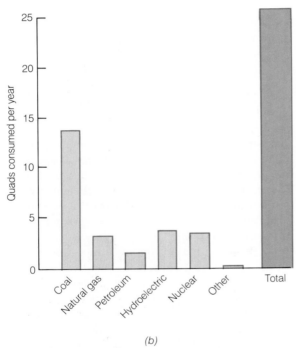

Figure 18.18 *The energy resources that were used to produce electricity in the United States in 1983 in percent (a) and in absolute amounts (b).* (Monthly Energy Review, *December, 1984. Energy Information Administration, Washington, D.C.*)

ceramic pellets. Then, the pellets are encased into fuel rods, which are placed in fuel-rod assemblies. Those steps entail virtually no health hazard to workers.

The assembled fuel rods are transported to nuclear power plants. A typical plant (which produces 1000 megawatts of electricity) requires approximately 30 metric tons of nuclear fuel per year. On the average, fuel rods remain in the reactor core (see Figure 18.21) for 3 years, and during that time, the uranium fuel undergoes *nuclear fission* (Figure 18.22) inside the reactor's core. That reaction, which splits uranium-235 atoms, releases heat and produces approximately 200 types of radioactive by-products, which are called *radioactive isotopes* (Chapter 13). Two of those isotopes are shown in Figure 18.22. Then, the heat that is released by the fission reactions in the core is cycled through a closed, high-pressure primary loop into a steam generator. And the pressurized steam that is formed is fed through a secondary loop to steam turbines. Because the primary loop is isolated from the secondary loop, radioactive materials that leak from the fuel rods or that form when the cycle water passes through the core are retained in the primary loop. From that point on, a nuclear-powered electricity-generating plant is similar to a coal-fired electricity plant. The steam produced in the secondary loop is used to drive steam turbines, which turn electrical generators.

Operating nuclear power plants release only a minute amount of radioactivity into the environment through the cooling water and into the atmosphere in the form of gases. The Nuclear Regulatory Commission (NRC) requires that those releases be kept at levels that are low enough to ensure persons who live at the perimeter of a nuclear power plant are not exposed (whole-body exposure) to more than 5 millirems of radiation per year. Table 18.3 shows the average level of radiation that is emitted from natural background radiation (such as from rocks and cosmic rays), medical diagnostic procedures, and the small increase in exposure from nuclear power emissions. The increased number of cases of cancer and genetic defects caused by exposure to those sources are also projected in Table 18.3. Natural background levels of radiation vary from 200 millirems in Wyoming and Colorado to 60 millirems in Florida.

After the fissionable uranium-235 that is contained in the fuel rods has decreased from 3 percent to about 1 percent, they are considered to be "spent"—that is, they can no longer be used to generate electricity. Then, the spent-fuel-rod assemblies, which contain high-level radioactive wastes, are removed from the reactor core and are stored at nuclear power plants (under water) in pools that absorb the heat that is released by decaying radioactive materials. That method of storage is used because no nuclear-

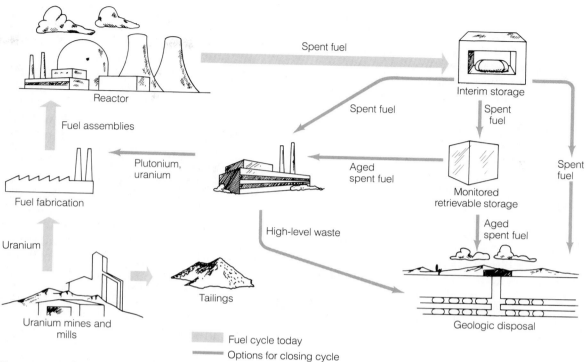

Fuel cycle today
Options for closing cycle

Figure 18.19 *The commercial nuclear fuel cycle that is used to produce electricity. In 1984, the reprocessing option was not being pursued, and the option of geologic disposal was meeting stiff political opposition from persons who did not want disposal sites near them.* [*Reprinted with permission from* Chemical and Engineering News *45:25 (July 18), copyright 1983 ACS.*]

Figure 18.20 *The coal cycle for the production of electricity. The coal cycle requires the transport of approximately 100 thousand times more fuel material than the nuclear cycle.*

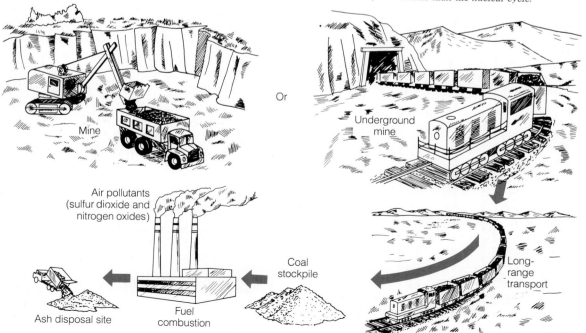

fuel reprocessing centers or permanent storage sites exist for those radioactive wastes. However, many nuclear power plants are approaching the limits of their storage capacity. The government has proposed to build temporary government storage facilities by 1995 or 1996. To date, however, the government has been hesitant to move ahead and build such facilities because of public opposition to the storage and transportation of the high-level radioactive wastes. The government would rather see the plants handle their own storage problems by expanding their temporary on-site storage facilities until a permanent repository is available. The Department of Energy promised to have a specific location for temporary storage of waste nuclear fuels picked by January 1986. (The technology for permanent high-level nuclear-waste disposal is discussed in Chapter 13.)

The reprocessing of spent nuclear fuels is a second potential way to handle nuclear wastes, because, in the process, valuable unfissioned uranium-235 could be recovered. Spent fuel rods, when they are removed, contain approximately 1 percent unfissioned uranium-235 as well as a similar amount of fissionable plutonium-239, which forms during operation of the reactor. In fact, the fissionable material that is recovered from spent fuel rods from three reactors would be sufficient to fuel a fourth reactor. Therefore, in essence, the operation of a nuclear reactor actually creates additional fissionable fuel. However, with reprocessing, the disposal of radioactive components would still be required. Currently, no reprocessing plants are licensed to recover fissionable materials from spent nuclear fuel. In fact, private companies that have attempted to carry out the complex reprocessing of nuclear wastes in the past encountered serious economic and technical difficulties, which caused them to shut down.

A major obstacle to reprocessing nuclear fuels is the isolation of plutonium-239, which is the primary ingredient of nuclear bombs. Thus procedures must be established to ensure that this material will not fall into the hands of terrorist groups, small nations that want to build their own atomic weapons, or deranged members of society before reprocessing plants can be licensed. In view of the complexities of these issues and the ponderous pace of governmental regulatory agencies, it is unlikely that nuclear wastes will be reprocessed in the United States

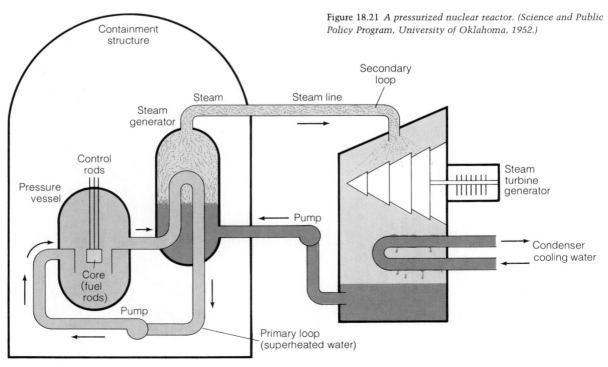

Figure 18.21 *A pressurized nuclear reactor. (Science and Public Policy Program, University of Oklahoma, 1952.)*

Pressurized water reactor

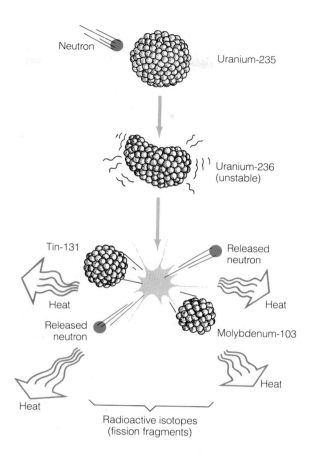

Neutron

Uranium-235

Uranium-236
(unstable)

Tin-131

Released
neutron

Heat

Heat

Released
neutron

Molybdenum-103

Heat

Heat

Heat

Radioactive isotopes
(fission fragments)

in the near future. And given the high capital costs that are involved plus the uncertainties concerning future governmental policies regarding the handling and disposal of nuclear waste, private industry is unwilling to accept the financial risks that are involved in reprocessing. Thus if the United States chooses to reprocess spent fuel in the future, it will probably take place at government-owned facilities, which is the case in Great Britain and France.

The Coal Cycle

Coal is the only fuel that can now compete with nuclear fuel as a means of generating enough electricity for our complex society. The coal cycle is more familiar to most persons. In that cycle, the coal is mined; it is loaded on trains, trucks, barges, or ships; and it is hauled long distances to power plants, where it is stockpiled. Before the coal can be converted to electricity, it is pulverized and then burned

Figure 18.22 *The process of nuclear fission is carried out in the core of a nuclear reactor. In the sequence shown here, uranium-235 is fissioned to form tin-131 and molybdenum-103. Two to three neutrons (an electrically neutral part of the nucleus of an atom) are released per fission event and continue the chain reaction. Fissionable materials yield tremendous amounts of heat per kilogram of material, relative to other fuels.*

Table 18.3 Annual radiation exposure of United States population

	DOSE		STATISTICAL PROJECTION	
	AVERAGE (mrem)	TOTAL (PERSON-rem)	CANCER	GENETIC DEFECT
Source				
Natural Background	100	21,700,000	3,050	193
Technology Enhanced*	5	1,000,000	150	10
Medical Diagnostics	85	18,500,000	2,600	164
Nuclear Power (General Public)†	0.03	6,000	1	0.05
Nuclear-Power (Workers)	600	33,000	5	0.3
Nuclear Weapons (Development and Fallout)	6	1,400,000	200	13
Consumer Products	0.03	6,500	1	0.06
Effects				
Total from All Radiation Sources			6,007	381
Total from All Causes, Known and Unknown			400,000	356,000
Percent from All Radiation Sources			1.5%	0.1%

Source: Radiation and human health. *Electric Power Research Institute Journal*, September, 1979, p. 7.
Note: Calculations by Ralph Lapp, based on estimates of the BEIR committee.
*Mainly from naturally occurring radionuclides redistributed by human activities, such as mining and milling of phosphate and burning coal.
†Assuming normal operation, normal exposure. The radiation from the Three Mile Island accident (50-mile radius) was 1.5 mrem, 3300 person-rem, expected to produce 1 cancer, 0.05 genetic defect.

to produce the steam that drives turbines. Although some of the air pollutants (most of the particulates) that are produced during combustion are removed during the process, sulfur dioxide, nitrogen oxides, and some sulfate particles are released into the atmosphere. And those emissions present a serious health hazard (see Chapter 10). Coal-fired power plants also create enormous waste-disposal problems. For example, each year, a 1000-megawatt power plant produces approximately 230,000 metric tons of fly ash in the course of burning 2.3 million metric tons of coal.

Risk Assessment: Nuclear Power versus Coal

Risk assessment of complex technology has only recently become a subject of investigation to help society choose appropriate technology. Although risk assessment of alternative technologies can help us to clarify relevant issues, it cannot resolve the trade-offs that must be made in terms of the impact on air, water, and land use as well as in terms of long- and short-term risks to human health. The types of data used in an assessment of energy sources are listed in Table 18.4.

Risk assessment can be thought of as a two-step process in which (1) scientific data are collected concerning specific aspects of a technology (for example, public and worker safety and health and environmental damage) to measure the risk, and then (2)

Table 18.4 Factors to be considered in choosing energy sources

Performance
    Efficiency—Overall Net Energy Balance
    Materials Availability
    Space Availability
    Reliability
    Adaptability to Another Energy Source
    Ease of Maintenance
    Long-Term Availability
Negative Impacts
    Surface Disruption
    Transportation and Distribution Disruptions
    Air Pollution
    Water Pollution
    Human Health and Safety Risks
    Effect on Ecosystems
    Waste Disposal or Recycling Problems
    Water Availability
Costs
    Initial Investment
    Operation and Maintenance Costs
    Dismantling Costs
    Availability of Capital and its Cost

a judgment is made, which is based on that information, concerning one's willingness to tolerate the risks in light of the perceived benefits. However, both steps of risk assessment have limitations that prevent it from providing final answers regarding acceptable risk levels.

It is difficult to collect data on illnesses, deaths, and disabilities that may be associated with a particular technology. For example, scientists know much more about the effects of exposure of organisms to low levels of radiation than they know about exposure to low levels of sulfur dioxide. In actual practice, it is much easier to measure exposure to radiation than it is to measure exposure to sulfur dioxide. Furthermore, risks can vary considerably from one place to another (for example, surface coal mining is much safer than underground mining). And some risks are not measurable, such as the long-term risks that are associated with nuclear power in making atomic weapons available to more nations.

Once the risks have been identified and, when possible, measured, persons can decide which risks are acceptable to them and which are not. However, two persons can examine the same information and come to opposite conclusions. In general, various segments of society do not always receive the same information, and sometimes the information is unreliable. Furthermore, most intuitive assessments of risk are often different from those that are based on statistics. For example, most persons are willing to take moderately high risks with familiar activities, such as driving a car, but they are unwilling to let technology, such as nuclear power plants, operate at the same level of risk. The reason seems to be that most persons believe that technology should be responsible for reducing risk, whereas they are generally unwilling to be accountable for the risks they place on themselves.

Many scientists have begun to do risk assessments of energy alternatives. For example, L. D. Hamilton of the Brookhaven National Laboratory has compared the health effects of generating electricity by using both coal and nuclear power. Those effects are compared for power plants that produce an equal amount of energy (1000 megawatts) in Figure 18.23. The area of each block that is shown is proportional to the number of deaths that occur within the public sector as well as for workers who are involved in mining, processing, transporting, and converting electricity for both fuel cycles. The estimated number of deaths that result are higher for

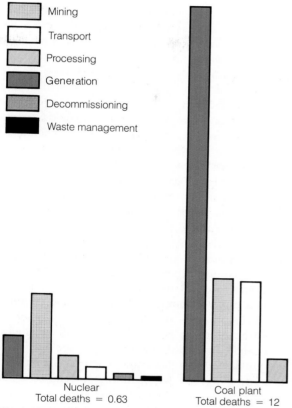

Mining
Transport
Processing
Generation
Decommissioning
Waste management

Nuclear
Total deaths = 0.63

Coal plant
Total deaths = 12

Figure 18.23 *Comparison of the public sector plus worker deaths for the nuclear cycle versus the coal cycle. The estimated number of deaths is expressed as number of deaths per 1000 megawatts of electricity production per year. The areas in the figure are proportional to the number of deaths. (L. D. Hamilton, Comparing the Health Impacts of Different Energy Sources. In Proceedings of International Atomic Energy Agency Series, Health Impacts of Different Sources of Energy, Vienna: International Atomic Energy Agency, 1982, p. 12.)*

workers who work with coal than for those who work with nuclear energy (in the generation, transportation, and mining segments of the cycle). However, this should not be surprising since coal requires the handling of far larger quantities of materials than those which must be handled in the nuclear cycle. And those steps all require more man-hours of work in relatively dangerous occupations.

However, nuclear energy has an additional risk—the risk of a catastrophic accident. For example, Norman Rasmussen of the Massachusetts Institute of Technology directed a $4 million study, which was completed in 1975, to assess the consequences and probability of core meltdown and a worst-case

possible accident. That study, which is known as the Rasmussen Report, estimated that a meltdown would occur once every 20,000 reactor operating years. (Such an accident would occur once every 50 years if all electricity were generated by nuclear power.) Thus if we averaged that probability over 50 years, the annual consequences of using nuclear power would be 0.2 deaths from acute radiation sickness and, eventually, 10 deaths from cancer.

The Rasmussen Report also makes estimates for a worst-case accident. Following a worst-case accident, 3500 deaths would occur from acute radiation illnesses and 45,000 deaths would occur from delayed onset of cancer. Evacuation and extensive cleanup costs for such a catastrophe were estimated at $14 billion. And the probability that such an accident could occur was estimated to be one chance in a million. Critics of nuclear power state that the Rasmussen Report underestimates the frequency of accidents and their severity by a factor of ten.

Since the Rasmussen Report, only one serious accident has occurred—at the Three-Mile Island plant on the Susquehanna River near Middletown, Pennsylvania. Because of the severity of that accident, President Carter set up a special commission to investigate it. The final report, which was called the Kemeny Report (after the President of Dartmouth College, who served as chairman of the commission), found that radiation exposure to the public was low and that it would have a negligible effect on human health including future effects. Mental stress was cited as the major health effect. The accident did not release as much radioactivity as was predicted by accident analysts. But it did cause severe damage to the generating facility, and costs to repair it are estimated to reach, ultimately, $1 to $2 billion if it can be put back into service. The costs will be much higher if the facility must be abandoned. In 1984, the cleanup was still incomplete, and the fate of the reactor was uncertain.

Scientists and the public remain divided on the nuclear power issue. For example, Herbert Inhaber, a nuclear engineer, has analyzed the safety of nuclear power and has concluded that it is 10 times safer, including catastrophic effects, than all other energy systems that he has investigated, with the exception of natural gas. He included solar and hydroelectric dams, which are discussed in Chapter 19, in his analysis.

John P. Holdren, Professor of Energy Resources at the University of California at Berkeley, who is equally concerned about the detrimental effects of coal and nuclear power, remains opposed to nuclear power, mainly for reasons that have to do with non-technical considerations. His primary concern is that nuclear power will speed up the proliferation of nuclear weapons among nations, because nuclear power plants create (breed) fissionable materials, such as plutonium-239, which can be separated out by using sophisticated technologies and used for the construction of nuclear weapons. Theft of materials for nuclear weapons is next in his order of personal concerns, followed by sabotage and accidents in the nuclear fuel cycle. Routine emissions and exposure to radiation are last on his list of concerns.

Costs: Nuclear Power versus Coal

The technology that is used to provide energy services is not selected solely on the basis of risk assessment. Economic factors also weigh heavily in decisions that are made both for or against using a particular technology. And the economics of nuclear power have changed considerably in the past decade, because the relative cost of nuclear power plant construction has been rising much faster than the cost of construction of coal-fired power plants. In fact, costs have risen so fast that the construction of several partially built nuclear power plants was halted between 1982 and 1984. Some of those plants, such as the Seabrook reactors in New Hampshire, already had $4 to $6 billion invested in their design and construction. One plant, the Zimmer plant in Ohio, will probably be converted to a coal-generating plant, while the others will remain idle monuments to an earlier economic era. Those plants were abandoned because their actual costs were exceeding the original estimates by a factor of ten, and the utilities found that the costs of borrowing the tremendous sums of money that were necessary to finance their completion were more than they could justify. Other reasons for the escalating costs include mismanagement of construction schedules and safety inspections, labor strikes, costly changes in design that are dictated, in part, by the Three-Mile Island reactor accident in 1979, and the financial liabilities that would occur in case of an accident. The Three-Mile Island accident is expected to cost the utility's owner approximately $1 to $2 billion to clean up, which is approximately double the plant's initial construction costs. Such costs are high enough to bankrupt many utilities.

The average generating cost for electricity generated by nuclear fuel or coal varies with geographical location largely because of differences in transportation costs for coal. In seven of the nine regions that have been studied, nuclear power has a small cost advantage over coal for plants that will begin to produce electricity in 1985 (Table 18.5). New designs and improvements in technology could change the future economics of nuclear power. For example, German scientists are constructing a nuclear power plant that uses a flow of uranium-containing pellets through the plant. That design avoids the possibility of core meltdown and, therefore, requires much less-expensive design features and safety equipment to prevent accidents. The plant is scheduled to begin operation in 1985. And in Fort St. Vrain, Colorado, the United States is experimenting with another design that avoids the possibility of meltdown. If those plants are successful, the economics of electric power will undergo yet another revolution in the coming decades.

Although some utilities have had an excellent safety and operational record, others have had serious problems with faulty components, fradulent record keeping, and lax operating procedures. The Kemeny Commission, in its investigation of the Three-Mile Island accident, found many questionable procedures in the operation of nuclear power plants. As a result, the industry has been required to implement new safety and evacuation proce-

Table 18.5 Annual costs of new nuclear and coal plants* by region (in 1982 cents per kilowatt-hour)

REGION	NEW NUCLEAR	NEW COAL
Northeast	3.97	4.39
Southeast	3.14	3.78
Great Lakes	3.28	3.92
North Central	3.28	3.78
Midwest	3.22	3.09
South Central	3.12	3.31
Mountain	3.20	2.85
Southwest	3.32	3.52
Pacific	3.47	4.37
Total United States	3.35	3.86

Source: "The Current Economics of Electric Generation from Coal in the U.S. and Western Europe," National Economic Research Associates Inc., January 1983, as reported in *Atomic Energy Clearinghouse*, Vol. 29, No. 5, January 31, 1983.
*Assuming 1985 on-line dates.

dures, to enhance the safety of their operation. Those procedures increase the cost of nuclear power.

Operation of nuclear power plants demands a commitment to excellence by a utility—a commitment that has not been made by all utilities in the past. Thus the problems that have arisen in the nuclear electrical utility industry have undermined the public's confidence in them, and now more than 50 percent of the population opposes them.

The unpredictability of costs and the erosion of public confidence has caused utility managers to be unwilling to plan new nuclear plant construction. In 1984, the Office of Technology Assessment evaluated the nuclear industry and concluded that "without significant changes in the technology, management, and levels of public acceptance, nuclear power in the United States is unlikely to be expanded in this century beyond the reactors already under construction." Not a single new order for a nuclear power plant has been placed since 1978 in the United States. As a result, in the foreseeable future, nuclear power will not play as significant a role in the production of electricity in the United States as was projected even 5 years earlier. Either conservation, coal, or some combination of both will probably take up most of the slack.

Conclusions

Our examination of past consumption and future oil supplies points out the unique "oil era" in which we live—a period during which the relative cost for energy (the proportion of our income that we spend for energy) has been small. Access to oil and its prod-

ucts and natural gas has been relatively easy and, consequently, inexpensive. But as the availability of those resources diminishes, prices will rise and other sources will compete with conventional sources. The challenge for society today is to evaluate which means of providing future energy services makes the most sense from economic and environmental viewpoints. In addition, it is important to keep in mind the time frame that will be necessary to convert from one type of technology to another. A strong emphasis on the all-out development of oil and natural gas resources now (without paying attention to how it is used) may increase our dependence on oil and natural gas and, eventually, intensify the pains associated with inevitable withdrawal.

The stage is set for change: petroleum reserves are declining worldwide, and in the United States nuclear power will generate less electrical power than was anticipated a few years earlier. In addition, coal has limited applications and increased use will further increase air-quality problems, such as acid rain. But government estimates of where our energy will come from in the year 2000 (Table 18.6) reveal that the energy picture will not be all that different from what is in place today. However, projected oil and natural gas reserves (Figure 18.10) illustrate that major changes must occur in energy services soon after the year 2000. We therefore must continue to assess the best means of supplying our society with the safest, cleanest, and most reliable sources of energy possible.

Since the United States is a net energy importer and a leader in the development of new technology, it has two critical tasks in the next decade. First, it must be a world leader in making the transition from

Table 18.6 Energy-use projections for year 2000 by source and use section (Quads)

	RESIDENTIAL	COMMERCIAL	INDUSTRIAL	TRANSPORTATION	ELECTRICITY	TOTAL
Coal	0.1	0.1	5.0	neg.	23.0	28.2
Biomass	1.0	0.5	1.9	0.2	0.3	3.9
Natural Gas	5.1	2.9	8.6	0.6	1.8	18.9
Oil	1.3	0.9	8.2	16.8	1.3	28.4
Hydro	—	—	neg.	—	3.7	3.8
Nuclear	—	—	—	—	7.9	7.9
Solar, Wind, and Geothermal	0.3	0.3	0.5	—	1.2	2.3
Total Direct Consumption	7.8	4.7	24.2	17.6	39.2	93.4
Electricity (Point of Use)	3.5	3.3	4.7	0.1		11.6

Source: U.S. Department of Energy, Office of Policy and Analysis, 1983.

an oil dependent economy to a technologic system based on renewable or essentially unlimited resources. Second, it can stabilize its own economy and improve its national security by reducing its vulnerability to interruptions of energy resources delivered from foreign countries.

In Chapter 19, we evaluate means of energy conservation and alternative means of supplying energy services. Consequences of these technologies are also evaluated so we can make wise decisions.

Summary Statements

During the past century, we have seen the rapid development of energy-consuming technology. That factor, along with the growth of our population, has resulted in a 15-fold increase in our nation's use of energy since 1900.

When the use of electricity is factored in, the residential–commercial, transportation, and industrial sectors of the United States' economy account for 35.3, 26.9, and 37.8 percent, respectively, of the 73.7 Quads of total energy consumed in 1984. Space heating is the major use of energy in the residential–commercial sector, while motor vehicles use most of the remaining energy that is consumed in the residential sector.

Crude oil is the raw material for a vast array of petroleum-based products. In 1984, oil supplied 42.1 percent of the total amount of energy that was needed in the United States, which continues to use oil at near the historic maximum rate, despite the fact that proven and undiscovered reserves would only last 36 years at current rates of consumption and importation. World oil supplies are estimated to last 60 years at present consumption rates. But United States' and world oil production rates are expected to begin to decrease in the early 1990s.

In contrast to oil, almost all the natural gas that is consumed in the United States is produced domestically. Estimates of the life expectancy of discovered and undiscovered natural gas resources range between 46 and 62 years based on present rates of consumption.

The use of coal has increased in the past decade, and in 1984, coal represented 23.3 percent of the total amount of energy that was used in the United States. Coal reserves are scattered throughout the United States, and they are large enough to last for several centuries. A greater reliance on coal for energy will result in more landscape disruption, air pollution, fly-ash disposal problems, and increased acid rain.

Both oil and natural gas supplies can be augmented through techniques for the gasification and liquefaction of coal and oil shale. Those techniques can be carried out after the resource is mined or *in situ* (in place). Massive subsidies by the United States government have resulted in only a few plants that produce minor amounts of synthetic fuels. Production of synfuels presently is not economical and weak world oil prices in the mid-1980s worsened the economic picture for synfuels.

Nuclear energy can be used to augment the production of electricity. Studies indicate that a nuclear fuel cycle is a safer means of generating electricity than is the coal cycle, but it has other social risks, such as the spreading of nuclear weapons to other countries. Management of the production of electricity at nuclear power plants by the various utilities ranges from excellent to inadequate. At some locations, construction and borrowing costs for nuclear power plants are greater than that for coal-fired power plants. Those costs and declining public confidence have resulted in the halting of construction of a number of nuclear power plants and no new plants are scheduled for construction. Therefore, nuclear power will not produce as much electricity as had been estimated earlier. New nuclear technology is being developed that precludes the possibility of meltdown and, therefore, the release of large amounts of radiation, which is the major concern of the public. Within the next decade, the viability of these plants will be known. Procedures for the permanent disposal of the high-level radioactive wastes that are generated at nuclear power plants are not yet complete.

Questions and Projects

1. In your own words, write a definition for each of the terms that are italicized in this chapter. Compare your definitions with those in the text.

2. Would an oil embargo today have a greater impact on the United States than the 1973 oil embargo did? Explain your answer.

3. Develop a list of the uses for oil, natural gas, coal, and nuclear fuel. State whether or not substitutes are available for each type of use if the resource is used up.

4. Is it possible for the United States to fulfill its energy demands without experiencing conflicts of interest with other countries?

5. We often encounter statements that claim, in effect, that we are going to run out of a fossil-fuel resource in X years. State reasons why such statements are oversimplifications.

6. List some of the side effects that the rapid increase in energy prices has had in our society.

7. Describe the rate at which you use toothpaste from a tube. Is your rate of use dependent on the size of the reserve in the tube? Does an analogous situation exist with respect to fossil-fuel reserves?

8. Why has such a tremendous amount of research and development been focused on synfuels?

9. Why are natural gas prices expected to climb faster than other sources of energy?

10. What are the environmental advantages and disadvantages of gasifying coal or oil shale *in situ*?

11. The exploitation of energy resources often creates conflicts with the management of other resources. Cite examples of such problems in your community and list possible ways of resolving them.

12. If your electric utility burns significant amounts of coal, where does it dispose of its fly ash? What is the impact of fly-ash disposal on the environment in your region?

13. Have a spokesperson from your electric utility come into your class to discuss its plans regarding future energy sources and their

impact on the environment. Ask about what efforts are being implemented by the utility to conserve energy.

14. Homes can be heated by natural gas, oil, coal, or electricity. Explain which source would require the most energy-efficient home and why.

15. Compare and contrast the hazards of the nuclear fuel cycle with those of the coal cycle.

16. If your electric utility were to increase or replace some of its generating capacity, where would you suggest that the power plant be located? What type of plant would you recommend? List the factors that go into your decision.

17. Is the quality of coal the same throughout the United States? What advantages does western coal have over eastern coal? What are its disadvantages?

18. Do you think that the owners of pipelines should be given the right of eminent domain? Defend your answer.

19. What strategies are available to enhance petroleum liquids and natural-gas supplies? Speculate on their relative environmental impacts.

20. Debate the pros and cons of government regulation of natural-gas prices.

21. Will the United States continue to increase its per-capita energy demand as has been the case for most of this century? Explain.

Selected Readings

BEEBE, G. W.: Ionizing radiation and health. *American Scientist* 70:35, 1982. A review of the health effects caused by exposure to radiation.

CARTER, L. J.: WIPP goes ahead, amid controversy. *Science* 222:1104, 1983. The status of the military's Waste Isolation Pilot Project in New Mexico is given.

COHEN, B. L.: Radiation pollution and cancer: Comparative risks and proof. *Cato Journal* 2:255, 1982. A comparison of the relative health and safety risks we take in implementing various technologies, including exposure to radiation.

COMMITTEE ON ENERGY AND NATURAL RESOURCES, UNITED STATES SENATE: *World Petroleum Outlook—1983.* Washington, D.C.: U.S. Government Printing Office, 1983. A government report that looks at various scenarios for energy sources and their environmental impacts.

DICK, R. A., AND WIMPFEN, S. P.: "Oil Mining." *Scientific American* 243(4):182, 1980. Discussion of technologies to mine oil that is not extractable by conventional technologies.

INTERNATIONAL ENERGY AGENCY: *World Energy Outlook.* Washington, D.C.: OECD Publications, 1982. Text with statistics and energy policy guidelines adopted by the International Energy Agency.

LANDSBERG, H. H.: *Energy: The Next Twenty Years.* Cambridge, Mass.: Ballinger, 1979. An excellent review of the energy resource problem, especially as it relates to international issues, produced by a panel of experts.

Managing Nuclear Wastes. In *Underground Space*, Vol. 6, Nos. 4–5. A special issue devoted entirely to nuclear waste disposal. New York: Pergammon Press, 1982.

PERRY, H.: Coal in the United States: A status report. *Science* 222:377, 1983. A review of the environmental and economic factors that affect the use of coal.

RUEDISILI, L. C., AND FIREBAUGH, M. W.: *Perspectives on Energy.* New York: Oxford University Press, 1982. A textbook with individual chapters written by experts on topics that deal with energy issues, technology, and environmental dilemmas.

SPENCER, D. F., GLUCKMAN, M. J., AND ALPERT, S. B.: Coal gasification for electric power generation. *Science* 215:1571, 1982. A review of the methods available for producing synthetic natural gas from coal.

THE NATIONAL RESEARCH COUNCIL: *Energy in Transition, 1985–2010.* New York: W. H. Freeman, 1980. The final report of the Committee on Nuclear and Alternative Energy Systems of the National Academy of Sciences to the Department of Energy, which covers a wide range of energy issues.

U.S. GEOLOGICAL SURVEY CIRCULAR 860: *Estimates of Undiscovered Recoverable Conventional Resources of Oil and Gas in the United States.* Washington, D.C.: U.S. Government Printing Office, 1981. An assessment of the size and location of gas resources in the United States.

U.S. GOVERNMENT PRINTING OFFICE: *Nuclear Power in an Age of Uncertainty.* Washington, D.C.: U. S. Government Printing Office, 1984. A discussion of the issues that affect the size and future of the nuclear power industry.

VICK, G. K., AND EPPERLY, W. R.: Status of the development of EDS coal liquefaction. *Science* 217: 311, 1982. Description of Exxon's process for producing liquid fuels from coal.

WASP, E. J.: Slurry pipelines. *Scientific American* 249(5):48–55, 1983. An assessment of pipelines as a means of transporting coal.

*Toyotas, fresh from Japan, being unloaded at the import dock at
Benicia, California, for distribution to Toyota dealers in Northern California. (© 1978 Barrie Rokeach.)*

CHAPTER 19 ENERGY ALTERNATIVES

Because deserts receive so much solar radiation, they are ideal sites for technologies that convert the sun's energy into other useful forms of energy, especially electricity. Thus researchers are now testing many different solar-energy systems at desert sites to evaluate their potential for energy production on a commercial scale. One such system is the California Edison Company's Vanguard I, which is a solar collector that is in the shape of a parabolic dish, which the company began testing at Rancho Mirage, California, in December of 1983 (Figure 19.1). Eventually, the company plans to construct a 30-megawatt generating station that will employ the Vanguard I design.

Vanguard I is especially promising because its cooling system requires very little water—a scarce commodity in desert regions, where solar technologies can operate at peak efficiency. Its dish-shaped parabolic mirror collects and concentrates the sun's rays at the mirror's focal point, where the temperature reaches 1100°C (2000°F). The focal point is composed of a metal surface that acts as a heat receiver, which heats hydrogen gas. Then, the heated gas is circulated through a closed-loop system, which connects the heat receiver with a device that is called a *Stirling engine*. And that engine, in turn, drives an electric generator.

The Stirling engine functions in much the same way as a gasoline-powered internal-combustion engine. Both types of engine are heat engines—that is, both convert heat energy into kinetic energy (energy of motion)—and both contain a hot, expanding "working fluid" that presses against pistons to cause them to move back and forth. However, internal-combustion engines generate heat internally by burning (combusting) gasoline in its cylinders, and the working fluid is a hot gas that is formed when the gasoline is burned. In contrast, the working fluid of a Stirling engine is heated externally—in this case, by solar radiation—and is pumped through the engine's cylinders, where it causes the pistons to move. Furthermore, in both types of engine, the moving pistons turn a drive shaft that transmits power to some other kind of mechanical device to perform useful work. In Vanguard I, the drive shaft transmits energy to a generator that produces electricity.

In a Stirling engine, the pistons move back and forth between hot and cool reservoirs, and the greater the temperature difference between those reservoirs, the greater is the engine's efficiency. Thus the cool reservoir must be kept as cool as possible, which is accomplished by a closed cooling system that employs a radiator (much like the type of radiator that cools an automobile engine). Because that cooling system uses the same water over and over again, relatively little water is needed and, because it is a closed system, very little water is lost by evaporation.

The design of the Stirling engine has several other advantages as well. It is relatively inexpensive to build; it is easy to maintain; and it can be designed with a movable dish that tracks the sun across the sky, which enables it to operate at near-maximum efficiency throughout the daylight hours. Also, it can be used in remote areas that lack other sources of electrical power, or it can feed electricity into an existing power grid. Its environmental impact is minimal (with the exception of a large space requirement for the parabolic dish). And it is quite efficient for a solar-power system. For example, in early tests, the Vanguard I version achieved a record sunlight-to-electricity conversion efficiency of 29.5 percent.

The Vanguard I with its Stirling engine is one example of an amazing array of choices for meeting our future energy needs that scientists and engineers have created during the past decade. Those choices fall into three categories: (1) energy conservation measures, (2) new technology, and (3) proven, but not yet fully developed, technologies. The parabolic-dish design that has been used in the Vanguard I and several other experimental solar-power designs is one of many promising technologies that are being evaluated for performance and economic viability as a means to augment our future energy supplies.

Because fossil-fuel resources are finite and some are already in short supply, and because nuclear energy has not been widely accepted, alternative means of furnishing energy services are now being investigated intensively. Alternative energy sources range in size from solar hot-water systems for homes to hydroelectric facilities that would inundate a small state. And some energy alternatives extend the useful lifetime of finite fossil-fuel reserves, whereas others exploit renewable energy resources or essentially undepletable ones. Some of those alternatives make environmental and economic sense now; others will require one or more decades of further development. We now examine the basis of those technologies, some of the major obstacles for their development, and their environmental effects.

Energy Conservation

Before alternative energy resources can provide a significant amount of energy, they will require large investments of capital, time, and materials. None of the alternatives described in this chapter are expected to add significantly to our conventional energy sources until the 1990s. In the meantime, we can extend the lifetime of our fossil fuels by implementing energy-conservation strategies. Energy-

conservation measures can, in fact, be viewed as a source of energy, whose contribution to our energy supplies is every bit as real as new sources of coal, oil, natural gas, or nuclear power. Simply put, conservation reduces the amount of energy that must be produced to meet our demands for energy services.

Strong motives exist for conserving energy. Conservation, of course, saves money, and efforts to conserve energy usually produce a cleaner environment. Also, energy that is saved by conservation reduces our reliance on foreign sources, which increases national security and decreases problems concerning the international balance of payments and international trade. (Between 1977 and 1983, the United States' net trade deficit for energy totaled $386 billion. The cost of imported oil was even more, but that cost was offset partially by coal exports.) Furthermore, most conservation measures are low-risk investments: they perform as advertized and have short pay-back periods—often less than 1 year. And some conservation measures are essentially cost-free, because they only involve a change in habits, such as turning off lights or changing thermostat settings. Instead of building new, large power plants, which

Figure 19.1 *The Vanguard I is a parabolic, solar-reflecting dish that is being evaluated as a means of generating electricity. The high temperatures produced at its focal point power an external combustion engine that produces electricity, which goes to Southern California Edison customers. Because the system requires very little cooling water, it is ideally suited for use in arid regions. (Southern California Edison Co.)*

take 6 or more years to plan and build, conservation measures can be implemented quickly—often with a series of small purchases instead of a single large investment.

There is no exact measure or definition of conserved energy. Typically, it is considered to be the difference between the amount of energy that is used before conservation measures are implemented and the amount of energy that is used after such procedures are installed. For example, the difference between the amount of energy that is used in a home before and after adding insulation would be a measure of conserved energy.

Residential–Commercial Measures

If you can identify the areas in which you consume large amounts of energy in your daily life, then you can identify the areas in which the largest gains can be achieved from conservation. Table 19.1 shows the uses for which the average American uses energy and the great amounts of energy that are used for residential space heating and cooling. For example, more than 60 million of this country's 80 million homes were constructed prior to 1970. Many of those homes are inadequately insulated, caulked, and weather-stripped, and most of them are equipped with single-glazed windows (that is, windows that have a single sheet of glass). Such houses not only lose heat during the winter, but they gain heat during summer. Thus poor weatherization places demands on both heating and cooling systems. Hence, it is often worthwhile to add insulation to homes that are inadequately insulated, because the pay-back pe-

riods are short—usually 1 to 2 years. Most utilities now offer free home inspections and advice on how much insulation to add, where to add it, its cost, and its pay-back time. And some utilities aggressively pursue conservation, because it is less expensive for them to continue to operate at existing levels of production than it is for them to increase their capacity to produce energy. For example, in 1984, some customers of the Jersey Central Power and Light Company could qualify to have a contractor install insulation and weather-stripping. That service was free to the customer, and the utility paid the contractor, depending on how much energy was conserved. The most generous offers for conservation come from utilities whose demand exceeds their capacity to produce energy. However, in regions where conservation efforts decrease the demand below production capacity, utilities may ask permission to raise their rates to preserve the rate of return for their owner-investors.

Homeowners can save energy in many other ways. For example, they can reduce water-heating costs by lowering the temperature settings on their hot-water heater, covering the hot-water heater with an insulation blanket, or replacing a worn-out water heater with a new, better-insulated model. Furnishing hot water accounts for approximately 18 percent of the energy use in a typical home.

In moderate climates, it is economical to replace electric-resistance space-heating systems with heat pumps. Heat pumps can extract a portion of the heat in cold outdoor air (outdoor air is cooled further) and the heat extracted is transferred from outdoor air to indoor air during the winter. In summer heat pumps function in reverse as an air conditioner. Heat pumps can save 25 to 50 percent over electrical-resistance heating if outdoor temperatures do not fall below approximately −7°C (20°F). Heat pumps were installed in 24 percent of all single-family homes that were built in this country during 1980. Also, space heating by natural gas can be accomplished more efficiently with new pulse-combustion furnaces, which operate at 90 to 95 percent efficiency as compared to the 60 to 65 percent efficiency of most conventional oil furnaces.

Builders and architects should be encouraged to incorporate into new homes such energy-saving features as overhangs (discussed later in this chapter), large windows that face the south, and movable, insulated window panels. Older homes can be fitted

Table 19.1 Energy used directly by United States' citizens, 1980

	QUADS PER YEAR	PERCENT
Space Heating*	5.054	25.2
Water Heating	1.734	8.6
Air Conditioning	0.363	1.8
Lighting, Cooking, Refrigeration, Drying, and Other Appliances	2.185	10.9
Transportation	10.750	53.5
Total	20.086*	100.0

Source: *World Petroleum Outlook, 1983.* Committee on Energy and Natural Resources, United States Senate.
*Excludes wood.

with some of the same energy-saving features, but the cost of those efforts can be considerable, since they involve extensive remodeling.

The electric energy that is used by household appliances accounts for approximately 8 percent of the total amount of energy that is used in the United States. Therefore, merely improving the operating efficiency of household appliances would reduce significantly the overall demand for electricity. (The energy consumption of home appliances and lighting fixtures is listed in Table 19.2). For example, frost-free refrigerator-freezer combinations typically use 1800 kilowatt-hours of electrical energy per year. According to federal estimates, refrigerators that contain more insulation, more efficient motors, and improved cooling systems could reduce the annual energy consumption of those appliances by 45 percent. If such refrigerators were in use in all the homes in the United States today, our total electrical demand would be lowered by approximately 3 percent. Other appliances that could be redesigned to be more energy efficient include air conditioners (a 22-percent energy savings is possible), television sets, and microwave ovens. Since energy prices are likely to rise in the future, the energy-efficient appliances are good long-term investments, even though their purchase price is greater than that for conventional, energy-hungry appliances.

Energy conservation measures that are available for the commercial sector are much the same as those for the residential sector. However, lighting which consumes 20 to 25 percent of the electricity in the United States, is of particular concern in businesses. Thus reductions in the amount of lighting that is used and improvements in lighting efficiency would save a great deal of energy. The amount of light that is emitted by an electrical lamp per unit of energy input varies greatly with different types of lamps. For example, fluorescent lights emit nearly four times as much light per unit of energy as ordinary incandescent light bulbs.

Conserving electric energy pays double dividends. On an energy-equivalent basis, electricity is the most expensive source of energy. Furthermore, every BTU of electric energy that is saved at the point of use reduces the nation's total consumption by approximately 3 BTUs. The reason for the savings multiplier is that power plants only convert approximately one-third of the energy that is in fossil or nuclear fuels to electricity. The remaining two-thirds

Table 19.2 Monthly energy consumption of home appliances

APPLIANCES	POWER IN WATTS	TIME USED PER MONTH IN HOURS	TOTAL KWH* PER MONTH
Air Conditioner (Window)	1,566	74	116
Blanket, Electric	177	73	13
Blender	350	1.5	0.5
Broiler	1,436	6	8.5
Clothes Dryer (Electric)	4,856	18	86
Clothes Dryer (Gas)	325	18	6.0
Coffee Pot	894	10	9
Dishwasher	1,200	25	30
Drill (¼ in. Elec.)	250	2	0.5
Fan (Attic)	370	65	24
Freezer (17 cu. ft.)	370	242	65
Freezer (17 cu. ft.), Frostless	360	400	144
Frying Pan	1,196	12	15
Garbage Disposal	445	6	3
Heat, Electric Baseboard, Avg. Size Home	10,000	160	1600
Iron	1,088	11	12
Light Bulb, 75-Watt	75	120	9
Light Bulb, 40-Watt	40	120	4.8
Light Bulb, 25-Watt	25	120	3
Oil Burner, ⅛ hp	250	64	16
Range	12,200	8	98
Record Player (Tube)	150	50	7.5
Record Player (Solid State)	60	50	3
Refrigerator-Freezer (14 cu. ft.)	326	290	95
Refrigerator-Freezer (14 cu. ft.), Frostless	615	250	152
Skill Saw	1,000	6	6
Sun Lamp	279	5.4	1.5
Television (B&W)	70	180	13
Television (Color)	90	180	16
Toaster	1,146	2.6	3
Typewriter	30	15	0.45
Vacuum Cleaner	630	6.4	4
Washing Machine (Auto)	512	17.6	9
Washing Machine (Wringer)	275	15	4
Water Heater	4,474	89	400
Water Pump	460	44	20

Source: Wisconsin Public Service Corporation, "Choosing a Home Wind Generator."
*KWH = kilowatt-hour, the basic measure of electricity. In 1983, the average cost of electricity in the United States was $0.063 per kilowatt-hour.

is unavoidably lost in the form of heat at the plant, and small losses occur along transmission lines.

Transportation Measures

During the 1930s, the automobile replaced the railroad train as the major means of personal transportation. The greater individual mobility that was provided by the automobile influenced the shape of our cities, increased the average commuting distances between home and work, and molded (to a great extent) the patterns of our individual lives. Today, in the United States, cars and trucks travel 2.4 trillion kilometers (1.35 trillion miles) each year and, in the process, burn up more than 431 billion liters (114 billion gallons) of fuel. Thus in the United States transportation uses up 85 percent of the liquid fuel used.

Five major ways exist to reduce energy consumption in the transportation sector: (1) by improving the efficiency of vehicles, (2) by operating equipment more efficiently, (3) by increasing the load factor, (4) by shifting travel from less-efficient to more-efficient modes, and (5) by reducing the total amount of travel.

Improving the fuel economy of automobiles is one of the goals of the Energy Policy and Conservation Act. That act requires automobile manufacturers to produce cars with an overall average fuel economy of 27.5 miles per gallon (mpg) by 1985 for passenger cars. Even though the Office of Technology Assessment estimates that the fuel economy for new cars should reach 45 miles per gallon by the year 2000 and could range as high as 80 miles per gallon, the federal government has not mandated any further improvements beyond the 1985 standards.

The amount of fuel that is used by motor vehicles is heavily dependent on the weight of the vehicles. Thus automobile manufacturers have been decreasing the size and weight of new cars to meet the mandated fuel economy. Unfortunately, smaller cars are more dangerous cars. Safety experts say that occupants of small cars are 3.4 to 8 times more likely to die in an accident than are occupants of a larger car. The success in decreasing fuel use by the United States' fleet depends on the size and mix of new cars that are purchased as well as further design improvements, such as computer-controlled operation. Box 19.1 describes several approaches that car manufacturers are taking to improve gasoline mileage.

Also, persons can operate their motor vehicles more efficiently if they adopt conservation measures, for example, by reducing speed on open highways, which lowers the amount of fuel that is used. The nationwide highway speed limit of 88-kilometers (55-miles) per hour, which was imposed in response to the 1973 Arab oil embargo, is estimated to save 200,000 barrels of oil per day. (It has also been credited with saving nearly 200,000 lives.) And lower speed limits would improve mileage even more, but they would be met by stiff opposition from the trucking industry and persons who live in the spacious western states who are unhappy with the current limit. Late in 1984, the National Academy of Sciences studied the questions of speed, fuel economy, and safety that surround the 88-kilometer per hour limit. They concluded that it should remain in force. Box 19.2 illustrates other measures all drivers can use to conserve fuel.

Trucks and aircraft can also be designed for greater fuel efficiency. For example, trucks that have diesel engines are more fuel-efficient than those that are equipped with gasoline engines. And the fuel efficiency of air transportation can be improved by using different types of airplanes for specific purposes. For example, wide-body aircraft are designed for long trips, but other designs are more efficient for short commuter routes.

Americans use their transportation systems inefficiently. For example, most persons in the United States drive to work alone (Figure 19.2), but in-

Figure 19.2 These statistics, which show how United States' citizens get to work, reveal the potential for car pooling and van pooling. (U.S. Census Bureau, compiled by American Demographics.)

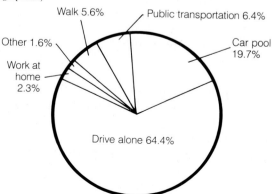

BOX 19.1

MORE MILES PER GALLON ...

Last September, when most American motorists had already begun seeing a decline in gasoline prices, Peugeot S.A. arrived in Detroit to show off its automotive engineering. Peugeot's demonstration made clear that, despite the swift advances made over the past 5 years in fuel efficiency, more progress is yet to come.

Peugeot's car, a five-passenger sedan, was driven at 55-miles per hour from Detroit to Knoxville, Tenn. It averaged 74 miles per gallon. At a slower speed—33 miles per hour along the same 500-mile route—it averaged 91 miles per gallon, or slightly shy of the 100-mile-per-gallon barrier that Peugeot's engineers had hoped to overcome.

"This isn't a futuristic test vehicle," asserted Larry Edwards, the national technical manager for Peugeot's United States operations. The Peugeot used in the test resembled a standard model already marketed in Europe, but achieved the greater economy through a small, supercharged diesel engine, flush glass and spoilers to improve wind resistance and the substitution of plastic and aluminum for several metal body parts to reduce its weight.

"These are refinements of technology that we've had for years," Mr. Edwards said, adding that he expects most major car manufacturers to "keep refining, and keep moving ahead."

Currently, two of the most active areas of research and development focus on the way fuels are burned in cylinders and the wind resistance of car bodies.

In both cases, engineers say electronic microprocessors may play an expanding role in controlling the car's performance.

This year, several manufacturers in this country and abroad have begun offering gasoline-powered engines that utilize a system known as the "fast burn." By making gasoline ignite in cylinders faster than it has in the past, this system generates enough compression to drive a car's pistons up and down using a minutely smaller amount of fuel in each rotation.

Different car companies have accomplished the fast burn through different methods, with some Japanese companies using more than one spark plug for each cylinder to increase the heat. The General Motors Corporation, meanwhile, has begun offering an engine in some of its Chevrolet models in which air, which must be mixed with gasoline in the cylinder, is injected through what the company calls a "swirling inlet."

Eudell Jacobsen, a General Motors engineer, said that the air in this system passes into the cylinders through a mechanism that makes it swirl like a tiny cyclone, thus creating turbulence inside the cylinder. Swept up in this air motion, gasoline molecules are ignited faster than if they are allowed to settle in a slow-moving air mixture. He estimated that the fast-burn system increases fuel efficiency by 2 percent.

Another improvement, which several companies are developing but none have introduced in passenger cars, may offer even greater efficiencies in diesel engines. Engineers call this system "direct injection" of fuel into cylinders.

A simple form of this system is already in use in most diesel-powered heavy-duty trucks. In these vehicles, the fuel is injected directly into cylinders, mixed with air and burned. But while diesel offers greater fuel efficiency, one of the disadvantages in trucks is that the fuel, when it cools, creates difficulties in ignition. As a result, commercial truckers often leave their engines running overnight when they are on the road in the winter, a practice that would not be practical for most other motorists.

In order to resolve ignition problems, as well as to overcome other problems with diesel fuel, the diesel engines in cars currently use a pre-chamber, or fuel cell, outside the cylinders from which the flow of fuel is controlled. Many diesel-powered cars are also equipped with electronic systems to heat up their engines on cold mornings.

The pre-chamber, however, complicates the injection system and wastes some fuel in the process. Thus many car companies are at work on fuel injector mechanisms that would be inside the cylinders. These systems, many engineers believe, would also use microprocessor-controlled heating systems to keep them warm. If a workable direct injection system can be brought to market, Mr. Jacobsen said, it could offer a 15 percent *reduction in diesel fuel consumption.*

Advancements in the designs of car bodies, meanwhile, have been getting a major share of the attention of automotive engineers. And in large part, this attention has resulted from the economics of the auto industry: because American companies tend to redesign their car bodies often, incorporating more aerodynamically efficient designs has proved less expensive than changing engines or drive trains.

At the Ford Motor Company, Larry B. Socha, manager of aerodynamics and flexibility engineering, said that aerodynamic improvements in his company's cars had improved fuel efficiency by an average of one mile a gallon since 1977, and that further improvements would add at least another one and a half miles a gallon by 1990.

The improvements at Ford, as well as at several other companies, have consisted largely of streamlining exterior body lines, sharpening the downward tilt of windshields and designing sleek appendages on the auto body, including fenders and rear view mirrors, that slice more efficiently through the air. These changes, although they do little to decrease fuel consumption in city driving, make cars more efficient at highway speeds.

At the same time, Ford says it has advanced aerodynamics a step further with the use of electronics. In its most advanced prototype, Ford is using a microprocessor-controlled system to tilt the front bumper area of the car at highway speeds in a way that resembles the lowering of the front end of a Concorde jet. The electronic system also lowers the entire body of the car from 6 inches off the ground, to 4 inches, when it increases speed to more than 10 miles per hour. And at highway speeds, the back end of the car rises hydraulically so it is higher than the front.

These improvements "are perfectly viable with the technology we have," Mr. Socha said. He added, "It always comes to a question of how much people will pay, but these things could be on the market in the very near future."

creased automobile occupancy and greater use of car pools, van pools, buses, and urban rail systems would diminish our fuel consumption. In van pooling, an employer makes a van available to approximately 10 employees, who use it exclusively for commuting to and from work. Those measures would also reduce congestion on highways, which, in turn, would conserve even more energy by reducing the amount of fuel that is consumed in stop-and-go, rush-hour traffic. Car pooling, van pooling, and the reservation of certain highway lanes for high-occupancy vehicles can reduce substantially the total amount of fuel that is used.

In choosing a mode of transportation, it is important to consider the amount of fuel that is used by a given system to transport one person a specified distance, usually 1 mile. Table 19.3 lists the energy efficiencies for different modes of travel for urban and intercity trips. Surprisingly, new rail systems, such as San Francisco's BART system and Washington D.C.'s METRO, are more inefficient than old rail systems. (The high cost of constructing rail transit systems make them the most heavily subsidized mode of travel per passenger mile. Furthermore, engineers and planners are finding that it is more costly to improve rail transportation than to expand bus service.) And the trend of improving fuel efficiency in automobiles means that, in the long run, intercity travel will be more energy efficient by car than by rail. However, buses will remain the most energy efficient mode of passenger transportation. Unfortunately, it appears that the quest for increased efficiency in terms of convenience and transporting large numbers of persons by mass transport systems is offset by increased crowding.

Industrial Measures

In 1984, industry in the United States used 27.9 Quads of energy. Manufacturing accounted for 75 percent of that consumption; mining, 12 percent; and agriculture and construction, each another 6 percent. In response to higher energy prices, between 1972 and 1981, energy efficiency in the four largest industries improved. Paper production accounted for a 25 percent increase in energy efficiency; petroleum, 20.8 percent; chemicals, 24.2 percent; and steel, 17 percent.

Improvements in energy efficiency fall into four general categories: (1) better housekeeping such as

---

BOX 19.2

AUTO ENERGY EFFICIENCY QUIZ

To assess your understanding of ways to improve the fuel efficiency of automobiles, answer each of these questions true or false.

1. *Automatic transmissions are more efficient than manual transmissions because the car is automatically kept in correct gear at all times.*
2. *Steel-belted radial tires can save as much as 3 percent on gasoline mileage over regular tires.*
3. *You use more fuel letting your car idle for 2 minutes than you do by turning off the engine and restarting when you are ready to drive again.*
4. *It is advisable from a fuel-saving aspect to warm your engine for at least 1 minute each morning before driving your car.*
5. *In order to achieve a quick, low-fuel start each time, it is a good idea to rev up the engine just before turning it off.*
6. *Because of the greater power, it is more efficient to use low gears while driving rather than the higher gears.*
7. *In late model cars, you will save fuel during warm weather by closing windows and turning on the air conditioner when driving over 30 mph.*
8. *Many drivers report they save from $300 to $1000 per year by riding to work or going shopping in a carpool rather than driving alone.*
9. *You can actually use less gasoline by taking a longer route with fewer stops and hills than a shorter route with stops and hills.*
10. *Underinflated tires will cause an auto to consume 5 percent more fuel than properly inflated tires.*

Answers: 1-F, 2-T, 3-T, 4-F, 5-F, 6-F, 7-T, 8-T, 9-T, 10-T.

Excerpted from "Auto Energy Efficiency Quiz," Wisconsin Energy News, February, 1984.

---

shutting off equipment that is not in use; (2) increased efficiency of equipment that has already been installed, such as the installation of computerized boiler controls; (3) shifts to new processes, such as the replacement of ingot casting by continuous casting in steel production; and (4) manufacture of less energy-intensive products (products whose manufacture consumes less energy), such as automobile parts that are made of plastic instead of metal. The first two measures usually have short pay-back periods, and most industries have already implemented them. In contrast, changes in procedures and the manufacture of new products take years to implement. Thus although industries can make considerable gains in energy efficiency during the next two decades by making such changes, enormous capital investments will be required in the process.

Also, industries and utilities can cooperate through a strategy that is called cogeneration. *Cogeneration* is the process whereby two useful forms of energy are produced from the same process. For example, the Anheuser-Busch brewery in St. Louis uses its steam plant to produce both beer and electricity. (See also Figure 13.7.)

The technology for the coproduction (cogeneration) of electricity and process heat is well established. Cogeneration processes can raise the overall efficiency of energy use from approximately 35 percent for electrical production to 60 to 76 percent when the waste heat from electricity production is used in an industrial process. The federal government is actively encouraging cogeneration under the Public Utility Regulatory Policy Act (PURPA) of 1978. That act provides a number of incentives to cogenerators, including federal tax breaks and exemption from the requirement that oil and natural gas not be used in power plants. Overall, the Department of Energy estimates that cogeneration could save 140,000 barrels of oil per day from 1980 to 1990.

Solar Power

Solar power is a very attractive option as an alternative energy source, because solar energy is free and it will last as long as the sun itself—several billion more years. However, not until late 1974 did Congress provide funds through the Solar Heating and Cooling Demonstration Act to encourage large-scale research and development of solar-powered

Table 19.3 Energy use by various urban and intercity travel modes

MODE	URBAN TRAVEL	INTERCITY TRAVEL	
	TOTAL ENERGY* BTUs PER PASSENGER MILE	BTUs PER PASSENGER MILE	BTUs PER SEAT MILE
Automobile (Compact)	—§	1900–2700	1000–1400
Automobile (Average)	14,000‡	2400–7600	1200–2000
Car pool	5400	—†	—†
Van pool	2400	—†	—†
Bus	3100	1100–1800	300–600
Heavy Rail	6500	1800–3700	400–1900
Rail (Commuter)	5000	1400–3200	700–1300
Light Rail (Streetcar)	5100	—†	—†
Aircraft (Wide Body)	—†	4800–6100	2000–4100
Aircraft (Average)	—†	5600–9600	2600–6100

Source: U.S. Department of Energy, *Reducing U.S. Oil Vulnerability*. Washington, D.C.: U.S. Government Printing Office, 1980.
*Includes all energy—construction, maintenance, and operation, including terminals over a lifetime.
†Generally not used for stated purposes.
‡The value for average occupancy is 10,200.
§Not available.

systems for heating, cooling, and electric power generation. Solar energy is manifested in many forms: radiation, organic products of photosynthesis, wind, running water, and ocean currents. A wide range of technologies can be used to convert those forms into electricity and other usable forms of energy. Solar systems are also gaining acceptance. It is estimated that 123,000 residences and commercial establishments in the United States installed solar systems (excluding wood burners) in 1981.

Radiation Collectors

The total amount of solar energy that falls on the United States each year is approximately 600 times greater than the total amount of energy that is consumed during the same period of time. Assuming that 100 percent of the energy that falls on the roof of a small house (93 square meters, or 1000 square feet) during the course of a year could be collected and sold at average rates for electricity, approximately $8000 worth of energy could be harnessed.

But the collection of solar radiation presents some problems because, unlike fossil or nuclear fuels, sunlight is diffuse and its availability varies with the time of day, the season, and cloud cover. In fact, the amount of solar radiation that reaches the ground across the United States annually averages only approximately 190 watts per square meter. Including inefficiency in collection, a square meter of a collection device would yield approximately 0.3 to 0.5 cents worth of electricity per hour. The average daily intensity that reaches the ground across the United States and Canada is presented in Figure 19.3 for December and June. The greatest potential for solar power is in the American Southwest, where cloudiness is minimal and sunlight is intense.

All technologies that are based on the collection of solar radiation are limited by the fact that the source is not continuous. Not only does the sun not shine at night, but its intensity varies with cloud cover as well as seasonally and geographically. For example, in some mid-latitude sites, where fair weather predominates, a higher average amount of annual insolation occurs than in equatorial regions that are plagued by persistent cloudiness. Seasonal changes include the length of day as well as the solar altitude (the angle at which solar radiation strikes a horizontal surface on earth). Seasonal contrasts in length of day are especially troublesome in middle and high latitudes. For example, in eastern Washington state, which is a relatively sunny location, the average amount of solar power that is available annually is 194 watts per square meter, but averages vary monthly from a low of 50 to a high of 343 watts per square meter, which is a sevenfold difference. However, in tropical latitudes, seasonal differentials in intensity are considerably lower.

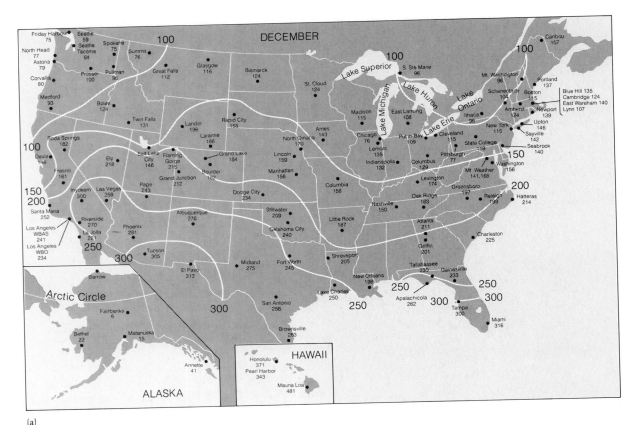

(a)

Figure 19.3 *Mean daily solar radiation at the earth's surface in the United States during (a) December—the month with the lowest levels— and (b) June—the month with the highest levels. (Units are in langleys: 1 langley = 1 calorie per square centimeter, or 3.69 BTUs per square foot.) (Federal Energy Administration.)*

The seasonal variability of solar-radiation intensity that is caused by changes in the solar altitude can be reduced by tilting solar collectors or equipping them with systems that enable the collectors to track the sun, so they remain perpendicular to the sun's beams. The seasonal variation in the amount of sunlight on a horizontal surface and vertical surfaces facing south and 45 degrees east or west of south are compared in Figure 19.4 for a location 43 degrees N latitude. (Note that the radiation data that are shown in Figure 19.3 were measured on a horizontal surface.) In general, sites at higher latitudes will experience greater seasonal variation, while sites further south experience less fluctuation. The advantage of tilted and tracking collectors over collectors that are fixed and horizontal is greatest in cloudy locations at high latitudes. And tilted solar collectors can nearly double the amount of energy that is collected during the winter months in mid-latitude regions.

Solar energy can be collected by passive systems or active systems. *Passive systems* use solar energy directly, without concentrating it or converting it into another form of energy. Examples of passive systems include a greenhouse, a solarium, or a large window that faces south. *Active systems* convert the sun's energy into another form, such as heated air, water, or electricity. Some active systems, including all those that produce electricity, focus (concentrate) the sun's energy before converting it into another form of energy. Active systems generally are attached to, or located next to, the system that uses the energy.

Solar Low-Temperature Systems Active solar systems using solar panels are frequently used for space and water heating in buildings, such as homes, apartments, schools, and businesses—places that can make use of the 65° to 100°C (150° to 212°F) tem-

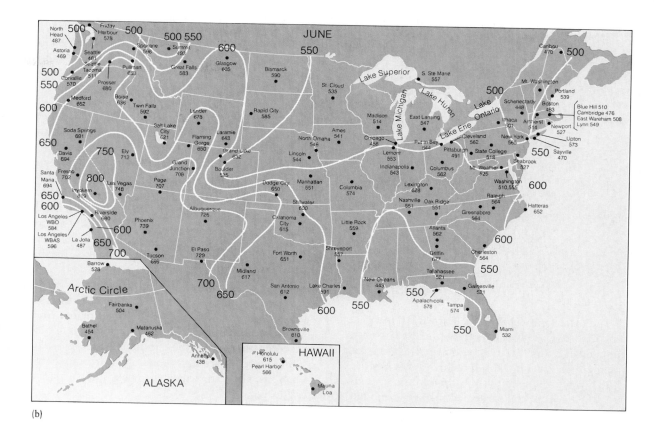

(b)

peratures that can be attained by flat-plate panels. However, because solar panels do not produce temperatures that are high enough to produce steam, those devices are not adequate for most industrial purposes.

Solar panels are framed panels of glass that trap solar energy. In most systems, the sunlight passes through two layers of glass that are separated by an insulating layer of still air before it is absorbed by a blackened metal plate. Then, the absorbed heat energy is transferred from the absorbing plate to a liquid that is in contact with the plate, and it is moved by a pump to wherever it is needed. Some systems use fans to remove heat from solar panels and thus eliminate the need for a liquid. Figure 19.5 shows a solar panel that is in use, and Figure 19.6 is a schematic diagram of a solar heating system. Typically, solar panels capture 30 to 50 percent of the solar energy that reaches the collector.

Solar panels frequently collect more heat during the day than can be used at that time. Thus the excess heat that is collected when the sun is shining brightly is usually stored in insulated water tanks or in compartments that are filled with rocks. In middle and high latitudes, conventional heating systems are needed as a backup, particularly during cold periods or prolonged periods of cloudiness.

Figure 19.4 *Typical seasonal availability of solar radiation on a surface in three different positions. These data are for Madison, Wisconsin, 43° N latitude. (D. R. Schram and D. M. Utizinger,* Passive Solar Heating for the Home. *Madison, Wisc.: University of Wisconsin Cooperative Extension Programs, 1983.)*

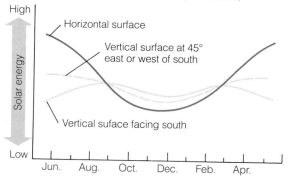

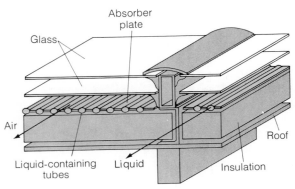

Figure 19.5 *A cutaway view of a solar panel. Heat that is absorbed by the blackened absorber plate is moved from the collector in the form of heated air or heated liquid. (Most actual systems would employ air or liquid, but not both.) A double layer of glass reduces heat loss.*

The most formidable obstacle to widespread and immediate use of solar energy for heating is cost. Although the "fuel" is free, the initial cost of the equipment and its installation is high. For example, in 1984, a typical solar system that would provide 40 to 60 percent of the space heating needs and hot water for a Wisconsin house would cost $15,000 if it were installed or $12,000 for do-it-yourselfers. State and federal tax credits lower the homeowner's cost to $8600 and $5600 respectively. However, because those systems would only provide 40 to 60 percent of a home's heating needs, a conventional heating unit also is required, which increases the capital cost even more.

Although the initial price for solar heating (and cooling) systems is much higher than for conventional systems, over the long term, solar heating systems promise to be more economical than systems that rely on conventional fuels that are becoming increasingly more expensive. But because many variables influence the economic feasibility of solar heating or cooling, people who contemplate such a venture should analyze their particular situations carefully. For example, unless a home is well insulated and has small windows on the north side (where more heat is lost than gained through windows, because the sun does not shine directly into windows that face the north), the investment in solar collectors that would be sufficient to heat the house will be high. Also, the climate of the site, the number of windows in the house, and the size and efficiency

of the system, including its storage component, should be considered. Then, once the total cost of the system has been computed, it should be weighed against the projected costs of more conventional systems. Because of federal and state tax incentives, solar systems for heating hot water are competitive with electrical hot-water heating in most places and may soon be competitive with natural gas (as its price continues to rise).

There is still considerable room for improvement in the reliability, durability, and cost of solar-panel systems that are used for space heating. Most solar-panel manufacturers are small businesses, many of which are not strong financially, and thus many fail. This fact, coupled with the fact that there has been no effort to standardize parts, makes repairs difficult.

Using solar energy to cool buildings is more complex than using it to heat them. Nevertheless, solar cooling is particularly attractive for the long term, since the peak electrical demand in most locations occurs during the summer. Although solar cooling systems are not economically competitive at present, they are expected to become so in the near future.

Another option for heating and cooling buildings is to use architectural design to capture some of the sun's energy during cold months and to protect buildings against excessive heating during the summer. Systems that are designed to be highly energy efficient architecturally are called passive solar systems. Passive solar systems use natural radiation, conduction, and convection of materials to collect and distribute heat in a building. An energy-efficient passive solar building incorporates five essential features: (1) good weatherization (including movable window insulation and plugging of air leaks), (2) large, transparent surfaces that face the south to admit solar radiation, (3) building mass (brick or concrete), which stores thermal energy during daylight, (4) provisions for heat distribution throughout the building, and (5) consideration of wind, shading, and the placement of shrubbery on the building site (see Figure 19.7).

Passive solar buildings can be designed to be heated by direct gain if the sunlight that enters a room is absorbed by the floor and furniture (Figure 19.8a). Since dark colors are the most efficient absorbers of sunlight, but are subject to fading, some passive-solar homes are designed with a concrete or

(a)

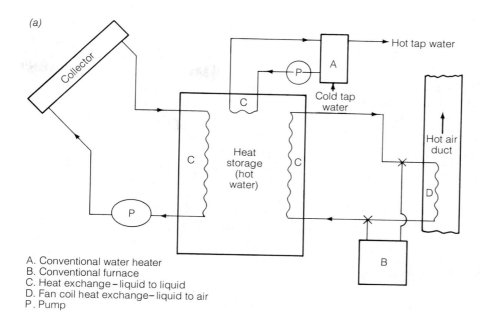

A. Conventional water heater
B. Conventional furnace
C. Heat exchange – liquid to liquid
D. Fan coil heat exchange – liquid to air
P. Pump

(b)

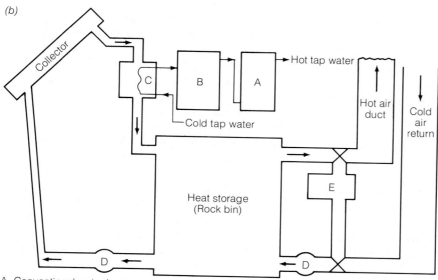

A. Conventional water heater
B. Solar hot water storage
C. Air to liquid heat exchanger
D. Blower
E. Conventional furnace

Figure 19.6 *Two solar heating systems: one that uses water to distribute heat (a), and one that uses air (b). (Federal Energy Administration.)*

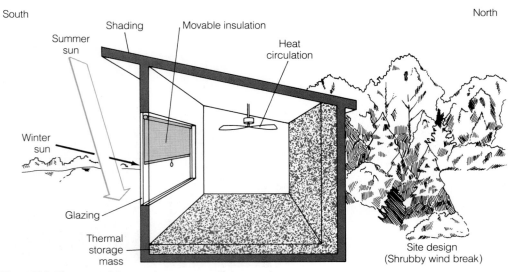

Figure 19.7 *The essential features for an energy-efficient passive solar building.*

brick wall inside the home close to a south-facing window (Figure 19.8*b*). These so-called Trombe walls store heat during the day and then release heat to the cooler surroundings at night. An attached sun space, such as a greenhouse or solarium, can also be attractively designed into a building and used to capture sunlight while being used to grow plants (Figure 19.8*c*). With both the Trombe wall and sun spaces, natural air circulation and fans are used to help circulate the heat throughout the building. Planting deciduous trees so that they shade windows in summer and allow sunlight to enter in winter is one landscaping technique available to reduce energy usage. Unfortunately, from the standpoint of energy conservation, passive solar designs have not been highly accepted, because the public and builders are unwilling to take the risks that are associated with modifying conventional designs for homes and buildings.

Solar High-Temperature Systems For some years now, scientists and engineers have been studying ways of using solar energy to produce temperatures that are higher than 100°C—hot enough to produce steam. In one system—which is called a *power tower system*—computer-controlled mirrors (which are called *heliostats*) track the sun and focus its energy on a single heat-collection point on a tower. In the systems that have been tested, the sunlight is con-

centrated approximately 200 times to produce temperatures as high as 480°C (900°F)—high enough to convert water into high-pressure steam for driving turbine generators. However, a large solar-powered system of this sort would require a large field that is filled with tracking mirrors. In fact, generation of 50 megawatts of electricity (which would be enough for 15,000 homes) would require approximately 1.6 square kilometers (1 square mile) of land, which would make the land unusable for most other purposes. For example, an experimental power tower, which is called *Solar One* (see Figure 19.9), is now operating at Barstow, California, in the Mojave Desert. That system (operated by Southern California Edison) consists of an array of 1818 heliostats that focus sunlight on a central collection tower. The system generates 10 megawatts of electricity, which is enough for 3000 homes. Also, in 1984, Southern California Edison was working on plans for a 100-megawatt solar-to-electricity power plant that would use two collection towers. Those solar-electricity plants are less costly to build and operate than coal-fired plants. Also, central-receiver systems, such as the power tower system, cause minimal pollution, but they have significant land-area requirements. Furthermore, the water that is required for cooling towers (to recondense spent steam) will present problems, because the climate that makes deserts an ideal place to locate such facilities also is responsible for the scarcity of water. However, other

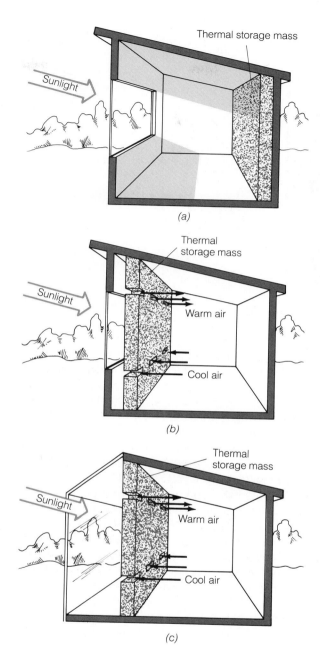

Figure 19.8 *Three design strategies for buildings heated with passive solar systems: (a) direct gain, (b) Trombe wall with natural or mechanical circulation, and (c) sunspace with provision for air circulation. (After D. R. Schram and D. M. Utzinger,* Passive Solar Heating for the Home. *Madison, Wisc.: University of Wisconsin Cooperative Extension Programs, 1983.)*

Figure 19.9 *The Solar I facility shown here, located at Barstow, California, is a 10-megawatt central receiver pilot plant. The facility uses an array of reflecting mirrors focused on a point to produce steam that is used for electricity generation. (Southern California Edison Company.)*

experimental systems, such as the Vanguard I (shown in Figure 19.1), eliminate the need for water cooling.

Solar collectors that concentrate the sun's energy can supply industry's need for temperatures that are higher than 100°C, which is the highest temperature that flat-plate solar collectors can produce. For example, a 1-megawatt power tower system is being used by Atlantic Richfield in California to produce steam for enhanced oil recovery. And other industries use parabolic mirrors in other arrangements. One system, which is shaped in the form of a trough, focuses and concentrates sunlight on tubes that are located along the mirror's focal length. Those collectors produce higher temperatures than flat-plate collectors (see Figure 19.10), which are usable for many industrial processes, including the pumping of irrigation water.

Photovoltaic Systems (Sunlight to Electricity) Solar *photovoltaic* systems have been called the ultimate energy technology, because they convert sunlight directly into electricity. Those systems, which have no moving parts, are fundamentally different from all other means of producing electricity and will have applications in all segments of society. At present, the only important applications of that technology are space satellites and earthbound devices that are located long distances from electricity

645

Figure 19.10 *This experimental trough-style tubular solar collector is used to produce temperatures above 100°C (212°F), which is the temperature that is required for many industrial processes. The system concentrates the sun's rays on a receiver tube that has a liquid inside to transfer energy. (Department of Energy.)*

grids. Although some watches, calculators, and other small electronic devices use this technology, their energy savings are inconsequential.

Photovoltaic systems use sunlight to create a *voltage*, or difference in electrical potential, in an electrical circuit (see Figure 19.11). In response to that voltage, an electric current (electricity) flows through the circuit. However, only certain materials (which are called *semiconductors*) develop the voltage necessary to produce a direct current when they are illuminated. Those materials are made of highly purified silicon, which has been treated with phosphorus and boron. The semiconductors are manufactured in the form of thin wafers that are placed perpendicular to incoming sunlight. Electricity moves through metal contact wires on the front and back sides of the wafer. Then, groups of wafers are wired together to form a photovoltaic panel.

The amount of electricity that can be produced by photovoltaic panels is a function of the intensity of sunlight and the efficiency of conversion (sunlight to electricity). Commercial flat-plate solar photo-

voltaic panels are 10- to 12-percent efficient, while laboratory models reach 15 percent. Since the maximum intensity of the sun is approximately 1200 watts per square meter, if a system has a 10 percent efficiency, the maximum output is 120 watts per square meter—just a little more than enough to run a 100-watt light bulb. In 1983, even in large quantities, the cost for a square meter of a photovoltaic panel was $500. Thus despite their simplicity, solar cells are extremely expensive because the manufacturing process for the purified silicon wafers is costly. A roof-top panel of solar cells that produces enough electricity for an average home (assuming a maximum output of 5000 watts at midday and 13 percent operating efficiency) would cost approximately $50,000. An array of photovoltaic panels is shown in Figure 19.12.

The goal of researchers who design and develop photovoltaic cells is to make them more cost competitive, so several alternatives are being explored. For example, mass-producing thin sheets of solar-grade silicon would decrease production costs, and the final cost for solar-voltaic cells would be lower if they were more efficient. Two types of more complex cells are being investigated (1) cells with lenses that intensify the radiation, which have a higher

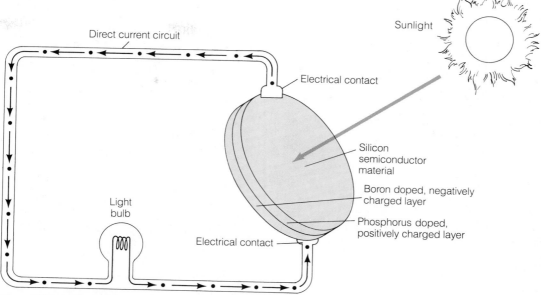

Figure 19.11 *A silicon photovoltaic cell, which converts sunlight directly into electricity. Sunlight frees electrons from the silicon atoms, producing an electric current that moves through the circuit. The process has an efficiency of approximately 15 percent. (Federal Energy Administration.)*

conversion efficiency, and (2) multilayer cells (several wafers sandwiched together), which are more complex but may attain conversion efficiencies that are as high as 30 percent. If those cost-reduction goals are achieved in the late 1980s or early 1990s, photovoltaic systems may begin to replace existing electricity grids that serve commercial and residential markets on a larger scale. It is estimated that by the year 2000, the worldwide installed capacity for photovoltaic electricity will be 5000 to 10,000 megawatts, which is enough electricity for 2 to 5 million Americans. Even though this is a relatively small amount of electricity, those installations will set the stage for rapid growth in the following decades if they are successful. How fast such systems are introduced depends on how fast the low-cost, large-scale manufacturing processes develop. But building large-scale manufacturing plants, which could decrease production costs of photovoltaic cells to approximately one-quarter the present cost, is a "catch-22" situation. Thus they will not be built until an enormous demand exists.

The large-scale use of photovoltaic systems would greatly reduce environmental problems that are associated with energy production. However, because of the relatively low intensity of sunlight, a great deal of space would be required for grids of photovoltaic panels. Also, the manufacture of those panels would produce hazardous wastes that will require treatment, and the disposal of worn-out panels (experts believe that the panels will function for approximately 10 years) will produce a large amount of solid wastes.

The effectiveness of solar panels (as well as that of other solar technologies) is limited by variations in sunshine due to cloudiness and nightfall. In response to that limitation, Peter Glaser, a solar-energy researcher, has drawn up a scheme to launch a 65-square-kilometer (25-square-mile) solar panel into a geosynchronous, or stationary, orbit that is 35,000 kilometers (22,000 miles) above the earth's surface. Such a system would receive about seven times more solar radiation than an earthbound collector of equivalent size, and it would generate 10,000 megawatts of power. That energy would be converted into microwaves and beamed in that form to earth, where it would be reconverted to electricity. An artist's conception of such a satellite panel is shown in Figure 19.13. That enormous system, which

Figure 19.12 *The world's largest array of photovoltaic panels tracks the sun across the sky in Victorville, California (northeast of Los Angeles). The experimental system produces 1 megawatt of electricity—enough for 400 homes.*

would be approximately 10,000 times larger than any satellite that has been launched so far, would cost approximately $2 trillion—a prohibitive figure. Furthermore, a vast array of other technical obstacles remain to be overcome.

Wind Power

For centuries, wind energy has been tapped by windmills to pump water and grind grain and, since the 1920s, to generate electricity on a small scale. Today, scientists are using the methods and materials of space-age technology to design and construct highly efficient windmills that convert the wind's energy into electrical energy. For example, large modern windmills, such as the one shown in Figure 19.14, are capable of generating up to 4 megawatts of power,

which is enough electricity for approximately 1200 homes. Small (5 to 40 kilowatt) and medium-sized (200 kilowatt) wind generators have also been developed.

In Chapter 9, we saw that winds are the result of vertical and horizontal imbalances between the amount of solar radiation absorbed and terrestrial radiation emitted by the earth's surface. And although only approximately 2 percent of the solar energy that reaches the earth is converted to the kinetic energy of wind, that amount of wind constitutes a tremendous quantity of energy. Theoretically, blades that are driven by the wind can convert a maximum of 60 percent of the wind's energy into mechanical energy. However, energy is lost when that mechanical energy is converted through a drive train into electricity, which decreases the practical efficiency. Typically, wind-generating systems extract only approximately 40 percent of the wind's energy and, on the average, transform just over one-third of their rated power output into electricity.

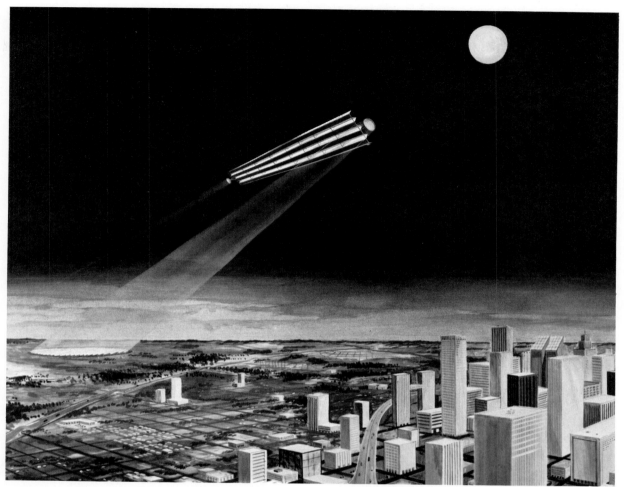

Figure 19.13 *Artist's conception of a satellite solar system. Energy that is collected by the satellite would be beamed to earth by means of microwaves and converted to electricity. The tremendous costs of the project have prevented its development. (NASA.)*

Also, average wind speeds must be at least 16-kilometers (10-miles) per hour before most electricity-generating systems can operate economically.

The power that a windmill can extract from the wind is directly proportional to air density, the area that is swept out by the windmill blade, and the cube of the wind speed (V^3). Thus wind speed is by far the most important consideration in evaluating the potential for wind energy in any region, because even small changes in wind speed translate into important changes in energy generation. For example,

a doubling of wind speed multiplies the available amount of wind energy by a factor of eight. Those factors make windy regions especially desirable as generation sites.

The most formidable obstacle to the development of wind-power potential is the variability of wind. As a result, the electrical output of wind turbines varies, so a wind-power system must contain some type of device to store the energy that is generated during gusty periods for use during lulls. For example, several thousand dollars' worth of storage batteries are required to store enough energy for a typical home if it is to function independently of other energy sources. Thus rather than using expensive collections of storage batteries, most wind-generation systems are connected to existing power grids.

Then, when excess electricity is produced, it is fed into the grid and the producer is credited by the utility company for the amount of electricity that is generated.

Economics suggest that centralized "wind farms" that serve entire communities are preferable to individual household wind turbines. Those "wind farms" would consist of 50 or more giant wind-power generators, each of which would be capable of producing several megawatts of electricity. Indeed, multimegawatt wind systems are currently being developed (Figure 19.14). In 1983, a shaft that supports the 100-ton rotors in one of those large machines cracked, which necessitated redesign and delays in the development of those large wind machines. In 1984, five 2-megawatt, or large, wind machines were operating in the United States. Some experts feel that the intermediate-sized machines (200 kilowatt) will be more economical to design and maintain.

Figure 19.14 The MOD-2 turbines have blades that are 100 meters (300 feet) from tip to tip and rest on towers 60 meters (200 feet) tall. The wind turbine generates 2.5 megawatts of electricity in winds between 14 and 45 miles per hour. (Department of Energy.)

Given current technical and economic limitations, wind power has its greatest immediate potential in those regions where winds are consistent in direction and relatively high (greater than 16 kilometers per hour) in average speed. In North America, such regions include the western high plains, the Pacific Northwest coast, the eastern Great Lakes, the south coast of Texas, and exposed summits and passes in the Rocky Mountains and the Appalachian Mountains. And small wind systems have strong potential in small, isolated communities and on individual farms and ranches. The impact of wind systems on the environment is minimal. Their drawbacks are that they may be somewhat noisy; they may detract from the beauty of the landscape; and they may kill some birds.

The estimated potential of wind energy in the United States, excluding off-shore areas, is equivalent to 10 to 20 Quads of energy per year. (In comparison, the consumption of energy in the United States from all sources in 1984 was 73.7 Quads.) Wind energy projects during the early 1980s had a production goal of 0.17 Quads of electric energy for 1985, which would require 600 large machines and 50,000 small machines. But that goal was not met because of limited government backing and technical problems with large machines. However, despite those setbacks, some experts believe that those problems will be solved and that 3 Quads of energy (3 percent of projected needs) could be contributed by wind power by the year 2000.

Biomass

Green plants are a continually growing source of materials that can be converted to usable forms of energy. Worldwide, some 50 Quads of energy are derived from biomass—mostly in the less-developed countries. Currently, the burning of biomass accounts for approximately 2.2 Quads, or approximately 3 percent, of the total amount of energy that is consumed in the United States. Thus David Pimentel of Cornell University estimates that approximately 0.6 billion metric tons of the 3.2 billion metric tons of biomass that is potentially available each year could be converted to energy. And that amount of energy would represent 11 Quads, or 11 percent, of the total projected amount of energy that will be consumed in the United States in the year 2000.

Wood has always been a desirable biofuel because it is the easiest to harvest, store, and handle. For example, in some of the northeastern states, wood furnishes 50 percent of the energy that is used for space heating. And other states that have large forested areas also use significant quantities of wood. In many small towns in those regions, wood stoves far outnumber oil or gas furnaces. Wood and forest residues (branches) are expected to be the largest source of fuels, as shown in Table 19.4. Wood wastes, bark, and sawdust have been used as an energy resource in the wood-products industry (lumber and paper) for a long time, and use of those wastes continues to increase. And some wood is used by a few utilities. For example, the largest wood-burning utility is a 50-megawatt plant in Burlington, Vermont.

Starches and sugars that are harvested from agricultural crops can be fermented to produce ethanol, which can be used as a fuel or fuel additive (for example, gasohol). Certain plant seeds, such as sunflower seeds, yield substantial amounts of oils that can be used as fuels after processing. And animal and human wastes can be digested anaerobically (see Chapter 8) to produce methane gas. Agricultural and forestry wastes (branches and bark) account for 80 percent of the total amount of biomass that is available, and municipal and industrial refuse and sewage sludge compose the remaining 20 percent.

However, implementing a much larger biomass energy program has serious implications for land use, water and air quality, human safety, and native flora and fauna. For example, the production of biomass that is used for energy will have to compete with other land uses. And the removal of wood, forest residues, and crop residues is expected to intensify erosion problems, which are already serious. Also, the diversion of animal wastes to fuel production would greatly reduce the organic content of soil. Currently, approximately 90 percent of animal wastes are returned to the land.

Harvesting agricultural crops and forests also removes nutrients (Table 19.5). For example, corn is the largest grain crop in the United States and nearly one-half of the nutrients in corn plants are in the residues. Thus farmers would have to use twice as much fertilizer on their corn fields if the residues are harvested for fuel. Although harvesting aquatic crops would help to alleviate eutrophication problems, we do not know how other aquatic organisms would be affected by that practice.

Planting forest "energy plantations" would alter the habitat for deer, birds, and other wildlife. And to obtain maximum energy yield, those plantations would have to be harvested in 2- to 10-year cycles. Furthermore, those monocultures would need applications of fertilizers and pesticides to maintain high production.

The use of biomass resources for energy also poses other problems. For example, serious air-pollution problems are occurring in many wood-burning regions. Studies that have been done by the Department of Energy indicate that wood smoke contains 18 different carcinogenic substances plus six other substances that increase mucus production in the respiratory system. And concern about rising carbon dioxide levels from the combustion of biomass and its contribution to modifying climate is also being watched by climatologists.

Also, the fermentation of grains (primarily corn) to produce ethanol (which is added to gasoline to produce gasohol) raises some serious ethical and social questions. For example, shunting corn grain to ethanol production reduces the supply of food that is available to feed animals in the United States, which causes prices to rise. And in light of the millions of hungry people in the world, many persons believe that it is unwise to burn food to run our automobiles. Moreover, David Pimentel estimates that in the United States, eight times more land (4.2 hectares) would be required to produce ethanol to power the average American automobile than to feed an average American (0.5 hectares). And that esti-

Table 19.4 Projected energy from biomass in the United States for the year 2000 (Quads)

RESOURCES	AUDUBON SOCIETY FORECAST	OFFICE OF TECHNOLOGY ASSESSMENT FORECAST
Wood	5.8	5–10
Grasses and Legumes	—	0–5
Grains	0.3	0–1
Crop Residues, Garbage, Animal Manure	2.8	0.9–1.5
Total	8.9	5.9–17.5

Source: *Side Effects of Renewable Energy Sources.* New York: National Audubon Society, December, 1982.

Table 19.5 Nutrient removal rates per hectare of harvest area

NUTRIENT	CORN*		FOREST (NORTHEAST)†	
	KILOGRAMS	(POUNDS)	KILOGRAMS	(POUNDS)
Nitrogen	224	(494)	600	(1300)
Phosphorus	37	(82)	63	(140)
Potassium	140	(309)	325	(716)
Calcium	6	(13)	800	(1760)

Source: D. Pimentel et al., Environmental and social costs of biomass energy. *Bioscience* 34:89, 1984.
*Corn yield: 125 bushels/acre.
†Forest yield, including residues: 500 metric tons/hectare.

mate assumes that no oil, gas, or coal would be used in the energy-intensive steps of producing alcohol. When one considers the net amount of energy that is derived from ethanol production, it appears to make economic sense only when organic wastes or waste heat from some other process (such as cogeneration) are used to produce it.

The Energy Security Act of 1980 set a goal to produce enough alcohol fuel by 1990 to supply 10 percent of the nation's gasoline consumption (60,000 barrels per day). However, changes in federal plans were made when it was feared that corn prices would rise unnecessarily. Thus the 1990 goal has been decreased to approximately 10,000 barrels per day—a level that is not expected to affect corn prices greatly. Also, because of current surplus grain production, various states have offered subsidies (tax incentives) for ethanol production. Without those subsidies, ethanol production would not be economically feasible, because ethanol that is produced (including subsidies) is calculated to cost $0.83 per liter ($3.14 per gallon), which is approximately three times the cost of a liter of gasoline.

Hydropower

The kinetic energy in falling water is a renewable source of energy that produces one-quarter of the world's electricity. Hydroelectric generating facilities range in size from water-wheel generators in mill pond outlets to Brazil's giant Itaipu Dam, a 12,600-megawatt facility on the Parana River along the Brazil–Paraguay border. Countries that have mountainous regions or rugged topography have the greatest hydropower potential. For example, Nor-

way obtains 99 percent of its electricity from this source, while Canada meets 67 percent of its demands and exports electricity to the United States. In the United States, hydropower accounts for about 14 percent (3.8 Quads) of the electricity that is generated.

Hydropower is normally generated by damming rivers. However, in Egypt, work has begun on a canal that will divert water from the Mediterranean Sea into an 18,000 square kilometer area of the Saharan Desert (the Qattara Depression) that is below sea level. The average efficiency of converting the energy in falling water to electricity is 75 to 80 percent, but in some newer installations, efficiencies are as high as 90 percent. Even though the United States has been building hydroelectric dams for decades, it has tapped only 42 percent of its hydroelectric potential, and worldwide, the total amount of developed potential is just 17 percent. The status of potential hydropower that can be developed in the United States is summarized in Figure 19.15.

The United States is building few dams for several reasons. Dams are expensive to construct and good dam sites are often located in areas with rugged topography, which are usually a great distance from the areas they serve. Thus long transmission lines are needed, which greatly increases both the cost of electrical transmission equipment and the amount of electricity that is lost in transmission. Also, the construction of transmission lines often runs into stiff opposition, because they cut unsightly paths across private and public lands. Furthermore, construction of hydropower dams on many of our rivers is prohibited by federal legislation, such as the Scenic and Wild Rivers Act, or by legislation that is

REGION	SITES WITH DAMS		UNDEVELOPED SITES	
	NUMBER	CAPACITY (MW)	NUMBER	CAPACITY (MW)
A	315	8,730	204	11,682
B	46	1,022	2	24
C	62	1,027	40	1,217
D	19	137	33	466
E	581	1,244	0	0
F	178	2,900	16	1,540
G	100	1,513	83	5,234
H	553	2,432	102	2,394
I	46	465	1	424
Alaska	10	17	49	3,510
Hawaii	7	9	4	29
Puerto Rico	13	35	7	24
Total	1,407	19,531	541	26,544

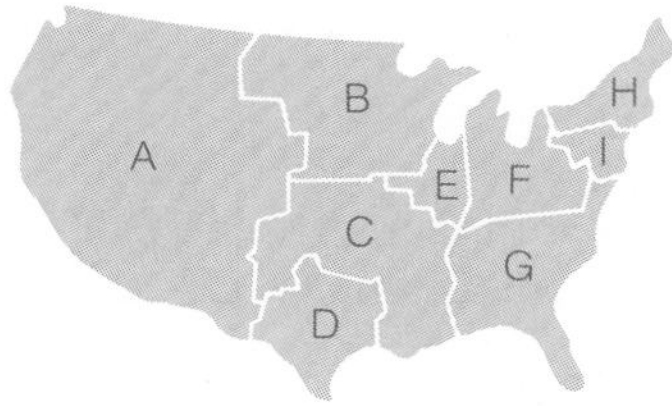

Figure 19.15 *Production of hydropower in the United States by region and the potential for new hydropower projects.* (Electric Power Research Institute Journal, 9:35, 1984.)

enacted by individual states. Thus few new, large hydroelectric facilities will be built in the United States—except perhaps, in Alaska, which has tremendous hydroelectric generating potential. However, worldwide, many large, expensive facilities are being planned.

In the United States, renewed interest is being expressed to put some of the 3000 generating facilities that currently exist at small dams back into service. Estimates of potential power production from small dams range between 6000 and 24,000 megawatts. For example, a 1980 study by the New England River Basin Commission identified 1750 unused small dams in that region, which have the potential to generate 1000 megawatts of power. And the commission estimated that 80 percent of that potential could be achieved if low-cost loans were available and if the electricity could be sold for at least $0.067 per kilowatt hour. (In 1983, the average cost of electricity in the United States was $0.063 per kilowatt hour.)

Also, renovation of hydroelectric facilities makes long-term sense, because once they are built, they provide inflation-resistant sources of energy that can last for centuries if they are maintained properly. Thus to encourage production from small plants, the United States government has provided low-interest loans and tax-depreciation benefits for such efforts. Furthermore, the Public Utilities Regulatory Policies Act of 1978 now requires utilities, in most cases, to buy power from small producers at the same prices they would pay for added coal or nuclear power.

Hydroelectric facilities may also be used to smooth out electrical production from other renewable energy sources by means of *pumped storage systems.* For example, excess power—which is essentially free power—from a wind generator or photovoltaic sysem can be used to pump water uphill into a reservoir. And since water that is impounded behind a dam is stored potential energy, it can be converted back into electric energy when it is needed by letting the water flow back downhill through turbines.

Ocean Thermal Energy Conversion

Another solar-energy conversion system that is currently being studied would exploit temperature differences between the cold bottom waters and the sun-warmed surface waters of tropical oceans. Such temperature differences are known to be as much as 22°C (40°F). And floating electric power plants could be anchored in those regions. In those systems, which are called *ocean thermal energy conversion* (OTEC) *systems,* liquid and gaseous ammonia would be analogous to water and steam in a conventional turbine electric generator. Thus warm surface water would be used to evaporate and pressurize ammonia, and the pressurized gas would drive the turbines. Then, the cold water that could be as deep as 900 meters (3000 feet) below the plant would be brought up to cool—and thereby condense—the ammonia, so that the evaporation–condensation–evaporation cycle could continue. Electric power from such systems could be transmitted to shore through underwater cables or could be used at sea to manufacture energy-intensive products, such as aluminum or ammonia.

However, ocean thermal energy conversion plants are inherently inefficient because of the relatively small temperature differences (compared to conventional power plants) within which they must operate. For example, the estimated operating efficiency for ocean thermal energy conversion systems is approximately 3 percent. And that low efficiency can only be compensated for by using truly enormous heat exchangers (cooling equipment) for condensation of the working gas (ammonia). Thus although the energy is free, the capital outlay for capturing the energy is enormous—so large, in fact, that ocean thermal energy conversion is, at present, an unlikely candidate to furnish future energy.

Tidal Power

Tidal energy is derived from the ceaseless motion of ocean tides. As a result of the gravitational attraction among the earth, moon, and sun, a rhythmic oscillation of ocean water levels occurs. And along coastlines, that pattern appears as a daily cycle of changing water levels, which usually differ by 1 to 10 meters (3 to 30 feet) in height. Thus tidal power facilities (dams) are filled when the tide rises, and the trapped (elevated) water is used to generate power when the tide levels fall. Although tidal energy is reliable, only one major tidal-powered electricity-generating plant is now in operation in the world. That plant, which is situated at the mouth of La Rance River in France, has been operating since 1966 and has a capacity of 160 megawatts of energy (Figure 19.16).

The only sites that are considered to be feasible for tidal power plants are regions where the difference in sea level is great between low and high tides and where a large bay or inlet exists to hold the incoming tide. However, because few such sites exist in the United States, tidal power will never be more than a minor energy source here. And recent

Figure 19.16 *The world's only major tidal-powered electricity-generating plant. This facility spans the estuary at the mouth of La Rance River in France. (Electrite de France, Michel Briguad; French Engineering Bureau.)*

proposals for the construction of tidal power plants on favorable sites (such as Passamaquoddy Bay, on the Maine coast) have been scrapped because of the high capital costs of construction. The environmental effects of operating tidal power plants have not been fully evaluated yet, but they could be considerable, because they would be located in estuaries.

Geothermal Power

Geothermal energy is the heat energy that is generated in the earth's interior. Most of that energy, which is the result of radioactive decay of certain elements that are located deep within the crust, is slowly conducted to the surface. In certain places, hot magma (molten subterranean rock) rises until it is close to the earth's surface and produces hot rock. In areas where groundwater contacts that rock (whose temperature is higher than 100°C) it becomes *su-*

perheated, that is, it remains liquid even though its temperature is higher than the normal boiling point of water. Regions in which that situation occurs are referred to as *wet steam fields*. In other locations—where the heated groundwater is less-tightly confined—it expands and turns into steam. Thus regions in which that situation occurs are called *dry steam fields*. In both types of steam fields, the heated groundwater is separated from the earth's surface by an impermeable rock layer, which is known as *caprock*.

The steam or superheated water is extracted by means of wells that are drilled through the caprock. In a dry steam field, the steam is used directly to drive a turbine, as shown in the left-hand drawing in Figure 19.17. But in a wet steam field, another step is required. When the wells are drilled in a wet steam field, the pressure is released and, as a result, a portion of the superheated well water turns immediately into steam. Then, the pressurized steam is separated from the hot water and is used to drive a turbine. In both systems, depressurized (spent) steam from the turbine is either released into the atmo-

Figure 19.17 *Somewhat different systems are employed in the two types of geothermal fields that are used to generate electric power. (After P. R. Ehrlich, A. H. Ehrlich, and J. P. Holdren,* Ecoscience: Population, Resources, Environment. *New York: W. H. Freeman, 1977.)*

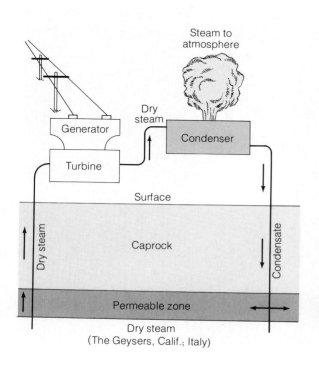

Dry steam
(The Geysers, Calif.; Italy)

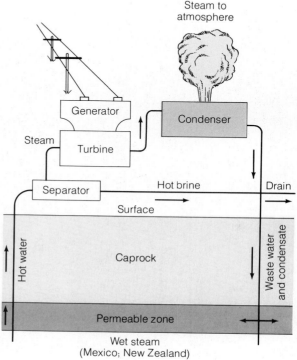

Wet steam
(Mexico; New Zealand)

sphere or condensed. When the spent steam is condensed, the electricity output of the system is increased.

One major problem with geothermal wells is the disposal of wastewater. In many instances, the wastewater contains at least trace amounts of sulfur dioxide, hydrogen sulfide, ammonia, and boron that can cause problems for plants and animals. And, frequently, the wastewater is extremely saline. For example, the yield of some wells in the Imperial Valley of California is only 20 percent steam; the rest is brine that is up to 30 times more saline than ocean water. Thus the saline water must be disposed of either at sea or by pumping it back underground.

However, those problems that exist because of contaminants in the steam can be solved by passing all the superheated water through a heat exchanger, where a hydrocarbon mixture is gasified and used to turn turbines. (Those systems are called *binary cycle systems*.) When that method is used, all the geothermal brines are pumped back into the ground. Binary cycle systems are expected to be more efficient (17 percent versus 12 percent) than *flash cycle systems*, in which superheated water is depressurized, which instantly transforms ("flashes") it into steam. Construction of a 70-megawatt, binary-cycle plant began in 1983 in California's Imperial Valley.

Another serious limitation of geothermal power is that many of the best sites for it are located in scenic areas and, therefore, would meet much public resistance if they were developed. Also, sites in national parks, such as Yellowstone National Park, are prohibited by law from being developed without an act of Congress.

Despite the limitations of geothermal energy, in certain areas, dry steam field operations can compete economically with other sources of electric power. For example, plant construction costs are generally two-thirds to three-fourths lower than for comparable fossil-fuel power plants, and geothermal operations have the added advantage of requiring less maintenance.

The largest known dry steam field is the Geysers facility (shown in Figure 19.18), which is located 145 kilometers (90 miles) north of San Francisco. Power production at the Geysers began in 1960 and has increased each year since then. By 1984, 1453 megawatts of power were being produced, and

Figure 19.18 *Part of the geothermal electricity-generating facility at the Geysers, California, northeast of San Francisco. Pictured are two of the eight units that produce power for Pacific Gas and Electric Company. (Pacific Gas and Electric Company.)*

the total operational plus planned capacity of that field is 2582 megawatts. And by 1992, planned additions will bring that figure up to just over 3000 megawatts, or 0.4 percent, of the projected electrical demand for that year. Although geothermal sources only make a small contribution to the total amount of energy that is generated on a nationwide basis, those contributions can be significant for local areas. Furthermore, geothermal sources can be used as a source of heat for buildings and commerce, such as for heating greenhouses in cold regions.

Nuclear Power

Breeder Fission Reactors

In Chapter 18, we saw that plutonium-239 (^{239}Pu) is one of the products that is formed in nuclear fission reactors. That readily fissionable material is formed from the nonfissionable isotope of uranium—uranium-238 (^{238}U) in the reactor core of a nuclear power plant. Because conventional reactors produce limited amounts of plutonium-239, three reactors are required to produce enough of it to fuel another reactor. However, by increasing the amount of plutonium or uranium-235 in the fuel and by using faster neutrons (the speed of neutrons in a conventional reactor is slowed with water to promote fission), a reactor can be made to create, or "breed," more plutonium-239 than it consumes. Therefore, reactors

that can breed plutonium-239 are called *breeder fission reactors*.

The sequence of nuclear reactions that is required for a breeder fission reactor to produce heat is shown in Figure 19.19. The continuous flow of energy that is released by the nuclear reactions is carried away from the reactor core by hot liquid (molten) sodium, as shown in Figure 19.20. And that molten sodium generates the steam that drives electric turbine generators.

The principal advantage of breeder reactors is that they can create fissionable fuel (plutonium-239) from the more abundant isotope of uranium (uranium-238, which is 99.3 percent of the uranium in natural ores), thus obtaining considerably more energy from every kilogram of uranium fuel. (Conventional reactors fission uranium-235 primarily, which constitutes only 0.7 percent of the total amount of uranium that is mined.) However, breeder reactors can utilize both uranium-235 and uranium-238—and mined uranium-238 is plentiful for breeder reactors because it is a by-product of nuclear-weapons pro-

Figure 19.19 *A breeder reactor converts the much more abundant but unfissionable uranium-238 into fissionable plutonium-239. Basically, the sequence of reactions in the core of a breeder reactor consists of three processes: (1) fission of the plutonium-239, which releases energy and two to three neutrons per fission; (2) a breeding step, in which plutonium-239 is formed from uranium-238; and (3) a control system that absorbs the excess neutrons so the reactor does not overheat.*

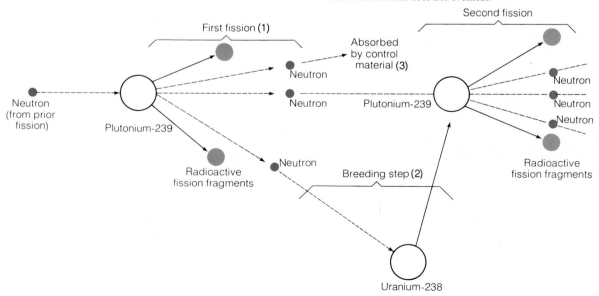

duction. In fact, current supplies of uranium-238 could fuel breeder reactors for 400, 1000-megawatt electricity-generating plants (the approximate number of plants that are necessary to meet our current rate of electrical consumption) for approximately 500 years. Thus if breeder technology were implemented, energy reserves in the United States would increase dramatically. As shown in Figure 19.21, even the energy in our tremendous coal reserves is small compared to the energy that could be released from existing supplies of uranium-238.

Breeder reactors pose the same safety problems that are created by conventional fission reactors as well as two additional, potential dangers. First, breeder reactors use plutonium-239, which is a highly toxic, alpha-emitting radioactive substance with a half-life of 24,000 years. Thus plutonium that could be released accidentally into the atmosphere could enter the body by inhalation, creating a risk of lung cancer. Plutonium also concentrates in liver and bone and is a potent carcinogen. And second, breeder reactors require nuclear weapons grade materials as fuel. Those fuels are recovered by reprocessing spent nuclear fuel to recover high-grade plutonium-239. It is only at reprocessing centers—and nowhere else in the nuclear fuel cycle—that the grade of plutonium that is required for nuclear weapons would be produced. At all other steps in the nuclear fuel cycle, the plutonium is mixed with other highly radioactive materials and would be extremely hazardous for potential bomb makers to handle and reprocess. Thus although stringent security measures can be taken to prevent theft, a chance always exists that terrorists could obtain plutonium-239 from a reprocessing center to construct a nuclear bomb and hold part of society hostage by threatening to detonate it. Finally, breeder reactors pose nuclear waste-disposal problems that are similar to those of conventional nuclear power plants (see Chapter 13).

The development of breeder reactors in this country was stalled by Congress in 1983. In 1972, Congress had authorized the construction of a 350-megawatt demonstration plant at a site near Clinch

Figure 19.20 *A breeder reactor, showing its heat-exchange system. Liquid sodium metal is used rather than water to transport the heat away from the reactor core, because liquid sodium does not slow the neutrons down as much as water. Fast-moving neutrons are necessary to promote the breeding step. (Atomic Industrial Forum.)*

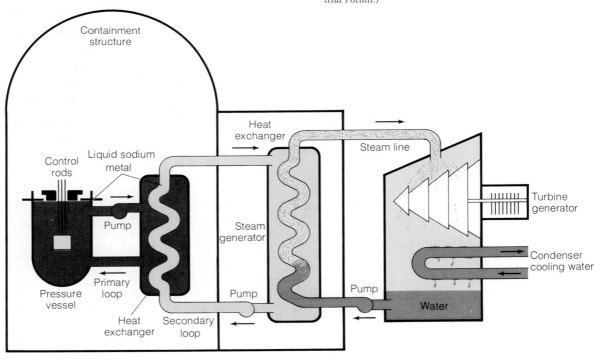

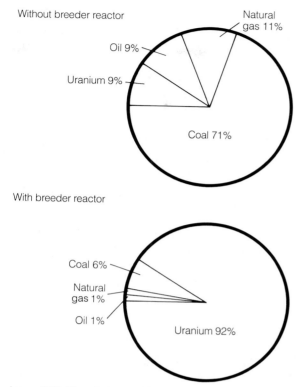

Without breeder reactor

Natural gas 11%

Oil 9%

Uranium 9%

Coal 71%

With breeder reactor

Coal 6%

Natural gas 1%

Oil 1%

Uranium 92%

Figure 19.21 *The relative size of energy reserves in the United States, with and without breeder reactors. The energy value of uranium resources increases 60-fold if it is used in a breeder reactor.* (The Electric Power Sourcebook, 1983. U.S. Committee for Energy Awareness, Washington, D.C.)

River, Tennessee. Design of the plant began soon thereafter, but in 1977, delays occurred because of the results of studies that were done on reprocessing spent fuel. As a result of those studies, attempts to define appropriate licensing procedures delayed the project further. Those delays held the project back an estimated 5 years. By 1983, $1.3 billion had been spent on research and development on the project and the total estimated costs for the breeder reactor had soared to $7.5 billion. Furthermore, parts of the design for the project were considered to be obsolete because of other research developments that had been made in breeder technology. Those costs and doubt that electricity needed to be generated by breeder reactors immediately led Congress to cancel funds for the project in 1983. The Department of Energy's best guess is that the generation of electricity from breeder reactors will not be economical for at least another 40 years. However, European countries, such as France, continue to develop breeder reactors.

Fusion Reactors

Another type of nuclear reaction that releases enormous quantities of energy—much more than such chemical reactions as the burning of fossil fuels produce—is *nuclear fusion*. In that process, the nuclei of light elements (elements of low atomic mass)—principally isotopes of hydrogen—are fused together to form heavier elements and, in the process, tremendous amounts of energy are released. For example, the energy-releasing reactions that occur within the sun and other stars are nuclear-fusion processes.

Physicists are trying to duplicate those processes on a small scale to learn how to achieve fusion. For example, a fusion reactor must maintain temperatures of 50- to 100-million degrees Celsius if the reactor's fuel—the heavy hydrogen isotopes, deuterium and tritium—are to slam into each other with enough force to overcome the natural forces that keep them separate. (The fusion reaction is shown as reaction 1 in Figure 19.22.) Since no materials can withstand those superhot temperatures, scientists must contain the fuel gases (actually, plasmas) in doughnut shaped "magnetic bottles" or use other very sophisticated technology to achieve those temperatures. Much research effort in the United States has gone into the design and building of such reactors (see Figure 19.23).

However, so far, experimental fusion reactors have failed to reach the 50- to 100-million °C temperature that is required, and those magnetic fusion reactors have only lasted for a few hundredths of a second. Scientists at the Princeton Plasma Physics Laboratory hope to achieve the required temperature by 1985. And by 1986, researchers hope to perform the first break-even experiment. The goal of that experiment is to demonstrate that it is possible to get as much energy out of a fusion reactor as goes into it.

The complexity and costs of developing fusion reactors is very high. Thus several different conceptual designs must be tested to determine which design is the simplest and least costly to develop. As a result, those persons or groups who are designing the experiments must compete for research funds,

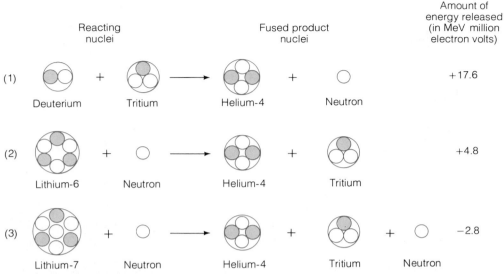

Figure 19.22 *The fusing of deuterium and tritium, which is a promising energy-releasing nuclear fusion reaction (reaction 1). Deuterium is abundant in seawater. The tritium that is required for reaction 1 would be produced from two common isotopes of lithium (lithium-6 and lithium-7) by means of reactions 2 and 3. A positive value for release of energy means that the reaction produces energy. Reaction 3 absorbs energy rather than releasing it, but its product, tritium, releases much energy in a subsequent step (reaction 1). Thus the net energy balance for reactions 1 and 3 combined is +14.8 MeV.*

which were in the range of $400 million to $500 million per year during the early 1980s.

If the technology of nuclear fusion is perfected, fusion reactors promise to offer three significant advantages over both fission and fossil-fuel power-generating systems. First, the ocean contains vast amounts of deuterium—and the fusion of the deuterium in just 1 cubic kilometer of seawater would yield an amount of energy that is nearly equivalent to that contained in all the earth's oil reserves. Second, the economic and environmental costs of extracting deuterium from seawater would be minimal. And third, fusion does not pose the hazards that nuclear fission does, because fusion-reaction products are inert (although some of the equipment that is associated with fusion reactors would become radioactive). However, the construction of fusion reactors would require the use of exotic (and, therefore, scarce) materials, and that factor alone could limit the implementation of nuclear fusion tech-

nology, even if it were perfected. The next decade of research on fusion technology will tell us much about its viability as a source of energy.

Choosing a Future Energy Strategy

Our discussion to this point has shown us that many potential alternatives are available to meet our future energy demands but, at this time, no single one of those alternatives stands out clearly as being the best.

The changeover to a different mix of basic energy resources and new energy technology will not occur overnight. It took decades to build today's existing energy-delivery systems, and many of those systems will be in operation for many more decades. Even if our energy alternatives were narrowed down to one "best choice" tomorrow, it would be prohibitively expensive to scrap most of our existing electric-generating plants and our petroleum-based transportation system. Thus it will take several decades to replace energy-delivery systems with new energy technology.

All methods of providing energy services—conservation, conventional sources (such as coal), and renewable sources (such as wind power)—have one or more drawbacks (for example, threats to health and safety, disruption of ecosystems, disruption of

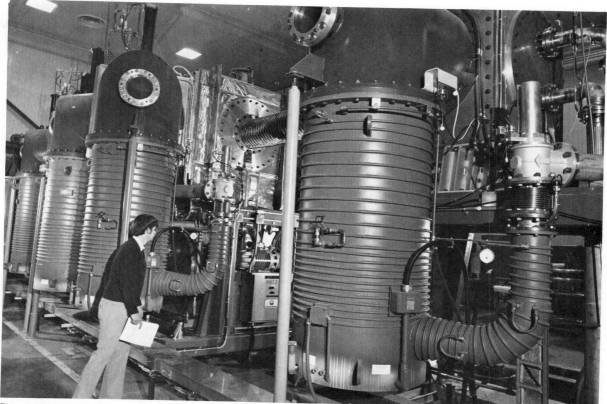

Figure 19.23 *A portion of an experimental nuclear fusion apparatus. Fusion's potential as a price-competitive, large-scale energy producer is at least 30 years away. (Department of Energy.)*

water supply, noise pollution, visual pollution, or psychological stress that is caused by the fear of the unknown effect of a technology—a particular problem with nuclear power). However, the negative impacts of energy technologies are much more difficult to document than their economic benefits.

We can gain an idea of the potential impacts of various energy alternatives by examining the amount of land, steel, concrete, and other resources that are required to produce an equivalent amount of energy. Table 19.6 compares the amount of those resources that are required for various energy alternatives compared to the nuclear fuel cycle, which produces the greatest amount of energy per unit of land area. The most striking feature in this comparison is that renewable resources, such as solar energy, have significantly larger requirements for land and struc-

Table 19.6 Amounts of resources required to produce an equivalent amount of energy using nuclear electricity generation as basis of comparison

	LAND	STEEL	CONCRETE
Electricity*			
Nuclear	1.0	1.0	1.0
Coal	1.7	1.0	0.5
Gas	1.1	1.0	0.5
Oil	1.4	1.0	0.5
Solar Energy			
Low Temperature	10	—	—
Central Receiver			
(Large Scale)‡	19	270	200
Photovoltaic	20	10	30†
Biomass	4500	1	0.5
Wind	110	240	95
Hydroelectric	100	60	570
Geothermal	30	—	—
Tidal	140	—§	—§

*Includes extraction, transportation, and conversion of fuel plus electricity distribution.
†Value is for materials (aluminum, copper, and glass) other than concrete.
‡Approximately 150 times as much other material required in addition.
§Estimated to be similar to hydroelectric power

tural materials than do the conventional methods of generating electricity. Also, energy systems that require large amounts of construction materials or fuel have the highest occupational health and safety risks associated with them on an energy-equivalent basis. They also have a high potential for negative impacts on the environment. The Audubon Society, which is an environmentally conscious organization, has recognized those risks. The high-risk side effects that they have identified for some of the alternatives to fossil fuels are identified in Table 19.7. And one of those alternatives—the harvesting of for-

est biomass (wood)—is, relatively speaking, very hazardous. On an energy-equivalent basis, it causes 14 times more injuries and illnesses than underground coal mining and 28 times more than oil and gas extraction. The basic reason for this high level of risk is the greater amount of labor that is required to harvest a unit of energy and the high rate of accidents that occurs as a result of logging activities. However, it may be possible to reduce those accident rates by making work procedures safer.

Threats that are posed by various technologies to the health and safety of the general public are

Table 19.7 High-risk side effects of renewable energy sources

ENERGY SOURCE	MATERIALS USE	ACCIDENTS AND ILLNESS	WATER USE	LAND USE	AIR POLLUTION	OTHER SIDE EFFECTS
Biomass	Increased Use of Chemicals to Make Fertilizers Materials Are Needed for Construction of Conversion facilities	Accident Rates Are High for Workers in Agriculture and Logging	Demand for Water Would Increase Runoff of Soil and Chemicals Could Be Large	Pressure Would Significantly Increase to Convert Cropland and Protected Lands Erosion Rates Could Increase if Proper Land Management Is Not Followed	Emissions from Boilers and Stills Could Be Significant—Especially if Using Nonrenewable Fuels Local Pollution from Wood Burning. Wind Erosion	Potential Loss of Some Critical Natural Habitats Possible Loss of Some Wildlife and Plant Species
Large-Scale Photovoltaic and Solar Thermal Energy Conversion Systems	Further Demand on Some Constituents of Photovoltaic Cells that Are in Short Supply Increased Use of Concrete, Steel, etc., in the Construction of Collector Farms	High Accident Rates for Construction of Large Tall Towers Worker Exposure to Toxics in Manufacture and Installation of Cells and Working Fluid System	Increased Water Use if Conversion System Has Water Cooling Improper Disposal of Cleaning Fluids, Working Fluids, and Used Cells Could Contaminate Water Supplies	Land Areas Are Required for Collector Farms		Military Applications of Solar Photovoltaic Systems Birds and Insect Kills Could Result from Improper Siting in Flight Paths Disruption of Natural Habitats Possible Local Climate Changes
Hydropower				Prime Natural Areas Might Be Flooded for Reservoirs		Radical Disruptions of Habitats by New Reservoirs Loss of natural free-flowing rivers
Wind Energy	Increased Use of Metals and Concrete for Construction of Wind Energy-Conversion Systems	Hazards from Working on Tall Structures		Some Land Required for Wind Energy-Conversion System Arrays and Possibly for New Transmission Lines (Multiple Uses of Land Are Possible Though)		Wind Energy-Conversion Arrays May Be Visually Unpleasing. Noise May Be Disturbing. Possible Radio and TV Interference

much harder to quantify than threats that are posed by occupational hazards, because it is very difficult to attribute illnesses and deaths among the general public to specific causes. Moreover, some of the new technologies undoubtedly will pose hazards that have yet to be identified. For example, synthetic fuel production (gas and liquids) from biomass or electricity from photovoltaic cells may well have high occupational and public-health impacts, because they use toxic materials. Also, some unforeseen environmental affects probably will become apparent once those new systems are used on a wider scale. For example,

more than a half century passed before the threat of acid rain received serious scientific scrutiny.

The movement toward the increased use of energy (conventional or renewable) requires an increased awareness of land use. For example, in general, each square meter of land that is used for residential, commercial, or industrial purposes requires one to two times the same amount of area elsewhere for the mines, generating plants, and so on, which are needed to support a conventional nuclear or coal energy cycle. Table 19.6 shows that renewable energy sources require even more land

Table 19.7 High-risk side effects of renewable energy sources (continued)

ENERGY SOURCE	MATERIALS USE	ACCIDENTS AND ILLNESS	WATER USE	LAND USE	AIR POLLUTION	OTHER SIDE EFFECTS
Geothermal Energy		Hazards from Hot, High-Pressure Fluids	Chemicals May Get into Groundwater—Especially Geopressurized Brine Systems	Possible Geologic Effects Might Lead to Subsidence or Earthquakes—Especially for Geopressurized Brines	Possible Emissions of Significant Quantities of Toxic Substances. Some Radioactive Substances Would Be Released	Possible Disruption of Scenic Areas
Ocean Thermal Energy Conversion			Possible Release of Chemicals and Waste Products into Water	Some Coastal Areas Would Be Used for Support Facilities	Possible Release of CO_2 from Deep Ocean Water Brought to the Surface	Disruption of Local Marine Ecosystem
Residential Photovoltaic	Increased Demand for Scarce Materials Needed for Manufacture of Photovoltaic Cells	Fire Hazards and Overheating of Rooftop Collectors Release of Toxics in Fires and Other Accidents	Possible Contamination from Disposal of Used Cells and Working Fluids			
Direct Solar Heating and Cooling	Construction Materials Needed (e.g., Glass, Metals, Concrete)	Accidents from Handling Working Fluids for Active Systems Glass Breakage Hazards	Disposal of Working Fluids Could Cause Water Pollution	Land Area Required for Storage Ponds		Collectors and Solar Architecture May Be Unpleasing to Some People
Conservation	Materials and Energy Are Needed for Manufacture of Insulation and Energy-Efficient Appliances	Possible Release of Toxics from Insulation During Fire Possible Long-term Effects from Radon Buildup Greater Severity of Accidents due to Increase in Small Cars			Possibility of Indoor Air Pollution if Houses Are too Tight	

Source: Larry Medsker, *Side Effects of Renewable Energy Sources*. New York: National Audubon Society, December, 1982.

than conventional sources. And, for many alternative sources of energy, the land would need to be located in urban areas, which already have high densities. For example, new supermarkets use approximately 50 percent of the energy they consume for refrigeration; 10 percent, for heating; and the remainder, for lighting, ventilation, and baking. However, even if the cost were not a factor, if those energy requirements were provided by adjacent solar collectors, the required panels would fill an area that would be five times the size of the store itself. And that estimate is for a supermarket that would be located in the Southwest, the most favorable solar location in the United States.

Other, indirect factors also must be considered in choosing an energy source. And, in many situations, conventional energy choices seem to pose the greatest risks. For example, burning fossil fuels adds to the carbon dioxide in the atmosphere and contributes to acid rain, which, in turn, may reduce the supplies and increase the costs of goods and services obtained from ecosystems over the long term. Nuclear fuels increase the risk of nuclear terrorism because of the accessibility of fissionable nuclear materials. And the struggle by nations for access to oil supplies increases the chance of war. In contrast, the renewable energy resources lack many of those indirect threats, and they place their negative impacts more directly on the persons who receive the benefit of the energy services. Those advantages could, in the long run, be the factors that tip the scales in favor of renewable energy sources.

However, in some cases, it may not be possible for us to use renewable resources as a substitute for fossil fuels. For example, the production of iron and steel requires the use of metallurgical grades of coal. And the most reasonable substitute for coal would be charcoal products from wood. But to meet current world demands for iron would require 4 million square kilometers of land, which would be devoted solely to wood production. That amount of land is equivalent to 20 percent of the world's forest production, and if agricultural land were used instead, it would require 30 percent of the world's farmland.

Conclusions

The Department of Energy has made estimates of the relative contribution of the various energy sources for the year 2000. Those estimates, given in Table 19.8, are for a midrange forecast. Earlier government forecasts have underestimated the potential of conservation and the ability of the consumer to adapt and respond to higher energy prices. But if those estimates are reasonably accurate, they show that conventional resources will still provide most of our energy needs in the year 2000. This fact alone means that we must take the necessary steps to minimize the negative environmental effects of these fuel cycles.

Nobody really knows the exact mix of fossil fuels and nonrenewable fuels that will be used in the future. Certainly oil and natural gas will be in short supply and will be replaced to some extent by coal

Table 19.8 Energy-use projections for year 2000 by source and use-section (Quads)

	RESIDENTIAL	COMMERCIAL	INDUSTRIAL	TRANSPORTATION	ELECTRICITY	TOTAL
Coal	0.1	0.1	5.0	neg.	23.0	28.2
Biomass	1.0	0.5	1.9	0.2	0.3	3.9
Natural Gas	5.1	2.9	8.6	0.6	1.8	18.9
Oil	1.3	0.9	8.2	16.8	1.3	28.4
Hydro	—	—	neg.	—	3.7	3.8
Nuclear	—	—	—	—	7.9	7.9
Solar, Wind, and Geothermal	0.3	0.3	0.5	—	1.2	2.3
Total Direct Consumption	7.8	4.7	24.2	17.6	39.2	93.4
Electricity (Point of Use)	3.5	3.3	4.7	0.1		11.6

Source: U.S. Department of Energy, Office of Policy and Analysis, 1983.

and renewable resources. Furthermore, we must gain more experience with renewable energy sources, so we can learn to minimize their negative effects on the environment. Which renewable energy sources will replace these fossil fuels will depend on which alternatives are available and which ones make the most economic and environmental sense for a specific geographical region.

Summary Statements

Existing fossil-fuel reserves can be extended through conservation. Energy conservation requires that we evaluate all our energy-consuming practices and learn which ones can be accomplished by using less energy.

Energy conservation in the residential sector means better weatherization of buildings, decreasing heat losses from hot-water heaters, and purchasing energy-efficient appliances.

Energy conservation in the transportation sector involves the purchase of smaller, fuel-efficient automobiles, using car pools or van pools to get to work, and careful planning of trips.

Future energy conservation measures in industry will proceed more slowly because it requires a great deal of capital to convert to more energy-efficient technologies. Industries can make the operation of power plants and industrial plants more energy efficient overall by using cogeneration—by making use of the waste heat or steam from an electricity-generating utility.

Solar energy is abundant, but it is also diffuse and expensive to collect. Both passive and active systems can be used to cut fossil-fuel use. Solar collectors are practical in many geographical areas for space heating and water heating. Large arrays of solar collectors can produce the high-temperature steam that is necessary to generate electricity. A 100-megawatt system is the largest system that has been designed so far.

Photovoltaic cells convert the sun's energy directly into electricity. Those systems are still very expensive, but mass-production technology promises to lower the costs substantially.

Growing plant materials (biofuels) as energy sources is possible, but tremendous amounts of land are needed to meet today's energy demands. Concern also exists that raising "energy crops" would divert land away from food production for a hungry world.

The generation of electric energy by wind has been feasible for decades. However, modern generators are still expensive and are best placed in particularly windy locations.

Electricity also is generated by hydroelectric dams. And some small dams are being returned to service because of the higher prices for electricity and government incentives. The United States has used approximately 42 percent of its favorable sites for dam construction. Construction of new dams is highly controversial because they cause social and environmental disruptions. Dams are also expensive to build.

Ocean thermal energy conversion (OTEC) exploits the temperature differences between warm surface ocean waters and cold bottom waters to transmit energy to floating electric power plants. Their inherent low efficiencies increase their size and cost.

Tidal power is possible only in a few locations, where tidal oscillations are particularly high. However, capital costs are prohibitive and environmental effects are poorly documented.

In some regions, hot rocks and hot subsurface water can be used to generate steam that can be employed to generate electricity. The western states hold the greatest potential. Some geothermal fields are difficult to extract energy from, and geothermal sites produce air and water pollutants that must be dealt with.

Breeder reactors would extend uranium resources by nearly 60 times. Major obstacles to implementing breeder reactors have been their high cost and legislator's unwillingness to fund a high-cost demonstration project at Clinch River, Tennessee. Questions of national security still remain concerning the handling of plutonium that can be used for nuclear weapons, which is also required for the operation of breeder reactors.

Nuclear fusion of light elements yields enormous quantities of energy. But the development of the technology that will permit the sustained generation of electricity by nuclear fusion remains questionable.

Government policies, especially tax incentives, play an important role in how fast new technologies can be demonstrated, and if successful, put into service.

The mix of fossil fuels, nuclear fuels, and renewable resources will change little by the year 2000. After that, the mix of coal and renewable resources used to begin to replace dwindling oil and natural gas supplies is less certain.

Questions and Projects

1. In your own words, write definitions for each of the terms that are italicized in this chapter. Compare your definitions with those in the glossary.

2. List six ways to conserve energy in your home or apartment. Rank them in order of pay-back time, size of savings, and the likelihood of implementation. An energy audit by your local utility will provide much of this information.

3. Cite reasons why car pools and van pools are not used more widely. Suggest some ways to try to overcome these barriers to their use.

4. Try to locate a cogeneration facility in your community, and determine how much energy is conserved by using cogeneration rather than having two single plants. What economic and regulatory incentives were important in the decision to use cogeneration?

5. Locate a solar home in your community and obtain some data on its performance. Report on changes that you would suggest for the building.

6. The government has given significant tax incentives for the construction of solar-energy devices. Do you think the solar industry should continue to be singled out as an energy industry that receives special tax benefits?

7. Using Figure 19.3, determine the amount of solar radiation that is available in your region. How does the amount of radiation that is available vary throughout the year. Is solar energy a viable alternative energy source in your region?

8. Identify the advantages and disadvantages of harvesting the sun's energy using the following methods: solar collectors, solar cells, windmills, hydroelectric dams, and biofuels. Which systems would be most useful for small applications (homes)? Which systems would be most useful for large-scale applications?

9. What use is made of organic wastes (see Chapter 13) to supplement energy sources in your region. Do not forget about industrial uses of organic wastes.

10. Do potential sites for the generation of hydropower exist in your region? If so, speculate on which groups would favor its development and which groups would oppose its development.

11. Are there any small, unused hydropower facilities in your region? Are there plans to restore those facilities? If so, find out what costs would be incurred.

12. Draw up a plan to provide electricity to your community. What mix of systems (hydro, nuclear, coal, solar, tidal, geothermal, or wind) would you use to meet those demands. Where would the facilities be located?

13. Cite the advantages and disadvantages of breeder reactors.

14. Distinguish between nuclear fusion and nuclear fission.

15. Most solar technologies require large amounts of land that are located near the point of use. What planning initiatives should be used to allow for greater reliance on solar energy?

16. Suggest changes that you could make to decrease the amount of fuel that you use for your personal transportation needs. Which ones would you be willing to initiate?

17. Suggest some energy-conservation ideas that are contrary to commonly held ideas.

18. Peak electrical demand in most communities occurs at about 4 P.M. each day. On an annual basis, peak demand occurs during the summer. What are the advantages of reducing peak demand? Suggest some policies that would lower the peak demand.

19. How can energy be conserved in the purchase and operation of an automobile?

20. More efficient energy utilization has occurred during the last decade as the result of substantially higher energy prices. If another rapid rise in energy prices were to occur, would it be possible to decrease the use of energy as fast as it was a decade ago?

Selected Readings

ANDERSON, B., AND WELLS, M.: *Passive Solar Energy: The Homeowners Guide to Natural Heating and Cooling.* Andover, Mass.: Brick House Publishing Co., 1981. A lay person's "how to" manual.

CARTER, J., ED.: *Solarizing Your Present Home.* Emmaus, Pa.: Rodale Press, 1982. A survey of ways to retrofit a home with active and passive solar systems.

CONN, R. W.: "The Engineering of Magnetic Fusion Reactors." *Scientific American* 249(4):60 (October), 1983. An explanation of the science and technology of magnetic fusion.

DEUDNEY, D.: *Rivers of Energy: The Hydropower Potential.* Washington, D.C.: Worldwatch Institute, Worldwatch Paper 44, 1981. A survey of the potential of hydropower on a worldwide basis.

DOUCETTE, D. B.: "The Embattled Breeder Reactor." *High Technology* 3:50 (July), 1983. A survey of the history of the development of the breeder reactor.

FLAVEN, C.: *Electricity from Sunlight: The Future of Photovoltaics.* Washington, D.C.: Worldwatch Institute, Worldwatch Paper 52, 1982. A discussion of the technology and economics of photovoltaic cells.

GLEICK, P. H., AND HOLDREN, J. P.: Assessing environmental risks of energy. *American Journal of Public Health* 71:1046, 1981. A discussion of the environmental impacts of conventional and alternate energy sources.

GRAY, C. L., AND VON HIPPEL, F.: "The Fuel Economy of Light Vehicles." *Scientific American* 244(5):48 (May), 1981. A survey of the strategies available for improving fuel economy.

HOLLANDER, E.: Federal study evaluates new hydro potential. *Electric Power Research Institute Journal* 9:34, 1984. A survey of the potential of small dams for producing electricity.

JOHANSSON, T. B., STEEN, P., BOGREN, E., AND FREDRIKSSON, R.: Sweden beyond oil: The efficient use of energy. *Science* 219:355, 1983. A discussion of how Sweden has attempted to become more energy efficient.

KERR, R. A.: Extracting geothermal energy can be hard. *Science* 218:668, 1982. A discussion of technical problems encountered in developing geothermal sources.

KREITH, F., AND MEYER, R. R.: Large-scale use of solar energy with central receivers. *American Scientist* 71:598, 1983. A survey of the development of the "power tower" solar to electricity systems.

MEDSKER, L.: *Side Effects of Renewable Energy Sources*. New York: National Audubon Society, 1982. Review of the possible environmental consequences of large-scale implementation of renewable energy technologies.

PIMENTEL, D., ET AL.: Environmental and social costs of biomass energy. *Bioscience* 34:90, 1984. A discussion of the potential of biomass as an energy source and the negative consequences of this technology.

RUEDISILI, L. C., AND FIREBAUGH, M. W., EDS.: *Perspectives on Energy*, 3rd ed. New York: Oxford University Press, 1982. The last two sections of this textbook deal with alternative energy sources, conservation, and energy policies, with articles written by experts in the field.

SHUPE, J. W.: Energy self-sufficiency for Hawaii. *Science* 216:1193, 1982. A discussion of strategies used in Hawaii to meet their energy needs.

SMIL, V.: On energy and land. *American Scientist* 71:15, 1984. A survey of land requirements for various energy technologies.

STRICKLER, D.: *Passive Solar Retrofit*. New York: Van Nostrand, Reinhold Co., 1982. Homeowners' guide to implementing passive solar features into an existing home.

THE CONSERVATION FOUNDATION: *State of the Environment, 1982*. Washington, D.C.: The Conservation Foundation, 1982. Chapter 5 gives an overview of energy issues.

THE ORGANIZATION FOR ECONOMIC COOPERATION AND DEVELOPMENT: *World Energy Outlook*. Washington, D.C.: OECD Publications, 1982. The Organization for Economic Cooperation and Development's International Energy Agency's forecasts to the year 2000 for all forms of energy.

UBELL, A.: *Energy-Saving Guide for Homeowners*. New York: Warner Books, 1980. Practical energy-saving tips for the homeowner.

U.S. SUPERINTENDENT OF DOCUMENTS: *Passive Solar Design Handbook*, Vol. 3. Washington, D.C.: U.S. Superintendent of Documents, 1983. Covers the design and use of passive solar systems as an energy source.

WAYNE, M.: Plugging cogeneration into the grid. *Electric Power Research Institute Journal* 6:6, 1981. Survey of how electric utilities are attempting to implement cogeneration technology.

WHITAKER, R.: New promise for photovoltaics. *Electric Power Research Institute Journal* 8:6, 1983. A survey of progress in photovoltaic cell research.

CHAPTER **20**

TOWARD A SUSTAINABLE ENVIRONMENT

Imagine that the world's population is reduced proportionately to equal the population of a town of 1000 persons. In this town, based on global statistics, 50 inhabitants would be American and 950 would be non-American. The 50 Americans would earn approximately half the total income of the town, would own 15 times the number of possessions of all others, and would use a highly disproportionate share of the electric power, fuel, steel, and other materials. The 50 Americans would produce nearly one-fifth of the town's food supply and eat approximately 70 percent more than their minimal needs require. Because many of the 950 non-Americans in the town would be hungry most of the time, they would harbor ill feelings toward the 50 Americans, who would appear to them to be enormously rich and ridiculously overfed. Of the 950 non-Americans, 200 would suffer at some time from malaria, cholera, typhus, or malnutrition. None of the 50 Americans would contract any of those diseases, and probably none would ever even worry about contracting them.

This scenario (based on figures developed by Dr. Henry Lepier) underscores the reality that, as Americans, we enjoy an enormous share of the world's advantages. However, the process of achieving and maintaining our affluent life-styles has imposed unexpected costs upon us, and at present we are plagued with numerous unforeseen problems associated with our way of life.

Many of the problems that are discussed in this book arise from our acquisition, use, and disposal of natural resources to support our affluent life-styles. For example, land is degraded as a result of coal mining; toxic substances enter the atmosphere as by-products of manufacturing; and groundwater quality is threatened by improper disposal of industrial wastes. At the same time, the supplies of many of the resources upon which we are so dependent are declining.

As we have pointed out so often in this textbook, because of the interrelationships that operate within the environment, disturbances in one sector are likely to have repercussions in others. Hence as we attempt to balance petroleum imports with food exports, we are making excessive demands on our valuable agricultural lands, and we are gradually draining them of their productivity. As demands for housing, food, petroleum, transportation, commercial goods, and recreation continue to grow, competing interests come into conflict over the use of our fixed supply of land—land that can satisfy few of the demands that are placed on it.

As per-capita demands for resources climb and per-capita supplies of resources fall, the economy's resource base begins to shrink. The resulting scarcity of goods elevates prices, aggravates inflation, increases unemployment, and inevitably leads to a decline in living standards. All these factors force our society to choose among undesirable alternatives—choices that are governed by availability rather than by quality. Simply put, then, our environmental problems can be viewed as matters of supply versus demand. In solving them, we must seek ways to attain a balance between supply and demand without sacrificing the quality of our environment.

Supply: Prospects and Strategies

We live on a planet whose supply of resources is finite. Probably the most important lesson that we learned from our journey to the moon and beyond is that the earth is our only home. We will not be establishing major colonies on other planets in the foreseeable future; nor will we be importing resources from other worlds. Hence the resources on earth are all we have. Already, as world population reaches the 5 billion mark, the world's per-capita output of some basic commodities is now declining. For example, the world's per-capita timber harvest peaked in 1980; the fish catch, in 1970; beef production, in 1976; grain production, in 1978; and oil, in 1979.

Some scientists, economists, and government officials believe that the concept of limited resources is misleading. They point out that the earth's crust and the oceans contain enormous quantities of minerals, albeit in very dilute concentrations. They argue, too, that abundant energy sources are available to us. For example, trillions of barrels of oil are locked up in oil shale. Those experts believe that, as economic conditions become more favorable and exploration techniques and mining technology are improved, copious supplies of many valuable resources will more than meet our needs far into the future (refer back to Box 12.1).

It is true that vast, untapped reserves of minerals and energy sources exist that are presently too expensive to exploit. However, for very low-grade mineral deposits, recovery may never prove to be feasible unless inexpensive fuel becomes available, which is an unlikely prospect for the future. In fact, all sources of energy are expected to grow increasingly more expensive. For example, as we have seen, even though solar energy is free, the cost of building a solar heating system is high. Granted, as the price of fossil fuel rises, solar energy is becoming more economically competitive. But the question remains concerning who will be able to afford either alternative: fossil fuels or solar-collection equipment. The minimum cost of even the cheapest fuel could conceivably outstrip the ability of those who have low and even moderate incomes to pay for it.

Many Americans possess an almost blind faith that technological innovation eventually will come to their rescue. Indeed, in a historical perspective, technology has helped to eradicate disease, expand our resource base, and raise our standard of living. And although past technological success may portend well for the future, our optimism must be tempered by the inherent limitations of technological research and development.

To formulate solutions, we must first understand the problems. However, most environmental problems are vastly more complex than they initially appear. Hence with respect to a particular problem, scientists and engineers may not possess sufficient understanding to focus their research efforts appropriately. For example, current research on alternative energy sources is proceeding in many directions, basically because researchers do not yet know which of their efforts will be most fruitful. Furthermore, some technological problems are more difficult to solve than others, assuming that solutions exist at all. Developing a commercially available nuclear fusion reactor, for instance, poses greater scientific challenges than did going to the moon.

Advances in technology usually involve complex efforts that must be expended over considerable periods of time and require the skills of many highly trained and creative personnel. In addition, applying new technological principles often employs rare and exotic—and, therefore, expensive—materials. The examples of nuclear-fission power plants illustrates the enormous investments that are involved in implementing technological advances. Although the direct costs of nuclear-generated electricity that are listed in a consumer's electrical bill are competitive with the costs of electricity from coal-fired power plants, such a comparison is actually simplistic and misleading. For approximately 20 years before the first nuclear power plant went into operation, consumers had been spending billions of their tax dollars on nuclear power through federal support of research by governmental agencies, universities, and private industry. Federal funds have also underwritten the costs of construction, fuel processing, health-hazard evaluation, and hazard insurance. Had tax-payers not subsidized those efforts, nuclear power would not be competitive with fossil-fuel power. In fact, without that support, nuclear power plants never would have left the drawing boards.

Development and implementation of any new technology takes time—perhaps too much time to help solve an immediate and pressing problem. Typically, a new technology must undergo lengthy testing and gradual scaling up from a pilot, or prototype, stage to full-scale operation. Some critics of nuclear power claim that many of the problems that are faced by that industry today can be traced to its deployment, which has been too rapid. In retrospect, they see the nation as having rushed the application of that new technology without taking enough time to develop adequate safeguards—especially with regard to the disposal of waste products.

Experience also tells us that every technological advance creates side effects that detract from the benefits it bestows. For example, any technique that we might develop for using low-grade deposits of minerals is certain to require substantial amounts of fuel and other resources (such as water) and to generate large quantities of wastes. And more waste products could, in turn, degrade other valuable resources (for example, by contaminating surface or groundwater, disrupting the landscape, or destroying wildlife habitats).

The mind is a fertile source of ideas and, undoubtedly, human ingenuity will continue to devise technological solutions to our resource supply problems. However, the promises of technology are limited. As we have noted, new technology is expensive; it typically requires abundant and inexpensive fuel for implementation; it takes time for development; and it may produce unacceptable side effects. Hence it is unwise to base our future solely on the belief that the earth is a vast storehouse of resources that inevitably will be unlocked by technological innovations. A more fruitful course of action would be to develop a greater reliance on technology that is energy efficient, suited to the real needs of society, and compatible with nature. Several strategies take us in that direction.

In addition to exploring for new deposits of oil, we should also work to improve the technology for enhanced oil recovery (see Chapter 12). Wherever possible, we should increase resource supplies through recycling, particularly for metals whose reserves are declining. For example, we have seen that by using scrap metal instead of virgin ore, we can conserve energy and that organic wastes can be converted into fuel. Also, recycling solid wastes saves energy, reduces pollution, and alleviates the disposal problem. In addition, we need to intensify our efforts to conserve energy—for example, by increasing the energy efficiencies and durability of household appliances, by making further improvements in automobile efficiency, and by building better insulated buildings.

Cropland is an indispensable resource that must be protected if we are to continue to produce sufficient supplies of food. Thus we must reduce soil erosion by wind and water and curb efforts to convert agricultural land to other less-productive uses. And we must reverse worldwide deforestation to ensure that we have adequate supplies of building materials, paper, wood fuel, and other wood products.

Where possible, we should develop or choose technologies that operate within the constraints that are imposed by the natural operation of ecosystems. For example, swamps, forests, and agricultural land can be used for tertiary treatment of wastewater effluents. By utilizing that natural cleansing process, we can reduce our dependency on energy-intensive wastewater treatment. Also, by using natural systems in this way, we can enhance the intrinsic value of biomes, which would give us more incentive to preserve them and the wildlife within them.

In summary, then, we question the point of view that inexpensive, abundant resources will be available in the immediate future. There is no doubt that new supplies will be discovered and exploited, but at what costs? The prudent path is to make better use of our resources by conservation and by reuse, recycling, and the recovery of the valuable materials in our wastes. But all this is not likely to be enough, particularly if our demands increase simultaneously. For this reason, we must seek to solve our problems by increasing the total supply of resources as well as by reducing our per-capita consumption.

Demand Strategies: Two Perspectives

Demand can be considered from two perspectives: (1) demand by the total population and (2) per-capita demand. The population growth rate has subsided greatly in the United States and its population may become stationary within 50 years. That projection portends well for the future. Because the rate of population growth in the United States has declined so markedly, Americans have a better opportunity to cope with problems of resource supply and demand.

However, per-capita consumption in the United States is quite another matter. Despite increasing costs, each succeeding generation of Americans has enjoyed an ever-rising level of affluence. For example, many American families now have air-conditioned homes that are equipped with at least two bathrooms, two television sets, two cars, a stereo, and a computer. Many persons own recreational vehicles, such as powerboats and campers. And recycling has yet to supplant planned obsolescence as a guiding principle of our economy. Our society is infatuated with disposal items; we believe that we can afford to use materials once or twice and then throw them away and buy more.

Our dependence on the automobile is one indicator of our level of affluence and our determination to maintain that level. For example, in the 1970s, we were faced with declining gasoline supplies and soaring prices, but few motorists voluntarily restricted their driving. As a result, fuel conservation was achieved only through legislative mandates, which set fuel-efficiency goals for automobile manufacturers and imposed a 55-mile per hour speed limit. However, legislated gains in fuel economy are offset to some extent by our preference for cars—even small cars—that are equipped with energy-consuming options, such as air conditioning and power steering. Furthermore, we now drive our cars greater distances than we did before the Arab oil embargo in 1973.

In recent years, some persons have, indeed, voluntarily reduced their level of fuel consumption. For example, more persons are riding bicycles and small motorcycles for transportation and placing greater emphasis on self-propelled sports, such as jogging, backpacking, canoeing, and downhill and cross-country skiing. There is no doubt that such activities consume fewer resources and do less environmental damage than powerboats and campers.

Also, a handful of persons have dramatically changed their life-styles and consumer habits by leaving the comforts of home and taking up a pioneer existence in such remote regions as Alaska or northern Maine. However, few of us really have any desire to undertake the hardships of the "good old days." We value the advantages that our society can provide in making our lives easier and, presumably, more enjoyable. Furthermore, few persons have the knowledge, skills, and time that is required to grow most of their own food, to build and maintain their own houses, to sew and mend their own clothes, and simply to survive on their own. Making a living without modern conveniences would be extremely difficult for most of us.

However, curbing our appetites or, for that matter, the appetites of other developed nations, is not going to solve the problems that are related to the demand for resources. Even though we are the most conspicuous consumers on earth, we are far from being alone. For example, in less-developed nations, the problem is not so much per-capita demand as it is the cumulative demands of a burgeoning population. Thus the most essential step in slowing demands for resources on a global scale is to stabilize the world's population. Indeed, even if all other efforts to increase resource supplies (such as energy conservation, reforestation, and cropland protection) are successful, unless population growth is brought under control, such efforts will be futile. The number of persons in the world eventually would outstrip the globe's ability to provide sufficient supplies of the basic necessities of life. Fortunately, the limits to the earth's carrying capacity are becoming increasingly clear to many national leaders (and individual couples), so birth rates are declining in some less-developed countries (see Chapter 16).

What You Can Do

What, if anything, can one person do about environmental issues? Actually, by reading this book, you have already taken the first step. You are now better informed about the roots and implications of environmental problems. We cannot anticipate tomorrow's new environmental problems, nor can we be sure which of today's problems will fade in importance. But by studying this book, you have a clearer idea of how the various segments of our society respond to environmental problems.

There are, of course, a great number of ways in which you could modify your life-style and thereby

The third section of this chapter is based on material provided by Professor Richard Tobin.

contribute to the solution of environmental problems. For example, you could decide to drive a smaller car that gets better mileage and emits fewer pollutants, or you could participate in a car pool to save gasoline and wear and tear on your automobile. You could improve energy conservation in your home by turning down your thermostat, adding insulation, and using air conditioners only when they are really needed. However, these are only a few examples; perhaps you have already decided on taking these actions or others.

Elect and Influence Policymakers

Personal efforts to conserve resources and reduce pollution are praiseworthy, but there are other effective ways to become involved in environmental decision making. One of those ways is to participate in the shaping of public policy by electing environmentally aware and responsive public officials. As we have seen, environmental issues are matters of public policy as well as matters of science and technology, because such issues entail important and often controversial questions that involve personal values, competing interests and goals, and the use of cost–benefit or risk–benefit analysis.

Progress in improving air and water quality has come about largely as a result of state and federal pollution control programs. And efforts to conserve scarce resources and clean up our environment are—at least in part—responses to government-imposed financial incentives or disincentives. For example, at least nine states have enacted laws that encourage recycling and discourage littering by requiring purchasers of canned or bottled beverages to pay a deposit along with the price of the beverages (the deposit being redeemable upon return of the beverage containers to the store or other collection facility). And real progress in the fight against soil erosion is likely only if we elect legislators who recognize the need for a national soil-conservation policy.

But how can we know which public officials are environmentally responsible? The League of Conservation Voters, located in Washington, D.C., can provide some help. The league is a nonpartisan, national campaign organization that promotes the election of environmentally aware public officials at the national level. The league publishes, before each presidential election, a thoroughly researched background study of each candidate's views on such is-

sues as air and water pollution, endangered species, pesticides, population growth, and nuclear power. The report is entitled *The Presidential Candidates: What They Say, What They Do on Energy and the Environment.* The league also publishes, once a year, the voting records of incumbent United States senators and representatives on environmental issues, and it assigns an environmental score to each member of congress on the basis of his or her voting record.

In the case of nonincumbents at the national level and all candidates at state and local levels, it is more difficult to determine who stands where on the issues, but it can be done. For example, one way is to check with the League of Women Voters, which frequently prepares information on the views of candidates for state and federal office. Another approach is to write or telephone a candidate's campaign office and get responses to questions about the candidate's views on environmental matters.

How can you influence government policymakers after election day? You can start by joining an environmental group, such as the National Wildlife Federation or the Sierra Club, that works to influence public policy by lobbying elected and appointed public officials. Such groups work to pass or defeat legislative bills and sometimes propose environmental legislation, and they monitor the implementation of environmental laws. Their agents meet with government policymakers to discuss environmental issues as well as pending environmental legislation. And they frequently report to their members on environmental issues (for example, the Sierra Club has its own monthly magazine). To locate such environmental groups and to learn more about them, consult the *Conservation Directory* (published every year by the National Wildlife Federation), which lists organizations, government agencies, and government officials concerned with natural resource use and management. That directory, which is a nominally priced paperback, contains the names and addresses of international, national, and state organizations and commissions; federal departments and agencies; state and territorial agencies; Canadian national citizens' groups; Canadian Provincial government departments and citizens' groups; and conservation–environment offices of foreign governments. For each organization, a brief description is given of what the group or agency does, its address, and a list of its publications, if any.

Knowledgeable citizens can also exert a more direct influence on government policymakers. For example, the most straightforward way is to write a well-informed letter to the appropriate policymakers on the environmental matter that concerns you. Such letters—even just a handful of them—can, and often do, make a difference. Elected officials are looking for cues: very few of them are well versed in environmental science, and they do not necessarily have a preformed opinion about every issue that comes before them. Moreover, astute politicians want to know both sides of an issue. When they get information from a lobbying group, they know that they are getting only one side of the story. As a result, even just a small number of well thought-out letters from their constituents can make a difference in how they see the issue in question—and how they vote on it.

However, caveats must be considered when writing to elected government officials: (1) avoid form letters that have been prepared for mass mailings; and (2) when writing to elected officials, be sure that you write to the ones who represent *you*. For example, if you are from New York, your letter is not likely to have much influence on a senator or representative from New Mexico or Florida.

Telephone calls as well as letters to the local offices of state legislators and to local officials (elected or appointed) can also be effective. (Calls to congressmen and bureaucrats in Washington are generally less effective than letters, because the message is less likely to get past the office staff. For example, a congressman might have 20 or more persons working for him. One of those staff members will take your call—but not all of them have regular, direct contact with their bosses. In contrast, a state legislator is likely to have only one or two assistants. Those aides see the legislator regularly, and they might well put you through directly to him or her.)

Another way to influence government decision makers is to attend and participate in public hearings on environmental matters. The federal Environmental Protection Agency and state environmental agencies, as well as their respective regional offices, can tell you when and where hearings will be held. They can also provide background material for the hearings. Be assured that any hearing about changes in rules or regulations that could adversely affect an industry will be attended by representatives from that industry. If such a hearing is not well publicized—and many are not—the public officials who preside over it may hear only the industry's side

of the issue unless persons or groups that are pro-environment have been particularly alert.

Also, your class or student organization can invite public officials to make presentations to classes or community groups. Most officials are usually quite willing to do this. The best format is a brief talk by the official about what he or she does or (if he or she is an elected official) what his or her views are, followed by a question-and-answer session. By means of such a program, you can not only ascertain the policymaker's views but also educate him or her. As we noted earlier, few policymakers are well versed in environmental science and environmental problems: they are looking for cues. Remember, though, that public officials rarely consider a problem solely in terms of its scientific aspects. The problem's economic and political aspects also weigh heavily.

Still another means of influencing public officials is to take carefully prepared, well-timed legal action. For example, it is possible to file civil law suits against the EPA for failure to perform many of its nondiscretionary duties. And nearly all the major federal environment laws, except the Superfund Law, allow *anyone* to file a suit to compel the EPA to fulfill its legal mandates. Several national environmental organizations have used those so-called citizen-suit provisions to force the EPA to issue standards for hazardous air pollutants. However, legal action can, of course, be expensive. But some environmental laws (such as the Clean Air Act Amendments) allow the courts to impose the plaintiff's litigation costs on the government if it loses the case.

Finally, you can influence other persons' views on the environment and contribute to environmental problem solving by your choice of a career. For example, you can enter business administration and help companies to formulate their policies on environmental quality and resource conservation. You can enter the political arena, where reforms and changes in priorities are clearly needed. You can enter research and development and seek new knowledge or develop new technological solutions. Or you can become a teacher, and seize the opportunity to expose your students to the value of a healthy environment.

Stay Informed

The more you know, the more you can do. Above all, then, you can become—and stay—as knowledgeable as possible about environmental matters. Every day, the electronic and print media bombard you with details about the latest environmental problems. You must interpret that information in the context of what you now understand about basic ecological principles. And if you are serious about keeping environmentally informed, you will seek out more information. The National Wildlife Federation's *Conservation Directory* lists not only organizations and officials who are concerned with the environment (as mentioned earlier), but also periodicals of interest, directories of interest, and sources of information that exist in audiovisual form. And it lists all of our national parks, monuments, wildlife refuges, and seashores.

The best single source of information about federal activities is the annual report of the Council on Environmental Quality (*Environmental Quality*). That report summarizes just about every major resource and environmental activity that the federal government has any interest in. Most college and university libraries are likely to have each year's report. Moreover, the Council on Environmental Quality usually provides a free copy to anyone who writes (722 Jackson Place, N.W., Washington, D.C. 20006).

For up-to-date information, seek out the *Environment Reporter: Current Developments*, which is published by the Bureau of National Affairs, a nongovernmental organization. That publication is, in effect, a weekly newspaper of environmental events at the state and federal levels. The Bureau publishes federal environmental laws and regulations and state laws in this same periodical. It also publishes a monthly *International Environment Reporter.*

Publications of federal agencies are another source of useful information on the environment. The Government Documents section of college and university libraries probably has (or can quickly get) most reports of such agencies as the EPA and the Department of Energy and the Interior. If you have not done so already, you might request a tour of the Government Documents section of your library. Librarians in such sections are generally quite willing to provide special tours that focus on particular topics. Note that government documents often are not included in the regular card catalog. If you do research without the benefit of government documents, you will often miss much relevant and current information.

Finally, if you are patient and persistent, you can use the Freedom of Information Act to get information from a federal agency that the agency has not provided in its publications. For example, suppose that you are researching air pollution in your town or county and that you want to know the emission levels of a nearby steel mill. The EPA occasionally publishes information concerning air quality for whole air-quality control regions, but it does not publish data for specific installations. However, by invoking the Freedom of Information Act, you can obtain the emission data for that steel mill if the EPA has it (which it probably does). Be sure to invoke the act when you write: "Under the provisions of the Freedom of Information Act, I request the following information. . . ." If you do not, the recipient agency may not feel obligated to provide the specific information you seek. Federal agencies are supposed to answer requests that are made under this act within 10 working days. If you do not receive an answer within a reasonable period of time, you can appeal through the same agency. If your appeal is unavailing, as a last resort, you can take the agency to court.

Do not overlook state agencies as sources of unpublished information; every state but Mississippi now has a law that is comparable to the federal Freedom of Information Act.

Conclusions

Some persons are confused, pessimistic, and even fatalistic about prospects for the human race. They feel that it is futile to try to do anything—that the problems which plague our planet are too numerous and too complex. Having never lived through a prolonged economic depression, and having known only prosperity and good times, many young Americans lack the experience and the confidence to face change. However, as many of us realize from trying to solve our own personal problems, running away or ignoring critical issues seldom solves them. And refusing to act can lead to despair, while searching for solutions focuses our energies in a way that could pay off. It is in our own self-interest to accept the challenge of change. Let us identify those environmental problems, tackle them—and solve them.

GLOSSARY

Abiotic Pertaining to nonliving components of the environment.

Acclimatization Adjustment of an organism to environmental change.

Acid Rain Rain that has become more acidic from falling through air pollutants and dissolving them, primarily sulfur dioxide.

Active Solar Collector A device that collects the sun's energy and converts it to another form, such as hot water or steam, that is used by a nearby system.

Activated Sludge Treatment A sewage treatment process in which a portion of the decomposer bacteria present in the waste is recycled to the beginning of the process, thus providing the system with acclimatized organisms.

Acute Exposure A single exposure to radiation or toxic substances that lasts seconds, minutes, or hours.

Adaptation A genetically controlled characteristic that enhances an organism's chances to survive and reproduce in its environment.

Aerobic Decomposers Microorganisms that require oxygen to break down organic wastes into carbon dioxide and water.

Aerosols Tiny solid and liquid particles suspended in the atmosphere.

A-horizon A layer of soil (topsoil), which contains most of the mineral nutrients and plant roots. Considerable amounts of soluble soil constituents are leached from this zone.

Air Mass A volume of air covering thousands of square kilometers that is relatively uniform in water vapor content and temperature.

Air Pollution Episode A period in which air pollution concentrations reach levels that are hazardous to human health.

Air Pressure The weight of the atmosphere over a unit area of the earth's surface.

Air Quality Control Region (AQCR) An area where at least two communities share common air pollution problems. Meteorological conditions and types of emission sources are relatively uniform within each AQCR.

Air Stability The relative buoyancy of air parcels in the atmosphere.

Albedo The fraction of received radiation that is reflected by a surface.

Alpha Particle A particle emitted from certain radioactive materials. One of the three major forms of radiation, it has the least penetrating power.

Ambient Air The air we breathe, the surrounding air

Anaerobic Decomposers Microorganisms that, in the absence of oxygen, break down certain organic wastes into methane, carbon dioxide, hydrogen sulfide, and water.

Antagonism An interaction in which the total effect is less than the sum of the effects taken independently.

Aphelion The time of year when the earth is farthest from the sun (July 3 or 4).

Appropriation Water Law The right to divert water from a water course for beneficial purposes. This legal system is used in most western states; ownership of water rights is not necessarily tied to land ownership.

Aquaculture   The growing and harvesting of fish and shellfish, usually in confinement.

Aquifer   A permeable layer of rock, soil or sediment that holds and can transmit water.

Area Landfill   A landfill formed by dumping wastes into valleys, canyons, or pits until full, then covering the wastes with earth.

Area Strip Mining   A mining method used to extract near-surface rock and mineral deposits (usually coal) in relatively flat terrain. Heavy earth-moving equipment removes the overburden and target material from a series of adjacent trenches.

Artesian Well   A free-flowing well tapped into a pressurized aquifer.

Asphyxiating Agents   Air pollutants that when inhaled deprive the blood of its oxygen supply; carbon monoxide is an example.

Assimilated Energy   The portion of ingested food energy that is absorbed by an animal's digestive system.

Atmospheric Nitrogen Fixation   The process by which the high temperatures associated with lightning cause nitrogen gas to combine with oxygen gas, eventually forming nitrate.

Atmospheric Windows   Infrared wavelength bands for which there is little or no absorption by components of the atmosphere.

Banking   An air quality management strategy whereby industries earn credits for reducing their air pollutant emissions to below maximum allowable levels. These credits may be bought and sold like any other commodity.

Base Flow   The dependable flow in a river, which results primarily from groundwater sources.

Belt of Soil Moisture   The top layer of soil from which plants obtain their moisture.

Bergeron Process   The process in cold clouds whereby ice crystals grow at the expense of supercooled water droplets.

Beta Particle   A particle with moderate penetrating power that is emitted from certain radioactive materials.

B-horizon   The soil layer beneath the A-horizon, consisting of weathered material and minerals leached from the A-horizon.

Bioaccumulation   The process whereby certain chemicals become more highly concentrated in organisms at the higher levels of a food web.

Biochemical Oxygen Demand (BOD)   The amount of oxygen required by microorganisms to decompose the organic wastes in a given volume of water.

Biodegradable   Pertaining to materials that can be decomposed by microorganisms.

Biological Control   The regulation of a pest population through the use of natural predators, parasites, or disease-causing bacteria or viruses.

Biological Nitrogen Fixation   The process by which blue-green algae and some bacteria convert atmospheric nitrogen gas into ammonia.

Biomass   The total weight, or mass, of organisms in a given area.

Biome   A major type of terrestrial ecosystem, such as the Arctic tundra or the Great Plains grasslands, that usually covers an extensive area.

Biotic   Pertaining to the living components (organisms) of the environment.

Breeder Fission Reactor   A type of nuclear reactor that produces slightly more fissionable material than it consumes.

British Thermal Unit (BTU)   The quantity of heat needed to raise the temperature of 1 pound of water 1 degree Fahrenheit.

Broad-Spectrum Pesticide   A chemical used to kill a wide variety of pests.

Bubbling   An industrial air quality management strategy that groups together emissions from many individual sources (i.e., smokestacks) in a given area under an imaginary "bubble" and considers them as a single aggregate source.

Carbon cycle   Circulation of carbon among various components of the environment. Major carbon reservoirs include the atmosphere, the oceans, and vegetation.

Carcinogen   A chemical substance or physical agent (such as radiation) capable of causing cancer.

Carnivore   An animal that feeds only on animals.

Carrying Capacity   The maximum population that a given area can sustain indefinitely.

Catalytic Converter   An automobile exhaust control device that chemically changes hydrocarbons and carbon monoxide in the exhaust into carbon dioxide and water vapor.

Cell   The basic structural unit of all organisms.

Chlorinated Hydrocarbons   A class of chlorine-containing chemicals, some of which are toxic and carcinogenic, and may bioaccumulate.

Chlorination   The addition of chlorine to drinking water or treated sewage treatment plant effluent for the purpose of disinfection.

Chloroorganic Compounds   Organic chemicals that contain chlorine as part of their molecular structure.

Chlorosis   A yellowing of plant leaves due to chlorophyll loss.

C-horizon   The bottom soil layer, a mineral made up of the partially decomposed, underlying bedrock.

Chromosomes   Cellular structures that contain the genetic (hereditary) information.

Chronic Exposure   Continuous or recurring exposure to radiation or toxic substances that lasts for days, months, or even years.

Cilia   Tiny undulating hairs lining the respiratory tract that protect the respiratory system from inhaled particulates.

Clarification   A water purification process whereby suspended materials are allowed to settle out in quiescent basins. Chemicals are often added to enhance the process.

Climate   The weather conditions of an area averaged over a period of time. Also, extremes in weather behavior that occurred during the same period.

Climatic Anomaly   Departure of temperature, precipitation, or other weather elements from long-term average values.

Climatic Destabilization   An increase in the frequency of extreme weather behavior (record temperatures and precipitation, for example).

Climax Stage   The assemblage of plants and animals at the endpoint of an ecological succession series.

Coal Gasification   One of several processes whereby coal is heated in the absence of oxygen and in the presence of steam to produce the combustible gases methane, carbon monoxide, and hydrogen.

Coal Liquefaction   One of several processes for manufacturing liquid fuels from coal.

Coevolution   The process by which two interacting populations—for example, a parasite and a host—adapt to each other to their mutual benefit.

Cogeneration   An energy conversion technology that produces two useful forms of energy simultaneously, usually electricity and hot water or low-pressure steam.

Coliform Bacteria   A type of bacteria that resides in the human intestine whose presence in water is used to indicate whether the water may be contaminated with disease organisms.

Collision-Coalescence Process   The process that occurs in warm clouds whereby large water droplets grow larger by colliding with smaller droplets.

Combined Sewer   A sewer system that transports both runoff and sanitary wastes through one large pipe to a sewage treatment plant.

Competition   Interactions of organisms to secure a resource in short supply.

Competitive Exclusion Principle   The generalization that two species having the same resource requirements cannot coexist indefinitely in the same area.

Composting   The production of a fertilizer or soil conditioner through the aerobic breakdown of cellulose wastes, such as grass clippings, leaves, and paper.

Compressional Warming   The temperature increase that occurs in a parcel of air when it compresses due to an increase in pressure.

Condensation   The process whereby water changes from the vapor state to the liquid state.

Condensation Nuclei   Tiny solid or liquid particles that initiate condensation.

Conduction   The transfer of heat from one object to another through direct physical contact.

Cone of Depression   A cone-shaped depression that forms in the water table when groundwater is pumped out more rapidly than it is recharged.

Consumers   Organisms that use other organisms as a food source.

Contour Farming   A soil conservation procedure in which a field is plowed parallel to land contours.

Contour Strip Mining   A mining method used to extract near-surface rock and mineral deposits (usually coal) in hilly or mountainous terrain. Earth-moving equipment removes the overburden and the deposit from a series of benches cut along contours of elevation.

Convection   A circulation pattern in which a relatively light substance (such as heated air) rises and a relatively heavy substance (such as cold air) sinks.

Convergent Evolution   The process whereby unrelated isolated populations become more similar to each other as a result of independent adaptation to similar environmental factors.

Convergent Plate Boundary   A boundary between lithospheric plates where one plate slides downward and under the adjacent plate, thereby bending the oceanic crust into a trench.

Cost–Benefit Analysis   A technique employed by economists in which all the gains or benefits of a project are compared with all the losses or costs of that project.

Cost Effectiveness Analysis   A technique employed by economists to determine how a particular goal can be achieved for the least cost.

Criteria Air Pollutants   Those air pollutants for which the Environmental Protection Agency has established national standards for ambient air.

Crude Birth Rate   The annual number of births per one-thousand people in a population.

Crude Death Rate The annual number of deaths per one-thousand people in a population.

Cultural Eutrophication A hastening of the natural nutrient enrichment of surface water by human activities such as agriculture and the discharge of nutrient-rich wastes.

Cyclone Collector A device that removes particulates from an industrial effluent air stream by inducing gravitational settling.

Decomposers Organisms such as bacteria, mushrooms, and maggots that feed on the remains of plants and animals.

Deep-Well Injection A method of disposing of liquid wastes by pumping them under pressure into deep subsurface cavities and pore spaces in bedrock.

Demographic Transition A pattern of change in birth and death rates in which a decline in the death rate is followed by a decline in the birth rate.

Denitrification The conversion of nitrates to nitrogen gas carried out by a specialized group of bacteria.

Density-Dependent Factors Interactions such as predation, parasitism, and competition whose influence on population size varies with the population density.

Density-Independent Factors Environmental factors such as weather whose influence on population size is little affected by the population density.

Deposition The process whereby water vapor changes directly to the solid (ice) state.

Desalination The process whereby the salt content of water is greatly reduced so that saline water can be used for human consumption or irrigation.

Desertification The degradation of terrestrial ecosystems as a result of deforestation, overgrazing, and poor soil and irrigation management.

Detritus Freshly dead or partially decomposed remains of plants and animals.

Detritus Feeders Organisms that feed on detritus.

Detritus Food Webs A food web based on decomposers feeding on detritus.

Diaphragm A rubber cup that covers the uterus and prevents entry of sperm.

Diffuse Insolation Solar radiation that is reflected or scattered to the earth's surface.

Dilution A reduction in the concentration of some substance, as when polluted water is mixed with clean water.

Direct Insolation Solar radiation that is transmitted directly through the atmosphere to the earth's surface.

Discharge The volume of water per unit time flowing past a fixed point in a river, stream, or pipe.

Dissolved Oxygen The oxygen dissolved in water. The amount is usually expressed in parts per million (ppm).

Divergent Plate Boundary A boundary between lithospheric plates where the adjacent plates drift apart and melted rock wells up from the earth's interior to fill the resulting gap.

Dose-Response Curve The response of a population to increasing doses of a toxic substance.

Doubling Time The time needed by a population to double in size when it is growing at a specific annual rate.

Drainage Basin The geographical region drained by a river or stream.

Dredging A process for mining streambed sands, gravel, and placer deposits through the use of chain buckets and drag lines.

Dry Deposition Removal of aerosols from the atmosphere by a combination of gravitational settling and impaction.

Dry Steam Field A region where the steam that is emitted from the earth's interior contains little condensed water.

Dust Dome A dome-shaped accumulation of particulates that often forms over urban-industrial complexes.

Dust Plume A dust dome elongated downwind from a city.

Ecological Efficiency The percentage of energy transferred from one trophic level to the next higher trophic level.

Ecological Succession The replacement of one ecosystem by another until an ecosystem is established that is best adapted to the environment.

Ecosystem A functional unit of the environment comprising the interactions of all organisms and physical components within a given area.

Ecotone A transition zone between two or more distinct ecosystems.

Ecotype A population of a species that is adapted to specific conditions found only in a part of the total range of the species.

Edge Effect The tendency toward increased species diversity in ecotones.

Electromagnetic Radiation Energy that can travel through a vacuum in the form of waves, for example, light.

Electromagnetic Spectrum The various forms of radiational energy arranged by wavelength, frequency, or both.

Electrostatic Precipitator A device that removes

particulates from an industrial effluent air stream by inducing an electric charge on the particulates. The particulates are then collected on plates of opposite charge.

Emigration   Migration out of a population.

Eminent Domain   The right of a government body to condemn land for public benefit.

Endangered Species   A species that is in immediate danger of extinction.

Energy   The ability to do work or to produce change.

Enhanced Oil Recovery   A variety of techniques for increasing the amount of oil extracted from both producing and abandoned wells.

Epilimnion   The warm, relatively less dense top layer of water in a stratified lake.

Estuary   A coastal ecosystem where fresh water and salt water meet.

Euphotic Zone   The surface layer of a body of water to which photosynthesis is confined.

Eutrophication   A natural process of nutrient enrichment that gradually makes lakes more productive.

Eutrophic Ecosystems   Ecosystems that are high in fertility and biological productivity.

Eutrophic Lake   A lake with a high rate of nutrient cycling and thus a high level of biological productivity.

Evaporation   Vaporization of water through a combination of transpiration by plants and direct evaporation from the surfaces.

Evapotranspiration   The combined vaporization of water from plants and land surfaces.

Expansional Cooling   The drop in temperature that accompanies the expansion of a gas. For example, air undergoes expansional cooling when it ascends and the pressure on the air decreases.

Exponential (Geometric) Growth   An increase by a constant percentage of the whole during a specific time period.

Externalities   Consequences of economic activities that are not reflected in market transactions. A neighborhood park increases property values (positive externality), while a garbage dump reduces nearby property values (negative externality).

Fall Turnover   A mixing process that occurs in fall in stratified lakes whereby the top waters mix with the bottom waters.

Family Planning   A voluntary program of fertility control used by a couple to achieve the family size of their choice.

Fertility Rate   The average number of births per woman in a population.

Floodplain   Land adjacent to a river that is covered by water when a river overflows its banks.

Flue-Gas Desulfurization (FGD)   A variety of methods for removing sulfur dioxide from industrial gaseous effluent. For example, when the effluent is channeled through a slurry of water and limestone, calcium in the limestone combines with sulfur to produce a calcium sulfite sludge, which is collected and disposed.

Fluorosis   The poisoning of animals, such as cattle, resulting from the ingestion of fluorides.

Food Chain   A sequence of organisms, such as green plants, herbivores, and carnivores, through which energy and materials move within an ecosystem.

Food Web   A network of interconnected food chains.

Food Web Accumulation   The process whereby certain chemicals become more highly concentrated in organisms at the higher levels of food webs.

Fossil Fuels   The remains of ancient plant and animal life that have been transformed into coal, oil, and natural gas.

Free Markets   Markets in which resources are allocated and prices established on the basis of individual, voluntary exchange among producers and consumers.

Gamete   A sexual reproductive cell, an egg or sperm.

Gamma Ray   A highly energetic form of radiation emitted by certain radioactive materials that results in whole-body exposure.

Gaseous Cycle   A pattern of material flow in which the atmosphere or the oceans provide the primary source of the substance for organisms. Examples include the carbon cycle and the oxygen cycle.

Gene   A portion of a chromosome that codes for a particular characteristic.

Gene Pool   All the genes of all the individual members in a given population.

Geopressurized Methane   Natural gas resources found deep in the earth's crust as free gas, gas dissolved in brines, and immobile gas trapped in sediments.

Geothermal Energy   Heat energy contained in the earth's interior.

Global-Scale Circulation   The largest scale of atmospheric motion, which includes the wind systems that circle the globe.

Gravitational Settling   Gravity-induced downward motion of particles suspended in air or water.

Grazing Food Web   A food web based on herbivores feeding on living plants.

"Greenhouse Effect"   The absorption and reradiation of terrestrial infrared radiation by atmospheric water vapor, carbon dioxide, and ozone.

Green Revolution A popular term for the dramatic increases in crop yields per hectare that sometimes accompany the introduction of new crop varieties and the improvement of soil, water, and pest management.

Ground Subsidence The sinking of the earth's surface that occurs when subsurface mineral deposits or fluids are withdrawn.

Groundwater Mining Withdrawal of groundwater at a rate that exceeds the natural recharge rate.

Groundwater Reservoir Extractable underground fresh water.

Haber Process An industrial process that uses natural gas as an energy source to synthesize ammonia from atmospheric nitrogen.

Habitat Island A restricted area of habitat that is surrounded by dissimilar habitats.

Habituation Learning to ignore unimportant stimuli.

Half-Life The amount of time required for the radioactivity emanating from a particular radioactive substance to be reduced by one-half.

Hazardous Pollutants Pollutants that pose a serious danger to human health even at extremely low concentrations.

Heat of Fusion The quantity of heat required to change 1 gram of a substance from the solid state to the liquid state.

Heat of Vaporization The quantity of heat required to change 1 gram of a substance from the liquid state to the vapor state.

Heavy Metals A group of elements whose compounds are toxic to humans when found in the environment; examples are cadmium, mercury, copper, nickel, chromium, lead, zinc, and arsenic.

Herbivore An animal that feeds only on plants.

Hibernation A state of inactivity exhibited by some animals during winter.

Homeostasis The tendency of a system to maintain nearly constant internal conditions in the face of a changing environment.

Humus Organic material that is slow to decay.

Hydrologic Cycle Circulation of water among various components of the environment. Major reservoirs include oceans and glaciers.

Hydropulping A technique used to recover paper fibers from paper wastes; the fibers are used to manufacture recycled paper.

Hydrothermal Vents Hot water produced when magma welling up within divergent plate zones comes in contact with sea water; associated with metallic sulfide ore deposits.

Hygroscopic Nuclei Condensation nuclei with a special affinity for water vapor.

Hypolimnion The cold, relatively dense bottom layer of water in a stratified lake.

Igneous Rock Rock formed by the crystallization and solidification of magma (or lava), either within the crust or on the earth's surface.

Immigration Migration into the population from elsewhere.

Impaction Removal of particulates from the air when they strike and adhere to buildings and other structures.

Inbreeding Mating between close relatives.

Incinerators Furnaces that are used to burn combustible wastes.

Incremental Decision Making A process used to formulate public policies; it involves small, incremental changes in existing policy.

Infiltration Seepage of surface water into soil and rock layers.

Inorganic Pertaining to materials composed of elements other than carbon, including water, oxygen, and minerals.

Inorganic Fertilizer Plant nutrients derived from mineral deposits (phosphate or potash) or from natural gas (ammonia).

Insecticide A chemical directed at killing insects.

Insight Learning The ability to respond correctly the first time to a new situation.

Integrated Pest Management A combination of pest control methods that is best suited for a particular pest and set of environmental and economic conditions.

Internal Crustal Processes Landscape sculpturing processes acting on the earth's crust from the earth's interior. Volcanic eruptions and earthquakes are examples.

Interspecific Competition Competition for a resource between populations of two or more species that reduces their well-being.

Intraspecific Competition Competition for a resource among members of the same species that reduces their well-being.

Intrauterine Device (IUD) A small metal or plastic device that is inserted into the uterus to prevent conception.

Isotopes Different forms of an element having different atomic mass. Isotopes of an element behave in a similar chemical manner, but have widely varying nuclear properties.

K-strategists Species that increase their probability of surviving to reproduce in the future rather than maximize their current rate of reproduction.

Kinetic Energy The energy attributable to something as a result of its motion.

Land Trust A community growth-control strategy in which private citizens donate undeveloped land to a nonprofit conservation agency for a specified time period with the stipulation that the land not be developed.

Landslide The gradual or abrupt downslope movement of rock, soil, mud, or a mixture of these materials.

Latent Heat Transfer The transport of heat energy from place to place as the result of changes in the phases of water.

Law of Energy Conservation Energy can be neither created nor destroyed, although it can change from one form to another.

Law of the Minimum The principle that the growth and well-being of an organism ultimately is limited by that essential resource that is in lowest supply relative to what the organism requires.

Law of Tolerance For each physical factor in an environment, a minimum and a maximum level exists beyond which no member of a particular species can exist.

LD$_{50}$ The quantity of a substance that will kill 50 percent of a test population when administered as a single dose.

Leaching The dissolving, transporting, and redepositing of materials by water seeping downward through soil or solid wastes.

Limiting Factor Any component of the environment that limits the ability of an organism to grow or reproduce.

Linear (Arithmetic) Growth An increase by a constant amount in a constant time period.

Liquefied Natural Gas Natural gas that has been liquefied by cooling it to $-162°C$ ($-259°F$), which reduces its volume to $1/600$ of normal.

Lithosphere The rigid uppermost portion of the mantle plus the overlying crust of the earth.

Litter Partially decomposed organic matter.

Load The sediments that a river transports in suspension and pushes along the bottom, plus the materials carried in solution.

Longshore Current A component of water motion parallel to the shoreline, the result of coastal wave action, that supplies sand to beaches.

Lower Tolerance Limit The minimum level of an environmental factor below which no member of a species can survive.

Macrophages Specialized, free-living cells in the air sacs of the lung that engulf and digest foreign particles.

Magnetic Separation A technique used to separate ferrous (iron-containing) metals from waste streams.

Manganese Nodules Deposits of manganese, copper, cobalt, and nickel found on the deep ocean bottom.

Marginal Cost The extra cost of producing one less unit of pollution.

Marine Sanctuaries Areas of the coastal zone set aside to preserve natural resources.

Marsh A treeless wetland that is dominated by grasses.

Mesoscale Systems Weather systems that are relatively small and localized, for example, sea breezes and thunderstorms.

Mesosphere The subdivision of the atmosphere situated between the stratosphere and an altitude of about 80 kilometers (50 miles); air temperature declines with altitude in the mesosphere.

Metamorphic Rock Rock formed when a preexisting rock is subjected to high temperatures, confining pressures, and chemically active fluids deep within the earth's crust.

Metamorphism A change in the form of a rock caused by high temperatures, high confining pressures, and chemically active fluids occurring deep within the earth's crust.

METROMEX Metropolitan Meteorological Experiment, a study that indicates an increase of precipitation downwind of St. Louis.

Microscale Circulation The smallest scale of atmospheric motion, which includes circulation patterns within meters of the ground and vegetation.

Mimicry An organism's presenting an appearance that resembles another organism, or an inanimate object. The mimic is often less subject to predation.

Mineral A solid characterized by an orderly internal arrangement of atoms, fixed chemical composition, and definite physical properties.

Minimum Tillage A farming procedure in which the surface soil is first loosened, and then planting, fertilizer application, and weed control are combined. This technique eliminates the more extensive seed bed preparation normally used and reduces soil erosion.

Minimum Viable Population The smallest number of individuals that is needed to ensure the survival of an isolated population.

Mixed-Market Economies Economic systems, such as that in the United States, that combine pri-

vate, competitive enterprise with some government involvement.

Mixing Depth Vertical distance between the earth's surface and the altitude to which convective currents extend.

Mixing Layer Surface layer of the atmosphere in which air is thoroughly mixed by convection.

Mixing Ratio The amount of water vapor mixed with other atmospheric gases. It is expressed in grams of water vapor per kilogram of dry air.

Model An appropriate representation or simulation of a real situation.

Mortality The death rate.

Mounded Landfill A landfill formed by piling wastes and covering them with earth to form a large mound. This technique is used where the water table is close to the surface.

Mutagen A chemical substance or physical agent (such as radiation) capable of producing a change in an individual's genetic makeup.

Mutation A random, inheritable change in chromosomes (the genetic material).

Mutualism A relationship between two species where the survival of each depends on the presence of the other.

Natality The rate of production of new individuals by birth, hatching, or germination.

National Forests Public lands set aside primarily to foster wise forest management and thereby insure sustained timber yield and watershed protection.

National Parks Lands regulated for the preservation of some natural feature as well as for the educational and recreational needs of the general public.

National Resource Lands (BLM lands) Public lands managed by the Bureau of Land Management and used primarily for grazing and mining.

National Wildlife Refuges Public lands set aside to protect the habitats of threatened or endangered species.

Natural Selection The differences in the survival and rate of reproduction of individuals in nature leading to an increase in the frequency of some characteristics and a decline in the frequency of others.

Negative Feedback A control mechanism that enables a system to adjust to an environmental change by counteracting the effects of the change.

Neritic Zone The region of the ocean that extends from the high tide line to the outer edge of the continental shelf.

Net Condensation The amount of water gained as the result of more water condensing into a liquid source from the vapor state than evaporating from the source.

Net Evaporation The amount of water lost as the result of more water evaporating from a source than returning to the source by condensation.

New Towns Planned communities designed to be efficient in operation, protective of the environment, and as self-sufficient as possible.

Nitrogen Cycle Circulation of nitrogen among various components of the environment. Major reservoirs include the atmosphere and oceans.

Nitrogen Fixation Processes by which atmospheric nitrogen gas is changed into forms that plants can use.

Nonbiodegradable Pertaining to materials resistant to degradation by microorganisms.

Nonpoint Sources Sources of pollutants in the landscape; for example, agricultural runoff.

Nonrenewable Resources Resources that once used are not regenerated.

No-Till Agriculture An agricultural practice whereby fields are seeded, fertilized, and applied with herbicides in one operation. This technique eliminates the more extensive seed bed preparation normally used and reduces erosion.

Nuclear Chain Reaction A self-perpetuating sequence of fission reactions within the nuclei of certain atoms such as uranium-235.

Nuclear Fission The fragmenting of an unstable nucleus of an element resulting in the release of energy and several neutrons, and the formation of two new nuclei, usually radioactive.

Nuclear Fusion The fusing together of atomic nuclei of light elements to form heavier elements, thereby releasing energy.

Obliquity of the Ecliptic The tilt of the earth's axis to the earth's orbital plane; responsible for the seasons.

Ocean Thermal Energy Conversion A process that produces energy by using temperature differences between surface and bottom waters of the ocean.

Oceanic Zone The region of open ocean that lies beyond the continental shelf.

O-horizon The surface layer of soil, composed of fresh or partially decomposed organic matter.

Oil Shale Shale deposits containing organic materials that yield oil-like products when the shale is heated to high temperatures.

Old-Field Succession Ecological succession following the abandonment of farmland.

Oligotrophic Ecosystems Ecosystems that are low in fertility and biological productivity.

Oligotrophic Lake A lake with a low rate of nutrient cycling and a low level of biological productivity.

Omnivore An animal that feeds on plants and other animals.

Onchocerciasis A waterborne disease, common in less developed tropical countries, caused by parasitic worms spread by black flies, that can result in blindness.

Open Pit Mining A surface mining method whereby the overburden is removed from a large area so that rock and mineral deposits can be excavated to considerable depth.

Optimum Concentration The level of an environmental factor that sustains the maximum population of a species.

Ore An economically important mineral deposit.

Organic Pertaining to materials produced by or derived from organisms or synthetic chemicals composed of carbon.

Organic Fertilizers Crop and animal residues that are used to improve soil fertility.

Orographic Lifting The rising motion of an air mass as a result of a topographic barrier.

Orographic Rainfall Rainfall triggered by an uplift of air caused by topographical features such as mountain ranges.

Oxygen Cycle Circulation of oxygen among various components of the environment. Major reservoirs include the atmosphere and oceans.

Ozone Shield A layer of ozone in the stratosphere that filters out potentially lethal intensities of ultraviolet radiation. Thus, organisms on earth are shielded by ozone.

Pangaea A supercontinent that existed 200 million years ago consisting of Eurasia, Africa, and the Americas.

Parasitism An interaction in which one organism (the parasite) obtains energy and nutrients by living within or upon another organism (the host).

Parts Per Million (ppm) The unit measure of the concentration of a component substance; for example, a 1 ppm concentration of arsenic is 1 part of arsenic to 999,999 parts of other material.

Passive Solar System Features designed into a building to reduce energy utilization for heating and cooling—for example, window placement minimizing solar capture in summer and maximizing solar capture in winter.

Perihelion The time of year when the earth is closest to the sun (January 3 or 4).

Permeability The capability of soil or rock to transmit water or air.

Persistent Term applied to substances that are not easily broken down by the metabolism of organisms, chemical reactions, or by physical forces that operate in the environment.

Pest An organism that humans consider to be undesirable. Often they are pioneer species.

Pesticide A chemical used to kill pests.

pH Scale A system to specify the degree of acidity and alkalinity; a pH of 7 is neutral, below 7 is acidic, and above 7 is alkaline.

Phosphorus Cycle Circulation of phosphorus among various components of the environment. Major reservoirs include the soil, sediments, and phosphate-bearing rock.

Photochemical Smog A noxious, hazy mixture of aerosols and gases formed by sunlight acting on oxides of nitrogen and hydrocarbons, common pollutants in urban-industrial air.

Photoperiod The relative length of day and night.

Photosynthesis The process by which green plants transform light energy into food (chemical) energy.

Photovoltaic Systems Devices that use semiconducting materials to convert sunlight directly to electricity.

Phytoplankton Free-floating, mostly microscopic aquatic plants.

The Pill An oral contraceptive consisting of a female hormone and a synthetic substance that is chemically similar to another female hormone.

Pioneer Species Organisms present in the initial stage in ecological succession.

Pioneer Stage The initial stage in ecological succession.

Placers Heavy mineral deposits found in sand and gravel layers in streambeds or in coastal areas.

Planetary Albedo The percentage of solar radiation reflected by the entire earth-atmosphere system; about 30 percent.

Plate Tectonics The slow movement of gigantic lithospheric plates over the surface of the earth.

Point Sources Discernible conduits, such as pipes, ditches, channels, sewers, tunnels, or vessels, from which pollutants are discharged.

Poleward Heat Transport The flow of heat from tropical latitudes into middle and high latitudes, primarily by the exchange of air masses.

Pollutant Standards Index (PSI) A measure of air quality based on the *one* criteria air pollutant exhibiting the highest concentration when compared with that pollutant's primary air quality standard.

Population A group of individuals of the same species occupying the same geographical region at the same time.

Population Momentum The tendency of a growing population to continue to grow for some time even if each couple, on the average, just replaces itself.

Positive Crankcase Ventilation A device that reduces automobile hydrocarbon emissions by channeling crankcase blow-by gases back through the engine.

Positive Feedback A control mechanism that reinforces rather than counteracts the adjustment of a system to an environmental change.

Power-Tower System A system containing mirrors that track the sun and focus reflected sunbeams on a tower where pressurized steam is produced to generate electricity.

Precipitation The return of water to the earth's surface in the form of rain, snow, ice pellets, and hail.

Predation An interaction in which one organism (the predator) kills and eats another organism (the prey).

Primary Air Pollutants Substances introduced into the atmosphere that, in sufficient concentrations, pose serious hazards to environmental quality.

Primary Air Quality Standards Maximum exposure levels of criteria pollutants in ambient air set by the Environmental Protection Agency and based on potential health effects on humans.

Primary Succession Ecological succession that begins in an area where no soil exists.

Primary Treatment The first stage in sewage treatment, which relies on the screening and settling out of insoluble water pollutants.

Producers Organisms, mainly green plants, that manufacture their food by utilizing raw materials from air and soil.

Productivity The accumulation of biomass during a specific period of time at a given trophic level.

Protective Coloration The physical appearance of an organism that blends unnoticed with its background.

Public Lands Lands held in trust by the federal government and subject to multiple use regulations, for example, national forests, national parks, and wildlife refuges.

Pyrolysis A conversion technique for decomposing organic substances such as garbage, wood, and coal by intense heat in the absence of oxygen. Usable products include combustible gases and liquids.

Quad An amount of energy equal to one quadrillion BTUs (1.0×10^{15} BTUs). 172 million million barrels of oil will yield approximately one Quad of energy.

Quarry A small-scale surface mine used for the recovery of rock.

Radiation The process by which energy flows from place to place as oscillating electromagnetic waves.

Radiational Cooling An air temperature drop, most pronounced at night, that is caused by emission of infrared radiation.

Radiational Temperature Inversion A temperature inversion formed by cooling of a surface air layer by emission of infrared radiation so that the coldest air is at the earth's surface and the air temperature increases with altitude.

Radioactivity The emission of highly energetic, dangerous rays or particles by certain substances.

Ranks of Coal Stages in the sequence of changes from peat to lignite to bituminous coal to anthracite coal.

Reclamation (land) Those activities that foster natural succession on land disturbed by mining.

Reclamation (resource) The separation of one or more components from a waste stream for recycling.

Recycling The recovery and reuse of resources.

Refuse-Derived Fuel Fuel obtained as a product of solid-waste conversion processes.

Replacement Rate The fertility rate at which a population has only the number of young necessary to replace itself.

Relative Humidity The ratio of the amount of water vapor present in air to the amount of water vapor in saturated air at that temperature, usually expressed as percent relative humidity.

Reserve That portion of a rock, mineral, or fuel deposit that can be extracted immediately both legally and economically.

Resource Recovery The recovery of materials, energy, or both from waste streams.

Respiration The liberation of energy from food within an organism.

Rhythm Method A method of birth control based on abstention from sexual intercourse during a woman's fertile period.

Riparian Doctrine A basic water law in most eastern states that gives the owner of land adjoining a stream or lake the right to receive the natural flow of a stream undiminished in quality and quantity.

Risk–Benefit Analysis An analytic approach employed by economists in which a product's or activity's risks are weighed against its benefits to determine if the product or activity should be allowed.

Rock Cycle The transformation of a rock into another type of rock as a consequence of change in the rock's geologic environment.

r-strategists Species that maximize their current rate of reproduction at the expense of their survival long enough to reproduce in the future.

Salinization An accumulation of salts at soil surface.

Saltwater Intrusion The movement of saline groundwater into a freshwater reservoir of underground water.

Sanitary Landfill A landfill consisting of compacted solid waste sealed between layers of clean earth.

Saturation The condition whereby a medium such as water contains its maximum possible concentration of some dissolved substance.

Saturation (Air) The condition whereby air contains its maximum concentration of water vapor.

Saturation Mixing Ratio The maximum concentration that water vapor can reach in air at a given temperature.

Saturation Vapor Pressure The pressure exerted by water vapor when air is saturated with it.

Scattering The dispersal of radiation in random or almost random directions when the radiation is intercepted by particles.

Scavenging (of pollutants) Removal of pollutants from air through washout by rainfall and snowfall.

Schistosomiasis A debilitating waterborne disease, common in less developed countries, caused by parasitic worms that live in calm water, especially irrigation canals.

Scientific Method A systematic form of inquiry that involves observation, speculation, and reasoning.

Scrubbing An industrial air pollution control strategy that removes soluble gases from effluents by spraying water through the effluent stream or by bubbling waste gases through water.

Secondary Air Pollutants Products of reactions among primary air pollutants; smog and acid rain are examples.

Secondary Air Quality Standards Maximum exposure levels of criteria pollutants in ambient air set by the Environmental Protection Agency, designed primarily to minimize damage to property and crops.

Secondary Succession Ecological succession that begins in an area where soil exists.

Secondary Treatment The removal of water pollutants from sewage through the growth and harvesting of bacterial cells.

Secure Landfill A landfill designed with rigid specifications for the storage of hazardous chemical wastes.

Sedimentary Cycle A pattern of material flow in which the soil or rocks provide the primary source of the substance for organisms. An example is the phosphorus cycle.

Sedimentary Rock Rock formed by the compaction and cementation of particles (sediments) of abiotic or biotic origin.

Seismic Gap Regions of quiescence within a seismically active belt.

Self-thinning The mortality associated with severe intraspecific competition among plants.

Sensible Heat Transfer The transport of heat energy from place to place as a result of conduction and convection.

Separated Sewer System A sewer system employing two pipes, one to transport surface runoff water and the other to transport sanitary wastes.

Sigmoid Growth Curve An S-shaped curve illustrating one pattern of growth that can result if an environmental factor begins to limit population growth.

Slash-and-burn Agriculture A farming method used in the tropics consisting of several steps; the clearing of overgrowth on a several-hectare plot, the burning of residue, the planting of crops for several years, and the abandoning of the site to allow the forest to reinvade it.

Sludge Suspended materials removed from water treatment processes and collected as a thick pasty ooze.

Social Hierarchy The pattern of dominant-subordinate relations among members of a population.

Soil Unconsolidated sediment consisting of both biotic and abiotic materials; a reservoir of nutrients and water that support vegetation.

Soil Horizon A layer of soil distinguished from layers above and below by characteristic physical properties and a distinct chemical composition.

Solar Constant Amount of radiational energy falling on a surface positioned at the top of the atmosphere and oriented perpendicular to the solar beam when the earth is at an average distance from the sun; about 2.00 calories per square centimeter per minute.

Solution Mining The removal of deep deposits of soluble minerals through the pumping of water down an injection well to dissolve the minerals and the retrieving of the solution via extraction wells.

Somatic Effect The direct effect of exposure to hazardous materials, such as radiation, on persons and often including the conversion of normal cells to cancer cells and other changes that reduce life expectancy.

Specific Heat The quantity of heat required to raise the temperature of 1 gram of a substance 1 degree Celsius.

Spring Turnover A mixing process that occurs in spring in temperate lakes whereby the bottom waters mix with the top waters.

Stable Air An air layer in which mixing of air is inhibited.

Stationary Population A population that is not changing in size.

Steering Winds Winds in the middle and upper troposphere that direct the path of storms and the exchange of air masses.

Sterilization A procedure, usually surgical, that renders an individual incapable of reproduction. Exposure to radiation and certain chemicals also may produce sterility.

Stratosphere The subdivision of the atmosphere that lies between the troposphere and an altitude of about 50 kilometers (30 miles); primary site of ozone formation.

Stream Channelization The straightening of meandering rivers for the purpose of transporting water more rapidly downstream and thereby reducing the threat of flooding.

Strip Cropping The planting technique used on sloped surfaces that reduces soil erosion through the alternation of a crop of low soil-anchoring ability with a crop of high soil-stabilizing ability.

Storm Sewer A large pipe network installed underground in cities to channel runoff to surface waters.

Subduction Zone Convergent boundaries of lithospheric plates where a downward moving plate enters the mantle.

Sublimation The process whereby water in the form of ice changes directly to water vapor without passing through the intervening liquid state.

Subsidence Temperature Inversion A temperature inversion formed by air sinking gradually over a wide area and warmed by compression; it forms at the top of the mixing layer.

Supercity A coalescence of adjacent cities and towns with a combined population of at least one million.

Supercooled Water Water that remains in the liquid state when cooled below its normal freezing point.

Superheated Water Water that remains in the liquid state when heated above its normal boiling temperature because of pressurization.

Supersaturation A relative humidity of more than 100 percent in the atmosphere.

Surface Crustal Processes Landscape sculpturing processes that encompass interactions between the earth's surface and wind, running water, glaciers, and organisms; weathering and erosion are examples.

Surface Mining The extraction of near-surface deposits of rock, minerals, or fuels after the removal of overburden.

Suspended Materials Substances that do not settle out of fluids because of their small size.

Swamp A wetland dominated by shrubs or trees.

Synergism An interaction in which the total effect is greater than the sum of the effects taken independently.

Synoptic-Scale Weather Circulation systems that influence the weather at the continental or oceanic scale, for example, air masses and high- and low-pressure centers.

Synthetic Fuels (Synfuels) Gaseous and liquid fuels produced primarily from coal and oil shale.

Tailings The waste residue of ore processing.

Temperature A measure of the molecular activity of a substance; the faster the molecular motion, the higher the temperature.

Temperature Inversion A temperature profile in the atmosphere characterized by an increase of temperature with altitude.

Ten-Percent Rule For ease of calculation, ecologists assume that only 10 percent of the energy at one trophic level is transferred to the next higher trophic level.

Teratogen A chemical substance or physical agent (such as radiation) capable of causing a birth defect.

Territorial Behavior Any activity whereby an individual actively defends an area against intruders of the same species.

Tertiary Treatment Any of several water treatment techniques used beyond conventional secondary treatment to improve water quality further.

Thermal Shock Interruption of the normal functioning of an organism because of a sudden change in the temperature of its environment. Fish are especially sensitive.

Thermocline The transition zone in a stratified lake between the upper warm layer and the lower cold layer, characterized by a rapid temperature decline as depth increases.

Thermosphere The uppermost subdivision of the atmosphere, characterized by a continuous increase of temperature with altitude.

Threatened Species A species that is still abundant in some parts of its range, but whose continued existence is in question because its numbers have declined significantly in other areas.

Threshold Limit Value The maximum concentration of a substance to which a person may be exposed for a long period of time without experiencing harmful effects.

Tidal Energy The energy associated with the tidal motion in the oceans, some of which can be recovered by hydroelectric generating stations located on oceanic tidal basins.

Tolerance Limits The range of an environmental factor beyond which no members of a particular species can survive.

Topography The physical relief of the landscape.

Toxic Substances Substances that cause serious illness or death in one dose or in low doses over a long time period.

Transfer Rates The rates at which energy and materials move among various components of the environment.

Transpiration The loss of water vapor from plants through the leaves.

Trench Landfill A landfill formed by digging wide trenches for waste disposal, then daily covering the waste with the earth that was excavated from the trench.

Trial-and-Error Learning Learning that takes place in response to prior mistakes.

Trihalomethanes A group of chlorine- and bromine-containing compounds, some of which are carcinogens, that are formed when drinking water or wastewater effluent containing organic chemicals is treated with chlorine.

Trophic Level The feeding position occupied by a given organism in a food chain, measured by the number of steps removed from the producers.

Troposphere The lowest subdivision of the atmosphere, and the site of most weather events.

Turbidity Reduction in light transmission through some medium (air or water) due to the presence of suspended particles.

Unstable Air Layer An air layer in which rising air continues to rise and sinking air continues to sink, resulting in mixing within the layer.

Upper Tolerance Limit The maximum level of an environmental factor above which no member of a species can survive.

Urban Heat Island A region of relatively warm air centered over an urban-industrial area.

Vapor Pressure The pressure exerted by a gas, for example, water vapor pressure in air.

Waste Exchange An organization that inventories waste chemicals so that they can be made available for reuse by other industries.

Water Cycle The ceaseless flow of water among the oceanic, atmospheric, and terrestrial reservoirs.

Watershed The geographical region drained by a river or stream and its tributaries.

Water Table The upper surface of the groundwater reservoir.

Weather The state of the atmosphere described in terms of such variables as temperature, cloudiness, and precipitation.

Weathering The chemical decomposition and mechanical disintegration of rock.

Weed A plant growing on a site where it is not wanted. Often they are pioneer species.

Wet Steam Field A region where steam is emitted from the earth's interior along with substantial quantities of hot water.

Wien's Displacement Law The higher the temperature of a radiating object, the shorter the wavelengths of its radiated energy.

Wilderness Areas Portions of national forests, national parks, and wildlife refuges where timbering, commercial activities, motor vehicles, and human-made structures are prohibited.

Zero Population Growth A state in which no population growth is occurring; synonymous with stationary population.

Zone of Aeration A layer of soil or rock in which pore spaces are partially filled with both air and water.

Zone of Saturation A layer of soil or rock in which the pore spaces are completely filled with water.

Zooplankton Weakly swimming, mostly microscopic aquatic animals found near the water surface.

A P P E N D I X I

CONVERSION FACTORS

	MULTIPLY	BY	TO OBTAIN
Length:	inches	2.540	centimeters
	feet	0.3048	meters
	statute miles	1.6093	kilometers
	nautical miles	1.852	kilometers
	centimeters	0.3937	inches
	meters	3.281	feet
	kilometers	0.6214	statute miles
	kilometers	0.5400	nautical miles
Mass and Weights:	ounces (avdp)	28.350	grams
	pounds	0.4536	kilograms
	tons	0.9072	metric tons
	grams	0.03527	ounces (avdp)
	kilograms	2.205	pounds
	metric tons	1.120	tons
Volume:	fluid ounces	0.02957	liters
	gallons	3.785	liters
	liters	0.2642	gallons
	liters	33.81	fluid ounces
Area:	square yards	0.8361	square meters
	acres	0.4047	hectares
	square miles	2.590	square kilometers
	square meters	1.196	square yards
	hectares	2.471	acres
	square kilometers	0.3861	square miles
	hectares	0.0100	square kilometers
	square kilometers	100.0	hectares
Pressure:	bars	0.9869	atmospheres
	inches mercury	25.40	millimeters mercury
	bars	1000.0	millibars

	MULTIPLY	BY	TO OBTAIN
Energy:	joules	0.2389	calories
	kilocalories	1000.0	calories
	joules	1.000	watt-seconds
	calories	0.003968	BTUs
	BTUs	252.0	calories
	kilowatt-hours	860,000.0	calories
	kilowatt-hours	3413.0	BTUs
	therms	2.52×10^7	calories
	therms	100,000.0	BTUs
	quads	1.0×10^{15}	BTUs
	barrels of crude oil	5.80×10^6	BTUs
	gallons of gasoline	1.25×10^5	BTUs
	tons (2000 lbs) coal	2.24×10^7	BTUs
	cubic feet natural gas	1020.0	BTUs
Power:	Horsepower	745.7	watts
	BTUs per minute	0.0176	kilowatts
	watts	0.057	BTUs per minute
	watts	0.00134	horsepower
	watts	0.0143	kilocalories per minute
	joules per second	1.00	watts

Temperature: Conversion Equations

$°F = 9/5°C + 32$

$°C = 5/9 (°F - 32)$

$K = °C + 273.15$

$K = 5/9 (°F + 459.7)$

A P P E N D I X **II** THE GEOLOGIC TIME SCALE

ERA	WHEN PERIOD BEGAN (MILLIONS OF YEARS AGO)	PERIOD	ANIMAL LIFE	PLANT LIFE	MAJOR GEOLOGIC EVENTS
Cenozoic	— 2 —	Quaternary	Rise of Civilizations	Increase in Number of Herbs and Grasses	Ice Age
	— 65 —	Tertiary	Appearance of First Men; Dominance on Land of Mammals, Birds, and Insects	Dominance of Land by Flowering Plants	
Mesozoic	— 135 —	Cretaceous		Dominance of Land by Conifers; First Flowering Plants Appear	
	— 180 —	Jurassic	Age of Dinosaurs		Building of the Rocky Mountains
	— 225 —	Triassic	First Birds		
	— 275 —	Permian	Expansion of Reptiles		Building of the Appalachian Mountains
	— 350 —	Carboniferous	Age of Amphibians	Formation of Great Coal Swamps	
Paleozoic	— 413 —	Devonian	Age of Fishes		
	— 430 —	Silurian	Invasion of Land by Invertebrates	Invasion of Land by Primitive Plants	
	— 500 —	Ordovician	Appearance of First Vertebrates (Fish)	Abundant Marine Algae	
	— 570 —	Cambrian	Abundant Marine Invertebrates	Appearance of Primitive Marine Algae	
Precambrian		—		Primitive Marine Life	

APPENDIX III   EXPRESSING NUMBERS AS POWERS OF TEN

Scientific notation—that is, expressions of numbers as powers of 10—is a means of writing very large or very small numbers without using the long string of zeros that follows or precedes them in conventional notation. The number 4,000,000, for example, has six zeros; writing them out every time the number is used could become awkward. But since 4,000,000 is obtained by multiplying 4 by 10 six times $(4 \times 10 \times 10 \times 10 \times 10 \times 10 \times 10)$, in scientific notation this number can be written as 4×10^6. On the other hand, a small number, such as 0.00004, is obained by dividing 4 by 10 five times:

$$\left(\frac{4}{10 \times 10 \times 10 \times 10 \times 10} \right)$$

In scientific notation, this number is written as 4×10^{-5}. A part of the system of scientific notation is shown below:

$$
\begin{aligned}
1,000,000 &= 10 \times 10 \times 10 \times 10 \times 10 \times 10 &&= 10^6 \\
100,000 &= 10 \times 10 \times 10 \times 10 \times 10 &&= 10^5 \\
10,000 &= 10 \times 10 \times 10 \times 10 &&= 10^4 \\
1,000 &= 10 \times 10 \times 10 &&= 10^3 \\
100 &= 10 \times 10 &&= 10^2 \\
10 &= 10 &&= 10^1 \\
1 &= 1 &&= 10^0 \\
0.1 &= 1/10 &&= 10^{-1} \\
0.01 &= 1/100 &&= 10^{-2} \\
0.001 &= 1/1,000 &&= 10^{-3} \\
0.0001 &= 1/10,000 &&= 10^{-4} \\
0.00001 &= 1/100,000 &&= 10^{-5}
\end{aligned}
$$

INDEX